Environmental Phosphorus Handbook

Environmental Phosphorus Handbook

Edited by

EDWARD J. GRIFFITH
Monsanto Company

ALFRED BEETON
Center for Great Lakes Studies

JEAN M. SPENCER
Baylor University

DEE T. MITCHELL
University of Arkansas

A Wiley-Interscience Publication

JOHN WILEY & SONS, New York . London . Sydney . Toronto

Library of Congress Cataloging in Publication Data:

Main entry under title:

Environmental phosphorus handbook.

"A Wiley-Interscience publication."
Includes bibliographical references.
1. Phosphorus—Addresses, essays, lectures.
I. Griffith, Edward J., ed.

QD181.P1E68 1973 546′.712 72-11574
ISBN 0-471-32779-4

Printed in the United States of America

10 9 8 7 6 5 4 3 2 1

AUTHORS

Z. S. Altschuler	U.S. Geological Survey Washington, D.C.
Walter E. Brown	Director American Dental Association Research Unit National Bureau of Standards Washington, D.C.
William Chamberlain	Indiana State University Terre Haute, Indiana
G. Donald Emigh	Director of Mining Monsanto Industrial Chemicals Company St. Louis, Missouri
Joseph Feder	New Enterprise Division Monsanto Company St. Louis, Missouri
L. W. Ferrara	International Minerals and Chemical Corporation Libertyville, Illinois
D. Jerome Fisher	Arizona State University Tempe, Arizona
H. L. Golterman	Limnological Institute Nieuwersluis The Netherlands
Edward J. Griffith	Monsanto Company St. Louis, Missouri
R. A. Gulbrandsen	U.S. Geological Survey Menlo Park, California
J. E. Harwood	National Institute for Water Research South African Council for Scientific and Industrial Research Pretoria, Republic of South Africa

W. H. J. Hattingh	National Institute for Water Research South African Council for Scientific and Industrial Research Pretoria, Republic of South Africa
P. R. Hesse	Senior Technical Officer Food and Agriculture Organization of the United Nations Rome, Italy
P. H. Hodson	Monsanto Company St. Louis, Missouri
Frank F. Hooper	Professor of Fisheries and Zoology The University of Michigan Ann Arbor, Michigan
Andre T. Jagendorf	Division of Biological Sciences Cornell University Ithaca, New York
Edward D. Jones III	Research Economist, Central Intelligence Agency Washington, D.C.
Emmett F. Kaelble	Monsanto Company St. Louis, Missouri
A. Fred Kerst	Director Research and Development Michigan Chemical Corporation St. Louis, Michigan
Richard A. Kimerle	Monsanto Company St. Louis, Missouri
John W. Lyons	Commercial Development Special Chemical Systems Monsanto Company St. Louis, Missouri
Lawrence J. Machlin	Life Science Department New Enterprise Division Monsanto Company St. Louis, Missouri
Kenneth M. Mackenthun	Director, Division of Applied Technology Environmental Protection Agency Washington, D.C.
Duncan McConnell	Ohio State University Columbus, Ohio
John F. McCullough	Division of Chemical Development Tennessee Valley Authority Muscle Shoals, Alabama

V. E. McKelvey — Director U.S. Geological Survey
Washington, D.C.

Carleton B. Moore — Professor of Geochemistry.
Director, Center for Meteorite Studies
Arizona State University
Tempe, Arizona

F. W. Morgan — Detergent and Fine Chemicals Division
Monsanto Industrial Chemicals Company
St. Louis, Missouri

John B. Nesbitt — Department of Civil Engineering
Pennsylvania State University
University Park, Pennsylvania

Shigeru Ohashi — Kyushu University
Fukuoka, Japan

Joseph C. O'Kelley — Biology Department
University of Alabama
University, Alabama

F. H. Rigler — Department of Zoology
University of Toronto

Charles E. Roberson — U.S. Geological Survey
Menlo Park, California

William Rorie — Industrial Testing Laboratories, Inc.
St. Louis, Missouri

Clair N. Sawyer — Consultant
Sun City, Arizona

Lewis G. Scharpf, Jr. — Monsanto Company
St. Louis, Missouri

Joseph Shapiro — University of Minnesota
Minneapolis, Minnesota

C. Y. Shen — Detergent and Fine Chemicals Division
Monsanto Industrial Chemicals Company
St. Louis, Missouri

Eugene A. Thomas — Zürich University and Cantonal Laboratory, Switzerland

John R. Van Wazer — Vanderbilt University
Nashville, Tennessee

Thomas P. Whaley — International Minerals and Chemical Corporation
Libertyville, Illinois

PREFACE

Environmental Phosphorus Handbook is a comprehensive study of phosphorus in the environment. The complexity of the subject is immediately apparent. *Environmental Phosphorus Handbook* is a compilation designed to furnish the serious student a "port of entry" into the expansive and intricate universe of phosphorus.

Phosphorus in the environment occurs in many forms and performs an all but limitless number of complex functions. Because of its complexity and the great number of places it is found, no single scientist nor scientific discipline is sufficiently diverse or inclusive to present a detailed profile of phosphorus. Currently the environment is under active investigation by the following scientific and professional areas: agronomy, astronomy, bacteriology, biology, botany, chemistry, ecology, economics, entomology, enzymology, geology, government, hydrology, ichthyology, limnology, mathematics, medicine, meteorology, mineralogy, nutrition, oceanology, phycology, physics, sanitary engineering, virology, zoology, and others too numerous to mention. The characteristics and properties of phosphorus become involved in many of these studies. *Environmental Phosphorus Handbook* is an organized compilation of articles written by recognized authorities from many diverse professional disciplines and international backgrounds, all focusing on environmental phosphorus as viewed from their particular vantage point.

The *Handbook* is organized in the following manner. Phosphorus is viewed telescopically beginning with studies of gigantic systems followed by smaller and smaller subsystems, until the details of the minute microscopic and submicroscopic systems are explored. The manifestations of phosphorus are presented as they appear in meteorites and lunar samples, as in the geological formation of the earth, as in living organisms, and as in the molecules and atoms that constitute these systems. With this background the behavior and problems of phosphorus in lakes, streams, and sewage are reviewed. Then as a change of pace, phosphorus is treated as a factor in the economy of the United States. Finally, a simplified editorial overview of phosphorus in the environment completes the study.

Environmental Phosphorus Handbook is a study in contrast. The reader's attention is directed to content and style variations in the subjects covered and the importance attached to different aspects of the "personality" of phosphorus. It is also called to the reader's attention that the ability to obtain reliable chemical analyses is the key to understanding for the majority of the work presented in this book. Without a detailed understanding of the quantity, forms, and properties phosphorus exhibits, little factual information can be gained about its functions and prediction becomes conjecture or fantasy.

EDWARD J. GRIFFITH

St. Louis, Missouri
June 1972

EDITORIAL NOTE

It was the editorial policy when *Environmental Phosphorus Handbook* was initially conceived that the phosphate controversy of the last two decades should not influence the contents of this volume. Rather than embarking upon the hopeless task of diverting opinion in areas ruled more by emotions than knowledge, it was desired that only the known scientific profile of phosphorus should be presented from a wide diversity of scientific disciplines. Once armed with as much of the scientific knowledge as could be presented in a single volume, the reader should become better prepared to arrive at his own unemotional evaluation of the role of phosphorus in our environment.

Although the original concept was idealistically sound, it proved to be unrealistic. Most authors with sufficient background and interest in phosphorus to contribute an article to the book do justifiably have an opinion based on their experiences and judgment. Consequently, the only defendable editorial stance that could be acceptable was one of neutrality. As a consequence, the opinions expressed in this volume are those of the authors and do not necessarily reflect the views of the editorial staff, the organizations that support this work, nor the publishers.

THE EDITORS

CONTENTS

Environmental Phosphorus Handbook

1

Phosphorus in Meteorites and Lunar Samples

CARLETON B. MOORE

Professor of Geochemistry, Director, Center for Meteorite Studies, Arizona State University, Tempe, Arizona

METEORITE ABUNDANCES

Phosphorus is one of the 20 most abundant elements in the solar system although as shown in Table 1, with the exception of chlorine, it has a lower abundance than the other elements in the second row of the periodic table. According to the rules of Oddo and Harkins the most abundant elements are those with a low atomic weight, but nuclei with odd mass numbers are considerably less abundant than their adjoining nuclides with even mass numbers. Natural phosphorus consists of the single nuclide ${}_{15}{}^{31}P$. This nuclide is apparently produced by the neutron capture *s*-process(4).

Analyses of meteorites show that phosphorus is a minor but ubiquitous element found in the many types of meteorites. Table 2 lists the contents in the stony and stony-iron meteorites. For comparison purposes the atomic abundances compared to 10^6 atoms of silicon have been calculated. Early determinations of phosphorus have not shown a great degree of precision. The values listed in Table 2 are taken from recent x-ray fluorescent or wet chemical colorimetric analyses. The values selected for the chondrites tend to be lower than those reported in earlier compilations. This accounts for the differences between the values in Table 2 and those in Table 1. In Table 1 Cameron (11) used the abundances in type I carbonaceous chondrites to calculate his cosmic abundances, whereas Goles (26) preferred the analyses of type II carbonaceous chondrites as being more representative of solar or cosmic abundances.

In stony meteorites phosphorus is reported by the analysts as P_2O_5, but schreibersite (iron–nickel phosphide) inclusions have been described in the metal phase of these meteorites; the actual state of the phosphorus shown by bulk chemical analysis is uncertain. General descriptions of the different meteorite classes have been reviewed by Mason (30). The chondrite nomenclature used in Table 2 is taken from Van Schmus and Wood (44). In general the phosphorus content of the chondrites appears to be fairly constant. There may be a higher abundance in the volatile (C, N, S, H) rich type I carbonaceous

TABLE 1 Relative Atomic Abundances of Second-Row Elements

Atomic Number	Element	Relative Cosmic Abundance	
		Cameron (11)	Goles (26)
10	Ne	2.36×10^6	4.4×10^6
11	Na	6.32×10^4	3.5×10^4
12	Mg	1.05×10^6	1.04×10^6
13	Al	8.51×10^4	8.4×10^4
14	Si	1.00×10^6	1.00×10^6
15	P	1.27×10^4	8.1×10^3
16	S	5.06×10^5	8.0×10^5
17	Cl	1.97×10^3	2.1×10^3
18	Ar	2.28×10^5	3.4×10^5

TABLE 2 Phosphorus (wt.% as P) in Stony and Stony-Iron Meteorites

Name	Number of Determinations	Range	Mean	Atoms per 10^6 Si	Ref.
Chondrites					
C-1 (Carbonaceous, type I)	1	—	0.08	7,000	45
C-2 (Carbonaceous, type II)	3	—	0.09	6,200	45
C-3 (Carbonaceous, type III)	4	0.10–0.11	0.11	6,400	45
H (Bronzite)	12	0.10–0.11	0.11	5,800	45
L (Hypersthene)	20	0.08–0.11	0.10	4,800	45
LL (Amphoterites)	11	0.08–0.18	0.11	5,300	33
E-4 (Enstatite, type I)	2	0.20–0.21	0.20	10,600	45
E-5 (Enstatite, intermediate)	1	—	0.19	9,100	45
E-6 (Enstatite, type II)	4	0.11–0.13	0.12	5,600	45
Calcium-poor achondrites					
Enstatite	1	—	0.008	260	45
Hypersthene	2	0.005–0.006	0.006	220	45
Chassignite	1	—	0.03	1,550	16
Ureilites	3	0.03–0.04	0.03	1,420	47
Calcium-rich achondrites					
Angrite	1	—	0.06	2,600	42
Nakhlite	1	—	0.05	2,000	27
Howardites	1	—	0.04	1,600	45
Eucrites	6	0.04–0.06	0.04	1,600	15
	2	0.04–0.05	0.04	1,600	45
Stony-irons					
Mesosiderites					
Silicate	9	0.11–0.64	0.36	—	37
Metal	9	—	0.21	—	9
Total	4	0.25–0.54	0.38	30,000	38
Pallasites	14	—	0.16	—	9
Metal	4	0.05–0.37	0.16	—	31
Olivine	1	0.08	0.08	—	31

chondrites and a depletion in the low iron L and LL group chondrites The high iron E-4 enstatite chondrites have a significantly higher phosphorus content than the E-6 enstatite chondrites. If both of these apparent trends are real, there appear to be two mechanisms controlling the distribution of phosphorus. The first is a concentration in the volatile-rich low-temperature phase of the carbonaceous chondrites and the second a separation in the metallic nickel–iron phase of the ordinary and enstatite chondrites.

Both the calcium-poor and calcium-rich achondrites have relatively low phosphorus contents. These meteorites appear to have undergone fractionation and crystallization by an igneous-type process that has separated the phosphorus from the silicate phase. On the other hand, the stony-iron mesosiderites have relatively high phosphorus contents in both the silicate and metal phases. This unusual phosphorus distribution will undoubtedly play a major role in solving the mode of origin of these significant meteorites. In the stony-iron pallasites the phosphorus is found primarily in the metallic phase, although in the Springwater pallasite phosphorus is found in the magnesium phosphate mineral farringtonite.

A determination of the phosphorus concentration is essential in a complete analysis of iron meteorites. In recent analyses of 100 iron meteorites by Moore *et al.* (35) the phosphorus contents reported ranged from 0.02 to 0.94 wt. % phosphorus. The phosphorus in iron meteorites is either dissolved interstitial phosphorus in the metallic phase or located in the iron–nickel phosphide mineral schreibersite. Doan and Goldstein (13) have pointed out that the total phosphorus in iron meteorites containing macroscopic schreibersite is usually underestimated. Samples of iron meteorites taken for chemical analysis generally avoid noticeable nonmetallic inclusions. They have estimated the phosphorus abundance in 11 selected meteorites by a point counting technique. Their results are compared with chemical data in Table 3.

Table 4 gives the phosphorus contents in iron meteorites arranged both by structural (Widmanstatten bandwidth) and chemical (gallium–germanium) classifications. Results for the structural classifications show little difference between individual groups. In the chemical classifications of iron meteorites developed by Wasson (46) low-phosphorus meteorites fall into groups IVa, IVb, and III, which also have low gallium and germanium contents. Within group IIIa there appears to be a positive correlation between nickel and phosphorus.

METEORITE PHASES

Phosphorus has an interesting distribution in the mineral phases found in meteorites. In iron meteorites it is found predominantly in the siderophilic

TABLE 3 Estimated Total Phosphorus in Selected Meteorites (13)

Name	Ga–Ge Group	Chemical P (wt. %)[a]	Vol. % P	Estimated Total P (wt. %)
Apoala	IIIa	0.19	6	0.9
Mount Edith	IIIb	—	5.5	0.8
Grant	IIIb	—	7	1.0
Breece	IIIb	—	5.5	0.8
Chupaderos	IIIb	—	8	1.2
Tieraco Creek	IIIb	—	8.8	1.3
Goose Lake	—	0.26–0.8	6	0.9
Rodeo	—	0.75	6	0.9
Wallapai	—	—	7	1.0
Carlton	—	0.13	5	0.75
Cowra	—	—	3.5	0.50

[a] Data taken from Moore *et al.* (35).

phosphide mineral schreibersite while in the stony meteorites it occurs primarily in lithophilic phosphates. In each group of meteorites, however, representatives of the other mineral types are sometimes found. The phosphorus-bearing minerals are listed in Table 5.

Ordinary chondrites contain the phosphates whitlockite and chlorapatite. Whitlockite is more often found in greater abundance. The remaining phosphates are found in other classes of meteorites which have evidently evolved under more variable chemical and physical conditions. The meteoritic mineral whitlockite was first recognized by Merrill (34) and characterized by Shannon and Larsen (41). They originally named it merrillite in honor of Merrill but later work by Fuchs (19) showed that it is identical to the better characterized terrestrial mineral whitlockite. A detailed electron microprobe study by Van Schmus and Ribbe (43) supported Fuch's conclusions. They indicated that the chlorapatite has essentially the composition $Ca_5(PO_4)_3(Cl_{0.8}F_{0.1}OH_{0.1})$ and the whitlockite may be represented as $2.55CaO{\cdot}0.28MgO{\cdot}0.15Na_2O{\cdot}0.02FeO{\cdot}P_2O_5$. Fuchs (21) gives similar analyses of whitlockite from a number of different meteorite types. It has a fairly uniform composition with only slight chemical variations indicating different chemical environments. Both whitlockite and chlorapatite are found as irregular grains with a nonuniform distribution. Grains up to 100 μ in diameter have been observed in the more recrystallized chondrites. In the chondrule-rich, less recrystallized ordinary and carbonaceous chondrites phosphate occurs in small difficultly recognizable granules. It is interesting to note that the

TABLE 4 Phosphorus (wt. % as P) in Iron Meteorites (35)[a]

Name		Number of Meteorites	Number of Analyses	Range	Mean
Structural classification					
H	Hexahedrites	8	16	0.13–0.29	0.26
Ogg	Coarsest octahedrites	4	8	0.16–0.94	0.38
Og	Coarse octahedrites	10	20	0.04–0.61	0.20
Om	Medium octahedrites	45	90	0.02–0.53	0.17
Of	Fine octahedrites	17	34	0.02–0.75	0.15
Off	Finest octahedrites	7	14	0.02–0.30	0.11
Dr	Ataxites	7	14	—	0.08
Chemical classification (Ga–Ge)					
IIa		7		0.39–0.46	0.44
IIb		2		0.18–0.24	0.21
I		6		0.15–0.34	0.20
IIIa		13		0.09–0.36	0.15
IIIb		2		0.20–0.63	0.41
IIIab		2		0.51–0.52	0.52
IVa		10		0.02–0.16	0.05
IVb		2		0.04–0.10	0.05
IIc		1		—	0.30
IId		2		0.61–0.65	0.63

[a] Data from Moore *et al.* (35).

TABLE 5 Phosphorus-Bearing Minerals in Meteorites

Mineral	Formula	Ref.
Schreibersite	$(Fe, Ni)_3P$	28, 30, 32
Barringerite	$(Fe, Ni)_2P$	10
Chlorapatite	$Ca_5(PO_4)_3Cl$	41
Whitlockite (merrillite)	$Ca_3(PO_4)_2$	19, 21
Graftonite	$(Fe, Mn)_3(PO_4)_2$	36
Sarcopside	$(Fe, Mn)_3(PO_4)_2$	36
Panethite	$Na_2Mg_2(PO_4)_2$	23
Brianite	$Na_2MgCa(PO_4)_2$	23
Stanfieldite	$Mg_3Ca_4Fe_2(PO_4)_6$	20
Farringtonite	$Mg_3(PO_4)_2$	14
Pe̊rryite	$(Ni, Fe)_x(Si, P)_y$	39, 17

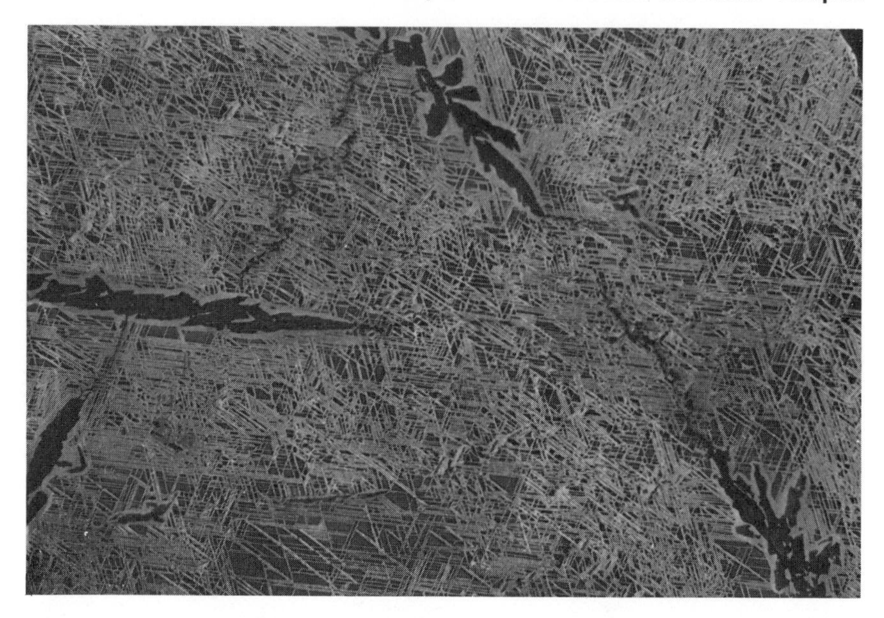

Figure 1. Large schreibersite crystals in a polished and etched section of the Carlton iron meteorite. The elongated crystals are surrounded by swathing kamacite in a fine-grained Widmanstatten pattern. Scale, 1 cm. Photo by J. Foster.

chlorapatites of meteorites differ from the predominantly fluorapatites of terrestrial rocks. No phosphate minerals have been recognized in the highly reduced enstatite chondrites. In the enstatite chondrites the phosphorus appears to be in schreibersite.

The more recently discovered rare phosphates are found in the metal-rich iron or stony-iron meteorites. The minerals graftonite, sarcopside, whitlockite, brianite, and panethite have been found associated with schreibersite in some troilite nodules in iron meteorites. Stanfieldite has been reported in both types of stony-iron meteorites, mesosiderites and pallasites. Pallasites have also been reported to contain farringtonite and whitfieldite. Consideration of the thermodynamic properties of these minerals indicates that the iron meteorites have a relative degree of oxidization about equal to the ordinary chondrites, which is higher than the highly reduced enstatite chondrites.

The role of reduced phosphorus in the iron meteorites and the ternary Fe–Ni–P system has been reviewed by Goldstein and Ogilvie (25), Buchwald (8), Axon (6), Comerford (12), Doan and Goldstein (13), and Reed (40). Schreibersite occurs in several distinct forms in iron meteorites. Macroscopic crystals may take the form of irregular hieroglyphic shapes or long thin blades known as Brezina lamellae (Fig. 1).

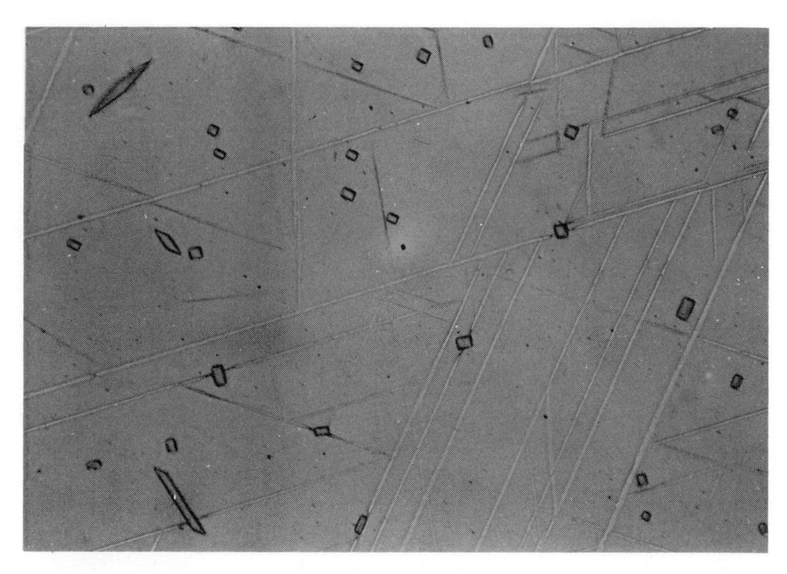

Figure 2. Microscopic rhombahedral schreibersite crystals (rhabdites) in a polished and etched section of the Bloody Basin iron meteorite. The long straight lines are Neumann bands or slippage planes produced by shock effects. Scale, 0.2 mm. Photo by P. Buseck.

These schreibersite crystals appear to have formed from liquid metal during the original crystallization of the meteorite parent bodies. Microscopic phosphides may form at the kamacite–taenite interface as the cooling meteorites enter the $\alpha + \gamma$ field of the iron–nickel phase diagram. Small rhombic crystals of schreibersite, commonly called rhabdites, often form from kamacite at low temperatures (Fig. 2). The phosphorus for these small crystals diffuses from the cooling kamacite crystals. Reed (40) has shown that the phosphorus content of kamacite may range from 0.02 to 0.2 wt. % phosphorus while that in adjacent taenite is lower by a factor of 3 or more. On the basis of their studies of the Fe–Ni–P phase diagram Doan and Goldstein (13) conclude that the phosphide structures can be explained in terms of the meteorite compositions, the Fe–Ni–P phase stability relations, and the diffusion of nickel and phosphorus in the solid state. The forms of phosphides found in iron meteorites yield information on the cooling history of the meteorite.

In addition to the common phosphide schreibersite $(Fe, Ni)_3P$, Buseck (10) has identified a new mineral, barringerite $(Fe, Ni)_2P$. This mineral was found in the pallasite Ollague. He concluded that this mineral, which occurs together with schreibersite, indicates that a state of local nonequilibrium exists in this meteorite. A nickel–iron silicide first recognized by Ramdohr

(39) in the Horse Creek iron meteorite was characterized by Fredriksson and Henderson (17). They named the mineral perryite and reported a composition of 81 wt. % Ni, 12% Si, 3% Fe, and 5% P.

An interesting calculation may be made on the contribution of extraterrestrial phosphorus to the hydrosphere of the earth. Current estimates of meteoritic influx are about 100 metric tons or 10^8 g per day. Taking 0.1% as the amount of phosphorus in this meteoritic material, the influx of extraterrestrial phosphorus to the earth would then be about 3.7×10^7 g per year. If about 70% of this material ends up in the oceans and the oceans have a mass of 1.4×10^{24} g, this would contribute 2.6×10^{-11} mg of phosphorus per liter of ocean water. This is a rather insignificant amount compared to the 0.07 mg/l of phosphorus currently present in ocean water and can hardly be considered a major source of phosphorus. If the figures above are integrated over the 4.6×10^9 year age of the earth, meteoritic phosphorus could contribute 0.12 mg/l to the ocean water, which is approximately what we see today. This figure, too, is probably insignificant compared to the amount of phosphorus that has been recycled through the oceans from weathered rocks and the biosphere over geologic time, but this is the only new source of phosphorus on earth.

LUNAR SAMPLES

The initial investigations of returned lunar samples from Apollo 11 indicate that phosphorus is present as a minor element in each of the analyzed specimens. Analyses reported by a number of investigators* did not show large variations between laboratories, analytical methods, or rock types. Type A or fine-grained vesicular basalts had phosphorus contents that ranged from 0.04 to 0.19 wt. % P_2O_5. Type B medium to coarse vuggy basalts showed 0.04 to 0.23% P_2O_5. Type C breccias had 0.08 to 0.15% P_2O_5, while the lunar regolith or fines contained from 0.06 to 0.32% P_2O_5. There are no apparent significant differences between the sample types. Individual analyses of glass fragments in the lunar regolith give similar results. On the basis of selected chemical analyses, Agrell *et al.* (2) calculated C.I.P.W. norms with 0.09% apatite in type A samples, 0.09% in type B samples, 0.17% in type C samples, and 0.12% in type D samples. Anderson *et al.* (5) interpret low P_2O_5/K_2O ratios in lunar glasses to indicate that the

* The results of the initial Apollo 11 investigations have been published in *Science*, **167**, 449–784 (1970) and *Proceedings of the Apollo 11 Lunar Science Conference*, Vols. 1–3, A. A. Levinson (Ed.), Pergamon Press, New York, 1970. Because of the number and similarity of the analyses, individual papers have not been referenced.

moon is more highly reduced than the earth and that part of the lunar phosphorus has been dissolved in the metal phase. Several investigators, including Adler *et al.* (1), Goldstein (24), and Keil *et al.* (29), have found phosphorus in high nickel–iron particles from breccias and fines. These particles are probably of meteoritic origin. No phosphorus or phosphides have been identified in the nickel-free metallic iron found in the lunar basalts. Fuchs (22), French (18), and Brown (7) have reported the occurrence of fluorapatite in lunar basalt. Albee and Chodos (3) have in addition characterized chlorofluorapatite and whitlockite in a sample of lunar fines.

The similarity of total phosphorus in the lunar fines and breccias as compared to the lunar basalts is undoubtedly due to the fact that the fines are primarily made up of crushed lunar basalt. Since meteoritic material has only slightly more phosphorus than the lunar basalts, a 1 to 2% meteoritic increment in the lunar fines will not make a recognizable contribution to the total phosphorus content. Since phosphorus is not a volatile element under lunar surface conditions, it, likewise, will not show a depletion due to shock effects as shown by volatile elements such as sulfur.

Subsequent studies* of Apollo 12 and 14 lunar samples have reported the presence of fragments with higher phosphorus contents than the Apollo 11 basalts. This primarily glassy material, suggested to be a highland basalt, has received the designation KREEP from its relatively high potassium, rare earth elements, and phosphorus contents. Electron microprobe analyses of this material indicates that it has P_2O_5 contents ranging from 0.2 to 1.5 wt. %. Some fines samples from Apollo 12 contain up to 50% of a KREEP component. It is estimated that the Apollo 11 fines may contain up to 10% of KREEP material.

REFERENCES

1. Adler, I., L. S. Walter, P. D. Lowman, B. P. Glass, B. M. French, J. A. Philpotts, K. J. F. Heinrich, and J. I. Goldstein, *Science*, **167**, 590 (1970).
2. Agrell, S. O., J. H. Scoon, I. D. Muir, J. V. P. Long, J. D. C. McConnell, and A. Peckett, *Science*, **167**, 583 (1970).
3. Albee, A. L., and A. A. Chodos, *Proceedings of the Apollo 11 Lunar Science Conference*, A. A. Levinson (Ed.), Pergamon Press, New York, 1970, p. 135.
4. Aller, L. H., *The Abundance of the Elements*, Interscience, New York, 1961.
5. Anderson, A. T., A. V. Crewe, J. R. Goldsmith, P. B. Moore, J. C. Newton, E. J. Olsen, J. V. Smith, and P. J. Wyllie, *Science*, **167**, 587 (1970).
6. Axon, H. J., *Prog. Mater. Sci.*, **13**, 184 (1968).

* Hubbard, N. J., and P. W. Gast, *Proceedings of the Second Lunar Science Conference*, A. A. Levinson (Ed.), MIT Press, Cambridge, Massachusetts, 1971, pp. 999–1021.

7. Brown, G. M., *Proceedings of the Apollo 11 Lunar Science Conference*, A. A. Levinson, (Ed.), Pergamon Press, New York, 1970, p. 195.
8. Buchwald, V. F., *Acta Polytech. Scand. Chem. Met. Ser.*, No. 51, (1966), 46 pages.
9. Buddhue, J. D., *Pop. Astron.* **54,** 149 (1946).
10. Buseck, P. R., *Science*, **165,** 169 (1969).
11. Cameron, A. G. W., *Origin and Distribution of the Elements*, Ahrens (Ed.), Pergamon Press, Oxford, 1968, p. 125.
12. Comerford, M. F., *Meteorite Research*, P. M. Millman (Ed.), Reidel, Holland, 1969, p. 780.
13. Doan, A. S., Jr. and J. I. Goldstein, *Meteorite Research*, P. M. Millman (Ed.), Reidel, Holland, 1969, p. 763.
14. Dufresne, E. R., and S. Roy, *Geochim. Cosmochim. Acta*, **24,** 198 (1961).
15. Duke, M. B., and L. T. Silver, *Geochim. Cosmochim. Acta*, **31,** 1637 (1967).
16. Dyakonova, M. I., and V. Y. Kharitonova, *Meteorit.*, **18,** 66 (1960).
17. Fredriksson, K., and E. P. Henderson, *Trans. Am. Geophys. Union*, **46,** 121 (1965).
18. French, B. M., *Proceedings of the Apollo 11 Lunar Science Conference*, A. A. Levinson (Ed.), Pergamon Press, New York, 1970, p. 433.
19. Fuchs, L. H., *Science*, **137,** 425 (1962).
20. Fuchs, L. H., *Science*, **158,** 910 (1967).
21. Fuchs, L. H., *Meteorite Research*, P. M. Millman (Ed.), Reidel, Holland, 1969, p. 683.
22. Fuchs, L. H., *Proceedings of the Apollo 11 Lunar Science Conference*, A. A. Levinson (Ed.), Pergamon Press, New York, 1970, p. 475.
23. Fuchs, L. H., E. Olsen, and E. P. Henderson, Abstracts, 29th Annual Meeting of the Meteoritical Society, 1966, p. 12.
24. Goldstein, J. I., *Proceedings of the Apollo 11 Lunar Science Conference*, A. A. Levinson (Ed.), Pergamon Press, New York, 1970, p. 499.
25. Goldstein, J. I., and R. E. Ogilvie, *Geochim. Cosmochim. Acta*, **27,** 623 (1963).
26. Goles, G., *Handbook of Geochemistry*, K. H. Wedepohl (Ed.), Springer-Verlag, New York, 1969, p. 116.
27. Greenland, L., Ph. D. Thesis, Australian National University, Canberra, 1964.
28. Haidinger, *Ann. Phys.*, **72,** 580 (1847).
29. Keil K., M. Prinz, and T. E. Bunch, *Science*, **167,** 597 (1970).
30. Mason, B., *Meteorites*, Wiley, New York, 1962.
31. Mason, B., *Am. Mus. Novit.*, No. 2163 (1963).
32. Mason, B., *Am. Mineralog.*, **52,** 307 (1967).
33. Mason, B., and H. Wiik, *Geochim. Cosmochim. Acta*, **28,** 533 (1964).
34. Merrill, G. P., *Am. J. Sci.*, **43,** 322 (1917).
35. Moore, C. B., C. F. Lewis, and D. Nava, *Meteorite Research*, P. M. Millman (Ed.), Reidel, Holland, 1969, p. 738.
36. Olsen, E., and K. Fredriksson, *Geochim. Cosmochim. Acta*, **30,** 459 (1966).
37. Powell, B. N., Ph. D. Thesis, Columbia University, New York, 1969.
38. Prior, G. T., *Miner. Mag.*, **18,** 151 (1918).
39. Ramdohr, P., *J. Geophys. Res.*, **68,** 2011 (1963).
40. Reed, S. J. B., *Meteorite Research*, P. M. Millman (Ed.), Reidel, Holland, 1969, p. 749.

41. Shannon, E. V., and E. S. Larsen, *Am. J. Sci.*, **9**, 250 (1925).
42. Urey, H. C., and H. Craig, *Geochim. Cosmochim. Acta*, **4**, 36 (1953).
43. Van Schmus, W. R., and P. H. Ribbe, *Geochim. Cosmochim. Acta*, **33**, 637 (1969).
44. Van Schmus, W. R., and J. A. Wood, *Geochim. Cosmochim. Acta*, **31**, 747 (1967).
45. Von Michaelis, H., L. Ahrens, and J. P. Willis, *Earth Planet. Sci. Lett.*, **5**, 387 (1969).
46. Wasson, J. T., *Geochim. Cosmochim. Acta*, **31**, 161 (1967).
47. Wiik, H. B., unpublished.

Abundance and Distribution of Phosphorus in the Lithosphere

2

V. E. McKELVEY

Director, U.S. Geological Survey, Washington, D.C.

Phosphorus is the eleventh most abundant element in the earth's crust. It forms only about 0.1% of the rocks that make up the bulk of the earth's crust, however, and is thus geochemically classed as a trace element. Even so, it is a rock-forming element, both in the sense that it is a constituent of most rocks, common enough to be generally reported in rock analyses, and in the sense that some rocks are composed chiefly of phosphorus-bearing minerals.

Although phosphorus may occur in both the trivalent negative and the quinquevalent positive state, in the lithosphere it occurs almost entirely as the quinquepositive ion, which is a part of the orthophosphate ion PO_4^{3-}. About 200 minerals are known that contain 1% or more P_2O_5. Most of the phosphorus in the earth's crust, however, occurs in species of the apatite group, $Ca_5(PO_4, CO_3)_3(F, Cl, OH)$; in igneous and metamorphic rocks the most common species is fluorapatite, $Ca_5(PO_4)_3F$, but in sedimentary phosphorites the typical mineral is carbonate fluorapatite (1, 38). Perhaps the second most important mode of occurrence of phosphorus in the lithosphere is in silicate minerals in igneous rocks, where it substitutes for silicon in SiO_4 tetrahedra; in rocks analyzed by Koritnig (35), about 2 to 26% of the total phosphorus occurred in this fashion.

Because phosphorus in the lithosphere occurs only in the quinquevalent state, its solubility is not affected by the redox potential of the environment. The solubility of apatite, however, is strongly influenced by pH; it is slowly soluble in neutral and alkaline waters below a pH of about 7 to 8, and its solubility increases with increasing acidity, decreasing hardness, and decreasing temperature. Apatite is less soluble than calcium carbonate and some other minerals with which it is associated in many occurrences, and it may form residual concentrates as other minerals are leached away during weathering. In subtropical or tropical climates, apatite itself is destroyed by weathering and while some phosphorus may be redeposited in the form of aluminum phosphate minerals such as wavellite $[Al_3(OH)_3(PO_4)_2 \cdot 5H_2O]$, most of it is leached out during advanced stages of weathering.

TABLE 1 Classification of Igneous Rocks and Their Average Phosphate Content [In Percent P_2O_5, as Reported by Nockolds (46)][a]

Other Essential Minerals	Essential Alkali Feldspar (Potassium Feldspar, or Albite, or Both)	Essential Potassium Feldspar and Lime-Bearing Plagioclase: Potassium Feldspar >60% of Total Feldspar Content	Essential Potassium Feldspar and Lime-Bearing Plagioclase: Potassium Feldspar between 60% and 40% of Total Feldspar Content	Essential Potassium Feldspar and Lime-Bearing Plagioclase: Potassium Feldspar between 40% and 10% of Total Feldspar Content	Essential Potassium Feldspar and Lime-Bearing Plagioclase: Potassium Feldspar <10% of Total Feldspar Content	No Essential Feldspar (i.e., <10% of Any Feldspar Present)
Quartz >10% of the rock	ALKALI GRANITE—0.14 (Alkali rhyolite[b])—0.07 PERALKALINE GRANITE—0.07 (Peralkaline rhyolite)—0.03	CALC-ALKALI GRANITE—0.18 (Calc-alkali rhyolite[b])—0.07	ADAMELLITE—0.20 (Dellenite[b])—0.12	GRANO-DIORITE—0.21 (Rhyodacite[b])—0.17	TONALITE—0.21 (Dacite[b])—0.17	
No essential quartz or feldspathoid (i.e., <10%)	ALKALI SYENITE—0.19 (Alkali trachyte)—0.18 PERALKALINE SYENITE—0.17 (Peralkaline trachyte)—0.12	CALC-ALKALI SYENITE: LEUCO-CRATIC—0.38 MELANO-CRATIC—0.70 (Calc-alkali trachyte)—0.20	MONZONITE—0.44 (Latite)—0.49	CALC-ALKALI MANGERITE—0.43 (Calc-alkali doreite)—0.33 ALKALI MANGERITE—0.63 (Alkali doreite)—0.66	DIORITE—0.35 (Andesite)—0.28 (Alkali andesite)—0.52 GABBRO—0.24 (Tholeitic basalt)—0.23 ALKALI GABBRO—0.44 (Alkali basalt)—0.39	PYROXENITE—0.09 ALKALI PYROX-ENITE—0.59 PERIDOTITE—0.05 ALKALI PERI-DOTITE—0.38
Feldspathoid >10% of the rock	NEPHELINE SYENITE—0.19 (Phonolite)—0.17 MALIGNITE—0.71		NEPHELINE MONZONITE—0.45 (Nepheline latite)—0.37	ESSEXITE—0.48 (Nepheline ordanchite)—0.60 GLENMUIRITE—0.66	THERALITE—0.31 (Nepheline tephrite)—0.35 TESCHENITE—0.50 (Analcite tephrite)—0.67	ULTRA-ALKALINE ROCKS—0.02–1.68 (Nephelinite, etc.)—0.35–1.02

[a] Names of plutonic rocks are in capital letters, names of corresponding volcanic types in parentheses.
[b] Includes obsidian of equivalent composition.

Phosphorus is an essential component of every living cell and the existence of plant and animal life is therefore contingent on its availability. Conversely, biologic processes influence the distribution of phosphorus and are responsible for its concentration in some deposits.

The distribution of phosphorus in the lithosphere is thus a function of its behavior in magmatic, weathering, sedimentary, and biologic processes. The following brief description of its distribution, however, focuses not so much on these processes as on their products as reflected in the abundance of phosphate in igneous, sedimentary, and metamorphic rocks, and in the deposits in which it is especially concentrated.

PHOSPHORUS IN ORDINARY ROCKS

Igneous Rocks

The distribution of phosphorus in igneous rocks is well shown in Nockolds' (46) fine compilation of average chemical analyses and normative mineral calculations of 2507 igneous rocks. Nockolds' classification of igneous rocks and his values for their average P_2O_5* contents are shown in Table 1. His average P_2O_5 contents of component plutonic rock types are shown on Fig. 1, which shows also the petrochemical fields of rock groups and individual rock types when considered in terms of their content of silica (SiO_2), alkalis ($Na_2O + K_2O$), and camafic constituents ($CaO + MgO + Fe_2O_3 + FeO$) alone.

In these analyses, the average P_2O_5 content of igneous rock types ranges from nil to 1.68%. Dunites, peridotites, pyroxenites, anorthosites, and some granites and syenites generally contain less than 0.1% P_2O_5. Alkalic rocks of intermediate, mafic, and ultramafic composition—such as alkali pyroxenite, alkali gabbro, alkali mangerite, and monzonite—generally contain more than 0.40% P_2O_5. Ultraalkalic rocks, such as turjaite, melteigite, ijolite, and malignite, contain the largest amounts—up to 1.68% P_2O_5 in the averages reported by Nockolds and as much as 3.5% in individual rocks (2).

In rocks containing less than about 10% combined normative calcium, magnesium, and iron silicates, the phosphate content increases, with few exceptions, with increasing camafic constituents, irrespective of the content of alkalis or normative feldspathoid minerals. In rocks containing more than about 10% combined camafic silicates, the P_2O_5 content generally increases with increasing alkalis and normative feldspathoid minerals.

* As is customary in rock analyses, phosphorus is reported in this paper as P_2O_5. To convert to P, multiply percentage P_2O_5 by 0.436.

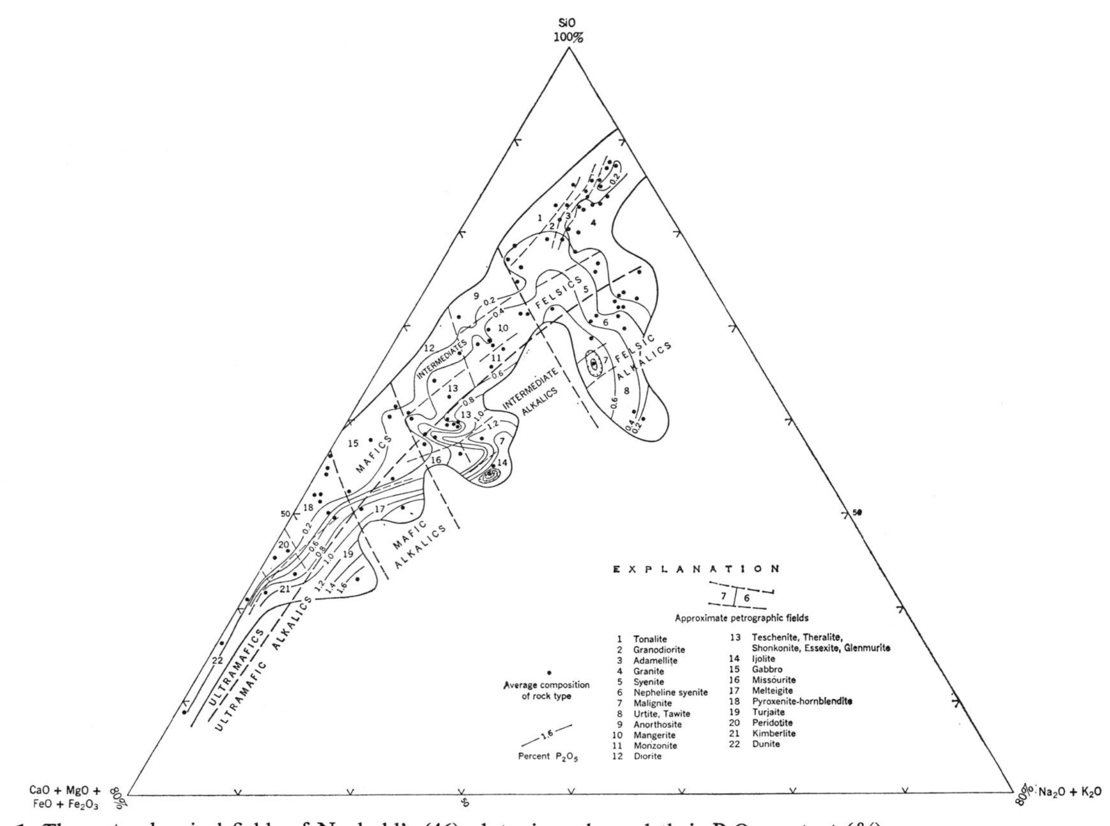

Figure 1. The petrochemical fields of Nockold's (46) plutonic rocks and their P_2O_5 content (%).

Normative ilmenite and magnetite closely parallel normative apatite in their distribution, the most notable exceptions being in those rocks containing more than about 2% apatite. As pointed out by Vogt (63) and Pecora (49), these elements are not geochemically coherent. Their parallel behavior, therefore, must reflect similar properties with respect to the behavior of their minerals in silicate melts.

Most plutonic rocks of silicic, intermediate, and subsilicic composition generally contain more phosphate than their volcanic equivalents. The reverse is true, however, for some alkalic and ultraalkalic varieties, such as monzonite, alkali mangerite, essexite, theralite, and teschenite.

In addition to its widespread distribution in ordinary igneous rocks, phosphate—also in the form of apatite—makes up the bulk of some rocks of igneous origin, generally in association with and genetically related to the phosphate-rich ultraalkalic rocks already mentioned. These rocks are discussed later in the section on igneous apatite deposits.

Igneous rocks thus range widely in their phosphate content, even when considered in terms of major groups (Table 2). Estimates of the average therefore depend on the abundance assumed for various rock types; and because silicic rocks are much more abundant in the upper continental crust than in the lower continental crust and in oceanic crust, such estimates also depend on which part of the earth's crust is being considered.

Approaching the problem from the standpoint of the composition of igneous rocks that would be required to match that of the sediments derived from them—mainly igneous rocks of the upper continental crust—Conway (8), Goldschmidt (22), Kuenen (36), Wickman (68), Taylor (61), Horn and Adams (27), and Brotzen (5) estimate the average igneous rocks to contain 0.27, 0.22, 0.28, 0.24, 0.25, and 0.27% P_2O_5, respectively. Weighting the most common types according to their observed abundance on various parts of the continents, however, Knopf (33), Vogt (63), Grout (23), Soloview [in Ronov and Yaroshevsky (55)], and Wedepohl (66)—using Nockolds' analyses shown in Table 2—estimated the average P_2O_5 content to be 0.26, 0.18, 0.16, 0.1, and 0.18%, respectively—with one exception, all in a lower range than the estimates derived from the composition of sediments.

From seismic evidence, Pakiser and Robinson (47) conclude that the lower continental crust is composed mainly of subsilicic rocks and that it makes up about 55 wt.% of the continental crust in the United States. Extrapolating these values to total continental crust and using Nockolds' analyses for silicic and subsilicic rocks, the igneous rocks of the continental crust contain an average of 0.23% P_2O_5.

In contrast to igneous rocks of the continental crust, those of oceanic crust are almost entirely of basaltic composition. Until recently, they were presumed to average about 0.3% P_2O_5, as for example, the composition

TABLE 2 Average Composition of Major Groups of Igneous Rocks (in Percent) (46)

	Silicic[a]	Intermediate[b]	Subsilicic[c]	Ultramafic[d]
SiO_2	68.9	54.6	48.4	43.8
TiO_2	0.5	1.5	1.8	1.7
Al_2O_3	14.5	16.4	15.5	6.1
Fe_2O_3	1.7	3.3	2.8	4.5
FeO	2.2	5.2	8.1	8.7
MnO	0.07	0.15	0.17	0.18
MgO	1.1	3.8	8.6	22.5
CaO	2.6	6.5	10.7	10.1
Na_2O	3.9	4.2	2.3	0.8
K_2O	3.8	3.2	0.7	0.7
H_2O+	0.6	0.7	0.7	0.6
P_2O_5	0.16	0.42	0.27	0.30
Percent of upper continental crust[e]	86	1	13	<1

[a] Granite, adamellite, granodiorite, tonalite, and volcanic equivalents.
[b] Syenite, monzonite, mangerite, diorite, and volcanic equivalents.
[c] Gabbro, norite, anorthosite, troctolite, and basalt.
[d] Pyroxenite, dunite, and peridotite.
[e] By volume, intrusive rocks only. Drawn from estimates of several authors by Wedepohl (66, p. 244). Volcanic rocks comprise only a few percent of all igneous rocks of the upper continental crust; of them, about 90% are tholeitic basalts and andesites (55).

assumed by Poldervaart (53). According to Engel and Engel (17), however, about 1% of oceanic crust is alkali basalt that averages about 0.48 P_2O_5, but the great bulk of it is a distinctive tholeitic basalt that averages only 0.13% P_2O_5.

Ronov and Yaroshevsky (55) have estimated the composition of the entire lithosphere and its major subdivisions on the basis of observed distributions of rocks in the upper crust and taking account of Engels' conclusions regarding the composition of oceanic crust. Although not calculated as such by Ronov and Yaroshevsky, their data yield a value for the phosphate content of average igneous rock of 0.22% P_2O_5.

Sedimentary Rocks

Because of the difficulty of classifying sedimentary rocks, which stems from the fact that many of them consist of mixtures of materials of diverse origin and manner of deposition, the average chemical composition of sedimentary rock varieties has not been reported in detail comparable to that for rocks of igneous origin. Data on the average phosphate content of sedimentary rocks are therefore mainly restricted to the major groups (Table 3).

Sandstone, shales, and carbonate rocks—the major sedimentary rock groups—contain an average of about 0.10, 0.17, and 0.07% P_2O_5, respectively, as judged from the convergence of the estimates shown in Table 3. Although these rocks are now exposed on the continents, most of them are of shallow-water marine origin. Deep-sea carbonate sediments contain about 0.1% P_2O_5 and deep-sea clays contain 0.3% or more (Table 3). The

TABLE 3 Average Phosphate Content of Common Sedimentary Rocks (Percent P_2O_5)

Ref.	Sandstone	Shale	Carbonate Rock	Deep-Sea Carbonate	Deep-Sea Clay	Deep-Sea Siliceous Ooze
(7, p. 30)	0.08	0.17	0.04	—	0.30[a]	—
(27, p. 285)	0.12	0.17	0.06	0.10	0.33	—
(55, p. 41)				0.11	0.21[a]	0.16
Platforms	0.16	0.11	0.07			
Geosynclines	0.11	0.15	0.10			
(36, p. 389)	0.08	0.17	0.04	0.18[b]	0.30[a] 0.21[c]	—
(34)	0.039	—	0.034	—	—	—
(16, p. 123)	—	—	—	0.15	0.14	0.27
(60, p. 373)[d]	—	—	—	0.3[b]	0.2[a] 0.2[c]	0.13[e]
(9)	—	—	—	0.08	0.34	—
Others[f]	0.1	0.16	0.067	—	0.24	—

[a] Red clay.
[b] Globigerina, pteropod, and coccolith ooze.
[c] Terrigenous clay.
[d] As recalculated by Poldervaart (53) to include CO_2.
[e] Combined from separate values for different siliceous oozes reported by Sujkowski.
[f] Sandstone estimated by Pettijohn (51, p. 515); shale by Wedepohl (67); carbonate rock by Gulbrandsen and Cremer (25, p. C125); deep-sea clay by Cronan (11).

differences in the phosphate content of these rocks reflect the chemical and mechanical separation of the constituents of parent rocks which takes place during their weathering, transport, and deposition.

Data on the average phosphate content of subgroups and varieties of sedimentary rocks may be briefly summarized. According to compilations of analyses of sandstones by Pettijohn (51), orthoquartzites contain less than 0.02% P_2O_5, arkoses and lithic arenites (subgraywackes) average 0.1%, and graywackes average 0.2%. The latter, which consist mainly of rock fragments, show less effects of weathering than ordinary detrital sediments, and are close to the composition of parent materials. This is true also of the tillites reported by Goldschmidt (21), which average 0.21% P_2O_5. Among modern oceanic sediments, the calcareous sands and oozes average 0.2% P_2O_5 and siliceous oozes average 0.3% [Rubey in Poldervaart (53)]. Most cherts contain less than 0.1% P_2O_5, but some of those containing abundant siliceous skeletal remains contain 0.2 to 0.8% or more (10). Marine bituminous black shales are generally richer in P_2O_5 than other shales, and black shales, cherts, and carbonate rocks associated with sedimentary phosphorites (which are described in a later section) contain admixed phosphorite grains in nearly all proportions.

Thin layers of phosphatic calcareous dolomitic siltstone have been found in the Eocene Green River Formation in Wyoming, generally associated with salt and trona [$Na_3H(CO_3)_2 \cdot 2H_2O$] beds of the Green River saline facies (37). Bradleyite ($Na_3PO_4 \cdot MgCO_3$) also occurs in some beds of the saline facies of the Green River in association with shortite ($Na_2CO_3 \cdot 2CaCO_2$) and trona (18), and NaH_2PO_4 in amounts of several percent has been reported from alkali encrustations (7). These occurrences apparently are the product of precipitation from highly saline alkaline lakes, and although other such lakes are known, they are rare. Most evaporite deposits, both lacustrine and marine, contain little or no phosphate (59).

Iron-rich sedimentary rocks of post-Precambrian age are characteristically enriched in phosphate, and this is true for all facies, including limonite, hematite, magnetite, chamosite, and siderite ores. Thus averages of analyses of these respective ores compiled by James (30) range from 0.91 to 1.78 P_2O_5 and the entire group of 87 determinations averages 1.43%. Basic slags made from the manufacture of steel from pig iron produced from such ores are used as phosphate fertilizer. Glauconites (greensands) are also generally phosphatic; the small group of analyses presented by James contains an average of 0.56% P_2O_5, and 35 analyses of New Jersey greensands listed by Mansfield (42) average about 1.0%. The phosphate in greensand occurs mainly in the form of pellets and skeletal fragments of apatite admixed in varying proportions and bearing no apparent direct chemical relationship to the glauconite. In some of the iron ores, the phosphate also occurs as apatite; in bog ores, however, it typically occurs as vivianite ($Fe_3P_2O_8 \cdot 8H_2O$) and in certain other iron ores it also occurs as an iron or aluminum phosphate mineral (31).

Precambrian sedimentary iron ores are generally only weakly phosphatic. The 41 analyses compiled by James average only 0.21 P_2O_5.

Estimates of the relative abundance of sedimentary rocks in the continental crust vary considerably. As Pettijohn has pointed out (50), estimates based on their abundance in measured sections differ considerably from those based on calculations of the amounts of the main rock groups required to match the composition of the average igneous rock from which they are derived. Thus estimates of the lithologic composition of sedimentary rocks based on field measurements range from 42 to 56% for shale, 14 to 40% for sandstone, and 18 to 29% for limestone, whereas estimates based on geochemical calculations range from 70 to 83, 8 to 12, and 5 to 14%, respectively, for the same rock types. Estimates of the average P_2O_5 content of sedimentary rocks of the continental crust reflect these differences in the weight assigned to various rock types. Thus, Clarke's (7) weighting—shale 80%, sandstone 15%, and limestone 5%—yields an average of 0.15% P_2O_5, and Poldervaart's weighting—shale 52%, sandstone 27%, and limestone 21%—yields an average of 0.1%.

Both sets of these average values are considerably less than those for the average phosphate content of igneous rock of the upper continental crust, suggesting that a substantial part of the phosphorus derived from the weathering of igneous rocks has been deposited in the deep-sea environment. Kuenen (36), has used this imbalance as one of the methods of estimating the total volume of deep-sea sediment. As shown in Table 3, deep-sea clays do have a higher phosphate content than sedimentary rocks of the continental crust, and while estimates of the abundance of various types of oceanic sediments and of their total volume differ considerably [see, for example, Wedepohl (67)], most indicate that the total phosphate in hemipelagic and pelagic sediments is much greater than that in sediments of the continental crust. This is illustrated in Table 4, drawn from Horn and Adams' (27) estimates of geochemical balances and element abundances.

Metamorphic Rocks

Metamorphic rocks are generally assumed to have essentially the same composition as their igneous and sedimentary equivalents. In view of the great chemical stability of apatite—the mineral in which phosphorus generally occurs in igneous, sedimentary, and metamorphic rocks—this assumption would be especially justified for phosphorus; Korzhinsky [in Mehnert (44)] ranks phosphorus next to aluminum and titanium in geochemical stability in

TABLE 4 Distribution of Phosphate in Sedimentary Domains, Water, and Igneous Source Rock (27)

	Mass (10^{14} metric tons of P_2O_5)	Percent P_2O_5 of Rock
Weathered Igneous Source Rock	51.4	0.25
Continent-shield sediments	3.44	0.13
Mobile belt-shelf sediments	11.12	0.14
Hemipelagic sediments	23.18	0.31
Pelagic sediments	13.76	0.30
Connate water	0.000237	—
Seawater	0.00225	—

metamorphism. The validity of the assumption concerning the equivalence of the composition of metamorphosed and unmetamorphosed rocks has been confirmed also by comparisons of observed and calculated compositions of rocks of the Baltic and Ukranian shields and the basement of the Russian platform by Ronov and Yaroshevsky (55). Because of their complexity in character and origin and the resulting difficulty of nomenclature and classification, the average composition has been estimated for only the main groups. As shown in Table 5 [see also Clarke (7)], the average phosphate

TABLE 5 Average Phosphate Content of Common Metamorphic Rocks (Percent P_2O_5) (53, p. 135)

Leptitites and halleflintas	0.1
Quartzfeldspathic gneiss	0.2
Mica schists	0.2
Slates	0.2
Phyllites	0.2
Amphibolites	0.3

content of these groups ranges from 0.1 to 0.3%, just as do the common igneous and sedimentary rocks.

Although phosphorus already within the rock is not much affected by metamorphic processes, it is commonly added in alkali metasomatism, including albitization and potassium feldspathization resulting from hydrothermal activity (44) as well as fenitization associated with carbonatite complexes (41, 19). The phosphate content of rocks so affected may be raised to the range of 0.4 to 0.6% P_2O_5.

Estimates of the abundance of metamorphic rocks based on measurements of mapped shield areas range from 17% in the Finnish and Canadian shields to 58% of the Baltic and Ukranian shields and basement of the Russian platform. Ronov and Yaroshevsky (55) estimate that of the volume of the upper continental crust, 37.6% is gneiss, 9.0% is crystalline schist, 1.5% is marble, and 9.8% is amphibolite; using Poldervaart's averages for these rocks (Table 5), they estimate that the average metamorphic rock contains 0.22% P_2O_5.

AVERAGE ABUNDANCE IN THE EARTH'S CRUST

Because sedimentary rocks make up only a few percent of the earth's crust—5, 6.2, and 7.9% according to the respective estimates of Clarke (7), Poldervaart (53), and Ronov and Yaroshevsky (55)—and because metamorphic rocks have essentially the same composition as their igneous and sedimentary counterparts, the composition of the earth's crust is nearly identical to that of igneous rocks. Most of the estimates of the average phosphate content of igneous rocks previously mentioned have been offered by their authors or interpreted by others as averages for the crust or component parts of it.

Worth mentioning here are a few additional estimates based on the observed abundance of crystalline rocks, including metamorphics, in shield areas, namely those of Sederholm (56), Grout (23), Shaw *et al.* (57), and Ronov and Yaroshevsky (55). The average P_2O_5 content of the upper continental crust in the areas studied by these authors is 0.11, 0.21, 0.15, and 0.2, respectively—similar as a group to those based on observed abundances of igneous rocks alone, and significantly lower than estimates derived from the composition of sediments.

For the crust as a whole, the average P_2O_5 content of 0.23% estimated by Ronov and Yaroshevsky (55) is probably the best available. Poldervaart's (53) estimate of 0.26% is made by similar methods but assumes a higher P_2O_5 content of oceanic crust than seems justified now.

From the uncertainties and the differences in observations and conclusions relating to the abundance of phosphorus in the earth's crust and its major subdivisions it is plain that the available estimates do not deserve great credence, especially as to the second significant figure. Perhaps their chief value is in indicating that in spite of the high phosphate content that may be found locally in both igneous and sedimentary rocks, the phosphate content of rocks in most areas is likely to be in the range of 0.15 to 0.2%.

PHOSPHATE DEPOSITS

In spite of the fact that ordinary rocks contain only 0.15 to 0.2% P_2O_5, they contain the great bulk of the phosphorus in the lithosphere. Only a fractional percentage of the total is concentrated in deposits consisting mainly of phosphate minerals, but these, of course, are the source of commercial phosphate production. Phosphate deposits may be classed into three broad groups: apatite deposits of igneous origin; sedimentary phosphorites; and guano and related deposits. Deposits that are secondarily enriched in phosphate form as the result of weathering of each of these major types, and as a group such enriched deposits supply a substantial part of commercial production.

Commercial phosphate deposits are described in Chapter 3 of this volume. The principal characteristics of the main groups, however, are described briefly in the following paragraphs.*

Apatite Deposits of Igneous Origin

Apatite deposits of igneous origin include pegmatites, veins, and disseminations principally associated with magnetite or ilmenite deposits of magmatic origin, and intrusive and metasomatized rocks associated with alkalic and ultraalkalic intrusions. The pegmatite deposits are generally of the order of hundreds to tens of thousands of tons in size and now rarely yield any commercial production. In some places apatite is recovered as a by-product of magnetite production or forms the basis for "Thomas meal," the phosphate-rich slag utilized in some places as a fertilizer, but neither the current nor potential production of phosphate from these deposits is significant.

The apatite deposits associated with alkalic intrusions, however, presently yield about 15% of the world's phosphate production, principally from the U.S.S.R., South Africa, Uganda, and Brazil. Individual deposits contain millions of tons of potentially recoverable apatite and a few may contain billions of tons. Most of them occur in intrusive complexes that have a ring-like structure and are associated with one or more of the following rocks—nepheline syenite, carbonatites, ultraalkaline rocks such as ijolite and urtite, and alkaline ultramafics such as alkali pyroxenite. Carbonatite commonly forms a central core in these complexes, but it is absent altogether in the largest igneous apatite deposit yet found—the Khibina nepheline syenite massif of the Kola Peninsula, U.S.S.R.

* For additional descriptions of individual deposits and of major types the reader is referred to the compilations of the United Nations (62) and the British Sulphur Corporation (3).

The apatite in the alkalic complexes occurs in several kinds of deposits (13, 14). Many of the carbonatites contain several percent apatite (49), and while this is not recoverable in unweathered rock, in some areas of deep weathering the apatite is concentrated in residual deposits that are minable. Apatite may also occur by itself, with magnetite, or in high concentrations in carbonatite, and in some deposits it occurs with olivine and magnetite in rocks called phoscorite. In some of these rocks it is recovered as a coproduct of magnetite. In the Khibina complex, the apatite is associated with nepheline in rock that averages about 27% P_2O_5.

The alkalic complexes typically contain a variety of commercially recoverable minerals, including niobium and rare earths (13), and have attracted much interest in exploration in recent years. Deans (14) estimates that some 200 carbonatite complexes are now known over the world, about 20 of which contain apatite deposits. Most of the carbonatites occur in shield areas, many of them associated with rift valley structures, and they range in age from Precambrian to Recent.

Marine Phosphorites

About 80% of the world's production of phosphate comes from marine phosphorites and their weathering derivatives, and they contain the bulk of the world's potential resources. Individual fields commonly contain scores or hundreds of millions of tons of minable reserves and billions of tons of potential resources.

The phosphorites are mainly formed in marine environments containing higher than average concentrations of phosphate. Such concentrations may result from the discharge of rivers draining humid highlands (6), from nutrient enrichment in estuaries (54, 52), and perhaps from submarine volcanic activity (43, 45, 20, 64). The most important concentrations, however, result from oceanic circulation (32, 40). Whereas warm surface waters of the ocean in most areas contain only 0.01 ppm PO_4 or less, deep cold waters contain nearly 0.3 ppm. Cold phosphate-rich water upwells to the surface in several environments (39), such as along the west coasts of continents in the trade wind belt (58), and leads to the growth of enormous quantities of organisms (4). Phosphate may be deposited in this environment by a number of inorganic and organic processes (24), including diagenetic precipitation and replacement (6, 12), and its concentration in rich deposits is enhanced by slow or nonaccumulation of diluting clastics or carbonates and by submarine reworking leading to the oxidation of organic matter and to the winnowing away of fine-grained impurities.

Phosphorites may contain one or more of a variety of textural components—pellets, nodules, oolites, or pisolites composed of microcrystalline particles of apatite; skeletal remains, mainly of parts that were phosphatic during life, such as teeth and bone, but also calcareous shells replaced by apatite; and crystalline apatite in cement or in aphanitic aggregates. Any one of these forms may be dominant, but the most common phosphorites consist of pellets, cement, and occasionally skeletal fragments. Unweathered phosphorites rarely contain more than 29 to 30% P_2O_5, and commonly contain much less because of carbonatic, siliceous, or carbonaceous impurities. Leached and oxidized rocks generally contain 32 to 35% P_2O_5, and phosphate is recovered by washing or flotation from some unconsolidated sands and weathered mudstones containing as little as 5 to 10% P_2O_5.

Marine phosphorites are commonly associated with carbonate rocks and quartz sandstones and the thickest and richest accumulations are generally associated also with black shales and cherts or diatomites.

Guano and Related Deposits

Most of the large accumulations of guano are formed at the surface by seafowl, but smaller quantities are formed by bats and to a lesser extent by other cave-dwelling mammals and birds (28). The bat guanos are most abundant in the cave districts of temperate and tropical regions. Although many bat-guano deposits have been found and mined, most of them have been measured in hundreds or thousands of tons, and only sporadic production is obtained from them now. Seafowl deposits are mainly confined to islands and coastal regions in low latitudes in the vicinity of the same areas of upwelling to which the marine phosphorites are related. The largest lie along the west coasts of Lower California, South America, and Africa, and on islands near the equatorial currents. Many used to be several hundred thousand tons in size, and although most of the fossil accumulations have been mined out, production from the current crop is continuing in some areas.

Fresh seafowl droppings contain about 22% N and 4% P_2O_5. Decomposition proceeds rapidly, and the phosphate content increases as the nitrogen (and total organic matter) decreases. Modern guano contains 10 to 12% P_2O_5, leached guano contains 20 to 32%. The mineralogy of guano is complex. Slightly decomposed deposits contain soluble ammonium and alkali oxalates, sulfates, and nitrates, and a variety of magnesium and ammonium–magnesium phosphates. Largely decomposed guano consists chiefly of calcium phosphates such as monetite ($HCaPO_4$) or whitlockite [β-$Ca_3(PO_4)_2$].

In areas where rainfall, however slight, is a normal event, the soluble phosphates of guano are carried to underlying rocks where they may be deposited as cavity fillings or replacements. Through this process, phosphate from guano has accumulated over long periods of geologic time. Most such deposits are of the order of only a few thousands or tens of thousands of tons in size, but a few formed on elevated islands are comparatively large deposits. For example, reserves on Nauru, an island in the equatorial region of the western Pacific Ocean, were originally about 90 million tons of rock averaging about 39% P_2O_5 (28), and those on Christmas Island in the Indian Ocean are 200 million tons (15). The mineralogy of phosphatized rocks derived from guano depends on the composition of the host rock (65). Where it is limestone, as on many of the coral atolls, the phosphate mineral is apatite; but where the underlying rock is a silicate, as in islands of volcanic origin, the phosphate minerals are calcium–aluminum or aluminum–iron phosphates.

CONCLUSION

Most rocks contain on the order of 0.15 to 0.2% P_2O_5, mainly in the form of apatite or in silicate minerals—both of which are relatively stable in the surface environment except under conditions of tropical weathering. Moreover, in most soils phosphorus added in the form of fertilizers tends to be fixed rather quickly into a relatively stable form—apatite in lime-rich soils or an iron or aluminum phosphate in less calcareous ones (29). The availability of phosphorus can be increased by acidifying the soil or by adding phosphate in relatively soluble form, but either effect is likely to be temporary.

Even though the availability of phosphates in rocks and soils is thus generally small, its local availability may be influenced by three factors. One is that rocks of much higher than average phosphate content may occur over large areas even though they form a negligible part of the earth's crust; this is true with respect to the alkalic intrusive rocks and certain phosphate-rich sediments where they are flat lying, but it is even more significant with respect to basaltic volcanic rocks, which cover tens of thousands of square miles in many areas. The second factor is that the phosphate in certain kinds of rocks may be more readily available than in others; Hutchinson (29) points out that phosphate is probably more easily liberated from sedimentary rocks than from igneous ones, presumably because of their greater porosity and permeability. This is particularly significant on a regional basis, for whereas sedimentary rocks make up only a few percent of the crust by volume, they cover a far larger proportion of the earth's surface—for example, 85% of the United States (26). The third factor that affects the availability of phosphate in local areas is the environment itself—the character of the climate, acidity of natural waters, and the presence of other minerals that may influence the liberation or fixation of phosphorus (29).

REFERENCES

1. Altschuler, Z. S., R. S. Clarke, Jr., and E. J. Young, *Geochemistry of Uranium in Apatite and Phosphorite*, U.S. Geological Survey Prof. Paper 314-D, 1958, pp. 45–90.
2. Barth, T. F. W., and I. B. Ramberg, "The Fen Circular Complex," in *Carbonatites*, O. F. Tuttle and J. Gittins (Eds.), Interscience, New York, 1966, pp. 225–257.
3. British Sulphur Corporation, *A World Survey of Phosphate Deposits*, 2nd ed., The British Sulphur Corporation, London, 1964, 206 pages.
4. Brongersma-Sanders, M., "Mass Mortality in the Sea," in *Treatise on Marine Ecology and Paleoecology*, Vol. 1, *Ecology*, J. W. Hedgpeth (Ed.), Geological Society of America Memoir 67, 1957, pp. 941–1010.
5. Brotzen, O., "The Average Igneous Rock and the Geochemical Balance," *Geochim. Cosmochim. Acta*, **30**, 863–868 (1966).
6. Bushinski, G. L., "Old Phosphorites of Asia and Their Genesis," *Akad. Nauk SSSR, Geol. Inst., Trans.*, No. 149, translated by Israel Program for Scientific Translations, Jerusalem, 1969 (1966), 206 pages.
7. Clarke, F. W., *The Data of Geochemistry*, U.S. Geological Survey Bulletin 770, 1924, 841 pages.
8. Conway, E. J., "Mean Geochemical Data in Relation to Oceanic Evolution," *Proc. Roy. Irish Acad., Sect. B*, **48**, 119–159 (1942).
9. Correns, C. W., *Die Sedimente des aquatorialen Atlantischen Ozeans. 2*, Wiss. Ergebnisse d. Duetschen Atlant. Expedition Meteor 1925–27, Vol. III, Berlin, Teil 3, 1937, 298 pages.
10. Cressman, E. R., "Nondetrital Siliceous Sediments," in *Data of Geochemistry*, 6th ed., U.S. Geological Survey Prof. Paper 440-T, 1962, 23 pages.
11. Cronan, D. W., "Average Abundances of Mn, Fe, Ni, Co, Cu, Pb, Mo, V, Cr, Ti, and P in Pacific Pelagic Clays," *Geochim. Cosmochim. Acta*, **33**, 1562–1565 (1969).
12. Cook, P. J., "Repeated Diagenetic Calcitization, Phosphatization, and Silicification in the Phosphoria Formation," *Geol. Soc. Am. Bull.*, **81**, 2107–2116 (1970).
13. Deans, T., "Economic Mineralogy of African Carbonatites," in *Carbonatites*, O. F. Tuttle and J. Gittins (Eds.), Interscience, New York, 1966, pp. 385–413.
14. Deans, T., "Exploration for Apatite Deposits Associated with Carbonatites and Pyroxenites," in *Proceedings of the Seminar on Sources of Mineral Raw Materials for the Fertilizer Industry in Asia and the Far East*, United Nations Mineral Resources Development Series, No. 32, 1968, pp. 109–119.
15. deKeyser, F., and I. R. McLeod, "Geological Environment of Australian Deposits of Phosphate and Gypsum," in *Proceedings Seminar on Sources of Mineral Raw Materials for the Fertilizer Industry in Asia and the Far East*, United Nations Mineral Resources Development Series, No. 32, 1968, pp. 165–170.
16. El Wakeel, S. K., and J. P. Riley, "Chemical and Mineralogical Studies of Deep-Sea Sediments," *Geochim. Cosmochim. Acta*, **25**, 110–146 (1961).
17. Engel, A. E. J., and C. G. Engel, "Rocks of the Ocean Floor," in *Fundamental Problems of Oceanology*, 2nd International Oceanographic Congress, Plenary Lectures, Izd, Nauka, Moscow, 1968, pp. 183–217 (in Russian).

18. Fahey, J. J., *Saline Minerals in the Green River Formation*, U.S. Geological Survey Prof. Paper 405, 1962, 50 pages.
19. Garson, M. W., "Carbonatites in Malawi," in *Carbonatites*, O. F. Tuttle and J. Gittins (Eds.), Interscience, New York, 1966, pp. 33–71.
20. Gibson, T. G., "Stratigraphy and Paleoenvironment of the Phosphatic Miocene Strata of North Carolina," *Geol. Soc. Am. Bull.*, **78**, 631–650 (1967).
21. Goldschmidt, V. M., "Grundlagen der quantitativen," *Geochem. Fortschr. Mineral. Krist. Petroger*, **17**, 112 (1933).
22. Goldschmidt, V. M., *Geochemistry*, Clarendon Press, Oxford, 1954, 730 pages.
23. Grout, F. F., "Petrographic and Chemical Data on the Canadian Shield," *J. Geol.* **46**, 486–504 (1938).
24. Gulbrandsen, R. A., "Physical and Chemical Factors in the Formation of Marine Apatite," *Econ. Geol.*, **64**, 365–382 (1969).
25. Guldbrandsen, R. A., and M. Cremer, *Coprecipitation of Carbonate and Phosphate from Seawater*, U.S. Geological Survey Prof. Paper 700-C, 1970, pp. C125–C126.
26. Higgs, D. V., "Quantitative Areal Geology of the United States," *Am. J. Sci.*, **247**, 575–583 (1949).
27. Horn, M. K., and J. A. S. Adams, "Computer-Derived Geochemical Balances and Element Abundances," *Geochim. Cosmochim. Acta*, **30**, 279–297 (1966).
28. Hutchinson, G. E., "The Biogeochemistry of Vertebrate Excretion," *Am. Mus. Nat. Hist. Bull.*, **96**, 554 pages (1950).
29. Hutchinson, G. E., "The Biogeochemistry of Phosphorus," in *The Biology of Phosphorus*, L. F. Wolterink (Ed.), Michigan State College Press, East Lansing, Michigan, 1952, pp. 1–35.
30. James, H. L., "Chemistry of the Iron-Rich Sedimentary Rocks," in *Data of Geochemistry*, 6th ed., U.S. Geological Survey Prof. Paper 440-W, 1966, 61 pages.
31. James, H. L., C. E. Dutton, F. J. Pettijohn, and K. L. Wier, *Geology and Ore Deposits of the Iron River-Crystal Falls District, Iron County, Michigan*, U.S. Geological Survey Prof. Paper 570, 1968, 134 pages.
32. Kazakov, A. V., *The Phosphorite Facies and the Genesis of Phosphorites*, USSR, Sci. Inst. Fertilizers and Insectofungicides Trans., No. 142, 1937, pp. 95–113.
33. Knopf, A., "Composition of the Average Igneous Rocks," *J. Geol.*, **24**, 620–622 (1916).
34. Koritnig, S., "Ein Beitrag zur Geochemie des Fluor," *Geochim. Cosmochim. Acta*, **1**, 89–116 (1951).
35. Koritnig, S., "Geochemistry of Phosphorus—I, The Replacement of Si^{4+} by P^{5+} in Rock-Forming Silicate Minerals," *Geochim. Cosmochim. Acta*, **29**, 361–371 (1965).
36. Kuenen, Ph. H., *Marine Geology*, Wiley, New York, 1950, 568 pages.
37. Love, J. D., *Uraniferous Phosphatic Lake Beds of Eocene Age in Intermontane Basins of Wyoming and Utah*, U.S. Geological Survey Prof. Paper 474-E, 1964, 66 pages.
38. McClellan, G. H., and J. R. Lehr, "Crystal Chemical Investigation of Natural Apatites," *Am. Mineral.*, **54**, 1374–1391 (1969).
39. McKelvey, V. E., *Phosphate Deposits*, U.S. Geological Survey Bulletin 1252-D, 1967, 21 pages.
40. McKelvey, V. E., R. W. Swanson, and R. P. Sheldon, "The Permian Phosphorite Deposits of Western United States," International Geological Congress, 19th Algiers, 1952, C. R., Sect. 11, Part 11, 1953, pp. 45–64.

41. McKie, D., "Fenitization," in *Carbonatites*, O. F. Tuttle and J. Gittins (Eds.), Interscience, New York, 1966, pp. 261–294.
42. Mansfield, G. R., *Potash in the Greensands of New Jersey*, U.S. Geological Survey Bulletin 727, 1922, 146 pages.
43. Mansfield, G. R., "The Role of Fluorine in Phosphate Deposition," *Am. J. Sci.*, **238**, 863–879 (1940).
44. Mehnert, K. R., "Composition and Abundance of Common Metamorphic Rock Types," in *Handbook of Geochemistry*, Vol. 1, K. H. Wedepohl (Ed.), Springer-Verlag, New York, 1969, pp. 272–296.
45. Miller, L. J., "The Origin of Sedimentary Phosphate Deposits," *Econ. Geol.*, **59**, 1619–1620 (1964).
46. Nockolds, S. R., "Average Chemical Compositions of Some Igneous Rocks," *Geol. Soc. Am. Bull.*, **65**, 1007–1032 (1954).
47. Pakiser, L. C., and R. Robinson, "Composition and Evolution of the Continental Crust as Suggested by Seismic Observations," *Tectonophys.*, **3**, 547–557 (1967).
48. Parker, R. L., "Composition of the Earth's Crust," in *Data of Geochemistry*, 6th ed., U.S. Geological Survey Prof. Paper 440-D, 1967, 19 pages.
49. Pecora, W. T., "Carbonatites—a Review," *Geol. Soc. Am. Bull.*, **67**, 1537–1556 (1956).
50. Pettijohn, F. J., *Sedimentary Rocks*, 2nd ed., Harper and Row, New York, 1957, 718 pages.
51. Pettijohn, F. J., "Chemical Composition of Sandstones—Excluding Carbonate and Volcanic Sands," in *Data of Geochemistry*, 6th ed., U.S. Geological Survey Prof. Paper 440-S, 1963, 21 pages.
52. Pevear, D. R., "The Estuarine Formation of the United States Atlantic Coastal Plain Phosphate," *Econ. Geol.*, **61**, 251–255 (1966).
53. Poldervaart, A., "Chemistry of the Earth's Crust," in *Crust of the Earth—a Symposium*, A. Poldervaart (Ed.), Geological Society of America Special Paper 62, 1955, pp. 119–144.
54. Redfield, A. C., B. H. Ketchum, and F. A. Richards, "The Influence of Organisms on the Composition of Seawater," in *The Sea*, Vol. 2, M. N. Hill (Ed.), Interscience, New York, 1963, pp. 26–77.
55. Ronov, A. B., and A. A. Yaroshevsky, "Chemical Composition of the Earth's Crust," in *The Earth's Crust and Upper Mantle*, P. J. Hart (Ed.), American Geophysics Union, Geophys. Mon. 13, 1969, pp. 37–57.
56. Sederholm, J. J., "The Average Composition of the Earth's Crust in Finland," *Finl. Comm. Geol. Bull.*, No. 70, 20 pages (1925).
57. Shaw, D. M., G. A. Reilly, J. R. Muysson, G. E. Pattenden, and F. E. Campbell, "An Estimate of the Chemical Composition of the Canadian Precambrian Shield," *Can. J. Earth Sci.*, **4**, 829–853 (1967).
58. Sheldon, R. P., "Paleolatitudinal and Paleogeographic Distribution of Phosphorite," in *Geological Survey Research*, 1964, U.S. Geological Survey Prof. Paper 501-C, 1964, pp. C106–C113.
59. Stewart, F. H., "Marine Evaporites," in *Data of Geochemistry*, 6th ed., U.S. Geological Survey Prof. Paper 440-Y, 1963, 53 pages.
60. Sujkowski, Zb. L., "Average Chemical Composition of the Sedimentary Rocks," *Am. J. Sci.*, **250**, 360–374 (1952).

61. Taylor, S. R., "Abundance of Chemical Elements in the Continental Crust: A New Table," *Geochim. Cosmochim. Acta*, **28,** 1273–1285 (1964).
62. United Nations, *Proceedings of the Seminar on Sources of Mineral Raw Materials for the Fertilizer Industry in Asia and the Far East*, United Nations Mineral Resources Development Series, No. 32, 1968, 392 pages.
63. Vogt, J. H. L., "On the Average Composition of the Earth's Crust, with Particular Reference to the Contents of Phosphoric and Titanic Acids," Skrifter, Norske Videnskaps-Akad., Oslo I, Mat. Nat. Kl., 1931, 7, pp. 1–48 (1931).
64. Warin, O. N., "Theories on the Genesis of Marine Phosphorites," in *Proceedings of the Seminar on Sources of Mineral Raw Materials for the Fertilizer Industry in Asia and the Far East*, United Nations Mineral Resources Development Series, No. 32, 1968, pp. 138–144.
65. Warin, O. N., "Deposits of Phosphate Rocks in Oceania," in *Proceedings of the Seminar on Sources of Mineral Raw Materials for the Fertilizer Industry in Asia and the Far East*, United Nations Mineral Resources Development Series, No. 32, 1968, pp. 124–132.
66. Wedepohl, K. H., "Composition and Abundance of Common Igneous Rocks," in *Handbook of Geochemistry*, Vol. 1, K. H. Wedepohl (Ed.), Springer-Verlag, New York, 1969, pp. 227–249.
67. Wedepohl, K. H., "Composition and Abundance of Common Sedimentary Rocks," in *Handbook of Geochemistry*, Vol. 1, K. H. Wedepohl (Ed.), Springer-Verlag, New York, 1969, pp. 250–271.
68. Wickman, F. E., "The 'Total' Amount of Sediments and the Composition of the 'Average Igneous Rock'," *Geochim. Cosmochim. Acta*, **5,** 97–110 (1954).

The Weathering of Phosphate Deposits—Geochemical and Environmental Aspects*

3

Z. S. ALTSCHULER

U.S. Geological Survey, Washington, D.C.

The roles of weathering are of fundamental importance in evaluating the pathways of phosphorus and a number of other critical elements in our environment. The dissolution of accessory igneous apatite is the primordial route through which phosphorus has been mobilized to the hydrosphere and made available for all subsequent biological activity and all sedimentary deposition. The weathering of phosphate deposits is more readily studied, however, and the subaerial regeneration of secondary phosphate minerals from these is a cause for the fixation, the concentration, or the dispersal of phosphate in the immediate environment of our regoliths and soils. The processes of alteration and regeneration of phosphate minerals also mediate the concentration or dispersion of many trace elements that fix in apatite or other phosphate minerals by solid solution or adsorption. To the extent that these elements fix preferentially in phosphate minerals, as do fluorine, strontium, or uranium, weathering may exert the major control on their distribution in phosphatic terrains.

This review surveys some of the effects of weathering through a presentation of chemical and mineralogical data on weathered, rich deposits of apatite. This restriction is realistic as a few varieties of apatite comprise the only widespread and commonplace primary phosphates in the lithosphere. In the form of fluorapatite [$Ca_{10}(PO_4)_6(F,OH)_2$], it is the accessory mineral of igneous rocks and the primary enriched phase of carbonatites. Francolite, the carbonate–fluor-bearing variety [$Ca_{10}(PO_4)_{6-x}(CO_3)_x(F,OH)_{2+x}$], comprises the primary marine phosphate in sedimentary rocks. Dahllite, the carbonate–hydroxylapatite [$Ca_{10}(PO_4)_{6-x}(CO_3)_x(OH,F)_{2+x}$], makes up the mineral matter of fossil bone, and hydroxylapatite [$Ca_{10}(PO_4)_6(OH)_2$] occurs in guano-altered limestones. Moreover, as fresh bone and teeth are related to either dahllite or hydrosylapatite, the alteration of geological apatites during weathering has a fundamental bearing on the physiology and etiology of bone.

* Publication authorized by the Director, U.S. Geological Survey.

This report also includes data on insular and coastal deposits of phosphate resulting from the alteration of igneous rocks and limestones by meteoric solutions of guano. Many hydrothermal or pegmatitic phosphate minerals undergo conversions or replacements through the combined effects of deuteric and meteoric solutions. Although these occurrences are mineralogically interesting (see Chapter 6) they are extremely limited volumetrically and areally, and therefore of little significance in the environment.

The conditions and the products of the weathering of phosphates may be understood to a limited extent from laboratory studies of calcium orthophosphates in aqueous media, and from the compositions and settings of apatite and unweathered phosphorites. These subjects are reviewed briefly before undertaking a critical analysis of the data on weathered deposits.

THE SYSTEM $CaO—P_2O_5—H_2O$ AT LOW TEMPERATURE

Critical reviews of the literature on the synthesis of calcium phosphates are presented in Eisenberger *et al.* (63), Van Wazer (216), and Mooney and Aia (157). The phase diagram of the system $CaO–P_2O_5–H_2O$ (Fig. 1) is that of Van Wazer (216) with the addition of octocalcium phosphate, as per Mooney and Aia (157). It is based principally on the early definitive studies of Bassett (23).

Although five phases are shown to be stable at 25°C in Fig. 1, of these only brushite ($CaHPO_4{\cdot}2H_2O$), monetite ($CaHPO_4$), and hydroxylapatite [$Ca_{10}(PO_4)_6(OH)_2$] occur as minerals. Pure $Ca_3(PO_4)_2$ (tricalcium phosphate) is not stable at 25°C or known as a mineral (70). The high-temperature mineral whitlockite, known from pegmatites (76) and chondrites (81), though often described as β-tricalcium phosphate, always contains minor amounts of magnesium and iron. The low-temperature subaerial variety, martinite, from guanitized limestones and dental calculi, contains carbonate as well (73). Tricalcium phosphate is apparently stabilized to form whitlockite and dental calculi by small substitutions of Ca^{2+} by Mg^{2+}, Mn^{2+}, or Fe^{2+} (207, 95). Rowles (180) precipitated whitlockite in the presence of magnesium and has shown that the whitlockite content of human calculi varies with magnesium. Keppler (112, 113) has proposed the formula $Ca_9(PO_4)_6(Ca,Mg,Fe)HPO_4$ to account for the compositional deviations of some occurrences as well as the symmetry requirements of the whitlockite structure. Nevertheless, a number of analyses conform to the $(Ca,Mg,Fe)_3$-$(PO_4)_2$ formulation for largely calcic whitlockite (82, 214) which is adhered to here in the absence of new structural information.

Octocalcium phosphate has been synthesized by hydrolysis of brushite in sodium acetate solutions (37, 165). It has been reported to be stable at 43°C

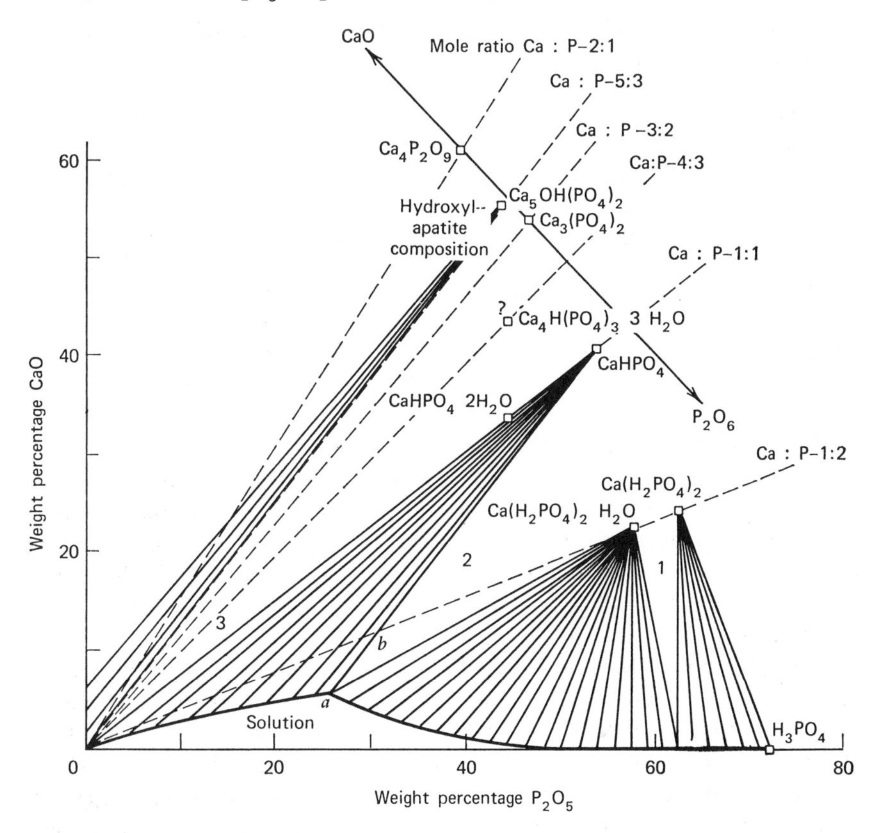

Figure 1. The system $CaO–H_2O–P_2O_5$ for the calcium orthophosphates at 25°C. From Van Wazer (216).

in phase studies by d'Ans and Knütter (13), and as a minor constituent of calculi (205). However, it is not established as an equilibrium phase at 25°C, or known as a mineral, although it should be looked for.

The naturally occurring calcium phosphates are given in Table 1, together with their thermodynamic solubility products (pK_{sp}) and those for β-tricalcium phosphate, octocalcium phosphate, and calcite.

The striking feature of the solubilities of the calcium phosphates is their pronounced decline with increasing basicity. The solubility of hydroxylapatite is a subject of controversy (164). Rootare and co-workers (178) have explained the compound's aqueous behavior by studies showing a surface-limited hydrolysis, as follows:

$$Ca_{10}(PO_4)_6(OH)_2 + 6H_2O = 4[Ca_2(HPO_4)(OH)_2] + 2Ca^{2+} + 2HPO_4^{2-}$$

$$4[Ca_2(HPO_4)(OH)_2] = 8Ca^{2+} + 4HPO_4^{2-} + 8(OH)^-$$

TABLE 1 Thermodynamic Solubility Products of Pertinent Calcium Phosphates

Mineral	pK_{sp} (25°C)	Ref.
Brushite $CaPHO_4 \cdot 2H_2O$	6.56	159
Monetite $CaHPO_4$	6.56	159
Whitlockite $(CaMg)_3(PO_4)_2$	Not known	
β-$Ca_3(PO_4)_2$	33.2	129
Octocalcium phosphate $Ca_8H_2(PO_4)_6 \cdot 3H_2O$	46.9	160
Hydroxylapatite $Ca_{10}(PO_4)_6(OH)_2$	115	50
Calcite $CaCO_3$	8.4	

These reactions have been confirmed, in essence, by LaMer (127), Deitz (59), and Corsaro *et al.* (54). As seen from these equations, and noted experimentally, the ratio of the sums of the calcium and phosphate ions at equilibrium equals the 10:6 ratio of apatite. Thus these intermediary reactions do not prevent the determination of a fixed solubility product for hydroxylapatite, and Clark's value, by both solution and precipitation studies, may be accepted.

The calcium phosphates that are more acid than apatite are all incongruently soluble. They undergo progressive hydrolysis to more basic compounds, liberating more phosphate than calcium to solution. This feature has been the basis of the low-temperature synthesis of whitlockite from $CaHPO_4$ (monetite) (37), apatite from $CaHPO_4$ (185), and apatite from $Ca_3(PO_4)_2$ (whitlockite) (99, 74). The precipitation of brushite is limited to acid solutions, even with excess CaO. Increase of pH above 7.0, or increase of CaO content (at equilibrium saturations of the respective phases), causes hydrolysis of brushite and tricalcium phosphate to hydroxylapatite. Kazakov (110) precipitated brushite below pH 6.4, a phase like whitlockite in the pH range 6.4 to 7.0, and hydroxylapatite above pH 7.0, by hydrolyses and also directly from solution.

These hydrolysis relations are indicated in Van Wazer's (216) phase diagram (Fig. 1) by the paths along which lines of constant Ca/P ratio converge to the origin in the apatite stability field. Those for compounds more acid than apatite (Ca/P = <1.67) cross progressively more basic regions ending in apatite, and those for more basic compounds cross more acid regions.

The stability relations of the calcium phosphates are altered in the presence of fluorine. That fluorapatite is markedly less soluble than hydroxylapatite is abundantly clear in synthesis, weathering, and physiological studies. Mere traces of fluoride ion cause apatite to precipitate at lower pH, and lower

CaO contents, than hydroxylapatite (110, 195, 156, 130). Fossil bone becomes increasingly more fluoridated with exposure to normal low-fluorine groundwater. The fluorine content of fossils was used to demonstrate relative geologic age and "intruded" fossils by Carnot (46) and to expose the Piltdown forgery (167, 221).

All marine apatites are francolites, containing integral carbonate, and fluorine in excess of the theoretical F/P_2O_5 ratio of 0.09 for fluorapatite. Kramer (117) has shown experimentally that francolite is more soluble than normal fluorapatite. Despite this, there is not a single documented example of the marine or subaerial precipitation of fluorapatite.

NATURE AND COMPOSITION OF PHOSPHORITES

Marine Phosphorite

The most widespread deposits of phosphorite, and the most important environmentally, are those of marine origin. Although they vary in occurrence, lithologic and faunal association, and degree of concentration, they are uniquely limited in essential mineralogy to carbonate-fluorapatite. This despite their diverse genetic natures—as oolites, granular detrital pellets, or fecal pellets; fragments of fossil bone, phosphatized shell or mold, or phosphatized limestone; concretions, or rare aphanitic layers of precipitated apatite (4).

The largest deposits are relatively rare basinal or platform accumulations, of blanket form, in which the phosphate reserves of individual fields range from hundreds of millions to billions of tons (see McKelvey *et al.* (140) and Emigh, Chapter 4, this volume).

In these deposits P_2O_5 may reach exceptional concentrations of 20 to 30% (54 to 80% apatite) in extensive zones 1 to 3 m thick. Such zones may be intercalated with phosphate-rich mudstones and dolomites (5 to 15% P_2O_5) as in the Phosphoria and Karatau Formations, or be separated by largely barren beds, as in Morocco or Israel. The Oulad-Abdoun basin, one of several phosphatic basins in Morocco, is roughly 90 by 60 km in size. Its phosphatic formation is 20 to 50 m thick, and contains five distinct and continuous beds of pelletal apatite, separated by lean or barren cherts, limestones, and marls. The phosphorites generally range from 1 to 2 m in thickness, and from 25 to 30% P_2O_5. The Bone Valley Formation of Florida lies about 10 ft below the surface almost continuously for more than 2000 mi^2. It is generally 25 to 35 ft thick and its lower 15 to 20 ft consists of clastic apatite averaging between 40 and 45% of the rock, and interlayered with quartz sand and montmorillonite clay.

Marine phosphate is much more common, however, in a variety of lean deposits. In many shales and limestones it occurs as dispersed, structureless, or spherulitic concretions or small nodules, commonly associated with organic matter, siderite, and pyrite, and apparently of diagenetic origin in reduced muds. In common sediments, P_2O_5 may be enriched to 1 to 5% in zones of coprolites, replaced foraminifera, phosphatic brachiopods, conodonts, and glauconite sand with apatite pellicles. Such zones may be several feet thick and may recur in repetitive sequences. Marine phosphate is concentrated also in exceptionally thin and extensive lag deposits of mixed detrital and biofragmental nature, often rich in sharks teeth and whale earbones, at diastems and unconformities (88, 38). These lean deposits may be upgraded to economic and environmentally significant proportions by weathering or reworking (6).

Marine phosphorite is characteristically limited in mode of occurrence, as in mineralogy. It occurs most commonly as granular, sand-size, water-worn ovules and oolites, in thin beds, which are persistent areally, in grade, and in pellet concentration. This is a consequence of its characteristic concentration in milieus and epochs of restricted detrital deposition. Deposits of rich phosphorite, or formations of lean phosphatic zones, underlie large areas and may be exposed for tens of miles along strike, exerting significant environmental influence. There are notable exceptions due to penecontemporaneous submarine reworking, or accumulation on undulating marine floors, causing fluctuations in grade, and lensing or basinal segregations. But even these deposits show a high degree of regularity, as illustrated in the succession of separate synclinal deposits in the Negev of Israel (25, 225), each individually small, but related stratigraphically and in mode of occurrence, and comprising a major field.

Synoptic discussions of the genesis and classification of phosphorites are given in Kazakov (111), McKelvey *et al.* (140), Himmelfarb (101), Bushinski (40), United Nations (212), and Gulbrandsen (92). Many detailed critical studies of important deposits are also reviewed and listed in United Nations (212) and Bushinski (39).

A dominant theme in phosphorite literature, first stressed by Kazakov (111), and expanded by McKelvey (138) and Sheldon (189), is the occurrence of major deposits in two broad tectonic settings, and corresponding lithofacies: those of the geosynclines, characterized by dark, organic-rich shales, cherts, and dolomites, with subordinate limestone, in sections up to several hundred feet thick (142, 39); and those of continental and coastal platforms, characterized by sequences of phosphatic limestones and marls, generally organic poor, quartzose, and clayey, in some cases, accompanied by cherts, and typically less than 100 ft thick (49). The subdivision may be controversial and forced when applied to individual deposits, as the lithologies are neither invariant nor exclusive, the platforms may be initially warped or folded later, or the locus of deposition may straddle or shift from the miogeosynclinal basin to the adjoining platform. Thus, the Phosphoria Formation, from Idaho to its platform facies in Wyoming, becomes thinner, coarser, more quartzose, less argillaceous, and calcareous rather than dolomitic (142, 189). The Moroccan deposits occur as very extensive flat-lying plates of intercalated limestones, marls, and limy phosphorites, with some interbedded cherts, but they are notably poor in shale, quartz sand, and organic matter (34). Whereas they now lie in broad basins west of the south and middle Atlas Mountains, they are classified here as platform deposits [see also Salvan (182) and Bushinski (40)].

Apatite Composition. The fine-grained, microcrystalline phosphate mineral of marine sediments is invariably carbonate-fluorapatite. In contrast to well-crystallized fluorapatite of igneous origin, this variety is generally deficient in P_2O_5 by 3 to 6 wt.%, has excess F and OH of 0.5 to 1.5 wt.%, and contains 2 to 3 wt.% of nonleachable carbonate (see analyses A, B, C, and D of Table 2). It has been called collophane and is essentially like the eucrystalline "francolites" and "staffelites" from Wheal Franco, Devon; Staffel, Germany; and Richtersveld, South Africa (89, 183, 60). Furthermore, it is characteristically smaller in unit-cell dimensions than fluorapatite (7). The structural difference must be related to the major and equally characteristic chemical deviations which are expressed in the structural formula

$$Ca_{10}(PO_4)_{6-x}(CO_3F)_x(F, OH)_2$$

in which x is generally close to 1. This formulation is based on the proposal of Borneman-Starynkevich and Belov (32) that $[FCO_3]^{3-}$ substitutes for $[PO_4]^{3-}$. It is in closest accord with the compositions of sedimentary apatites of very widespread provenance (146, 8, 91, 196, 222). Some marine apatites like that of the Phosphoria Formation (Analysis B, Table 2) contain notable substitution of sulfate for phosphate, apparently coupled with the replacement of calcium by sodium (91).

From the synthesis of apatites, as well as structural and compositional studies of natural materials, it is known that apatites can sustain a great variety of substitutions in the general formula $A_{10}(XO_4)_6Z_2$, as follows:

A = Ca, Sr, Mn, Pb, Mg, Ba, Zn, Cd, and possibly Fe and Cu among divalent elements,
= Na, K, Rb, and possibly Ag among monovalent elements,
= Sc, Y, R.E. (rare earths), Bi, and possibly Ti among trivalent elements,
= U, and probably Th and Zr among quadrivalent elements.

XO_4 = PO_4, SiO_4, SO_4, AsO_4, VO_4, CrO_4, BeO_4, probably CO_3F and CO_3OH, BO_4, and possibly GeO_4, AlO_4, and $Fe^{III}O_4$.

Z = F, OH, Cl, Br, I, and possibly O in defect apatites.

TABLE 2 Sedimentary Apatites of Primary (Marine) and Secondary (Continental) Derivations

		A	B	C	D	E	F
	Fluorapatite: Theoretical	Bone Valley Formation, Fort Meade, Florida[a]	Phosphoria Formation, Phosphoria Gulch, Idaho[a]	Morocco	Francolite: Staffel, Germany	Dahllite: Mouillac, France	Phosphorite: Dunnellon, Florida
CaO	55.5	52.7	52.29	53.00	54.88	55.16	51.08
MgO		—	0.06	0.16	0.31	—	0.05
SrO		0.03	0.05	—	—	—	—
Na_2O		0.2	0.77	1.19	—	0.77	0.10
K_2O		0.2	0.14	0.16	—	0.28	0.06
P_2O_5	42.3	37.5	36.82	35.11	37.71	38.57	35.99
V_2O_5			0.14	0.00	0.24	—	0.01
CO_2 total		—	—	4.12	3.36	4.46	2.19
CO_2 nonleached		1.9	1.66	—	—		
SO_3		0.1	1.98	1.40	0.00	0.05	0.01
SiO_2 total		—	—	0.86	—	0.40	5.01
SiO_2 soluble		1.1	0.03	—	—	—	—
Al_2O_3		1.1	—	0.45	—	0.44	0.52

Fe_2O_3[b]		0.9	—	0.12	—	0.34	0.69
F	3.8	4.0	4.05	4.24	4.11	0.19	3.91
Cl	—	—	—	<0.005	0.00	0.02	0.008
H_2O+	—	1.8	1.13	1.63[c]	1.14	0.72	1.67[c]
H_2O-	—	0.7	0.78		0.04	0.48	—
Other		—	—	0.07	—	—	0.057
	101.6	102.2	99.90	102.50	101.79	101.88	101.36
Less							
$O = (F_2, Cl_2, S)$	1.6	1.6	1.70	1.78	1.73	0.08	1.65
Total	100.0	100.6	98.20	100.72	100.06	101.80	99.71

[a] Corrected analysis excluding insoluble residue and carbonate soluble in 0.5 *M* triammonium citrate.
[b] Total iron determined as Fe_2O_3.
[c] Includes organic carbon and nitrogen.

A Apatite pellet composite, marine. Altschuler *et al.* (8, p. 49). Analyst: R. S. Clarke, Jr.

B Apatite pellet composite, marine. U.S. Geological Survey, unpublished. Sample from R. A. Gulbrandsen. Analysts: L. Jenkins and J. Budinsky.

C Phosphate rock, marine, commercial concentrate; probably from Kouribga. Jacob *et al.* (104, p. 23). Analysts: authors. (Other represents TiO_2, 0.025; MnO, 0.003; Cr_2O_3, 0.045.)

D Secondary druse in Nassau, Germany phosphorite. Gruner and McConnell (89). Analyst: R. B. Ellestad.

E Cave phosphorite. McConnell (132). Analyst: Ledoux and Company.

F Florida "hardrock phosphate". Subaerial limestone replacement. Jacob *et al.* (104, p. 20). Analysts: authors. (Other represents TiO_2, 0.039; MnO, 0.014; Cr_2O_3, 0.004.)

These are not to be understood as simultaneous or equally permissible. Thus, chlorapatite is structurally different than fluorapatite; Pb, VO_4, and AsO_4 are larger and not completely isomorphous with Ca and PO_4. Many other substitutions are coupled: RE^{3+} and SiO_4^{4-}, for Ca^{2+}, and PO_4^{3-} in britholites, and SO_4^{2-} and SiO_4^{4-} for $2PO_4^{3-}$ in the wilkeite-ellestedite series are examples. [For apatite structures see Naray-Szabo (163), Mehmel (147), Beevers and McIntyre (24), Kay *et al.* (108); for recent developments see Elliot (64); for synthetic replacements see Jaffe (105), Spencer and Uhler (198), Trommel and Eitel (208), Merker and Wondratschelz (149); for natural varieties see McConnell (131, 132), Vasileva (218), Palache *et al.* (172), and for trace elements in purified sedimentary apatites see Altschuler *et al.* (8).]

Phosphorite Composition. The major element composition of phosphorites is illustrated by Table 3, and is best comprehended by comparing Tables 2 and 3. The silica is largely present as quartz, with minor feldspar and accessory minerals. The alumina and the potassium oxide are mainly in clays, and the iron in iron oxide minerals. The shales of the Phosphoria Formation are illitic and kaolinitic (90), but montmorillonite, attapulgite, kaolinite, and chlorite clays are known in phosphorites of both facies (6, 39, 192), as are minor amounts of glauconite pellets, pyrite, and accessions of volcanic material (49). Individual beds or formations of the platform, however, tend to be simple mineralogically and chemically, and obviously calcitic rather than dolomitic.

The analyses in Table 4 portray the trace metal contents determined in comprehensive studies of a geosynclinal and a platform deposit. Although traces of most of these elements have been found in apatites, they may be additionally enriched in phosphorites due to organic matter, sulfide minerals, and accessories like zircon (Zr, Th) or tourmaline (B). Krauskopf (118) and Gulbrandsen (91) have shown unusual concentrations of As, Mo, Ni, V, Zn, Cr, Cu, Sb, and Se in organic-rich shaly phosphorites, presumably tied to organic matter. Borisenko (31) and Blokh and Kochenov (30) have shown atypically high scandium and rare earths in the pyritic and organic-rich bone beds. Some of the contrasts between the geosynclinal and platform assemblages in Table 4 appear to be explainable on this basis. Swaine (200) has compiled data from all sources, and Kholodov (114) and Tooms *et al.* (204) have critically reviewed trace element enrichment (relative to crustal abundances) in phosphorites as a class. Tooms *et al.* (204), using Swain's data for phosphorites of all types, find Ag, As, Cd, Cr, I, La, Mo, Pb, Sb, Se, Sn, Sr, U, and Y enriched by a factor of more than 2 over crustal abundances. Kholodov lists B, Ba, Be, Co, Li, Pb, Ni, all RE, V, Zn, and Zr, in addition, as enriched over crustal abundances.

TABLE 3 Compositions of Representative Primary Phosphorites

	Geosynclinal Facies			Platform Facies	
	Phosphoria Formation (Permian), Coal Canyon, Wyoming			Lower Cambrian, China	Leipers Limestone (Ordovician), Tennessee
	A	B	C	D	E
SiO_2	17.7	10.0	58.1	19.74	2.43
TiO_2	0.1	0.1	0.4	0.10	0.04
Al_2O_3	2.5	0.15	9.6	3.12	0.39
Fe_2O_3	1.1	1.0	3.4	1.90	1.27
CaO	28.9	43.9	7.0	41.98	52.08
MgO	9.9	5	1.7	0.21	0.55
Na_2O	0.8	1.0	0.7	0.54	0.28
K_2O	0.9	0.6	3.4	N.D.	0.18
P_2O_5	2.6	29.5	2.6	23.41	11.68
CO_2	30.3	3.0	3.4	5.52	28.22
SO_3	1.0	2.9	2.3	N.D.	1.55
V_2O_5	0.1	0.1	0.1	N.D.	—
H_2O^+	N.D.	N.D.	N.D.	0.24	0.65[b]
H_2O^-	0.6	0.8	1.0	0.33	—
F	0.3	3.1	0.4	2.72	1.12
Cl	—	—	—	0.67	0.01
C	—	—	—	0.62	—
Organic	2.9[a]	4.4[a]	6.1[a]	—	—
Other	—	—	—	0.73	—

[a] Determined as L.O.I.—(CO_2 + H_2O^-); thus includes H_2O^+.
[b] Includes organic C and N.

A Average of 15 quartzose and phosphatic dolomite and limestones layers, Meade Peak Phosphatic Shale Member. Gulbrandsen (90). Analysts: U.S. Geological Survey.

B Average of 15 black phosphorite layers, Meade Peak Phosphatic Shale Member. Gulbrandsen (90). Analysts: U.S. Geological Survey.

C Average of 15 dark, phosphatic mudstone layers, Meade Peak Phosphatic Shale Member. Gulbrandsen (90). Analysts: U.S. Geological Survey.

D Black, calcareous phosphorite, Chinese platform. Bushinski (English Translation) (39). Analysts: Academy of Science, USSR. (Other represents FeO, 0.21; BaO, 0.47; S, 0.05.)

E Phosphatic limestone, blue-rock mine, Gordonsburg, Tennessee. Jacob *et al.* (104, p. 21). Analysts: authors.

TABLE 4 Minor Elements in Representative Primary High-Grade Phosphorites (ppm)

	A Geosynclinal	B Platform
	Phosphoria Formation	Cretaceous, Uzbekistan
Ag	3	2
As	40[a]	—
Ba	100	200
Be	—	2
Co	—	7
Cr	1000	7
Cu	100	2
Ca	—	7
La	300	200
Mn	30	200
Mo	30	2
Nd	300	—
Ni	100	7
Pb	—	7
Sb	7[a]	—
Sc	10	—
Se	10[a]	—
Sr	1000	70
U	90[a]	—
V	300	7
Y	300	20
Yb	10	—
Zn	300	70
Zr	30	20

[a] Arithmetic averages.

A Modal values of spectrographically detected metals in many analyses of phosphatic shales. Gulbrandsen (91, p. 774). Analysts: U.S. Geological Survey.

B Modal values of spectrographically detected metals in many analyses of calcareous phosphorites of Kizyl-Kum. Kapustyanski (107).

Analytical data on the individual rare earths in purified marine apatite are given in Table 5. Altschuler *et al.* (9) have shown that the rare earths inherently accumulated in marine apatite reflect their distribution in seawater, in being relatively depleted in Ce and slightly enriched in the heavier lanthanons. Similar but not identical distributions have been shown for deep-sea nodules (62) and fishbones (17). Semenov *et al.* (187) and Alexiev and

TABLE 5 Spectrographic Analyses of Rare Earths, Yttrium, and Scandium in Primary Apatites and Secondary Phosphates (in ppm)

	Primary		Weathered		
	A	B	C	D	E
La	150	1500	0.8	0.5	400
Ce	120	2700	2.0	1.2	100
Pr	30	500	0.11	0.07	—
Nd	70	1500	1.0	0.45	5
Sm	30	430	<0.3	<0.3	—
Eu	4	20	<0.3	<0.3	—
Gd	14	330	0.3	0.3	5
Tb	4	40	0.17	<0.15	—
Dy	16	300	0.16	0.12	—
Ho	4	50	0.07	0.05	—
Er	21	220	0.12	0.08	—
Tm	2	—	<0.07	0.06	—
Yb	8	100	0.12	0.07	50
Lu	3	N.D.	<0.07	<0.06	—
Y	110	1750	1.25	1.05	600
Sc	3	70[a]	N.D.	N.D.	2

[a] Average of 10 samples.

A Average of three apatite composites, Bone Valley Formation, Florida. Altschuler *et al.* (9, 1967, Table 1). Analysts: S. Berman and F. Cuttitta.

B Average of 67 pyritic and organic-rich fossil fishbone fragments, Maikop and Mangyshlak deposits, U.S.S.R. Kochenov and Zimovieff (115).

C Carbonate–fluorapatite, supergene limestone replacement, Florida "hardrock", Dunnellon, Florida. Altschuler, unpublished. Analysts: S. Berman and F. Cuttitta.

D Hydroxylapatite, guanitized limestone, Mona Island, Puerto Rico. Altschuler, unpublished. Analysts: S. Berman and F. Cuttitta.

E Lateritoid aluminum phosphate, Bone Valley Formation, Clear Springs, Florida. Sample has major quartz, millisite, and crandallite; minor wavellite. Altschuler, unpublished. Analysts: C. Waring and H. Worthing.

Arnaudov (2) give many complete analyses of rare earths in phosphorites showing a range in total $(R.E.)_2O_3$ from 0.01 to 0.16%, and an average content of 0.08%. Their data are in relative rather than absolute percents and therefore not compared in Table 5. The marine phosphorites (in contrast to all phosphorites) are clearly enriched in all the rare earths. Being impure they generally do not show the cerium depletion, and are more comparable to marine shales, thereby reflecting the other mineral constituents present in addition to apatite [see comparisons in Altschuler *et al.* (9)].

Uranium is a characteristic syngenetic trace element in all marine phosphorites. It occurs in the apatite mineral, presumably as a proxy for calcium, as in the most recent and least weathered apatites it occurs principally as tetravalent uranium, a form in which it is almost identical in size to the calcium ion (8, 51). Comprehensive studies of uranium in phosphate deposits are given in Davidson and Atkin (55), Thompson (202, 203), McKelvey *et al.* (141), Cathcart (47), Clarke and Altschuler (51), Altschuler *et al.* (8), and Sheldon (188). Uranium typically makes up 0.005 to 0.02% of marine apatite (Table 6). Uranium is also readily emplaced postdepositionally in apatite from circulating groundwater, as in bones and isolated pebbles and concretions, which may attain 0.*X*% uranium (55, 1, 29), or by re-exposure to seawater, as in marine reworking, which may enrich pebbles to as much as 0.1% uranium. Uranium(IV) is readily oxidized to uranium(VI) by heating, oxidative weathering, and by autoxidation during radioactive decay (51, 8), which facilitates leaching of uranium from apatite (116) during weathering.

Phosphate Deposits in Carbonatites

Massive accumulations of igneous apatite occur in carbonatites of volcanic complexes and in their shallow subvolcanic igneous rocks. [Pecora (173), Deans (57), and papers in Tuttle and Gittins (211) give critical summaries and extensive bibliographies.] The apatite-rich masses form cone sheets, dikes, and plugs, or arcuate lenses and sheet-like segregations within these structures, in the volcanic carbonatite complexes of Africa and South America (57), and similar structures in the plutons of the Kibina syenite massif of the Kola Peninsula. Apatite is a commonplace accessory in most rocks of such complexes (173), but its unusual accumulation is most common in nepheline and magnetite-bearing syenites, pyroxenites, ijolites, and calcitic sövites, often rich in pyrochlore (57, 84) or, as in the Kola and Synnir complexes, rich in biotite, sphene, and olivine. In many carbonatites sövitic rocks have been intruded and secondarily mineralized by rare earth- and strontium-rich ankeritic phases to form heterogeneous and complex lithologies greatly enriched in many exotic and rarely juxtaposed minerals

TABLE 6 Total and Tetravalent Uranium Content of Sedimentary Apatites (8)

Sample	Material	Total U (%)	U(IV) (%)	U(IV)/U Percentage
Bone Valley Formation (Pliocene)				
B.L.-1	Black pebble	0.0011	0.0010	91
B.L.-2	Dark pebble	0.0089	0.0056	63
B.L.-3	Fine pebble composite	0.016	0.013	81
Ho-14	4–8 mm pebble composite	0.032	0.016	50
Ho-15	4–8 mm pebble composite	0.022	0.011	50
Ho-16	4–8 mm pebble composite	0.015	0.007	42
Ho-17	4–8 mm pebble composite	0.021	0.010	48
Va-7	Apatite pellets	0.011	0.005	45
Va-7a	Apatite pellets	0.009	0.006	67
WA-10	Apatite pellets	0.0075	0.003	40
PV-5	Apatite pellets	0.007	0.003	43
Hawthorn Formation (Miocene, recently reworked on Gulf Coast beaches)				
GP-1	Apatite pellets	0.0067	0.0039	58
GP-2	Apatite pellets	0.0059	0.0037	63
Tennessee phosphates, Bigby Limestone (Ordovician)				
Ak-2	Limestone, phosphatic	0.00004	0.00001	25
Ak-1	Brown-rock phosphate (Cenozoic weathering)	0.00074	0.00002	3
Moroccan phosphates (Senonian)				
Mor-3	A daily production average	0.012	0.0018	15
Mor-5	A daily production average	0.012	0.0015	13
Mor-11	A daily production average	0.008	0.0006	8
Phosphoria Formation (Permian)				
Conda-300	Apatite pellets	0.0063	0.0015	24
RAH-185	Apatite pellets	0.017	0.003	18
Pacific Ocean, Southern California (post-Miocene)				
69	Phosphate nodules	0.0089	0.0061	69
106	Phosphate nodules	0.0068	0.0040	59
127	Phosphate nodules	0.0041	0.0028	68
158	Phosphate nodules	0.0081	0.0050	62
162	Phosphate nodules	0.0125	0.0093	74
183	Phosphate nodules	0.0051	0.0028	55
Fossil bone				
1	Hells Creek Formation	0.015	0.0004	3
RW-4468	Taxpayer sample	{ 0.078	0.054 }	70–76
		{ 0.071	0.050 }	
W-3841	Taxpayer sample	0.85	0.40[a]	—
		0.82	0.50	61
Phosphatic arkose				
	Gas Hills, Wyoming	0.74	0.14[a]	—
			0.17	23

[a] Determination from HCl solution; uncorrected for 20% deficiency in recovery.

and elements (173). The unusual nature of these source rocks for weathering is documented in Garson (84). (See Table 7.)

The apatites of carbonatites are generally fluorapatites (see Table 8), somewhat enriched in rare earths and strontium. However, unusual varieties of strontian apatite (analysis 4), rare earth–strontian apatite (belovite, analysis 3), cerian silicate–apatite (britholite), and yttrian silicate–apatite (abukumalite) are not uncommon in smaller vein and dike occurrences within these terrains.

In its common accessory form the apatite composes but a few percent of the alkaline igneous rocks. Individual pipes and sheets containing 5 to 25% apatite are common, however, and richer bodies are known in many carbonatite bodies. In the Tundulu complex, Garson (84) has documented individual cone sheets with 90% apatite. In the Synnir pluton apatite composes as much as 80% of many zones pyroxenites (15).

Individual apatitic masses in carbonatites range from square yards to perhaps several square miles (Kola Peninsula) in outcrop area. But the larger parent bodies and terrains so enriched may extend over tens of square miles, and the volumes of apatite in such complexes may aggregate to tens of millions of tons, and in some to hundreds of millions (211). In the unique Khibina massif, deposits averaging 65 (high grade) and 45% (low grade) apatite and aggregating to reserves of 2 billion tons occupy an arcuate belt 10 mi long.

WEATHERING OF PHOSPHATE DEPOSITS

We may consider the geochemistry of weathered phosphate under a simple scheme based on the two parameters of parent lithology and weathering intensity. Primary sedimentary phosphates may be grouped into calcareous, argillaceous, marly, and siliceous types. However, as chert and quartz are relatively insoluble compared to calcite and apatite our examination reduces essentially to a study of two lithologies, calcareous and argillaceous phosphates, as they are affected by variations in weathering intensity. One of the fundamental notes that emerges is that the mineralogies and even the textures and morphologies of the weathered deposits are similar, under comparable degrees of weathering in chemically similar lithologies, whether igneous or sedimentary, insular or terrestrial, and regardless of the climate!

In calcareous phosphorites, groundwater alteration is controlled by bicarbonate equilibria, the regime is alkaline to neutral, and the weathering products are residual, precipitated, or metasomatic apatites, depending on the intensity or duration of weathering (see Fig. 2). In argillaceous terrain, acid meteoric waters are not as strongly buffered, and acid groundwater

TABLE 7 Chemical Analyses and Modes of Apatite-Rich Rocks from Carbonatites in Malawi (84)

	1	2
SiO_2	1.40	17.30
TiO_2	0.00	0.49
Al_2O_3	0.25	2.65
Fe_2O_3	3.97	14.10
FeO	—	—
MnO	1.80[b]	2.52
MgO	9.24	0.40
CaO	17.46	28.65
Na_2O	N.D.	0.60
K_2O	N.D.	1.05
SrO	18.24	0.33
BaO	2.72	0.41
Rare earth oxides + ThO_2	8.00	2.43
Nb_2O_5	Trace	0.91
H_2O+	0.00	2.85
H_2O-	—	0.82
P_2O_5	5.15	20.80
CO_2	29.80	1.90
F	N.D.	2.27
SO_3[a]	1.18	0.22
Zn	0.62	N.D.
		100.70
Less ($O = F_2$)		0.95
Total	99.83	99.75
Apatite	2.18	48.91
Quartz	1.40	9.36
Fe and Mn oxides and goethite	—	19.50
Feldspar	0.0	11.91
Calcite, dolomite, ankerite	55.25	2.34
Strontianite	25.36	0.44
Monazite	7.15	—
Bastaesite	—	3.29
Florencite	5.00	—
Barite	3.01	0.70
Fluorite	—	0.70
Sphalerite	0.92	—
Anatase	—	0.49
Pyrochlore	Trace	1.36
Clay minerals	—	1.00

[a] Total S as SO_3.
[b] MnO_2.

1 Strontianite-rich sövite, Kangankunda. Analysts: R. Pickup and N. Cogger.
2 Apatite rock, northeast slope Nathace Hill. Analyst: R. Pickup.

Table 8 Apatites from Carbonatites

	Primary			Residual			Replacement
	1	2	3	4	5	6	7
CaO	54.28	54.20	5.23	51.04	55.0	54.60	55.35
MgO	0.10	—	0.16	0.10	—	—	0.14
BaO	0.00	N.D.	0.96	—	—	—	—
SrO	1.60	0.62	33.60	4.72	(0.8[a])	(0.41[a])	—
Na_2O	0.15	0.26	3.60	—	—	—	—
K_2O	0.04	0.01	0.20	—	—	—	—
RE_2O_3	0.68	0.84	24.20	0.48	—	0.32	—
P_2O_5	40.67	41.80	28.88	40.20	41.3	41.85	39.55
V_2O_5	0.15	—	—	—	—	—	—
SiO_2	0.04	—	0.20	—	0.9	—	—
Al_2O_3	0.33	—	0.00	—	—	0.19	—
Fe_2O_3	0.20	—	0.60	0.45	—	0.28	0.11
SO_3	—	—	1.12	—	—	—	—
CO_2	N.D.	0.20	—	N.D.	0.2	0.10	1.80
F	2.79	1.70	0.00	3.40	4.1	2.13	3.79
Cl	0.00	0.10	—	0.05	0.05	—	—
H_2O+	0.13	0.10	0.89	0.59	0.12	0.20	0.53
H_2O-	—	—	—	—	—	—	0.30
Insoluble	—	—	—	0.10	—	—	0.05
	101.16	99.83		101.13	101.67	99.67	101.62
Less ($O = F_2, Cl_2$)	1.18	0.74		1.44	1.8	0.90	1.60
Total	99.98	99.09	99.64	99.69	100.87	99.77	100.72

[a] Later analysis, not included in total.

1 Apatite, Kukisvumchorr, Kola Peninsula. Vladovetz (220, p. 74). Analyst: V. I. Vladovetz.
2 Apatite, Panda Hill, Tanganyika. Van der Veen (213, p. 81), as in Deans (57, p. 393).
3 Belovite (Sr–R.E. apatite), Kola Peninsula. Borodin and Kazakova (33, p. 613). Analyst: M. F. Kazakova.
4 Strontian apatite, residuum on Nkombwa Hill, Zambia. Deans (57, p. 393). Analyst: R. Pickup.
5 Apatite, limonitic weathered zone, on Lueshe carbonatite, Kivu, Congo. Brasseur *et al.* (36, p. 76). Analysts: authors. (Contains Na, 0.1–0.2; Mg, Ba, Al, Fe, each 0.02–0.04.)
6 Apatite concentrate, residual soil, Sukulu, Uganda. Fleming and Robinson, 1960, p. 971, as in Deans (57, p. 393).
7 Francolite, phosphatized zone, Busumbu, Uganda. Davies (56, p. 143). Analyst: author.

alteration takes place, creating lateritic weathering zones of alkaline earth aluminophosphates (crandallite and millisite) and, eventually, simple aluminum phosphates (wavellite and augelite). In marly terrains, or in formations containing intercalated limestones (or dolomites) and shales, mild or short-term weathering causes residual enrichment of apatite, as has been noted on surficial exposures of the Phosphoria Formation (139, 67) and the Kara Tau phosphorites (39). Once decalcified, marly terrains may be replaced by aluminum phosphates, and the superimposition of an extensive aluminum phosphate zone over a deeper complex of residual and replacement apatites has been described in the Bolshie Dzhebarty phosphorites of the Eastern Sayan district in Siberia (28), the Bellona Island deposits (224), and areas within the hardrock field of peninsula Florida (69).

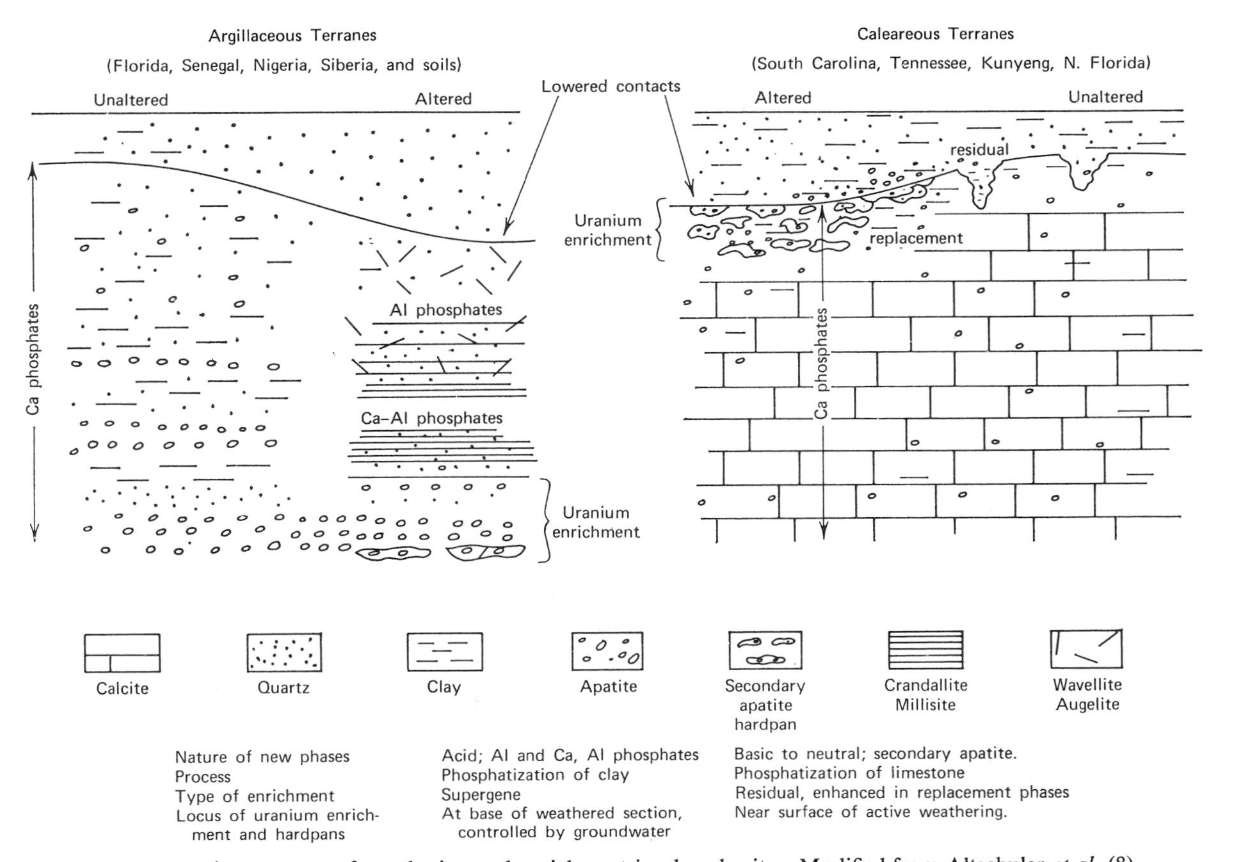

Figure 2. Contrasting patterns of weathering and enrichment in phosphorites. Modified from Altschuler *et al.* (8).

The two major groups of weathered calcareous phosphorites and lateritic aluminous phosphorites are described below. Insular phosphorites, though sharing many of the features of the preceding classes, are treated separately due to their relations to deposits of guano.

The literature on weathered phosphate deposits is quite large. The papers of Smith and Whitlatch (197), Hutchinson (103), McKelvey *et al.* (140), Malde (145), White and Warin (223), Altschuler *et al.* (6), Deans (57), United Nations (212), and Zanin (228) contain extensive reports on specific areas and comprehensive bibliographies.

Weathered Calcareous Phosphorites

Marine Phosphorites. The action of acid rainwater on limestone terrains produces a bicarbonate-rich, moderately alkaline groundwater (83, 119). The presence of apatite in the limestone would not greatly modify this consequence in view of the much greater solubility of calcite and the stability of basic calcium phosphates in alkaline solutions. Analyses of Florida groundwater in phosphatic carbonate terrains not rich in sulfate show a dominance of $[HCO_3]^{-1}$, and a pH range that is commonly 7.5 to 8.0 (21, 93). In such terrains apatite appears not to dissolve in incipient or mild weathering, but concentrates residually on exposed surfaces and in places of enhanced drainage, due to the differential removal of calcite. Such residual concentration is noticeable over all outcrop surfaces. Beds in the Phosphoria Formation which contain 29% P_2O_5 at depth, contain 32 to 34% P_2O_5 on surface exposures (139). Residual deposits encircle the outcrops on hillsides as in the "rim" and "collar" deposits of Tennessee, and fill joints enlarged by solution as in the "cutters" of the Tennessee or the karrenfelds (basins of solution-enlarged joints and intervening pinnacles) of the Florida hardrock fields (197, 219, 69).

In both the Tennessee brown-rock and the Florida hardrock fields secondary deposits are related to post-Miocene uplift of the limestone plateaus. Such uplift induces a lowering of the groundwater table and a renewal of stream incision, causing an acceleration of infiltration and of underground limestone solution. The influence of tectonic control is marked in the distribution of the residual deposits. The major deposits in Tennessee occur as discontinuous blankets in the major river valleys and their tributaries. They decline in grade, thickness, and continuity in the interfluve and upland areas; in headwater areas rim deposits are more prominent (197).

Indicative of the lack of solution of apatite, residual deposits are characteristically clastic, being composed of the "primary" apatite pellets and pebbles inherited from the parent limestone and dolomite. Analyses of purely residual phosphorites show little chemical change other than the removal of $CaCO_3$ or $CaMg(CO_3)_2$ and relative increase in the other constituents (197).

In extensively weathered calcareous phosphorites patchy discontinuous deposits of massive and aphanitic apatite underlie the purely residual mantles, or impregnate their lower zones (see Fig. 2). The secondary textures are reflected in the terminology of the mining districts where such deposits are called "lump rock" or "plate rock" or "nodular" phosphate. These masses generally lack distinct lower boundaries and grade downward into unaltered limestone. Varieties of banded, precipitated apatite also occur, commonly overlying or plastered upon the patches of massive apatite or the limestone. These deposits are collectively an instance of the solution of apatite in the upper soil horizon and its supergene transport and fixation at depth, by metasomatic replacement of limestone or by direct precipitation. They are considered collectively as replacement deposits, as even the precipitated varieties are derived from the interaction of phosphate-enriched groundwater with the solution products of limestone.

The following sequence of chemical processes is inferred by the writer from the zonation and from the textural varieties of residual and replacement deposits.

1. Apatite-rich mantles accumulate upon progressive decalcification of the upper soil zones.
2. Acid meteoric water is thus no longer substantially buffered by calcite and the groundwater may be slightly acid to slightly alkaline (pH 6.5 to 7.5).
3. Meteoric and groundwaters dissolve apatite and become enriched in P_2O_5 in descending through residually accumulated apatite.
4. Limestone encountered at depth dissolves, enriching groundwater in Ca^{2+} and HCO_3^-, and raising the pH to the probable range of 7.5 to 8.5.
5. Due to the increased concentration of Ca^{2+} and the increase of pH, apatite precipitates at the locus of reaction with limestone, resulting in the metasomatic replacements known as lump rock or plate rock. Where precipitation is from the ambient sclutions it results in the banded coatings.

The underground solution of limestone throughout this process enhances collapse and brecciation and causes an intermingling of precipitated and replaced varieties, and an infiltration of the overlying clastic residues. This is seen in descriptions of fractured and recemented blocks of "plate rock" and in the pockets of secondarily cemented clastic apatite, in which the cements are younger generations of apatite [see Smith and Whitlatch (197) for Tennessee; literature and authors summarized in Malde (145) for South Carolina; Espenshade and Spencer (69) and Vernon (219) for northern Florida; and Bliskovskiy (28) for the Sayan, U.S.S.R.].

As both residual and replacement phases of secondary phosphate adjoin each other and are mined together within any one field, there are unfortunately little data on these separate phases. Analyses of bulk material from the Florida hardrock deposits (analysis F, Table 2) reveal that a virtually monomineralic composition of carbonate-fluorapatite may be the end product of massive replacement. Note the similarity to primary apatite (analysis A, Table 2). Metasomatic apatite has been identified as carbonate-fluorapatite in South Carolina (145) and Florida (69).

Where weathering and leaching are pronounced or long continued, extensive solution of apatite and replacement of limestone occur. Thus even in the Tennessee brown-rock ores, so commonly cited as an example of residual phosphorites, Smith and Whitlatch (197) find metasomatic apatite to be a major component of the secondary phosphates. This is equally true in the Florida hardrock field where most of the secondary phosphate was mined from depressions in the Ocala Limestone (Eocene) or the Suwannee Limestone (Oligocene), though derived by supergene drainage through the overlying Hawthorn Formation (Miocene).

Table 9 presents chemical data for a profile through a representative deposit of replacement apatite. The most notable changes are the concentration of apatite and sparsely soluble minerals at the top of the section, and the complementary increase of carbonate downward.

Table 10 displays the effects on uranium content of different patterns of enrichment. In purely residual concentration little solution of the apatite occurs and the relations between U and P_2O_5 are the same before and after weathering. In replacement deposits there is an enhanced enrichment of uranium. This would require the solution of appreciable apatite in the overlying rocks, liberating uranium to solution, prior to its selective uptake from percolating groundwater by apatite re-forming at depth. This type of enhanced enrichment prevails in extensive zones of hardpan or concretions, at relatively shallow subsoil depths. Secondary apatite deposits with uranium in the range of $0.0X\%$ to $0.X$ have been noted in the Cooper Marl, South Carolina (up to 0.2% U), in arkosic sandstone of the Gas Hills, Wyoming (0.74% U) (8), and in arkosic sandstones of Bavaria (up to 0.47% U) (1).

Thus both residual and replacement processes cause zones of unusual concentration of P_2O_5, F, U, and to some extent, Sr and Mn. Moreover in residual accumulation the enrichments are created in surficial materials and have more immediate and pronounced environmental effects. However, even the replacement deposits are sufficiently shallow to contribute significantly to the radioactivity and to the fluorine, phosphate, and uranium contents of the vadose waters and the soils.

Carbonatites. Weathering imposes major economic and environmental changes on the apatite deposits of carbonatites. The apatitic sövites and

TABLE 9 Chemical and Mineralogical Analyses of Profile through Top of Cooper Marl, Charleston, South Carolina, Showing Variation in Supergene Phosphatization with Depth[a]

	1	2	3	4	5	6	7	8	9
SiO_2	57.7	40.8	31.5	28.9	28.9	27.7	23.0	24.1	21.4
Al_2O_3	5.4	5.2	4.2	4.2	5.4	4.3	3.6	3.6	3.6
Total Fe as Fe_2O_3	1.6	2.1	1.0	1.0	1.2	0.81	0.61	0.82	1.0
MnO	0.02	0.04	0.04	0.04	0.02	0.02	0.02	0.02	0.02
MgO	1.0	6.6	9.2	10.1	7.0	4.2	3.0	2.8	2.1
CaO	15.5	18.0	21.7	21.6	24.0	29.7	34.6	34.4	37.2
Na_2O	1.0	0.89	0.72	0.68	0.65	0.76	0.71	0.71	0.58
K_2O	1.3	1.0	0.83	0.82	0.93	0.82	0.69	0.70	0.66
TiO_2	0.35	0.31	0.26	0.23	0.27	0.24	0.17	0.20	0.18
F	1.06	0.64	0.40	0.44	0.30	0.28	0.23	0.30	0.25
Cl									
SrO[b]	0.07	0.03	0.02	0.01	0.02	0.02	0.03	0.05	0.03
P_2O_5	9.6	5.2	3.8	3.2	2.8	2.4	2.1	2.4	2.1
CO_2	1.8	14.8	21.9	23.6	22.0	24.3	27.3	26.6	28.0
Total S as SO_3	0.65	0.54	0.54	0.43	0.59	0.57	0.38	0.52	0.56
H_2O	3.8	3.8	3.3	3.3	4.7	3.1	2.5	2.5	2.5
Total	100.85	99.95	99.41	98.55	98.78	99.22	98.94	99.72	100.18
Less O = F_2, Cl_2, and S	0.45	0.27	0.17	0.18	0.13	0.12	0.10	0.13	0.10
Corrected total	100.40	99.68	99.24	98.37	98.65	99.10	98.84	99.59	100.08
Normative values for apatite, calcite, and dolomite.									
$Ca_{10}(PO_4)_6F_2$[c]	23.1	12.7	9.1	7.9	6.7	5.8	5.1	5.8	5.1
$CaCO_3$	4.5	5.0	9.6	7.5	22.2	39.2	51.3	50.0	59.0
$CaMg(CO_3)_2$	—	26.3	37.0	42.5	25.6	14.7	9.9	9.0	4.2

[a] From Malde (145). Analyst: U.S. Geological Survey. Samples at 0.5 to 1.0 ft intervals, No. 1 at 23.5 ft and No. 9 at 31 ft depths; ground elevation, 35 ft.
[b] Determined spectrographically as Sr.
[c] Present as francolite.

TABLE 10 Uranium Enrichment in Residual (Tennessee) and Replacement (South Carolina) Phosphorites. From Altschuler *et al.* (8)

Columbia, Tennessee Residual Enrichment	Percent P_2O_5	Percent U
(*a*) Limestone wall of enlarged joint, Bigby Limestone	1.60	0.00004
(*b*) "Brown rock" phosphate at contact	27.2	0.00074
(*c*) Enrichment factor (*b*/*a*)	16	18
Cooper Marl, South Carolina Enhanced Enrichment		
(*a*) Unaltered marl, Ashley River bank	4.0	0.005
(*b*) Overlying phosphatic hardpan	25.9	0.074
(*c*) Enrichment factor (*b*/*a*)	6	15
Effect of Continued Drainage on Selective Fixation of Uranium in Phosphatic Hardpan		
Stream bank, Ninemile station		
Dark fragment	27.0	0.035
Light fragment	26.9	0.11
Roadcut, south of Lambs		
Dark fragment	27.2	0.064
Light fragment	25.5	0.12

beforsites are similar to marine phosphatic limestones in their massive carbonate contents and display similar alteration sequences. Under mild weathering the dominant effect is decalcification, which produces residual soils greatly enriched in the minerals less soluble than calcite, and forms potentially valuable deposits of phosphate, iron, niobium, and rare earths. As the parent rocks are steeply dipping, the enrichment in their residual soils is pronounced. As the carbonatites generally compose the central, elevated plugs or cores of eroded volcanic centers, their residual soils are shed widely, forming blankets on the pediment slopes and filling the valleys to all sides with talus and alluvium of unusual composition (56, 57) (for chemical and mineralogical analyses see Tables 7 and 8).

In Jacupiranga, Brazil, the unaltered apatitic sövites contain 5 to 15% apatite as isolated round grains or aggregates of prisms in schlieren and veinlets. The solution of carbonates has produced a residium 1 to 10 m thick over these rocks, averaging 22% P_2O_5 (55% apatite) and 26% Fe_2O_3. The red soils over the Sukula Hills complex of Uganda contain apatite reserves of 200 million tons, representing the insoluble residue of billions of tons of carbonatite (58). These soils range from 50 to 220 ft in thickness and contain "about 32% apatite, 28% magnetite, 30% goethite and hematite, 7% quartz, and minor accessories including 0.4% pyrochlore" (58, p. 115). Of the nine carbonatite complexes mined for apatite six are exploited as residual soils (57). At Dorowe (Rhodesia) volcanic pipes containing calcite, apatite, magnetite, and vermiculite are decalcified to a depth of 200 ft and average 15 to 20% apatite (106).

Partial analyses of several Indian carbonatites and the soils overlying them portray the extent of residual concentration for a number of the elements characteristically enriched in carbonatites (Table 11).

As shown in Table 8 and in numerous chemical and mineralogical studies (56, 84, 148, 57), the primary apatite of the carbonatites is fluorapatite and the apatite from the large residual blankets shows virtually identical composition and has obviously not been dissolved or altered during the weathering in which it was concentrated. In fact residual apatite has the characteristic ovoid shape inherited from the parent sövites, in both Brazilian (148) and East African (57) occurrences.

Notable deposits of reprecipitated and metasomatic apatite also occur in several carbonatites. As in the sedimentary phosphorite, they represent a later or more intense stage of weathering than the purely residual deposits, and they are distinguished from the latter compositionally and texturally. As is seen in Table 8 the secondary apatite is carbonate-fluorapatite. At Busumbu fibrous bands of reprecipitated francolite surround fluorapatite grains (56). In the residual soils of the Bukusu complex, secondary phosphate occurs below 50 ft as a francolite cement, indurating masses of granular residual apatite (58). The secondary apatite generally occurs as concretionary and nodular masses in cavernous rock (56, 58).

In the Waterburg district of West Transvaal, decalcification and enrichment were accompanied by collapse. Subsequent solution of the apatite in the decalcified zone has caused a recementation of the ferruginous and apatitic carbonate breccia (57). At Glenover (South Africa), a phosphatic breccia forms a bowl-shaped body 500 ft deep in the center of the carbonatite mass, and consists of metasomatized beforsite cemented by drusy apatite (86).

The minerals accompanying apatite also suggest the extent and the intensity of weathering. Where secondary apatite is prominent, goethite and limonite rather than olivine are prominent. Alteration beyond pure decalcification is also seen in the supergene occurrences of gerasimovskite [$NbTi(OH)_9$] as a replacement of epistolite, smirnovskite ($Th_{1-x}PO_4 \cdot H_2O$), and hydrated silicorhabdophanes as replacements of belovite and steenstrupine, and in

TABLE 11 Analyses of Indian Carbonatites and Residual Soils (58)

	Carbonatites[a]					Residual Soils[b]				
Region	Ba	Sr	Nb	Ce	P_2O_5	Ba	Sr	Nb	Ce	P_2O_5
Amba Dongar, Gujarat	0.60	0.42	0.03	0.2	N.A.	1.0	0.12	0.05	0.8	2.33
	4.7	0.76	0.13	0.92	1.32	9.2	0.27	0.08	1.1	1.35
Newania, Rajasthan	Trace	1.2	0.06	0.05	2.50	Trace	0.12	0.035	0.05	5.24
	0.02	0.9	0.005	0.02	N.A.	0.03	0.06	0.06	0.1	6.25
Sevathur, Madras	0.12	0.91	N.D.	0.04	1.38	0.06	0.1	0.006	0.07	8.18
	0.09	1.10	N.D.	0.05	3.09	0.12	0.2	0.02	0.1	6.38

[a] N.A., not analyzed; N.D., not detected; approximate limits of detection: Ba 50, Sr 20, Nb 10, Ce 100 ppm.
[b] Grab samples, not necessarily the weathered products of the samples shown on the same line.

the hydration of pyrochlore and the development of niobian leucoxene, in the chemically altered soils over carbonatites at the Lovozero and Kiya massifs, U.S.S.R. (186), and in Brazil and Africa (213).

The principal effects of the weathering of apatitic carbonatites are to convert indurated rocks into residuums sufficiently enriched in phosphate to be extensively mined, and sufficiently soft and loose to facilitate mining, as well as natural dispersal by stream erosion and slope wash. This leads to a notable areal enhancement, in soils, of P_2O_5, F, Sr, R.E., and Th. Under conditions of advanced decalcification and weathering these exotic soils are thus sources of groundwater enrichment as well as of new rare minerals of supergene origin. [See discussion below and Fischer (72).]

Where apatite is not sufficiently enriched to be mined for itself, it may be a valuable by-product of the mining of niobium or rare earths. [See Table 7 and also Heinrich (97) and Deans (58).]

Weathered Argillaceous Phosphates

Lateritic Aluminum Phosphates. In areas of intense or prolonged groundwater drainage, argillaceous phosphorites are irregularly transgressed by extensive light-colored zones of leaching and alteration composed of aluminum phosphates. These zones are essentially lateritic. This is evident chemically in the vertical changes from a calcic and silicate-rich rock at the base of the zone to one in which all bases and silicates (other than quartz) are removed, and physically, in the extremely porous and vesicular rock in which the open sponge-like texture is indurated and maintained by the crystallization of secondary (aluminum phosphate) minerals (26, 11).

Zones of aluminum phosphate alteration have been described from Nigeria (181), the Florida land pebble (5, 11), and hardrock fields (69), Senegal (42, 194), and Siberia (227). Except for the Siberian occurrences, these are all products of Quaternary weathering. They are all supergene in origin and they all show the same paragenetic sequence which is essentially the progressive replacement of apatite and clay by the calcium aluminum phosphates crandallite [$CaAl_3(PO_4)_2(OH)_5 \cdot H_2O$] and millisite [$(Na, K)CaAl_6(PO_4)_4(OH)_9 \cdot 3H_2O$] and eventually wavellite [$Al_3(OH)_3(PO_4)_2 \cdot 5H_2O$] or, as in Senegal (42), augelite [$Al_2(PO_4)(OH)_3$]. Moreover, early in the leaching history, the three-layer clays, montmorillonite or illite, are transformed to kaolinite (11, 43, 228). In Florida this clay transformation is a widespread consequence of the regional weathering brought about through intracrystalline leaching of silica and epitaxial nucleation of new kaolinite on the residual two-layer sheet (10). The massive occurrences of millisite, with its essential sodium, may be confined to areas where montmorillonite had not been entirely transformed to kaolinite (171).

The nature of the lateritic alteration in supergene aluminum phosphate replacement is shown in compilations of analyzed profiles through the altered sections of the Bone Valley Formation, Florida (Table 12) and the Belka deposits of Siberia (Table 13). SiO_2 is higher on the average in the upper part of the section where it reflects almost entirely quartz. CaO, CO_2, and F, all reflecting apatite, and Na_2O, partly from apatite, increase sharply from top to bottom. The relative changes are shown in Fig. 3 where the major oxides for the Bone Valley profile are plotted on a silica-free basis. As little clay remains in this section, this is equivalent to removal of quartz, which remains inert during the alteration (11). CaO and Al_2O_3 have a complementary relationship in the aluminum phosphate alteration, and P_2O_5 has a slight and steady gain toward the bottom of the section, reflecting a general loss of some phosphate minerals during the leaching and alteration. Except for original bedding irregularities (a thick clay at sample Ho-22) the general slopes of curves for CaO and Al_2O_3, and their ratios to P_2O_5, correspond to a gradual replacement of apatite (3.3CaO to $1P_2O_5$) by crandallite and wavellite (both $3Al_2O_3$ to $2P_2O_5$). Thus, Al_2O_3 increases half as rapidly as CaO decreases.

Capdecomme (42) reports augelite in the upper part of the alteration zone in Senegal, and as its Al/PO_4 ratio is even higher than that of wavellite it may represent a more terminal facies of alteration; however, the paragenetic relations are not known.

Iron is mobilized to form zones of concretions and hardpans in the lateritized phosphates. Limonite and goethite are the principal minerals and are precipitated from downward-moving humate-rich groundwater in the Florida deposits (6). Due to interaction with phosphatic solutions these secondary iron deposits may be partly replaced by vivianite [$Fe_3(PO_4)_2 \cdot 8H_2O$] concretions, and by beraunite [$FeFe_4(PO_4)_3(OH)_5 \cdot 3H_2O$] and cacoxenite [$Fe_4(PO_4)_3(OH)_3 \cdot 12H_2O$] (27). Van Tassel (215) reports supergene development of cacoxenite, associated with strengite ($FePO_4 \cdot 2H_2O$) and secondary apatite, all replacing a breccia in sinks in Paleozoic limestone. Moore (158) has elucidated the structural relations among the basic iron phosphates and has noted the subaerial succession of dufrenite by beraunite in phosphatized iron deposits.

Both ferric and aluminum phosphates are stable relative to apatite under conditions of acid alteration; however, iron phosphates are not common and are volumetrically unimportant in sedimentary phosphates compared to the ubiquitous aluminum phosphates. In the lateritized Florida phosphates, iron phosphates occur solely as scattered concretions. In the Senegalese lateritized phosphates, iron occurs principally as goethite and as a substituent in millisite [cf. variety pallite (44)]. The relative paucity of iron phosphates may be due to the prominence of humic accumulations in the soils of highly humid regions, to the reducing effects of organic acids in waters draining these, and to the relatively high solubility of the ferrous phosphate compounds that would tend to form (102).

Uranium is significantly enriched in the lateritoid aluminum phosphates, and somewhat differentiated within these deposits. It is associated intimately and almost entirely with the major phosphate minerals. Only rare traces of uranium minerals have been reported for phosphorites, autunite [$Ca(UO_2)_2(PO_4)_2 \cdot 12H_2O$] from Florida (5) and torbernite [$Cu(UO_2)_2(PO_4)_2 \cdot 12H_2O$] in Morocco (14). In the Bone Valley Formation, for example, the unaltered phosphorite averages 0.008% in uranium and contains 10 to 15% P_2O_5. In contrast, typical aluminum phosphate sections have 0.012% uranium and 8 to 12% P_2O_5 (11). Moreover, as the aluminum phosphate zone in Florida has 60 to 70% quartz and minor kaolinite, both low in uranium, the zone's phosphate fraction has sustained a twofold to fourfold enrichment in uranium, supplied from an originally thicker section (12).

The distribution of uranium within the aluminum phosphate zone is shown in the graphs of Fig. 3. Uranium content increases sharply in the basal units, declines in the middle, and is relatively low in the upper half. Moreover, uranium parallels calcium in its distribution and associates most strongly with the more calcic phases. Thus, apatite pebbles at the base of the zone are the most enriched component, commonly having from 0.1 to 0.3% uranium, whereas pebbles from the equivalent unaltered strata contain only 0.02% uranium. Concentrates of crandallite and millisite from the middle of the zone contain about 0.03 to 0.05% uranium, and wavellite from the upper part of the zone has only 0.002 to 0.004% uranium (45, 8).

The pattern of uranium distribution accords well with the picture of supergene alteration for the aluminum phosphate zones. Uranium is liberated during the solution of apatite in the upper part of the zone, is kept soluble as uranyl and uranous ions in the descending acid groundwater, and is secondarily enriched in crandallite and in the porous, partly leached apatite pebbles, at the base, by structural replacement of U(IV) for calcium and probably by adsorption, and surface reaction of U(VI) (8). The progressive development of wavellite causes a continual exclusion of uranium and a remobilization and downward migration into those zones where crandallite and apatite still exist.

Scandium and the rare earths may also become unusually concentrated in secondary phosphates. The data are meager but compelling. Scandium may be anomalously elevated in diagenetically altered phosphorites, reaching 150 ppm in the organic- and pyrite-rich Maikop bone beds (31). More typical, primary, marine apatite contains only 1 to 10 ppm Sc (9) and unaltered phosphorite from the Phosphoria Formation has about 10 ppm Sc (91). Nevertheless secondary aluminum phosphate minerals derived from

TABLE 12 Chemical and Mineralogical Analyses of Profile through Lateritically Altered Phosphate Zone. Bone Valley Formation, Homeland, Florida (11)

Constituent	Ho-20	Ho-20A	Ho-21	Ho-22	Ho-23	Ho-24	Ho-25
SiO_2	51.48	62.60	40.92	57.24	69.46	68.08	51.32
Al_2O_3	8.26	5.98	12.48	14.17	8.16	9.40	14.91
Fe_2O_3[a]	2.76	2.86	2.61	2.17	1.32	1.33	2.19
MnO	0.70	0.53	2.14	0.16	0.16	0.14	0.13
MgO	0.01	0.01	0.01	0.01	0.01	0.01	0.01
CaO	8.98	8.25	8.00	3.10	0.90	0.20	1.20
Na_2O	0.23	0.13	0.12	0.15	0.06	0.04	0.03
K_2O	0.00	0.00	0.00	0.00	0.00	0.00	0.00
TiO_2	0.37	0.40	0.63	0.63	0.31	0.42	0.65
P_2O_5	19.72	13.60	20.79	12.79	11.61	11.32	16.35
Loss on ignition[b]	6.76	6.09	11.72	9.96	7.91	9.18	13.24
CO_2	0.80	0.24	0.26	0.05	0.02	0.02	0.05
F	1.15	0.52	0.63	0.59	0.51	0.70	0.53
Cl	0.03	0.02	0.04	0.01	0.03	0.02	0.01
SO_3[c]	0.01	0.01	0.01	0.01	0.01	0.01	0.01

Cr_2O_3	0.01	0.01	0.01	0.01	0.01	0.01	0.01
V_2O_5	0.00	0.00	0.00	0.00	0.00	0.00	0.00
U[d]	0.01	0.03	0.02	0.01	0.01	0.01	0.01
Total	101.28	101.28	100.39	101.06	100.49	100.89	100.65
Less O = F_2, Cl_2	0.49	0.23	0.28	0.25	0.23	0.31	0.22
Corrected total	100.79	101.05	100.11	100.81	100.26	100.58	100.43
H_2O—(110°C)	0.35	0.49	1.13	0.64	0.73	0.39	0.88

Major minerals listed in decreasing order of abundance[e].

Quartz	Quartz	Quartz	Quartz	Quartz	Quartz	Quartz
Apatite	Crandallite	Crandallite	Crandallite	Wavellite	Wavellite	Wavellite
Crandallite	Apatite	Wavellite	Wavellite	Crandallite	Goethite	Crandallite (tr)
Wavellite	Wavellite	Kaolinite	Kaolinite	Goethite	Crandallite (tr)	Goethite (tr)
Kaolinite	Kaolinite	Apatite (tr)	Goethite	Kaolinite		
Goethite	Goethite	Goethite				

[a] This represents total iron, some of which may be present as ferrous iron.
[b] The figures for loss on ignition include adsorbed water (H_2O^-) and exclude CO_2.
[c] This represents total sulfur; no sulfides were found.
[d] Uranium is reported as a metal, as its valence state was not determined.
[e] tr = trace = < 2%.

TABLE 13 Chemical Analyses of Profile through Lateritically Altered Aluminum Phosphate Zone, Belka, Gornaya Shoriya, U.S.S.R. (228)

Depth (m)	SiO_2	Al_2O_3	Fe_2O_3	P_2O_5	CaO	MgO	MnO	TiO_2	K_2O	Na_2O	CO_2	L.O.I.	H_2O^-	F	Sum	O=F	Total
1.5	11.60	31.52	20.00	10.26	2.66	1.45	0.92	1.85	0.41	0.14	0.27	18.31	0.46	0.08	99.93	0.03	99.90
3.5	18.14	32.19	18.81	7.78	2.58	0.50	0.33	2.70	0.65	0.14	0.00	16.26	0.50	0.012	100.59	0.005	100.58
7.0	8.17	22.34	23.60	14.08	4.80	4.80	2.16	2.05	0.90	0.20	0.00	15.42	1.42	0.30	100.24	0.12	100.12
14.5	18.45	26.76	20.20	11.88	4.92	0.00	0.09	2.95	0.64	0.11	0.00	13.46	0.58	0.08	100.12	0.03	100.09
20.0	14.53	18.22	10.60	20.72	22.16	0.52	0.58	1.05	0.53	0.20	1.40	8.48	0.20	1.00	100.19	0.42	99.77
25.0	18.59	18.62	7.70	18.52	24.70	0.00	0.27	1.15	0.43	0.18	0.80	9.06	0.24	0.88	101.14	0.37	100.77
30.0	19.80	13.83	7.35	23.35	26.65	0.00	0.31	0.85	0.38	0.16	2.24	5.18	0.16	1.13	101.38	0.47	100.91
40.0	14.82	18.62	10.94	20.08	22.16	0.54	0.57	1.05	0.55	0.28	0.00	10.04	0.30	0.88	100.83	0.37	100.46

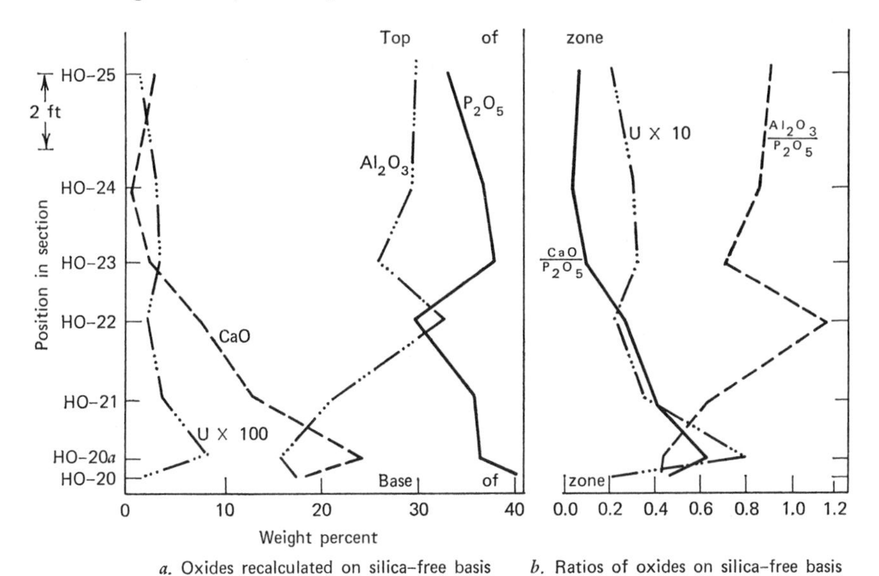

Figure 3. Chemical differences in the aluminum phosphate zone related to stratigraphy: (*a*) oxides recalculated on silica-free basis; (*b*) ratios of oxides on silica-free basis. From Altschuler *et al.* (11).

the Phosphoria Formation in Utah and Nevada, notably crandallite, millisite, goyazite, and variscite, attain from 10 to 1000 ppm Sc (80). As much as 0.8% Sc is found in crandallite from Fairfield, Utah (80) where scandium also accumulates sufficiently through supergene alteration to form a separate phosphate mineral, sterrettite $ScPO_4{\cdot}2H_2O$ (Mrose and Wappner, 1959). Generally, however, the massive secondary aluminum phosphate deposits of Senegal and Florida contain less than 10 ppm Sc (80).

Plumbogummite Group Minerals and Lateritized Carbonatites. Crandallite and other isostructural compounds of the plumbogummite group (Table 14) have the alunite type of structure in which the simple cations are held at large open sites with twelvefold coordination (98). This makes crandallite particularly suitable as a host for large cations, and Frondel (79) has described the scavenging of Sr by crandallite in the lateritic aluminum phosphates. The studies of Capdecomme and Orliac (44) show extensive uptake of Sr and Ba by solid solution in the crandallite from Thies, Senegal, and from Brazil (sample K, Table 15). Their analyses and data are given in Table 15. Two of the samples plotted are really goyazites. A possible instance of the same crystallochemical control is seen in the partial analysis of rare earths in a mixed millisite-crandallite rock from the Bone Valley Formation

TABLE 14 Aluminum Phosphate Minerals of Alunite Structure Found in Lateritized Phosphate Deposits

Plumbogummite	$PbAl_3(PO_4)_2(OH)_5H_2O$
Crandallite	$CaAl_3(PO_4)_2(OH)_5H_2O$
Gorceixite	$BaAl_3(PO_4)_2(OH)_5H_2O$
Goyazite	$SrAl_3(PO_4)_2(OH)_5H_2O$
Florencite	$CeAl_3(PO_4)_2(OH)_6$
Woodhouseite	$CaAl_3(PO_4)(SO_4)(OH)_6$
Svanbergite	$SrAl_3(PO_4)(SO_4)(OH)_6$

(analysis E, Table 5) in which there is very extensive enrichment of the large lanthanum ion relative to cerium, in comparison to the primary apatite sources of these elements (compare analysis A, Table 5).

The plumbogummite group minerals develop also as secondary products of intense weathering over carbonatites (143). Monazite and florencite of hydrothermal origin survive unaltered along with strontium- and rare earth-rich apatite, barite, and strontianite in the partly decalcified, residually enriched soils at Kangankunde Hill (57). Under more intense weathering these minerals and the rock-forming aluminosilicates are altered and transformed as apatite dissolves, and phosphoric acid is liberated after complete decalcification. Thus McKie (143) and Deans (57) record florencite and rare earth-rich goyazite and gorceixite, accompanied by earthy secondary monazite, as supergene products over African carbonatites. Secondary earthy monazite is also found as a weathering product along with probable florencite at Magnet Cove, Arkansas (179). An impressive supergene deposit, consisting of early varieties of secondary monazite, barite, gorceixite, goethite, psilomelane, and niobium and tantalum minerals, overlies the deeply weathered and completely decalcified Mrima carbonatite in Kenya (52). These secondary replacement and reprecipitated minerals occur in soils of great thickness, and are devoid of calcite or apatite, though developed on sovites.

Secondary gorceixite nodules are found at several localities in weathered marl of the Bashi Formation, a fossiliferous glauconitic sand in Alabama (154). Although called gorceixite, the mineral is unusually rich in strontium, rare earths, and calcium, and is essentially $(Ba, Ca, Sr, Ce)Al_3(PO_4)_2(OH)_5 \cdot H_2O$ (154). Though their origin is described as a problem the nodules are probably a secondary alteration product of the glauconitic marl, as glauconites are notably accompanied by apatite pellets or coated by apatite pellicles.

An unusual though volumetrically insignificant occurrence of crandallite and hydroxylapatite, as secondary alteration products of the continued

TABLE 15 Analyses of Crandallites from Thies, Senegal (44)

	E		F		G	
Sample	wt. %	mol. ratio	wt. %	mol. ratio	wt. %	mol. ratio
Na_2O	0.08	0.001	0.09	0.0015	0.07	0.001
K_2O	0.04	0.0004	0.04	0.0004	0.11	0.001
CaO	11.22	0.200	11.87	0.212	4.65	0.083
BaO	Traces		Traces		2.25	0.015
SrO	2.91	0.028	0.70	0.007	11.26	0.108
MgO	0.20	0.318	0.20	0.329	0.20	0.312
Al_2O_3	32.44	0.015	33.58	0.024	31.86	0.011
Fe_2O_3	2.37	0.200	3.84	0.207	1.71	0.204
P_2O_5	28.45	0.937	29.38	0.928	28.91	0.806
H_2O-	1.47	0.046	0.75	0.023	0.49	0.112
H_2O+	16.86		16.71		14.50	
HF insoluble	0.93		0.46		2.24	
Fe_2O_3	0.18		0.10		0.10	
TiO_2	1.08		0.49		0.15	
SiO_2	1.93		1.88		1.40	
Total	100.16[a]		100.09[a]		99 90[a]	

[a] Published values were wrong; they are corrected here.

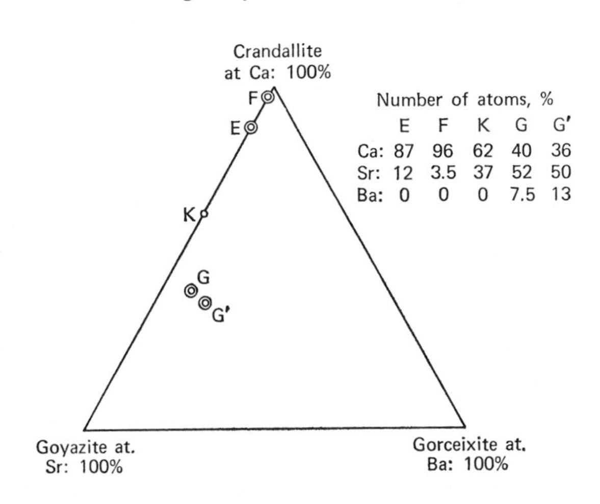

leaching and weathering of supergene variscite ($AlPO_4 \cdot 2H_2O$), has been described from the well-known Fairfield, Utah, locality by Larsen (128). The material results from the continued attack on earlier variscite nodules by supergene phosphate-rich solutions, and the crandallite is itself replaced by millisite and wardite, and all of these are later coated or filled with englishite, gordonite, overite, montgomeryite, and sterrettite. This occurrence is important, along with a few others, in revealing that variscite is not the stable form of subaerial aluminophosphate. Woodhouseite and svanbergite occur in phosphatized Cambrian shales of Kazakhstan, accompanying extensive uraniferous gorceixite (231).

WEATHERING RELATED TO GUANO

Conditions of Guano Accumulation

Accumulations of excrement form the only other important category of phosphate deposits. The subject has been treated in masterful detail by Hutchinson (103)* who noted that massive fecal concentration occurs only when large colonies of "organisms [which] are large, feed over a wide [and highly productive] tropophoric field, and return to a limited site for rest and reproduction" (103, p. 4). For persistent accumulation the site must be relatively free of predators and of frequent or heavy rain. Only the roosts of birds and bats fulfill these requirements to any geochemically important degree.

The requirements for substantial deposits of phosphatic bat guano (chiropterite) are met best in dry caves of karst terrains, in warm, fertile regions. The cave areas of the West Indies, Indonesia, and Thailand are notable past and present-day examples. Many caverns in the limestone belts of Bohemia, Appalachia, and Texas and New Mexico still support bat colonies, but their guano, for the most part, was deposited in pre-Holocene and, presumably, warmer periods. The requirements for bird guanos are fulfilled on arid islands and, subordinately, on coastal headlands adjoining highly productive regions of the sea.

Thus, most insular guano is found on small islands, in belts of the world ocean marked by major divergences in circulation, where colder, nutrient-rich, subsurface water is brought to the surface by pronounced upwelling. Such fertile belts, flourishing in plankton and fish, are notable on the west sides of the continents in tropical and trade-wind latitudes, at the inner edges of the Peru, California, Benguela, and West Australian currents. The Peruvian guano islands, the deposits of the Mejillones Peninsula in

* This discussion leans heavily on his work.

Chile, and of Saldanha Bay in South Africa are well-known examples of such geophysical control. Fertile belts of upwelling occur in the boundary zones of the westward-flowing north and south equatorial currents with equatorial countercurrents. Cold water wells up and flows transversely poleward from the divergence. Many major Pacific island phosphate deposits owe their existence to this current configuration. Thus Nauru, Ocean, Howland, Baker, and Jarvis Islands all lie close to 0° latitude in the South Pacific equatorial divergence.

Nutrient enrichment may be prominent in other locales as well. For physical, chemical, and biological oceanography see the comprehensive treatises of Sverdrup and co-workers (199), Hill (100), and Hedgepeth (96). For comprehensive and critical biogeochemical data and interrelations, and many cultural aspects, Hutchinson (103) is indispensable; for cogent summaries and interpretations see McKelvey (138) and Sheldon (190), who presents a paleotemporal interpretation. White and Warin (223) give a full review and extensive new observations of the geology and distribution of insular phosphate in the southwest Pacific, and, with Hutchinson (103), full bibliographies.

The semisolid excretion of birds is a highly aqueous mixture of nitrogenous organic compounds, rich in uric acid or urates, oxalates, purines, and containing moderate amounts of phosphate, sulfate, Na, K, Ca, Mg, Cl, and keratin. Uric acid compounds are decomposed readily by bacteria, producing ammonia, oxalic acid, and CO_2. Fresh guano quickly loses urates, ammonia, and organic matter, and eventually, the soluble oxalates, and alkali and alkaline earth chlorides, sulfates, and some phosphates (see Table 16). With aging and persistent leaching, the residual material, essentially a weathered guano, is a concentrated deposit of inorganic calcium phosphates, which may also have substantial calcium sulfate. At intermediate stages, many highly hydrated and readily soluble, ammoniacal, and alkaline phosphate, sulfate, and oxalate minerals are formed in the altering guano. These soluble leachates and minerals are largely evanescent, but the phosphate may be fixed by metasomatic alteration of the underlying rocks to produce a variety of new deposits. Thus guano is both a subject and a source of weathering.

Following Hutchinson (103) and White and Warin (223) we consider here three general categories: nitrogenous guano, phosphatic guano, and phosphatized rock.

Nitrogenous Guano

Table 16 portrays the compositional nature of old nitrogenous guano from the world's principal deposits, the quantity and nature of its solutes, and

TABLE 16 Old Nitrogenous Guano, Chincha Islands, Peru

	I				II			
	Soluble in Cold H_2O (%)	Soluble in Hot H_2O (%)	Insoluble in H_2O (%)	Total (%)	Soluble in Cold H_2O (%)	Soluble in Boiling H_2O (%)	Insoluble in H_2O (%)	Total (%)
H_2O-	22.20	—	—		10.67	—	8.06	18.73
H_2O+		1.33	3.46	28.48				
Organic matter	1.50				0.24	1.00	2.97	4.21
Humus	—	—	2.64	2.64	—	—		
CaO	—	0.10	11.87	11.97	0	0.61	29.85	30.47
MgO	—	0.08	0.73	0.80	—	—	1.09	1.09
NH_3	5.21	1.48	—	6.68	0.14	—	—	0.14
Na_2O	?Trace	?0.06	—	?0.06	1.91	—	—	1.91
K_2O	4.37	—	—	4.37				
SO_3	3.64	—	—	3.64	0.68	—	—	0.68
P_2O_5	4.35	0.42	10.20	14.92	0.64	0.53	38.56	39.73
Oxalic acid	5.03	—	1.44	6.47	Trace	—	—	Trace
Uric acid	—	16.52	—	16.52	—	0	—	—
Cl	1.70	—	—	1.70	0.87	—	—	0.87
Sand	—	—	1.56	1.56	—	—	2.04	2.04

I Normal guano. Analysis given in Hutchinson (103) from Denham Smith (1845).
II "Concrete" guano. Differentiated lumps in normal guano. Analysis given in Hutchinson (103) from Denham Smith (1845).

the progressive mineralization and chemical change incident to decomposition and leaching. [For discussion of fresh droppings and modern guano see Hutchinson (103) whose compilation documents nitrogenous organic matter as high as 80%.] The "concrete" guano (analysis II) is a cemented differentiate in the normal guano (analysis I) and the composition of its insoluble phase is virtually that of brushite, on the assumption that much of the water must be assigned to it. The less leached, normal guano, by contrast, still retains appreciable oxalate, water, and soluble phosphate and sulfate. Furthermore, the proportions of insoluble phosphate and calcium reveal the presence of noncalcic phosphate.

Deposits of nitrogenous guano are stratified by annual accretion and occur as thick caps surmounting their insular bases by as much as 44 m, as at South Chincha (mean depth 28.5 m). The islands are small and the deposits have no broad environmental significance, and are miniscule in terms of global phosphate geochemistry [see calculations in Hutchinson (103)]. Nevertheless they have been important economically. An aggregate of roughly 12 to 15 million tons of guano may have been present on the three Chincha Islands (103). Their local environmental influence may be highly significant, however, in view of their enrichment of the ambient waters by highly nutrient leachates, and the capacity of these to generate red tides and massive fishkills.

Phosphatic Guanos and Phosphatized Limestone

The calcium phosphates derived from guanos may be purely residual accumulations. They also occur as mixtures of phosphatized, wave-washed, calcareous debris, cemented by both precipitated and residual calcium phosphate. These occur as continuous, thin crusts over hundreds of small islands in the Pacific, the Caribbean, and the seas of southeast Asia, and are called "crust guanos" or atoll phosphates. Reflecting a component of residual guano, these deposits are collectively termed phosphatic guano. Lastly, the calcium phosphates may consist largely of massive varieties of precipitated and metasomatic (or replacement) deposits mixed to varying degree with finer, and less cohesive, particulate calcium phosphate, and subdivided texturally into "coherent" and "incoherent" (169). These are no longer guanos. The large economically exploited insular deposits of Nauru, Ocean, Makatea, Christmas, and Curaçao are of this type. [Geologic descriptions and classifications of insular phosphates are given in Elschner (65, 66), Hutchinson (103), Nugent (166), Owen (169, 170), Power (175, 176), Rodgers (177), and White and Warin (223).]

The crust or atoll phosphates occur on smaller low-lying atolls or saucer-shaped islands, forming relatively continuous skins, commonly less than 1 m thick. The larger metasomatic deposits, in contrast, occur on uplifted islands, and share many of the features of residual and replacement deposits on continental limestone plateaus. They form thick caps over etched surfaces, filling solution cavities and karrenfelds of steep-walled gullies and intervening pinnacles.

As may be expected in this geomorphic regime, and as noted by Hutchinson (103) and White and Warin (223), the textural and petrographic varieties grade into one another, and the available literature, based largely on bulk economic sampling, shows few geochemical distinctions (albeit these rocks have been classified interminably and may display significant and subtle differentiation if studied in detail). We shall therefore consider the compositions of three groups: leached or residual guano, crust guanos ("crust phosphates" of White and Warin), which together make up the phosphatic guanos, and the massive metasomatic calcium phosphates of phosphatized limestone.

Phosphatic Guano. Analyses of phosphatic guanos are given in Table 17. In contrast to nitrogenous guano (cf. analysis I, Table 16), these analyses are greatly enriched, as a class, in calcium phosphate and greatly reduced in total nitrogen content. After subtraction of normative gypsum and calcite, all of the analyses are too low in calcium to contain principally apatite or whitlockite, and brushite or monetite must be a major phase. The material from the Mejillones Peninsula and the chiropterite from the Drachenhohle are both essentially residual or leached guanos. The Mejillones Peninsula is notably fogbound and washed by sea spray, giving its deposits a high MgO content and unusual mineralogy. The minerals bobierite [$Mg_3(PO_4)_2 \cdot 8H_2O$] and newberyite [$MgH(PO_4) \cdot 3H_2O$] which have been reported from Mejillones guano (120, 103), in addition to halite, gypsum, and the major brushite, are obvious reaction products of sea salts with the soluble phosphate of the aging guano. The only other notable occurrences of these minerals are in the guanos and related metasomatized rock of basalt caves. MacIvor (136, 137) found bobierite and newberyite with struvite [$Mg(NH_4)(PO_4) \cdot 6H_2O$], dittmarite [$Mg_5(NH_4)H_4(PO_4)_5 \cdot 8H_2O$], and schertelite

$$[Mg(NH_4)_2H_2(PO_4)_2 \cdot 4H_2O]$$

in the leached chiropterite of the Skipton basalt caves, Victoria, and Lacroix (122) found newberyite after olivine in Tunnel des Cormorants, Réunion.

The chiropterite from Drachenhohle contains hydroxylapatite and brushite. Schadler (184) reports these as collophane and brushite, (see his analyses). Machatschki (144), in related deposits, finds brushite, "collophane," and, in rockfall material, variscite.

The crust guano of Jarvis Island is interstratified with gypsum, apparently the result of episodic marine flooding of the low-lying island's central

TABLE 17 Phosphatic Guanos

	I	II	III	IV
CaO	30.7	37.36	34.84	28.22
MgO	7.9	1.76	0.57	0.59
K_2O	0.5	5.54[c] (K₂O + Na₂O)	0.46	1.40
Na_2O	1.5		0.33	0.36
Fe_2O_3	0.2	0.50	0.09	2.39
P_2O_5	35.9	34.33	17.68	22.63
SO_3	1.6	—	27.02	0.23[f]
CO_2	1.6	0.46	—	2.81
N	0.7	1.04	0.65[d]	0.69
Org.	6.5[a]	12.72[b]	5.99[e]	4.99[g]
H_2O+	—	—	12.12 (H_2O+ and H_2O-)	5.12
H_2O-	7.7	4.83		—
Cl	2.3	—	0.20	—
SiO_2	2.5 (SiO_2 + Insol)	1.69 (SiO_2 + Insol)	—	21.93
Insol			0.27	—
Other	—	0.81	—	6.10

[a] Includes NH_3 and H_2O+.
[b] Includes H_2O+ and presumably, N compounds.
[c] Given as alkali salts, presumably includes Cl.
[d] $NO_3 = 0.32$, $NH_3 = 0.04$.
[e] Presumably includes N compounds, and some H_2O+.
[f] Equals S.
[g] Equals C.

I Leached guano, residual, Mejillones Peninsula, Chile. From Krull (1894) as given in Hutchinson (103). Analyst: Vohl.

II Crust guano, Raza Island, Gulf of California. From Voelcker (1876) as given in Hutchinson (103). Other = Al_2O_3.

III Crust guano interstratified with gypsum, Jarvis Island, Pacific Ocean. From Liebig (1860) as given in Hutchinson (103).

IV Chiropterite, residual, Drachenhohle cave, Mixnitz, Austria. Schadler (184). Other = TiO_2 0.24, Al_2O_3 5.78, MnO 0.08.

depression. Subtraction of gypsum from the analysis leaves a composition like that of pure monetite or brushite. The crust of Raza Island (analysis II, Table 17) was originally described as a continuous carapace of tricalcium phosphate [see summary in Hutchinson (103)] totaling 20,000 tons of ore. Its mineralogy is not known, but major monetite is reported in deposits on nearby Pedro Maritr Island (103) and from its analysis it may be a mixture of monetite and martinite.

Phosphatized Limestone. Analyses of commercial and thus representative samples of three major insular phosphorites and one sample from Nueva Leon, Mexico, are given in Table 18. The latter is an apparent cave filling consisting of red earthy martinite with traces of banded apatite and interstitial amorphous materials, and thus of probable bat-guano origin (41). The materials are all notable in their low contents of organic matter, K_2O, Al_2O_3, and Fe_2O_3, and thus show terminal metamorphosis of the parent guano, as well as reaction with and replacement of relatively pure limestone. Although the elevated islands have been typically dolomitized prior to guano accumulation (169) it is only when appreciable martinite is present that appreciable MgO persists, as at Curaçao. At Nueva Leon both MgO and ZnO are prominent constituents stabilizing the low-temperature whitlockite.

Although described originally as collophane, nauruite, ornithite, and a host of other names, the massive insular phosphorites tend to be dahllitic (analysis 3, Table 18) and notably francolitic (analysis 4) in composition (85, 77, 210). The fluorine contents of Angaur (2.96), Makatea (3.42), Ocean (2.97), and Nauru (2.48) apatites are all elevated (104), yet do not show the excess fluorine (higher than 3.8) characteristic of sedimentary marine francolite deposition (8). There seems little reason to doubt that the fluorine is principally of postdepositional origin, although the question of how and when it was acquired in such quantity is open (103). Studies of hydroxylapatite of bat-guano origin show related postdepositional uptake of both fluorine and uranium from probable marine sea spray and wash (8, 3). The apparent origin of the massive insular phosphate caps as a succession of deposits over newly uplifted terraces (103) invokes a suggestion of early postdepositional replacement of hydroxyl by marine fluoride from storm waves.

The dominant characteristic of the large insular calcic phosphates is their basic nature. As examples the deposits of Angaur, Nauru, Ocean, and Makatea Islands are principally, if not entirely, apatitic. The various calcic phosphates of cave and insular origins may, however, display an evolutionary pattern in which the mineralogy and composition are determined by the incongruent solution of the more acid calcium phosphates, and their progressive recrystallization to apatite. With more intensive or prolonged leaching, deposits would be transformed from brushite-monetite to whitlockite to apatite—a sequence accompanied by a parallel decline in organic, nitrogenous, and ammoniacal compounds. This evolution is possibly displayed in the sequence of chemical changes (Tables 16 through 18) from decomposed guano, through residual and crust guanos, to the Curaçao deposits, and finally to the deposits of the massive insular caps.

TABLE 18 Phosphatized Rock: Massive Calcium Phosphates from Guanitized Terrains

	1	2	3	4
	Nueva Leon, Mexico	Curacao Island, Caribbean	Christmas Island, Malaysia	Ocean Island, Pacific
CaO	46.35	49.96	52.50	54.08
MgO	2.17	1.23	0.10	0.00
SrO	—	—	—	—
Na_2O	—	0.88	0.53	0.49
K_2O	0.05	0.09	0.05	Trace
P_2O_5	44.28	38.59	39.46	40.32
V_2O_5	0.02	0.01	—	—
CO_2	—	3.90	2.28	1.06
SO_3	0.73	0.61	0.00	0.00
SiO_2	2.80	0.39	0.60	0.40
Al_2O_3	—	0.40	0.80	—
Fe_2O_3[a]	0.11	0.61	—	0.20
F	0.02	0.70	1.32	2.97
Cl	—	0.16	0.02	0.01
H_2O^+	1.63	2.47[b]	2.05	1.88[c]
H_2O^-	0.10	—	—	—
Other	1.99	0.024	—	—
		100.03	99.71	101.41
Less $O = (F, Cl)_2$	—	0.33	0.56	1.25
Total	100.25	99.70	99.15	100.16

[a] Total Fe as Fe_2O_3.

[b] Includes organic carbon and nitrogen.

[c] Loss on ignition less CO_2.

1. Commercial phosphate rock, martinite, in cavernous limestone. Cady *et al.* (41, p. 182). Analyst: Mrs. E. L. Hufschmidt. Other = ZnO.

2. Commercial phosphate rock, martinite with minor apatite and gypsum, overlying and replacing limestone terrace. Jacob *et al.* (104, p. 23). Analysts: authors. Others = TiO_2 0.024, MnO < 0.003, $Cr_2O_3 < 0.003$.

3. Commercial phosphorite from insular cap over dolomitic limestone. Dahllite. Jacob *et al.* (104, p. 23). Analysts: authors.

4. Commercial phosphorite from insular cap over dolomitic limestone. Francolite. Jacob *et al.* (104, p. 23). Analysts: authors.

Phosphatized Silicate Rocks—Deposits of Iron and Aluminum Phosphate

The accumulation of guano on silicate rocks produces another major category of metasomatic phosphates, those of iron and aluminum. Most of the occurrences may be grouped as follows: (*a*) isolated volcanic islands or uplifted atolls with exposed volcanic cores, such as Clipperton (168), Malpelo (134), and Redonda (121); (*b*) coastal regions of crystalline rocks like Saldanha Bay, South Africa (61), Sugar-loaves, New Zealand (191), and the Mejillones Peninsula, Chile (103); (*c*) caves in basalt [Réunion Island (123, 124)] or other igneous rocks [Bhomi Hills, Liberia (19)]; (*d*) residual or fluviatile soils in caves as in the Drachenhohle (144) or the Jenolan Caves of New South Wales (155); (*e*) residual soils over volcanic islands as on Kita-Daito-Jima (226) or on limestone islands with volcanic ash as on Bellona Island (223) and Christmas Island (209).

Deposits of Al–Fe phosphates are probably quite common but generally very small, and of little economic or environmental significance. Exceptional deposits occur at Kito-Daito-Jima Island, on which over 10 million tons of mixed Al–Fe phosphates were formed by the action of guano solutions on terra rossas, weathered pumice, and hypersthene andesite (18); the Saldanha Bay deposits of South Africa, estimated to contain "several hundreds of thousands of tons of largely aluminous phosphate" replacing granite blocks and boulders (61); and those of Grand Connetable, Christmas, and Trauhira Islands (103).

In the reaction of acid and ammoniacal guano or its leachates with igneous rocks, the products are most commonly the simple Al or Fe phosphates of the orthorhombic isomorphous series: variscite ($AlPO_4{\cdot}2H_2O$), barrandite [$(Al,Fe)PO_4{\cdot}{_2}H_2O$], and strengite ($FePO_4{\cdot}{_2}2H_2O$); or the parallel series of monoclinic dimorphs: metavariscite, metabarrandite (clinobarrandite), and metastrengite (phosphoriderite) (133). The more complex minerals, taranakite (palmerite, minervite; analysis 1, Table 19)

$$[(K,NH_4)_2Al_6(PO_4)_6(OH)_2{\cdot}18H_2O]$$

and leucophosphite [$\dot{2}(K,NH_4)Fe_2(PO_4)_2OH{\cdot}5H_2O$], occur more rarely, generally in caves, appearing to require excess alkali for their generation and a humid atmosphere to persist (103) (Table 19). Dufrenite

$$[FeFe_4(PO_4)_3(OH)_5 2H_2O],$$

rockbridgeite [$FeFe_4(PO_4)_3(OH)_5$], and newberyite ($MgHPO_4{\cdot}3H_2O$) are also rare occurrences. The acid ammonium phosphates of magnesium, hannayite and schertelite are incongruently soluble, transforming to newberyite with leaching (75).

TABLE 19 Representative Metasomatic Al–Fe Phosphates

	1	2	3	4
	Taranakite	Leucophosphite	Ferroan Variscite[a]	Aluminous Strengite
Al_2O_3	22.89	0.88	28.75	4.11
Fe_2O_3	1.17	36.85	3.45	33.40
CaO	Trace	0.16	0.00	—
MgO	—	0.02	0.24	—
Na_2O	—	0.22	0.05	—
K_2O	8.04	7.86	0.05	—
$(NH_4)_2O$	0.90	1.99	—	—
P_2O_5	37.10	33.46	42.46	33.74
SiO_2	—	4.32	1.42	4.31
H_2O+	29.16	11.67	1.57	14.90
H_2O-	—	1.23	22.58	3.80
Other	0.38	0.85	0.27	5.50
Totals	99.64	99.51	100.84	99.76

[a] Analysis recalculated to include H_2O-, as variscite loses hydration water below 110°C.

1 Cave deposit, "Palmerite" Monte Alburno, Italy. Casoria (1904) as given in Palache *et al.* (172). Other = insoluble.

2 Cave deposit, replaced hematitic gneiss, accompanied by strengite, Bomi Hills, Liberia. Axelrod *et al.* (19). Other = TiO_2 0.41, FeO 0.14, C 0.30.

3 Phosphatized gneiss and diabase, Connetable Islands. Jacob *et al.* (104). Other = TiO_2 0.12, SO_3 0.08, F 0.05, Cr_2O_3 0.013, V_2O_3 0.004.

4 Phosphatized augite-andesite, Malpelo Island. McConnell (134). Other = TiO_2 5.42, MnO 0.08. Mineralogy: strengite and metastrengite 87.2%, ilmenite and leucoxene 6.9%, quartz and opal 4.3% (134).

Initially these secondary minerals occur as pseudomorphs of feldspar or other phenocrysts closely adjoined by liberated silica in the form of opal or chalcedony, and the compositions of the metasomatic phases reflect the parent minerals. Thus at Réunion, Lacroix (124) records newberyite after olivine, and barrandite after pyroxene, in basalt. At Clipperton, sanidine is replaced by colloidal aluminum phosphate and quartz. With enhanced reaction the entire groundmass is replaced with less fidelity, resulting in a mixture of variscite and silica (168). With massive replacement several Al–Fe phosphates commonly develop together. At Malpelo Island augite-andesite is replaced by strengite, metastrengite, and quartz, and variscite and metavariscite together replace feldspathic amygdaloid (134). At Bomi

Hills, Liberia, leucophosphite, strengite, and rockbridgeite replace hematitic gneiss (19). At Lake Wheelamby, Australia, barrandite and leucophosphite replace serpentinite (191). More basic rocks are expectably replaced by iron-rich members of the variscite-strengite isodimorphous series, and conversely variscite is more prominent in granitic rocks, although the mixed Al–Fe phosphate barrandite is the most common replacement. It occurs in granites at Saldanha Bay, South Africa, and the Mejillones Peninsula (103), in diabase as at Redonda Island (135), and even replacing serpentine (191) and limburgite (209). Where the substrate is grossly differentiated the replacements are likewise zoned. Thus, at Trauhira Island, Brazil, soil development on a diabase stock has caused zonal segregation of iron and aluminum and there is a corresponding zonation of the replacing dufrenite in the iron zone and variscite in the aluminous rock (16).

PARAGENESIS OF ALUMINUM PHOSPHATES IN WEATHERING

The phosphatization of argillaceous material leads to the generation of the crandallite-wavellite assemblage of minerals in contrast to the simple Al–Fe phosphates of the variscite-strengite dimorphous series which develops on fresh igneous rocks. On limestone islands containing clay derived from weathered volcanic ash, crandallite composes the "phosphatic clay" that has accumulated in the central depressions, as on Bellona (224), Niue (71), and Cook (230) Islands. The distinction is vivid on Christmas Island where both fresh and lateritized igneous rocks are present and crandallite replaces the kaolinitic laterite, whereas barrandite replaces the fresh rock (209). The paragenesis of the crandallite-wavellite assemblage in guanitized terrains is virtually the same as that described for purely meteoric, supergene alteration of argillaceous phosphorites. It is essentially a succession of replacements of apatite by hydrated calcic and alkali calcic aluminum phosphates (crandallite and millisite), which are in turn replaced by the hydrated aluminum phosphate, wavellite (11). Moreover in both domains the replacing phases may be derived from two directions, phosphatization of clay or the attack of aluminum-bearing solutions on apatite. Both replacements have been observed in the Bone Valley Formation (11), and on Christmas (209), and Bellona (224) Islands.

Most of the iron and aluminum phosphates are unstable and undergo progressive alteration with continued weathering. Several tendencies are observed: change to greater hydration, change to higher Al–Fe/P ratios, and, in the case of iron phosphates, a change to greater oxidation (78, 158). Contrary to early, incomplete, phase studies tending to show that variscite and strengite are the stable and expectable forms of soil phosphate (53, 229), the evidence is overwhelming that the stable phases are crandallite and millisite if alkalis are present, and wavellite and augelite without alkalis. In extensive and massive developments of aluminophosphates (e.g., Liberia, Florida, Senegal), variscite is not recorded and crandallite is the dominant phase. Recent solubility studies are in accord (20). Only in the notably metamorphosed deposits of Sarysai is massive variscite of pimary (sic), weathering, and metamorphic origin reported, along with crandallite (174). However, the original relations are obscured by profound recrystallization and the variscite may be an example of retrograde alteration.

In the presence of excess alkali, the minerals formed initially are minyulite, taranakite, or leucophosphite. [See Cole and Jackson (53) and Haseman *et al.* (94) for laboratory studies and Bannister and Hutchinson (22), Axelrod *et al.* (19), and Murray and Dietrich (162) for field examples.] With continued leaching these are replaced under acid conditions by simple Al–Fe phosphates of the variscite-strengite group. The phases are comparable in Al–Fe/P ratios and the reactions are reversible and dependent on pH and alkali contents (53). If the ion activity product of $AlPO_4$ is high, taranakite dissolves incongruently to amorphous aluminum phosphate (201). In nature taranakite is replaced by barrandite (126) and leucophosphite appears to be succeeded by phosphosiderite and strengite (19). Furthermore, barrandite is replaced by crandallite on Christmas Island (209), variscite is replaced by crandallite at Fairfield, Utah (128), and at Bomi Hills some strengite is apparently succeeded by crandallite (19). Thus, the paragenetic sequences shown in Table 20 appear to obtain among secondary phosphates under

TABLE 20 Paragenetic Sequences in Aluminophosphates during Weathering

Parent Material and Agent	Products of Alteration
Igneous rock or clay + guano	Leucophosphite → strengite → crandallite
	Taranakite → barrandite → crandallite
Igneous rock + guano	SiO_2, Barrandite → crandallite + millisite → wavellite
Clay + guano	SiO_2, Crandallite + millisite → wavellite → augelite
Clay + guano	SiO_2, Wavellite
Apatite + groundwater	Crandallite + millisite → wavellite
Apatite + groundwater	Wavellite
Clay + groundwater	SiO_2, Crandallite + millisite → wavellite $\xrightarrow{?}$ augelite
Clay + groundwater	SiO_2, Wavellite

continued weathering. The paragenetic sequences will obviously vary according to the composition of the replaced rock. Thus in iron-deficient rocks a highly aluminous taranakite may form and be succeeded by variscite rather than barrandite. Where clays are highly leached prior to phosphatization, montmorillonite may be transformed to kaolinite (10), sodium may be depleted, and millisite may not form with crandallite (11, 171). Augelite appears to form only with intense or long-term leaching. Though shown in Table 20 as a possible phase, augelite is rare.

SOME MAJOR ENVIRONMENTAL EFFECTS OF THE WEATHERING OF PHOSPHATES

Several environmental consequences of phosphate weathering merit additional mention: the development of new patterns of mineral zonation and distribution related to the drainage; the redistribution of many elements, and the enhanced reconcentration of uranium and radioactivity through such zonation; and the upgrading of lean deposits to economic deposits.

Studies of the regional occurrence of the aluminum phosphate zone in Florida (48) show that the thickness and grade contours conform to the topography of the river valleys. The thickest sections and the greatest uranium enrichment generally occur under the outer slopes of the flood plains. Changes in grade and thickness migrate upstream in the tributaries, and toward the master streams on the interfluves (see Fig. 4). The relations of the lateritic alteration to the groundwater table and to the stream valley suggest that the leaching is influenced by stream incision and a lowering of the groundwater table caused by regional uplift during the Quaternary. Similar conclusions were reached in studies of the weathering and residual enrichment of the Tennessee phosphates in the Nashville Basin (197).

Thus some of the most important effects of weathering are the creation of new areal patterns of major element distribution, at shallow depth in the subsoil and within drainage basins, and hence of fundamental importance in the regional habitation. Appreciable calcium, phosphorus, and fluorine are lost from the soil zones and released to groundwater in lateritic aluminum phosphate development (Tables 12 and 13). Moreover the remaining phosphorus is fixed in less soluble aluminum phosphate. The new mineral hosts effect drastic redistribution and important fixation of minor elements as well. Rare earths, barium, strontium, and scandium become enriched over large areas rich in crandallite, and become sufficiently concentrated to form structurally related, independent minerals like florencite, gorceixite, and goyazite. The rare earths may also precipitate as earthy monazite and hydrous rhabdophane. Scandium forms sterrettite ($ScPO_4{\cdot}2H_2O$), an analog of variscite.

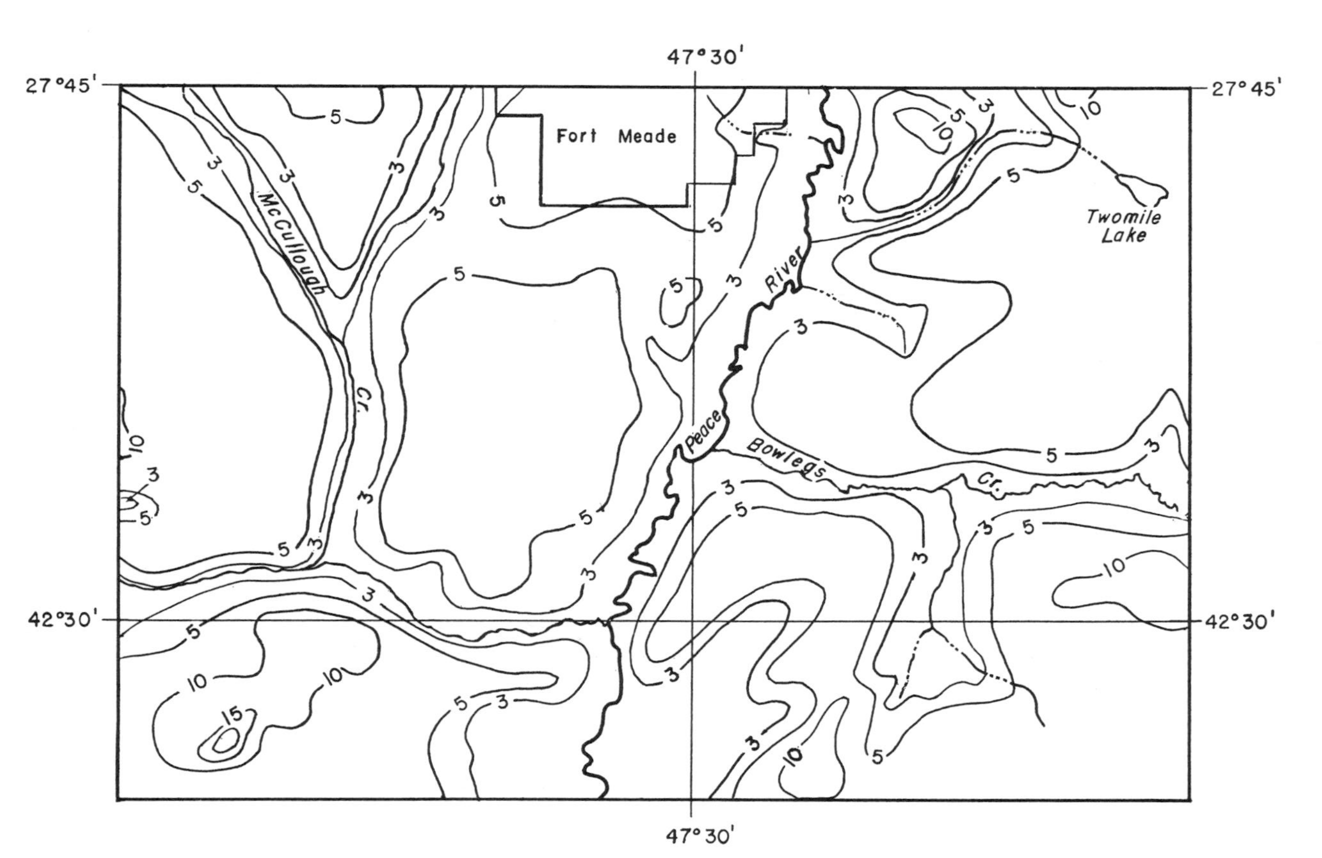

Figure 4. Isopach map of the aluminum phosphate deposits in the Peace River Valley south of Fort Meade, Florida. From Cathcart (48).

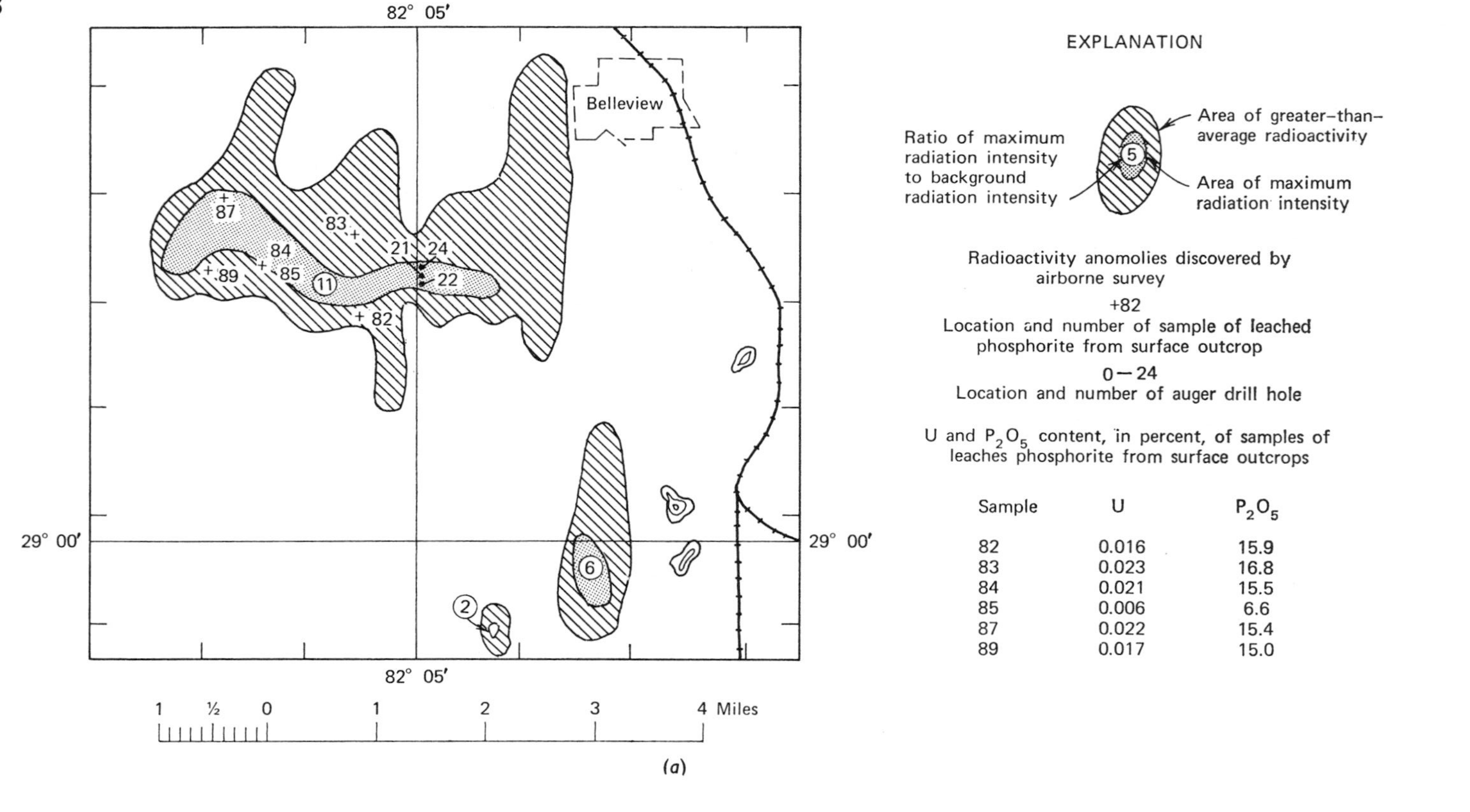

Sample	U	P_2O_5
82	0.016	15.9
83	0.023	16.8
84	0.021	15.5
85	0.006	6.6
87	0.022	15.4
89	0.017	15.0

(*a*)

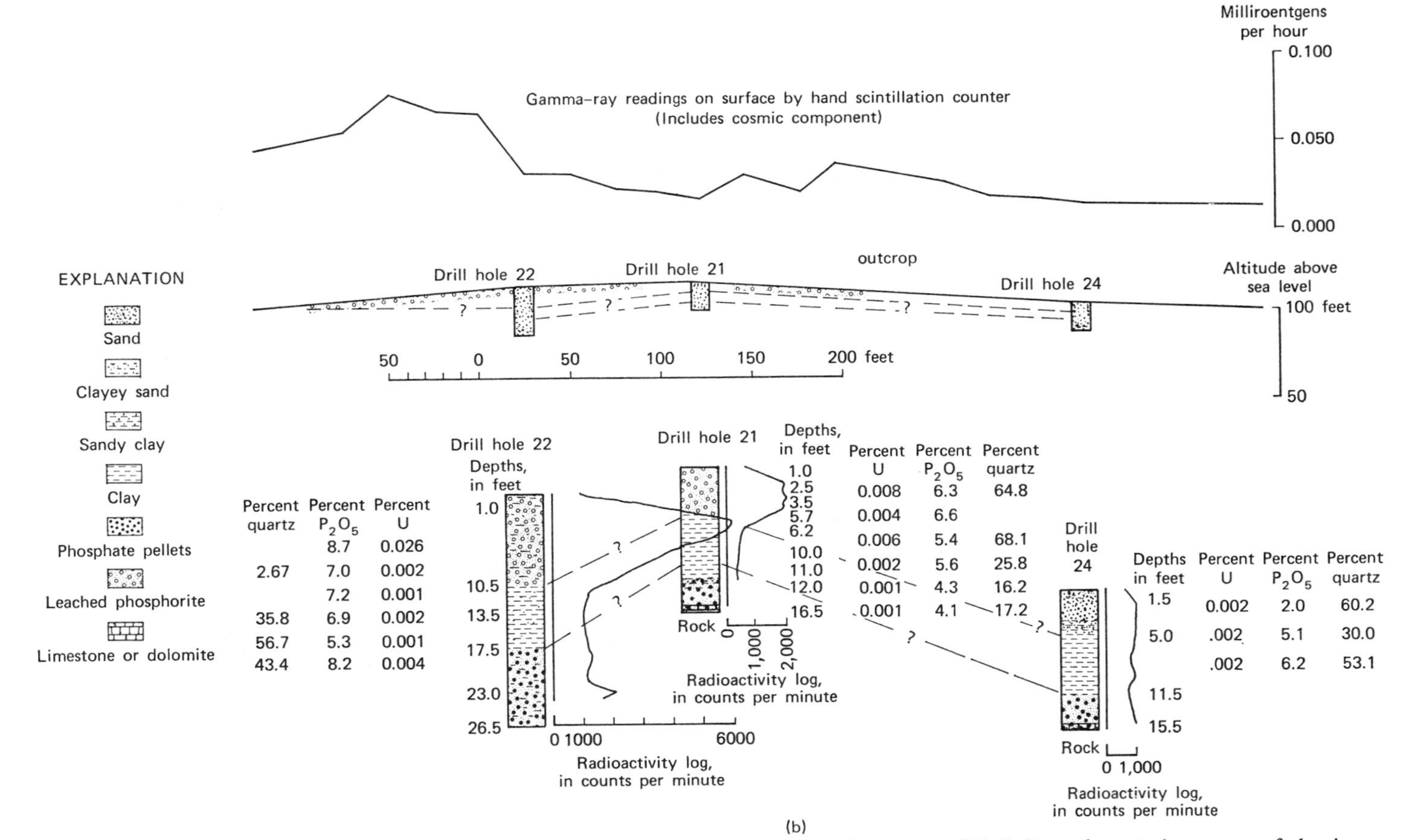

Figure 5. (*a*) Map of radiation intensity, and (*b*) profiles for lithology, radioactivity, and contents of U, P_2O_5, and quartz in an area of aluminum phosphate deposits south of Ocala, Florida. Assembled from Espenshade (68).

Of particular consequence is the pronounced basal enrichment in uranium that characterizes lateritic aluminum phosphate zonation (see Ref. 11 and Fig. 3) as well as the supergene, metasomatic apatites of calcareous terrains as in the Charleston, South Carolina area (Table 10). Areal radioactivity surveys over lateritically altered outliers of the phosphatic Hawthorn Formation in central Florida (161, 68) reveal radioactivity anomalies as great as 11 times the normal background that are readily detectable at elevations 500 ft above ground, despite the fact that the zone of maximum uranium enrichment may be covered by 5 to 10 ft of surficial material. The high secondary porosities and open textures of the lateritized aluminum phosphates and their overlying alluviated sand mantles permit the outgassing of radon, a major source in the generation of radioactivity. Figure 5 illustrates the relations of such anomalous radioactivity to the occurrence of uraniferous aluminum phosphate rock. Radioactivity of comparable magnitude has been reported over metasomatic apatite deposits and river pebble deposits in the Charleston and Edisto Island areas of South Carolina (152) and the river pebble deposits of the lower Peace River, Florida (150, 151).

The generation of enriched phosphate deposits from low-grade shelf sediments in the southeastern United States has occurred

1. "wherever subaerial weathering has leached the underlying sparsely phosphatic Hawthorn limestone to produce sinks, with pinnacles of limestone (karrenfelds) and residual pockets of the less soluble phosphate nodules;
2. "where fluviatile reworking has further concentrated weathering residues to form 'river pebble' deposits;
3. "where marine transgression has reworked residuum into extensive bedded deposits known as 'land-pebble' phosphate; or
4. "where leaching through a thin cover of Hawthorn has phosphatized subjacent limestone, to form 'hardrock' phosphate" (6).

These processes and products of concentration are of fundamental importance ecologically, even where economic proportions are not attained.

REFERENCES

1. Abele, G., K. Berger, and M. Salger, "Die Uranvorkommen im Burgsandstein Mittelfrankens," *Geol. Bavarica*, **49,** 3–90 (1962).
2. Alexiev, E., and B. Arnaudov, "Rare Earths, Uranium, and Thorium in Certain Bulgarian Phosphorites," *Tr. Vurkhu Geol. Bulg.*, **5,** 69–78 (1965) (in Bulgarian).
3. Altschuler, Z. S., "Petrography of the Phosphorites," in *Geology of Isla Mona, Puerto Rico*, C. A. Kaye, U.S. Geological Survey Prof. Paper 317-C, 1959, pp. 157–162.

4. Altschuler, Z. S., *Precipitation and Recycling of Phosphate in the Florida Land-Pebble Phosphate Deposits*, U.S. Geological Survey Prof. Paper 525-B, 1965, pp. B91–B95.
5. Altschuler, Z. S., and C. E. Boudreau, *A Mineralogical and Chemical Study of the Leached Zone of the Bone Valley Formation—A Progress Report*, U.S. Geological Survey Trace Elements Inv. 102, 1949, 67 pages.
6. Altschuler, Z. S., J. B. Cathcart, and E. J. Young, *Geology and Geochemistry of the Bone Valley Formation and Its Phosphate Deposits, West-Central Florida*, Geological Society of America Guidebook (Field Trip No. 6), Ann. Meeting Miami Beach, Florida, 1964, 68 pages.
7. Altschuler, Z. S., E. A. Cisney, and I. H. Barlow, "X-Ray Evidence of the Nature of Carbonate-Apatite," (abstr.), *Geol. Soc. Am. Bull.*, **63**, 1230–1231 (1952).
8. Altschuler, Z. S., R. S. Clarke, and E. J. Young, *Geochemistry of Uranium in Apatite and Phosphorite*, U.S. Geological Survey Prof. Paper 314-D, 1958, pp. 45–90.
9. Altschuler, Z. S., F. Cuttitta, and S. Berman, *Rare Earths in Phosphorites—Geochemistry and Economic Potential*, U.S. Geological Survey Prof. Paper 575-B, 1967, pp. B1–B9.
10. Altschuler, Z. S., E. J. Dwornik, and H. Kramer, "The Transformation of Montmorillonite to Kaolinite During Weathering," *Science*, **141**, 148–152 (1963).
11. Altschuler, Z. S., E. R. Jaffe, and F. Cuttitta, *The Aluminum Phosphate Zone of the Bone Valley Formation, Florida, and Its Uranium Deposits*, U.S. Geological Survey Prof. Paper 300, 1956, pp. 495–504.
12. Altschuler, Z. S., and E. J. Young, *Residual Origin of the "Pleistocene" Marine Sand Mantle in Central Florida Uplands and Its Bearing on Terraces and Cenozoic Uplift*, U.S. Geological Survey Prof. Paper 400-B, 1960, pp. B202–B207.
13. d'Ans, J., and R. Knütter, "Zur Abgrenzung der Existenzgebiete der Dicalciumphosphate und des Apatites im System H_2O—$Ca(OH)_2$—H_3PO_4 bei 25°C," *Angew. Chem.*, **65**, 578–581 (1953).
14. Arambourg, C., and J. Orcel, "Observations préliminaires sur a présence d'un vanadate d'urane dans les gisements de phosphate du Maroc," *Acad. Sci. Paris C. R.*, **233** (25), 1635–1636 (1951).
15. Arckhangel'skaya, V. V., "Synnyr Pluton of Alkalic Rocks and Its Apatite," *Dokl. Acad. Sci. USSR, Earth Sci. Sect.*, **158**, 114–117 (1965, originally 1964) (English translation).
16. Arimori, T., and N. Yoshida, "Mineralogical Study on the Brazilian Aluminum Phosphate Ore," *Mem. Fac. Ind. Arts, Kyoto Tech. Univ. Sci. Tech.*, No. 12, 41–47 (1963).
17. Arrhenius, G., and E. Bonatti, "Neptunism and Volcanism in the Ocean," in *Progress in Geochemistry*, F. Koczy and M. Sears (Eds.), Pergamon Press, New York, 1964, pp. 1–16.
18. Aso, Y., *Phosphates*, Maruzen Company, Tokyo, 1940, 209 pages, U.S. Navy Translation No. 399, April 22, 1946 (cited in Hutchinson, 1953).
19. Axelrod, J. M., M. K. Carron, C. Milton, and T. P. Thayer, "Phosphate Mineralization at Bomi Hills and Bambuta, Liberia," *Am. Mineral.*, **37**, 883–909 (1952).
20. Bache, B. W., "Aluminum and Iron Phosphate Studies Related to Soils, 1. Solution and Hydrolysis of Variscite and Strengite," *J. Soil Sci.*, **14**, 113–123 (1963).
21. Back, W., "Preliminary Results of a Study of Calcium Carbonate Saturation of Ground Water in Central Florida," *Int. Assoc. Sci. Hydrol. Publ.*, **8**, 43–51 (1963).

22. Bannister, F. A., and G. E. Hutchinson, "The Identity of Minervite and Palmerite with Taranakite," *Mineral. Mag.*, **28,** 31–35 (1947).

23. Bassett, H., "The System CaO—P_2O_5—H_2O at 25°," *J. Chem. Soc.*, **111,** 620–642 (1917).

24. Beevers, C. A., and D. B. McIntyre, "The Atomic Structure of Fluorapatite and Its Relation to Tooth and Bone Material," *Mineral. Mag.*, **27** (194), 254–257 (1946).

25. Bentor, Y. K., "Relations entre la tectonique et les Depots de phosphates dans le Neguev israelien," International Geological Congress, 19th, Algiers, 1952, C. R., Sect. 11, Part 11, 1953, pp. 93–101.

26. Bergendahl, M. H., "Wavellite Spherulites in the Bone Valley Formation of Central Florida," *Am. Mineral.*, **40,** 497–504 (1955).

27. Blanchard, F. N., and S. A. Denahan, "Cacoxenite and Beraunite from Florida," *Am. Mineral.*, **53,** 2096–2101 (1968).

28. Bliskovskiy, V. Z., "Phosphorite-Bearing Weathered Crust in the Bol'shie Dzhebarty (Eastern Sayan)," *Lithol. Miner. Deposits*, No. 4, 33–42 (1967) (English translation).

29. Bliskovskiy, V. Z., and A. I. Smirnov, "Radioactivity of Phosphorites," *Geokhim.*, No. 6, 744–747 (1966) (in Russian).

30. Blokh, A. M., and A. B. Kochenov, "Element Admixture in Bones of Fossil Fish," *Geol. Mestorozhd. Redk. Elem.*, **24,** 1–100 (1964) (in Russian).

31. Borisenko, L. F., "Occurrence of Scandium in Bone Remnants of Fishes of Tertiary Age," *Tr. Inst. Mineral. Geokym. Kristallokhim. Redk. Elem.*, No. 7, 65–70 (1961) (in Russian).

32. Borneman-Starynkevich, I. D., and N. V. Belov, "Isomorphic Substitutions in Carbonate Apatite," *C. R. Acad. Sci. USSR*, **26,** 804–806 (1940).

33. Borodin, L. S., and M. F. Kazakova, "Belovite, a New Mineral from Alkali Pegmatites," *Dokl. Akad. Nauk SSSR*, **96** (3) (1954).

34. Boujo, A., "Nouvelles données sur le Crétacé et l'Eocène phosphaté du gisement des Gantour," *Notes Serv. Géol. Maroc*, **28,** 7–16 (1968).

35. Bowie, S. H. U., and D. Atkin, "An Unusually Radioactive Fossil Fish from Thurso, Scotland," *Nature*, **177** (4506), 487–488 (1956).

36. Brasseur, H., P. Herman, and A. Hubaux, "Apatites de l'est du Congo et du Ruanda," *Ann. Soc. Géol. Belg.*, **85,** 61–85 (1961).

37. Brown, W. E., J. Lehr, J. P. Smith, and A. W. Frazier, "Crystallography of Octocalcium Phosphate," *J. Am. Chem. Soc.*, **79,** 5318–5319 (1957).

38. Bushinski, G. I., "On Shallow Water Origin of Phosphorite Sediments," in *Delta and Shallow Marine Deposits*, van Straaten (Ed.), Elsevier, Amsterdam, 1964, pp. 62–70.

39. Bushinski, G. I., "Ancient Phosphorites of Asia and Their Formation," *Tr. Geol. Inst. SSSR*, **149,** 192 (1966).

40. Bushinski, G. I., "The Origin of Marine Phosphorites," *Lithol. Miner. Resour.*, No. 3, 292–311 (1966).

41. Cady, J. G., W. L. Hill, E. V. Miller, and R. M. Magness, "Occurrence of Beta Tricalcium Phosphate in Northern Mexico," *Am. Mineral.*, **37,** 180–183 (1952).

42. Capdecomme, L., "Étude mineralogique des gites dephosphates alumineux de la region de Thiès (Sénégal)," International Geological Congress, 19th, Algiers, 1952, C. R., Sect. 11, Fasc. 11, 1953, pp. 103–117.

43. Capdecomme, L., and G. Kulbicki, "Argiles des gîtes de phosphates alumineux de la région de Thiès, Senegal," *Soc. Fr. Min. Crist. Bull.*, **77,** 500–518 (1954).
44. Capdecomme, L., and M. Orliac, "Sur les caractères chimiques et thermiques des phosphates alumineux de la région de Thiès (Sénégal)," *Coloq. Int. Phosphates Miner. Solides, Toulouse*, **2,** 45–55 (1968).
45. Capdecomme, L., and R. Pulou, "Sur la radioactivité des phosphate de la region de Thiès (Sénégal)," *Acad. Sci. Paris C. R.*, **239,** 288–290 (1954).
46. Carnot, A., "Recherches sur la composition générale et la teneur en fluor des os modernes et des os fossils des différents âges," *Annales des mines*, Ser. 9, Vol. 3, 155–195 (1893).
47. Cathcart, J. B., "Distribution and Occurrence of Uranium in the Calcium Phosphate Zone of the Land-Pebble Phosphate District of Florida," *Proceedings of the International Conference on the Peaceful Uses of Atomic Energy*, **6,** 514–519, Geneva, 1955; U.S. Geological Survey Prof. Paper 300, 1956, pp. 489–494.
48. Cathcart, J. B., *Economic Geology of the Fort Meade Quadrangle, Polk and Hardee Counties, Florida*, U.S. Geological Survey Bulletin 1207, 1966, 97 pages.
49. Cathcart, J. B., "Florida-Type Phosphate Deposits of the United States," *Proceedings of the Seminar on Sources of Mineral Raw Materials for the Fertilizer Industry of Asia*, United Nations, New York, 1968, pp. 178–186.
50. Clark, J. S., "Solubility Criteria for the Existence of Hydroxyapatite," *Can. J. Chem.*, **33,** 1696–1700 (1955).
51. Clarke, R. S., Jr., and Z. S. Altschuler, "Determination of the Oxidation State of Uranium in Apatite and Phosphorite Deposits," *Geochim. Cosmochim. Acta*, **13** (2 and 3), 127–142 (1958).
52. Coetzee, G. L., and C. B. Edwards, "The Mrima Hill Carbonatite, Coast Province, Kenya," *Geol. Soc. S. Africa, Trans.*, **62,** 373–397 (1959).
53. Cole, C. V., and M. L. Jackson, "Solubility Equilibrium Constant of Dihydroxy-aluminum Dihydrogen Phosphate Relating to a Mechanism of Phosphate Fixation in Soils," *Soil Sci. Proc.*, **15,** 84–89 (1950).
54. Corsaro, G., P. Lauderbach, and H. Schwantje, "Formation and Behavior of Hydroxyapatite," *J. Am. Water Works Assoc.*, **56,** 347–357 (1966).
55. Davidson, C. F., and D. Atkin, "On the Occurrence of Uranium in Phosphate Rock," International Geological Congress, 19th, Algiers, 1952, C. R., Sect. 11, Fasc. 11, 1953, pp. 13–31.
56. Davies, K. A., "The Phosphate Deposits of the Eastern Province, Uganda," *Econ. Geol.*, **42,** 137–146 (1947).
57. Deans, T., "The Economic Mineralogy of African Carbonatites," in *Carbonatites*, O. F. Tuttle and J. Gittins (Eds.), Interscience, New York, 1966.
58. Deans, T., "Exploration for Phosphate Deposits Associated with Carbonatites and Pyroxenites," *Proceedings of the Seminar on Sources of Raw Materials for the Fertilizer Industry of Asia*, United Nations, New York, 1968, pp. 109–118.
59. Deitz, V. R., H. M. Rootare, and F. E. Carpenter, "The Surface Composition of Hydroxylapatite Derived from Solution Behavior of Aqueous Suspensions," *J. Colloid Sci.*, **19,** 89–101 (1964).
60. deVilliers, J., "Francolite from Richtersveld, South Africa," *Am. J. Sci.*, **240,** 443–447 (1942).

61. Du Toit, A. L., *The Phosphates of Suldanha Bay*, South African Geological Survey Memoir 10, 1917, 34 pages.
62. Ehrlich, A. M., "Rare-Earth Abundances in Manganese Nodules," Ph.D. Thesis, Massachusetts Institute of Technology, Cambridge, Massachusetts, 1968, 255 pages.
63. Eisenberger, S., A. Lehrman, and W. D. Turner, "The Basic Calcium Phosphates and Related Systems. Some Theoretical and Practical Aspects," *Chem. Rev.*, **26,** 257–296 (1940).
64. Elliot, J. C., "Recent Progress in the Chemistry, Crystal Chemistry, and Structure of the Apatites," *Calc. Tiss. Res.*, **3,** 293–307 (1969).
65. Elschner, C., *Corallogen Phosphat-Inseln Austral-Oceaniens und ihre Produkte*, Lubeck, Max Schmidt, 1913.
66. Elschner, C., "Beitrage zur Kenntnis der Koralleninseln des Stillen Ozeans," *Z. Prakt. Geol.*, **31,** 69–73 (1923).
67. Emigh, G. D., "Petrology and Origin of Phosphates," in *Anatomy of the Western Phosphate Field*, L. A. Hale (Ed.), Intermountain Association of Geologists, Salt Lake City, Utah, 1967, pp. 103–114.
68. Espenshade, G. H., *Geologic Features of Areas of Abnormal Radioactivity South of Ocala, Marion Co., Florida*, U.S. Geological Survey Bulletin 1046-J, 1958, pp. 205–219.
69. Espenshade, G. H., and C. Spencer, *Geology of Phosphate Deposits of Northern Peninsular Florida*, U.S. Geological Survey Bulletin 1118, 1963, 115 pages.
70. Farr, T. D., "Phosphorus, Properties of the Element and Some of Its Compounds," Tennessee Valley Authority Chemical Engineering Report No. 8, TVA, Wilson Dam, Alabama, 1950, 99 pages.
71. Fieldes, M., G. Bealing, G. G. Claridge, N. Wells, and N. H. Taylor, "Mineralogy and Radioactivity of Niue Island Soils," *N. Z. J. Sci.*, **3,** 658–675 (1960).
72. Fischer, D. J., "Pegmatite Phosphates and Their Problems," *Am. Mineral.*, **43,** 181–207 (1958).
73. Fleischer, M., "Discussion of Nomenclature," *Am. Mineral.*, **44,** 326 (1944).
74. Fouretier, G., "La précipitation du phosphate tricalcique et l'hydroxyapatite," *C. R.*, **205,** 413–415 (1937).
75. Frazier, A. W., J. R. Lehr, and J. P. Smith, "The Magnesium Phosphates Hannayite, Schertelite, and Bobierite," *Am. Mineral.*, **48,** 635–641 (1963).
76. Frondel, C., "Whitlockite: A New Calcium Phosphate $Ca_3(PO_4)_2$," *Am. Mineral.*, **26,** 145–152 (1941).
77. Frondel, C., "Mineralogy of the Calcium Phosphates in Insular Phosphate Rock," *Am. Mineral.*, **28,** 215–232 (1943).
78. Frondel, C., "The Dufrenite Problem," *Am. Mineral.*, **34,** 513–540 (1949).
79. Frondel, C., "Geochemical Scavenging of Strontium," *Science*, **128,** 1623–1624 (1958).
80. Frondel, C., J. Ito, and A. Montgomery, "Scandium Content of Some Aluminum Phosphates," *Am. Mineral.*, **53,** 1223–1231 (1968).
81. Fuchs, L. H., "Occurrence of Whitlockite in Meteorites," *Science*, **137,** 425–426 (1962).
82. Fuchs, L. H., E. Olsen, and E. P. Henderson, "On the Occurrence of Brianite and Panethite, Two New Phosphate Minerals from the Dayton Meteorite," *Geochim. Cosmochim. Acta*, **31,** 1711–1719 (1967).

83. Garrels, R. M., *Mineral Equilibria*, Harper and Row, New York, 1960, 254 pages.
84. Garson, M. S., "Carbonatites in Malawi," in *Carbonatites*, O. F. Tuttle and J. Gittins (Eds.), Interscience, New York, 1966, pp. 33–71.
85. Geiger, T., "Carbonate Apatites," *Schweiz. Min. Petrog. Mitt.*, **30,** 161–181 (1950).
86. Gittins, J., "Summaries and Bibliographies of Carbonatite Complexes," in *Carbonatites*, O. F. Tuttle and J. Gittins (Eds.), Interscience, New York, 1966, pp. 417–570.
87. Goldberg, E. D., and R. H. Parker, "Phosphatized Wood from the Pacific Sea Floor," *Geol. Soc. Am. Bull.*, **71,** 631–632 (1960).
88. Goldman, M. I., "Basal Glauconite and Phosphate Beds," *Science*, New Series, **56,** 171–173 (1922).
89. Gruner, J. W., and D. McConnell, "The Problem of the Carbonate Apatites. The Structure of Francolite," *Z. Kristallogr.*, **97,** 208–215 (1937).
90. Gulbrandsen, R. A., *Petrology of the Meade Peak Phosphatic Shale Member of the Phosphoria Formation at Coal Canyon, Wyoming*, U.S. Geological Survey Bulletin 1111-C, 1960, pp. C71–C146.
91. Gulbrandsen, R. A., "Chemical Composition of Phosphorites of the Phosphoria Formation," *Geochim. Cosmochim. Acta*, **30** (8), 769–778 (1966).
92. Gulbrandsen, R. A., "Physical and Chemical Factors in the Formation of Marine Apatite," *Econ. Geol.*, **64,** 365–382 (1969).
93. Hanshaw, B. B., W. Back, and M. Rubin, "Carbonate Equilibria and Radiocarbon Distribution Related to Groundwater Flow in the Florida Limestone Aquifer, U.S.A.," Int. Assoc. Sci. Hydrology Symposium of Dubrovnik, 1965, pp. 601–614.
94. Haseman, J. F., J. R. Lehr, Jr., and J. P. Smith, "Mineralogical Character of Some Iron and Aluminum Phosphates Containing Potassium and Ammonium," *Soil Sci. Soc. Am. Proc.*, **15,** 76–84 (1950).
95. Hayek, E., and H. Newesley, "Uber die Existenz von Tricalciumphosphat in Wassriger Lösings," *Monatsh. Chem.*, **89,** 88–95 (1958).
96. Hedgepeth, J., Ed., *Treatise of Marine Ecology and Paleoecology*, Vol. 1, *Ecology*, Geological Society of America Memoir 67, 1967, 1296 pages.
97. Heinrich, E. W., *The Geology of Carbonatites*, Rand McNally, Chicago, Illinois, 1971, 607 pages.
98. Hendricks, S. B., "The Crystal Structure of Alunite and the Jarosites," *Am. Mineral.*, **22,** 773–780 (1937).
99. Hendricks, S., M. E. Jefferson, and V. M. Mosely, "The Crystal Structure of Some Natural and Synthetic Apatite-Like Substances," *Z. Kristallogr.*, **81,** 352–369 (1932).
100. Hill, M. N., Ed., *The Seas*, Vol. I, *Ideas and Observations*, Vol. II, *The Composition of Sea Water*, Wiley, New York, 1962, 1963.
101. Himmelfarb, B. M., "Principal Geological Patterns of Distribution of Phosphorite Deposits in the U.S.S.R.," in *Geology of Deposits of Rock-Forming Minerals*, Goskhimizdat, 1959 (in Russian).
102. Hsu, P. H., and M. L. Jackson, "Inorganic Phosphate Transformation by Chemical Weathering in Soils as Influenced by pH," *Soil Sci.*, **90,** 16–24 (1960).
103. Hutchinson, G. E., "Survey of Contemporary Knowledge of Biogeochemistry: 3. The Biogeochemistry of Vertebrate Excretion," *Am. Mus. Nat. Hist. Bull.*, **96,** 554 pages (1950).

104. Jacob, K. D., W. L. Hill, H. L. Marshall, and D. S. Reynolds, *The Composition and Distribution of Phosphate Rock with Special Reference to the United States*, U.S. Department of Agriculture Technical Bulletin 364, 1933, 90 pages.
105. Jaffe, E. B., *Abstracts of the Literature on Synthesis of Apatites and Some Related Phosphates*, U.S. Geological Survey Circular 135, 1951, 78 pages.
106. Johnson, R. L., "The Shawa and Dorowa Carbonatite Complexes, Rhodesia," in *Carbonatites*, O. F. Tuttle and J. Gittins (Eds.), Interscience, New York, 1966, pp. 205–224.
107. Kapustyanski, I. D., "Rare Elements in the Cretaceous Phosphorites and Phosphatized Formations in the Kul'dzuk-Tau Mts. (Kyzyl-Kum)," *Nauchn. Tr. Tashk. Gos. Unn.*, No. 249, Geol. No. 21, 230–239 (1964).
108. Kay, M. I., R. A. Young, and A. S. Posner, "Crystal Structure of Hydroxyapatite," *Nature*, **204,** 1050–1052 (1964).
109. Kaye, C. A., *Geology of Isla Mona, Puerto Rico*, U.S. Geological Survey Prof. Paper 317-C, 1959, pp. 141–178.
110. Kazakov, A. V., "The System $CaO—P_2O_5—H_2O$ in the Field of Low Concentration (Synthesis of Tricalcium Phosphate and Hydroxylapatite)," *Sci. Inst. Fertilizers and Insectofungicides Trans.* (*Leningrad*), **139,** 3–73 (1937).
111. Kazakov, A. V., "The Phosphorite Facies and the Genesis of Phosphorites," in "Geological Investigations of Agricultural Ores USSR," *Sci. Inst. Fertilizers and Insectofungicides Trans.* (*USSR*), No. 142, 95–113 (1937), special issue in English for the 17th International Geological Congress.
112. Keppler, U., "Zum Whitlockite-Problem," *Neues Jahrb. Miner., Monatsh.*, 171–176 (1965).
113. Keppler, U., "Structural Investigation of 'Calcium Phosphate' and Isotypic Structures," *Colloq. Int. Phosphates Minér. Solides, Toulouse*, **1,** 128–131 (1968).
114. Kholodov, V. N., "On Rare and Radioactive Elements in Phosphorites," *Akad. Nauk Inst. Mineral. Geokgm. Kristallokhim. Redk. Elem., Tr.*, **17,** 67–108 (1963) (in Russian).
115. Kochenov, A. V., and V. V. Zinovieff, "Distribution of Rare Earth Elements in Phosphatic Fish Rests from the Maikop Sediments," *Geochem.*, **8,** 714–725 (1960).
116. Kolodny, Y., and I. R. Kaplan, "Uranium Isotopes in the Sea-Floor Phosphorites," *Geochim. Cosmochim. Acta*, **34,** 3–24 (1940).
117. Kramer, J. R., "Sea Water—Saturation with Apatites and Carbonates," *Science*, **146** (3644) 637–638 (1964).
118. Krauskopf, K. B., "Sedimentary Deposits of Rare Metals," *Econ. Geol., 50th Ann. Vol.*, 411–463 (1955).
119. Krauskopf, K. B., *Introduction to Geochemistry*, McGraw-Hill, New York, 1967, 721 pages.
120. Lacroix, A., "Sur la bobierrite," *C. R. Acad. Sci., Paris*, **106,** 631–633 (1888).
121. Lacroix, A., "Sur un gisement de redondite à la Martinique," *Bull. Soc. Fr. Mineral.*, **28,** 13–16 (1905).
122. Lacroix, A., "Sur la transformation des roches volcaniques en phosphate d'alumine sous l'influence de produits d'origine physiologique," *C. R. Acad. Sci., Paris*, **143,** 661–664 (1906).
123. Lacroix, A., "Sur l'existence d'une variété de minervite à la Réunion," *Bull. Soc. Fr. Mineral.*, **33,** 34–39 (1910).

124. Lacroix, A., "Sur les minéraux du guano de la Réunion," *Bull. Soc. Fr. Mineral.*, **35,** 114–119 (1912).

125. Lacroix, A., "Le gisement phosphate de l'ile Juan de Nova," *Bull. Soc. Fr. Mineral.* **41,** 100–103 (1918).

126. Lacroix, A., *Le volcan actif de l'ile de la Réunion et ses produits*, Gauthier-Villars, Paris, 1936, 297 pages.

127. LaMer, V. K., "The Solubility Behavior of Hydroxyapatite," *J. Phys. Chem.*, **66,** 973 (1962).

128. Larsen, E. S., 3rd, "The Mineralogy and Paragenesis of the Variscite Nodule from Near Fairfield, Utah," *Am. Mineral.*, **27,** 281–300, 350–372, 441–451 (1942).

129. Lerman, A., "Identification and Analysis of Phosphate Minerals," in *Analytical Chemistry of Phosphorous Compounds*, M. M. Hajmann (Ed.), Wiley, New York, 1970.

130. McCann, H. G., and F. A. Bullock, "Reactions of the Fluoride Ion with Powered Enamel and Dentine," *J. Dent. Res.*, **34,** 59–67 (1955).

131. McConnell, D., "The Substitution of SiO_4 and SO_4 Groups for PO_4 Groups in the Apatite Structure, Ellestadite End Member," *Am. Mineral.*, **22,** 977–986 (1937).

132. McConnell, D., "A Structural Investigation of the Isomorphism of the Apatite Group," *Am. Mineral.*, **23,** 1–19 (1938).

133. McConnell, D., "Clinobarrandite and the Isomorphous Series Variscite-Metavariscite," *Am. Mineral.*, **25,** 719–725 (1940).

134. McConnell, D., "Phosphatization at Malpelo Island, Columbia," *Geol. Soc. Am. Bull.*, **54,** 707–716 (1943).

135. McConnell, D., "The Petrography of Rock Phosphates," *J. Geol.*, **58,** 16–23 (1950).

136. MacIvor, R. W. E., "On Minerals Occurring in Australian Bat Guano," *Chem. News*, **85,** 181–182 (1902).

137. MacIvor, R. W. E., "Further Note on Minerals Occurring in Australian Bat Guano," *Chem. News*, **85,** 217 (1902).

138. McKelvey, V. E., "Successful New Techniques in Prospecting for Phosphate Deposits," in *Natural Resources*, Vol. II of U.S. Papers for the United Nations Conference on Applications of Science and Technology for Less Developed Areas, 1963, 355 pages.

139. McKelvey, V. E., and L. D. Carswell, "Uranium in Phosphoria Formation," *Proceedings of the International Conference on the Peaceful Uses of Atomic Energy*, Vol. 6, Geneva, 1955, pp. 503–506; U.S. Geological Survey Prof. Paper 300, 1956, pp. 483–487.

140. McKelvey, V. E., J. B. Cathcart, Z. S. Altschuler, R. Swanson, and K. L. Buck, "Domestic Phosphate Deposits," in *Soil and Fertilizer Phosphorus in Crop Nutrition: Agronomy Monograph IV*, Academic Press, New York, 1953, pp. 347–376.

141. McKelvey, V. E., D. L. Everhart, and R. M. Garrels, "Origin of Uranium Deposits," *Econ. Geol. 50th Ann. Vol. 1905–1955*, Part 1, 464–533 (1955).

142. McKelvey, V. E., J. S. Williams, R. P. Sheldon, E. R. Cressman, T. M. Cheney, and R. W. Swanson, *The Phosphoria, Park City, and Shedhorn Formations in the Western Phosphate Field*, U.S. Geological Survey Prof. Paper 313-A, 1959, 47 pages.

143. McKie, D., "Goyazite and Florencite from Two African Carbonatites," *Mineral. Mag.*, **33,** 281–297 (1962).

144. Machatschki, F., "Beiträge zur Kenntnis der Ablagerungen," in *Die Drachenhöhle bei Mixnitz*, Vols. 7 and 8, O. Abel and G. Kyrle (Eds.), Vienna, (Verlag) Österr. Staatsdruckerei, Spelaeol. Monogr., 1931, pp. 225–245.
145. Malde, H. E., *Geology of the Charleston Phosphate Area, South Carolina*, U.S. Geological Survey Bulletin 1079, 1959, 105 pages.
146. Maslennikov, B. M., and F. A. Kavitskaya, "The Phosphate Substance of Phosphorites," *Dokl. Akad. Nauk SSSR*, **109**, 990–992 (1956).
147. Mehmel, M., "Über die Structur des Apatits," *Z. Kristallogr.*, **75**, 323 (1930).
148. Melcher, G. C., "The Carbonatites of Jacupiranga, Sao Paulo, Brazil," in *Carbonatites*, O. F. Tuttle and J. Gittins (Eds), Interscience, New York, 1966, pp. 169–181.
149. Merker, L., and H. Wondratschek, "Uber Lücken in derApatitstruktur," *Fortschr. Mineral.*, **36**, 73–74 (1958).
150. Meuschke, J. L., *Airborne Radioactivity Survey of the Ft. Meyers Areas, Charlotte and Lee Counties, Florida*, U.S. Geological Survey Geophys. Inv. Map GP 122, 1955.
151. Meuschke, J. L., *Airborne Radioactivity Survey of the Gardner Area, De Soto, Hardee, Manatee, and Lee Counties, Florida*, U.S. Geological Survey Geophys. Inv. Map GP 122, 1955.
152. Meuschke, J. L., *Airborne Radioactivity Survey of the Edisto Island Area, Berkeley, Charleston, Colleton, and Dorchester Counties, South Carolina*, U.S. Geological Survey Geophys. Inv. Map 123, 1955.
153. Meuschke, J. L., R. M. Moxham, and T. E. Bortner, *Airborne Radioactivity Survey of Parts of the Atlantic Ocean Beach, North and South Carolina*, U.S. Geological Survey Open-File Report, 1953.
154. Milton, C., J. M. Axelrod, M. K. Carron, and F. S. MacNeil, "Gorceixite from Dale Co., Alabama," *Am. Mineral.*, **43**, 688–694 (1958).
155. Mingaye, J. C. H., "On the Occurrence of Phosphatic Deposits in the Jenolan Caves, New South Wales," *Rep. Australas. Assoc. Adv. Sci.*, **7**, 327–331 (1898).
156. Montel, G., and G. Chaudron, "Reactions du fluorure de sodium dissous sur le phosphate tricalcique hydraté et l'hydroxyapatite précipitee," *C. R. Acad. Sci.*, **242**, 1182–1184 (1956).
157. Mooney, R. W., and M. A. Aia, "Alkaline Earth Phosphates," *Chem. Rev.*, 433–462 (1961).
158. Moore, P. B., "Crystal Chemistry of the Basic Iron Phosphates," *Am. Mineral.*, **55**, 135–169 (1970).
159. Moreno, E. C., W. E. Brown, and G. Osborn, "Solubility of Dicalcium Phosphate Dihydrate in Aqueous Systems," *Soil Sci. Soc. Am. Proc.*, **24**, 94–98 (1960).
160. Moreno, E. C., W. E. Brown, and G. Osborn, "Stability of Dicalcium Phosphate Dihydrate in Aqueous Solutions and Solubility of Octocalcium Phosphate," *Soil Sci. Soc. Am. Proc.*, **24**, 99–102 (1960).
161. Moxham, R. M., *Airborne Radioactivity Surveys for Phosphate in Florida*, U.S. Geological Survey Circular 230, 1954, 4 pages.
162. Murray, J. W., and R. V. Dietrich, "Brushite and Taranakite from Pig Hole Cave, Giles, Co. Virginia," *Am. Mineral.*, **41**, 616–626 (1956).
163. Naray-Szabo, S., "The Structure of Apatite $(CaF)Ca_4(PO_4)_3$," *Z. Kristallogr.*, **75**, 387 (1930).

164. Neuman, W. F., and M. W. Neuman, *The Chemical Dynamics of Bone Mineral*, University of Chicago Press, Chicago, Illinois, 1958, 209 pages.

165. Newesley, H., "Die Realstruktur der Oktocalcium-phosphat," *Monatsh. Chem.*, **95**, 94–101 (1963).

166. Nugent, L. E., "Elevated Phosphate Islands of Micronesia," *Geol. Soc. Am. Bull.*, **59**, 977–994 (1948).

167. Oakley, K. P., "The Composition of the Piltdown Hominoid Remains," *Br. Mus. Nat. Hist. Bull., Geol.* **2** (6), 254–265 (1955).

168. Obermuller, A. G., "Contribution a l'étude geologique et minerale de l'ile Clipperton," in *Recherche Geologique et Minerale en Polynesie Francaise*, Paris, Insp. Gen. Min. Geol., 1959.

169. Owen, L., "Notes on the Phosphate Deposits of Ocean Island; with Remarks on the Phosphates of the Equatorial Belt of the Pacific Ocean," *Q. J. Geol. Soc. London*, **79**, 1–15 (1923).

170. Owen, L., "The Phosphate Deposits on Ocean Island," *Econ. Geol.*, **22**, 632–634 (1927).

171. Owens, J. P., Z. S. Altschuler, and R. Berman, "Millisite in Phorphorite from Homeland, Florida," *Am. Mineral.*, **45**, 547–561 (1959).

172. Palache, C., H. Berman, and C. Frondel, *Dana's System of Mineralogy*, 7th ed., Vol. 2, Wiley, New York, 1951, 1224 pages.

173. Pecora, W. T., "Carbonatites—a Review," *Geol. Soc. Am. Bull.*, **67**, 1537–1556 (1956).

174. Popova, Z. D., "Variscite, the Principal Ore-Forming Mineral of the Sarysai Deposits," *Izv. Alzad. Nauk Kaz. SSR, Ser. Geol. 1963*, No. 1 (1963) (in Russian).

175. Power, F. D., "Phosphate Deposits of Ocean and Pleasant (Nauru) Islands," *Trans. Aust. Inst. Mineral. Eng.*, **10**, 213 (1905).

176. Power, F. D., "Phosphate Deposits of the Pacific Ocean," *Econ. Geol.*, **20**, 266 (1925).

177. Rodgers, J., "Phosphate Deposits of Former Japanese Islands," *Econ. Geol.*, **43**, 400–407 (1948).

178. Rootare, H. M., V. R. Deitz, and F. G. Carpenter, "Solubility Product Phenomena of Hydroxyapatite–Water Systems," *J. Colloid Sci.*, **17**, 179 (1962).

179. Rose, H. J., L. V. Blade, and M. Ross, "Earthy Monazite at Magnet Cove, Ark.," *Am. Mineral.*, **43**, 995–997 (1958).

180. Rowles, S. L., "The Precipitation of Whitlockite from Aqueous Solution," *Colloq. Int. Phosphates Miner. Solides, Toulouse*, **1**, 151–155 (1967).

181. Russ. W., *The Phosphate Deposits of Abeokuta Province*, Nigeria Geological Survey Bulletin 7, 1924, pp. 9–38.

182. Salvan, H., "Les phosphates de chaux sédimentaires du Maroc, leurs caractéristiques et leurs problèmes (essai de synthèse)," *Notes Maroc.*, No. 14, 7–20 (1960) (in French).

183. Sandell, E., M. Hey, and D. McConnell, "The Composition of Francolite," *Mineral. Mag.*, **25**, 395–401 (1939).

184. Schadler, J., "Die Ablagerungen," in *Die Drachenhöhle bei Mixnitz*, Vols. 7 and 8, O. Abel and G. Kyrle, Vienna, (Verlag) Osterr. Staatsdrucherei, Spelaeol. Monogr., 1931, pp. 169–224.

185. Schleede, A., W. Schmidt, and H. Kindt, "Zur Kenntnis der Calciumphosphate und Apatite," *Z. Elektrochem.*, **38** (8a) (1932).

186. Semenov, E. I., "The Mineralogy of Niobium and Tantalum in the Crust of Weathering," *Tr. Inst. Miner. Geokhym., Kristallokhym. Red. Elem.*, 30–35 (1967).

187. Semenov, E. I., V. N. Kholodov, and R. L. Barinskii, "Rare Earths in Phosphorites," *Geochem.*, **5**, 501–507 (1962).

188. Sheldon, R. P., *Geochemistry of Uranium in Phosphorites and Black Shales of the Phosphoria Formation*, U.S. Geological Survey Bulletin 1084-D, 1959, pp. D83–D115.

189. Sheldon, R. P., *Physical Stratigraphy and Mineral Resources of Permian Rocks in Western Wyoming*, U.S. Geological Survey Prof. Paper 313-B, 1963, pp. B49–B271.

190. Sheldon, R. P., *Paleolatitudinal and Paleogeographic Distribution of Phosphorite*, U.S. Geological Survey Prof. Paper 501-C, 1964, pp. 106–113.

191. Simpson, E. S., "Contributions to the Mineralogy of Western Australia," *J. Roy. Soc. West. Aust. Sect.*, **7**, 18, 61–74 (1932).

192. Slansky, M., *Contribution a l'étude géologique du bassin sédimentaire côtier du Dahomey et du Togo*, Mém. No. 11, Bur. Rech. Géol. Miner., 1962, 270 pages, (in French).

193. Slansky, M., "Généralités sur la sédimentation phosphatée et la recherche du phosphate," *Bur. Rech. Géol. Miner. Bull.*, No. 1, 43–61 (1964).

194. Slansky, M., A. Lallemand, and G. Millot, "La sedimentation et l'altération lateritique des formations phosphatées du gisement de Taiba (Republique du Sénégal)," *Bull. Serv. Carte Géol. Alsace et Lorraine*, **4,** N 14 (1964).

195. Smirnov, A. I., R. B. Ivnitskaya, and T. P. Zalavina, "Experimental Data on the Possibility of the Chemical Precipitation of Phosphate from Sea Water," *Gos. Nauchnoizsled. Inst. Gorno-Khim. Syr'ya Tr.* (translated by State Sci.-Research Inst. for Chem. Raw Materials), No. 7, 289–302 (1962).

196. Smith, J. P., and J. R. Lehr, "An X-Ray Investigation of Carbonate Apatites," *J. Agric. Food Chem.*, **14** (4), 342–349 (1966).

197. Smith, R. W., and G. I. Whitlatch, *The Phosphate Resources of Tennessee*, Tennessee Division of Geology Bulletin 48, 1940.

198. Spencer, M. J., and K. Uhler, *The Structure, Composition, and Growths of Bone, a Bibliography*, Armed Forces Medical Library, Washington, D.C., 1955, 190 pages.

199. Sverdrup, H. W., M. W. Johnson, and R. H. Fleming, *The Oceans, Their Physics, Chemistry, and General Biology*, Prentice-Hall, Englewood Cliffs, N. J., 1942, 1087 pages.

200. Swaine, D. J., *The Trace Element Content of Fertilizers*, Commonwealth Bureau of Soils, Technical Communication, No. 52, Herpenden, England, 1962, 306 pages.

201. Taylor, A. W., and E. L. Gurney, "Solubilities of Potassium and Ammonium Taranakites," *J. Phys. Chem.*, **65**, 1613–1616 (1961).

202. Thompson, M. E., *Distribution of Uranium in Rich Phosphate Beds of the Phosphoria Formation*, U.S. Geological Survey Bulletin 988-D, 1953, pp. D45–D67.

203. Thompson, M. E., *Further Studies of the Distribution of Uranium in Rich Phosphate Beds of the Phosphoria Formation*, U.S. Geological Survey Bulletin 1009-D, 1954, pp. D107–D123.

204. Tooms, J. S., C. R. Summerhayes, and D. S. Cronan, "Geochemistry of Marine Phosphate and Manganese Deposits," *Oceanogr. Mar. Biol. Ann. Rev.*, **7**, 49–100 (1969).

205. Tovberg-Jensen, A., and K. G. Hansen, "Tetracalcium Hydrogen Triphosphate Trihydrate, a Constituent of Dental Calculus," *Experientia*, **XIII** (8), 311 (1957).

206. Trautz, O. R., E. Fessendon, and M. G. Newton, "Magnesian Whitlockite," *J. Dent. Res.*, **33,** 687 (1954).

207. Trautz, O. R., and E. Fessenden, "Formation and Stability of Whitlockite and Octocalcium Phosphate," *J. Dent. Res.*, **37,** 78 (abs.) (1958).

208. Trommel, G., and Eitel, W., "Synthesis of Silicate Apatites of the Britholite-Abukumalite Group," *Z. Kristallogr.*, **109,** 271–279 (1957).

209. Trueman, N. A., "The Phosphate, Volcanic, and Carbonate Rocks of Christmas Island, Indian Ocean," *J. Geol. Soc. Aust.*, **12** (2), 261–284 (1965).

210. Trueman, N. A., "Substitutions for Phosphate Ion in Apatites," *Nature*, **210,** 937–938 (1966).

211. Tuttle, O. F., and J. Gittins, Eds. *Carbonatites*, Interscience, New York, 1966, 610 pages.

212. United Nations, *Proceedings of the Seminar on the Sources of Mineral Raw Materials for the Fertilizer Industry in Asia and the Far East*, United Nations Mineral Resources Development Series, No. 32, 1968, 392 pages.

213. Van der Veen, A. H., "A Study of Pyrochlore," *Verh. Ned. Geol. Ges.*, **22,** 1–188 (1963).

214. Van Schmuss, W. R., and P. H. Ribbe, "Composition of Phosphate Minerals in Ordinary Chondrites," *Geochim. Cosmochim. Acta*, **33,** 637–640 (1969).

215. Van Tassel, R., "Strengite, phosphosidérite, cacoxénite et apatite fibroradiée de Richelle," *Bull. Soc. Belge Géol.*, **68,** 360–368 (1959).

216. Van Wazer, J. R., *Phosphorous and Its Compounds*, Vol. I, Interscience Publishers, New York, 1958.

217. Van Wazer, J. R., *Phosphorous and Its Compounds*, Vol. II, Interscience Publishers, New York, 1961.

218. Vasileva, Z. V., "Sulphur-Bearing Apatites," *Geochem.*, 464–471 (1958) (English translation).

219. Vernon, R. O., *Geology of Citrus and Levy Counties, Florida*, Florida Geological Survey Bulletin 33, 1951, 256 pages.

220. Vladovetz, V. I., "On Two Apatites of the Kukisvumchorr Deposit in the Khibin Tundra," *Trans. Arct. Inst., Leningrad*, Vol. XII, 71–100 (1933).

221. Weiner, J. S., K. P. Oakley, and W. E. Le Gross Clark, "The Solution of the Piltdown Problem," *Br. Mus. (Nat. Hist.) Bull., Geol.*, **2** (3), 141–146 (1953).

222. Whippo, R. E., and B. L. Murowchick, "The Crystal Chemistry of Some Apatites," *Trans. Soc. Mineral. Eng.*, 257–263 (1967).

223. White, W. C., and O. N. Warin, *A Survey of Phosphate Deposits in the Southwest Pacific and Australian Waters*, Bureau of Mineral Resources of Australia, Bulletin 69, 1964.

224. White, W. C., and O. N. Warin, "Bellona Island Phosphate Deposits," Report No. 35 in *Br. Solomon Isl. Geol. Rec.*, **2** (1959–62), 72–88 (1965).

225. Wurzberger, U. S., "A Survey of Phosphate Deposits in Israel," *Proceedings of the Seminar on Sources of Mineral Raw Materials for the Fertilizer Industry in Asia*, United Nations, New York, 1968, pp. 152–164.

226. Yamanari, H., *Aluminium Phosphate Deposits in Kita-Daito-Jima*, Tohoku Imp. Univ. Res. Contrib., Inst. Palaeont. 15, 1935 (in Japanese).

227. Zanin, Y. N., "Classification of Phosphate Deposits Related to Weathering," *Dokl. Akad. Nauk SSSR*, **176,** 689–692 (1967) (English translation).

228. Zanin, Y. N., "Geology of the Phosphate Crusts of Weathering and Associated Phosphate Deposits," *Tr. Inst. Geol. Geofiz., Akad. Nauk SSSR, Sibirsk. Otd.*,Vyp. 85, 158 pages (1969) (in Russian).

229. Kittrick, J. A., and M. L. Jackson, "Application of Solubility Product Principles to the Variscite-Kaolinite System, *Soil Sci. Soc. Amer. Proc.*, **19,** 455–7 (1955).

230. Norrish, K., "Some Phosphate Minerals of Soils," *Proc. 2nd Austral. Conf. Soil Sci.*, **1** (17), 1–15 (1957).

231. Maulenov, A. M., and G. G. Bel'skii, "The Forms of Bonding and the Conditions of the Concentration of Some Microelements in Phosphates from Black Shales," *Izvest. Akad. Nauk Kazakh. SSR*, Ser. Geol., No. 6, 42–50 (1970) (in Russian).

4

Economic Phosphate Deposits

G. Donald Emigh

Director of Mining, Monsanto Industrial Chemicals Company, St. Louis, Missouri

Just what constitutes an "economic phosphate deposit" is difficult to convey in an article as short as this, because the economics of a deposit is subject to many forces including politics, nationalism, grade, and location, to mention some of the more important. Later in this article brief descriptions of some present phosphate rock-producing operations are presented to assist the reader in a better understanding of what it takes to make an "economic phosphate deposit."

Reserves of phosphate rock are mentioned to better orient the reader; however, I speak of reserves with reluctance. Some reserve estimates are good but some are not. Contributing to this latter type of estimate are such factors as lack of adequate prospecting results and inadequate background of those making calculations. Another problem with quoting published reserve figures is that commonly they are given without reference to that all-important factor—their economics.

DEFINITIONS

The purpose of definitions is to enable better understanding of this article and other literature on the subject.

1. *Phosphate rock*, commonly called "rock" and with igneous phosphate commonly called "apatite." This is the expression commonly used for mined, or mined and beneficiated, calcium phosphate which is then the raw material for the next stage of manufacturing. Most phosphate rock, whether mined and used directly or mined and beneficiated into a concentrate, is a fine-grained material. The expression *phosphate rock* has no relation to its phosphate content. Thus, the phosphate rock of Idaho used in the production of phosphorus contains about 24% P_2O_5, whereas some of the phosphate rock of Morocco contains 36.6% P_2O_5.

2. *Phosphorite.* A deposit of phosphate, directly or indirectly of sedimentary origin, which is of economic interest.

3. *Grade of phosphate rock.* The calcium phosphate content of phosphate rock is expressed in different world areas by one of the following terms:

BPL (bone phosphate of lime)	P_2O_5 (phosphorus pentoxide)
TPL (triphosphate of lime)	P (phosphorus—not commonly used).

An illustration of relationship is as follows:

$$80\% \text{ BPL} = 80\% \text{ TPL} = 36.6\% \text{ } P_2O_5 = 16\% \text{ P}$$

4. *Units of weight*

tonne or m.t.	= metric ton (2205 lb)
ton or l.t. (outside the United States)	= long ton (2240 lb)
ton (in the United States)	= short ton (2000 lb)
s. ton or s.t.	= short ton (2000 lb)

5. *Ore or matrix.* Used to denote naturally occurring phosphatic material to be upgraded into phosphate rock.

6. *Wet or green acid.* Phosphoric acid produced by treating phosphate rock with sulfuric acid.

Reference to weights and grades in this article are generally those expressions in most common usage for any one particular area.

USES OF PHOSPHATE ROCK

There are two broad uses for phosphate rock—in fertilizers and in industrial chemicals. About 22% of the rock used in the United States goes into industrial chemicals—the remainder into fertilizer. Worldwide, rock is used in fertilizer in a higher proportion than in the United States.

In the making of fertilizer nearly all rock is processed first with sulfuric acid to produce phosphoric acid. Any phosphate rock can be used to make wet acid; however, different rock has different economic advantages or disadvantages. Thus everything else being equal phosphorite rock is preferred over apatite rock because the pellets are more porous and acidulation takes place more quickly. Certain impurities in rock are undesirable and can determine whether or not a rock can be marketed. For example, iron, aluminum, and magnesium can set up undesirable sludging conditions in the production of wet acid; hence there are limits set up in specifications for these three metals. Calcite and dolomite in rock are undesirable because they consume sulfuric acid and thus impose another discipline on what constitutes an "economic phosphate deposit."

The hydrocarbon content of some phosphorite rock sets up undesirable foaming conditions in the making of wet acid and such rock may have to absorb the cost of removing hydrocarbons by heat treatment.

Such phosphate-containing compounds as detergents, toothpaste, pharmaceuticals, and food-treating compounds are included among industrial phosphates. Their preparation requires a high-purity phosphoric acid. Wet acid contains too many trace impurities to be easily used, such as iron, arsenic, sulfate, magnesium, manganese, vanadium, calcium, aluminum, and chromium. Some industrial chemicals, with lower specifications on impurities, are produced by cleaning up wet acid; however, the great majority are produced by using the high-purity acid produced from elemental phosphorus. This phosphorus is obtained by treating a blend of prepared rock, coke, and silica in large electric furnaces.

As in the production of wet acid, certain impurities are undesirable in rock used to make phosphorus. For example, calcium carbonates are heavy consumers of electricity, and iron is undesirable because it results in the production of the uneconomic alloy ferrophosphorus.

The parameters that determine an "economic phosphate deposit" for the production of phosphorus thus include the availability of a large supply of low-cost electricity, low-cost silica as gravel or crushed quartzite, reasonably priced coke, low-cost rock, and low-cost transportation for the phosphorus produced.

TYPES OF PHOSPHATE DEPOSITS

There are two types of phosphate deposits—sedimentary and igneous. About 85% of the world production of rock in 1969 was from the sedimentary type, and 15% from the igneous type. Whether of sedimentary or igneous origin, the phosphate mineral is basically the same. It is calcium phosphate containing fluorine. In sedimentary deposits I call the phosphate mineral francolite. In igneous deposits it is the mineral apatite. The ratio of fluorine to P_2O_5 is about 1:10. In other words, rock containing 30% P_2O_5 contains about 3% F. Guano contains low fluorine as do the secondary guano deposits.

Sedimentary Types of Phosphate Deposit

The great economic phosphate deposits of the world are of sedimentary origin. My proposed classification for these deposits is shown in Table 1 (3). The phosphorites can be divided into guano and pellet types; however, the guano type is insignificant in quantity compared to the pellet type. Pellets are of diverse origin and are believed to result from the original aragonite (calcium carbonate) of the pellet being replaced by phosphate in seawater. Pellets are generally ovoid in shape. Sizes commonly differ greatly between different beds but are of an order of magnitude of about 0.25 mm × 0.35 mm.

TABLE 1 Classification of the World's Great Economic Sedimentary Phosphate Deposits. From Hale (3)

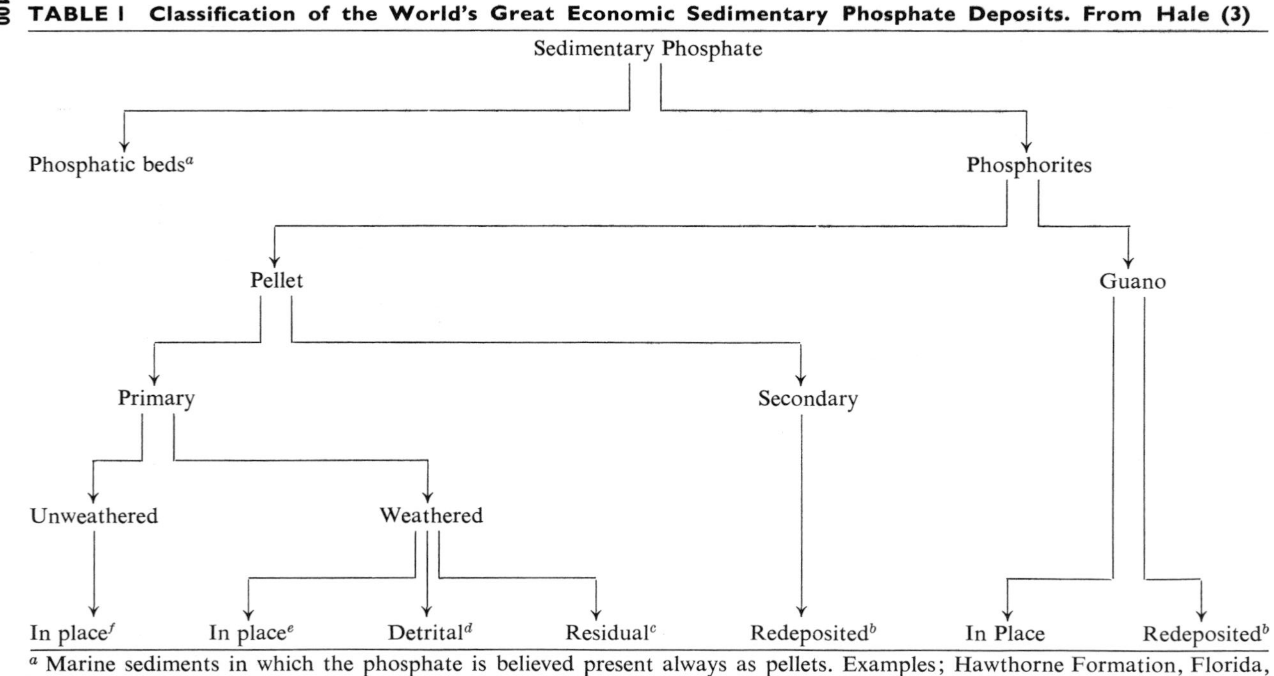

[a] Marine sediments in which the phosphate is believed present always as pellets. Examples; Hawthorne Formation, Florida, and Bigby Limestone, Tennessee.

[b] Phosphatized rocks. Examples: white rock of Tennessee, hardrock of Florida.

[c] Derived from phosphatic beds. Examples: brown rock of Tennessee.

[d] Examples: pebble field of Florida and other similar fields in the southeastern United States.

[e] Surface and near-surface weathered areas of all marine phosphorites. Examples: Phosphoria Formation, phosphorites in Zacatecas, Mexico, presently worked phosphorite beds of North Africa (one exception).

[f] Unweathered areas of all marine phosphorites. Examples: Phosphoria Formation, Zacatecas area in Mexico, lower unworked beds in Morocco.

The pellet sedimentary phosphorites are generally of two types: One is the original (primary) marine bed—the other, marine beds exposed to surface conditions of weathering and subsequent reconcentration of the phosphate pellets into unconsolidated alluvial deposits. Examples of the original marine-bed type are the Phosphoria Formation of the western United States, and deposits in Jordan, Israel, Egypt, Tunisia, Morocco, Spanish Sahara, Australia, and elsewhere. Examples of the surface-reworked sedimentary rock deposits are those of Florida and Tennessee. Phosphorites in any one region of the earth commonly have widespread lateral extent and may have originally covered many hundreds and even thousands of square miles.

Surface (and near-surface) weathering of phosphatic beds is of vital importance in creating deposits of economic importance. In the case of primary beds such weathering removes cementing calcium carbonate thereby removing this deleterious material and leaving the bed softer, easier to mine, and higher grade in phosphate. In the case of unconsolidated phosphatic alluvial deposits weathering is all important in removing the original cementing calcium carbonate thereby releasing the phosphate pellets for redistribution and concentration on the surface.

When weathering is carried to an extreme it can be harmful. Such weathering, both in primary phosphorite beds and in reworked alluvial beds, can cause calcium in the calcium phosphate pellets to be partially or wholly replaced by aluminum derived from admixed clay. Phosphate rock containing any significant quantity of aluminum phosphate cannot be used economically in the standard process of producing phosphoric acid from rock; consequently, the high aluminum phosphate areas of rock deposits must be removed in mining and cast to waste. This is especially true in the Florida mining fields where the aluminum phosphate zone is called the "leached zone."

Igneous Types of Phosphate Deposit

Phosphate deposits of igneous rock origin, although accounting for only about 15% of present world production of phosphate rock, are nevertheless of significant importance. This is especially true because such deposits in the Kola Peninsula of the U.S.S.R. contain large reserves and today account for most of the world's production of igneous phosphate rock.

At many places in the world crystalline apatite is associated with intrusive complexes of alkalic rocks. The rock types are varied and characterized generally by those minerals deficient in silica with respect to alkalis and aluminum (4). The silicate rocks occurring in the alkalic complexes show a wide variety of texture and minerals including as the principal ones:

mafic types—pyroxene (augite), olivine, and biotite
felsic types—potash feldspars and nepheline.

The large alkalic complex of the Kola Peninsula appears to be a marginal differentiate containing unusually high concentrations of apatite in an intrusive composed largely of nepheline. Elsewhere in the world there are two hundred known alkalic complexes. These are generally circular stocks or plugs of limited areal extent—1 to 20 mi^2. Often associated with them, and commonly in the center of the circular complexes as the last stages of the intruded complex, are carbonate rocks emplaced either in a molten condition or, as believed by Pecora (4), by hot aqueous solutions containing CO_2. The most common carbonate is calcite followed by dolomite. Siderite and ankerite can also be present. Such carbonate deposits are called carbonatites. Some carbonatites contain high enough concentrations of certain minerals to be of economic interest. Principal among these, now being recovered or having potential importance, are apatite, pyrochlore (a mineral containing columbium), copper minerals, vermiculite, and calcite.

Pecora (4) suggests that carbonatites may be classed in two broad types: (*a*) those with apatite and magnetite—generically associated with feldspathoid-bearing rocks, and (*b*) those with rare earth minerals (the batsnaesite of Mt. Pass, California).

DESCRIPTIONS OF THE MAJOR OPERATING "ECONOMIC PHOSPHATE DEPOSITS"

Sedimentary

The following description of the principal pellet phosphorites in the United States includes brief comments on the origin and evolution of the deposits; this is apropos to a better understanding of an "economic phosphate deposit." One reason for describing the United States deposits is that they pretty well encompass all types of occurrences. Another reason is that I am personally better acquainted with them than with many of those in other parts of the world.

Following the description of United States deposits brief resumes will be given of some of the other important world pellet phosphorites with reference to the types described in the United States.

Southeastern United States—Central Florida. The world's largest rock-producing area is in central Florida, east of the port city of Tampa. In 1971, using only about 90% of capacity, central Florida produced 32 million short tons of rock. This was 35% of the world production.

The phosphate occurs in the Bone Valley Formation of Pliocene age. The unconsolidated formation is made up largely of silica sand, clay, and phosphate pellets. The Bone Valley Formation is lying on the Hawthorne limestone-dolomite formation of Miocene age which contains phosphate pellets.

The components of the Bone Valley Formation were largely derived from weathering of the Hawthorne Formation which resulted in removal of carbonates and the leaving of the less soluble minerals such as phosphate pellets. The less soluble minerals were later worked by surface waters to form the Bone Valley Formation. This formation is now covered with fine-grained silica sand—the present relatively flat land surface. Only the lower part of the Bone Valley Formation is mined for its phosphate by open-pit mining using large draglines. These first cast aside the sand overburden and upper part of the Bone Valley Formation, then recover the phosphate matrix. In the upper part of the matrix zone weathering has caused chemical reactions between aluminum of the clays and calcium phosphate of the pellets resulting in the partial, or complete, replacement of calcium in the pellets by aluminum of the clays. The resulting white, pulverous aluminum phosphate is the mineral wavellite. The zone of wavellite is called the "leached zone." It is too erratic in occurrence, and too low grade, to have economic importance and is, therefore, cast to waste by the draglines.

Cathcart (2) describes part of the Bone Valley district including its geology and some of its mines.

This Bone Valley district of central Florida is roughly circular and 50 mi in diameter. At present nine companies produce rock. Operations are confined to the northern half of the 50-mi-diameter areas because of better mining characteristics and higher grades of rock; however, the southern part of the field has received serious attention and will someday be placed in production.

Matrix of economic interest underlies only a part of the general areas described here. Elsewhere it is too thin, too low grade, or has been removed by dissecting streams.

In the producing northern district the overburden on the matrix is 5 to 50 ft thick, averaging perhaps 22 ft. The matrix is 0 to 40 ft thick, averaging about 15 ft. An "economic phosphate deposit" here is one falling within certain parameters that include, among other things, grade of the rock, thickness of overburden, and thickness and grade of the matrix. Current practice is generally not to exceed a stripping ratio of 2 yd^3 of overburden to 1 yd^3 of matrix and 4 yd^3 of matrix to 1 ton of rock produced.

Large draglines move material for a few cents a cubic yard. As higher overburden ratios are encountered still larger draglines, with their lower operating costs, become necessary.

The matrix is pumped as a water slurry at 35% solids, from the mines to the beneficiation plants through 20-in. steel pipe for distances of up to 5 mi. This is not low-cost transportation but it has the following advantages: It can handle the wet and sticky matrix; agitation in the pipelines breaks up clay and otherwise helps liberate the individual phosphate pellets; and pipelines can be laid across swampy ground.

The 20-odd beneficiation plants in the district in general process the matrix in a similar manner. This consists of coarse screening to remove trash and waste rock, finer screening to produce a phosphate pellet rock called "pebble," desliming to remove silt-size material, and flotation to produce high-grade rock. Desliming is at 150 mesh. Flotation is on two-sized feeds. The size of the coarser one differs between plants but falls within the following mesh parameters: 14–28/28–48. Pebble rock is that rock coarser than the 14 to 28 mesh.

The deslimed matrix sent to the flotation circuits is called "flotation feed" or "heads." It is running as low as 14% BPL in the central Florida district although the average in the field is about 20 to 25% BPL. In prospecting, drill holes not only determine the thickness of overburden and matrix but laboratory washing tests are run on the matrix samples to determine quantity and quality of pebble, grade of flotation feed, and quantity and quality of flotation rock. An economic phosphate deposit is generally one that will produce at least 2500 tons of rock per acre.

Pebble rock is generally several percentage points lower in grade than rock produced by flotation; however, it is an important market item. In the district the quantity of pebble rock produced per ton of total product varies greatly but will average about 30%. Pebble rock is considerably less costly to produce than flotation rock.

The quantity of slimes in the matrix also varies widely from place to place but will average about 33% for the district. The slimes are made up of silt particles of about one-third silica, one-third clay, and one-third phosphate.

Reserves in Florida are stated in terms of short tons of recoverable rock. In central Florida reserves now economic, or on the verge of being so, are about 1 billion tons. Average grade will be less than past production and will be about 68% BPL. In addition, there are approximately 1 billion tons of reserves not quite economic today.

Southeastern United States—Northern Florida. The phosphatic Hawthorne Formation described in central Florida extends north in Florida 200 mi to the Florida-Georgia line and then into southern Georgia. Matrix similar to that described in central Florida is present in places and is especially prevalent in north Florida in Hamilton and Columbia Counties. The one rock-producing operation started in 1965. The matrix in this area is somewhat thinner than in central Florida and there is less pebble; however, there is less of the troublesome slime.

Presently known economic reserves in north Florida are several hundred million tons of rock.

The extensive matrix reserves of southern Georgia are not economic now but will be someday. They represent a potentially very large reserve of rock.

For the geology, and other information including reserves, of southern Georgia and northern Florida refer to Sever *et al.* (5)

Southeastern United States—North Carolina. An extensive phosphorite deposit in Beaufort County, North Carolina, covering many hundreds of square miles has received considerable attention over the past 15 years. The western edge of the bed is thin but it increases in thickness eastward attaining a thickness of over 100 ft. It is Miocene in age.

The phosphorite consists mostly of unconsolidated quartz sand, small calcite shells and fragments, clay, and phosphate pellets. Interspersed through the phosphorite are thin hard beds of phosphatic limestone, dolomite, and silty claystones and coquinas. The grade of the phosphorite is about 16% P_2O_5.

The phosphorite beds are overlain by the Miocene Yorktown Formation and it in turn by a thin layer of recent sediments. The Yorktown Formation consists for the most part of unconsolidated quartz sand with some clay. Its thickness over the phosphorite is 40 to 200 ft.

The one operator in the field started its full-scale mine-flotation plant in 1966. Large draglines strip overburden and mine the matrix. Currently about 96 ft of overburden are removed and about 38 ft of phosphorite mined. For details on this operation as well as comments on general geology see Caldwell (1).

Rock reserves in this North Carolina field have been estimated at 1.5 to 10 billion short tons. About 500 million short tons may be available by surface mining. The remainder probably will be available only by pumping a slurry of the phosphorite through holes drilled from the surface. Several companies with holdings in the area have been experimenting with this approach but the results have not been published. Initial efforts along these lines met with only limited success.

Tennessee. There has been a long-established phosphate-mining industry in southern middle Tennessee. Most mining has been concentrated in Maury County and the counties surrounding it.

Up to the mid-1930s most of the phosphate was mined for the fertilizer industry but since then it is used locally in electric furnaces producing phosphorus. The area is the largest phosphorus producer in the world.

Nearly all of the phosphate matrix is "brown rock." It is the residual product of the leaching away of underlying phosphatic limestone leaving, at the surface, the more insoluble components of the limestone, that is, clay, detrital quartz grains, and phosphate pellets. The brown color of the matrix largely results from iron oxide derived from the original finely dispersed pyrite in the limestone.

Most of the brown rock comes from the Bigby Limestone of Ordovician age. The phosphate in the unweathered limestone is present as originally deposited pellets. The grade of the limestone, in the mining areas, is as much as 7% P (35% BPL) but the average is probably 3% P (15% BPL) or less.

The sedimentary formations are flat lying. Topography is gently rolling hills with vertical relief of generally less than 200 ft interspersed with broad areas of very little vertical relief.

Where the phosphatic limestone is at the surface of a broad, flat area, brown-rock deposits may cover several hundreds of acres. Where the limestone is in a hilly area, only a narrow rim of brown rock develops along the outcrop. This rim of brown rock is more or less 100 ft wide.

In places the matrix rests on the limestone as a fairly even-thickness layer, but commonly water moving along the top of the limestone will follow joints, thus deepening and widening the joints and resulting in accumulation of brown rock in elongated buried troughs called "cutters." A cutter may be only a few feet wide and deep but it may be 30 ft wide and 40 ft deep and up to several hundred feet long.

The matrix is now covered by brown-colored soil with thicknesses of a few inches to 40 ft. Because the dimensions of residual deposits are so erratic it is not possible to describe them accurately. An oversimplification would be to say that an average matrix deposit would be 5 ft thick and covered with 10 ft of soil.

Small draglines, 1.5 to 3.5 yd^3, have been found best in stripping the overburden and loading matrix into trucks. They are especially required to dig matrix out of narrow cutters.

Matrix as mined contains about 33% sticky clay with the remainder being largely phosphate pellets and detrital quartz grains. Grades are 8.5 to 11% P. Most matrix is washed to reduce clay and iron oxide content. Both materials are undesirable in electric furnaces as both consume electricity—the clay in going into the slag and the iron oxide in forming ferrophosphorus alloy.

Matrix reserves of the region are limited.

Western United States. The phosphatic member of the Phosphoria Formation, Permian in age, originally covered 135,000 mi^2 in Montana, eastern Idaho, western Wyoming, northeastern Nevada, and northern Utah. After deposition it was covered with great thicknesses of sediments which were later lithified. Subsequent severe regional folding and faulting left the Phosphoria Formation in all altitudes—from flat lying to vertical. Like all pelletal phosphorites it is marine in origin.

The present stage of erosion exposes many miles of phosphorite outcrop extending along a 450-mi line roughly southeastwardly from Garrison, Montana to Vernal, Utah.

The lithology of the Phosphoria Formation, with its contained phosphorites, varies greatly along the above-mentioned line. At each end of the line there is only one phosphorite bed, whereas near the middle of the 450-mi line, in eastern Idaho, four phosphorite beds are mined. At the north end of the line one bed is mined 3.5 to 5 ft thick at 31.50% P_2O_5. One bed is also mined near Vernal, Utah, 450 mi away. It is 19 ft thick and contains 19% P_2O_5. In the eastern Idaho area the four beds have an aggregate thickness of about 50 ft across a formation width of 185 ft. The grades of the four beds range from 31 to 18% P_2O_5. Combined they average 24 to 28% P_2O_5.

At present phosphate is mined at eight locations—one in Montana, two in Utah, one in Wyoming, and four in Idaho. In 1971 yearly production of rock was 4.2 million tons.

Thirty-five percent of present production goes into fertilizer. This is both beneficiated and unbeneficiated rock. The other 65% goes into the production of elemental phosphorus by electric furnaces.

Rock used to produce phosphorus is not beneficiated. Most of this rock has a content of 24 to 28% P_2O_5.

About 5% of present production is from underground mining, the balance by surface mining.

With a great variation of phosphorite thicknesses and grades now being mined in the west the factors governing an "economic phosphate deposit" are many. Among them are the following: availability of large quantities of electrical power for making phosphorus; by-product sulfuric acid available from base metal smelters in Idaho, Utah, and British Columbia, Canada, for making fertilizer; adequate railroad transportation; and availability of weathered phosphorite for surface mining in Idaho.

Surface and near-surface weathering of the phosphorites of the Phosphoria Formation removes hydrocarbons and cementing calcium carbonates. This upgrades the ore several percentage points of P_2O_5 and softens the material so it is readily mined without blasting and, if beneficiated, requires no grinding. Such weathering can proceed down the dip several hundred feet or even further if surface ground slopes are such as to maintain a shallow overburden over the phosphorite.

Beds of chert (cryptocrystalline quartz) are commonly associated with pellet phosphorite. In southeastern Idaho the main chert lies stratigraphically on top of the 185-ft shale member of the Phosphoria Formation containing the phosphorite beds. It is the upper member of the Phosphoria Formation and is about 200 ft thick. It contains phosphate pellets throughout, but not in concentrations of economic interest. The Rex Chert changes lithologically both laterally and vertically. This Rex Chert is commonly moved as overburden in open-pit mining. As with the phosphorite, near-surface weathering is important in the Rex Chert because it softens the rock, thus facilitating stripping. This is especially true where the chert facies is more a cherty limestone than a pure chert.

Over the years there have been many estimates made on western phosphate ore reserves. These show many billions of tons. All of these estimates have good basis for what is in the ground; however, considering the present economics the phosphate ore reserves are probably less than 1 billion tons.

The reader is referred to Hale (3) for an exhaustive description of western phosphate and its industry; it covers geology, origin of phosphate, mineralogy, mining, beneficiation, and use among its subjects. Much of this publication's information is applicable to all pellet phosphorites.

From the description above of pellet phosphorite in the United States, the following type deposits can be defined—and these encompass most deposits elsewhere.

Type A. Those developed by weathering in place of highly phosphatic beds. Weathering removes calcium carbonates and hydrocarbons leaving a more or less soft, unconsolidated bed higher in phosphate than the original bed. An example is the surface and near-surface outcrops in much of the western United States Phosphoria Formation.

Type B. The original highly phosphatic bed is hard because of cementing calcium carbonates. Infrequently some of the cementing material is cryptocrystalline silica. Again, an example is the unweathered part of the Phosphoria Formation.

Type C. Those formed by release of phosphate pellets from a calcium carbonate host rock by leaching away of the calcium carbonate followed by surface waters reworking them into their present position. The deposits in Florida are an example of this type.

Type D. Those represented by residual material left in place after weathering and leaching away of the calcium carbonate host rock, such as Tennessee brown-rock deposits.

In world pellet phosphorites, type *A* has been the most important producer to date and will continue to be for some time. Within type *A* are all degrees of weathering. Generally the only completely weathered material is at or near the surface; however, special conditions can change this. For example, faults can allow deeper circulation of surface water and therefore deeper weathering and leaching.

Morocco. For many years Morocco has been one of the world's largest rock exporters with ocean shipments made through the ports of Casa Blanca and Safi. In recent years, Morocco has maintained yearly shipments at about 10 million metric tons and has set a goal of increased production to reach a rate of 18 million metric tons a year by 1972.

There are three main rock-producing areas, which can be referred to as Khouribga, Youssoufia, and Ben Guerir. They can be covered by a rectangle about 35 mi by 125 mi located south of Casa Blanca and east of Safi.

Phosphorite beds occur in flat-lying beds of phosphates associated with beds of phosphatic limestone, chert, and marl. Dips are 0 to 20°. Geologic ages are Late Cretaceous and Eocene. The number of phosphorite beds in any one area is one to six. Thicknesses range from 1 to 9 m with an average of perhaps 2 m.

Until the start of the first open-pit mining in the Khouribga area in 1961 all mining was underground. The phosphorite worked underground was the upper weathered bed (type *A*). In places weathering was intense enough to permit working the upper two and, in places, three beds by open pit; it was at such a place that the 1961 open-pit operation began.

Operations consist of

1. Khouribga—four open-pit mines, four underground mines, capacity 13 million metric tons per year.
2. Youssoufia—three underground mines, capacity 3.4 million metric tons per year.
3. Ben Guerir—one open-pit mine, 1970 capacity 650,000 tons per acre. Production from this area is scheduled at 2 million metric tons per year in 1972, and 10 million metric tons per year in 1980.

Mining of the soft type *A* beds in Morocco has consistently yielded a high-quality high-grade rock with grades of 70 to 82% TPL.

Over the years estimates of rock reserves have been all the way from less than a billion metric tons to 40 billion metric tons. Forty billion metric tons is the most recent figure. The bulk of this reserve is probably unweathered material with a TPL content of 55 to 70% (type *B*). Much of the diluent is calcite that flotation, to date, has not been effective in removing. Presumably upgrading through calcining may be the treatment eventually used.

All rock operations in Morocco, and sales, are by the government concern O.C.P. (*Office Cherifien des Phosphates*) with headquarters in Rabat.

Maintenance and growth of its phosphate rock industry are important to Morocco. There are about 18,000 men directly employed. Production is a significant part of total exports and contributes to nearly half the revenue for the national railway system. Like all rock producers, O.C.P. is thoroughly aware of the need to increase efficiency. Their initiating open-pit operations during the past decade is one indication of this awareness.

Tunisia. The phosphate rock-producing industry of Tunisia is operated by the government, which acquired the majority of ownership of the three former French companies in the early 1960s. Operations are through GAFSA (*Compagnie des Phosphates et du Chemin de fer Gafsa*).

Five underground mines and six beneficiation plants now have a rock capacity of over 4 million metric tons a year. These are in two areas. One mine is at Kalaa Djerda in the west-central part of the country, near the Algerian border, the other four are near the town of Gafsa in southwestern Tunisia. By 1976 three additional mines will raise total rock capacity to 5.4 million metric tons per year. Rock from Kalaa Djerda is taken by railroad northeast to the port of La Goulette (near Tunis). The others are serviced by a railroad extending east to the ports of Sfax and Gabes.

The phosphorites are Eocene in age and there are nine units named Units 0 to VIII. Current production is largely from Unit II which has a thickness of 1.5 to 4 m. The ore is type *A* except Kalaa Djerda which is type *B*. Grade of ore is about 62% TPL. Upgrading of the soft ore is done in wet washing plants and dry plants using air separation. Grades of concentrates (rock) are 65 to 68% TPL. A higher grade product of 70 to 73% TPL, referred to as epure, is also produced by blending phosphoric acid with rock.

The grade of Tunisian rock is on the low side compared to competitive rock from other countries. This imposes a restriction on export markets. To overcome this Tunisia has an active study underway to raise the grade, perhaps by calcination, and to improve the relatively low mine productivity by mechanization. Possibly open-pit mining can be developed to lower operating costs further. In order to better utilize low grade Tunisian rock a new phosphoric acid plant was completed in Gabes in 1971. The phosphoric acid will be exported as is triple superphosphate manufactured in plants in Sfax.

As with Morocco the rock industry of Tunisia is highly important to the country's economy. Nearly 12,000 persons are directly connected with operations. Tunisia's known rock reserves are estimated to be about 500 million metric tons.

Spanish Sahara. In 1962 the Spanish government company ENMINSA (*Empresa Nacional Minera del Sahara*) began a search in Spanish Sahara for phosphate deposits of economic interest. Careful field work found new deposits after exploring 20,000 k^2. Drilling, trenching, and shaft sinking in the most favorable section, named Bu-Craa, has developed 1.7 billion metric tons of phosphate ore averaging 3.6 m thick and about 70% TPL. It is equivalent in age to the Moroccan phosphate.

The phosphorite beds dip at a low angle from their outcrop and in part, at least, are amenable to surface mining. Specimens indicate the ore to be type *A*. The formations with which the phosphorites are associated contain clay, marl, and chert.

After unsuccessfully endeavoring to interest private foreign companies as partners, ENMINSA decided to proceed on its own with a rock-producing operation. A 61-mile belt conveyor was installed from the ore deposit at Bu-Craa to the coast at El Aauin. Port and beneficiation facilities were built at El Aauin and will be operational in 1972. Initial rock capacity will be 3.3 million metric tons per year from 5 million tons of ore. Grade of the concentrate (rock) will be 75 and 80% TPL. It is hoped markets will develop so that rock capacity can be tripled by 1975. Operations are under Fosfatos de Bu-Craa (FOSBUCRAA), controlled by ENMINSA.

Known rock reserves are 1700 million metric tons.

Australia. A large phosphorite field of Middle Cambrian age was found in northwestern Queensland in 1966. Associated rock formations are cherts, black siltstones, and siliceous shales and limestone.

Several companies have acquired extensive land holdings and have carried on field work—largely drilling. Thickness of the pellet phosphorite has been reported as 15 to 63 ft and grades of 16 to 30% P_2O_5. One company reported reserves of 2 billion tons averaging 17% P_2O_5; another, 500 million tons. Because of its isolated location it appears doubtful whether production will take place before 1980 at the earliest. Specimens from the outcrops indicate type *A* ore.

Other Important World Rock-Producing Phosphorites*

Location	1971 Rock Production (thousand metric tons)	Reserves Ore or Rock (million metric tons)
U.A.R.	600	Rock: 350
Algeria	550	Ore: 500† Grades: 53–61% TPL
Syria	50	Ore: 800 Grades: 24–29% P_2O_5
Israel	900	Ore: 200 Grades: 24–33% P_2O_5
Jordan	500	Rock: 300
Senegal	1400 (Taiba)	Rock: (pellet) 80
	150 (Thies)	Rock: (wavellite) 100
Togo	1700	Rock: 50
U.S.S.R.	8500 (1968)	Rock: 3000 Grades 20–30% P_2O_5
Nauru	1900	Rock: 55
Christmas Island	1000	Rock: 200
Ocean Island	600	Rock: 5

* These are all pellet phosphates except possibly Nauru, Ocean Island, and Christmas Island The phosphate from these three islands may have had a guano origin.

† In the Djebel Onng Mine area with an equal amount elsewhere in Algeria.

Igneous

Russia. The large apatite-bearing alkalic rock deposits of the Kola Peninsula were discovered in 1923 and production began a few years later. The intrusive is a sheet up to 600 ft thick composed largely of nepheline, and is rich in apatite. The mining area is near the town of Kirovsk and is about 100 mi south of the seaport of Murmansk. The area is north of the Arctic Circle and subjected to extremely severe weather conditions.

There are four mines near Kirovsk—one worked by open pit, the others largely by underground mining. Beneficiation plants upgrade the ore by flotation. Apatite rock, grade 39.5% P_2O_5, is shipped by rail north to Murmansk for export and south for internal use.

Planned production for 1975 is 14.50 million metric tons of apatite to be recovered from 33.6 million metric tons of ore. These figures indicate that the apatite rock recovered as a flotation concentrate represents a recovery of 17% P_2O_5 from the ore.

Production in 1970 was 11.33 million metric tons of apatite rock. By-product nepheline, 29% Al_2O_3, is shipped 700 mi south to a plant near Leningrad where it is converted to aluminum metal, soda-potash salts, and cement.

It is now believed that apatite-nepheline ore containing less than 12% P_2O_5 can be processed. This will double the ore reserves, making them sufficient to yield 17 million metric tons of rock a year for 75 years or a total reserve of 1,275,000,000 metric tons of apatite rock.

South Africa. The Palabora alkalic complex, intruded into granite, is located in the eastern Transvaal of South Africa. Within the complex there are now three separate open-pit mining operations. One produces copper and magnetite, another phosphate and some copper, and the third vermiculite.

The surface area of the complex is 2 mi by 5 mi and is oval in shape. After the emplacement of the original alkalic intrusive several subsidiary pipes were intruded. One of these has as its center a carbonatite core with copper sulfide minerals. Surface area of the carbonatite is 2500 by 5200 ft. This carbonatite is now a large-scale producer of copper and by-product magnetite which started open-pit operations in 1966.

The rock in the subsidiary pipe surrounding the carbonatite is pyroxnite containing diopside, phlogopite, apatite, and magnetite. It is called foskorite. It contains 5 to 10% P_2O_5 in the form of apatite and is the material now being mined open pit for apatite. Current production of apatite rock, averaging 37% P_2O_5, is 1 million metric tons a year. Together with other much smaller rock production this now makes South Africa self-sufficient in phosphate rock.

Although Palabora phosphate was first produced in the period of 1932 to 1934, it has been only since the establishment by the South African government of the present operating company in the 1950s, Phosphate Corporation of South Africa (FOSKOR), that the present large-scale mine-mill plant has been built and placed in operation.

Four million tons of the foskorite is reported to yield 600,000 tons of apatite rock containing 37% P_2O_5. This then represents a recovery of 5.5% P_2O_5 from the foskorite ore. Reserves of foskorite are stated as "hundreds of millions of tons" and "enough for hundreds of years."

Rhodesia. Dorowa and Shawa are the names of two alkalic complexes located 100 mi south of Salisbury. Dorowa is the only one now being mined.

The Dorowa complex is 2 mi long and horseshoe shaped. Syenites around the periphery grade to nepheline-rich rocks toward the center. Carbonatite occurs as a few small plugs and dikes. The development of the alkalic rocks around the carbonatite is referred to as "fenitization." Included in such rocks are syenites and pyroxenites. Apatite-magnetite and apatite-vermiculite rocks occur as lenticular bodies in the alkalic intrusive.

An altered yellow-colored decomposed area of "syenitic fenites" near the north rim is richest in apatite and magnetite and is the location of present mining operations. The area considered as ore covers 4 mi². The soft weathered rock is mined by rippers and scrapers and upgraded by flotation. About 7½ tons of ore is required to produce a ton of apatite rock. Reserves and grade of ore are 37 million tons containing 8% P_2O_5 as apatite. Present operations, started in 1965, produce 100,000 tons of apatite rock a year, which makes Rhodesia self-sufficient for phosphate fertilizer.

Statistics on Other Igneous Phosphate Deposits

Location	1971 Rock Production (metric tons)	Reserves Ore or Rock (million metric tons)	Comments
Uganda			
Sukulu	16,000	Ore: 200, Grade: 13% P_2O_5, 0.20% Cb_2O_5	Residual soils (surface mining)
Brazil			
Jacupriangа	150,000	Ore, Grade: 5% P_2O_5	Unweathered carbonatite (surface mining)
Araxa	40,000	Rock (ore): 90 Grade: 21–30% P_2O_5	
Canada			
Multi-Minerals Limited	0	Ore: 42, Grade: 8.6% P_2O_5 0.20% Cb_2O_5	Unweathered carbonatite (underground)

MARKETPLACE PROBLEMS OF THE PHOSPHATE ROCK INDUSTRY

Phosphate rock is produced in over 30 countries. In size they are all the way from tiny Naura to the USSR.

World production of rock in 1971 was 83.6 million metric tons, of which 77% was from the non-Communist world. Of the non-Communist production about 59% (38 million metric tons) was produced by private company operations and 41% (26.2 million metric tons) from governmental owned or controlled operations. Non-Communist rock exported into world markets in 1971 was 40 million metric tons, of which 65% was from countries where operations are either completely government owned or are government controlled.

In 1971 nearly half the quantity of that year's rock production was imported by countries who have limited or no domestic rock production. The table (p.115) summarizes this importation of rock.

The tabulation does not take into account the movement of finished phosphate fertilizer; however, it does highlight the important parts of the world which depend on imported phosphate.

Phosphate rock is a relatively low-value commodity. Whether or not a phosphate deposit is economic depends on two factors:

1. cost delivered at the marketplace, and
2. quality.

Competition from other rock producers, both from within a country and from abroad, determines the rock price. There are some exceptions to the statements above where political and other factors become involved such as barter and a country's desire to be self-sufficient for internal use. Barter deals involve such materials as coal, tea, and equipment.

Disregarding artificial cost factors such as taxes and royalties, the cost to deliver rock at the market is governed by the cost to mine, mill, and transport the finished rock. Even where low-wage conditions prevail, underground mining is not cheap. Consequently, producers are making every effort to utilize less expensive, and more efficient, surface-mining methods. Volume is a factor in all costs so producers naturally push for highest possible production.

The cost to develop and equip a large new phosphate rock-producing facility is high. For example, the first phase of the new operation in Spanish Sahara will cost $100 million; the final phase for 10 million metric tons a year is $200 million. The cost to increase production in Morocco will be $150 million. Such high costs necessitate large production.

Imports of Phosphate Rock in 1971

Area	Number of Countries	Million Metric Tons	Percent of Total Imports
Western Europe	17	18.93	47.3
Eastern Europe	8	7.42	18.5
Asia	9	6.12	15.3
Africa	1	0.00	0
North America	2	2.47	6.2
Latin America	9	1.82	4.6
Australia, New Zealand	2	3.22	8.1
	48	39.98	100.0

The largest importers (in million metric tons) in the areas above were as follows.

Western Europe		Eastern Europe		Asia	
France	3.7	Poland	2.2	Japan	2.9
West Germany	2.8	East Germany	1.4	India	0.9
United Kingdom	1.7	Czechoslovakia	1.0	China	1.1
Italy	1.9	Yugoslavia	0.9	South Korea	0.5
Belgium	2.0				
Spain	1.6				
Netherlands	1.6				

North America		Latin America		Australia–New Zealand	
Canada	2.4	Mexico	1.0	Australia	2.1
		Brazil	0.6	New Zealand	1.1

The wide variation between field conditions for competing rock is exemplified by comparing Florida and Morocco. The exported rock from Florida may be produced from matrix (ore) containing only 9% P_2O_5 but must compete with rock from Morocco much of which requires little or no beneficiation, contains 34% P_2O_5, and is considerably closer to the large western European market than Florida. Florida is able to compete because of favorable factors including the applicability of large mechanized mining methods, availability of adequate water, and geographic location close to ocean shipping. Conversely, in the past at least, Morocco has been faced with such conditions as costly underground mining and the need to provide for outside services and facilities for its employees.

An "economic phosphate deposit" is constantly faced with changing outside conditions challenging its ability to survive. On the local level in the United States, at least, in recent years increasing public interest has resulted in strict controls on air and water pollution thereby increasing costs. The same is true of public pressure for mined land reclamation.

On the export level an "economic phosphate deposit" faces the threat of better deposits being found and developed elsewhere in the world.

FUTURE OF THE ROCK INDUSTRY

As noted in this article, there are many, and changing, parameters that establish an "economic phosphate deposit."

Phosphate rock is man's only source of vitally needed phosphate and will always be produced. Fortunately, the earth's resources of phosphate rock are for all practical purposes unlimited (6). The known reserves of rock available under more or less existing cost parameters is extremely large and will take care of man's requirements for any foreseeable future. New phosphate deposits will continue to be found. Eventually man can turn to the lower grade deposits and here the cost parameters will probably be considerably different than those of today.

REFERENCES*

1. Caldwell, A. B., "Lee Creek Open-Pit Mine and Fertilizer Plants," *Eng. Min. J.*, **169,** No. 1, 59–83 (Jan. 1968).
2. Cathcart, J. B., *Economic Geology of the Keysville Quadrangle, Florida*, U.S. Geological Survey Bulletin 1128, 1963, 82 pages.
3. Hale, L. A., Ed., *Anatomy of the Western Phosphate Field*, Intermountain Association of Geologists (Utah Geological Survey), Salt Lake City, Utah, 1967, 287 pages.
4. Pecora, W. T., "Carbonatites, a Review," *Geol. Soc. Am. Bull.*, **67,** 1537–1555 (1956).
5. Sever, C. W., J. B. Cathcart, and S. H. Patterson, *Phosphate Deposits of South-Central Georgia and North-Central Peninsular Florida*, Georgia State Division of Conservation, Department of Mines, Mining, and Geology, 1957, 62 pages.
6. Emigh, G. D., "World Phosphate Reserves—Are There Really Enough?" *Eng. Min. J.*, **173** (4), 90–95 (April 1972).

* Considerable material in this article comes from publications of The British Sulphur Corp., London, that is, bimonthly issues of *Phosphorus and Potassium* and *World Survey of Phosphate Deposits*, 3rd Edition, 1971, 180 pages.

Inorganic Phosphorus in Seawater*

5

R. A. GULBRANDSEN and CHARLES E. ROBERSON

U.S. Geological Survey, Menlo Park, California

Phosphorus was known to be present in seawater by 1865 (9), and by 1878 many quantitative determinations of phosphorus had been made on samples of seawater from various parts of the oceans and seas of the world (34). The concentration of phosphorus in seawater is very small, an average of about 2.3 μg atoms P/l (0.07 mg P/l) (25). This is inorganic phosphorus, as contrasted with organic forms, and is the dominant mode of phosphorus occurrence in seawater. The inorganic phosphorus occurs nearly entirely as orthophosphate ions; traces of polyphosphates have been reported (39) and are considered as a possible indicator of pollution. Because phosphorus is one of the elements essential to all forms of life, it is one of the basic nutrient elements, the so-called nonconservative elements in seawater whose proportions with respect to such major constituents as sodium and chlorine are not constant. Interocean mixing, estimated to occur within a time period of about 1000 years (4), accounts for the nearly constant proportions of the conservative elements in seawater.

Most chemical determinations of phosphorus in seawater are of inorganic orthophosphate; however, total phosphorus determinations, which include the organic forms, have been made frequently in recent years. Concentrations of phosphorus are given here as μg atoms P/l (microgram atoms of phosphorus per liter). Conversion of μg atoms P/l to other familiar concentration units is as follows:

$$\mu\text{g atoms P/l} \times 30.974 \times 10^{-3} = \text{mg P/l}$$
$$\mu\text{g atoms P/l} \times 94.971 \times 10^{-3} = \text{mg PO}_4\text{/l}$$
$$\mu\text{g atoms P/l} \times 70.972 \times 10^{-3} = \text{mg P}_2\text{O}_5\text{/l}$$
$$\mu\text{g atoms P/l} = \text{mg atoms P/m}^3$$

* Publication authorized by the Director, U.S. Geological Survey.

Summaries on phosphorus in the world's oceans by Sverdrup *et al.* (41), Graham and Moberg (12), Barnes (2), Redfield (25), Redfield *et al.* (26), and Armstrong (1) have been especially helpful in the preparation of this chapter and should be referred to for additional information. Riley (28) discusses the analytical chemistry of phosphorus in seawater.

SOURCE OF PHOSPHORUS AND RESIDENCE TIME

The phosphorus in the oceans is derived principally from the continents by river runoff. Although igneous rocks are the original source of all the phosphorus, much of the phosphorus carried in present-day runoff is recycled; that is, it is derived from previously formed marine sedimentary rocks.

The total amount of phosphorus in the oceans, 9.8×10^{10} metric tons (15), is small compared to the total of 22.4×10^{14} metric tons calculated by Horn and Adams to have been weathered throughout geologic time from igneous rocks, even though the total weathered includes both dissolved phosphorus and the phosphorus in detrital solid phases. The amount in the oceans is small also when compared with the total of 7.3×10^{11} metric tons of phosphorus estimated by McKelvey and co-workers (20) to have been deposited in a single phosphatic rock unit of marine origin, the Permian Phosphoria Formation of the western United States. These comparisons suggest that some control mechanism prevents the buildup of the phosphorus content in seawater and that possibly a balance exists between the amount added to the oceans and the amount subtracted in a unit of time.

Barth (3) assumed such a steady state and proposed a calculation that gives an average time for the passage of an element through the seawater cycle. The average time τ, now commonly called residence time (11), is obtained by dividing the total amount of the element in the oceans by the amount supplied per year. Estimating the amount of phosphorus supplied per year to the oceans as about 2×10^6 metric tons (20) and the value cited above, 9.8×10^{10} metric tons, for the total in the oceans, the residence time is

$$\tau = \frac{9.8 \times 10^{10}}{2 \times 10^6} = 4.9 \times 10^4 \text{ years}$$

This figure is only useful as an order-of-magnitude estimate because of the few data available for making an estimate of the amount of phosphorus carried to the oceans in river water. It does tell us, however, that the residence time is geologically short and that some mechanism is preventing a buildup of phosphorus in the oceans. The phosphorus residence time is intermediate in magnitude in comparison with the residence times for 15 other elements reported by Goldberg and Arrhenius (11). It is similar to silicon and barium but very short as compared to sodium, which has a residence time of 2.6×10^8 years.

PROCESSES REMOVING PHOSPHORUS FROM SEAWATER

Phosphorus is extracted from seawater by organisms as an essential constituent of life, by precipitation of carbonate fluorapatite, the marine variety of apatite, and probably by adsorptive processes to some extent.

Most of the phosphorus is continuously and rapidly recycled. In a cycle it may be used by a long chain of living forms before being returned to solution upon the decomposition of organic matter. Some of the phosphorus, however, is regularly lost as organic-matter sediment, including the phosphatic animal hard parts—teeth, bones, and shells. The concentration of phosphorus in most sedimentary rocks is very small, generally measured in hundredths or tenths of a percent, but the total amount of this derivation is very large. The removal of phosphorus as organic-matter sediment, however, is not inherently a control mechanism and does not explain the level of phosphorus content in seawater.

Very large amounts of phosphorus have been deposited since late Precambrian time in the mineral carbonate fluorapatite, which is the phosphate mineral of phosphorites and phosphatic sedimentary rocks. Geologically old phosphatic animal hard parts also are composed of this mineral and are abundant in some rocks. Carbonate fluorapatite is the ubiquitous and primary mineral of the many marine phosphate deposits known in the world (18, 19). It appears likely that the equilibrium relationship of this mineral with seawater is the basic control of the phosphorus content observed in seawater.

The relationship, however, is complex and not fully known. Solubility experiments by Roberson (30) indicate that seawater in general is somewhat supersaturated with phosphate, whereas precipitation experiments by Smirnov and co-workers (38) suggest that seawater is somewhat undersaturated. In a discussion of the factors involved in the formation of marine apatite, Gulbrandsen (13) noted that the apatite equilibrium is dominated by the calcium carbonate equilibrium, which is the principal control of the calcium concentration in seawater. A high concentration of calcium is compatible with the solubility of calcium carbonate, but to maintain simultaneous equilibrium with respect to apatite requires that the phosphorus concentration be very low.

The main consideration to note here, even though the physical chemical parameters are only approximately known, is that the low phosphorus content of present-day seawater is consistent with general equilibrium considerations and that the phosphorus content of seawater throughout much of earth history may have been controlled by equilibrium with the solid phase, carbonate fluorapatite.

IONIC SPECIES OF INORGANIC PHOSPHORUS

The species of inorganic phosphorus in seawater are orthophosphate ions, represented by phosphoric acid (H_3PO_4) and its dissociation products, $H_2PO_4^-$, HPO_4^{2-}, and PO_4^{3-}, and ion complexes of these ions with some of the other constituents of seawater.

Phosphoric acid is a triprotic acid which dissociates in water as follows:

$$H_3PO_4 \rightleftarrows H^+ + H_2PO_4^- \tag{1}$$

$$H_2PO_4^- \rightleftarrows H^+ + HPO_4^{2-} \tag{2}$$

$$HPO_4^{2-} \rightleftarrows H^+ + PO_4^{3-} \tag{3}$$

The equilibrium constants at 25°C and 1 atm pressure are

$$\frac{\alpha_{H^+}\,\alpha_{H_2PO_4^-}}{\alpha_{H_3PO_4}} = K_1 = 10^{-2.16} \tag{1a}$$

$$\frac{\alpha_{H^+}\,\alpha_{HPO_4^{2-}}}{\alpha_{H_2PO_4^-}} = K_2 = 10^{-7.21} \tag{2a}$$

$$\frac{\alpha_{H^+}\,\alpha_{PO_4^{3-}}}{\alpha_{HPO_4^{2-}}} = K_3 = 10^{-12.32} \tag{3a}$$

In Eqs. (1a), (2a), and (3a), K_1, K_2, and K_3 are thermodynamic dissociation constants [derived from association constant of Sillén and Martell (35, p. 180)]. Also $\alpha_i = \gamma_i m_i$ where α_i, γ_i, and m_i are activity, individual ion activity coefficient, and molality, respectively, of species i. For ionic strength equal to zero, γ_i values are unity; for ionic strength greater than zero, γ_i must be known. The concentration term m_i represents the amount of free ion and differs from the total of the species by the amount that occurs as ion complexes or ion pairs. The magnitude of ion complexing for the major ions in seawater has been shown by Garrels and Thompson (10) to be significant. Data on complexing of the phosphate species are presented in Table 1. The amount of complexing appears to be greatest for the highest valent species, PO_4^{3-}, as is also indicated by Kester and Pytkowicz's (17) work in determining the apparent dissociation constants in seawater.

TABLE I Possible Ion Pairs or Complexes Involving Phosphate Species in Seawater. Constants Shown Are for Association Reactions at 25°C, Ionic Strength near 0, and 1 atm

	Log of Formation Equilibrium Constant for			
Ligand	Ca^{2+}	Mg^{2+}	Na^{+}	K^{+}
$H_2PO_4^-$	1.41[a]	[b]	[b]	[b]
HPO_4^{2-}	2.74[a]	2.5[c]	1.15[d]	1.04[d]
PO_4^{3-}	6.46[a]	[e]	[f]	[b]

[a] Chughtai *et al.* (7).
[b] No data found for this species.
[c] Sillén and Martell (35, p. 182, Ref. 40G).
[d] Extrapolation of values for $I = 0.2$ found in Sillén and Martell (35, p. 182, Ref. 56S).
[e] By analogy with $CaPO_4^-$ one would expect the constant for this species to be nearly as large.
[f] No data found, except that Chughtai *et al.* (7) reported no evidence of complexing of phosphate with sodium. This, of course, would be applicable to the ligands $H_2PO_4^-$ and HPO_4^{2-} with sodium although at least one value is reported for HPO_4^{2-} (this table).

Apparent Dissociation Constants

Because of the difficulty of obtaining values for γ_i in a complex electrolyte like seawater and because of the largely unknown parameters for calculating the extent of complexing between phosphate anions and metal ions, it is sometimes useful to employ apparent constants. These, based on experiments, allow one to calculate the total concentration of a given species. This value includes the concentrations of the free uncomplexed species and those bound to cations in some sort of association.

The apparent dissociation constants are defined by

$$\frac{\alpha_{H^*}[H_2PO_4^-]_T}{[H_3PO_4]_T} = K_1' \quad (1b)$$

$$\frac{\alpha_{H^*}[HPO_4^{2-}]_T}{[H_2PO_4^-]_T} = K_2' \quad (2b)$$

$$\frac{\alpha_{H^*}[PO_4^{3-}]_T}{[HPO_4^{2-}]_T} = K_3' \quad (3b)$$

The brackets denote molar concentrations of the various species. For example, $[PO_4^{3-}]_T$ includes uncomplexed PO_4^{3-} and that bound in ion complexes with various cations other than H^+. The term α_{H*} is defined by $pH = -\log \alpha_{H*}$ where α_{H*} is commonly measured with a glass electrode and a reference electrode with liquid junction.

Apparent dissociation constants that furnish information about the phosphate species in seawater are presented in Table 2 along with the thermodynamic dissociation constants. The general effect of other ions in solution (ionic strength) is shown by noting the large difference between all the apparent dissociation constants and their respective thermodynamic ones. A comparison of the apparent constants provides some measure of the magnitude of ion complexing of phosphate species in seawater. The similar values of K_1' for artificial seawater and the one calculated for the case in which no complexing occurs indicate essentially no complexing of $H_2PO_4^-$.

Table 2 Dissociation Constants of Phosphoric Acid in Some Types of Solutions

Medium	Ionic Strength	Dissociation Constant (K)	Log K
Pure water[a]	0	K_1	−2.16
	0	K_2	−7.21
	0	K_3	−12.32
Hypothetical noncomplexing[b]	0.68	$K'_{1(0)}$	−1.68
	0.68	$K'_{2(0)}$	−6.74
	0.68	$K'_{3(0)}$	−11.76
NaCl[c]	0.68	$K'_{1(NaCl}$	−1.55
	0.68	$K'_{2(NaCl)}$	−6.39
	0.68	$K'_{3(NaCl)}$	−11.00
Artificial seawater[d]	0.68	K_1'	−1.63
	0.68	K_2'	−6.06
	0.68	K_3'	−8.86

[a] These are thermodynamic constants. For 25°C, $I = 0$. Calculated from formation constants reported in compilation of data by Sillén and Martell (35, p. 180, Ref. 29B).

[b] These constants (25°C), which assume no complexing, are obtained from thermodynamic values above, using individual ion activity coefficients, calculated from mean activity coefficients ($\gamma\pm$) for pure salt solutions (31). In addition, it is necessary to use the assumption that $\gamma\pm_{KCl} = \gamma_{K^+}$. [See Truesdell and Jones (42).] The value for $\gamma_{H_3PO_4^\circ}$ was assumed to be 1.13 by analogy with $\gamma_{H_2CO_3^\circ}$ (10).

[c] Experimental values (20°C) measured in 0.68 M NaCl solution (17).

[d] Experimental values (20°C) for 33‰ salinity (17).

A comparison of K_2' with both the noncomplexing case and the sodium chloride solution indicates appreciably more complexing of HPO_4^{2-} than $H_2PO_4^-$, and a similar comparison for K_3' reveals a very large magnitude of PO_4^{3-} complexing. Kester and Pytkowicz (17) estimate that only 56% of the HPO_4^{2-} ions is free and only 0.4% of the PO_4^{3-} ions is free.

The values of K_1', K_2', and K_3' in seawater for a range of salinity and temperature have been measured by Kester and Pytkowicz (17) and are presented in Tables 3, 4, and 5.

Species Distribution

The distribution of the species of phosphate ions in seawater cannot yet be determined accurately, but the data that have been presented make approximations of the distribution possible. As an example, we use the following parameters for seawater:

$$\begin{aligned} \text{pH} &= 8.0 \\ \text{salinity} &= 33 \text{ parts per thousand} \\ \text{temperature} &= 20°\text{C} \\ \text{pressure} &= 1 \text{ atm} \\ \sum[\text{P}]_T &= 2.3\ \mu\text{g atoms P/l} \end{aligned}$$

Taking the values for the apparent dissociation constants (K_1', K_2', and K_3') given in Table 2 and the pH value above of 8, the equations for the apparent dissociation constants [(1b), (2b), and (3b) of p. 207] can be expressed as follows:

$$\frac{[H_2PO_4^-]_T}{[H_3PO_4]_T} = \frac{10^{-1.63}}{10^{-8.0}} = 2.3 \times 10^6 \tag{1c}$$

$$\frac{[HPO_4^{2-}]_T}{[H_2PO_4^-]_T} = \frac{10^{-6.06}}{10^{-8.0}} = 87 \tag{2c}$$

$$\frac{[PO_4^{3-}]_T}{[HPO_4^{2-}]_T} = \frac{10^{-8.86}}{10^{-8.0}} = 0.14 \tag{3c}$$

Equation (1c) shows that undissociated H_3PO_4 is negligible in amount relative to $H_2PO_4^-$; Eq. (2c) shows that HPO_4^{2-} is 87 times more abundant than $H_2PO_4^-$; and Eq. (3c) shows that HPO_4^{2-} is also more abundant, 7.2 times, than PO_4^{3-}. These comparisons therefore suggest that HPO_4^{2-} is the major phosphate ion in seawater with the above indicated parameters, and the ion retains this apparent dominance, as shown below, even after considering the effects of ion complexing.

TABLE 3 Values of K_1' for Phosphoric Acid ($\times 10^2$). From Kester and Pytkowicz (17)

Temperature (°C)	Salinity (‰)								
	29	30	31	32	33	34	35	36	37
0	2.84	2.87	2.87	2.86	2.85	2.84	2.83	2.82	2.81
1	2.80	2.83	2.83	2.82	2.82	2.82	2.81	2.81	2.80
2	2.74	2.77	2.77	2.78	2.78	2.78	2.79	2.79	2.80
3	2.69	2.72	2.73	2.74	2.75	2.76	2.77	2.78	2.78
4	2.63	2.66	2.68	2.69	2.71	2.72	2.74	2.76	2.77
5	2.58	2.61	2.63	2.66	2.68	2.70	2.73	2.75	2.77
6	2.53	2.56	2.58	2.61	2.64	2.66	2.69	2.73	2.75
7	2.46	2.50	2.54	2.58	2.61	2.64	2.68	2.72	2.75
8	2.40	2.44	2.49	2.54	2.58	2.62	2.66	2.70	2.74
9	2.34	2.38	2.42	2.48	2.54	2.59	2.64	2.68	2.73
10	2.30	2.35	2.40	2.46	2.51	2.56	2.62	2.67	2.72
11	2.34	2.39	2.44	2.49	2.53	2.58	2.64	2.70	2.76
12	2.44	2.48	2.52	2.56	2.61	2.66	2.72	2.79	2.87
13	2.57	2.60	2.63	2.67	2.72	2.77	2.84	2.92	3.10
14	2.67	2.69	2.72	2.76	2.81	2.87	2.93	3.01	3.13
15	2.70	2.72	2.75	2.78	2.83	2.89	2.96	3.05	3.15
16	2.70	2.71	2.74	2.77	2.81	2.85	2.90	3.00	3.08
17	2.70	2.70	2.70	2.71	2.72	2.73	2.74	2.75	2.80
18	2.70	2.66	2.64	2.63	2.58	2.56	2.54	2.52	2.50
19	2.69	2.62	2.55	2.49	2.42	2.39	2.37	2.30	2.32
20	2.68	2.58	2.49	2.41	2.35	2.31	2.28	2.26	2.25
21	2.60	2.54	2.48	2.45	2.37	2.33	2.31	2.27	2.23
22	2.52	2.50	2.47	2.46	2.40	2.36	2.33	2.30	2.24
23	2.42	2.45	2.46	2.47	2.44	2.40	2.36	2.33	2.26
24	2.34	2.41	2.45	2.48	2.48	2.46	2.41	2.35	2.27
25	2.26	2.36	2.44	2.50	2.52	2.50	2.45	2.37	2.28
26	2.22	2.32	2.40	2.51	2.55	2.52	2.46	2.39	2.30
27	2.18	2.28	2.39	2.52	2.59	2.56	2.50	2.42	2.32
28	2.15	2.23	2.38	2.53	2.62	2.59	2.53	2.44	2.36
29	2.13	2.19	2.37	2.54	2.65	2.62	2.55	2.46	2.38
30	2.10	2.15	2.36	2.55	2.68	2.65	2.57	2.48	2.40

TABLE 4 Values of K_2' for Phosphoric Acid ($\times 10^6$). From Kester and Pytkowicz (17)

Temperature (°C)	Salinity (‰)								
	29	30	31	32	33	34	35	36	37
0	0.48	0.50	0.52	0.54	0.56	0.58	0.59	0.60	0.62
1	0.49	0.51	0.53	0.55	0.57	0.59	0.61	0.62	0.64
2	0.50	0.52	0.54	0.56	0.58	0.60	0.62	0.63	0.65
3	0.51	0.53	0.55	0.57	0.60	0.62	0.63	0.64	0.66
4	0.53	0.55	0.57	0.59	0.61	0.63	0.65	0.66	0.68
5	0.54	0.56	0.58	0.60	0.62	0.64	0.66	0.67	0.69
6	0.55	0.57	0.59	0.61	0.63	0.65	0.67	0.68	0.70
7	0.56	0.58	0.60	0.62	0.64	0.66	0.68	0.69	0.71
8	0.57	0.59	0.61	0.63	0.66	0.68	0.69	0.71	0.72
9	0.59	0.61	0.63	0.65	0.67	0.69	0.70	0.72	0.73
10	0.60	0.62	0.64	0.66	0.68	0.70	0.71	0.73	0.73
11	0.61	0.63	0.65	0.67	0.69	0.71	0.72	0.73	0.74
12	0.62	0.64	0.66	0.68	0.70	0.71	0.72	0.74	0.74
13	0.63	0.65	0.67	0.69	0.70	0.72	0.73	0.74	0.75
14	0.63	0.65	0.67	0.69	0.71	0.72	0.73	0.75	0.76
15	0.64	0.66	0.68	0.70	0.72	0.73	0.74	0.75	0.76
16	0.65	0.67	0.69	0.72	0.75	0.76	0.77	0.78	0.79
17	0.65	0.68	0.72	0.76	0.79	0.81	0.83	0.85	0.87
18	0.65	0.69	0.74	0.79	0.83	0.86	0.90	0.94	0.98
19	0.65	0.70	0.76	0.81	0.86	0.91	0.96	1.01	1.05
20	0.65	0.71	0.77	0.82	0.88	0.93	0.98	1.03	1.07
21	0.68	0.72	0.76	0.81	0.87	0.92	0.97	1.01	1.05
22	0.70	0.72	0.75	0.80	0.85	0.90	0.95	0.99	1.03
23	0.71	0.73	0.75	0.79	0.83	0.88	0.93	0.97	1.02
24	0.72	0.73	0.75	0.78	0.81	0.85	0.90	0.95	1.01
25	0.73	0.74	0.75	0.76	0.79	0.83	0.87	0.93	1.00
26	0.73	0.74	0.75	0.76	0.77	0.81	0.85	0.91	0.98
27	0.74	0.75	0.75	0.75	0.75	0.79	0·83	0.89	0.96
28	0.75	0.75	0.75	0.74	0.73	0.77	0.81	0.87	0.94
29	0.76	0.76	0.74	0.73	0.71	0.75	0.79	0.85	0.92
30	0.78	0.77	0.74	0.71	0.69	0.73	0.77	0.83	0.90

TABLE 5 Values of K_3' for Phosphoric Acid ($\times 10^9$). From Kester and Pytkowicz (17)

Temperature (°C)	Salinity (‰)								
	29	30	31	32	33	34	35	36	37
0	0.16	0.20	0.26	0.30	0.34	0.30	0.26	0.20	0.16
1	0.17	0.21	0.27	0.32	0.36	0.33	0.29	0.24	0.20
2	0.18	0.22	0.28	0.34	0.38	0.35	0.31	0.28	0.25
3	0.19	0.23	0.29	0.36	0.41	0.38	0.35	0.32	0.29
4	0.20	0.24	0.30	0.38	0.43	0.41	0.38	0.36	0.34
5	0.20	0.25	0.32	0.40	0.46	0.44	0.42	0.40	0.38
6	0.21	0.26	0.33	0.42	0.49	0.48	0.46	0.44	0.42
7	0.22	0.28	0.36	0.45	0.52	0.51	0.51	0.48	0.46
8	0.23	0.29	0.38	0.47	0.54	0.53	0.52	0.51	0.49
9	0.23	0.31	0.41	0.51	0.58	0.59	0.57	0.54	0.51
10	0.24	0.34	0.44	0.54	0.61	0.60	0.59	0.58	0.57
11	0.25	0.36	0.48	0.58	0.67	0.65	0.62	0.59	0.56
12	0.26	0.39	0.55	0.67	0.73	0.69	0.65	0.61	0.55
13	0.26	0.43	0.60	0.71	0.79	0.74	0.69	0.62	0.54
14	0.27	0.47	0.65	0.77	0.86	0.81	0.73	0.64	0.53
15	0.27	0.52	0.76	0.90	0.95	0.92	0.82	0.66	0.49
16	0.42	0.62	0.80	0.95	1.01	0.94	0.83	0.70	0.58
17	0.60	0.75	0.90	1.04	1.09	1.01	0.90	0.77	0.64
18	0.84	0.94	1.02	1.11	1.17	1.10	0.99	0.86	0.73
19	1.04	1.11	1.17	1.22	1.26	1.21	1.12	1.00	0.88
20	1.18	1.26	1.31	1.35	1.37	1.34	1.28	1.20	1.11
21	1.31	1.37	1.43	1.49	1.55	1.55	1.54	1.54	1.53
22	1.37	1.47	1.58	1.69	1.80	1.82	1.84	1.85	1.86
23	1.43	1.58	1.73	1.94	2.05	2.09	2.13	2.16	2.19
24	1.48	1.68	1.88	2.08	2.28	2.35	2.42	2.48	2.54
25	1.56	1.79	2.03	2.29	2.50	2.63	2.73	2.80	2.86
26	1.59	1.89	2.19	2.48	2.74	2.87	2.99	3.11	3.23
27	1.69	1.99	2.30	2.65	2.97	3.12	3.27	3.42	3.57
28	1.80	2.09	2.44	2.88	3.20	3.34	3.48	3.73	3.87
29	1.90	2.19	2.62	3.08	3.44	3.64	3.84	4.04	4.24
30	2.00	2.29	2.70	3.17	3.67	3.90	4.13	4.35	4.58

TABLE 6 Distribution of Phosphate Species

Species	μg atoms P/l. Total	"Free"
H_3PO_4	Negligible	—
$H_2PO_4^-$	0.023	ca. 0.023
HPO_4^{2-}	2.0	1.1
PO_4^{3-}	0.276	0.001
$\Sigma[P]$	2.3	

The distribution of the phosphate ions in terms of concentrations is calculated from the equations above and the mass-balance equation,

$$\sum[P]_T = [PO_4^{3-}]_T + [HPO_4^{2-}]_T + [H_2PO_4^-]_T + [H_3PO_4]_T \quad (4)$$

$\sum[P]$ is the total phosphorus in seawater, in this case 2.3 μg atoms P/l. Because undissociated H_3PO_4 is negligible in amount, only Eq. (2c), (3c), and (4) are utilized, and the last term of Eq. (4) is dropped. Solving the simultaneous equations with the values presented above yields concentrations (total) of the individual species which are listed in Table 6.

Kester and Pytkowicz (17) found that the magnitude of ion complexing among the phosphate species is greatest for PO_4^{3-} and that it decreases with decreasing ionic charge. They estimated that only 0.4% of PO_4^{3-} is free and that 56% of the HPO_4^{2-} is free. Their data indicate that most of the $H_2PO_4^-$ is probably free, but they were not able to estimate the amount. From these percentages the concentrations of HPO_4^{2-} and PO_4^{3-} that are considered free are presented in Table 6. Although the amount of ion complexing is large for two of the phosphate species, HPO_4^{2-} is clearly the major phosphate ion in seawater.

DISTRIBUTION OF PHOSPHORUS IN OCEANS

The average inorganic phosphorus content of seawater is about 2.3 μg atoms P/l and ranges from essentially none in some surface waters to 12 μg atoms P/l in the region of the Andaman Islands in the Bay of Bengal (24). The common range is from a few tenths to about 3.5 μg atoms P/l. Other high concentrations, 5.2 and 4.4, have been observed in the bottom waters of the Santa Monica and Santa Barbara basins, respectively (29). Much of the deep water of the Black Sea has a high phosphorus content, up to 7.5 μg atoms P/l (37).

High concentrations of phosphorus around the Andaman Islands are of special interest because they are associated with milky-white turbidity that may be due to phosphate precipitation (24). The precipitate is not carbonate, nor has it been verified as phosphate. If phosphate is being precipitated there, the reported phosphorus concentrations would be the first natural observation of a limiting concentration of phosphorus in seawater. Reddy and co-workers (24) also note that the Andaman Island region supports the highest production of phytoplankton in the northern Indian Ocean, which would be expected on the basis of the high phosphorus concentrations.

The phosphorus content of the oceans can be characterized in general by its distribution in vertical profiles. A typical profile shows the lowest concentration at the surface, a rapid increase in concentration to a maximum that commonly occurs at a depth of about 1000 m, and a small decrease in concentration that varies only slightly with increasing depth. Examples of vertical profiles for parts of the oceans and seas are shown in Figs. 1, 2, and 3, and the data are presented in Tables 7, 8, and 9. The shape of the profiles

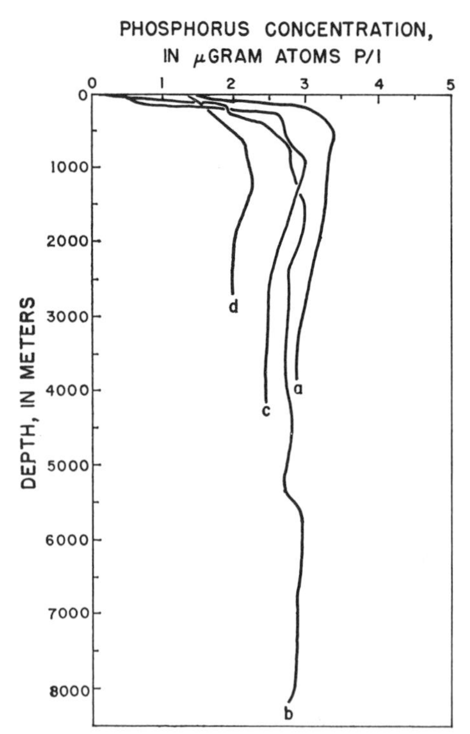

Figure 1. Vertical profiles of the distribution of phosphorus in the Pacific Ocean. The lettered curves correspond to the data in Table 7.

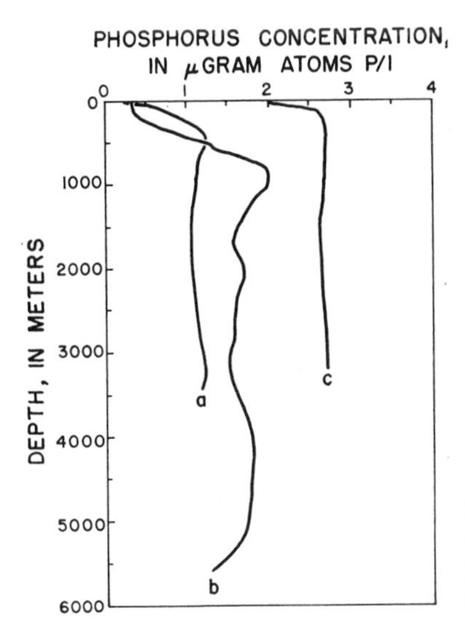

Figure 2. Vertical profiles of the distribution of phosphorus in the Atlantic Ocean. The lettered curves correspond to the data in Table 8.

reflects in general the role played by phosphorus in the life cycle. The thin surface layer is where photosynthesis takes place and is the zone of phosphorus consumption. Below is the zone of rapid phosphorus regeneration where the phosphorus maximum is attained; it is where phosphorus is returned to solution upon the dissolution and oxidation of organic excrement and dead organisms. Below the maximum is the deep zone which is also involved in the life cycle of phosphorus but on a longer time basis than the upper

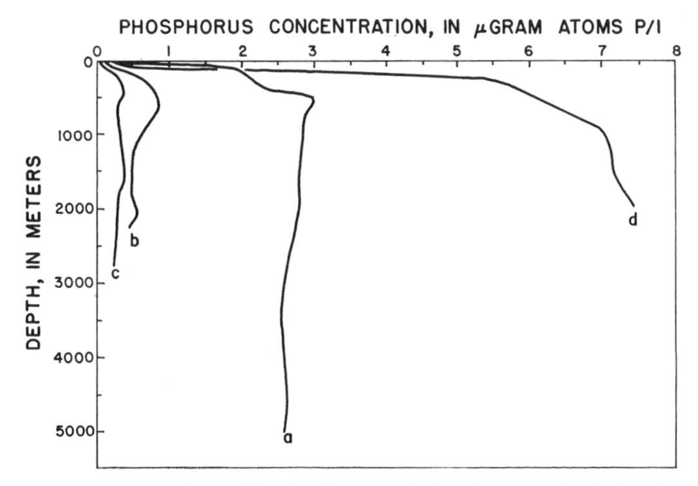

Figure 3. Vertical profiles of the distribution of phosphorus in the Indian Ocean, Red Sea, Mediterranean Sea, and Black Sea. The lettered curves correspond to the data in Table 9.

TABLE 7 Vertical Distribution of Phosphorus in the Pacific Ocean

North Pacific[a]		North Pacific[b]		South Pacific[c]		South Pacific[d]	
Depth (m)	P (μg atoms/l)	Depth (m)	P (μg atoms/l)	Depth (m)	P (μg atoms/l)	Depth (m)	P (μg atoms/l)
0	1.41	0	0.19	0	0.49	0	1.27
10	1.46	24	0.19	10	0.49	9	1.27
19	1.44	48	0.35	33	0.44	18	1.24
29	1.52	72	0.45	53	0.44	27	1.24
49	1.61	95	0.93	86	0.46	45	1.24
73	1.77	143	1.64	129	0.57	68	1.34
97	1.83	197	1.94	149	0.72	90	1.38
146	2.90	247	1.90	168	0.91	112	1.67
192	2.93	294	1.94	191	1.34	134	1.61
288	3.18	297	1.87	212	1.48	179	1.53
482	3.38	387	2.36	259	2.15	224	1.68
720	3.35	480	2.54	316	2.67	269	1.68
954	3.32	493	2.39	421	2.64	350	1.86
1430	3.30	589	2.58	525	2.66	526	2.03
1911	3.21	780	2.81	630	2.61	701	2.18
2366	3.10	971	2.81	735	2.82	876	2.15
2392	3.07	1410	2.96	840	2.98	1052	2.25
2856	2.96	1895	2.96	946	2.94	1316	2.25
2861	3.00	2380	2.74	1052	2.93	1766	2.09
3347	2.85	2865	2.78	1262	2.91	2228	2.00
3352	2.89	3350	2.71	1574	2.85	2702	2.00
3837	2.89	3835	2.71	1882	2.76		
3842	2.85	4320	2.81	2212	2.58		
		4805	2.78	2679	2.49		
		5290	2.64	3020	2.48		
		5771	2.94	3306	2.48		
		6252	2.94	3881	2.46		
		6733	2.87	4173	2.46		
		7214	2.87				
		7695	2.87				
		8175	2.78				

[a] North Pacific, 50°58′ N., 145°29′ W., September, Station 110-61. Norpac Committee (23).
[b] North Pacific, Mindanao Trench, 5°23′ N., 127°48′ E., January, Station 162. Bruneau *et al.* (5).
[c] South Pacific, 23°45′ S., 81°7′ W., November, Station 37. Scripps Institution of Oceanography, University of California, (33).
[d] South Pacific, 57°22.3′ S., 170°52.2′ W., January, Station UmA-266. Ishino *et al.* (16).

TABLE 8 Vertical Distribution of Phosphorus in the Atlantic Ocean

North Atlantic[a]		North Atlantic[b]		South Atlantic[c]	
Depth (m)	P (μg atoms/l)	Depth (m)	P (μg atoms/l)	Depth (m)	P (μg atoms/l)
6	0.26	1	0.35	0	2.0
53	0.59	50	0.25	20	2.1
102	0.74	100	0.35	40	2.0
199	0.88	199	0.40	60	2.4
296	1.05	299	0.65	80	2.4
393	1.22	398	1.00	100	2.5
490	1.23	498	1.30	150	2.7
587	1.22	597	1.35	200	2.7
685	1.12	697	1.70	400	2.7
782	1.17	796	2.05	800	2.7
878	1.09	896	2.05	1500	2.6
976	1.14	995	2.05	2500	2·7
1073	1.14	1194	1.80	3200	2.7
1268	1.12	1393	1.70		
1351	1.05	1667	1.55		
1549	1.10	1966	1.70		
1746	1.06	2365	1.60		
1944	1.09	2764	1.60		
2242	1.05	3164	1.50		
2540	1.14	3563	1.65		
2838	1.14	3962	1.80		
3136	1.22	4361	1.80		
3434	1.15	4760	1.75		
		5160	1.70		
		5559	1.25		

[a] North Atlantic, 53°15′ N., 24°58′ W., August, Station 3854. Worthington (44).
[b] North Atlantic, 24°31′ N., 31°38′ W., October, Station 3599. Worthington (43).
[c] South Atlantic, 57°36′ S., 30° W., April–June. Harvey (14).

zones. The deep zone receives regenerated phosphorus from the organic matter that passes through the zone of phosphorus maximum and constitutes a large reservoir of phosphorus that is brought to the surface for reuse in the life cycle by the major mixing processes of the oceans.

Phosphorus in Surface Waters

The phosphorus content of surface waters varies seasonally over most parts of the oceans. When conditions favor the growth of organisms, the phosphorus

TABLE 9 Vertical Distribution of Phosphorus in the Indian Ocean, Red Sea, Mediterranean Sea, and Black Sea

Indian Ocean[a]		Red Sea[b]		Mediterranean Sea[c]		Black Sea[d]	
Depth (m)	P (μg atoms/l)	Depth (m)	P (μg atoms/l)	Depth (m)	P (μg atoms/l)	Depth (m)	P (μg atoms/l)
0	0.29	1	0.07	1	0.04	10	0.12
23	0.26	25	0.07	97	0.11	50	0.65
45	0.36	49	0.13	194	0.29	100	1.03
68	0.61	98	0.52	290	0.32	150	3.42
90	1.26	197	0.30	387	0.38	200	4.33
134	1.94	295	0.73	484	0.41	300	5.55
183	2.07	590	0.90	600	0.28	500	6.07
282	2.10	984	0.68	800	0.31	1000	7.00
380	2.20	1251	0.56	999	0.34	1500	7.17
479	2.36	1448	0.51	1195	0.37	1750	7.30
577	3.00	1646	0.54	1393	0.40	1930–2050	7.46
772	2.87	1843	0.52	1588	0.42		
968	2.87	2040	0.60	1746	0.31		
1539	2.81	2139	0.57	1945	0.32		
2019	2.81	2188	0.52	2141	0.30		
2499	2.71	2218	0.57	2338	0.30		
3458	2.55	2227	0.51	2537	0.27		
4418	2.65	2237	0.46	2732	0.27		
4956	2.55						
4985	2.55						

[a] Indian Ocean, 4°47′ S., 88°18′ E., March, Station 205. Bruneau *et al.* (5).
[b] Red Sea, 25°21.7′ N., 36°6.5′ E., June, Station 5644. Neumann and Densmore (21).
[c] Mediterranean Sea, 38°45.7′ N., 6°54.0′ E., July, Station 5653. Neumann and Densmore (21).
[d] Black Sea, Average of 11 stations, August–September. Skopintsev *et al.* (37).

content of water in the photosynthesis zone is depleted. Replenishment occurs at times unfavorable for the growth of organisms by diffusion and current mixing. Graham and Moberg (12) report that a concentration of 0.61 μg atoms P/l measured in May in the middle North Atlantic was reduced to 0.12 μg atoms P/l by August. Essentially all the phosphorus may be utilized by the end of the growing season. Very high surface concentrations, up to 2.5 μg atoms P/l (1), support a large production of phytoplankton in the seas around Antarctica, but the concentration there has never been observed to be depleted beyond about 0.5 μg atoms P/l (8). Tropical areas of the oceans do not generally show the range of concentrations observed in higher latitudes.

Areas of upwelling provide a continuous supply of phosphorus to the surface and constitute the major areas of high organic productivity in the oceans. They are the result of major ocean current divergences and occur principally along the west coasts of continents at subtropical latitudes, such as off Peru, California, and northwest and southwest Africa. Upwelling off Arabia and Somaliland is seasonal and related to the monsoon. Upwelling takes place on a large scale around the Antarctic continent, especially in the southern part of the Atlantic Ocean (41). Divergences that take place between the north and south equatorial currents with the equatorial countercurrent in the Pacific Ocean also produce upwelling and are areas of high productivity (41). Phosphorus concentrations in upwelling waters are as high as 2 μg atoms P/l off the Peruvian coast (6) and, as noted above, 2.5 μg atoms P/l in the Antarctic region. A plot of the surface concentration in the Pacific Ocean (Fig. 4) shows high phosphorus in regions of horizontal divergence and low phosphorus in regions of convergence (27). Ryther (32) estimates that the areas of major upwelling comprise only 0.1% of the world ocean area but are the highest in mean primary organic productivity and yield about one-half the world's fish supply.

The Sargasso Sea of the North Atlantic appears to be unusually low in phosphorus. Data of Graham and Moberg (12, p. 13) indicate that surface phosphorus concentration is below 0.1 μg atom P/l in the winter as well as in the summer. It is an area of convergence and sinking water.

Zone of Maximum Phosphorus

Whereas phosphorus is extracted from seawater in the surface zone of photosynthesis, it is returned or regenerated in large part in the underlying waters where a maximum concentration with respect to a vertical profile is characteristically attained. The maximum is due in general to the dissolution and oxidation of dead organisms sinking from the surface. In the Antarctic region Clowes (8) found the greatest concentration of phosphorus where there was the greatest decomposition of zooplankton. An oxygen minimum generally corresponds roughly with the phosphorus maximum (26). The maximum usually occurs in the depth range from 250 to 2000 m (12, 41, 2, 1), and commonly is located near 1000 m. The maximum tends to be at shallower depths in the equatorial regions than at midlatitudes. It may not be pronounced in high latitudes in winter when vertical variations are small.

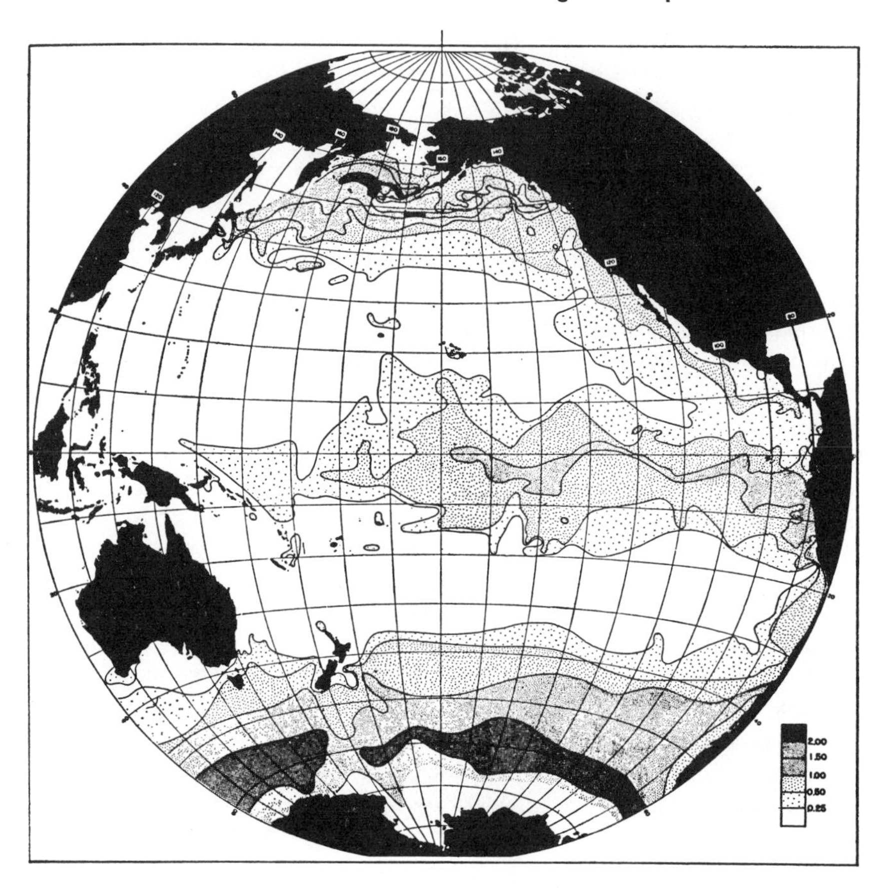

Figure 4. Distribution of P, in μg atoms/l, at the surface of the Pacific Ocean. From Reid (27, p. 293). Reproduced with permission of *Limnology and Oceanography*.

Phosphorus Content of the Deep Zone and Comparison of Oceans

Below the maximum which generally is at depths of about 1000 m, the phosphorus content changes only little, though decreasing slightly in most profiles and showing some variation due to the inherent characteristics of some deep layers or currents of seawater. Because the deep zone comprises the greater part of the total oceanic mass and because it does not show the rapid variations, both vertically and laterally, that are characteristic of the upper shallower waters, phosphorus determinations of the deep water are used to show regional trends in phosphorus concentration and to compare the phosphorus concentration of the major oceans and seas.

In a comparison of the phosphorus concentration of Pacific and North Atlantic waters at 2000 m depth, Graham and Moberg (12) found that the average concentration at 117 Pacific stations was 2.5 μg atoms P/l compared to an average for 22 stations in the North Atlantic of 1.1 μg atoms P/l. They cite other data to show that the concentration in the Arctic Ocean is low, as it is in the North Atlantic, and at all depths is nearly everywhere less than 1 μg atom P/l.

The distribution of phosphorus at a depth of 2000 m in the oceans of the world is shown by Redfield (25) and is presented here in Fig. 5. The anomalously low concentration of phosphorus in the North Atlantic Ocean and adjacent Arctic Ocean as compared to the other oceans and seas of the world is evident. The phosphorus distribution for the entire Atlantic Ocean reveals the circulation pattern which carries deep Antarctic water north across the equator and into the North Atlantic. Because the Arctic water that flows southward into the North Atlantic is low in phosphorus, there is no concentration gradient like that in the South Atlantic, and the North Atlantic water is poor in phosphorus. The oceans extending south of the equator join in a common sea around Antarctica that has a high phosphorus concentration, and the similar phosphorus concentrations of the

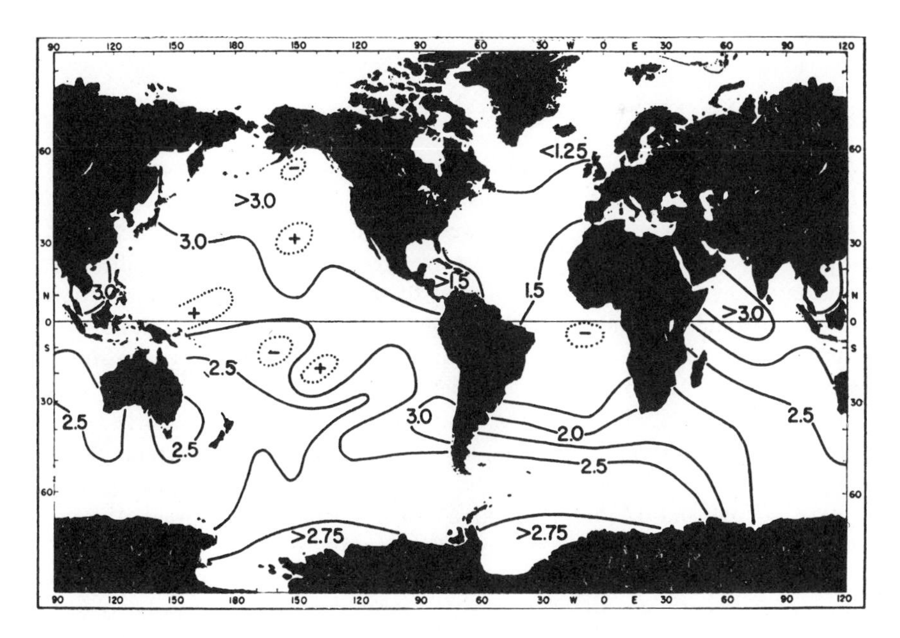

Figure 5. Distribution of phosphorus at a depth of 2000 m in the oceans of the world. Contour interval is 0.25 μg atoms/l. Redfield (25, p. 209). Reprinted with permission of *American Scientist*.

South Pacific, South Atlantic, and Indian Oceans are the result of this common source.

The North Pacific is the richest in phosphorus of all the oceans, a feature inherent unto itself as there is little interchange of its water with the South Pacific and only a small connection with the Arctic. The northern part of the Indian Ocean has also developed high concentrations like the North Pacific; this feature is apparently due to factors of local origin.

The Mediterranean and Red Seas are noteworthy for the low phosphorus concentrations that they exhibit, generally less than 1 μg atom P/l (14, 21, 22). Both seas have sources of inflowing water that are low in phosphorus. Surface waters depleted in phosphorus flow into the Mediterranean through the Strait of Gibraltar and into the Red Sea through Bab-el-Mandeb. The return flow from the seas to the oceans goes out the entrances over sills and under the inflowing waters. The return water is denser, concentrated by evaporation, and richer in phosphorus than the inflowing water. These conditions therefore tend to deplete continuously or at least work to maintain low phosphorus contents of the greater parts of both seas.

In the Black Sea, other factors are at work that bring about an enrichment of phosphorus in the deep water. The high concentrations are due to stagnation, hydrogen sulfide generation, and lower pH values than in regular seawater.

RELATIONS OF PHOSPHORUS TO OTHER NUTRIENTS

Many elements are necessary to the growth of marine organisms; some of the major nutrients are phosphorus, nitrogen, carbon, oxygen, and sulfur. It is phosphorus and nitrogen, however, whose quantities appear to limit organic development (25). The others are all present in excess. Redfield (25) considers that the relative constancy of the atom ratio of nitrogen to phosphorus in plankton and seawater, about 15, is of special significance. Although observed ratios in some seas and parts of the major oceans show a large variation, 1.2 to 30.2 (1), Redfield believes that the ratio of 15 holds for the oceans as a whole and the small range of variation, 12 to 19, found by Chow and Mantyla (6) for deep ocean waters between 2000 and 7000 m, supports his view. From many values of phosphorus and nitrogen in Pacific waters, Chow and Mantyla found a linear correlation between the constituents and calculated a line in terms of μg atom/l. of $N = [(17.5 \pm 4.0)P] - 6.5$. This equation agrees reasonably well with the one calculated by Stefánsson and Richards (40) of $N = 15.82P - 9.95$ for values obtained only on Pacific waters of the Washington and Oregon coastal region.

Silicon shows a seasonal variation in concentration in surface waters similar to that observed for phosphorus and nitrogen (41). It is not required by all organisms, however, and is utilized mainly by diatoms. The vertical distribution of silicon is markedly different from phosphorus (12). It increases in concentration with depth less abruptly than phosphorus and continues to increase to much greater depths than phosphorus. Chow and Mantyla (6) found that in the deep zone between 2000 and 6000 m the atom ratio of silicon to phosphorus ranged from 55 to 65 in Pacific Ocean waters, from 40 to 50 in the Indian Ocean, and from 20 to 40 in the Atlantic Ocean.

Carbon dioxide and oxygen concentrations in seawater are related to phosphorus concentrations. The general relationship is one in which carbon dioxide and phosphorus are consumed by phytoplankton in the near-surface zone of photosynthesis and oxygen is liberated (36). Below this zone the process is reversed. Oxygen is utilized in the oxidation of dead organisms and regenerates carbon dioxide, and phosphorus is released in the decomposition process. In vertical profile, the phosphorus maximum corresponds approximately with a total carbon dioxide high and an oxygen low. Graham and Moberg (12) found a remarkable inverse correlation of pH, which is closely related to total carbon dioxide content, with phosphorus in some vertical profiles.

Redfield (25) describes a reaction under anaerobic conditions that relates sulfur, carbon, and oxygen utilization with the concentration of phosphorus. He gives the following equation for sulfate reduction at depth under anaerobic conditions:

$$SO_4^{2-} + 2C \rightarrow 2CO_2 + S^{2-}$$

where C represents the reduced carbon in organic matter. The decomposition of organic matter by sulfate-reducing bacteria in this reaction also liberates phosphorus. His data for phosphorus content plotted against the combined utilization of oxygen and sulfate in the anaerobic waters of the Black Sea and Cariaco trench show a good correlation. He suggests that the phosphorus content is therefore an indication of the quantity of organic matter that has been decomposed.

REFERENCES

1. Armstrong, F. A. J., "Phosphorus," in *Chemical Oceanography*, Vol. 1, J. P. Riley and G. Skirrow (Eds.), Academic Press, New York-London, 1965, pp. 323–364.
2. Barnes, H., "Nutrient Elements," in *Treatise on Marine Ecology and Paleoecology*, Vol. 1, *Ecology*, J. W. Hedgpeth (Ed.), Geological Society of America Memoir 67, 1957, pp. 297–344.
3. Barth, T. F. W., *Theoretical Petrology*, Wiley, New York, 1952, 387 pages.
4. Broecker, W. S., R. D. Gerard, M. Ewing, and B. C. Heezen, "Geochemistry and Physics of Ocean Circulation," in *Oceanography*, M. Sears (Ed.), American Association for the Advancement of Science, Publ. 67, 1961, pp. 301–322.

5. Bruneau, L., N. G. Jerlov, and F. F. Koczy, "Physical and Chemical Methods, Appendix, Table I, Physical and Chemical Data," in *Swedish Deep-Sea Expedition Reports, 1947–48*, Vol. 3, No. 4, 1953, pp. 99–112.
6. Chow, T. J., and A. W. Mantyla, "Inorganic Nutrient Anions in Deep Ocean Waters," *Nature*, **206**, No. 4982, 383–385 (1965).
7. Chughtai, A., R. Marshall, and G. H. Nancollas, "Complexes in Calcium Phosphate Solutions," *J. Phys. Chem.*, **72**, 208–211 (1968).
8. Clowes, A. J., "Phosphate and Silicate in the Southern Ocean," *Discovery Rep.*, **19**, 1–120 (1938).
9. Forchhammer, G., "On the Composition of Sea Water in the Different Parts of the Ocean," *Roy. Soc. London Philos. Trans.*, **155**, 203–262 (1865).
10. Garrels, R. M., and M. E. Thompson, "A Chemical Model for Sea Water at 25°C and One Atmosphere Total Pressure," *Am. J. Sci.*, **260**, 57–66 (1962).
11. Goldberg, E. D., and G. O. S. Arrhenius, "Chemistry of Pacific Pelagic Sediments," *Geochim. Cosmochim. Acta*, **13**, 153–212 (1958).
12. Graham, H. W., and E. G. Moberg, *Chemical Results of the Last Cruise of the Carnegie, Chemistry—1*, Carnegie Institute Washington, Publ. 562, 1944, pp. 1–58.
13. Gulbrandsen, R. A., "Physical and Chemical Factors in the Formation of Marine Apatite," *Econ. Geol.*, **64**, 365–382 (1969).
14. Harvey, H. W., *The Chemistry and Fertility of Sea Water*, Cambridge University Press, New York, 1955, 224 pages.
15. Horn, M. K., and J. A. S. Adams, "Computer-Derived Geochemical Balances and Element Abundances," *Geochim. Cosmochim. Acta*, **30**, 279–298 (1966).
16. Ishino, M., K. Nasu, Y. Morita, and M. Hamada, "Oceanographic Conditions in the West Pacific Southern Ocean in Summer of 1964–1965," *Tokyo Univ. J. Fish.*, **9**, 115–208 (1968).
17. Kester, D. R., and R. M. Pytkowicz, "Determination of the Apparent Dissociation Constants of Phosphoric Acid in Sea Water," *Limnol. Oceanogr.*, **12**, 243–252 (1967).
18. Lehr, J. R., G. H. McClellan, J. P. Smith, and A. W. Frazier, "Characterization of Apatites in Commercial Phosphate Rocks," in *Phosphates naturels—Phosphates dans l'agriculture*, Colloque International sur les Phosphates Mineraux Solides, Toulouse, May 1967, Vol. 2, 1968, pp. 27–44.
19. McClellan, G. H., and J. R. Lehr, "Crystal Chemical Investigation of Natural Apatites," *Am. Mineral.*, **54**, 1374–3191 (1969).
20. McKelvey, V. E., R. W. Swanson, and R. P. Sheldon, "The Permian Phosphorite Deposits of Western United States," International Geological Congress, 19th, Algiers, 1952, C. R., Sect. 11, Part 11, 1953, pp. 45–64.
21. Neumann, A. C., and D. Densmore, *Oceanographic Data from Mediterranean Sea, Red Sea, Gulf of Aden, and Indian Ocean, Atlantis Cruise 242 for the International Geophysical Year 1957–58*, Woods Hole Oceanographic Institute, Ref. 60–2, 1959, 44 pages.
22. Neumann, A. C., and D. A. McGill, "Circulation of the Red Sea in Early Summer," *Deep-Sea Res.*, **8**, 223–235 (1962).
23. Norpac Committee, *Oceanic Observations of the Pacific, 1955, the Norpac Data*, University of California Press, Berkeley and Los Angeles, University of Tokyo Press, Tokyo, 1960, 532 pages.
24. Reddy, C. V. G., P. S. N. Murty, and V. N. Sankaranarayanan, "An Incidence of

Very High Phosphate Concentrations in the Waters around Andaman Islands," *Curr. Sci.*, **37**, 17–19 (1968).

25. Redfield, A. C., "The Biological Control of Chemical Factors in the Environment," *Am. Sci.*, **46**, 205–221 (1958).
26. Redfield, A. C., B. H. Ketchum, and F. A. Richards, "The Influence of Organisms on the Composition of Sea-Water," in *The Sea*, M. N. Hill (Ed.), Interscience, New York, 1963, pp. 26–77.
27. Reid, J. L., Jr., "On Circulation, Phosphate–Phosphorus Content, and Zooplankton Volumes in the Upper Part of the Pacific Ocean," *Limnol. Oceanogr.*, **7**, 287–306 (1962).
28. Riley, J. P., "Analytical Chemistry of Sea Water," in *Chemical Oceanography*, Vol. 2, J. P. Riley, and G. Skirrow, (Eds.), Academic Press, New York–London, 1965, pp. 295–424.
29. Rittenberg, S. C., K. O. Emery, and W. L. Orr, "Regeneration of Nutrients in Sediments of Marine Basins," *Deep-Sea Res.*, **3**, 23–45 (1955).
30. Roberson, C. E., "Solubility Implications of Apatite in Sea Water," in *Geological Survey Research 1966*, U.S. Geological Survey Prof. Paper 550-D, 1966, pp. D178–D185.
31. Robinson, R. A., and R. H. Stokes, *Electrolyte Solutions*, Butterworth, London, 1968, 571 pages.
32. Ryther, J. H., "Photosynthesis and Fish Production in the Sea," *Science*, **166**, No. 3901, 72–76 (1969).
33. Scripps Institution of Oceanography, Preliminary Report, Step-I Expedition, Sept. 15 to Dec. 14, 1960, Part I, Physical and Chemical Data, SIO Ref. 61-9, University of Calif., La Jolla, 1961, 47 pages.
34. Schmidt, Carl, "Hydrologische Untersuchungen," *Mélanges Phys. Chim.*, **10**, Part 3, 525–628 (1878).
35. Sillén, L. G., and A. E. Martell, *Stability Constants of Metal-Ion Complexes*, Chemical Society of London, Special Publ. 17, 1964, 754 pages.
36. Skirrow, G., "The Dissolved Gases—Carbon Dioxide," in *Chemical Oceanography*, Vol. 1, J. P. Riley, and G. Skirrow (Eds.), Academic Press, New York-London, 1965, pp. 227–322.
37. Skopintsev, B. A., S. N. Timofeyeva, A. F. Danilenko, and M. V. Sokolova, "Organic Carbon, Nitrogen, Phosphorus, and Their Mineralization Products in Black Sea Water," *Oceanology*, **7**, No. 3, 353–363 (1967).
38. Smirnov, A. I., R. B. Ivnitskaya, and T. P. Zalavina, "Experimental Data on the Possibility of the Chemical Precipitation of Phosphate from Sea Water," Gos. Nauchn.-Issled. Inst. Gorno-Khim Syr'ya Tr. (translated by State Sci.-Research Inst. for Mining Chem. Raw Materials), No. 7, 1962, pp. 289–302.
39. Solórzano, L., and J. D. H. Strickland, "Polyphosphate in Sea Water," *Limnol. Oceanogr.*, **13**, pp. 515–518 (1968).
40. Stefánsson, U., and F. A. Richards, "Processes Contributing to the Nutrient Distributions off the Columbia River and Strait of Juan de Fuca," *Limnol. Oceanogr.*, **8**, 394–410 (1963).
41. Sverdrup, H. U., M. W. Johnson, and R. H. Fleming, *The Oceans, Their Physics, Chemistry and General Biology*, Prentice-Hall, Englewood Cliffs, N.J., 1942, 1087 pages.

42. Truesdell, A. H., and B. F. Jones, "Ion Association in Natural Brines," *Chem. Geol.*, **4**, 51–62 (1969).

43. Worthington, L. V., *Oceanographic Data from the R.R.S. Discovery II*, International Geophysical Year Cruises 1 and 2, 1957, Natl. Inst. Oceanography, Surrey, England, and Woods Hole Oceanographic Inst., Woods Hole, Mass., W.H.O.I. Ref. No. 58–30, 1958, 84 pages.

44. Worthington, L. V., *Oceanographic Data from the R.R.S. Discovery II*, International Geophysical Year Cruise 3, 1958, Natl. Inst. Oceanography, Surrey, England, and Woods Hole Oceanographic Inst., Woods Hole, Mass., W.H.O.I. Ref. No. 59-54, 1959, 75 pages.

Geochemistry of Minerals Containing Phosphorus

6

D. JEROME FISHER

Arizona State University, Tempe, Arizona

Practically all natural mineral phosphorus occurs as anisodesmic phosphates built up of tetrahedral $(PO_4)^{3-}$ radicals. In Table 1 are listed 205 such minerals, along with one phosphide (schreibersite). For the sake of ready reference the minerals are arranged here in alphabetical order. Since this tends to hide close relationships among groups or series of minerals that are isostructural or closely similar in chemical composition, each mineral is assigned a number, and if it is a member of such a grouping the number carries a letter suffix. Table 2 lists 39 such groups, each carrying two to eight species, totaling 138 minerals. The remaining 67 phosphate minerals appear in Table 1 as isolated units without any letter suffix to their numbers. The chemical formulas of 206 minerals are given in Table 1, along with the crystal system. In addition there are three columns of data that serve as an index to the tables in Chapter 7.

What cations are combined with the (PO_4) radical to form these minerals? It is seen from Table 3 that more than 30 are so involved, and this figure neglects hydrogen as well as OH and H_2O. Thus 95 minerals carry iron (51 Fe^{2+}; 44 Fe^{3+}), 60 aluminum, 56 calcium, and 45 manganese. These figures are based on the formulas shown in Table 1; and since these list only the major cations as a rule, one must consider these numbers in this light. In fact there are not many iron phosphates that are entirely free from manganese, and practically all manganese phosphates carry at least some iron. As the Fe–Mn phosphates weather in the oxidized zone, the Fe^{2+} goes to the Fe^{3+} state; it is only in ferric phosphates (with all the iron oxidized) that the Mn^{2+} may then be converted to Mn^{3+}. Table 3 is helpful if one wishes to determine the names of phosphate minerals containing a given cation, since the numbers given in this table may be interpreted by means of Table 1. Table 3 is especially useful for determining the names of phosphate minerals carrying two or three specified cations, since under these conditions many numbers may be eliminated before referring to Table 1.

TABLE I Alphabetical List of Mineral Phosphates

Number	Name	Composition	Strongest Line A	Mean Index of Refraction	Birefringence	Crystal* System
1a	Alluaudite	$Na_5Ca_2(Mn, Fe'')_9Fe_3'''(PO_4)_{12}$	2.72	1.74	0.02	M
2a	Amblygonite	$(Li, Na)Al(PO_4)(F, OH)$	3.158	1.605	0.022	Tr
3	Anapaite	$Ca_2Fe''(PO_4)_2 \cdot 4H_2O$	3.14	1.613	0.047	Tr
4	Andrewsite	$(Cu, Fe)Fe_3'''(PO_4)_3(OH)_2$	3.22	1.820	0.017	O
5a	Apatite	$Ca_5(PO_4)_3F$	2.80	1.633	0.003	H
26b	Arrojadite	$Na_2(Fe'', Mn)_5(PO_4)_4$	3.04	1.669	0.010	M
6	Arthurite	$(Cu_2Fe_4)(As, P, S)_4O_{16}(O, OH)_4 \cdot 8H_2O$	4.28	1.78	0.015 ±	M
69c	Attakolite	$(Ca, Mn)Al_2[(P, Si)O_4]_2 \cdot (OH)$	3.09	1.664	0.020	O
8	Augelite	$Al_2PO_4(OH)_3$	3.34	1.576	0.014	M
94b	Autunite	$Ca(UO_2 \mid PO_4)_2 \cdot 10H_2O$	10.4	1.577	0.023	Te
9	Azovskite	$Fe_3'''(PO_4)_2(OH)_3 \cdot nH_2O$	—	1.758	W	M
10	Babefphite	$BaBePO_4F$	3.19	1.629	0.003	O
51b	Barbosalite	$Fe''Fe_2'''(PO_4)_2(OH)_2$	3.36	1.79 +	0.065	M
94c	Bassetite	$Fe(UO_2 \mid PO_4)_2 \cdot 10\text{–}12H_2O$	—	1.574	0.020	M ?
5b	Belovite	$(Sr, R.E.)_5(PO_4)_3(OH)$	2.90	1.660	0.020	H
11	Beraunite	$Fe''Fe_5'''(PO_4)_4(OH)_5 \cdot 6H_2O$	10.37	1.786	0.040	M
73b	Bergenite	$Ba(UO_2)_4(PO_4)_2(OH)_4 \cdot 8H_2O$	3.08	1.690	0.038	O
12	Berlinite	$AlPO_4$	3.37	1.524	0.006	H
13	Bermanite	$Mn''Mn_2'''(PO_4)_2(OH)_2 \cdot 4H_2O$	9.65	1.729	0.063	M
69b	Bertossaite	$Li_2CaAl_4(PO_4)_4 \cdot (OH)_4$	3.06	1.636	0.018	O
14	Beryllonite	$NaBePO_4$	2.84	1.558	0.009	M
40b	Beusite	$(Mn, Fe'', Ca)_3(PO_4)_2$	3.50	1.703	0.020	M
103b	Bobierrite	$Mg_3(PO_4)_2 \cdot 8H_2O$	6.96	1.520	0.033	M
15	Bøggildite	$Na_2Sr_2Al_2(PO_4)F_9$	3.16	1.466	0.007	M
30b	Bolivarite	$Al_2PO_4(OH)_3 \cdot 5H_2O$	—	1.506	W	—

16	Bradleyite	$Na_3Mg(PO_4)(CO_3)$	3.32	1.525	0.07 +	?
17	Brazilianite	$NaAl_3(PO_4)_2(OH)_4$	5.05	1.609	0.020	M
18a	Brianite	$Na_2MgCa(PO_4)_2$	2.63	1.605	0.010	O
77b	Brockite	$CaTh(PO_4)_2 \cdot H_2O$	3.03	1.680	0.015	H
19a	Brushite	$CaHPO_4 \cdot 2H_2O$	7.62	1.546	0.012	M
20	Cacoxenite	$Fe'''_9(PO_4)_4(OH)_{15} \cdot 18H_2O$	11.9	1.600	0.08	H
21a	Calcioferrite	$Ca_2Fe'''_2(PO_4)_3OH \cdot 7H_2O$	5.10?	1.575	0.005 +	M
5c	Carbonate apatite	$Ca_5(PO_4, CO_3OH)_3 \cdot F$	2.82	1.628	0.009	H
32b	Cassidyite	$Ca_2(Ni, Mg)(PO_4)_2 \cdot 2H_2O$	2.70	1.66	0.025	Tr
98b	Chalcosiderite	$CuFe'''_6(PO_4)_4(OH)_8 \cdot 4H_2O$	3.77	1.84	0.069	Tr
61b	Cheralite	$(R.E., Ca, Th)PO_4$	3.07	1.78	0.037	M
22a	Childrenite	$(Fe'', Mn)AlPO_4(OH)_2 \cdot H_2O$	2.81	1.683	0.042	O
5d	Chlorapatite	$Ca_5(PO_4)_3Cl$	2.78	1.668	0.001	H
98c	Coeruleolactite	$CaAl_6(PO_4)_4(OH)_8 \cdot 4H_2O$	3.71	1.582	0.008	Tr?
32c	Collinsite	$Ca_2(Mg, Fe)(PO_4)_2 \cdot 2H_2O$	2.69	1.642	0.025	Tr
90b	Corkite	$PbFe'''_3(PO_4)(SO_4)(OH)_6$	3.03	1.93	W	H
23	Cornetite	$Cu_3(PO_4)(OH)_3$	3.04	1.81	0.055	O
39b	Crandallite	$CaAl_3H(PO_4)_2(OH)_6$	2.97	1.619	0.006	H
104b	Cyrilovite	$NaFe'''_3(PO_4)_2(OH)_4 \cdot 2H_2O$	8.17	1.804	0.03	Te
24	Delvauxite	$Fe'''_4(PO_4)_2(OH)_6 \cdot 7H_2O$	—	1.72	0.00	—
73c	Dewindtite	$Pb(UO_2)_4(PO_4)_2(OH)_4 \cdot 8H_2O$	3.07	1.767	0.006	O
25	Diadochite	$Fe'''_2(PO_4)(SO_4)(OH) \cdot 5H_2O$	4.35	1.62	0.050	Tr?
26a	Dickinsonite	$Na_2(Mn, Fe'')_5(PO_4)_4$	3.037	1.66	0.014	M
27	Dufrenite	$CaFe''_2Fe'''_{10}(PO_4)_8(OH)_{12} \cdot 4H_2O$	3.15	1.84	0.06	M
28	Dumontite	$Pb_2(UO_2)_3(PO_4)_2(OH)_4 \cdot 3H_2O$	4.27	1.87	0.05?	M
29	Englishite	$K_2Ca_4Al_8(PO_4)_8(OH)_{10} \cdot 9H_2O$	9.30	1.572	0.002	M?
22b	Eosphorite	$(Mn, Fe'')AlPO_4(OH)_2 \cdot H_2O$	2.83	1.648	0.029	O
30a	Evansite	$Al_3(PO_4)(OH)_6 \cdot 6H_2O$	—	1.485	0.00	—

* Abbreviations in the last column are for *I*sometric, *H*exagonal, *T*etragonal, *O*rthorhombic, *M*onoclinic, *T*riclinic.

TABLE I (continued)

Number	Name	Composition	Strongest Line A	Mean Index of Refraction	Birefringence	Crystal* System
31	Faheyite	$MnBe_2Fe'''_2(PO_4)_4 \cdot 6H_2O$	5.72	1.631	0.021	H
32a	Fairfieldite	$Ca_2(Mn, Fe)(PO_4)_2 \cdot 2H_2O$	3.22	1.644	0.018	Tr
33	Farringtonite	$Mg_3(PO_4)_2$	3.83	1.544	0.019	M
98d	Faustite	$ZnAl_6(PO_4)_4(OH)_8 \cdot 4H_2O$	3.68	1.613	0.04	Tr
5e	Fermorite	$(Ca, Sr)_5[(P, As)O_4]_3(F, OH)$	2.86	1.66	W	H
1b	Ferrialluaudite	$Na_2Ca(Mn, Fe'')_4Fe'''_8(PO_4)_{12}$	3.06	1.805	0.06	M
95b	Ferrisicklerite	$Li_1(Fe''', Mn'')PO_4$	2.95	1.778	0.035	O
1c	Ferroalluaudite	$Na_6Ca(Fe'', Mn)_8Fe'''_4(PO_4)_{12}$	2.71	1.73	0.02	M
34	Fillowite	$Na_2(Mn, Fe'', Ca)_8(PO_4)_6 \cdot H_2O$	2.81	1.672	0.005	H
39c	Florencite	$CeAl_3(PO_4)_2(OH)_6$	2.95	1.695	0.010	H
35	Fluellite	$AlPO_4F_2(OH) \cdot 7H_2O$	6.60	1.490	0.032	O
36a	Frondelite	$(Mn, Fe'')Fe'''_4(PO_4)_3(OH)_5$	3.20	1.880	0.033	O
37a	Glucine	$Ba_2Be_8(PO_4)_4(OH)_8 \cdot H_2O$	10.8	1.56?	0.024	—
39d	Gorceixite	$BaAl_3H(PO_4)_2(OH)_6$	2.92	1.618	0.007	H
38a	Gordonite	$MgAl_2(PO_4)_2(OH)_2 \cdot 8H_2O$	9.78	1.543	0.024	Tr
39a	Goyazite	$SrAl_3H(PO_4)_2(OH)_6$	2.95	1.63	0.010	H
40a	Graftonite	$(Fe'', Mn, Ca)_3(PO_4)_2$	2.86	1.713	0.017	M
41	Griphite	$(Mn, Ca, Na, Fe'')_3(Fe''', Al)_2[PO_3(OH, F)]_3$	2.74	1.64 ±	0.00	I
42a	Hannayite †	$Mg_3(NH_4)_2H_4(PO_4)_4 \cdot 8H_2O$	6.96	1.522	0.035	Tr
43a	Herderite	$CaBe(PO_4)(OH, F)$	3.14	1.611	0.028	M
95c	Heterosite	$(Fe''', Mn''')PO_4$	3.48	1.89	0.05	O
90c	Hinsdalite	$PbAl_3(PO_4)SO_4(OH)_6$	2.78	1.671	0.018	H
44a	Hopeite	$Zn_3(PO_4)_2 \cdot 4H_2O$	9.10	1.598	0.010	O

45	Hureaulite	$(Mn, Fe'')_5H_2(PO_4)_4 \cdot 4H_2O$	3.14	1.655	0.010	M
46	Hurlbutite	$CaBe_2(PO_4)_2$	3.67	1.601	0.007	O
94d	Hydrogen autunite	$H_2(UO_2)_2(PO_4)_2 \cdot 8H_2O$	9.03	1.583	0.014	Te
5f	Hydroxylapatite	$Ca_5(PO_4)_3 \cdot OH$	2.81	1.651	0.007	H
47	Isokite	$CaMgPO_4 \cdot F$	3.18	1.601	0.025	M
48a	Kehoeite	$(Zn, Ca)(AlPH_3)_2O_{12} \cdot 4H_2O$	3.35	1.53	0.00	I
105b	Kingite	$(AlOH)_3(PO_4)_2 \cdot 9H_2O$	9.10	1.514	W	Tr
49	Koninckite	$Fe'''PO_4 \cdot 3H_2O$	8.42	1.660	0.012	Te
76d	Kryzhanovskite	$Fe_2'''Mn''(PO_4)_2(OH)_2 \cdot H_2O$	3.14	1.80	0.03	M?
76b	Landesite	$Mn_9''Fe_3'''(PO_4)_8 \cdot (OH)_3 \cdot 9H_2O$	3.20	1.728	0.015	O
50	Laubmannite	$(Fe'', Mn)_3Fe_6'''(PO_4)_4(OH)_{12}$	5.16	1.847	0.052	O
38b	Laueite	$Mn''Fe_2'''(PO_4)_2 \cdot (OH)_2 \cdot 8H_2O$	9.91	1.656	0.097	Tr
51a	Lazulite	$(Mg, Fe'')Al_2(PO_4)_2(OH)_2$	3.07	1.626	0.033	M
52	Leucophosphite	$K(Fe''', Al)_2(PO_4)_2(OH) \cdot 2H_2O$	6.79	1.721	0.032	M
53	Libethenite	$Cu_2PO_4(OH)$	4.81	1.745	0.086	O
51c	Lipscombite	$(Fe'', Mn)Fe_2'''(PO_4)_2(OH)_2$	3.3	—	—	Te
95d	Lithiophilite	$Li(Mn, Fe'')PO_4$	2.53	1.679	0.011	O
54	Lithiophosphatite	Li_3PO_4	3.98	1.557	0.017	O
55a	Ludlamite	$Fe_3''(PO_4)_2 \cdot 4H_2O$	3.96	1.675	0.044	M
56	Lueneburgite	$Mg_3(PO_4)_2[B(OH)_3]_2 \cdot 6H_2O$	—	1.541	0.026	M
39e	Lusungite	$(Sr, Pb)Fe_3'''H(PO_4)_2(OH)_6$	2.98	1.8 ±	0.035	H
96b	Magnesiotriplite	$(Mg, Fe, Mn)_2(PO_4)F$	3.33	1.649	0.020	M
57	Melkovite	$CaFe'''H_6(MoO_4)_4(PO_4) \cdot 6H_2O$	2.92	1.838	W?	—
32d	Messelite	$Ca_2(Fe, Mn)(PO_4)_2 \cdot 2H_2O$	6.34	1.653	0.040	Tr
58b	Metaankoleite	$K_2(UO_2)_2(PO_4)_2 \cdot 6H_2O$	8.92	1.580	0.003	Te
58c	Metaautunite	$(Na_2, Ca, Sr)(UO_2)_2(PO_4)_2 \cdot nH_2O$	8.47	1.604 ±	0.012	Te
58d	Metabassetite	$Fe''(UO_2)_2(PO_4)_2 \cdot 8H_2O$	4.89	1.61	0.014	M
105c	Metakingite	$(AlOH)_3(PO_4)_2 \cdot 4H_2O$	7.40	—	—	—

* Shown in 1971 to be a mixture of dittmarite [$(NH_4)MgPO_4 \cdot H_2O$] and newberyite.

TABLE I (continued)

Number	Name	Composition	Strongest Line A	Mean Index of Refraction	Birefringence	Crystal System
83b	Metaschoderite	$Al_2(PO_4)(VO_4)\cdot 6H_2O$	7.50	1.604	0.028	M
58a	Metatorbernite	$Cu(UO_2)_2(PO_4)_2\cdot 8H_2O$	3.69	1.626	0.002	Te
58e	Metauranocircite	$Ba(UO_2)_2(PO_4)_2\cdot 8H_2O$	8.93	1.622	0.013	O?
72b	Metavariscite	$AlPO_4\cdot 2H_2O$	2.71	1.558	0.031	M
101b	Metavauxite	$Fe''Al_2(PO_4)_2(OH)_2\cdot 8H_2O$	2.75	1.561	0.027	M
104c	Millisite	$Na_2CaAl_{12}(PO_4)_8\cdot 8H_2O$	4.77	1.602	0.018	Te
59	Minyulite	$KAl_2(PO_4)_2(OH, F)\cdot 4H_2O$	5.60	1.534	0.007	O
60	Mitridatite	$Ca_2Fe_3'''(PO_4)_3(OH)_4\cdot nH_2O$	2.73	1.765	0.01?	M?
61a	Monazite	$(R.E.)PO_4$	3.09	1.791	0.054	M
62	Monetite	$CaHPO_4$	3.38	1.615	0.053	Tr
2b	Montebrasite	$(Li, Na)Al(PO_4)(\mathit{OH}, F)$	3.153	1.613	0.022	Tr
21b	Montgomeryite	$Ca_2Al_2(PO_4)_3OH\cdot 7H_2O$	5.11	1.578	0.010	M
63	Moraesite	$Be_2(PO_4)OH\cdot 4H_2O$	7.00	1.482	0.028	M
64	Morinite	$NaCa_2Al_2(PO_4)_2(F, OH)_5\cdot 2H_2O$	2.94	1.563	0.014	M
2c	Natromontebrasite	$(Na, Li)Al(PO_4)(OH, F)$	—	1.603	0.021	Tr
95e	Natrophilite	$Na(Mn, Fe'')PO_4$	2.61	1.674	0.013	O
65	Newberyite	$MgHPO_4\cdot 3H_2O$	5.34	1.517	0.019	O
66	Ningyoite	$(Ca, R.E., U)PO_4\cdot H_2O$	3.02	1.64	W	O
67	Nissonite	$CuMg(PO_4)OH\cdot 2\frac{1}{2}H_2O$	11.14	1.62	0.037	M
68	Overite	$Ca_3Al_8(PO_4)_8(OH)_6\cdot 15H_2O$	2.83	1.574	0.012	O
69a	Palermoite	$Li_2SrAl_4(PO_4)_4\cdot(OH)_4$	3.09	1.642	0.017	O
18b	Panethite	$NaMgPO_4$	3.01	1.576	0.012	M

44b	Parahopeite	$Zn_3(PO_4)_2 \cdot 4H_2O$	7.56	1625	0.023	Tr
38c	Paravauxite	$Fe''Al_2(PO_4)_2(OH)_2 \cdot 8H_2O$	9.82	1.558	0.019	Tr
70	Parsonsite	$Pb_2UO_2(PO_4)_2 \cdot 2H_2O$	4.20	1.85 +	0.012	Tr
76c	Phosphoferrite	$(Fe'', Mn)_3(PO_4)_2 \cdot 3H_2O$	3.18	1.680	0.028	O
55b	Phosphophyllite	$Zn_2Fe''(PO_4)_2 \cdot 4H_2O$	2.83	1.614	0.021	M
71	Phosphorroesslerite	$MgHPO_4 \cdot 7H_2O$	4.22	1.485	0.009	M
72a	Phosphosiderite	$Fe'''PO_4 \cdot 2H_2O$	2.78	1.725	0.046	M
73a	Phosphuranylite	$Ca(UO_2)_4(PO_4)_2(OH)_4 \cdot 8H_2O$	7.91	1.710	0.042	O
39f	Plumbogummite	$PbAl_3H(PO_4)_2(OH)_6$	2.97	1.653	0.022	H
58f	Przhevalskite	$Pb(UO_2)_2(PO_4)_2 \cdot 4H_2O$	3.61	1.749	0.013	O
74	Pseudoautunite	$Ca_2(UO_2)_2(PO_4)_4(H_3O)_4 \cdot 5H_2O$	6.20	1.568	0.029	H
101c	Pseudolaueite	$MnFe_2'''(PO_4)_2(OH)_2 \cdot 8H_2O$	9.93	1.650	0.060	M
75	Pseudomalachite	$Cu_5(PO_4)_2(OH)_4$	4.49	1.835	0.056	M
95f	Purpurite	$(Mn''', Fe''')PO_4$	2.95	1.86	0.07	O
5g	Pyromorphite	$Pb_5(PO_4)_3 \cdot Cl$	2.96	2.06	0.01	H
76a	Reddingite	$(Mn, Fe'')_3(PO_4)_2 \cdot 3H_2O$	3.20	1.656	0.032	O
73d	Renardite	$Pb(UO_2)_4(PO_4)_2(OH)_4 \cdot 7H_2O$	7.97	1.736	0.024	O
77a	Rhabdophane	R.E. $(PO_4) \cdot \frac{1}{2}H_2O$	3.02	1.654	0.049	H
36b	Rockbridgeite	$(Fe'', Mn)Fe_4'''(PO_4)_3(OH)_5$	3.19	1.880	0.022	O
78	Roscherite	$(Ca, Mn, Fe)BePO_4(OH) \cdot \frac{2}{3}H_2O$	9.59	1.641	0.015	M
94e	Sabugalite	$HAl(UO_2)_4(PO_4)_4 \cdot 16H_2O$	9.61	1.583	0.019	Te
94f	Saléeite	$Mg(UO_2)_2(PO_4)_2 \cdot 10H_2O$	9.85	1.574	0.015	Te
79	Salmonsite	$Mn_9''Fe_2'''(PO_4)_8 \cdot 14H_2O$	9.42	1.660	0.015	M?
80	Sampleite	$NaCaCu_5(PO_4)_4Cl \cdot 5H_2O$	9.60	1.677	0.050	O
81	Sanjuanite	$Al_2(PO_4)(SO_4)OH \cdot 9H_2O$	10.77	1.49 ±	0.015 +	M?
95g	Sarcopside	$Fe_3''(PO_4)_2$	3.03	1.728	0.062	M
82	Schertelite	$(NH_4)_2MgH_2(PO_4)_2 \cdot 4H_2O$	5.94	1.515	0.015	O
83a	Schoderite	$Al_2(PO_4)(VO_4) \cdot 8H_2O$	7.9	1.548	0.024	M

TABLE I (continued)

Number	Name	Composition	Strongest Line A	Mean Index of Refraction	Birefringence	Crystal System
84	Scholzite	$Zn_2Ca(PO_4)_2 \cdot 2H_2O$	8.5	1.593	0.014	O
85	Schreibersite *	$(Fe, Ni)_3P$	2.19	Opaque		Te
51d	Scorzalite	$(Fe'', Mg)Al_2(PO_4)_2 \cdot (OH)_2$	3.24	1.670	0.041	M
86	Seamanite	$Mn_3''(PO_4)(BO_3) \cdot 3H_2O$	2.846	1.663	0.025	O
95h	Sicklerite	$Li_{1-}(Mn'', Fe''')PO_4$	3.01	1.735	0.030	O
38d	Sigloite	$(Fe''', Fe'')Al_2(PO_4)_2(O, OH) \cdot 8H_2O$	9.69	1.586	0.056	Tr
87	Souzalite	$(Mg, Fe'')_3Al_4(PO_4)_4(OH)_6 \cdot 2H_2O$	2.69	1.642	0.034	M
88	Spencerite	$Zn_4(PO_4)_2(OH)_2 \cdot 3H_2O$	9.40	1.602	0.020	M
18c	Stanfieldite	$Ca_3(Mg_3Fe_2''Ca)(PO_4)_6$	2.82	1.622	0.012	M
42b	Stercorite	$Na(NH_4)H(PO_4) \cdot 4H_2O$	6.60	1.442	0.030	Tr
72c	Sterrettite	$ScPO_4 \cdot 2H_2O$	4.88	1.590	0.029	M
101d	Stewartite	$Mn''Fe_2'''(PO_4)_2(OH)_2 \cdot 8H_2O$	9.98	1.655	0.069	Tr
99b	Strengite	$Fe_3'''PO_4 \cdot 2H_2O$	4.38	1.719	0.030	O
76e	Strunzite	$Mn''Fe_2'''(PO_4)_2(OH)_2 \cdot 8H_2O$	9.02	1.67	0.10	M
89	Struvite	$Mg(NH_4)(PO_4) \cdot 6H_2O$	4.26	1.496	0.009	O
90a	Svanbergite	$SrAl_3(PO_4)(SO_4)(OH)_6$	2.98	1.633	0.014	H
55c	Switzerite	$Mn_3(PO_4)_2 \cdot 4H_2O$	8.55	1.628	0.030	M
91	Taranakite	$K_3Al_5H_6(PO_4)_8 \cdot 18H_2O$	15.5	1.508	0.006	H
92	Tarbuttite	$Zn_2PO_4(OH)$	2.78	1.705	0.053	Tr
2d	Tavorite	$LiFe'''PO_4(OH)$	3.05	1.81 ±	?	Tr
93	Tinticite	$Fe_3'''(PO_4)_2 \cdot (OH)_3 \cdot 3H_2O$	3.91	1.745	?	O ?
94a	Torbernite	$Cu(UO_2)_2(PO_4)_2 \cdot 10H_2O$	10.3	1.592	0.010	Te
95a	Triphylite	$Li(Fe'', Mn)PO_4$	2.53	1.695	0.008	O

96a	Triplite	$(Mn, Fe'')_2(PO_4)F$	2.86	1.658	0.022	M
96c	Triploidite	$(Mn, Fe'')_2(PO_4)OH$	3.04	1.726	0.005	M
51e	Trolleite	$Al_4(PO_4)_3(OH)_3$	3.20	1.639	0.024	M?
97	Tsumebite	$Pb_2CuPO_4(OH)_3 \cdot 3H_2O$	4.71	1.920	0.071	M
98a	Turquoise	$CuAl_6(PO_4)_4(OH)_8 \cdot 4H_2O$	3.68	1.62	0.04	Tr
51f	Unnamed	$Fe_3''Fe'''(PO_4)_3$	3.31	—	—	—
37b	Uralolite	$CaBe_3(PO_4)_2(OH)_2 \cdot 4H_2O$	3.56	1.525	0.026	M
94g	Uramphite	$(NH_4)(UO_2)PO_4 \cdot 3H_2O$	3.79	1.585	0.021	Te
94h	Uranocircite	$Ba(UO_2)_2(PO_4)_2 \cdot 10H_2O$	5.10	1.583	0.014	Te
99a	Variscite	$AlPO_4 \cdot 2H_2O$	5.37	1.588	0.031	O
100	Vauquelinite	$Pb_2Cu(CrO_4)(PO_4)OH$	3.31	2.22	0.11	M
101a	Vauxite	$Fe''Al_2(PO_4)_2(OH)_2 \cdot 6H_2O$	5.45	1.555	0.011	Tr
43b	Vaeyrynenite	$MnBe(PO_4)OH$	3.45	1.662	0.027	M
102	Veszelyite	$(Cu, Zn)_3PO_4(OH)_3 \cdot 2H_2O$	3.64	1.658	0.055	M
48b	Viséite	$NaCa_5Al_{10}Si_3P_5O_{30}(OH)_{18} \cdot 8H_2O$	2.92	1.53	0.00	I
103a	Vivianite	$Fe_3''(PO_4)_2 \cdot 8H_2O$	6.70	1.603	0.053	M
96d	Wagnerite	$Mg_2(PO_4)F$	2.99	1.572	0.014	M
104a	Wardite	$NaAl_3(PO_4)_2(OH)_4 2H_2O$	4.74	1.588	0.009	Te
105a	Wavellite	$Al_3(PO_4)_2(OH)_3 \cdot 5H_2O$	3.44	1.535	0.025	O
39g	Waylandite	$(Bi, Ca)Al_3(PO_4)_2(OH)_6$	2.93	—	—	H
19b	Weinschenkite	$(Y, Er)PO_4 \cdot 2H_2O$	4.21	1.612	0.040	M
106	Whitlockite	$Ca_3(PO_4)_2$	2.85	1.629	0.003	H
5h	Wilkeite	$Ca_5(PO_4, SO_4, SiO_4)_3(F, O)$	2.80	1.645	0.004	H
96e	Wolfeite	$(Fe'', Mn)_2PO_4(OH)$	2.93	1.742	0.005	M
90d	Woodhouseite	$CaAl_3(PO_4)(SO_4)(OH)_6$	2.94	1.636	0.011	H
107	Xanthoxenite	$Ca_4Fe_2'''(PO_4)_4(OH)_2 \cdot 3H_2O$	3.06	1.715	0.020	M?
61c	Xenotime	$Y(PO_4)$	3.44	1.720	0.10	Te
36c	Zinc rockbridgeite	$ZnFe_4'''(PO_4)_3(OH)_5$	3.31	1.83	0.06	O
96f	Zwieselite	$(Fe'', Mn)_2(PO_4)F$	2.87	1.693	0.019	M
	*Barringerite is $(Fe,Ni)_2P$		2.23	Opaque		H

TABLE 2 Phosphate Mineral Groups

With More Than Two Members			With But Two Members	
Group Number	Group Name	Number of Minerals	Group Number	Group Name
1	Alluaudite	3	19	Brushite
2	Amblygonite	4	21	Calcioferrite
5	Apatite	8	22	Childrenite
18	Brianite	3	26	Dickinsonite
32	Fairfieldite	4	30	Evansite
36	Frondelite	3	37	Glucine
38	Gordonite	4	40	Graftonite
39	Goyazite	7	42	Hannayite
51	Lazulite	6	43	Herderite
55	Ludlamite	3	44	Hopeite
58	Metatorbernite	6	48	Kehoeite
61	Monazite	3	69	Palermoite
72	Phosphosiderite	3	77	Rhabdophane
73	Phosphuranylite	4	83	Schoderite
76	Reddingite	5	99	Variscite
90	Svanbergite	4	103	Vivianite
94	Torbernite	8		
95	Triphylite	8		
96	Triplite	6		
98	Turquoise	4		
101	Vauxite	4		
104	Wardite	3		
105	Wavellite	3		
		106		
	Other column	32		
	Total	138		

TABLE 3 Cation Components of Mineral Phosphates (Omitting H, OH, H_2O)

Cation	Mineral Numbers (as in Table 1)[a]
Al	2a–c; 7, *8*, *12*, 15, 17, 21b; 22a–b; 29, *30a–b*; *35*, 38a, c–d; 39a–d, f–g; 41, 42a; 47, 48a–b; *51*a, d–*e*; 52, 59, 64, 68, 69a–b; *72b*; *81*, *83a–b*; 87, 90a, c–d; 91, 94e; 98a, e–d; *99a*; 101a–b; 104a, c; *105a–c*
As	5e; 6
Ba	10; 37a; 39d; 58e; 73b; 94h
Be	10, 14, 31, 37a–b; 43a–b; 46, *63*, 78
Bi	39g
B	56, 86
Ca	1a–c; 3, *5a*, *c–f*, *h*; 7, 18a, c; *19a*; 21a–b; 27, 29, 32a–d; 34, 37b; 39b, g; 40a–b; 41, 43a; 46, 47, 48a–b; 57, 58c; 60, 61b, *62*, 64, 66, 68, 69b; 73a; 74, 77b; 78, 80, 84, 90d; 94b; 98c; 104c; 105a; *106*, 107
C	5c, g; 16
Cl	5d; 80
Cr	100
Cu	4, 6, *23*, *53*, 58a; 67, *75*, 80, 94a; 97, 98a–b; 100, 102
Fe″	1, 3, 4, 6, 11, 18c; 22a–b; 26a–b; 27, 32a, c–d; 34, 36a–b; 38c–d; 40a–b; 41, 45, 50, 51a–d, f; *55a*–b; 58d; 76a, c; 78, 85, 87, 94c; *95*a, d–e, *g*; 96a–c, e–f; 101a–b; *103a*
Fe‴	1, 2d; 4, *9*, 11, *20*, 21a; *24*, *25*, 27, 31, 36a–c; 38b, d; 39e; 41, *49*, 50, 51b–c, f; 52, 57, 60, *72a*; 76b, d–e; 79, 90b; *93*, 95b–c, f, h; 98b; *99b*; 101c–d; 104b; 107
F	2a–c; 5a, c, e, h; 10, 15, 35, 41, 43a; 47, 59, 64, 96a–b, d, f
K	29, 52, 58b; 59, 91
Li	2a–d; *54*, 69a–b; 95a–b, d, h
Mg	16, 18a–c; 32b–c; *33*, 38a; 42a; 47, 51a, d; 56, *65*, 67, *71*, 82, 87, 89, 94f; *96*b, *d*; *103b*
Mn″	1, 7, 13, 22a–b; 26a–b; 31, 32a, d; 34, 36a–b; 38b; 40a–b; 41, 43b; 45, 50, 51c; *55c*; 76a–e; 78, 79, *86*, 95a–b, d–e, h; 96a–c, e–f; 101c–d
Mn‴	13, 95c, f
Mo	57
Na	1a–c; 2a–c; 14, 15, 16, 17, 18a–b; 26a–b; 34, 41, 42b; 48b; 58c; 64, 80, 95e; 104a–c
NH_4	42a–b; 82, 89, 94g
Ni	32b; 85
Pb	5g; 28, 39e–f; 58f; 70, 73c–d; 90b–c; 97, 100
R.E.	5b; 19b; 39c; *61a*–b; *77a*
Sc	*72c*
Si	5h; 7, 48b
S	5h; 6, 25, 81, 90a–d
Sr	5b, e; 15, 39a, e; 58c; 69a; 90a
Th	61b; 77b
U	28, 58a–f; 66, 70, 73a–d; 74, 94a–*d*–h
V	83a–b
Y	19b; *61c*
Zn	36c; *44a–b*; 48a; 55b; 84, *88*, *92*, 98d; 102

[a] *Note.* The italicized numbers refer to phosphates that carry only the single metal cation under which they are listed. Anions listed include Cl, F, S.

Many of the minerals listed in Table 1 are rare, being known only as very minor ingredients of a single or at most a few deposits. Apatite (fluorapatite) is the chief phosphate mineral in igneous rocks. This as well as carbonate and hydroxylapatite are found in sediments, both primary and residual. The Florida phosphorites are rich in crandallite and millisite, minerals which a generation ago were considered rare, as well as wavellite and carbonate fluorapatite. A number of the phosphates with weak birefringence, other than the apatites, are typically very fine grained and thus difficult to identify—see the list in Table 4, Chapter 7. Future studies may show some of them to be of considerable economic importance. Wavellite, vivianite, and even turquoise are fairly common phosphate minerals in certain types of surficial deposits. But a larger number of phosphate species is found in the pegmatites, where they may be "primary," as amblygonites, triphylites, alluaudites, lazulites, monazite, xenotime, and so on, or as later ("hydrothermal") minerals or as supergene alteration products. Geologic occurrences are given in Table 2, Chapter 7.

Identification of Phosphorus-Bearing Minerals

7

D. JEROME FISHER

Arizona State University, Tempe, Arizona

Once a given mineral is known to contain phosphorus, its identity within the limits of the present status of our knowledge may be established by powder x-ray diffraction (Table 1), or by determining indices of refraction using oil-immersion techniques with the polarizing microscope (Table 2).

The chemist in making use of these tables should be warned that few minerals are pure chemical compounds of fixed composition. With the presence of vicarious components one must expect some intensity (as well as minor spacing) variation (Table 1), and often quite pronounced variation in *n* values (Table 2). Where one is dealing with a series or group containing two to eight members (Table 2, Chapter 6), the trends in such variations can be evaluated. Thus if we consider the eight members of the triphylite group (95a–h) arranged in order of increasing values of index of refraction (α) as shown in Table 3, the following facts emerge. Omitting the first two minerals of this table, it is clear that we are dealing with three pairs of minerals, each of which shows solid solution variation of the Fe/Mn ratio. The first series (d, a) containing bivalent Fe carries a "normal" content of Li. But as the iron suffers oxidation there is a concomitant loss of Li, resulting in the second series (h, b). When all the iron reaches the ferric state, further oxidation produces trivalent manganese, the Li is completely removed, and the result is a third series (f, c). Along with these changes an increase in indices of refraction occurs, but in any one series the figures may vary between the values shown; in extreme cases results somewhat outside the ranges given may be attained.

All the members of the triphylite group as shown in Table 3 are orthorhombic, except sarcopside which is monoclinic with angle β essentially 90° and unit-cell dimensions like those of the others. It should be emphasized that some of the groupings as listed in Table 2, Chapter 6, contain members assigned to them in a tentative way. This is a field of active research, and in the future no doubt some modifications will be required.

TABLE I Mineral Phosphate Powder Diffraction Data

Strongest Line	Next Stronger Lines		Name
2.19	2.11/7	1.97/7	Schreibersite
2.53	3.01/9	3.47/8	Lithiophilite
2.53	3.02/9+	3.49/9	Triphylite
2.61	2.59/10	2.87/9	Natrophilite
2.63	2.73/9	2.68/9	Brianite
2.69	3.04/8	1.67/6	Collinsite
2.69	3.79/9	5.35/8	Souzalite
2.70	3.03/9+	2.67/8	Cassidyite
2.71	4.25/9	4.58/8	Metavariscite
2.71	6.25/6	2.53/5	Ferroalluaudite
2.72	3.11/6	6.29/5	Alluaudite
2.73	5.57/8	8.81/7	Mitridatite
2.74	1.64/6	3.07/4	Griphite
2.75	4.67/9	4.32/9−	Metavauxite
2.78	2.86/6	1.84/2	Chlorapatite
2.78	2.96/8	5.59/7	Hinsdalite
2.78	4.69/8	4.37/8	Phosphosiderite
2.78	6.12/9	3.70/9	Tarbuttite
2.80	2.70/9	2.24/8	Wilkeite
2.80	2.70/6	2.77/6−	Apatite
2.81	2.78/6	2.72/6	Hydroxylapatite
2.81	3.02/7	3.64/7−	Fillowite
2.81	5.27/4	2.42/4	Childrenite
2.82	2.72/9	3.45/7	Carbonate apatite
2.82	3.75/8	2.51/8	Stanfieldite
2.83	2.42/6	5.23/5	Eosphorite
2.83	4.40/9	8.84/8	Phosphophyllite
2.83	9.40/8	5.29/6	Overite
2.84	3.65/9	2.28/7	Beryllonite
2.846	2.51/8	2.29/6	Seamanite
2.85	2.58/8	1.71/7	Whitlockite
2.86	2.75/6	3.49/5	Fermorite
2.86	3.03/7	3.24/6	Triplite
2.86	3.50/9	2.72/7	Graftonite
2.87	3.06/9	3.26/7	Zwieselite
2.90	3.17/7	2.79/7	Belovite
2.92	1.74/6	3.46/5	Viséite
2.92	3.54/8	8.42/7	Melkovite
2.92	5.66/8	2.17/8	Gorceixite
2.93	3.09/9	3.18/8	Wolfeite

TABLE I (continued)

Strongest Line	Next Stronger Lines		Name
2.93	5.66/7	3.48/5	Waylandite
2.94	1.89/9	2.19/8	Woodhouseite
2.94	3.47/8	4.70/7	Morinite
2.95	2.45/10	4.37/7	Purpurite
2.95	2.49/10	5.01/9	Ferrisicklerite
2.95	5.65/9	2.18/8	Florencite
2.95	5.71/7	3.50/7	Goyazite
2.96	2.99/8	2.07/7	Pyromorphite
2.97	2.18/5	5.70/4	Crandallite
2.97	5.70/8	3.45/6	Plumbogummite
2.98	2.22/6	5.74/4	Svanbergite
2.98	5.77/9	3.53/6	Lusungite
2.99	2.84/10	3.15/9	Wagnerite
3.01	2.53/10	4.32/9	Sicklerite
3.01	2.71/7	5.10/6	Panethite
3.02	2.81/8	2.13/8	Ningyoite
3.02	4.40/8	2.83/8	Rhabdophane
3.03	3.54/8	6.06/5	Sarcopside
3.03	4.37/7	2.83/7	Brockite
3.03	5.86/7	2.24/6	Corkite
3.037	2.71/9	3.21/7	Dickinsonite
3.04	2.72/8	3.22/6	Arrojadite
3.04	2.92/10	3.19/8	Triploidite
3.04	4.29/9	3.17/8	Cornetite
3.05	3.29/9	4.99/5	Tavorite
3.06	3.50/8	2.69/8	Xanthoxenite
3.06	2.72/6	6.26/5	Ferrialluaudite
3.06	3.29/7	3.10/7	Bertossaite
3.07	3.14/9	6.15/7	Lazulite
3.07	3.26/7	2.86/7	Cheralite
3.07	3.87/9	4.39/6	Dewindtite
3.08	3.83/7	2.88/7	Bergenite
3.09	2.87/7	3.30/5	Monazite
3.09	3.13/8	4.34/7	Attakolite
3.09	4.36/7	3.13/6	Palermoite
3.14	1.85/10	1.82/10	Kryzhanovskite
3.14	2.86/8	2.20/7	Herderite
3.14	2.87/7	3.72/6	Anapaite
3.14	8.01/7	2.98/7	Hureaulite
3.15	12.0/9	5.01/9	Dufrenite

TABLE I (continued)

Strongest Line	Next Stronger Lines		Name
3.153	2.97/9	4.66/8	Montebrasite
3.158	4.64/9	2.96/8	Amblygonite
3.16	3.89/8	3.96/7 –	Bøggildite
3.18	2.72/8	4.25/7	Phosphoferrite
3.18	3.02/9	2.63/8	Isokite
3.19	2.16/10	1.52/10	Babefphite
3.19	3.39/5	1.59/5	Rockbridgeite
3.20	3.09/8	2.51/5	Trolleite
3.20	3.38/5	1.60/5	Frondelite
3.20	5.05/9	2.75/8	Reddingite
3.20	5.10/8	4.28/8	Landesite
3.22	2.12/8	5.01/5	Andrewsite
3.22	3.01/8	6.33/7	Fairfieldite
3.24	3.20/10	3.14/8	Scorzalite
3.31	1.60/9	3.21/7	Unnamed
3.31	3.20/9	1.60/7	Zinc Rockbridgeite
3.31	4.73/7	2.89/6	Vauquelinite
3.32	2.66/10	2.57/8	Bradleyite
3.33	1.60/6	2.06/6 –	Lipscombite
3.33	3.14/10	2.97/10	Magnesiotriplite
3.34	3.51/9	4.00/8	Augelite
3.35	3.13/10	1.92/7	Kehoeite
3.36	3.31/8	4.84/6	Barbosalite
3.37	4.28/3 –	1.84/2 –	Berlinite
3.38	2.99/9	2.76/8	Monetite
3.44	1.77/8	2.15/7	Xenotime
3.44	3.26/10	8.58/8	Wavellite
3.45	7.25/9 –	2.89/9 –	Vaerynenite
3.48	4.29/8 –	2.73/8 –	Heterosite
3.50	2.86/10	2.71/6	Beusite
3.56	2.65/10	3.07/9	Uralolite
3.61	9.08/9	1.62/6	Przhevalskite
3.64	6.96/4	4.49/3	Veszelyite
3.67	3.03/9	2.78/9	Hurlbutite
3.68	2.89/8	6.70/7	Faustite
3.68	2.91/8	6.17/7	Turquoise
3.69	8.66/9	3.24/8	Metatorbernite
3.71	2.88/9	6.81/8	Coeruleolactite
3.77	3.39/7	3.02/6	Chalcosiderite
3.79	2.23/9	1.70/8	Uramphite

TABLE I (continued)

Strongest Line	Next Stronger Lines		Name
3.83	3.41/10	2.39/8	Farringtonite
3.91	3.28/9 +	3.01/8 +	Tinticite
3.96	2.77/10	2.54/10	Ludlamite
3.98	3.79/10 –	2.67/8 +	Lithiophosphatite
4.20	3.39/8	3.23/8	Parsonsite
4.21	7.50/9	3.02/9	Weinschenkite
4.22	12.63/9	6.35/8	Phosphorroesslerite
4.26	5.60/6	2.92/6 –	Struvite
4.27	3.00/9 –	2.95/9 –	Dumontite
4.28	4.81/9 ?	6.97/9 ?	Arthurite
4.35	8.30/9	8.70/8	Diadochite
4.38	5.50/8	3.11/8	Strengite
4.49	2.39/7	2.44/6	Pseudomalachite
4.71	3.23/10	2.90/9	Tsumebite
4.74	2.99/7	2.59/7	Wardite
4.77	2.98/9	2.81/8	Millisite
4.81	2.63/10	2.93/8	Libethenite
4.88	4.51/9	2.90/8	Sterrettite
4.89	3.46/10	8.59/6	Metabassetite
5.05	2.99/8	2.74/8	Brazilianite
5.10	2.04/10	10.1/6	Uranocircite
5.10 ?	12.0/9(?)	2.9/7 ?	Calcioferrite
5.11	2.89/9	12.0/8	Montgomeryite
5.16	15.3/9	3.45/7	Laubmannite
5.34	4.71/6	3.46/4	Newberyite
5.37	4.26/10	3.04/10	Variscite
5.45	5.97/8	4.94/6	Vauxite
5.60	3.37/10	6.80/6	Minyulite
5.72	7.28/9	3.24/6	Faheyite
5.94	2.97/5 –	5.21/4	Schertelite
6.20	3.25/10	1.92/9	Pseudoautunite
6.34	3.17/10	3.02/8	Messelite
6.60	2.89/10	4.60/5	Stercorite
6.60	3.29/8	2.66/8	Fluellite
6.70	2.97/6	2.70/6	Vivianite
6.79	5.99/7	3.06/7	Leucophosphite
6.96	2.94/3	8.04/2	Bobierite
6.96	3.46/5	5.15/3	Hannayite
7.00	3.28/9	4.24/6	Moraesite
7.40	5.02/9	3.19/8	Metakingite

TABLE I (continued)

Strongest Line	Next Stronger Lines		Name
7.50	14.9/6	11.1/4	Metaschoderite
7.56	2.99/9	4.48/7	Parahopeite
7.62	3.80/3	3.06/1	Brushite
7.9	15.8/4	11.1/2	Schoderite
7.91	3.96/6	3.15/6	Phosuranylite
7.97	3.99/9	5.83/8	Renardite
8.17	4.08/8	4.14/8	Cyrilovite
8.42	3.77/3	2.98/2	Koninckite
8.47	3.61/9 –	2.11/7	Metaautunite
8.50	2.80/6	4.27/4	Scholzite
8.55	2.59/6	7.13/4	Switzerite
8.92	3.73/7	3.25/6	Metaankoleite
8.93	3.73/8	5.48/7	Metauranocircite
9.02	5.32/8	4.35/6	Strunzite
9.03	3.80/9	3.27/8	Hydrogen autunite
9.10	3.48/8	4.53/6	Hopeite
9.10	3.45/8	3.48/6 +	Kingite
9.30	2.86/7	1.72/6	Englishite
9.40	3.50/6 –	2.33/3	Spencerite
9.42	2.85/10 –	3.16/9 +	Salmonsite
9.59	5.93/9	3.17/9	Roscherite
9.60	3.04/10	4.30/8	Sampleite
9.61	3.48/5	4.93/4	Sabugalite
9.65	4.82/5	5.09/4	Bermanite
9.69	6.46/9	4.86/9	Sigloite
9.78	3.17/8	2.83/7	Gordonite
9.82	6.38/9	4.20/9	Paravauxite
9.85	3.49/9	4.95/8	Saléeite
9.91	3.28/9	4.95/8	Laueite
9.93	5.87/7	3.47/4	Pseudolaueite
9.98	3.93/4	2.99/4	Stewartite
10.3	4.94/9	3.58/9	Torbernite
10.37	3.06/7	4.79/4	Beraunite
10.4	5.19/5	3.58/5 –	Autunite
10.77	4.13/6 –	5.28/4	Sanjuanite
10.8	2.41/10	1.39/9	Glucine
11.14	2.79/3 –	4.37/2	Nissonite
11.94	22.02/8	3.34/5	Cacoxenite
15.5	7.90/8	3.83/8	Taranakite

TABLE 2 Optical and Other Data on Mineral Phosphates

ω or β	Sign	Crystal System	α	(Birefringence)	γ	$2V$	Number	Name	(Specific) Gravity	Clea-vage	Geologic Occurrence
1.442	+	Tr	1.439		1.469	35	42b	Stercorite	1.57	—	S
1.466	+	M	1.462		1.469	79	15	Bøggildite	3.66	—	P
1.482	−	M	1.462		1.490	65	63	Moraesite	1.81	—	P
1.485	−	M	1.477		1.486	38	71	Phosphorroesslerite	1.73	—	W
1.485		I		00			30a	Evansite	1.9	—	P, W
1.490	+	O	1.478		1.510	L	35	Fluellite	2.18	P	P
		M?	1.484−		1.499+		81	Sanjuanite	1.94	—	W
1.496	+	O	1.495		1.504	37	89	Struvite	1.71	G	S
1.506				W			30b	Bolivarite	2.04	—	W
1.508	−	H	1.502			O	91	Taranakite	2.12	—	S
1.514		Tr		W			105b	Kingite	2.2	—	S
1.515	+	O	1.508		1.523	L	82	Schertelite	1.8	—	S
1.517	+	O	1.514		1.533	45	65	Newberyite	2.1	E	W
1.520	+	M	1.510		1.543	71	103b	Bobierrite	2.2	E	S
1.522	−	Tr	1.504		1.539	L	42a	Hannayite	1.9	E	S
1.524	+	H			1.530	O	12	Berlinite	2.64	—	H
1.525±			1.49		1.56		16	Bradleyite	2.73	—	S
1.525	−	M	1.510		1.536	L	37b	Uralolite	2.1		W
1.53±		I		00			48a	Kehoeite	2.34	—	(H, W)?
1.53		I		00			48b	Viséite	2.2	—	W
1.534	+	O	1.531		1.538	L	59	Minyulite	2.45	?	W
1.535	+	O	1.525		1.550	L	105a	Wavellite	2.36	E	W(P, H)
1.541	−	M	1.522		1.548	62	56	Lueneburgite	2.1	P	S
1.543	+	Tr	1.534		1.558	75	38a	Gordonite	2.2	E	W
1.544	+	M	1.540		1.559	55	33	Farringtonite	2.80	F	Met.
1.546	+	M	1.539		1.551	86	19a	Brushite	2.3	E	S
1.548	+	M	1.542		1.566	61	83a	Schoderite	1.88	—	W

TABLE 2 (continued)

ω or β	Sign	Crystal System	α	(Birefringence)	γ	$2V$	Number	Name	(Specific) Gravity	Clea-vage	Geologic Occurrence
1.555	+	Tr	1.551		1.562	32	101a	Vauxite	2.40	—	W
1.557	+	O	1.550		1.567	69	54	Lithiophosphatite	2.46		P
1.558	+	Tr	1.554		1.573	M	38c	Paravauxite	2.4	E	W
1.558	+	M	1.551		1.582	62	72b	Metavariscite	2.5	G	P, W, S
1.558	−	M	1.552		1.561	68	14	Beryllonite	2.8	E	P
1.56?			1.547−		1.571+		37a	Glucine	2.3	—	H?
1.561	+	M	1.550		1.577	L	101b	Metavauxite	2.34	—	W
1.563	−	M	1.551		1.565	40	64	Morinite	2.94	G	P
1.568	−	H	1.541		1.570	32	74	Pseudoautunite	3.28	E	H
1.572	−	M?	1.570		1.572	S	29	Englishite	2.65	E	S
1.572	+	M	1.568		1.582	28	96d	Wagnerite	3.15	P	H?
1.574	−	M?	1.56		1.580	62	94c	Bassetite	3.1	E	W
1.574	−	O	1.568		1.580	75	68	Overite	2.5	E	S
1.574	−	Te	1.559			O	94f	Saléeite	3.3	E	W, P
1.575	−	M	1.570			O?	21a	Calcioferrite	2.5	E	S
1.576	+	M	1.574		1.588	50	8	Augelite	2.7	E	P, H
1.576	−	M	1.567		1.579	51	18b	Panethite	3.0		Met.
1.577	−	Te	1.554		1.577	20±	94b	Autunite	3.15	E	W
1.578	−	M	1.572		1.582	75	21b	Montgomeryite	2.53	E	S
1.580	−	Te			1.583	O	58b	Metaankoleite	3.54	E	P
1.582	+	Tr	1.580		1.588	40	98c	Coeruleolactite	2.6	—	W
1.583	−	Te	1.569			O	94d	Hydrogen autunite	3.4	E	W
1.583	−	Te	1.564			M	94e	Sabugalite	3.2	E	W
1.583	−	Te	1.574		1.588	70	94h	Uranocircite	3.5?	E	W
1.585	−	Te	1.564		1.585	O	94g	Uramphite	3.7	G	W
1.586	+	Tr	1.563		1.619	76	38d	Sigloite	2.35	E	W
1.588	−	O	1.563		1.594	M	99a	Variscite	2.6	G	P?, S

1.588	+	Te			1.597	O	104a	Wardite	2.85	E	P, S
1.590	−	M	1.572		1.601	60	72c	Sterrettite	2.4	G	W, S
1.592	−	Te	1.582			O	94a	Torbernite	3.22	E	P, W
1.593	+	O	1.588		1.602	65	84	Scholzite	3.1		P, H?
1.598	−	O	1.589		1.599	37	44a	Hopeite	3.1	E	W
1.600	+	H			1.680	O	20	Cacoxenite	2.26	—	P, W
1.601	−	O	1.595		1.602	70	46	Hurlbutite	2.88	—	P
1.601	+	M	1.598		1.623	39	47	Isokite	3.3	G	H
1.602	+	Te	1.584			M	59	Millisite	2.85	?	S
1.602	−	M	1.586		1.606	49	88	Spencerite	3.1	E	W
1.603	+	Tr	1.594		1.615	L	2c	Natromontebrasite	3.04	E	P
1.603	+	M	1.580		1.633	83	103a	Vivianite	2.7	E	P, H, W, S
1.604±	−	Te	1.592±			S	58c	Metaautunite	3.6	E	W
1.604	+	M	1.598		1.626	59	83b	Metaschoderite			W
1.605	−	Tr	1.591		1.613	L	2a	Amblygonite	3.08	E	P, H
1.605	−	O	1.598		1.608	65	18a	Brianite	3.17		Met.
1.609	+	M	1.602		1.622	73	17	Brazilianite	3.0	G	P
1.610	−	M	1.603		1.617	L	58d	Metabassetite	3.4±	E	W
1.611	−	M	1.591		1.619	75	43a	Herderite	3.01	P	P
1.612	+	M	1.605		1.645	M	19b	Weinschenkite	3.3	E	P, W
1.613	+	Tr	1.602		1.649	54	3	Anapaite	2.81	E	H, S
1.613	+	Tr	1.602		1.624	74	2b	Montebrasite	3.06	E	P
1.613		Tr		0.04			98d	Faustite	2.9		
1.614	−	M	1.595		1.616	45	55b	Phosphophyllite	3.1	E	H
1.615	−	Tr	1.587		1.640	L	62	Monetite	2.9	P	S
1.618	+	H			1.625	O	39d	Gorceixite	3.1	—	W
1.619	+	H			1.625	O	39b	Crandallite	2.78	E	S
1.620	−	M	1.584		1.621	19	67	Nissonite	2.73		W
1.62	+	Tr?	1.618		1.67	S	25	Diadochite	2.2	—	W
1.62	+	Tr	1.61		1.65	40	98a	Turquoise	2.8	—	W
1.622	−	O	1.61		1.623	S	58e	Metauranocircite	4.1	E	H?, W

TABLE 2 (continued)

ω or β	Sign	Crystal System	α	(Birefringence)	γ	$2V$	Number	Name	(Specific) Gravity	Clea-vage	Geologic Occurrence
1.622	+	M	1.619		1.631	50	18c	Stanfieldite	3.15		Met.
1.625	+	Tr	1.614		1.637	L	44b	Parahopeite	3.3	E	P–W
1.626	−	M	1.604		1.637	70	51a	Lazulite	3.1	—	P, H
1.626	+	Te	1.624			O	58a	Metatorbernite	3.7	E	P, H, W
1.628	−	H	1.619			O	5c	Carbonate apatite	3.0		S
1.628	−	M	1.602		1.632	42	55c	Switzerite	3.0	E	P–W
1.629	+	O			1.632	O	10	Babefphite	4.3		H
1.629	−	H	1.626			O	106	Whitlockite	3.1	—	P, S, Met.
1.63	+	H			1.64	O	39a	Goyazite	3.26	E	W, S
1.631	+	H			1.652	O?	31	Faheyite	2.66	E	P
1.633	−	H	1.629			O	5a	Apatite	3.15	P	All
1.633	+	H			1.647	O	90a	Svanbergite	3.2	F	M, W
1.636	−	O	1.624		1.642	53	69b	Bertossaite	3.10	G	P
1.636	+	H			1.647	20	90d	Woodhouseite	3.0	E	H
1.639	−	M?	1.619		1.643	49	51e	Trolleite	3.1	P	W
1.64±		I		00		O	41	Griphite	3.4	—	P–W
1.64		O		W			66	Ningyoite			H?
1.641	+	M	1.636		1.651	L	78	Roscherite	2.92	G	P
1.642	+	Tr	1.632		1.657	80	32c	Collinsite	3.0	F	W?
1.642	−	O	1.627		1.644	20	69a	Palermoite	3.2	E	P
1.642	−	M	1.618		1.652	68	87	Souzalite	3.1	G	P
1.644	+	Tr	1.636		1.654	L	32a	Fairfieldite	3.08	E	P
1.645	−	H	1.641			O	5h	Wilkeite	3.1	P	H?
1.648	−	O	1.628		1.657	50	22b	Eosphorite	3.06	P	P–H
1.649	+	M	1.641		1.661	60	96b	Magnesiotriplite	3.57		
1.650	+	M	1.626		1.686	80	101c	Pseudolaueite	2.46	—	P–H

1.651	−	H	1.644			O	5f	Hydroxylapatite	3.0	P	H, M
1.653	+	Tr	1.640		1.680	M	32d	Messelite	3.0	E	P–W
1.653	+	H			1.675	O	39f	Plumbogummite	4.0	—	W
1.654	+	H			1.703	O	77a	Rhabdophane	4.0	—	P–W
1.655	−	M	1.649		1.659	76	45	Hureaulite	3.2	G	P–W
1.655	−	Tr	1.614		1.683	77	101d	Stewartite	2.5	—	P–W
1.656	−	Tr	1.591		1.688	M	38b	Laueite	2.5	E	P
1.656	+	O	1.651		1.683	41	76a	Reddingite	3.0	P	P–H
1.658	+	M	1.650		1.672	82	96a	Triplite	3.79	G	P–H
1.658	+	M	1.640		1.695	71	102	Veszelyite	3.5	G?	W
1.660	−	H	1.640			O	5b	Belovite	4.19	P	P
1.660	−	Te	1.648			S	49	Koninckite	2.40	F?	P–W
1.660	+	M?	1.655		1.670	L	79	Salmonsite	2.88	F?	P–W
1.66±		Tr	1.647±		1.672±		32b	Cassidyite			W
1.66	+	M	1.654		1.668	L	26a	Dickinsonite	3.4	E	P
1.66	−	H		W		O	5e	Fermorite	3.52	—	W
1.662	−	M	1.640		1.667	46	43b	Vaeyrynenite	3.18	E	P–H
1.663	−	O	1.640		1.665	40	86	Seamanite	3.1	F	W
1.664	+	O	1.655		1.675	84	7	Attakolite	3.23	—	H?
1.6684	−	H	1.6675			O	5d	Chlorapatite	3.17	P	H, Met
1.669	−	M	1.663		1.673	83	26b	Arrojadite	3.55	E	P
1.670	−	M	1.639		1.680	58	51d	Scorzalite	3.4	—	P, M, H
1.67	−	M	1.62		1.72	42	76e	Strunzite	2.5	—	P–W, S
1.671	+	H			1.689	O	90c	Hinsdalite	3.65	E	H?
1.672	+	H	1.671		1.676	30−	34	Fillowite	3.4	E	P
1.674	+	O	1.671		1.684	75	95e	Natrophilite	3.4	G	P
1.675	+	M	1.653		1.697	82	55a	Ludlamite	3.2	E	P–H
1.677	−	O	1.629		1.679	22	80	Sampleite	3.20	E	W
1.679	+	O	1.676		1.687	63	95d	Lithiophilite	3.5	E	P
1.680	+	H			1.695	O	77b	Brockite	3.9	—	W
1.680	+	O	1.672		1.700	68	76c	Phosphoferrite	3.2	P	P–H
1.683	−	O	1.649		1.691	40	22a	Childrenite	3.25	P	P–H

TABLE 2 (continued)

ω or β	Sign	Crystal System	α	(Birefringence)	γ	$2V$	Number	Name	(Specific) Gravity	Clea-vage	Geologic Occurrence
1.690	−	O	1.660		1.698	45+	73b	Bergenite	4.1	E?	W
1.693	+?	M	1.684		1.703	87	96f	Zwieselite	3.93	G	P–H
1.695	+	H			1.705	O	39c	Florencite	3.7	G	P, H?, S
1.695	+	O	1.694		1.702	62	95a	Triphylite	3.5	E	P
1.703	+	M	1.702		1.722	25	40b	Beusite	3.70	G	P
1.705	−	Tr	1.660		1.713	50	92	Tarbuttite	4.1	E	W
1.710	−	O	1.668		1.710	15	73a	Phosuranylite	4.1	E	P–W
1.713	+	M	1.711		1.728	40	40a	Graftonite	3.7	G	P, Met.
1.715	−	M?	1.704		1.724	L	107	Xanthoxenite	3.0	E	P–H
1.719	+	O	1.711		1.741	S	99b	Strengite	2.9	G	P–W
1.720	+	Te			1.820	O	61c	Xenotime	4.3	E	I, P, M, S
1.72		I					24	Delvauxite	1.9	—	W
1.721	+	M	1.707		1.739	84	52	Leucophosphite	2.9		P–W, S
1.725	−	M	1.692		1.738	62	72a	Phosphosiderite	2.7	G	P–W
1.726	+	M	1.725		1.730	M	96b	Triploidite	3.7	G	P
1.728	−	O	1.720		1.735	L	76b	Landesite	3.0	G	P–W
1.728	−	M	1.670		1.732	28	95g	Sarcopside	3.8	G	P, Met.
1.729	−	M	1.687		1.750	75	13	Bermanite	2.85	E	P–W
1.73		M	1.72		1.74	L	1c	Ferroalluaudite	3.6	G	P
1.735	−	O	1.715		1.745	L	95h	Sicklerite	3.3	G	P–W
1.736	−	O	1.715		1.739	41	73d	Renardite	4.3	E	W
1.74	+	M	1.73		1.75	L	1a	Alluaudite	3.58	G	P
1.742	+	M	1.741		1.746	M	96e	Wolfeite	3.8	G	P
1.745	−	O	1.703		1.789	L	53	Libethenite	3.9	—	W
1.745		O?					93	Tinticite	2.8	—	S
1.749	−	O	1.739		1.752	30	58f	Przhevalskite		G	W
1.758		M		W		M	9	Azovskite	2.6	—	W, S

1.765+	–	M?	1.762–		1.770+		60	Mitridatite	2.8±	—	P–W
1.767	–	O	1.762		1.768	M	73c	Dewindtite	5.0	E	W
1.778	–	O	1.757		1.792	M	95b	Ferrisicklerite	3.3	G	P–W
1.78		M		W–M			6	Arthurite	3.2		W
1.780	+	M	1.779		1.816	18	61b	Cheralite	5.3		P–W
1.786	+	M	1.775		1.815	M/L	11	Beraunite	2.9	G	P–W
1.791	+	M	1.790		1.844	15	61a	Monazite	5.1	G	I, P, M, S
1.79+	+?	M	1.77		1.835	L	51b	Barbosalite	3.60		P
1.8±	+?	H		0.03–0.04		O	39e	Lusungite			P
1.80?	+?	M	1.79		1.82	40	76d	Kryzhanovskite	3.31	G?	P–W
1.804	–	Te	1.773±			O	104b	Cyrilovite	3.08	E?	P–W
1.805	+	M	1.78		1.84	79	1b	Ferrialluaudite	3.6	G	P
1.81	–	O	1.765		1.82		23	Cornetite	4.1	—	W
1.807	+					75	2d	Tavorite	3.29		P–W
1.820	+	O	1.813		1.830	L	4	Andrewsite	3.50		W
1.83	+	O	1.82		1.88	M	36c	Zinc rockbridgeite	3.5	E	P–W
1.835	–	M	1.789		1.845	50	75	Pseudomalachite	4.2	F	W
1.838				W			57	Melkovite	2.97	E	W
1.840	–	Tr	1.775		1.844	22	98b	Chalcosiderite	3.2	—	W
1.84	+	M	1.83		1.89		27	Dufrenite	3.34	E	W
1.847	+	O	1.840		1.892	M	50	Laubmannite	3.3	F?	W
1.85+	–	Tr	1.85		1.862		70	Parsonsite	5.7	—	P–W
1.86	+	O	1.85		1.92	M	95f	Purpurite	3.3	G	P–W
1.87	+	M	1.85		1.90	L	28	Dumontite	5.65		W
1.880	–	O	1.860		1.893	M	36a	Frondelite	3.4	E	P–W
1.880	+	O	1.874		1.896	M	36b	Rockbridgeite	3.4	E	P–W
1.89	–	O	1.86		1.91	L	95c	Heterosite	3.3	G	P–W
1.920	+	M	1.885		1.956	L	97	Tsumebite	6.1	—	W
1.93	–	H		W		O	90b	Corkite	4.3	E	W
2.06	–	H	2.05			O	5g	Pyromorphite	7.0	P	W
2.22	–	M	2.11		2.22	O	100	Vauquelinite	6.0	—	W

TABLE 3 Minerals of the Triphylite Group

Number	Name	Composition	α	β	γ	Strongest Line	Next Stronger Lines	
95g	Sarcopside	$Fe''_3(PO_4)_2$	1.670	1.728	1.732	3.03	3.54/8	6.06/5
95e	Natrophilite	$Na(Mn, Fe'')PO_4$	1.671	1.674	1.684	2.61	2.59/10	2.87/9
95d	Lithiophilite	$Li(Mn, Fe'')PO_4$	1.676	1.679	1.687	2.53	3.01/9	3.47/8
95a	Triphylite	$Li(Fe'', Mn)PO_4$	1.694	1.695	1.702	2.53	3.02/9	3.49/9
95h	Sicklerite	$Li_{1-}(Mn'', Fe''')PO_4$	1.715	1.735	1.745	3.01	2.53/10	4.32/9
95b	Ferrisicklerite	$Li_{1-}(Fe''', Mn'')PO_4$	1.757	1.778	1.792	2.95	2.49/10	5.01/9
95f	Purpurite	$(Mn''', Fe''')PO_4$	1.85	1.86	1.92	2.95	2.45/10	4.37/7
95c	Heterosite	$(Fe''', Mn''')PO_4$	1.86	1.89	1.91	3.48	4.29/8 –	2.73/8 –

TABLE 4 Mineral Phosphates with Birefringence Less Than 0.005

n	Name	*n*	Name
1.485	Evansite[a]	1.64	Ningyoite
1.506	Bolivarite[a]	1.64 ±	Griphite
1.514	Kingite	1.645	Wilkeite
1.53	Kehoeite	1.66	Fermorite
1.53	Viséite	1.668	Chlorapatite
1.572	Englishite	1.72	Delvauxite[a]
1.580	Metaankoleite	1.758	Azovskite[a]
1.626	Metatorbernite	1.8 ±	Lusungite
1.629	Babefphite	1.838	Melkovite
1.629	Whitlockite	1.93	Corkite
1.633	Apatite		
Unknown birefringence			
1.745	Tinticite	?	Metakingite
1.81 ±	Tavorite	?	Unnamed
		?	Waylandite
Opaque			
—	Lipscombite	—	Schreibersite

[a] Powder diffraction data not available. They are also lacking for bassetite, lueneburgite, and natromontebrasite.

Mineral phosphates having zero or very low birefringence are given in order of increasing index of refraction in Table 4, which also contains a few additional optical data. A number of other mineral phosphates that are isotropic or with low birefringence have been listed in the older literature, but these are regarded as being too poorly characterized to warrant inclusion here.

In working with x-ray powder diffraction films or diffractograms it is recommended that one record the spacing for all (up to six) fairly strong lines. In succession each of these may be taken as the strongest line when comparing the results with the data in Table 1. One can then quickly determine whether or not the other two lines listed in the table agree with any of those observed. This technique obviates the necessity of repeating all the data of the table in every possible order. For a final check one can compare his results with the more extensive data available on the applicable x-ray powder diffraction card of the American Society for Testing and Materials. In Table 1 the strongest line is assumed to be of intensity 10; the next two stronger lines are listed in order of decreasing intensity; the intensity values are given following the virgule on a scale of 1 to 10.

If one is able to identify a mineral from its powder diffraction film, and if it turns out to belong to a solid solution series, its approximate composition can then be obtained as a rule if reasonably accurate indices of refraction values are measured. Thus the mineralogist finds a combination of these two techniques to be of maximum value in many cases.

The number appearing in the first column of Table 2 represents β for a biaxial mineral; in those cases in which it corresponds to ω the mineral is uniaxial. Under these circumstances the $2V$ value is 0° unless the crystal is under strain. If uniaxial, the n value listed under α is really ε in the case of an optically negative mineral (no n value being listed under the γ column). For a positive uniaxial mineral no n value appears under the α column, and the number given under the γ column represents the ε value. In the third column I represents either isotropic or isometric with birefringence of zero. In the next two columns if the indices of refraction values are not known for an anisotropic mineral, the birefringence may be listed as W (weak), or in a few cases a figure appears. Where the numerical value of the optic axial angle ($2V$) for a biaxial mineral is unknown, the sixth column may contain a letter S (small), M (medium), or L (large). The quality of any cleavage (if present) is indicated in the penultimate column by a letter; thus E (excellent), G (good), F (fair), and P (poor). A dash (—) indicates absence of cleavage; a blank means no knowledge, but probable absence. The last column gives some hint of the geologic occurrence, where

H hydrothermal deposit
I igneous rock
M metamorphic rock
P pegmatite
S sedimentary rock
W product of meteoric (cool ground) waters or weathering.
Met meteorites.

If a dash separates two letters, such as P–W, this refers to a weathering product of a pegmatite mineral.

8

The Compounds of Phosphorus

JOHN R. VAN WAZER

Vanderbilt University, Nashville, Tennessee

SCOPE OF PHOSPHORUS CHEMISTRY

Everyone recognizes the vast number and variety of molecular structures making up organic chemistry, which is the chemistry of compounds based on carbon atoms. However, even many chemists do not realize that a similar great variation and wide extent is to be found in the chemistry of compounds based on phosphorus. During the last 25 years, this chemistry has been the subject of extensive study throughout the world and, as a result, a host of new phosphorus compounds have been described. It is obvious from what is already known about phosphorus chemistry that vast numbers of additional new compounds based on phosphorus are simply waiting to be prepared by some investigator in the future.

As a result of the investigations done to date, there are a number of reference works devoted solely to the inorganic (1–2) or the organic (3–6) chemistry of phosphorus as well as a book (7, 8) in which the entire subject of pure and applied phosphorus chemistry is viewed as a whole. In addition, there is a series of books (9) devoted solely to surveys of phosphorus chemistry and a journal (10) on this subject has recently appeared.

In the systematization of organic chemistry, it is common practice to classify compounds in terms of homologous series in which the molecular weight is increased by addition of a fixed assemblage of atoms (generally the CH_2 group) in which carbon plays a central role. Similarly, the chemistry of phosphorus compounds is fruitfully treated (7, 8) in terms of series of compounds ranging from simple structures in which there is only a single phosphorus atom to more complex ones in which there are many. Because of their occurrence in nature, ease of preparation, and relatively good chemical stability, the family of hydrocarbons represents the key reference series of compounds in organic chemistry. A similar situation is not found in phosphorus chemistry, since the family of phosphorus hydrides (the phosphorus analogs to the hydrocarbons) is made up of compounds that are often hard to prepare, not very stable, quite reactive, and highly toxic. However, the phosphate family represents structures that are chemically quite stable, naturally occurring, and readily handled in the laboratory so that the phosphates (ortho- and polyphosphates) play about the same role (except for nomenclature) in the chemistry of phosphorus that the hydrocarbons do in the chemistry of carbon.

Nearly all of the phosphorus in nature (8) and in commerce (11) is present in the form of phosphate salts and various organic and inorganic derivatives of them.

STRUCTURAL CHEMISTRY OF PHOSPHORUS

In all of its compounds, except perhaps for the alkali and alkaline earth metal phosphides (7) phosphorus is covalently bonded to its neighboring atoms. With five electrons in its valence shell, phosphorus commonly shares electrons with three or four neighboring atoms (i.e., coordination number of 3 or 4) as indicated below, where the symbol Z stands for the nearest-neighbor atom donating one electron, $\cdot$, to the two-electron bond and the crosses, $\times$, stands for the valence electrons of the phosphorus.

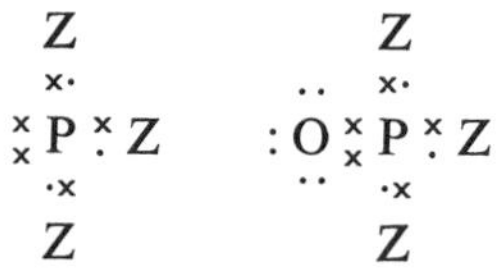

$$
\begin{array}{ccc}
 & Z & \\
 & \times\cdot & \\
\overset{\times}{\underset{\times}{}} & P & \overset{\times}{\underset{\cdot}{}}\,Z \\
 & \cdot\times & \\
 & Z &
\end{array}
\qquad
\begin{array}{ccc}
 & & Z \\
 & \cdot\cdot & \times\cdot \\
: & O\,\overset{\times}{\underset{\times}{\cdot}} & P\,\overset{\times}{\underset{\cdot}{}}\,Z \\
 & \cdot\cdot & \cdot\times \\
 & & Z
\end{array}
$$

However, phosphorus is known to exhibit all coordination numbers ranging from 1 through 6. These coordination numbers are exemplified by the following set of compounds in which unshared pairs of electrons on the phosphorus are denoted by a doublet of electron dots: HCP:; $R_3P\underset{\cdot\cdot}{\ddot{P}}R$, :$PCl_3$; $OPCl_3$; PCl_5; and PF_6^-, where R stands for an alkyl group. For coordination numbers of 4 or less, only *s* and *p* electrons are needed for a minimum-basis-set description of the ground state of the molecule but, with coordination numbers 5 and 6, it is necessary to invoke *d* orbitals.

From the viewpoint of the environmental scientist, all compounds in which phosphorus exhibits coordination numbers 1, 2, and 5 are probably of no importance; and, of the structures with coordination number 6, only the hexafluorophosphate anion PF_6^- seems to be of interest, because of its reasonable stability in aqueous solution. There are some compounds in which phosphorus exhibits coordination number 3 which may persist for a long enough time in water or soil so as to be of marginal interest. Thus, under anaerobic conditions the trialkylphosphines (R_3P) may persist for quite a while; but, in the presence of air, they are rather rapidly oxidized to the respective phosphine oxides (R_3PO). Other structures such as normal esters of phosphorous acid [$(RO)_3P$] undergo hydrolytic degradation rather rapidly and will be persistent only under special conditions.

Most phosphorus compounds of interest to the environmental scientist are based on phosphorus having a coordination number of 4. Chief among these are the phosphates; but, because of the high stability of the P—C bond, such structures as the phosphine oxides (R_3PO), the phosphinic acids [$R_2PO(OH)$] and their derivatives, and the phosphonic acids [$RPO(OH)_2$] and their derivatives must also be considered. The synthesis of phosphonates (particularly 2-aminoethylphosphonic acid) by biological systems has been known for over a decade (12); and it has been shown (13) that *Escherichia coli* can derive phosphate values by scission of the C–P bond of methyl- or ethylphosphonic acid when these were the sole source of nutrient phosphorus. However, it is thought that compounds exhibiting C–P bonds are not readily split by microorganisms to give orthophosphates, in the usual situation where other phosphates are available.

It also must be remembered that any highly insoluble species that are present as sizable chunks or even as reasonably large particulates may persist for rather long periods of time, even in such unlikely places as lake bottoms, and so on. Thus, iridium phosphide (Ir_2P), which has been used as a tip on fountain pens, is surely sufficiently stable to persist for many years in, say, a stream bed. An example of a highly persistent insoluble material that is of great practical interest to environmental scientists is the very slowly decomposable organic residues that are found in all forms of soils, including the organic detritus in swamps and eutrophic water bottoms. This slowly decomposable organic matter is rather high in nitrogen but also contains phosphorus. A typical C/N/P atomic ratio (14) for such material is in the order of 100:10:1. Environmental scientists should always be on the alert for insoluble persistent forms of phosphorus, either organic or inorganic in nature.

Phosphate Structures (7)

Phosphates are defined as being those chemical structures in which each phosphorus atom is more or less tetrahedrally surrounded by four oxygen atoms. If one to three corners of such PO_4 tetrahedra are shared with other PO_4 tetrahedra, through an oxygen atom common to just one pair of tetrahedra so as to produce P—O—P bonds, the resulting phosphates are called condensed phosphates.

The lone PO_4 group corresponding to the orthophosphate moiety may occur as a triply charged anion or as structures in which there are covalent bonds from one or more of the four oxygens to atoms other than a phosphorus atom. For example, the monohydrogen phosphate anion HPO_4^{2-} exhibits an O—H bond; and phosphate esters, such as trimethylphosphate $[(CH_3O)_3PO]$, have C—O bonds as do mixed anhydrides with organic acids, such as acetyl phosphate $[CH_3C(O)OPO_3H^-]$. There are many other kinds of structures such as tris(trimethylsilyl)phosphate $([(CH_3)_3SiO]_3PO)$ which are also classed as orthophosphates.

The examples of orthophosphate derivatives given in the preceding paragraph correspond to the commonly discussed situation in which up to three of the oxygens of a PO_4 group can form covalent bonds with other atoms through sharing of electrons. However, the fourth oxygen may also act as a bridge between the phosphorus and other atoms. For example, trimethyl phosphate forms a complex with europium(III) to give the cation $[(CH_3O)_3PO]_6Eu^{3+}$ in which six phosphoryl oxygens are bonded to each europium atom (15). In addition, each oxygen may have as many as four nearest neighbor atoms bonded to it so that a given PO_4 group may exhibit a functionality as high as 12. The well-known phosphovanadic and phosphotungstic acids represent a situation in which a number of VO_6 or WO_6 interlinked octahedra surround a PO_4 tetrahedra, with much sharing of oxygen atoms. In these structures each of the four oxygens tetrahedrally surrounding the phosphorus atom are joined to three different vanadium or tungsten atoms (16)!

In the condensed phosphates a given PO_4 group may share up to three of its oxygen atoms with neighboring PO_4 groups, allowing only one P—O—P bridge between a given pair of phosphorus atoms. The simplest condensed or polyphosphate is the pyrophosphate (represented by the $P_2O_7^{4-}$ anion) in which two PO_4 groups are bonded together by the sharing of one oxygen atom. The tripolyphosphate (also known as triphosphate) corresponds to three PO_4 groups connected together in a row by the sharing of oxygens so as to give a P—O—P—O—P molecular backbone. The triphosphate anion has the formula $P_3O_{10}^{5-}$. Increasingly larger P—O—P chain molecules are readily prepared. Indeed, straight-chain polyphosphates exhibiting hundreds of phosphorus atoms are well known. In the case of the longer-chain phosphates, it is commonly found that the various preparative methods lead to molecular size distributions rather than to polyphosphate chains all having the same length. A common physical form for the longer-chain polyphosphates (including a small amount of cyclic phosphate) is a glass; hence the phrase "vitreous phosphates." The term "sodium hexametaphosphate" has been incorrectly applied to the class of vitreous phosphates exhibiting average chain lengths of 10 to 25 phosphorus atoms per molecule.

Simple cyclic phosphates, consisting of rings of alternating phosphorus and oxygen atoms, are also well known and are generally called cyclic metaphosphates. There has been some recent research on phosphates having bird-cage structures (17). As yet, there has been no acceptable description in the literature of a branched-chain phosphate.

Commercial interest and much of the research activity on condensed phosphates have been centered on the unbranched chain structures known as the polyphosphates which exhibit the generic formula $M_{n+2}P_nO_{3n+1}$, where M stands for one equivalent of a metal ion, hydrogen, and so on. For $n = 1$, this generic formula represents the orthophosphate; for $n = 2$, the pyrophosphate; and for $n = 3$, the tripolyphosphate. The cyclic metaphosphates, which are rare in both commerce and nature, exhibit the generic formula $M_nP_nO_{3n}$; and the most common representative of these compounds is sodium trimetaphosphate, for which $n = 3$.

PHOSPHATE PROPERTIES (7, 8)

The phosphate chemical and physical properties that are of especial interest to the environmental scientist are treated individually elsewhere in this book. However, we shall touch upon them briefly here. From the viewpoint of environmental science, the key properties are (*a*) hydrolytic degradation of esters and of condensed phosphates, (*b*) precipitation to form materials of low solubility, (*c*) sorption on surfaces, and (*d*) formation of soluble complexes with metal ions (sequestration).

Presumably, any compound when exposed to the combined action of air and moisture over a sufficiently long length of time will be transformed into an orthophosphate. Of all of the processes involved in such conversions, the hydrolytic degradation of condensed phosphates as mediated by enzymes is of great practical importance. Although the straight-chain phosphates undergo hydrolysis to the orthophosphate in sterile systems, this rate is approximately 10^{-3} to 10^{-4} times slower at normal temperatures than the enzymatic hydrolysis in dilute aqueous solution—a process which has a half-life of only a few hours. An interesting feature of this enzymatic hydrolysis is found in the fact that no induction period is needed for microorganisms to become acclimated to a polyphosphate substrate since the metabolic processes of all forms of life involve P—O—P splitting (as well as C—O—P splitting) enzymes so that the appropriate degradative enzymatic system is already present and operative.

Since the ortho- and polyphosphate anions are multiply charged negative species they react readily with multiply charged positive ions to give precipitates. Many of these precipitates are crystalline, but perhaps even a greater number are amorphous. As indicated by the Ostwald–Gay-Lussac step rule (according to which an unstable form is usually produced before the stable form appears), phosphates tend to precipitate first as an amorphous mass of variable composition—a mass of material which slowly converts to one or more stable crystalline species. These amorphous precipitates exhibit extremely complicated chemical structures, generally being in the form of inorganic network polymers incorporating occasional terminal groups, such as the hydroxyl group. These amorphous materials, although difficult to characterize, probably play an important role in practical environmental chemistry.

Both ortho- and polyphosphates are known to sorb strongly on various surfaces. The chemistry of such sorption processes is extremely complex and has been unraveled in only a few instances. The sorption is often related to the precipitation processes discussed above, in that electrostatic forces between the negatively charged phosphate and a positively charged surface are often involved. The sorptive characteristics of the polyphosphates are strikingly evidenced in the ability of a small amount of a chain phosphate to convert a plastic clay mass to a thin, soupy fluid. This comes about through the fact that the edges of the clay particles are positively charged, whereas the faces have negative sites so that there is an edge-to-face interaction in plastic clay masses which causes the individual clay platelets to form a "house-of-cards" structure. Sorption of the polyphosphate on the positive sites on the edges of the clay platelets destroys this structure and hence greatly increases the fluidity of the system. Another important phosphate interaction that has received some study in the case of polyphosphates involves the sorption of the negative phosphate ions on the positive sites of proteins, thereby causing various more or less striking rheological and other physical and chemical changes.

Considerable work has been carried out by soil scientists on the sorption of orthophosphates on siliceous particulate matter, either in natural mixtures or as pure clay minerals (18–20). Not surprisingly, the sorption is found to be very complex with some of the sorbed phosphate being subject to reversible sorption-desorption while another portion is found to be fixed and not subject to exchange with dissolved phosphates. This sorption has been profitably treated (21) in terms of a competition between the phosphate and other anions for the sorption sites. These sites are related to the presence of fourfold-coordinated aluminum which introduces a negative charge into a gel exhibiting a tetrahedral silica-type structure. These complicated and important sorption processes in which orthophosphates become bound to the surface of soil particles or organic detritus are fundamental to a thorough understanding of the behavior of orthophosphates in natural systems and are worthy of much more study than has thus far been accorded them.

Both ortho- and polyphosphates form soluble complexes with metal ions. The orthophosphate complexes are very weak for the alkali and alkaline earth metals and only become important for transition metals such as iron. On the other hand, the polyphosphates form relatively strong complexes with the alkali and alkaline earth metals as well as with transition metals. Indeed, one of the main reasons polyphosphate builders are used in commerce is their ability to sequester calcium ions and thereby to reduce the effective water hardness by formation of polyphosphatocalcium complex ions. Although these complexes should not be ignored, it appears that, at the phosphate concentrations corresponding to most natural waters (even though somewhat polluted), the processes of surface sorption and precipitation dominate the process of forming soluble complexes.

In summary, it is probably safe to say that the emphasis in the environmental chemistry of phosphorus in freshwaters will be on the orthophosphate ion and its sorption on the surface of colloidal particles, either suspended or making up bottom muds. Further, polyphosphates and phosphate esters will undergo such rapid enzymatic degradation that their persistence time will be in the range of only a few hours. For seawater, there is the problem of persistent phosphonates. Indeed, it has been suggested (22) that, because of the great chemical stability of the C—P bond, the aminoalkylphosphonates may comprise a significant fraction of the dissolved organic phosphorus in the oceans.

ANALYTICAL TECHNIQUES (23, 24)

At the present time, there are really no good, general analytical methods of tested reliability for the differential assay of phosphorus in the majority of samples of interest to the environmental scientist. The various analytical methods for total phosphorus content are usually reliable; but the investigator should rightly be suspicious of analytical results in which chemical labels are placed on various cuts from, say, solubility or density fractionations. Continuing work is needed on the separation of the various organic and inorganic phosphorus species dissolved in water, when the total phosphorus concentration is of the order of 1 ppm. Probably the effort here should be concentrated on nondestructive methods (such as freeze-drying or ion-exchange sorption) for concentrating the sample to be followed by elution chromatography for separations, using a column. The new technique (25) of applying Fourier-transform technology to ^{31}P nuclear-magnetic-resonance spectroscopy may also be valuable in environmental studies, if a method for nondestructive concentration precedes it.

The problem of determining what phosphorus-containing species are present in either organic or inorganic solids that are amorphous or poorly crystalline is extremely vexing, since extractive or degradative procedures have a good probability of causing drastic chemical changes in the sample. At the present state of the art, it seems doubtful that general procedures for such differential phosphorus assays will be developed. At best, it seems that one can only hope for some type of a systematic classification that is only vaguely relatable to chemical structure. However, these kinds of studies may be of great practical importance and should be vigorously pursued.

REFERENCES

1. *Gmelins Handbuch der Anorganischen Chemie, Phosphor*, 8th ed., Verlag Chemie, Weinheim, 1965, System Number 16, Parts A through C.
2. Palache, C., H. Berman, and C. Frondel, *Dana's System of Mineralogy*, Vol. II, Wiley, New York, 1951, pp. 660–1016.
3. Kirby, A. J., and S. G. Warren, *The Organic Chemistry of Phosphorus*, Elsevier, Amsterdam, 1967.
4. Hudson, R. F., *Structure and Mechanism in Organo and Phosphorus Chemistry*, Academic Press, New York, 1966.
5. Purdela, D., and R. Vilceanu, *Chimia and Compusilor Organici Al Fosforului Si Al Acizilor Lui*, Editura Academici Republicii Socialiste Romania, 1965.
6. Sasse, K., *Houben-Weyl Methoden der Organischen Chemie, Phosphor-Verbindungen*, Vol. XII, George Thieme Verlag, Stuttgart, 1963–1964, Parts I and II.
7. Van Wazer, J. R., *Phosphorus and Its Compounds*, Vol. I, Interscience, New York, 1958.
8. Van Wazer, J. R., *Phosphorus and Its Compounds*, Vol. II, Interscience, New York, 1961.
9. *Topics in Phosphorus Chemistry*, M. Grayson and E. J. Griffith (Eds.), Vols. 1–6, Interscience, New York, 1964–1969.
10. Grayson, M., *Phosphorus and the Heavier Group Va Elements*, Gordon and Breach, London, six issues per year starting July 1971.
11. Van Wazer, J. R., *Industrial Chemistry and Technology of Phosphorus and Phosphorus Compounds—a Survey*, Interscience, New York, 1968.
12. Quin, L. D., in *Topics in Phosphorus Chemistry*, Vol. 4, M. Grayson and E. J. Griffith (Eds.), Interscience, 1967, pp. 23–48.
13. Zeleznick, L. D., T. C. Myers, and E. B. Titchener, *Biochem. Biophys. Acta*, **78**, 546 (1963).
14. Kafkafi, U., in *Analytical Chemistry of Phosphorus Compounds*, M. Halmann (Ed.), Interscience, New York 1972.
15. Graham, H. P., and M. D. Joesten, *J. Inorg. Nucl. Chem.*, **32**, 531 (1970).
16. Van Wazer, J. R., *J. Chem. Doc.* **4**, 84 (1964).
17. Glonek, T., T. C. Myers, and J. R. Van Wazer, *J. Am. Chem. Soc.*, **92**, 7214 (1970).
18. Muljadi, D., A. M. Posner, and J. P. Quirk, *J. Soil Sci.*, **17**, 212 (1966).
19. Hingston, F. J., R. J. Atkinson, A. M. Posner, and J. P. Quirk, *Nature*, **215**, 1459 (1967).

20. Cloos, P., A. Herbillon, and J. Echeverria, *Proc. 9th Int. Soil Sci. Soc. Conf.*, Adelaide, **2**, 733 (1968).
21. Hingston, F. J., R. J. Atkinson, A. M. Posner, and J. P. Quirk, *Proc. 9th Int. Soil Sci. Soc. Conf.*, Adelaide, **1**, 669 (1968).
22. Kittredge, J. S., M. Horiguchi, and P. M. Williams, *Comp. Biochem. Physiol.*, **29**, 859 (1969).
23. Halmann, M., *Analytical Chemistry of Phosphorus Compounds*, Interscience, New York, 1972.
24. Riemann, W., III, and J. Beaukenkamp, in *Treatise on Analytical Chemistry*, Vol. V, I. M. Kolthoff, P. J. Elving, and E. B. Sandell (Eds.), Interscience, New York, 1961, Part 2, pp. 317—402.
25. Van Wazer, J. R., in *Analytical Chemistry of Phosphorus Compounds*, M. Halmann (Ed.). Interscience, New York, 1972.

Origin and Fate of Organic Phosphorus Compounds in Aquatic Systems

9

FRANK F. HOOPER

Professor of Fisheries and Zoology, The University of Michigan, Ann Arbor, Michigan

Although inorganic orthophosphate is the form of phosphorus that is biologically mobile and the principal state in which phosphorus is interchanged among various biological components in water systems, a major share of the phosphorus resources of the earth's surfaee waters is in organic form. The principal pools or storage sites of organic phosphorus in the hydrosphere are (*a*) the organic compounds of living and dead particulate suspended matter (seston), (*b*) a variety of filterable organic compounds usually termed "dissolved," (*c*) the organic compounds of rooted and encrusting plants of the bottom, (*d*) the phosphorus of free-swimming animals, and (*e*) phosphorus present in bottom sediments. As phosphorus is cycled in aquatic systems, the size of these organic pools is constantly changing as the biota of the system responds to changing physical conditions and to inputs of phosphorus and other materials from the surrounding watershed.

The significance of suspended and dissolved phosphorus fractions has been fully recognized only during the last one or two decades. Because of the early discovery that inorganic phosphate (orthophosphate) is the principal form of phosphorus utilized in plant nutrition, orthophosphate was the form first measured in field studies. Early chemical surveys of natural waters (12, 65) reported only this form of phosphorus. In fact, many recent reports, some of which have set policy dealing with water quality problems [Federal Water Pollution Control Administration (19)], have considered only this form of phosphorus even though there is an abundance of data now available that indicate the inorganic phosphorus compounds often make up less than 10% of the total phosphorus of many systems (40).

The first extensive reports of organic fractions in freshwaters were the surveys of Wisconsin lakes by Juday *et al.* (45) and Juday and Birge (44). These reports and a series of studies that followed (38, 17, 74) demonstrated that large quantities of organic phosphorus compounds existed within lakes and in the ocean. With study of the dynamics of lake and marine phytoplankton and the phosphorus requirements of phytoplankton, it became clear that the supply of inorganic phosphate present at any one time was completely inadequate to supply the needs of cells during blooms and periods of rapid increase (38, 74). Tracer technology provided the means of investigating the dynamics of phosphorus exchanges in lakes and streams (39, 30, 75, 88, 4). These studies demonstrated that inorganic phosphate must always be considered a transient state and that there is rapid cycling of phosphorus atoms between the several large pools or compartments of phosphorus within lake and stream systems. It was demonstrated that there may be large fluxes of phosphorus through a bacterial cell in a matter of minutes (31, 75) and that a complete replacement of phosphorus within the epilimnion of a lake may occur within a week or less (30). It also became clear that the organic phosphorus components of both lake and marine systems are important in the cycling of phosphorus, and that measurement of organic phosphorus is necessary in the assessment of phosphorus resources that are utilized in biological production. Perhaps the most important concept which emerged from these studies is that neither particulate nor dissolved organic phosphorus compounds are rapidly lost by sedimentation, but that major fractions are regenerated by bacterial action and are reutilized in the production process.

In the last decade it has become clear that little progress can be made toward solution of the practical problems of control of algae and other aquatic nuisances and toward the assessment of the fate of phosphorus compounds introduced as waste into surface waters, without an adequate understanding of the nature of the organic phosphorus residues and their fate. This seemingly obvious relationship has for many years been overlooked by engineers and technicians concerned with waste-disposal problems.

MEASUREMENT OF ORGANIC PHOSPHORUS FRACTIONS

The various suspended and dissolved phosphorus fractions found in freshwaters have been defined by limnologists and oceanographers on a more or less operational or methodological basis. Phosphorus fractions designated as "dissolved" are ordinarily those which pass through a fine filter which retains bacteria (0.45 μ). Suspended particles of smaller size containing phosphorus are included in this category along with ionic phosphorus compounds and dissolved macromolecules. The dissolved fraction is in turn separated into two components depending on reactivity with the reagents used in the colorimetric determination of phosphate (15). Compounds reacting with molybdate to form heteropoly blue without digestion or hydrolysis, and at the pH of the Denigès test, are termed soluble phosphate phosphorus, orthophosphate, or soluble reactive phosphate phosphorus (83). In the heteropoly blue method and its modifications, water samples are acidified. The soluble phosphate fraction will therefore also contain acid-soluble colloidal precipitates as well as organic compounds that can be hydrolyzed to orthophosphate at low pH. Suspended phosphorus compounds removed by filtration but which are dissolved at the pH of the reaction are called soluble sestonic phosphorus and include such compounds as calcium and iron phosphate. Living and dead particulate organic phosphorus material is ordinarily termed sestonic organic phosphorus or particulate unreactive phosphorus. Dissolved organic fractions that do not react with molybdate, but require additional hydrolysis and/or digestion, ordinarily are termed dissolved organic phosphorus or dissolved unreactive phosphorus. Several different systems of terminology have been used to describe the above-mentioned phosphorus fractions. Most of the older literature uses the designations given by Hutchinson (40). Strickland and Parsons (83) have proposed a system in which the phosphorus fractions reacting with molybdate without digestion or hydrolysis are termed reactive as contrasted to those requiring chemical treatment, termed unreactive. A more comprehensive system of classification has been proposed by Olsen (66) in which he takes into account states of hydrolysis and also includes additional size categories of particulate phosphorus. Readers are referred to Olsen's paper for synonomy of terms and also for review of methodology.

The categories of phosphorus compounds discussed above are the fractions most widely reported in limnological and oceanographic studies. Olsen points out that there are wide differences in procedures used among investigators. Earlier investigations (44, 45) did not undertake filtration of samples before performing the Denigès reaction or before carrying out oxidation of organic fractions. Thus the oxidized matter included the bacteria and plankton components. In these studies organic phosphorus values were obtained by subtracting the amount of phosphorus reacting with molybdate without oxidation from the amount reacting after oxidation was complete. The procedure above combines particulate or sestonic phosphorus with that termed soluble organic.

Olsen (66) correctly points out that the fraction usually labeled soluble organic may include polyphosphates and pyrophosphate compounds which should properly be called inorganic. He also notes that the Denigès technique is not specific for orthophosphate and that at higher temperature and at the pH level used in this procedure short-chain polyphosphate may be hydrolyzed and thus may be included in the soluble phosphate fraction. The interest in detergent additions to freshwaters has stimulated a large number of analyses of polyphosphates in waters receiving domestic waste, but there have been few attempts to isolate the individual compounds containing phosphorus from unpolluted waters.

A number of biochemical procedures enable further separation of the dissolved organic category into additional fractions. These procedures are discussed in detail later in this paper.

THE QUANTITY AND DISTRIBUTION OF ORGANIC PHOSPHORUS COMPOUNDS IN AQUATIC SYSTEMS

It is beyond the scope of this chapter to review all existing data on the occurrence of organic phosphorus compounds in water, except within broad limits. I shall only attempt to summarize data from a selected series of studies that are representative of a broad spectrum of unpolluted waters. The highest values of organic phosphorus have been reported from urban areas and in areas with intensive agricultural development (18). Forested, undeveloped watersheds in general have lakes and streams of lower organic phosphorus content (84).

The data from the 479 northeastern Wisconsin lakes reported by Juday and Birge (44) are among the most complete sets of data ever gathered for a single geographic area. They are perhaps typical of the range of variation to be found in an area of relatively uniform topography and geology. Lakes of this area are of low to moderate hardness. The range was from 1 to 25.9 mg/l, fixed carbon dioxide. The range of organic carbon of the plankton of these lakes was from 0.2 to 8.86 mg/l; however, all but three of the values obtained were below 3.0 mg/l. The range of total organic phosphorus in this series of lakes was from 8 to 103 mg/m^3. However about 60% of the values fell within the range of 10 to 20 mg/m^3. The mean total organic phosphorus for the 479 lakes was 20.3 mg/m^3. This was seven times the mean concentration of soluble inorganic phosphorus in this series of lakes. The ratio of soluble inorganic phosphorus to total organic phosphorus ranged from 1:1 to 1:89. The smaller ratios were rare except in the bottom waters of stratified lakes.

The report discussed above also provides an account of the vertical distribution of organic compounds within stratified lakes. The surface and bottom waters of the majority of lakes studied did not differ in the amounts of organic and inorganic phosphorus in the early spring, but the quantity of both types increased below the thermocline as the summer progressed. Increases of this sort have been noticed in the hypolimnia of lakes throughout the world and arise from the settling of plankton organisms from the upper water, followed by their decay and conversion from the organic state into inorganic phosphate.

Juday *et al.* (45) demonstrated that there is an irregular but positive correlation between the amount of organic carbon and the amount of organic phosphorus. Although there was wide variation in the organic carbon content for a given set of phosphorus values, there was a well-defined trend in the mean values between 12 and 28 mg/m³ organic phosphorus. Since many of these waters varied in color and many were deeply stained with organic residues from bogs, it is not surprising that there was a low correlation between these two measurements.

In most lakes, the sestonic component of the organic phosphorus appears to be much larger than the dissolved organic fraction. The sestonic phosphorus of a series of Ontario lakes averaged 65.4% of all phosphorus (76). This was compared to 28.7% for dissolved organic phosphorus and 5.9% for inorganic phosphate. Figures given for Connecticut lakes by Hutchinson (40) agree rather closely with the values above (62% for sestonic phosphorus, 28.5% for dissolved organic, and 9.5% for soluble inorganic). For the series of Wisconsin lakes studied by Juday and Birge discussed above, the percentage of these three fractions can be computed by using these authors' figures for average phosphorus content of the seston. Estimated in this way, sestonic phosphorus becomes 67% of the total, soluble organic is 21%, and soluble inorganic is 12% of the total phosphorus. In Lake Superior, Putnam and Olson (73) found that the organic phosphorus content was from 72 to 100% of the total phosphorus, but they did not separate dissolved organic from the sestonic phosphorus component. In Lake Erie, Chandler and Weeks (9) found that an average of 61% of the total phosphorus was in an organic state, but again the percentage in dissolved and sestonic components was not determined. In polluted waters the percentage of inorganic phosphorus may be higher than the values above. In the offshore surface waters of Lake Ontario 77% of the phosphorus was organic, but inshore water receiving urban waste appears to have a much larger percentage of orthophosphate [International Lake Ontario-St. Lawrence River Water Pollution Board (41)].

It is also of value to consider the percentages of phosphorus contained per unit of dry weight of various components of the biota in natural waters. Bowen (8) gives figures of 0.425% of dry weight for the phosphorus of marine plankton, 3.0% for bacteria, 2.8% for brown algae, and 2.3% for angiosperm plants. Birge and Juday (7) give the percentage phosphorus for a variety of plant and animal plankton in Lake Mendota. For the blue-green *Microcystis* and *Anabaena*, the figures are 0.52 and 0.53%, respectively. For the green alga *Cladophora* they report a value of 1.4%. The aquatic macrophyte *Myriophyllum* had 0.52% phosphorus on a dry-weight basis. Figures for planktonic cladocera range from 1.02% for *Cyclops* to 1.60% for *Daphnia pulex*. Insects and other aquatic vertebrates range from 0.64 to

0.93%. For mixed centrifuge plankton from northeastern Wisconsin lakes, Birge and Juday report percentages of phosphorus ranging from 0.18% in Clear Lake to 0.39% in Lower Greshan Lake. The percentage of phosphorus in catches of zooplankton made with a tow-net ranges from 0.173% in Turtle Lake to 0.604% in Crystal Lake.

CELLULAR PHOSPHORUS COMPOUNDS AND THEIR METABOLISM

The particulate phosphorus found in natural waters consists chiefly of cellular phosphorus compounds within a variety of microorganisms including planktonic algae and bacteria, and also nonliving detritus particles undergoing decomposition. During the degradation process within natural waters, many of these organic compounds may be released at least temporarily into the water in a free state. In their "free" state, the cellular products constitute a part of the "dissolved" fraction. To discuss adequately the role of organic phosphorus in natural waters, it is of value to describe first the various intracellular phosphorus compounds, consider their function in the cell, and review the present state of our knowledge regarding their metabolism.

The vital role of phosphorus in the growth and reproduction of all living matter is well documented. Phosphorus is involved in the entire range of metabolic processes. It is connected in an essential way with virtually all energy transformation systems in living cells. It enters the cell as inorganic phosphate by means of an active transport process, and once inside the cell it becomes incorporated (*a*) into a variety of organic phosphorus compounds, and (*b*) into condensed inorganic phosphates (polyphosphates). In algae the disposition of phosphorus into these compounds arises from reactions involved in two major energy-yielding chains present in all living plant cells. These are (*a*) the process of photophosphorylation within chloroplasts whereby light energy is used for esterification of inorganic phosphate into ATP (adenosine triphosphate) according to the equation:

$$\text{inorganic phosphate} + \text{ADP (adenosine diphosphate)} + \text{light} \longrightarrow \text{ATP (adenosine triphosphate)}$$

and (*b*) phosphorylation coupled with respiration which is associated with mitochondria (oxidative phosphorylation):

$$\text{ADP} + \text{inorganic phosphate} + \text{energy from oxidation} \longrightarrow \text{ATP}$$

In all organisms, most of the energy-consuming reactions are coupled with the hydrolysis of the energy-rich phosphate bond of ATP. In these reactions, phosphate groups are actively transferred by the action of enzymes.

The ATP arising from photophosphorylation is used by the plant cells to support growth and all energy-requiring metabolism such as CO_2 assimilation, uptake of various organic and inorganic compounds, synthesis of amino acids, synthesis of proteins and nucleic acids, and synthesis of inorganic polyphosphates. This close coupling of phosphorus with cell energetics and photosynthesis makes photosynthetic production dependent on available phosphates. It is clear that a phosphate deficiency in algae soon reduces photosynthesis and other metabolic reactions (70), and may lead to unusual accumulations of fat, starch, and other cell-wall substances. This, in turn, suggests a secondary interference with the nitrogen metabolism of the plant (6).

The phosphorus content of cells grown in laboratory culture varies within a wide range. In *Asterionella formosa*, phosphorus varied from 6×10^{-8} to 4×10^{-6} μg P/cell (54). The minimum amount for this species was found to be 5×10^{-8}. μg P/cell (56). For *Chlorella pyrenoidosa* it varied from a minimum of 1×10^{-7} to a maximum of 1.5×10^{-6} μg P/cell (1). There have been extensive investigations aimed at determining the phosphorus necessary to sustain growth of various species of algae in laboratory cultures (10). Rodhe (78) was able to group species of algae into ranges of minimum phosphorus tolerance.

In nature and also in the laboratory it has been demonstrated that planktonic algae take up phosphorus in excess of their actual needs when it is in adequate supply. With this surplus they appeared to be able to grow even when the external sources are depleted (20). This acts as a luxury phosphorus uptake and appears to have no effect upon growth (22), and appears to be stored in a loosely bound, readily exchangeable state (23).

Polyphosphates

Of the numerous intracellular phosphorus compounds of bacteria and algae, polyphosphates have received a great deal of attention because for many years they were believed to be the compounds used for energy storage within the cell (34). Polyphosphates are one of three types of condensed inorganic phosphate compounds containing pentavalent phosphorus, in which various numbers of tetrahedral PO_4 groups are linked together by oxygen bridges. Condensations of PO_4 groups giving cyclic structures and having the elemental formula $M_nP_nO_{3n}$ in which M is a metal are properly called metaphosphates. Cross-link condensation chains are usually termed ultraphosphates. The term polyphosphates is usually reserved for linear condensed molecules having the formula $M_{(n+2)}P_nO_{(3n+1)}$. The chain length of polyphosphates ranges from 2 units (pyrophosphates) to the insoluble Maddrell's salts of chain length up to 10^4. Only the molecules of lower chain length have been identified as individual compounds. Nearly all of the polyphosphates from living cells are regarded as mixtures of various proportions of polyphosphates of different molecular size. Polyphosphates of varying chain lengths have been extracted from yeast cells using a variety of solvents and pH ranges. Fractions have varied in chain length from 4 to 260, and molecular weights range from 530 to 30,700 (52). Column and paper chromatography have been used to separate polyphosphates and metaphosphates of shorter chain lengths from living cellular materials.

Polyphosphates appear to be characteristic of microorganisms and have been isolated from a wide variety of bacteria, fungi, protozoa, and all major groups of algae (28, 51). In addition, they have been identified from some higher plants and animals (61, 55).

Some controversy exists regarding the chemical state of polyphosphates within cells. Since some polyphosphates cannot be extracted with acid, for many years they were believed to be bound with protein or ribonucleoprotein complexes (46). Harold (28) now favors the view that complexes do not exist in vivo but are formed during the extraction process and that polyphosphates exist free in living cells. In bacterial cells much of the polyphosphate may be localized in granules that stain metachromatically with basic dyes. These granules can be identified in cells with either the light or electron microscope and have been called volutin granules. In addition to polyphosphates, volutin granules may contain RNA, lipid, protein, and Mg^{2+} (89). In yeast and in *Chlamydomonas* phosphates may be localized in vacuoles.

The various enzymatic transformations of polyphosphates within bacterial and algal cells have been extensively investigated. The biosynthesis of polyphosphate is accomplished in the living cell by the addition of the terminal phosphoryl group of ATP to the chain according to the following reaction:

$$\text{ATP} + (\text{polyphosphate})_n \rightleftarrows \text{ADP} + (\text{polyphosphate})_{n+1}$$

The discovery by Kornberg (49) of the polyphosphate kinase catalyzing the reaction above led to the hypothesis that polyphosphate was in some way related to the energy storage processes within the cell. The reaction above appears to be reversible at low ATP–ADP ratios. Other enzymes have been isolated from cellular material which bring about the transfer of phosphate from polyphosphate to adenosine monophosphate (91) and which bring about the phosphorylation of glucose and fructose (85, 86). A number of enzymes which have been isolated from living tissues will hydrolyze pyrophosphates, tripolyphosphates, tetrapolyphosphates, metaphosphates, and polyphosphates of high molecular weight to orthophosphate (25, 64). At this time the evidence is not clear whether hydrolysis is direct to orthophosphate or through intermediates such as ATP.

The polyphosphate content of living cells has been shown to fluctuate within a wide range. Polyphosphates usually cannot be detected in phosphorus-starved cells. They are low or undetectable during the exponential growth phase of cells. On the other hand, they accumulate to very high levels in cells with nutritional deficiencies (90), up to 20% of the dry weight of yeast cells (53). When sulfur, carbon, or nitrogen is limiting, *Nitrosomonas* converts nucleic acid into polyphosphate. When the limiting element is restored, polyphosphate is converted back into nucleic acid (87).

The cycle of polyphosphate in *Aerobacter aerogenes* and its genetic control has been outlined by Harold (26, 27) and Harold and Harold (29). Briefly, it appears that in growing cells, the synthesis of nucleic acid inhibits polyphosphate synthesis and stimulates the degradation of polyphosphate. Consequently little or no polyphosphate is deposited. If growth and nucleic acid synthesis stop because of the exhaustion of a nutrient, net polyphosphate synthesis is promoted. This is presumed to occur because competition for ATP is relieved and polyphosphate accumulates at a rate determined by the level of polyphosphate kinase. It has also been shown that cells subject to prior phosphate starvation contain elevated levels of polyphosphate kinase and are thus capable of very rapid polyphosphate synthesis when orthophosphate is provided. Thus in *Aerobacter*, under favorable conditions for growth, net polyphosphate synthesis occurs at a low rate and nucleic acid synthesis proceeds, whereas under unfavorable environmental circumstances, polyphosphate synthesis takes place and rapid accumulation occurs.

In algae Kuhl (50) found that there is rapid incorporation of orthophosphate into polyphosphate in the presence of light. In the dark, incorporation is depressed but not abolished. Thus in algae it appears that the formation of polyphosphate is stimulated by light but is not totally dependent on it. In algae as in other microorganisms the polyphosphate content is a function of growth phase and is lowest in the exponential phase and highest in old cultures. There also appears to be a reciprocal relationship between RNA and the polyphosphate content (81).

Harold (28) reviews theories on the function of polyphosphate in living cells. His review raises doubts as to the validity of the earlier hypothesis that polyphosphates are stores of phosphate-bound energy from which phosphoryl groups may be transferred to ADP from ATP. In *Aerobacter aerogenese* polyphosphates do not break down in cells when energy generation is limited or blocked by chemicals. Polyphosphate degradation is hydrolytic and results in dissipation of energy-rich phosphate bonds. Thus Harold feels that the hypothesis that polyphosphate acts as an energy storage device is no longer universally tenable. Polyphosphate accumulation accounts for only a small fraction of the ATP generated by cells.

Harold (28) provides evidence that polyphosphates serve as a phosphate reserve and storage center for phosphorus within cells from which rapid biosynthesis of nucleic acid and phospholipids can take place. This leads to the general concept of polyphosphates acting as regulatory agents of cellular phosphorus. On the one hand, there is the cyclic regeneration of ATP and its participation in numerous reactions in which metabolic energy is produced and utilized. On the other hand, in growing cells ATP is continually drawn into nucleic acids and phospholipids. There appear to be valid biochemical reasons for believing cells must control the steady-state concentrations of orthophosphates, ADP, and ATP. Polyphosphate appears to be the compound regulating these levels by serving, in effect, as a metabolic buffer.

THE QUANTITY OF PHOSPHORUS FRACTIONS IN LIVING CELLS

The percentages of the various phosphorus fractions within bacterial and algal cells fluctuate considerably depending on light conditions, concentration of the phosphorus in the environment, pH, and temperature. Overbeck (68) provides data on four of the above fractions in *Scenedesmus quadricauda*. The TCA (trichloroacetic acid) soluble fraction made up 27% of the phosphorus of normal *Scenedesmus* cells. About half of this fraction was orthophosphate and half was 7-min hydrolyzable phosphate (acid-soluble polyphosphate). The TCA-insoluble phosphorus made up the remaining 73% of the cell phosphorus. This consisted of a fraction hydrolyzable by the 7-min hydrolysis (50.6% of cell phosphorus) and stable organic phosphorus (22.4%)

Correll (1965)* fractionated particulate material of two size categories collected from Antarctic waters into five chemical fractions. The large-size particles consisted chiefly of zooplankters while the smaller category were chiefly diatoms. The larger size particles contained much higher quantities of RNA and polyphosphate (27 to 55% of total phosphorus) than the smaller size particles. On the other hand, the smaller category of particle was richer in phospholipids (13 to 29% of total phosphorus). Both classes of particle contain more orthophosphate, oligopolynucleotides, and oligophosphates than healthy laboratory plankton, which indicated that enzymatic degradation of cells had occurred. By incubating aliquots of these two samples with ^{32}P, rates of incorporation of the label into various fractions were observed. As expected, the smaller particles took up radioactivity about six times faster than the larger size fraction. The RNA–polyphosphate fraction was labeled more rapidly than any of the other four organic fractions separated. However, phospholipids were also labeled rapidly and reached a high specific activity.

* D. Correll, "Pelagic Phosphorus Metabolism in Antarctic Waters," *Limnol. Oceanogr.*, **10**, 364–370 (1965).

REGENERATION OF PHOSPHORUS FROM ORGANIC COMPOUNDS

Demonstration that the stores of soluble phosphate measured in the open water of the photosynthetic zones of lakes and the ocean were inadequate to provide the necessary phosphorus for production of phytoplankton for any extended period (38, 45, 74) made it necessary to hypothesize that (*a*) phosphorus was being brought into the open water from greater depths and/or bottom sediments, or (*b*) that major stores of phosphorus were regenerated from the phosphorus of the phytoplankton, zooplankton, or the dissolved organic phosphorus. It is now clear that recycling of phosphorus may be rapid and that regeneration can account for a major share of the needs of the system.

In the sea and in lakes it is difficult or impossible to determine the origin of soluble phosphorus compounds that appear in the open water since these compounds may arise by (*a*) in situ decomposition of the phytoplankton and zooplankton organisms themselves, (*b*) excretion by the plankton organisms, (*c*) regeneration from bottom sediments and transport to the photosynthetic areas, or (*d*) in situ release of dissolved organic compounds by algae and bacteria, and subsequent breakdown of the organic compounds into soluble phosphate. All of these processes may be operating simultaneously or one process may predominate to the exclusion of others. In shallow freshwater and estuarine ecosystems, the influence of land drainage and/or the littoral sediments may be predominant in providing the needed phosphorus resources for the phytoplankton. In circumstances in which water is freely circulating over the substrata, it is difficult to separate the sediment- and water-regeneration processes. Phosphorus from the mud is carried into the open water so that there is seldom a measurable increase in concentration in the layer of water immediately adjacent to the sediment.

Perhaps the most successful approach to the study of cycling of phosphorus in shallow systems where there are several regeneration processes operating simultaneously has been made by the use of radiophosphorus (^{32}P). Kinetic analyses of the movement of the label from the water into the various "pools" or "compartments" within the system have shown a logarithmic decline followed by decrease at a much lower rate as the tracer is returned to the open water from pools. From measurements of the rates of disappearance and reappearance of ^{32}P and the rates of loss of stable phosphorus added as fertilizer to lake systems, Hayes *et al.* (30) have estimated turnover times for the participating phosphorus of lake systems. Their calculations suggest that the turnover time is long for rich eutrophic systems and short for oligotrophic lakes and bog lake ecosystems. Pomeroy (71) reviewed data on the residence time and turnover rate from ^{32}P experiments in water systems and found the residence time range to be from 0.05 to 200 hr. He suggests that a short residence time indicates either depleted phosphorus resources, or high metabolic activity, or a combination of these two conditions. He found that turnover rates fall between 0.1 and 1.0 μg P/(m^3)/(hr) regardless of the phosphate concentration of the water except in biologically active systems where values between 1.0 and 20 μg P/(m^3)/(hr) are encountered. Turnover time was considered more important than phosphate concentration in maintaining high productivity.

In many unpolluted freshwater systems, rapid regeneration within the water is believed to take place because the concentration of orthophosphates is frequently 1 mg/m^3 or lower and variations in quantity of orthophosphate may be small despite large changes in plankton biomass (45). The transient state of orthophosphate can be demonstrated in a striking way by simple aquarium experiments in which radiophosphorus is added as a tracer (31, 69). Aquarium experiments as well as in situ lake measurements by Rigler (75) suggest that the rate of turnover in bacterial cells is a matter of a few minutes and that turnover of the dissolved organic phosphorus pool in the water may be a matter of a few hours.

There can be little doubt that the major share of the phosphorus of the hypolimnia of stratified lakes comes from regeneration of the phosphorus from plankton settling below the thermocline. Radiophosphorus added to the epilimnion promptly appears as soluble phosphorus in bottom strata (39, 60). However, even in the deepest layers of lakes regeneration is a complex problem since in this zone phosphorus may also come from sediments. The classical study by Mortimer (63) in Esthwaite water demonstrates that regeneration of phosphate from bottom mud proceeds when the surface layer of the mud becomes chemically reduced and an iron colloid layer is broken down. Mortimer's studies indicate that the dissolved phosphate which moves from the mud into water may be transported vertically in the hypolimnion by currents arising from internal waves.

Large-scale changes in the phosphorus content in the deeper zones of the ocean, however, can hardly arise from processes other than those taking place in the open water. In the Gulf of Maine, the large seasonal changes in the relative proportions of particulate phosphorus, dissolved organic phosphorus, and soluble phosphorus indicated that regeneration and recycling are taking place (74). In this study it appeared that 25% of the phosphorus came from regenerative processes. In the open ocean a major share of the phytoplankton cells produced in the euphotic zone is consumed by herbivorous zooplankters living in this zone, and is not lost by sedimentation (47). After digestion and assimilation by zooplankton it is subsequently excreted into the water. Ketchum and Corwin (48) recorded changes in the phosphorus fractions within the euphotic zone during a plankton bloom in the Gulf of Maine. The magnitude of the regeneration process can be deduced from their field measurements. During the 10-day period in which the bloom was studied, 14.2 μg atoms P/m^2 was removed from the 0 to 50 m stratum. This phosphorus gave an observed increase of 7.58 μg atoms/m^2 of new particulate phosphorus (plankton) and 1.05 μg atoms/m^2 increase in dissolved organic phosphorus. All of the organic phosphorus was liberated in the upper 25 m. Below the euphotic zone (137 m) 2.7 μg atoms/m^2 of inorganic phosphorus appeared in the water, presumably from regeneration of particulate matter sinking to this zone. The overall rate of regeneration of phosphorus in this study was 4.0×10^{-3} μg atoms/m^3 per day. This rate under bloom conditions can be compared to rates in the Gulf of Maine under nonbloom situations of 2.0×10^{-3} μg atoms/m^3 per day, and the average, 0.01×10^{-3}, and the maximum 0.30×10^{-3} μg atoms/m^3 per day, for the North Atlantic Ocean given by Riley (77).

In field studies such as those discussed above, it is seldom possible to identify the source of the organic material being decomposed. Ketchum and Corwin (48) point out that the dissolved and organic phosphorus which appear in their study may have come from excretion of phytoplankton as well as zooplankton. Cultures of phytoplankton cells may liberate up to 25% of the organic carbon fixed during the 24-hr period (32). Carbon excretion presumably would be accompanied by sizable loss in organic phosphorus. However, autolysis and bacterial degradation of phytoplankton and zooplankton together with excretion of living zooplankton organisms are believed to account for a major share of regeneration in the sea. The relative importance of autolysis as compared to bacterial decomposition of phytoplankton and zooplankton cells has not been adequately evaluated. There are several laboratory studies in which the breakdown of plankton has been followed. Hoffman (33) found that 20 to 25% of the inorganic phosphorus and 30 to 40% of the organic phosphorus is liberated shortly after death, and between 80 and 90% is liberated within 24 hr. Marshall and Orr (58) found a complete breakdown of the phosphorus of dead marine copepods of the genus *Calanus* within 2 days. The rapidity of decomposition led the above-mentioned authors to conclude that autolysis, rather than bacterial action or the action of free enzymes, was the most important regenerative process. Grill and Richards (24) followed the regeneration of phosphorus from a culture of phytoplankton (chiefly a centric diatom) which had been incubated in seawater and then stored in the dark. During the first day they noted that a sizable fraction of the inorganic phosphorus passed into particulate form, presumably uptake by bacteria. The increase in particulate phosphorus continued for approximately 8 days, at which time there was a sudden increase in the dissolved organic phosphorus and a decrease in particulate phosphorus. This change, they believe, was a result of autolytic release similar to that described by Hoffman (33). The dissolved organic phosphorus liberated was quickly acted upon by the bacteria population which was rapidly increasing so that after 17 days it was almost completely reassimilated into particulate matter. After 4 weeks there was again decomposition of particulate phosphorus and an increase in inorganic phosphorus. This continued until the end of the experiment. By plotting the decay of various components, Grill and Richards developed a simple kinetic model of the mineralization process. The process was approximated by the resultant of the first-order decay rates of two labile organic fractions plus a refractory fraction which did not change during the experiment. They believed the initial decay rate represented organic phosphorus released from diatoms. Decay of this fraction gave a second fraction which appeared to be the phosphorus contained within the bacterial cells. This fraction, in turn, gave rise to (*a*) inorganic phosphorus and (*b*) refractory organic products of bacterial metabolism and the undecomposed particulates of bacterial cells.

Cooper (14) observed the release of phosphorus from the decomposition of a mixed culture of marine zooplankters and from a mixed culture of marine diatoms. He observed a large release of inorganic phosphorus from the zooplankton within the first 12 hr. An amount of phosphorus equivalent to that in the plankton was set free within a period of 6 days. Regeneration of phosphorus from the diatom culture was somewhat delayed. Very little phosphate appeared within the first 3 days but thereafter rapid liberation took place, reaching a maximum amount in approximately 1 month. After this maximum there was a slow decrease in phosphate for a 4-month period.

Although there are little or no data on excretion rates of the zooplankters under field conditions, there are data from laboratory studies and from in situ field measurements in closed systems. The excretion rates of herbivores depend on feeding rates, density of the herbivores, density and quality of the food, temperature, and other variables. A wide variation in regeneration rates is encountered. Although most attention has been paid both in marine and freshwater studies to the larger herbivorous zooplankters, the microzooplankton may be of great significance.

Johannes (42) demonstrated with a series of marine animals of varying sizes that the rate of excretion of dissolved phosphorus per unit of body weight increased as the body weight decreased. This suggests that the smaller short-lived species may play a very important role in mineralization. The large numbers of protozoa, rotifers, and other small metazoa inhabiting the mud-water interface as well as the open water are capable of liberating large quantities of phosphorus. Excretion of phosphorus is not confined to forms grazing upon living cells. Many of the microzooplankters as well as benthic filter feeders may ingest both living and dead cells (16). Dead cells together with their covering of epiphytic bacteria may constitute a large fraction of the organic material in many freshwater ecosystems. However, the microplankton may release inorganic phosphorus in the absence of bacteria. Hooper and Elliott (36) demonstrated the release of inorganic phosphorus from bacteria-free cultures of the ciliate *Tetrahymena pyriformes* that contained autoclaved extracts of the organic matter from lake mud. Margalef (57) observed release of inorganic phosphorus from cultures containing cladocera and the alga *Scenedesmus*. Release of phosphorus increased as the number of cells of *Scenedesmus* increased. Margalef believed that the algae stimulated phosphotase activity in the digestive tract of the cladocera. Rigler (76) verified Margalef's finding that living crustacea secrete phosphatases into the medium, and Overbeck (67) found that a bacterial suspension releases phosphate enzymatically from sodium glycerophosphate and makes this phosphorus available to the alga *Scenedesmus quadricauda*.

The early experiments of Gardiner (21), who used mixed cultures of the larger marine zooplankters, demonstrated that large stores of inorganic phosphate are excreted by these animals within a 3-hr test period. These and later studies by Marshall and Orr (58) provide estimates of excretion rates of cultured animals under different conditions of feeding. More refined measures of excretion rates come from studies in which diatoms labeled with ^{32}P were fed to the marine copepod *Calanus* and the rate of loss of metabolized label from the medium was measured (59). In these studies the biological half-life of phosphorus was estimated to be 14 days. A kinetic analysis of Marshall and Orr's data by Conover (13) indicated that the phosphorus of *Calanus* existed in two pools, one of labile phosphorus with a short biological half-life (approximately 0.375 days), and a second pool of more stable phosphorus with a half-life of approximately 13 days. It was estimated that between 94 and 99% of the total phosphorus in *Calanus* was in the stable form and that the mean turnover rate from both pools was about 10% of the body phosphorus per day. Ketchum (47) summarizes unpublished data of Conover on excretion rates of a variety of marine organisms. These rates vary from 0.03 to 6.9% of the total phosphorus content of the organisms per day.

Pomeroy *et al.* (72) found much higher excretion rates for estuarine plankton than those reported above. Net plankton excreted an amount of phosphorus nearly equal to its total phosphorus content in a day. Slightly more than half of the excreted phosphorus was phosphate; the remainder was insoluble organic compounds. Rigler (76) found that *Daphnia magna* excreted inorganic phosphorus into the medium at the rate of 8.4×10^{-3} μg per animal hour. This loss was independent of epizootic bacteria, and was independent of the production of feces. He was not able to detect loss of organic phosphorus. His excretion rates for *Daphnia* appear to be much higher than those reported by Gardiner (21) and are slightly higher than those reported by Marshall and Orr for *Calanus*. Barlow and Bishop (5) followed the regeneration of phosphorus from zooplankton by collecting net plankton from the epilimnion (5 m) and from the hypolimnion (45 m) of Cayuga Lake and re-suspending the plankton at 100 times its original concentration in bottles of water taken from the depth from which the plankton was collected. Excretion rates ranged from 30 to 120% of body phosphorus per day and were considerably higher than those reported by Ketchum (47) but within the range reported by Pomeroy *et al.* (72). The epilimnetic zooplankton population of Cayuga Lake consisted chiefly of cladocera while the hypolimnetic plankton consisted chiefly of copepods. The latter were, on the average, of larger size. Excretion rates were consistently lower for the epilimnetic zooplankton as compared to the hypolimnetic forms of the same body size.

The fecal pellets produced by herbivorous zooplankton have been suggested as a source of debris that may be converted into phosphorus compounds. Marshall and Orr (59) found that *Calanus* produced pellets at the rate of one pellet every 5 to 6 min. However, these authors believe the phosphorus content of pellets is low and their studies on digestion indicate that the phosphorus assimilation efficiency of *Calanus* is high. Thus the role of fecal pellets in the regeneration cycle of phosphorus is not clear.

The role of bacteria in the regeneration process appears to be complex. Both the tracer studies on aquatic microcosms (31, 69) and the direct measurements of decomposition in seawater (24) indicate that living bacteria not only incorporate inorganic phosphorus rapidly but following uptake much of the phosphorus is released as dissolved organic phosphorus. Although most of the direct evidence for bacterial production of dissolved organic phosphorus comes from experiments with laboratory microcosms, much of the dissolved organic phosphate in nature may be of bacterial origin.

In following the uptake of ^{32}P by bacteria, Phillips (69) noted that normally there is not a release of dissolved organic phosphorus during the growth phase of cells and that release is usually associated with a decline in cell numbers. He concluded that release was associated with death of microorganisms. Bacteria also appeared to be the agents responsible for the conversion of dissolved organic phosphorus to inorganic phosphate. Phillips (69) added labeled dissolved organic phosphorus to seawater under sterile and nonsterile conditions. With bacteria present, labeled dissolved organic phosphorus declined rapidly and the label entered the particulate phase (bacteria). Later, inorganic phosphorus appeared. On the other hand, the label remained in organic form in sterile cultures. From these experiments, Phillips concluded that there was no conversion of dissolved organic phosphorus to inorganic phosphate without bacteria.

Johannes (43) found that marine bacteria utilized 80% of the dissolved organic phosphorus released by an amphipod. He also found that a large fraction of this organic phosphorus was hydrolyzed in a sterile medium. He noted a release of organic phosphorus by bacteria-free diatoms after their growth phase had ceased, but was unable to report regeneration of dissolved inorganic phosphorus from dissolved organic phosphorus in the presence of bacteria; contrary to the findings in freshwater systems, he believes that marine bacteria (living or dead) release little dissolved organic phosphorus.

Under anoxic conditions, cultures of bacteria and other organisms have been shown to release large quantities of orthophosphate very rapidly (80). When oxygen is restored to the medium, much of this phosphorus is reabsorbed. The phosphorus liberated appears to have been cellular, since release was inhibited by mercuric chloride and by 2,4-dinitrophenol. The fractions of the phosphorus compounds in the systems at various times during the release and uptake process indicated that the phosphorus liberated during the early stage of release was phosphorus that could be extracted by acid. Later in the release process RNA and DNA fractions were produced. There did not appear to be a release from phospholipids or phosphoproteins. Experiments by Chu (11) in which a variety of organic phosphorus compounds were added to algae cultures indicated that certain compounds (glycerophosphoric acid, sodium nucleinate, and lecithin) could be utilized directly by diatom cells. However, these compounds were also broken down into orthophosphate by bacteria in the presence of seawater. Thus in the presence of high bacterial populations, direct uptake of organic phosphorus compounds by algal cells may be of minor importance.

IDENTIFICATION OF DISSOLVED ORGANIC COMPOUNDS

Despite the wealth of data on the quantity of dissolved organic phosphorus in marine and freshwater systems, only fragmentary and isolated efforts have been made toward the identification of these compounds. More important, little or no effort has gone toward linking these organic fractions with the metabolic activities of bacterial and plant cells from which many of these materials must arise. Other than suggestions that the dissolved organic phosphorus represents the more refractory compounds that decompose slowly (47, 24) and that they are frequently liberated by bacterial cells, there are few guides as to their role in the biochemical cycles of either marine or freshwater systems.

Phillips (69) separated the dissolved organic phosphorus of seawater that had been equilibrated with ^{32}P into six fractions. Three were identified as nucleotides or polynucleotides on the basis of adsorption on Norit A charcoal. Three other chromatographic peaks were not adsorbed and were believed to represent phosphorylated hydrocarbons. The fractionation of dissolved organic phosphorus of bog water samples (37) showed that in the aphotic zone, between 0.3 and 4.2% of the total phosphorus was adsorbed by charcoal and presumably was nucleic acid. This fraction, however, was not present in the euphotic zone. From 1.7 to 12.3% of the total phosphorus was labile inorganic phosphorus, presumably polyphosphate or pyrophosphate. This fraction was present at all depths. The identified fractions, however, represented a small portion of the dissolved organic phosphorus since 35 to 47% of the organic phosphorus remained unidentified.

Since algal and bacterial cells are known to accumulate large quantities of polyphosphates, the question arises whether or not these fractions exist free in the euphotic zone since they would be expected to be liberated by the lysis of cellular material. Solórzano and Strickland (82) and Armstrong and Tibbitts (3) found only small quantities of polyphosphates in unpolluted coastal waters. Solórzano and Strickland suggest that polyphosphate does not accumulate to any extent in unpolluted situations since it appears to be taken up rapidly by algal cells even in the presence of bacteria. However, since it accumulates when there are deficiencies of other nutrients (28) and under adverse environmental conditions, it may occur in tropholytic waters and its presence in the euphotic zone may be an indication of a limiting micronutrient.

Because little of the organic phosphorus of seawater is easily hydrolyzed by enzyme phosphate esters Solórzano and Strickland (82) suggest that the organic phosphorus isolated from seawater is mainly nucleic acid. Particulate DNA has been isolated from seawater by Holm-Hansen and co-workers (35). However, Armstrong *et al.* (2) found that nucleic acids were easily degradable by ultraviolet light, and thus might not be expected in the surface water but might appear in the aphotic layers, as reported above for bog lakes (37).

A proper assessment of the role of dissolved organic phosphorus compounds cannot be made until more effort is expended toward isolation and identification of compounds from the open water of freshwater and marine ecosystems. To enable proper interpretation of such data, there must be concurrent studies of the composition of algal and bacterial cells and their nutritional status. Assuming fractions can be identified and their source determined, efforts should be made to assess the turnover times of various fractions so that their significance in the dynamic processes of the phosphorus cycles can be evaluated.

REFERENCES

1. Al Kholy, A. A., "On the Assimilation of Phosphorus in *Chlorella pyrenoidosa*", *Physiol. Plant.*, **9**, 137–143 (1956).
2. Armstrong, F. A. J., P. M. Williams, and J. D. H. Strickland, "Photooxidation of Organic Matter in Sea Water by Ultraviolet Radiation, Analytical and Other Applications," *Nature*, **211**, 481–483 (1966).
3. Armstrong, F. A. J., and S. Tibbitts, "Photochemical Combustion of Organic Matter in Sea Water, for Nitrogen, Phosphorus, and Carbon Determination," *J. Mar. Biol. Assoc. U.K.*, **48**, 143–152 (1968).
4. Ball, R. C., and F. F. Hooper, "Translocation of Phosphorus in a Trout Stream Ecosystem," *Proceedings of the First Symposium on Radioecology*, Reinhold, New York, 1963, pp. 217–218.
5. Barlow, J. P., and J. W. Bishop, "Phosphate Regeneration by Zooplankton in Cayuga Lake," *Limnol. Oceanogr. Suppl.*, **10**, R15–R24 (1965).
6. Bergmann, L., "Stoffwechsel und Mineralsalzernährung einzelliger Grünalgen. II. Vergleichende Untersuchungen über den Einfluss mineralischer Faktoren bei heterotropher und mixotropher Ernährung," *Flora (Jena)*, **142**, 493–539 (1955).
7. Birge, E. A., and C. Juday, "The Inland Lakes of Wisconsin. The Plankton. I. Its Quantity and Chemical Composition," Wisconsin Geological and Natural History Survey, Bulletin No. 64, Scientific Series No. 13, 1922, pp. 1–222.
8. Bowen, H. J. M., *Trace Elements in Biochemistry*, Academic Press, London-New York, 1966.
9. Chandler, D. C., and O. B. Weeks, "Limnological Studies of Western Lake Erie, V. Relations of Limnological and Meteorological Conditions to the Production of Phytoplankton in 1942," *Ecol. Monogr.*, **15**, 436–456 (1945).
10. Chu, S. P., "The Influence of the Mineral Composition of the Medium on the Growth of Planktonic Algae, II. The Influence of the Concentration of Inorganic Nitrogen and Phosphate Phosphorus," *J. Ecol.*, **31**, 109–148 (1943).
11. Chu, S. P., "The Utilization of Organic Phosphorus by Phytoplankton," *J. Mar. Biol. Assoc. U. K.*, **26**, 285–295 (1946).
12. Clarke, F. W., *Data of Geochemistry*, 5th ed., U.S. Geological Survey, Bulletin 770, 1924, pp. 63–121.
13. Conover, R. J., "The Turnover of Phosphorus by *Calanus finmarchicus*". Addendum to S. M. Marshall and A. P. Orr, "On the Biology of *Calanus finmarchicus*," (1961), *J. Mar. Biol. Assoc. U. K.*, **41**, 484–488 (1961).
14. Cooper, L. H. N., "The Rate of Liberation of Phosphate in Sea Water by the Breakdown of Plankton Organisms", *J. Mar. Biol. Assoc. U. K.*, **20**, 197–200 (1935).
15. Denigès, G., "Réaction de Coloration Extrêmement Sensible des Phosphates et des Arséniates," *C. R. Acad. Sci., Paris*, **171**, 802–804 (1920).
16. Edmonson, W. T., "Trophic Relations of the Zooplankton," *Trans. Am. Microsc. Soc.*, **76**, 225–245 (1957).
17. Einsele, W., "Die Umsetzung von zugeführtem anorganischen Phosphat im eutrophen See und ihre Rückwirkung auf seinen Gesamthaushalt," *Z. Fisch.*, **39**, 407–488 (1941).

18. Engelbrecht, R. S., and J. J. Morgan, "Land Drainage as a Source of Phosphorus in Illinois Surface Waters," Algae and Metropolitan Wastes, U.S. Public Health Service, SEC TR W61-3, 1961, pp. 74–79.
19. Federal Water Pollution Control Administration, "Physical and Chemical Quality Conditions. Water Quality Investigations. Lake Michigan Basin," U.S.D.I., 1968, 81 pages.
20. Franzew, A. W., "Ein Versuch der physiologischen Erforschung der Produktionsfähigkeit des Moskauflusswassers," *Microbiol.*, **1**, 122–130 (1932).
21. Gardiner, A. C., "Phosphate Production by Planktonic Animals," *J. Const. Int. Explor. Mer.*, **12**, 144–146 (1937).
22. Gest, H., and M. D. Kamen, "Studies on the Phosphorus Metabolism of Green Algae and Purple Bacteria in Relation to Photosynthesis," *J. Biol. Chem.*, **176**, 299–318 (1948).
23. Goldberg, E. D., T. J. Walker, and A. Whisenand, "Phosphate Utilization by Diatoms," *Biol. Bull.*, **101**, 274–284 (1951).
24. Grill, E. V., and F. A. Richards, "Nutrient Regeneration from Phytoplankton Decomposing in Seawater," *J. Mar. Res.*, **22**, 51–69 (1964).
25. Grossman, D., and K. Lang, "Inorganic Poly- and Metaphosphatases as well as Polyphosphates in the Animal Cell Nucleus," *Biochem. Z.*, **336**, 351–370 (1962).
26. Harold, F. M., "Enzymic and Genetic Control of Polyphosphate Accumulation in *Aerobacter aerogenes*," *J. Gen. Microbiol.*, **35**, 81–90 (1964).
27. Harold, F. M., "Regulatory Mechanisms in the Metabolism of Inorganic Polyphosphate in *Aerobacter aerogenes*," *Colloq. Int. Centre Natl. Rech. Sci.* (*Paris*), **124**, 307–315 (1965).
28. Harold, F. M., "Inorganic Polyphosphates in Biology: Structure, Metabolism, and Function," *Bacteriol. Rev.*, **30**, 772–794 (1966).
29. Harold, F. M., and R. L. Harold, "Degradation of Inorganic Polyphosphate in Mutants of *Aerobacter aerogenes*," *J. Bacteriol.*, **89**, 1262–1270 (1965).
30. Hayes, F. R., J. A. McCarter, M. L. Cameron, and D. A. Livingstone, "On the Kinetics of Phosphorus Exchange in Lakes," *J. Ecol.*, **40**, 202–216 (1952).
31. Hayes, F. R., and J. E. Phillips, "Lake Water and Sediments: IV. Radiophosphorus Equilibrium with Mud, Plants, and Bacteria under Oxidized and Reduced Conditions," *Limnol. Oceanogr.*, **3**, 459–475 (1958).
32. Hellebust, J. A., "Excretion of Some Organic Compounds by Marine Phytoplankton," *Limnol. Oceanogr.*, **10**, 192–206 (1965).
33. Hoffman, C., "Untersuchungen über die Remineralisation des Phosphors im Plankton," *Kieler Meeresforsch., Bd.*, **12**, 25–36 (1956).
34. Hoffman-Ostenhof, O., "Some Biological Functions of the Polyphosphates," *Colloq. Int. Centre Natl. Rech. Sci.* (*Paris*), **106**, 640–650 (1962).
35. Holm-Hansen, O., W. H. Sutcliffe, Jr., and J. Sharp, "Measurement of Deoxyribonucleic Acid in the Ocean and Its Ecological Significance," *Limnol. Oceanogr.*, **13**, 507–514 (1968).
36. Hooper, F. F., and A. M. Elliott, "Release of Inorganic Phosphorus from Extracts of Lake Mud by Protozoa," *Trans. Am. Microsc. Soc.*, **72**, 276–281 (1953).
37. Hooper, F. F., "Nutrient Cycling and Productivity of Dystrophic Lake-Bog Systems," Tech. Prog. Rep., submitted to A.E.C., Nov. 1969.

38. Hutchinson, G. E., "Limnological Studies in Connecticut. IV. The Mechanism of Intermediary Metabolism in Stratified Lakes," *Ecol. Monogr.*, **11**, 21–60 (1941).

39. Hutchinson, G. E., and V. T. Bowen, "Limnological Studies in Connecticut: IX. A Quantitative Radiochemical Study of the Phosphorus in Linsley Pond," *Ecol.*, **31**, 194–203 (1950).

40. Hutchinson, G. E., *A Treatise on Limnology*, Wiley, New York, 1957, 1015 pages.

41. International Lake Ontario–St. Lawrence River Water Pollution Board, *Pollution of Lake Ontario and the International Section of the St. Lawrence River*, Vol. 3, Reports to the International Joint Commission, Water Pollution Boards, 1969, 329 pages.

42. Johannes, R. E., "Phosphorus Excretion and Body Size in Marine Animals: Microzooplankton and Nutrient Regeneration," *Science*, **146**, 923–924 (1964).

43. Johannes, R. E., "Uptake and Release of Dissolved Organic Phosphorus by Representatives of a Coastal Marine Ecosystem," *Limnol. Oceanogr.*, **9**, 224–234 (1964).

44. Juday, C., and E. A. Birge, "A Second Report on the Phosphorus Content of Wisconsin Lake Waters," *Trans. Wis. Acad. Sci., Arts, Lett.*, **26**, 353-382 (1931).

45. Juday, C., E. A. Birge, G. I. Kemmerer, and R. J. Robinson, "Phosphorus Content of Lake Waters of Northeastern Wisconsin," *Trans. Wis. Acad. Sci., Arts, Lett.*, **23**, 233–248 (1928).

46. Katchman, B. J., and J. R. Van Wazer, "The 'Soluble' and 'Insoluble' Polyphosphates in Yeast," *Biochim. Biophys. Acta*, **14**, 445–446 (1954).

47. Ketchum, B. H., "Regeneration of Nutrients by Zooplankton," International Council for Exploration of the Sea, Rapports et Procés, Verbaux des Reunions, **153**, 142–147 (1962).

48. Ketchum, B. H., and N. Corwin, "The Cycle of Phosphorus in a Plankton Bloom in the Gulf of Maine," *Limnol. Oceanogr., Suppl.*, **10**, R148–R161 (1965).

49. Kornberg, S. R., "Adenosine Triphosphate Synthesis from Polyphosphate by an Enzyme from *Escherichia coli*," *Biochim. Biophys. Acta*, **26**, 294–300 (1957).

50. Kuhl, A., "Die Biologie der kondensierten anorganischen Phosphate," in *Ergebnisse der Biologie*, Vol. 23, H. Autrum (Ed.), Springer, Berlin, 1960, pp. 144–186.

51. Kuhl, A., "Inorganic Phosphorus Uptake and Metabolism," in *Physiology and Biochemistry of Algae*, R. A. Lewin (Ed.), Academic Press, New York, 1962, pp. 211–229.

52. Langen, P., E. Liss, and K. Lohmann, "Art, Bildung und Umsatz der Polyphosphate der Hefe," *Colloq. Int. Centre Natl. Rech. Sci. (Paris)*, **106**, 603–612 (1962).

53. Liss, E., and P. Langen, "Versuche zur Polyphosphat-Überkompensation in Hefezellen nach Phosphat-verarmung," *Arch. Mikrobiol.*, **41**, 383–392 (1962).

54. Lund, J. W. G., "Studies on *Asterionella formosa* Hass. II. Nutrient Depletion and the Spring Maximum, part I, Observations on Windermere, Esthwaite Water and Blelham Tarn, Part II, Discussion," *J. Ecol.*, **38**, 1–14, 15–35 (1950).

55. Lynn, W. S., and R. H. Brown, "Synthesis of Polyphosphate by Rat Liver Mitochondria," *Biochem. Biophys. Res. Commun.*, **11**, 367–371 (1963).

56. Mackereth, F. J., "Phosphorus Utilization by *Asterionella formosa* Hass," *J. Expt. Bot.*, **4**, 296–313 (1953).

57. Margalef, R., "Rôle des Entomostracés dans la régéneration des phosphates," *Int. Assoc. Theor. Appl. Limnol.*, **11**, 246–247 (1961).

58. Marshall, S. M., and A. P. Orr, "On the Biology of *Calanus finmarchicus*, VIII, Food Uptake, Assimilation and Excretion in Adult and Stage V *Calanus*," *J. Mar. Biol. Assoc. U. K.*, **34**, 495–529 (1955).
59. Marshall, S. M., and A. P. Orr, "On the Biology of *Calanus finmarchicus*, XII, The Phosphorus Cycle: Excretion, Egg Production, Autolysis," *J. Mar. Biol. Assoc. U. K.*, **41**, 463–488 (1961).
60. McCarter, J. A., F. E. Hayes, L. H. Jodney, and M. L. Cameron, "Movement of Materials in the Hypolimnion of a Lake as Studied by the Addition of Radioactive Phosphorus," *Can. J. Zool.*, **30**, 128–133 (1952).
61. Miyachi, S., "Inorganic Polyphosphate in Spinach Leaves," *J. Biochem.* (*Tokyo*), **50**, 367–371 (1961).
62. Miyachi, S., and H. Tamiya, "Distribution and Turnover of Phosphate Compounds in Growing *Chlorella* Cells," *Plant Cell. Physiol.* (*Tokyo*), **2**, 405–414 (1961).
63. Mortimer, C. H., "The Exchange of Dissolved Substances between Mud and Water in Lakes, I and II," *J. Ecol.*, **29**, 280–329 (1941); **30**, 147–201 (1942).
64. Muhammed, A., "Studies on Biosynthesis of Polymetaphosphate by an Enzyme from *Corynebacterium xerosis*," *Biochim. Biophys. Acta*, **54**, 121–132 (1961).
65. Ohle, W., "Chemische und physikalische Untersuchungen norddeutscher Seen," *Arch. Hydrobiol.*, **26**, 386–464, 584–658 (1934).
66. Olsen, S., "Recent Trends in the Determination of Orthophosphate in Water," in, *Chemical Environment in the Aquatic Habitat*, H. L. Golterman and R. S. Clymo (Eds.), Proceedings of the I.B.P. Symposium, Noord-Hollandsche Uitgevers Maatschappij, Amsterdam, 1967, pp. 63–105.
67. Overbeck, J., "Die Phosphatasen von *Scenedesmus quadricauda* und ihre ökologische Bedeutung," *Verh. Int. Verein. Limnol.*, **14**, 226–231 (1961).
68. Overbeck, J., "Untersuchungen zum Phosphathaushalt von Grünalgen VI, Ein Beitrag zum Polyphosphatstoffwechsel des Phytoplanktons," *Ber Deutch Bot. Gesell*, **76** (8), 276–286 (1963).
69. Phillips, J. E., "The Ecological Role of Phosphorus in Waters with Special Reference to Microorganisms," in *Principles and Applications in Aquatic Microbiology*, H. Heukelekian and N. C. Dendero (Eds.), Wiley, New York, 1964, pp. 61–81.
70. Pirson, A., "Functional Aspects in Mineral Nutrition of Green Plants," *Ann. Rev. Plant Physiol.*, **6**, 71–114 (1955).
71. Pomeroy, L. R., "Residence Time of Dissolved PO_4 in Natural Waters," *Science*, **131**, 1731–32 (1960).
72. Pomeroy, L. R., H. M. Mathews, and H. S. Min, "Excretion of Phosphate and Soluble Organic Phosphorus Compounds by Zooplankton," *Limnol. Oceanogr.*, **8**, 50–55 (1963).
73. Putnam, H. D., and T. A. Olson, "A Preliminary Investigation of Nutrients in Western Lake Superior 1958–1959," School of Public Health, University of Minnesota, Minneapolis, 1959, 32 pages (mimeographed).
74. Redfield, A. C., H. P. Smith, and B. H. Ketchum, "The Cycle of Organic Phosphorus in the Gulf of Maine," *Biol. Bull.*, **73**, 421–443 (1937).
75. Rigler, F. H., "A Tracer Study of the Phosphorus Cycle in Lake Waters," *Ecol.*, **37**, 550–562 (1956).
76. Rigler, F. H., "The Uptake and Release of Inorganic Phosphorus by *Daphnia Magna Straus*," *Limnol. Oceanogr.*, **6**, 165–174 (1961).

77. Riley, G. A., "Oxygen, Phosphate, and Nitrate in the Atlantic Ocean," *Bull. Bingham Oceanogr. Collect.*, **13,** 1–126 (1951).
78. Rodhe, W., "Environmental Requirements of Freshwater Plankton Algae," *Symb. Bot. Ups.*, **10,** 1–149 (1948).
79. Schmidt, G., and S. J. Thannhauser, "A Method for the Determination of Desoxyribonucleic acid, Ribonucleic acid, and Phosphoproteins in Animal Tissues," *J. Biol. Chem.*, **161,** 83–89 (1945).
80. Shapiro, J., "Induced Rapid Release and Uptake of Phosphate by Microorganisms," *Science*, **155,** 1269–1271 (1967).
81. Smillie, R. M., and C. Krotkov, "Phosphorus-Containing Compounds in *Euglena gracilis* Grown under Different Conditions", *Arch. Biochem. Biophys.*, **89,** 83–90 (1960).
82. Solórzano, L., and J. D. H. Strickland, "Polyphosphate in Seawater," *Limnol. Oceanogr.*, **13,** 515–518 (1968).
83. Strickland, J. D. H., and T. R. Parsons, "A Manual of Sea Water Analysis," *Fish. Res. Board of Can., Bull.*, **125,** 37–53 (1965).
84. Sylvester, R. O., "Nutrient Content of Drainage Water from Forested, Urban and Agricultural Areas," Algae and Metropolitan Wastes, U.S. Public Health Service SEC TR W61-3, 1961, pp. 80–87.
85. Szymona, M., O. Szymona, and S. Kulesza, "On the Occurrence of Inorganic Polyphosphate Hexokinase in Some Microorganisms," *Acta Microbiol. Pol.*, **11,** 287–300 (1962).
86. Szymona, O., and T. Szumilo, "Adenosine Triphosphate and Inorganic Polyphosphate Fructokinases of *Mycobacterium phlei*," *Acta Biochim. Pol.*, **17,** 129–144 (1966).
87. Terry, K. R., and A. B. Hooper, "Polyphosphate and Orthophosphate Content of *Nitrosomonas europaea* as a Function of Growth," *J. Bacteriol.*, **99,** 103–109 (1970).
88. Whittaker, R. H., "Experiments with Radiophosphorus Tracer in Aquarium Microcosms," *Ecol. Monogr.*, **31,** 157–198 (1957).
89. Widra, A., "Metachromatic Granules of Microorganisms," *J. Bacteriol.*, **78,** 664–670 (1959).
90. Wilkinson, J. F., and J. P. Duguid, "The Influences of Cultural Conditions on Bacterial Cytology," *Int. Rev. Cytol.*, **9,** 1–76 (1960).
91. Winder, F. G., and J. M. Denneny, "Utilization of Metaphosphate for Phosphorylation by Cell-Free Extracts of *Mycobacterium smegmatis*," *Nature*, **175,** 636 (1955).

Solubilities of Phosphates and Other Sparingly Soluble Compounds

WALTER E. BROWN

Director, American Dental Association Research Unit, National Bureau of Standards, Washington, D.C.

Rapid increases in phosphate levels in lakes and rivers near heavily populated areas have made the understanding of factors that control these concentrations a matter of great social importance. Phosphate is thought to be the nutrient that is the limiting factor in the growth of algae in most waters (48); it is possible that carbonate, under some conditions of heavy pollution, may also be a limiting substance. Solubilities of salts of these two anions have been proposed as having considerable control over their concentrations in lake waters. The same considerations do not apply to nitrogen, however, because all of its salts are quite soluble. Even though metabolic processes such as growth and decay of organisms are probably of primary importance in establishing the concentrations of dissolved orthophosphate and carbonate ion at any instant, particularly during warm periods, it appears reasonable that physicochemical processes such as dissolution of suspended particles and sediments, leaching of rocks and soil, and precipitation of insoluble salts have major roles in establishing overall, long-term levels.

The observation that continual entry of phosphate into a lake does not increase the phosphate level in proportion to the amount that has entered (48) clearly indicates that at least some of the phosphate is being removed by a precipitation process. The high levels of calcium frequently found in lake waters suggest that a calcium phosphate may be one of the forms by which it is removed. This alone is justification for the treatment given here on the solubilities of calcium phosphates. The emphasis here is on thermodynamic and operational points of view and on making the treatment sufficiently general so that it can be applied to sparingly soluble phosphates of cations other than calcium and to salts of other weak acids.

Much of the information described here was developed in studies (27, 36–38, 42–44) of calcium phosphates for application to systems containing tooth, bone, and dental calculus; the method of approach, however, is equally applicable to other complex, multicomponent systems.

Solubilities of sparingly soluble salts comprise a subject that has been basic to chemistry over the years, and more recently has been treated extensively as applied specifically to minerals and limnetic solutions (53, 26). One might wonder what more can be said on the subject. However, the treatment given here differs from others in the emphasis that is placed on how to interpret data for a multicomponent system in terms of a four-component system. This should be useful in the interpretation of data collected under natural conditions. Also, stress is placed on considerations that are vital for making measurements on solubilities of other salts, such as those of iron which are thought to limit the phosphate concentrations under acidic conditions. There is urgent need, on the one hand, for more solubility and thermodynamic data on a fairly large variety of salts and minerals. On the other hand, it should be recognized that the data must be quite accurate if one is to avoid serious misinterpretations. For example, a relatively small error in the solubility product of one compound may have a dramatic effect because it may change the identity of the compound that is thought to be most stable for some solution compositions.

The application of solubility principles even to the simple ternary system $Ca(OH)_2$—H_3PO_4—H_2O is far more complex than is usually thought. Much of this complexity goes back to the fact that orthophosphoric acid is both a weak acid and a polybasic acid; but there are other complicating factors, including those caused by the presence of ions other than calcium and phosphate, by the presence of more than one solid phase, and by the formation of ion pairs and complexes. It is possible to treat each of these factors provided it is adequately understood, but we have found it desirable, both conceptually and experimentally, to start out with the simplest system first and then to add complicating factors. Some of these complicating factors are described next as a preliminary to examining the system in greater detail.

When the solubility of calcium phosphates is thought to be the factor that limits the concentration of a solution, attention is usually fixed on hydroxyapatite, $Ca_5(PO_4)_3OH$, as the saturating solid phase because it is the least soluble of the several calcium phosphates under the situations encountered most frequently. Actually, directing one's attention solely to this salt is an oversimplification because hydroxyapatite is only one of several salts with which such a solution may be supersaturated; under these circumstances it becomes difficult to predict which will precipitate. The other sparingly soluble calcium phosphates include $CaHPO_4 \cdot 2H_2O$, $CaHPO_4$, $Ca_8H_2(PO_4)_6 \cdot 5H_2O$, and β-$Ca_3(PO_4)_2$; two others, tetracalcium phosphate, $Ca_4O(PO_4)_2$, and α-tricalcium phosphate, α-$Ca_3(PO_4)_2$, do not persist when in contact with aqueous solutions. Experience has shown that the two hydrated salts,

$CaHPO_4{\cdot}2H_2O$ and $Ca_8H_2(PO_4)_6{\cdot}5H_2O$, are the ones that precipitate most easily, particularly at ambient temperatures, provided the degree of supersaturation with respect to hydroxyapatite is sufficiently high so that their solubilities also have been exceeded. Subsequently, these salts tend to convert to the more stable forms, sometimes going through a sequence of phases of increasing stability, but this can be a slow process. Under circumstances where one of the other salts has formed, it may not be possible to describe the composition of the solution from solubility considerations unless that salt is taken into account. This is true even though the salt is metastable with respect to hydroxyapatite and one or more of the other calcium phosphates (provided the rate of its dissolution is great compared to the rates of precipitation of the more stable salts).

Ions in the solution, other than calcium or phosphate, affect the solubilities of the solid phases in several ways.

1. The most important factor is probably the acidity or basicity of these components. This comes about because the concentration of the PO_4^{3-} ion is strongly repressed by relatively low concentrations of hydrogen ions. Thus, as described later, the calcium and phosphate concentrations in a solution saturated with respect to hydroxyapatite at a pH of about 4 could be as high as 10^{-1} *M*. As is shown later, when several other components are present it is the net acidity (i.e., surplus of acid over base) or the net basicity of these components that is one of the dominant factors in setting the solubility of any calcium phosphate in a given solution.

2. Both calcium and phosphate ions enter into the formation of ion pairs or complexes, between themselves as well as with other possible components. As noted later, uncertainties in the theory of ionic solutions make it difficult to be certain about the presence or absence of specific calcium–phosphate ion pairs. This uncertainty comes about because the concentrations of these ion pairs are usually quite low (less than 10 to 20% of the total calcium or orthophosphate in solution) and because it is always necessary to make extrathermodynamic calculations to arrive at ionic activities. However, we can now have some confidence that these ions do form pairs because extensive studies (26, 36–38, 42, 44, 53) of the ternary system $Ca(OH)_2$—H_3PO_4—H_2O at 5, 15, 25, and 37°C have shown that the solubilities of four of the above-mentioned calcium phosphates can be described accurately if it is assumed that the ion pairs $CaHPO_4^0$ and $CaH_2PO_4^+$ are present in solution. In the pH range above 7 or 8, the ion pair $CaPO_4^-$ may be present in significant concentrations (12). Stable complexes are known to form between phosphate ions and a variety of multivalent cations found in natural waters (e.g., iron and aluminum); calcium ions form complexes with a rather limited number of anions found in natural bodies of water, primarily organic ions, and now, possibly, polyphosphate ions. Carbonate

ions, because of their ubiquitous presence in all natural waters, must be given special consideration in this respect because there is still some uncertainty about the stability of calcium–carbonate ion pairs (30, 32).

3. The presence of other ions also increases the ionic strength of a solution, thereby affecting the solubilities, but usually in a less spectacular way than the first two effects. A high ionic strength has two effects on the equilibria: (*a*) the solubility is enhanced by decreases in the ionic activity coefficients, and (*b*) the distribution of phosphate ions among the various species is altered because the activity coefficient of a multivalent ion such as PO_4^{3-} should be more affected than that of, say, $H_2PO_4^-$. Evidence of the latter effect can be sought in the way the ionic strength affects relative solubility of an acid calcium phosphate (e.g., $CaHPO_4$) as compared to a more basic calcium phosphate [e.g., $Ca_5(PO_3)OH$] and in how it affects the apparent electroneutrality unbalance [see Eq. (12)]. Neither of these effects is large and, thus, would require very accurate data.

Finally, nonthermodynamic factors, such as rates of dissolution and precipitation, barriers to nucleation or crystal growth, ionic transport rates, and conversion of metastable solid phases into more stable solid phases, serve to make these systems difficult to handle experimentally and necessitate caution in the interpretation of results.

Recitation of these difficulties is made not to discourage the reader but instead to stress the need for use of a valid thermodynamic treatment, a chemically and physically appropriate model, and a systematic approach that will help to clarify the concepts and avoid some of the pitfalls. This is the major objective of this chapter.

SOLID PHASES

As noted earlier there are five solid calcium phosphates that are of importance in terms of being sparingly soluble salts that are relatively stable in aqueous systems. It is generally considered that hydroxyapatite, $Ca_5(PO_4)_3OH$, is the least soluble of these salts. The other salts in order of increasing solubility under normal conditions are whitlockite or β-tricalcium phosphate, β-$Ca_3(PO_4)_2$; octacalcium phosphate, $Ca_8H_2(PO_4)_6 \cdot 5H_2O$; anhydrous dicalcium phosphate or monetite, $CaHPO_4$; and dicalcium phosphate dihydrate or brushite, $CaHPO_4 \cdot 2H_2O$. Some investigators have tended, more or less, to disregard the fact that each of these salts has individual characteristics and that each, therefore, would have its own solubility for a given solution composition, temperature, and pressure. Thus, although $Ca_5(PO_4)_3OH$ is the most stable salt under most conditions found in nature, it becomes less stable (i.e., more soluble) than $CaHPO_4$ and $CaHPO_4 \cdot 2H_2O$ if the solution is sufficiently acid. As is described below, the pH at which the relative stability of two salts is reversed (i.e., the singular point) depends on the concentrations of other components. The apparent solubility behavior of $Ca_5(PO_4)_3OH$ in a solution more acid than the singular point, or which at some time had been more acid than the singular point, may be quite different from what one would expect for $Ca_5(PO_4)_3OH$; the formation of the more acid calcium phosphates, especially if formed as surface coatings, can lead to apparent high solubilities. This effect is accentuated by the facts that (*a*) the solutions tend to be so dilute that a large volume of aqueous phase is required to dissolve a relatively small amount of the more acid calcium phosphate phase, and (*b*) hydroxyapatite crystal growth appears to be a slow process so that the elimination of the acid calcium phosphate by conversion to hydroxyapatite may not occur to a significant extent.

These facts have not always been taken adequately into account; the resulting apparent anomalous behavior may have led some investigators to the conclusion that the properties of calcium phosphates in general, and $Ca_5(PO_4)_3OH$ in particular, are rather variable (45, 16, 49, 31). Actually when working with well-purified materials, it was possible to show from solubility measurements (26, 36–38, 42, 44, 53) that each behaves in a quite normal fashion in terms of its solubility and thermodynamic properties.

The order of increasing solubilities of these salts listed above is true only for a limited range of pH. This is shown in Fig. 1 where the negative logarithm of the calcium concentration, pCa, at saturation is plotted against pH for the five salts mentioned above (36–38, 42, 44, 53, 41). A similar diagram is obtained when the negative logarithm of the total orthophosphate concentration, pP, is plotted against pH. It is important to keep in mind that these plots apply only to the ternary system (i.e., solutions which contain only calcium, phosphate, and water ions). The relative solubilities (at 25°C) of the salts at a given pH are easily seen from this graph. At pH 7, the order of increasing solubility is $Ca_5(PO_4)_3OH$, β-$Ca_3(PO_4)_2$, $Ca_8H_2(PO_4)_6 \cdot 5H_2O$, $CaHPO_4$, and $CaHPO_4 \cdot 2H_2O$; below pH 4.8 the most stable salt is $CaHPO_4$, and even $CaHPO_4 \cdot 2H_2O$ is more stable than $Ca_5(PO_4)_3OH$ below pH 4.3. The pH values for various singular points for the ternary system are defined by the broken vertical lines. If one were to extend this plot to high concentrations, four other singular points would be found where the solubilities of the two monocalcium phosphates, $Ca(H_2PO_4)_2$ and $Ca(H_2PO_4)_2 \cdot H_2O$, which are found only under very acid conditions, become less than those of $CaHPO_4$ and $CaHPO_4 \cdot 2H_2O$ (7, 22).

Hydroxyapatite, $Ca_5(PO_4)_3OH$

The compound of greatest interest in most physiological and other natural circumstances is hydroxyapatite because this compound is a prototype of the principal crystalline material in tooth, bone, and many minerals. It is

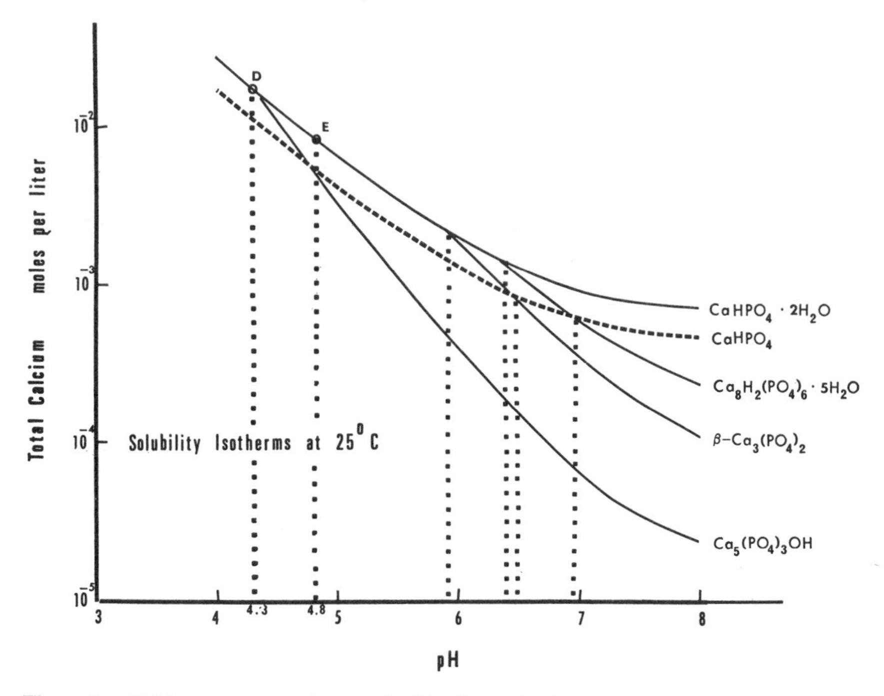

Figure 1. Calcium concentrations and pH values of solutions saturated with respect to various calcium phosphates in the ternary system $Ca(OH)_2$–H_3PO_4–H_2O at 25°C. (Reproduced with the permission of *Clinical Orthopaedics*, No. 44, 1966.)

the ultimate alteration product of fertilizers in soils, and undoubtedly is a factor in the precipitation of phosphates under limnetic circumstances and in sewage and waste-treatment facilities. Because of the relationship between its structure and its chemical properties, the crystal structure of hydroxyapatite (29, 55–57) is shown projected on the *c* face in Fig. 2 and in a stereoscopic illustration in Fig. 3. Some of the atoms in Fig. 2 are shaded to show their relationship to $Ca_8H_2(PO_4)_6{\cdot}5H_2O$, described later. The stereoscopic illustration has the advantage of showing the groups of ions in three-dimensional perspective. The ellipsoid representing each atom depicts its apparent thermal motions. The unit cell contains two formula weights and consists of the following groups of ions: (*a*) two OH^- ions lying on the hexagonal axis at the center of the diagram. The oxygen of the OH^- ion lies about 0.3 Å above or below the plane formed by three nearby calcium ions, and the hydrogen attached to this oxygen has a strong vibrational motion and is even farther away from the calcium triangle; (*b*) six calciums in the form of two triangles separated by half a unit-cell translation plus a 60° rotation; (*c*) six PO_4 groups in two triangles similar to those of the calciums; and

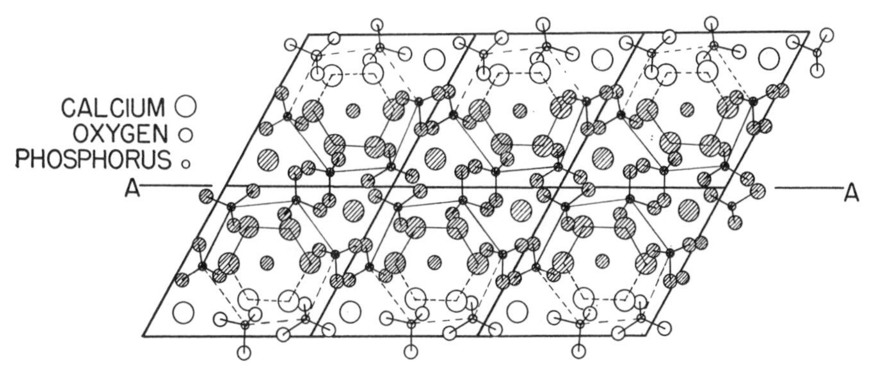

Figure 2. The structure of hydroxyapatite, $Ca_5(PO_4)_3OH$, projected into the *c* face. Six unit cells are shown. The line A–A defines a plane of compatibility with the structure of octacalcium phosphates. (Reproduced with the permission of *Clinical Orthopaedics*, No. 44, 1966.)

(*d*) four calciums made up of pairs located on the trigonal axes near the corners of the cell depicted in Figs. 2 and 3. By keeping in mind these four sets of atoms, one can easily visualize the structure of hydroxyapatite. A closely related mineral of importance to these considerations is fluorapatite, $Ca_5(PO_4)_3F$. It differs from $Ca_5(PO_4)_3OH$ in that the OH^- ions have been replaced by F^- ions situated at the centers (i.e., in the planes) of the calcium triangles instead of being 0.3 Å away. The displacement of OH^- ions from the centers of the calcium triangles in $Ca_5(PO_4)_3OH$ introduces the possibility of disorder in the three-dimensional arrangement of these ions. This disorder probably affects the solubility of hydroxyapatite through its effects on entropy and internal energy. The disorder may be enhanced by partial replacement of OH^- by F^- or O^{2-} ions or by vacancies.

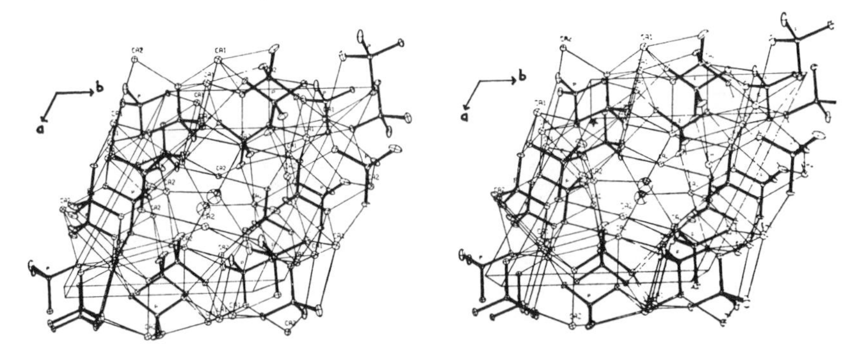

Figure 3. A stereoscopic drawing of the structure of hydroxyapatite as viewed approximately down the *c* axis. The atoms are denoted by their thermal ellipsoids so that the larger ellipsoids near the center of the diagrams (which have large librational motions) represent hydrogens and the smaller ones are the oxygens of the OH group.

Normally one might expect all the OH^- groups in a given column to point in one direction because this gives maximum separation of protons that are on adjacent OH^- groups. On the other hand, an F^- or O^{2-} ion located in the plane of the calcium triangle can act as a center of inversion by being hydrogen bonded to the OH^- groups above and below it. This arrangement could decrease the enthalpy and increase the entropy over what might be expected from a simple solid solution, thereby decreasing the solubility significantly. Several studies on the solubilities of laboratory preparations of hydroxyapatite have shown that it is possible to describe its solubility in a straightforward manner. However, evidence is accumulating showing that enamel, which is the most crystalline of the apatites produced in vivo, has a variable solubility even though fluoride is present in minor amounts. This may be caused by the presence of crystal imperfections and impurities (such as carbonate), by the presence of other calcium phosphates, or by particle-size effects. Furthermore, the effects of F^- ion on the solubility of fluorapatite–hydroxyapatite mixtures are poorly understood beyond being able to say that such mixtures tend to form in the presence of F^- ion and are, therefore, more stable than hydroxyapatite alone. These uncertainties create major problems in the application of solubility principles to limnetic solutions, as they do for tooth and bone chemistry.

Octacalcium Phosphate, $Ca_8H_2(PO_4)_6 \cdot 5H_2O$

The structure of this salt (5) bears a special relationship to that of hydroxyapatite and is shown in projection in Fig. 4 and as a stereoscopic illustration

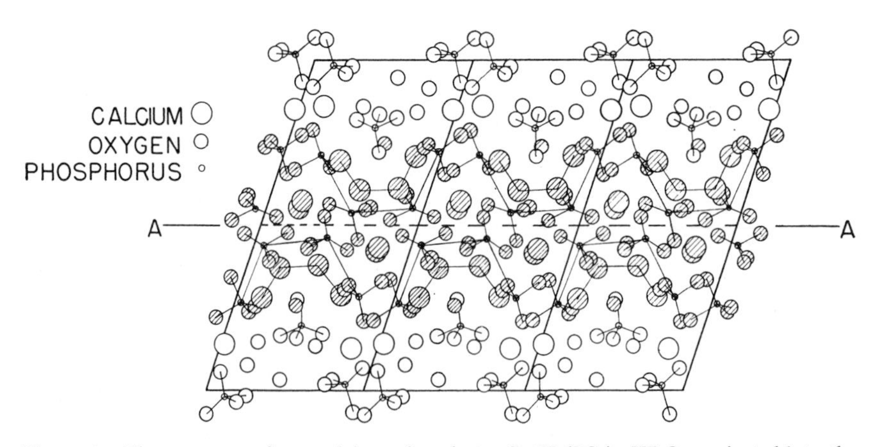

Figure 4. The structure of octacalcium phosphate, $Ca_8H_2(PO_4)_6 \cdot 5H_2O$, projected into the *c* face. Three unit cells are shown. The line A–A defines a plane of compatibility with the structure of hydroxyapatite.

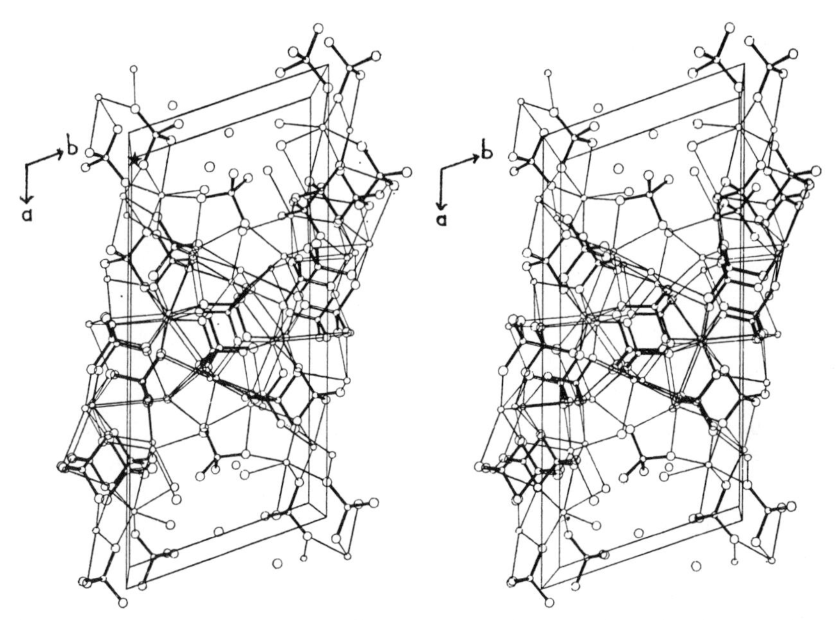

Figure 5. A stereoscopic drawing of the structure of octacalcium phosphate as viewed approximately down the *c* axis. The PO_4 groups are defined by the heavy lines and the calciums by circles that are smaller than those for the oxygens.

in Fig. 5. The unit cell contains two formula weights. The shaded atoms in Fig. 4 correspond in position very closely to those of hydroxyapatite which are shaded in Fig. 2. The presence of this layer in $Ca_8H_2(PO_4)_6{\cdot}5H_2O$ permits an epitaxial relationship with hydroxyapatite (8, 9). The resultant formation of surface layers and intracrystalline mixtures is believed to be a major factor that has led to confusion about the composition and properties of hydroxyapatite. The unshaded atoms in Fig. 4 have essentially no relationship to the structure of hydroxyapatite. The water molecule [O(43) in Ref. 25] is of chemical interest in that this site may be only partially occupied, thus introducing possible variability into the chemical properties of octacalcium phosphate.

As a result of the similarity in the two structures, it appears that $Ca_8H_2(PO_4)_6{\cdot}5H_2O$ may participate in the nucleation and growth as a precursor to hydroxyapatite; it may act, also, as a transition layer between the hydroxyapatite crystals and the aqueous phase. It is thought (6) that a hydrated structure such as octacalcium phosphate will have a lower interfacial energy as compared to that of an anhydrous ionic structure such as hydroxyapatite. Thus, octacalcium phosphate may have an important bearing on other properties of hydroxyapatite such as (*a*) influencing the morphology of the crystallite, (*b*) participating in the reactions with fluoride ions, and (*c*) possibly stabilizing hydroxyapatite so that it can exist as very fine crystallites. The existence of hydroxyapatite in this finely divided form in most of its occurrences is one of its most important properties, giving it the characteristics of a colloid.

Additional insight into the manner in which octacalcium phosphate may complicate the chemistry of hydroxyapatite can be gained if we consider the the different processes by which hydroxyapatite can be formed by precipitation from an aqueous solution. It was proposed (6) that hydroxyapatites may be formed through three mechanisms.

1. One of these is easily carried out in the laboratory. Octacalcium phosphate is precipitated first (e.g., by hydrolysis of $CaHPO_4 \cdot 2H_2O$ or by titration of a calcium solution with a phosphate solution at room temperature and approximately neutral pH). Once formed, the octacalcium phosphate can be hydrolyzed to hydroxyapatite by boiling or by treating with an alkaline solution. In this process the larger crystals tend to become coated so that complete hydrolysis is difficult, and the gross composition tends to be intermediate to the two salts. An important aspect of this process is that it tends to occur in situ so that the resulting hydroxyapatite crystals have the platy or blade-like morphology of the original octacalcium phosphate. This is an important indication of the previous history of such crystals.
2. A second and the most obvious process would be the growth of hydroxyapatite directly without an intermediary phase such as octacalcium phosphate. Under many conditions, this route is very slow or ineffectual. Many possible constituents of lakes and rivers, such as carbonate, magnesium, and polyphosphate ions, seem to inhibit precipitation of hydroxyapatite.
3. A third process might be a combination of these two and would involve the successive steps of precipitation of, say, a unit-cell-thick layer of octacalcium phosphate on the (100) face of hydroxyapatite followed by its hydrolysis to hydroxyapatite; this process would then be repeated, each time converting a unit-cell thickness of octacalcium phosphate to a double unit-cell thickness of hydroxyapatite. Such a mechanism was proposed (6) because of the structural similarity described above and because direct precipitation of hydroxyapatite seems to be very difficult.

Fluoride ions would play an important role in the first and third of these processes in that they tend to accelerate the hydrolysis of octacalcium phosphate to an apatite. This process was proposed (6) as a partial explanation of why extremely small amounts of fluoride in tooth mineral can have such profound effects in reducing caries. Limnetic conditions are such that they, too, could facilitate the formation of octacalcium phosphate as an intermediary in the formation of hydroxyapatite since the growth of octacalcium phosphates seems to be preferred at lower temperatures. This mechanism and the proposed role of F^- ions could, therefore, be important factors in the kinetics of precipitation in natural bodies of water.

Whitlockite, β-$Ca_3(PO_4)_2$

Like hydroxyapatite, β-$Ca_3(PO_4)_2$ is a mineral (24, 25). It is somewhat unusual in that cations smaller than calcium (Fe^{2+} and Mg^{2+}, in particular) seem either to make it more stable or to facilitate its formation in aqueous media. For a long time the composition of this salt was in doubt because of a discrepancy between the unit-cell constants and the apparent symmetry (35). Recent determinations of the structure are resolving both the uncertainty in the structure and the reason for the stabilizing effect of small cations.

Recent solubility measurements on the pure calcium salt (53) have shown that above about pH 6 (see Fig. 1) this salt is second only to hydroxyapatite in its insolubility. Therefore, under suitable circumstances one might expect that it would form at least as an intermediary phase to hydroxyapatite. Special considerations should be given to this compound in those modifications of sewage or industrial-waste treatment processes in which the presence of Fe^{2+} or Mg^{2+} ions may facilitate formation of β-$Ca_3(PO_4)_2$.

Brushite, $CaHPO_4 \cdot 2H_2O$

Although under most circumstances this salt is the most soluble of the five that we are discussing here, it deserves special consideration; because of crystal-growth kinetics it appears to form very readily. Therefore, it is found under many circumstances where it is metastable with respect to the other four salts. Another feature that seems to be important to both this compound and octacalcium phosphate is the presence of waters of hydration. It seems that the presence of these waters may reduce the interfacial energy of these salts when they are in aqueous systems. This would enhance the relative stability of the hydrated salts when the particles are extremely small, that is, during the nucleation and initial growth stages. Thus, it is possible that one of these salts is actually more stable than hydroxyapatite during the incipient stage of formation. In any event, it is a common experience in the laboratory to find that brushite (or octacalcium phosphate) forms initially and the more stable anhydrous salts form subsequently, either by hydrolysis in situ, as in the case of transition from octacalcium phosphate to hydroxyapatite, or through a dissolution step followed by a precipitation step, as seems to be the case usually in the conversion of brushite into one of the other more insoluble forms.

Monetite, $CaHPO_4$

Although this salt is always more stable than brushite, except possibly at very high pressures and temperatures below about 5°C, it is encountered less frequently than brushite under most circumstances. This is in part due to the fact that the salt is somewhat difficult to detect by x-ray methods. Petrographic microscopy should be used when its presence is suspected because its birefringence is relatively high (52). It has a relatively low probability of occurrence under limnetic conditions for two reasons: (*a*) the pH range in which it is the most stable salt (about 5 to 1) is rarely encountered in lakes and rivers; and (*b*) its metastability with respect to several other calcium phosphates is not compensated by a rapid rate of crystal growth as is the case with brushite.

Amorphous Calcium Phosphate

In recent years considerable emphasis has been placed on the observation that precipitated calcium phosphates frequently display no crystalline characteristics. These products are also thought (19–21) to have a composition closely approximating that of $Ca_3(PO_4)_2$ except that they contain considerable water. These materials cannot be treated according to the solubility product principles described later even though they might have a fairly fixed composition, because it is unlikely that they fulfill the necessary condition of having a fixed standard free energy of dissolution per mole. There is no reason to believe that a material as ill defined as amorphous calcium phosphate would have fixed thermodynamic properties. Whether such materials ever have significant effects on the compositions of limnetic solutions is difficult to say. It seems probable that they would nearly always be metastable with respect to all five of the sparingly soluble calcium phosphates described above, and can therefore be formed only when (*a*) the solution is so highly supersaturated that the reactions are too rapid for crystal growth of any of the well-defined crystalline phases or (*b*) the solution is only moderately supersaturated, but crystal growth poisons are present that prevent the formation of the other phases.

Calcium Carbonates

Calcite and aragonite are the two most commonly occurring anhydrous calcium carbonates. Several other anhydrous forms, including vaterite, several high-pressure modifications of calcite (14, 15), and a multiple twinned type of aragonite structure, should be kept in mind as having an outside chance of being involved in some way in growth from water solutions.

The two hydrated salts, $CaCO_3 \cdot H_2O$ (51, 39, 33) and $CaCO_3 \cdot 6H_2O$ (3, 18), are more likely participants in nucleation in crystal growth; they would tend to be favored by (*a*) properties such as interfacial energy and kinetics of growth, and (*b*) conditions such as low temperatures, high pressures, and growth poisons (Mg^{2+} and polyphosphates) that poison the growth of the anhydrous salts more than that of the hydrated salts.

THE LIQUID PHASE

Definition of Components

For the sake of convenience in defining the activities or chemical potentials of the components, we have chosen to describe the ternary system with the components $Ca(OH)_2$, H_3PO_4, and H_2O. For many years the components that were chosen to describe this system were CaO, P_2O_5, and H_2O. However, in dilute solutions, the activities and chemical potentials of the oxides cannot be visualized or expressed as easily as those of $Ca(OH)_2$ and H_3PO_4, which can be defined as follows:

$$A_{Ca(OH)_2} = (Ca^{2+})(OH^-)^2, \quad (1)$$

$$A_{H_3PO_4} = (H^+)^3(PO_4^{3-}) \quad (2)$$

$$A_{H_2O} = (H^+)(OH^-) \quad (3)$$

$$\mu_{Ca(OH)_2} = \mu^{\circ}_{Ca(OH)_2} + RT \ln (Ca^{2+})(OH^-)^2 \quad (4)$$

$$\mu_{H_3PO_4} = \mu^{\circ}_{H_3PO_4} + RT \ln (H^+)^3(PO_4^{3-}) \quad (5)$$

and

$$\mu_{H_2O} = \mu^{\circ}_{H_2O} + RT \ln (H^+)(OH^-) \quad (6)$$

In this chapter, parentheses denote activities of ions, and square brackets denote concentrations.

The very useful fact that the chemical potential of any component, whether it is a weak or a strong electrolyte, can be defined in terms of an ionic product is frequently overlooked. As a consequence of this definition, the chemical potential of an acidic or basic component rises or falls as the pH is changed even though the concentration of the component is held constant. A graphic illustration of this effect is described later where the chemical potential of $Ca(OH)_2$ is shown to decrease along an isotherm even though the concentration of calcium ion is increased over a thousandfold. For this reason it is important to distinguish clearly between the concentration of a component and its chemical potential. Although these effects are generally recognized in the treatment of inorganic processes, they should be given more consideration than presently in treating metabolic processes where the driving forces for diffusion and chemical reactions also involve differences in chemical potentials.

In the ternary system, the chemical potentials are relatively easily definable quantities because the calcium and phosphorus concentrations and pH are measured directly. One then needs only to know the ionic strength to be able to use an empirical relationship such as the Debye-Hückel or the Davies equation (37) to calculate the activities of the individual ions. In these calculations, one must also use equilibrium constants for the dissociation of H_3PO_4 and for the calcium–phosphate ion pairs that may be present.

The application of the equilibrium constants obtained from the study of the ternary system to those that contain additional components is described in a later section. It is noted here, however, that in natural waters there are exceptional difficulties in defining the concentrations of the components. Phosphate, for example, occurs in various other forms such as (*a*) organically bound phosphate, (*b*) particulate material, (*c*) adsorbed on particulate material, (*d*) complexed ions, and (*e*) possibly as polyphosphates or metaphosphates. It is only the "free" orthophosphate and the orthophosphate ion paired to calcium (which can usually be calculated) that should be counted as affecting the solubility of a calcium phosphate salt.

The task of defining the components to be used in describing a system should not be taken lightly. Although neutral salts and nonionizing components tend to have secondary effects on solubility and frequently can be ignored, some components may introduce new solid phases even when present in very low concentrations, which would then dominate the nature of the equilibrium. HF is an example of such a component because it forms highly insoluble fluorapatite (or fluorapatite–hydroxyapatite solid solutions). In general, it is advantageous to choose all other components as acids or bases because it is the acidity or basicity that is usually the primary factor affecting the solubility of a phosphate or carbonate salt because of its effect on the degree of dissociation of H_3PO_4 or H_2CO_3. The exception to this rule would be a constituent that forms soluble complexes or insoluble precipitates. As is shown later, the selection of the additional components as acids and bases makes it easier to treat their effects by combining them into a single variable which has been called the "electroneutrality unbalance."

Solubility Isotherms

We have already mentioned some of the complications that are introduced into understanding the solubilities of the calcium phosphates because of the fact that phosphoric acid is a weak polybasic acid. For this reason, we approach the subject from a basic position. The Gibbs phase rule, when applied to a system containing three components and in which a solid phase

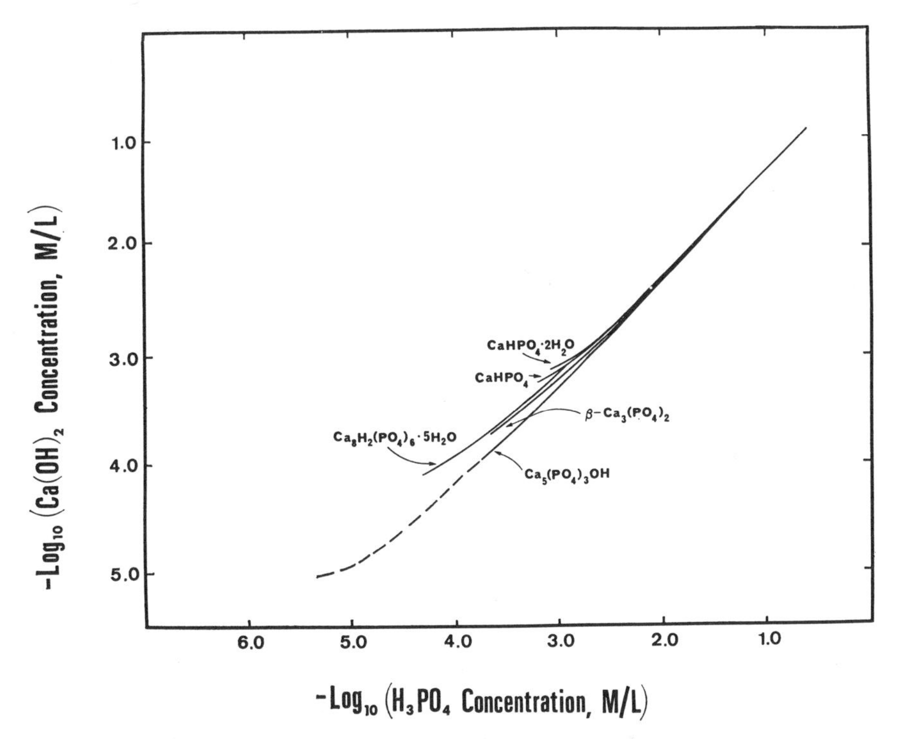

Figure 6. The phase diagram for the ternary system, $Ca(OH)_2$–H_3PO_4–H_2O at 25°C showing solubility isotherms for the salts $Ca_5(PO_4)_3OH$, β-$Ca_3(PO_4)_2$, $Ca_8H_2(PO_4)_6 \cdot 5H_2O$, $CaHPO_4$, and $CaHPO_4 \cdot 2H_2O$. The coordinates are the negative logarithms of the concentrations of $Ca(OH)_2$ and H_3PO_4. Broken lines represent calculated points.

and a liquid phase are in equilibrium at fixed temperature and pressure, allows only one degree of freedom. This fact is easily seen from Fig. 6, a logarithmic plot of the concentration of calcium hydroxide against that of phosphoric acid. The compositions of the solutions that are saturated with respect to hydroxyapatite are given by the line designated $Ca_5(PO_4)_3OH$. The fact that the equilibrium compositions are described by a line (the "isotherm") is a consequence of the single degree of freedom. Thus, if the $[Ca(OH)_2]$, $[H_3PO_4]$, or pH of a saturated solution is specified, then the composition of the solution is completely fixed by the point where that composition or pH line crosses the isotherm. Along this isotherm the chemical potential of hydroxyapatite is constant even though the concentrations of the components $Ca(OH)_2$ and H_3PO_4 vary over three orders of magnitude. This is true because saturation requires that

$$\Delta F^{\circ}_{HA} + 9\,\mu_{H_2O} = 5\,\mu_{Ca(OH)_2} + 3\,\mu_{H_3PO_4} \tag{7}$$

where ΔF°_{HA} is the standard free energy of dissolution of hydroxyapatite. This relationship is a direct consequence of the assumption that the solution is saturated with respect to hydroxyapatite and, therefore, the reaction

$$Ca_5(PO_4)_3OH(s) + 9H_2O(l) = 5Ca(OH)_2(aq) + 3H_3PO_4(aq) \quad (8)$$

takes place without a change in free energy. In the dilute solutions under consideration μ_{H_2O}, the chemical potential of water, is essentially constant.

One can now visualize a hypothetical process in which the composition is changed toward the region of high concentrations and low pH's (toward the upper right along the isotherm), at all times maintaining equilibrium. In this process, five times the decrease in chemical potential of $Ca(OH)_2$ is exactly compensated by three times the increase in chemical potential of H_3PO_4. The chemical potential of $Ca(OH)_2$ is decreased during this process because the pH has a greater effect on the chemical potential of $Ca(OH)_2$ than does the concentration.

The isotherm is the line of demarcation between the region of supersaturation with respect to hydroxyapatite (above and to the left of the isotherm in Fig. 6) and the region of undersaturation. In the latter region, the reaction defined by Eq. (8) would be accompanied by a decrease in free energy. It must be kept in mind that Eq. (8) is not the only reaction that can be written. Isotherms are shown in Fig. 6 for other calcium phosphates which define the solution compositions where there would be no change in free energy for the dissolution-precipitation reaction for each of these salts. The fact that these isotherms have regions of positive slope in the range of concentrations of primary interest to us is an important characteristic of phosphatic systems and is a consequence of phosphoric acid being a weak electrolyte. The same situation applies to the solubilities of calcium carbonates. It accounts for the well-known phenomenon that calcium carbonate dissolves when the partial pressure of CO_2 is increased. On the other hand, the slope of an isotherm for a salt formed from a strong acid and a strong base should be monotonically negative over the entire range (in the absence of complex formation). The slope of the isotherm for hydroxyapatite becomes negative when the pH reaches a value in the vicinity of pK_3 of phosphoric acid, but the phosphate concentrations are then so low and the pH's so high that this is not a range of significance except possibly when waste water is being treated with lime.

With the exception of the isotherm for octacalcium phosphate, which was calculated by use of a tentative solubility constant, the other isotherms given in Fig. 6 represent experimental measurements. The isotherms for all the salts shown in Fig. 6 are nearly superimposed in the range of higher concentrations and lower pH's. This is why it is necessary to use a plot of the type shown in Fig. 1 to separate the isotherms for the various salts. The fact that these isotherms are nearly superimposed in Fig. 6 relates to a property of phosphoric acid which can be used to give additional insight into these systems, as described in the section on the "electroneutrality surface." Next, however, we consider some of the consequences of having two solid phases in equilibrium with a liquid phase.

Singular Points

As noted above, when a single equilibrating solid phase is present, application of the phase rule shows that the ternary system has one degree of freedom; the presence of a second equilibrating solid phase reduces the degrees of freedom by one so that the system becomes invariant (at a given pressure and temperature). This is represented in Fig. 1 as the point of intersection where the isotherms of two solids intersect; the composition at the point of intersection is known as the "singular point." A total of six singular points are shown in Fig. 1. The singular points, although present, cannot be seen in Fig. 6 because the isotherms are so nearly superimposed. One can approximate three additional singular points by extrapolating the isotherms of $Ca_5(PO_4)_3OH$, $Ca_8H_2(PO_4)_6{\cdot}5H_2O$, and β-$Ca_3(PO_4)_2$, Fig. 1, to the points where they intersect. However, the metastability of these compositions with respect to $CaHPO_4{\cdot}2H_2O$ and $CaHPO_4$ (and the ease with which $CaHPO_4{\cdot}2H_2O$ precipitates) prevents them from having more than theoretical significance. The invariance of the calcium and phosphorus concentrations and pH at the singular points is destroyed when the system contains more than three components. It is shown next that the variability in the concentrations of these components does not signify that their chemical potentials are variable at the singular point.

At the singular point, saturation with respect to two solids requires that the solubility products for both are satisfied. Thus at the singular point of $Ca_5(PO_4)_3OH$ and $CaHPO_4$, for example, we have

$$(Ca^{2+})^5(PO_4^{3-})^3(OH) = K_{HA} \tag{9}$$

and

$$(Ca^{2+})(HPO_4^{2-}) = K_m \tag{10}$$

or

$$(Ca^{2+})(H^+)(PO_4^{3-}) = K_m K_3, \tag{11}$$

where K_{HA} and K_m are the solubility products for hydroxyapatite and monetite, respectively, and K_3 is the third dissociation constant for phosphoric acid. Simply by dividing Eq. (9) by the third power of Eq. (11) and substituting K_w for $(H^+)(OH^-)$, we have

$$(Ca^{2+})(OH^-)^2 = K_{HA}^{1/2}\left(\frac{K_w}{K_m K_3}\right)^{3/2} \tag{12}$$

and by dividing Eq. (9) by the fifth power of Eq. (10), we have

$$(H^+)^3(PO_4^{3-}) = (K_m K_3)^{5/2}\left(\frac{K_w}{K_{HA}}\right)^{1/2} \tag{13}$$

These equations reveal that the chemical potentials of $Ca(OH)_2$ and H_3PO_4 are fixed at the singular point at a given temperature or pressure because the addition of other components will not change the quantities on the right-hand sides of Eqs. (12) and (13).

The singular points are important features of the system because they represent compositions with relatively high probability. This is a direct result of the phenomenon of metastability. An example illustrates this point. Consider a solution which is saturated with respect to hydroxyapatite and which has a pH less than that of its singular point with brushite (i.e., to the left of pH = 4.3 in Fig. 1). The solution will be supersaturated with respect to brushite, which, therefore, can precipitate, making the solution undersaturated with respect to hydroxyapatite. Dissolution of the basic salt, hydroxyapatite, and precipitation of the acid salt, brushite, will cause the pH of the solution to rise. This process would continue until the pH and solution composition reached the singular point. Conversely, brushite in a solution with a pH above that of the singular point would tend to cause the precipitation of hydroxyapatite; subsequent dissolution of brushite would drive the pH down; again, these processes would continue until the singular point was reached.

Relationships similar to (12) and (13) are obtainable at the singular point for any pair of calcium phosphates. The chemical potential of $Ca(OH)_2$ or H_3PO_4 would again be constant, although for each pair it would have a different value. According to Eq. (12), the activity of the calcium ion at the singular point would increase a hundredfold with a tenfold increase in hydrogen ion activity. These increases can be brought about only by the presence of other components (e.g., HCl, H_2CO_3, H_2SO_4) in the solution; concomitantly, the PO_4^{3-} ion activity would be decreased by the presence of these other anions. One can see these relationships better by keeping in mind that the charge of the foreign anion is balanced by a greater calcium concentration. The consequent greater calcium ion activity results in a lower PO_4^{3-} ion activity because of Eqs. (9) and (11). Similarly, according to Eq. (13), the activity of PO_4^{3-} would be increased, and the activities of the calcium and hydrogen ions would be decreased, by the presence of components such as NaOH, KOH, and $Mg(OH)_2$. Since the composition and pH of the singular point solution is affected by the presence of other ions, Eqs. (12) and (13) are especially useful for estimating the pH of the singular point of a solution in which the phosphate concentration is greatly exceeded by those of other anions such as carbonate and sulfate, as is the case in many natural waters. Note the direction of these effects: An excess of added anion increases the calcium and hydrogen ion activities, and an excess of added cation increases the PO_4^{3-} activity and the pH.

Since $Ca(OH)_2$ is a legitimate solid in the ternary system, it could coexist with hydroxyapatite to form a singular point that could be of great interest in the treatment of sewage with lime. However, this region of the system has not been studied adequately because the phosphate concentrations are too low to measure conveniently.

The "Electroneutrality Surface"

Knowledge of the dissociation constants of phosphoric acid and the stability constants for calcium-phosphate ion pairs permits one to calculate the pH of any dilute solution for which the concentrations $[Ca(OH)_2]$ and $[H_3PO_4]$ are given. Furthermore, one can define a surface in three-dimensional space by plotting $-\log[Ca(OH)_2]$ against $-\log[H_3PO_4]$ in the plane of the paper, as was done in Fig. 6, and the pH in the third direction. This surface can be depicted as in Fig. 7 by lines of constant pH which define the contours of the surface. Although this surface is more properly called a "pH surface," it was first called (4) the electroneutrality surface for the reason that it is necessary to introduce the condition of electroneutrality,

$$[H_2PO_4^-] + 2[HPO_4^{2-}] + 3[PO_4^{3-}] + [OH]^- - [H^+] - 2[Ca^{2+}] - [CaH_2PO_4^+] = 0 \quad (14)$$

to be able to calculate the pH at any given point. More details about this surface are given elsewhere (4).

The important feature about this surface is the presence of two steep faces where the pH contours fall very close to one another in the range of pH from about 4 to 10. The existence of these steep faces is the reason why the isotherms in Fig. 6 are nearly superimposed over much of their lengths. These isotherms, like the points for all electrically neutral compositions, must lie in the electroneutrality surface. In those regions where the surface is very steep, and when the surface is viewed from above as in Fig. 6, the isotherms will appear nearly superimposed. When the isotherms are projected into a face that contains the pH axis, as is the case in Fig. 1, the isotherms appear well separated. Diagrams similar to Fig. 6 may be calculated for fixed compositions of a fourth component and for other temperatures and pressures through use of the solubility and ion-pair formation constants that have been derived from studies of the ternary system.

The two steep faces, in the pH ranges 4 to 6 and 8 to 9, are separated by a slightly flattened region at pH 7. Since pK_2 for phosphoric acid is 7.20, it is apparent that the flattening of the surface in this region corresponds to the

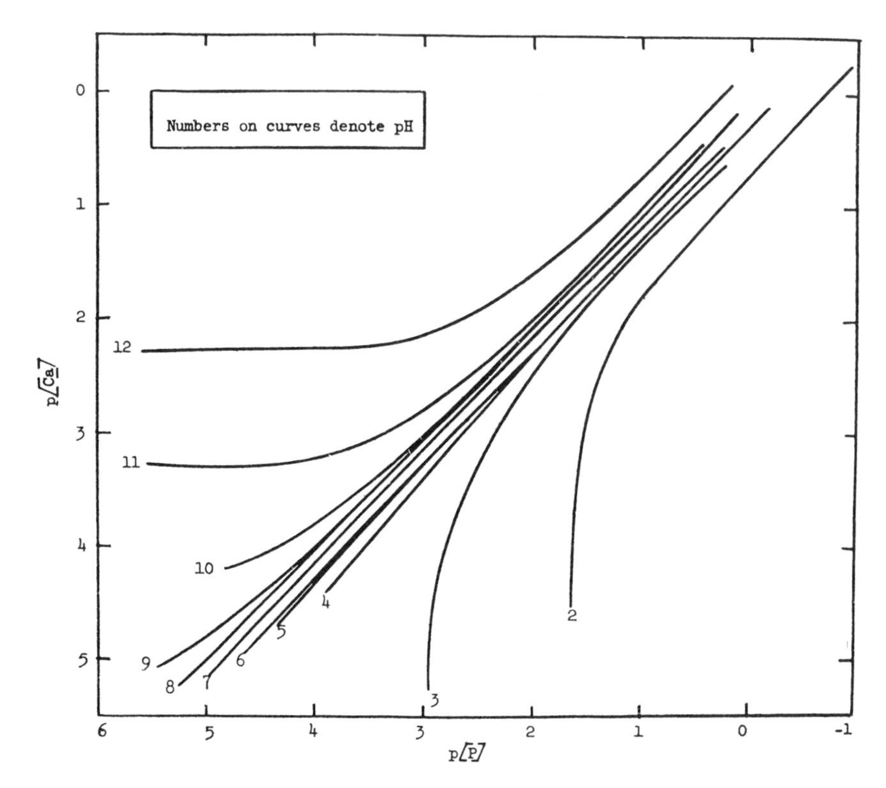

Figure 7. The electroneutrality surface for the ternary system $Ca(OH)_2$–H_3PO_4–H_2O. The lines of constant pH represent contours of a surface in three-dimensional space in which pH is the vertical coordinate.

pH where the two ions $H_2PO_4^-$ and HPO_4^{2-} have equal activities and the system has a maximum in its buffer capacity. The flat surfaces below pH of 4 and above 10 are regions that are buffered by the combinations H_3PO_4/$H_2PO_4^-$ and HPO_4^{2-}/PO_4^{3-}, respectively. Thus the steep faces occur at pH values that are intermediate to the pK values for the acid. Accordingly, the system $Ca(OH)_2$—H_2CO_3—H_2O should have a steep face at a pH of about 8.3.

The presence of the steep faces in Fig. 7 is very valuable for interpreting experimental data or for designing experiments. For example, it is difficult to tell whether a solution is saturated with respect to any specific calcium phosphate from a knowledge of the calcium and phosphate concentrations alone; it is necessary to know the pH as well. In other words, in the steep faces the calcium and phosphate concentrations are slowly changing variables, whereas the pH is the most sensitive variable, and it is the one that should be chosen if the approach to equilibrium is to be monitored. This is not true, of course, in a multicomponent system that contains a high concentration of buffer.

Another use for the type of diagram shown in Fig. 7 comes from the fact that the solutions, regardless of their state of saturation with respect to the various calcium phosphate salts, are restricted in their compositions to a rather narrow region. Most solutions of interest fall in the pH range 5 to 9. Thus this is a region of fairly high probability from the standpoint of finding solution compositions, and one can use it for purposes of approximation in much the same way as one would use an isotherm. The presence of other components will displace the steep surface in a manner that is described in a later section. However, before doing this it is necessary to consider the effects of additional components in a more general way.

Multicomponent Systems

Isotherms describing the solubility of hydroxyapatite in the presence of constant amounts of a fourth component are shown in Fig. 8. For reasons given later, the amount of the fourth component is given in terms of the negative logarithm of the concentration, $U(\pm)$, of this component, where $U(-)$ represents an acid and $U(+)$ a base. The isotherm labeled $\mathrm{p}U(0)$ is the same as the one given for hydroxyapatite in the ternary phase diagram in Fig. 6. In accord with the phase rule, the addition of a fourth component introduces a second degree of freedom, but by restricting the concentration of that component to a given value, the degrees of freedom are again reduced to one. Thus, all solutions containing 10^{-4} mole/l of, say, HCl would have compositions defined by the line $\mathrm{p}U(-) = 4$ when saturated with respect to hydroxyapatite; solutions containing 10^{-3} mole/l of HCl would correspond to the isotherm labeled $\mathrm{p}U(-) = 3$. These isotherms lie in the region which, in the absence of HCl, would be supersaturated with respect to hydroxyapatite, thus showing that the effect of the acid, as one would expect, is to increase the solubility of hydroxyapatite. The latter two isotherms were calculated on the basis of equilibrium constants obtained from the ternary system. The position of the isotherm should not be shifted very much if, say, HCl were replaced by HNO_3 or by any other strong acid of the same normality that did not have specific interactions with calcium or phosphate ions. A similar set of isotherms, shown below that of $\mathrm{p}U(0)$ in Fig. 8, is obtained for a series of solutions with constant concentrations of a base such as NaOH or KOH. Examples with $\mathrm{p}U(+) = 3$ and 4 are shown in Fig. 8. In the range of higher pH's and lower calcium and phosphate concentrations, the curves with positive values of $\mathrm{p}U(\pm)$ acquire negative slopes as do the curves with negative values of $U(\pm)$. However, this portion of the

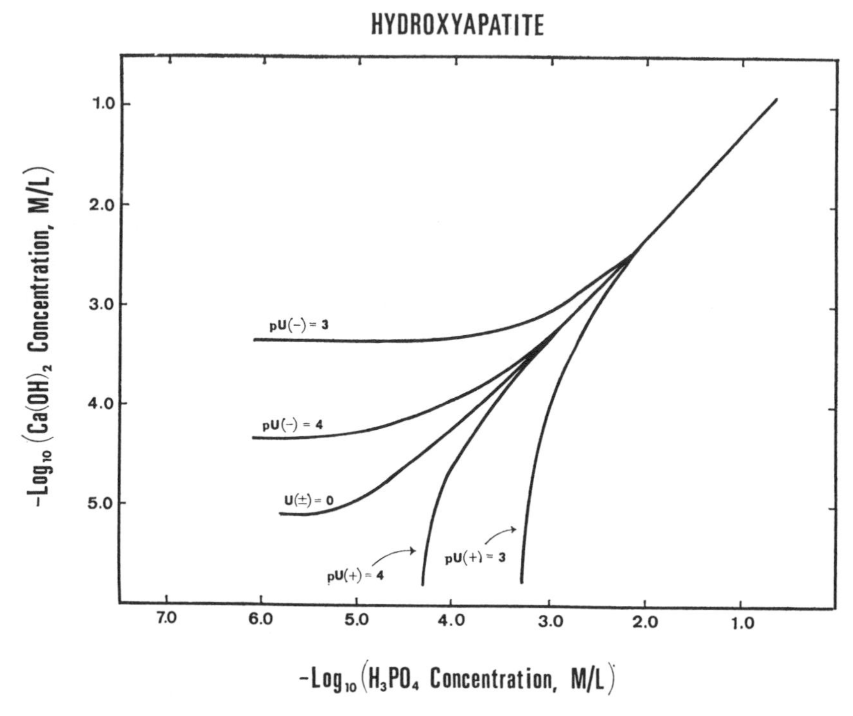

Figure 8. Calculated solubility isotherms for hydroxyapatite at 25°C in the "generalized four-component system," $Ca(OH_2–H_3PO_4–H_2O–U(\pm)$. The concentration of acid or base, is defined by Eqs. (15) and (16).

phase diagram is in the range of such low phosphate concentrations that it does not have much practical significance.

According to the view presented here, it is the acidity or basicity of the other components that is important as long as they do not form significant concentrations of ion pairs and complexes. Furthermore, in the presence of several of such components, it is the excess of base over acid, or vice versa, that determines the position of the isotherm. This leads to the definition of the quantity $U(\pm)$,

$$U(\pm) = \sum_i \nu_i C_i \tag{15}$$

where ν_i is the valence and C_i is the concentration of the ith ion, and the sum is taken over all species of ions other than calcium, phosphate, and water ions, and calcium–phosphate ion pairs. Because the solution is electrically neutral, an alternative definition of $U(\pm)$ is given by

$$U(\pm) = -\sum_j \nu_j C_j \tag{16}$$

where the sum is taken over calcium, phosphate, and water ions (i.e., the ions present in the ternary system). This is why the quantity $U(\pm)$ has been called (4) the "electroneutrality unbalance." In essence, $U(\pm)$ is the measure of the excess of basic over acidic components, but it takes into account only the dissociated portions of these acids and bases. Ion pairs and complexes are counted as acids or bases depending on whether their valence charges are negative or positive, respectively. For these reasons, the quantity $U(\pm)$ can be considered to be the fourth component, which when positive is a base and when negative is an acid. Its use partially ignores the effects of ionic strength on activity coefficients, but for the dilute solutions under consideration here, this would not be a major source of error. Ion pairs and complexes (other than the calcium–phosphate ion pairs) must be taken into account explicitly, and the concentrations of calcium or phosphate in these forms must be subtracted from the quantities $[Ca(OH)_2]$ and $[H_3PO_4]$ used as the coordinates in Fig. 8.

The isotherms for the various values of $pU(\pm)$ can be viewed as forming a three-dimensional surface in a way somewhat similar to that of Fig. 7 except that the coordinate in the vertical direction would be $pU(\pm)$ instead of pH. Since each point on an isotherm has a specific pH, one can define lines of constant pH on this surface. These are shown, in projection, in Fig. 9, which is a duplicate of Fig. 8 except that it shows lines of constant pH.

Figure 9 displays at a glance considerable information about how solubility of hydroxyapatite is affected by other components over the normal range of limnetic solutions. Compositions with pH's less than 4 are not shown because hydroxyapatite is no longer the most stable phase; above pH = 9, the orthophosphate concentrations are too low to be of much significance compared to other forms of phosphate found in natural waters. Figure 9 shows that a wide range of calcium and phosphate concentrations can be saturated with respect to hydroxyapatite at a given pH. The breadth of this range (i.e., the fanning out of the isotherms) is much greater in the dilute region of the diagram. Figure 9 also shows that the calcium and phosphate concentrations of saturated solutions are quite high in the acid region. As noted previously, the phosphate concentration at a pH of 4 is remarkably high, being of the order 0.1 *M* over a considerable range of values for $U(\pm)$.

The lines of constant pH can be used to determine whether or not the solution is saturated even when one does not know the value of $U(\pm)$ for that solution. Thus, the lines of constant pH can be used as "isotherms" when $U(\pm)$ is not known or is a variable. For example, a solution undersaturated with respect to hydroxyapatite would have a pH below that shown in Fig. 9 for a point defined by its calcium and phosphorus concentrations. A disadvantage of this method (as with any other method) is that either the concentrations of ion pairs and complexes must be negligible or they must be taken into account explicitly in the manner noted previously.

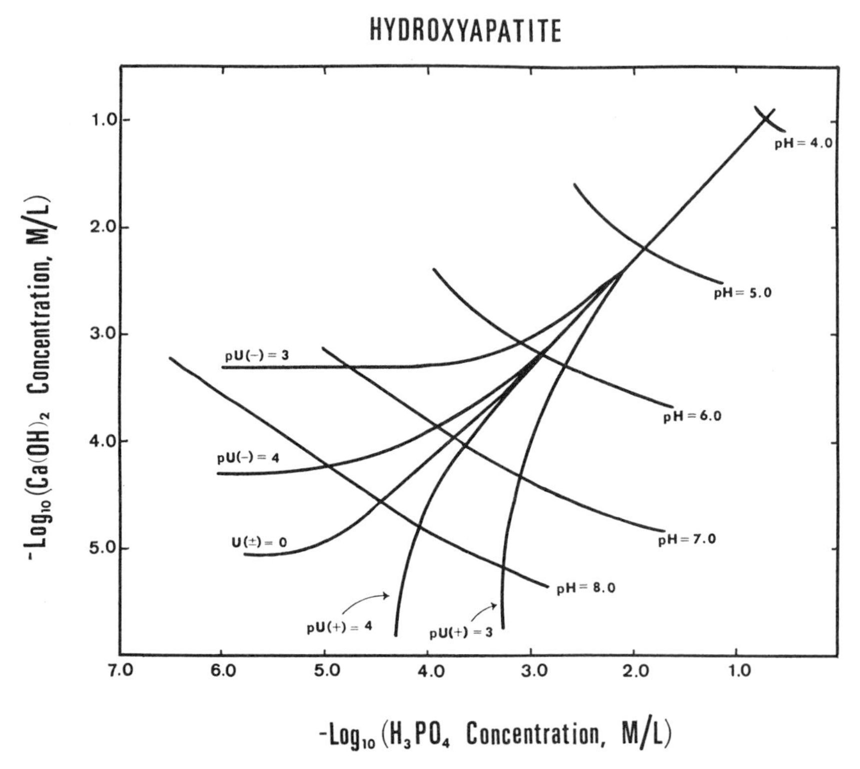

Figure 9. A repeat of Fig. 8 showing lines of constant pH.

Effects of Dilution

The positive slopes of the isotherms in Figs. 7 through 9 have an interesting effect on what happens when a saturated solution is diluted, say, by rainwater which is relatively free of solutes. The slope of an isotherm over much of its length is so nearly 1 that even after a severalfold dilution the composition point is not far from the isotherm. The degree of undersaturation of the new solution, as measured by the ion-activity product, would probably be quite low; however, as long as the composition lies on a "steep face" of the electroneutrality surface (i.e., in the pH range 4 to 10) it is likely to be near the isotherm. This is true even when the value of $U(\pm)$ of the initial solution is quite large because the new composition and the newly applicable isotherm both lie in the "steep face" for solutions with the new value of $U(\pm)$. This phenomenon can be explained on a physical basis. Dilution of the solution causes a decrease in the calcium concentration, but this is largely compensated by an increase in the OH^- and PO_4^{3-} ion concentrations so that the ion-activity product is not reduced as much as one might expect. On the other hand, dilution of a saturated solution with one that has a positive value of $U(\pm)$ is likely to lead to a final solution that is supersaturated with respect to the salt. The reason for this is that the isotherm applicable for the final solution will have been shifted farther to the right than would be the case if $U(\pm)$ of the diluent were zero or negative, and the composition of the mixture will tend to lie to the left and above the applicable isotherm (i.e., in the region of supersaturation). This type of consideration exemplifies the use of the "generalized four-component phase diagram." Detailed calculations, based on the initial and final compositions of the solutions, would also permit one to analyze the effects of dilution for any given situation, but this would be much more tedious than the use of the phase diagram, and it would be nearly impossible except when the initial and final pH's were known. These pH's would have to be known in any case where the solutions contain weak electrolytes because their contributions to $U(\pm)$ vary with pH and dilution. Carbonate, of course, is such an electrolyte; sulfate and chloride, the other principal anions contributing to $U(\pm)$ in most waters, are not affected by pH in the ranges found in most lake waters.

Application to Natural Waters

No effort is made here to test the application of Fig. 9 to specific examples of lake water because this would require knowledge that the solution was saturated with respect to hydroxyapatite. However, it should be noted that the calcium concentration usually exceeds that of phosphate by one to two orders of magnitude. This is consistent with a large negative value of $U(\pm)$ which would be supplied by the carbonate, chloride, and sulfate ions. The latter provide most of the negative charge needed to counterbalance the positive charge of the calcium ions. Large negative values of $U(\pm)$ lead to high calcium and low phosphate concentrations as seen in Fig. 9 when the pH is above about 6.

Other Calcium Phosphates

A diagram similar to Fig. 9 can be calculated for each of the other four calcium phosphates for which solubility constants are available. Because of the way the isotherms tend to superimpose, the positions of the isotherms in these diagrams would be quite similar to those in Figs. 8 and 9. They would differ considerably, however, in the locations of the pH lines in Fig. 9. The greater the relative solubility of a given salt, the more its lines of constant pH would be shifted toward the upper right in the diagram. Increasing the ionic strength with neutral salts, which tend to increase the solubility of the salt, would shift the pH lines in the same direction. Similarly, factors that enhance solubility of a given salt, such as small crystallite size, crystal impurities, or structural defects, would shift the positions of the pH lines, but they would shift the positions of the isotherms only slightly. The effects of these structural factors on the solubility of hydroxyapatite are poorly understood, but there is reason to believe that they may be quite significant (46).

Other Electroneutrality Surfaces

The electroneutrality surface shown in Fig. 7 for the system $Ca(OH)_2$—H_3-PO_4—H_2O was calculated taking into account the ion pairs of $CaHPO_4^0$ and $CaH_2PO_4^+$, but these are of relatively minor importance over most of the achievable range of concentrations. A system such as $Fe(OH)_3$—H_3PO_4—H_2O, where complexing between the cation and the phosphate is strong, would produce a grossly different surface. The shape of this surface can be calculated provided the stability constants for the complexes are known. The electroneutrality surface for such a system may also have steep faces, but these are likely to be displaced toward lower pH's. The reason for this displacement would relate to the facts that (*a*) the complexing enhances the dissociation of the protons on the phosphate groups, and (*b*) the inflections in the steep faces correspond approximately to the pH values half-way between the first and second and the second and third dissociation constants of the acid. As noted above, the system $Ca(OH)_2$—H_2CO_3—H_2O is also likely to have an electroneutrality surface with a steep face at a pH of about 8.3. The location and slope of this face may be of considerable value in describing calcium carbonate solubility. As with the phosphate system, other ions will affect the shape of the surface for the carbonate system. The presence of other ions might be handled, as in Figs. 8 and 9 for each of the calcium carbonates, by defining a quantity similar to $U(\pm)$ given here for the phosphate system. On this diagram, the lines of constant chemical potential of the component H_2CO_3 would correspond to lines of constant partial pressure of CO_2. It can be easily shown that the chemical potential of $Ca(OH)_2$ is also a constant along these lines.

Chemical Potential Diagrams

Figures 1 and 6 have considerable utility in that the quantities that are plotted are directly measurable. However, the presence of other components destroys their utility and it becomes necessary to use a phase diagram such as Fig. 9 where one can compare the directly measured quantities (calcium and phosphate concentrations and pH) with functions (isotherms and lines of

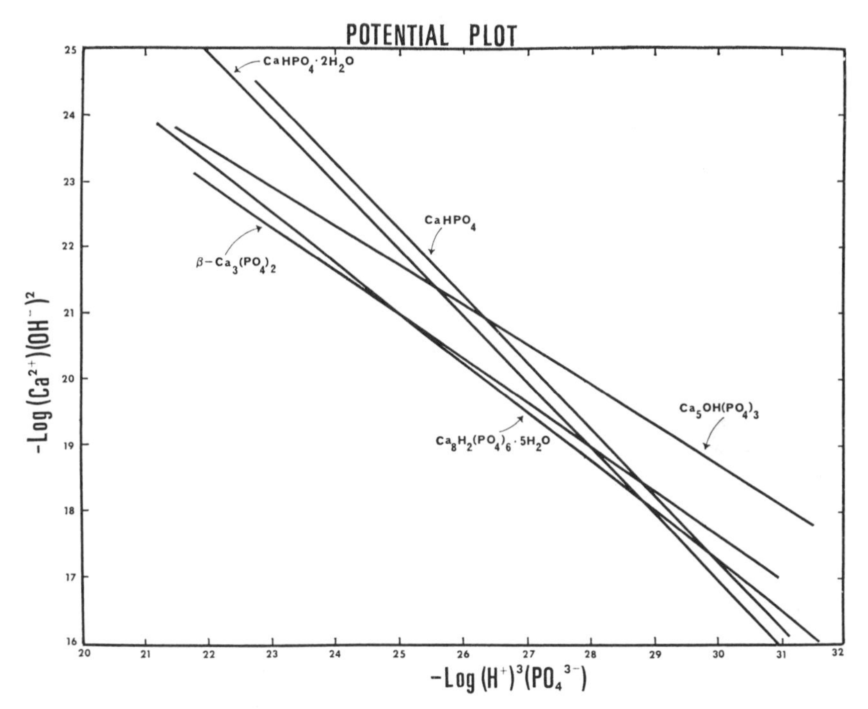

Figure 10. A "potential diagram" for calcium phosphates. The coordinates, the negative logarithms of the activities of the components $Ca(OH)_2$ and H_3PO_4, are proportional to the chemical potentials of these components.

constant pH) that have been derived from theoretical considerations. There is another type of derived diagram (Fig. 10), usually called a "potential diagram," which is quite useful in multicomponent systems. Equation (7), which is valid regardless of the number of components, can be written

$$\mu_{Ca(OH)_2} = -\tfrac{3}{5}\mu_{H_3PO_4} + \tfrac{1}{5}(\Delta F^0_{HA} + 9\mu_{H_2O}) \tag{17}$$

Substituting Eqs. (4) and (5) into (17) gives

$$\log(Ca^{2+})(OH^-)^2 = -\tfrac{3}{5}\log(H^+)^3(PO_4{}^{3-}) + K \tag{18}$$

It can be seen from Eq. (18) that the chemical potential of $Ca(OH)_2$ is linearly related to that of H_3PO_4, and that the negative inverse of the slope of the line, $\frac{5}{3}$, is the Ca/P ratio of hydroxyapatite, the equilibrating phase. Similar relationships have been derived for the other calcium phosphates (34) in a more general way. The slope and position of the line for each solid are characterized by the Ca/P ratio and its solubility product (i.e., its standard free energy of dissolution). Brushite and monetite would both have a line with a slope of 1, but the line for brushite would be displaced in the direction of higher chemical potentials.

Potential diagrams (34) have proved quite useful in attempting to identify the phase that may be saturating a given solution or a series of solutions. In addition to the usual difficulties related to uncertainties as to the formation of ion pairs and complexes and whether equilibrium has been attained, this type of plot has two shortcomings: (*a*) it is strictly a comparison of *derived* quantities (involving ion activities which can be calculated only through the use of arbitrary conventions), and (*b*) the concentrations are not shown explicitly, so that two solutions differing greatly in composition may fall on the same point in the potential plot. In spite of these shortcomings, potential plots are useful adjuncts to the other types of plots.

It is apparent in Fig. 10 that the point where two lines intersect (the singular point) has constant values for the chemical potentials of $Ca(OH)_2$ and H_3PO_4 regardless of what other components may be present. This is a direct result of Eqs. (12) and (13).

DESIGN OF SOLUBILITY EXPERIMENTS

Some rules for the design of valid solubility experiments are given in this section. One needs to make only a cursory review of the literature on the solubilities of sparingly soluble calcium phosphates to see that the results of different investigators can be in considerable disagreement. When properly carried out, solubility measurements are tedious and costly. It is important, therefore, to design the experiments carefully to avoid introducing a weakness that will destroy the value of the results. The following principles should help in this respect.

Phase Rule Considerations. The phase rule provides minimum criteria which, although frequently ignored, must be fulfilled before equilibrium experiments can be valid. The first step in the design of the experiments should be to analyze the number of degrees of freedom in terms of the components to be used in the equilibrations. Our experience has been that it is best to minimize the number of components so that the number of degrees of freedom is limited to one. In this way it is possible to plot the experimental data directly in the forms given in Figs. 1 and 6. Experimental errors can frequently be detected as points that do not fall on a smooth curve. Buffers are frequently used to control the pH, but there is another way of controlling the pH range (described later) which does not introduce additional components with the concomitant uncertainty of having specific interactions with the calcium and phosphate ions. The addition of a constant amount of another component to all of the equilibrations in a given series does not, in effect, increase the number of degrees of freedom, but it may prevent the use of the electroneutrality balance as an internal check.

Solid Phases. The first requirement is that the identity and pertinent properties of all the solid phases that can form in the system should be known. X-Ray diffraction patterns and chemical analyses are necessary but insufficient criteria for establishing the purity of a given preparation. Both are insensitive to the presence of minor phases and coatings that could markedly alter the apparent solubility of the major phase. Examination with a petrographic microscope is essential for establishing the absence of impurities and for ascertaining that the crystals are suitable for the measurements.

Criteria for Equilibrium. Strict criteria must be used in ascertaining that equilibrium is achieved in the solubility experiments. Whenever possible equilibrium should be approached from both supersaturation and undersaturation. The approach from supersaturation can usually be achieved by preequilibrating at a different temperature or by the use of a salt that is more soluble than the one under study. However, the approach to equilibrium from supersaturation does not always go smoothly; intervening precipitation of a metastable salt or a slow rate of crystal growth may preclude equilibration from supersaturation. Both of these factors appear to apply to solutions that are supersaturated with respect to hydroxyapatite. Knowledge of this type can be of practical value, however, in understanding the behavior of the system under natural conditions.

Monitoring Approach to Equilibrium. One should use the most sensitive variable to monitor the approach to equilibrium. As noted previously, changes in pH can be detected more readily than changes in calcium and phosphate concentration when the composition is on the "steep faces" of the electroneutrality surface. The same would not be true in the ranges of pH below about 4 and above about 10. In the range 4 to 10, however, relatively large changes in pH have been found to occur even when no change in calcium or phosphorus concentration could be detected. An electroneutrality surface is so easily calculated by the use of computers that it should be calculated prior to design of the experiments so that one can decide on the most sensitive variable for the purpose of monitoring. A disadvantage of the use of pH for this purpose is that it cannot be measured continuously on the solution being equilibrated because of possible contamination from the salt bridge. Continuous monitoring can be achieved, however, through the use of conductivity (1).

Equilibration Times. Constancy in the monitored quantity should be achieved over a large range of equilibration times. It is sometimes possible to get very short equilibration times by the use of leaching columns. The comparison of these data with those obtained from long-term end-over-end rotations can provide confidence that equilibrium has been achieved. It should be recognized, however, that an approach to equilibrium that requires dissolution of one or more solid phases and precipitation of another, particularly when the liquid is in the dilute range, can be exceedingly slow and can give steady-state conditions that mimic equilibrium. Lack of equilibrium can sometimes be detected by plots of the experimental data, such as Figs. 1 and 6 above, or by use of the adjustment procedure, described later, for treatment of the data.

Condition of Electroneutrality. Since this is a condition that must be fulfilled by any chemical solution, it should be used to test the validity of the data. A nonzero value indicates either (*a*) an error in analysis, (*b*) presence of impurities, or (*c*) an incorrect calculation of the phosphate ion distribution because of the use of inappropriate phosphoric acid dissociation constants, ion-pair formation constants, or expressions for calculating ion-activity coefficients.

Solid-to-Solution Ratio. If the ratio, weight of the solid to the volume of aqueous phase, is too small, the rate of approach to equilibrium may be too slow for it to be achieved in a reasonable length of time. This problem becomes especially important if difficulties with preventing growth of microorganisms are likely to limit the equilibration time. On the other hand, high solid-to-solution ratios are likely to lead to two types of difficulties: (*a*) small amounts of impurities—either as foreign ions or as secondary solid phases—may exert a disproportionate effect on the equilibrium composition, and (*b*) the amounts of desorption of calcium and phosphate ions from the surface may be large compared to that obtained from dissolution of the crystalline portion of the sample. Although in theory desorption and resorption of ions from surfaces should not affect the equilibrium constants derived from the experiment, these processes do alter the apparent stoichiometry of the reaction. These effects can be of relatively large magnitude when working with insoluble salts in the dilute region of the phase diagram.

Stoichiometry of the Reaction. Another internal check on the validity of the experimental results is the requirement that the solid should dissolve stoichiometrically. Desorption reactions, as noted above, can invalidate the use of this as a check, but these can be minimized by a preliminary equilibration with the solution to be used in the final equilibration or by the use of the leaching-type technique (26). Hydrolysis (i.e., conversion of a metastable salt under study to one that is more stable) will alter the apparent stoichiometry of the reaction. However, hydrolysis does not necessarily vitiate the results. The composition of the solution may still be adequately representative of equilibrium with respect to the metastable phase provided the latter phase is present in preponderant amounts so that its rate of dissolution exceeds greatly that of precipitation of the more stable phase.

Metastability. For the reason cited above, it is possible to study metastable equilibria provided favorable equilibration conditions are chosen and other precautions cited here are taken into account to verify that the results represent equilibria. In many phase-diagram studies it is possible to mix the components together in almost any form that will give the desired final composition and then to wait long enough for final equilibrium to be attained. Equilibria involving dilute solutions, however, tend to be slow, and one must start with the solid phase to be studied, whether it is the stable or the metastable one, and use experimental techniques that facilitate rapid equilibrium. Sufficiently large solid-to-solution ratios and continuous stirring are particularly important.

Range of Concentrations Studied. The range of compositions studied should be as large as possible within the limitations provided by the accuracy in the analyses, on the dilute side, and uncertainties in treating the data (i.e., calculation of activity coefficients) on the concentrated side. Another possible limiting factor is metastability with respect to other salts in the system. When the supersaturation with respect to the more stable salt becomes too great, then its rate of formation becomes excessive.

A broad range of concentrations is necessary if one is to distinguish between the effects related to the solubility product of the salt and those due to ion-pair formation. This can be seen by considering a potential plot such as that given in Fig. 1. Unless ion-pair formation is properly taken into account, the line would not be linear. This lack of linearity would not be detected if the range of chemical potentials which is covered by the experimental data is too narrow. It is fortunate that ion pairs such as $CaHPO_4^0$ and $CaH_2PO_4^+$, which differ by the dissociation of a hydrogen ion, will usually have their most significant concentrations in different pH ranges of the phase diagram, thus making it possible to detect their presence if the pH range is sufficiently broad. In the system $Ca(OH)_2$—H_3PO_4—H_2O it is possible to achieve a broad range by equilibrating the solid with various initial concentrations of H_3PO_4. In this way, concentration ranges of more than three orders of magnitude can be covered (37, 38, 44). In a multicomponent, buffered system, a similar range of chemical potentials can be achieved by use of a large range of Ca/P ratios in the solutions, and in the system $Ca(OH)_2$—H_2CO_3—H_2O, varying the partial pressure of the CO_2 achieves the same purpose as varying the initial concentration of H_3PO_4.

The apparent concentration of the ion pair $CaH_2PO_4^+$ is greatest in the low pH range where no difficulties are encountered with metastability or measurement of the calcium and phosphate concentrations, but the high ionic strengths of the solutions make calculation of ionic activities uncertain. The concentration of $CaHPO_4^0$ in solutions saturated with respect to $CaHPO_4 \cdot 2H_2O$ or $CaHPO_4$ is assumed to be constant throughout the range. However, its concentration becomes significant only in the higher part of the pH range when the total calcium and phosphorus concentrations are small because of the reduced solubility of the salt. Analytical errors and metastability with respect to the more basic calcium phosphates tend to be the limiting factors in this range.

The Equilibrium Model. The model to be used for treatment of the data entails selection of (*a*) dissociation constants for the acid, (*b*) stability constants for the ion pairs or complexes, (*c*) an expression for calculating individual ion activities, and (*d*) standard values for the thermodynamic properties of ions and components and their variations with temperature. Decisions have to be made as to whether the reaction is taken to be stoichiometric, how to weight the measured variables, and whether the condition of electroneutrality is to be used as a condition function.

The greatest difficulty is encountered with the choice of types of ion associations to be used in the model. The following species have been included (2, 37, 38, 44, 10–12, 23) in descriptions of calcium–phosphate equilibria: $CaHPO_4^0$, $CaH_2PO_4^+$, $CaPO_4^-$, $H_5(PO_4)_2^-$, $H_4(PO_4)_2^{2-}$, $H_3(PO_4)_2^{3-}$, $CaH_3(PO_4)_2^-$, $CaH_2(PO_4)_2^{2-}$, $CaH(PO_4)_2^{3-}$, $Ca(PO_4)_2^{4-}$, and $Ca_2H_2(PO_4)_2^0$. The relatively low concentrations of the ion pairs and other ionic associations in this system make it difficult to select the appropriate species to be used in the model. The same appears to be true of the system $Ca(OH)_2$—H_2CO_3—H_2O. The situation can best be described by some examples. The maximum ion-pair formation found (37) in the solutions saturated with respect to $CaHPO_4 \cdot 2H_2O$ in the pH range 3.6 to 7.7 was about 25% of the calcium. Hydroxyapatite (36), which is the most insoluble salt over much of this pH range, provides an example of the other extreme. Generally, formation of the ion pairs $CaHPO_4^0$ and $CaH_2PO_4^+$ in solutions saturated with respect to hydroxyapatite is not very significant, never exceeding 3.5% of the calcium. There are two qualifications: (*a*) In the low pH range (below 5), the solubility of hydroxyapatite approaches that of $CaHPO_4 \cdot 2H_2O$ and, therefore, these solutions may contain significant concentrations of the $CaHPO_4^+$ pair. (*b*) When the Ca/P ratio of the solution is high, as is the case with many natural waters, the fraction of the phosphate in the form of ion pairs may be quite large even though their absolute concentrations are low.

Possible errors in the dissociation constants for phosphoric acid and uncertainties in the use of expressions for calculating ionic activity coefficients further confound the problem of selecting ion pairs for the model because one can never be certain that some of the apparent formation of ion pairs is not attributable to errors from these sources. It should be kept in mind, also, that equilibrium constants such as those for dissociation of phosphoric acid and ion pairs were derived from systems that are likely to differ grossly in composition from the one being studied, and a totally different mathematical treatment may have been used in deriving these constants. Although some consolation may be taken from the fact that the constants are sometimes "limiting law" values, frequently they are not; and at other times the extrapolations are so great that specific ionic interactions may limit their accuracy for use in other systems. Consideration, therefore, should be given to refining the values for such constants using the solubility data from the system being studied and keeping in mind the empirical character of the calculations.

In general one should avoid the temptation provided by the fact that the fit of experimental data to a mathematical model can always be improved by including more equilibrium species in the model. However, the number of such equilibria should be kept to a minimum to be realistic. In spite of the large number of species (cited earlier) which have been suggested as having sensible concentrations in the dilute range of the system $Ca(OH)_2$—H_3PO_4—H_2O, it is found (26, 36–38, 42, 44, 53) that the pairs $CaHPO_4^0$ and $CaH_2PO_4^+$ sufficed to describe the data in the pH range from about 4 to 7. The pair $CaPO_4^-$ (12) may have to be included in the higher pH range. Considerable help in reaching the decision concerning the minimum number of ion pairs was provided by the adjustment procedure for treating data described below. There are specific reasons for questioning the reliability of constants derived for ion associations involving more than one calcium and one phosphate.

Solubility studies and potentiometric (pH) titrations are the two methods that are most commonly used to determine stability constants for ion pairs and complex ions. Reference is made later to the advantage given by having the additional constraints provided by the conditions of saturation and stoichiometric dissolution. Titration studies (11) used in studying calcium–phosphate ion association have led to the conclusion that multiple ion groups were present. However, calculations show that some of the solutions used in these studies were highly supersaturated. Before the stability constants for multiple groupings such as $CaH_3(PO_4)_2^-$ and $Ca_2H_2(PO_4)_2^0$ derived from these studies can be fully accepted, it is necessary to show that the supersaturation did not lead to formation of a colloidal precipitate that went undetected. Supersaturation in a titration experiment is likely to give more difficulty than is given by the same phenomenon in the study of the solubility of a metastable solid; the presence of a large amount of equilibrating solid tends to overcome the effects due to precipitation of small amounts of the more stable phase, and the longer equilibration times used in solubility measurements tend to reveal any changes taking place in the solution. In selecting an expression for activity coefficients, one should be aware of the deficiencies in the two that are most often used. On the one hand, there are thermodynamic strictures (28) against the use of the Debye-Hückel equation to calculate activity coefficients of ions of different valencies; on the other hand, there is an apparent physical unreality in the Davies equation (13) which requires the use of the same activity coefficient for all ions of a given valence. As a result, one should keep in mind the more conservative view that the primary objective is to treat the data in a way such that the derived constants have transferal value from a laboratory system to the one in which they are going to be applied. That is, one should seek to achieve maximum cancellation of errors through the use of a uniform treatment and compatible constants. A step has been taken in this direction by assembling an internally consistent set of thermodynamic values for the calcium phosphates which is based on solubility measurements (40).

Treatment of Data. Now that the use of computers permits sophisticated treatment of large blocks of data, there is little justification for point-by-point calculation of the equilibrium constant, followed by calculating an average value and a standard deviation of the average from the individual constants. We have found most satisfactory (37, 47) the use of a generalized least-squares procedure in which the various equilibrium expressions, the electroneutrality, and the stoichiometry of the reaction (when possible) for the model are introduced as constraints, the variations of which are to be minimized. Estimates of the experimental errors enter into the weighting scheme as part of the calculation of the errors of the derived constants. One of the most useful features of the output is the list of adjustments that would have to be made in the experimental data to minimize the deviations in the constraints imposed by the model. Consistently large adjustments reveal that the model is inadequate or that systematic errors, possibly due to lack of equilibrium, are present in the data.

One might anticipate intuitively that the more restrictive an experimental model, the more reliable are the constants that one can derive from the data. It was possible to test this hypothesis (37) through the use of the adjustment procedure. Incorporation of the condition functions based on electroneutrality and congruent dissolution into the mathematical model significantly decreased the standard deviation of the adjustment and thus increased the reliability of the derived constants over those obtained when these condition functions were omitted. Similarly, the use of the solubility data as compared to titration data has the advantage of introducing one or two extra constraints which should increase the reliability of the dissociation or stability constants derived from the data. The only constraints applicable to titration data are the dissociation equations and the electroneutrality expression.

A computer program has been reported (50) for handling titration-type data for the calculation of ion associations. This program was not designed to take into consideration the condition functions related to saturation and stoichiometric dissolution, nor was it explicitly stated whether the condition function for electroneutrality had been included.

Reporting of Results. Although it should not be necessary to make this admonition, the fact is that two types of deficiencies frequently appear in the reports on solubilities of sparingly soluble salts. The primary data, concentrations of all components and pH, are not always given along with the derived equilibrium constants and/or graphs. The absence of primary data implies a lack of esteem for the data, and it prevents the reevaluation of the data at a later time when a more appropriate model may be available. Another questionable practice is to list the logarithms of the constants for a given set of experiments rather than the constants themselves. It is not always recognized by the reader that constancy in pK is far less significant than constancy in K. Finally, some estimate should be given of the variation in the derived product that may be caused by errors or by junction potentials in the pH measurements. Very frequently the error in pH, which is a logarithmic quantity, is the greatest source of variation in the derived constants. In the adjustment procedure described above, this error is taken into account automatically, otherwise it may show up only in the actual variations of the derived equilibrium constants.

ACKNOWLEDGMENTS

This investigation was supported in part by research grant DE-00572 to the American Dental Association from the National Institute of Dental Research and is part of the dental research program conducted by the National Bureau of Standards, in cooperation with the American Dental Association; the United States Army Medical Research and Development Command; the Dental Sciences Division of the School of Aerospace Medicine, USAF; the National Institute of Dental Research; and the Veterans Administration.

REFERENCES

1. Avnimelech, Y., *Israel J. Chem.*, **6**, 375 (1968).
2. Bjerrum, N., *K. Dan. Vidensk. Selsk. Mat.-fys. Meddr.*, **31**, 1 (1958).

3. Brooks, R., L. M. Clark, and E. F. Thurston, *Philos. Trans. Roy. Soc.*, **A243,** 145 (1951).
4. Brown, W. E., *Soil Sci.*, **90,** 51 (1960).
5. Brown, W. E., *Nature*, **196,** 1048 (1962).
6. Brown, W. E., *Clin. Orthop.*, **44,** 205 (1966).
7. Brown, W. E., and J. R. Lehr, *Soil Sci. Soc. Am. Proc.*, **23,** 7 (1959).
8. Brown, W. E., J. R. Lehr, J. P. Smith, and A. W. Frazier, *J. Am. Chem. Soc.*, **79,** 5318 (1957).
9. Brown, W. E., J. P. Smith, J. R. Lehr, and A. W. Frazier, *Nature*, **196,** 1050 (1962).
10. Childs, C. W., *J. Phys. Chem.*, **73,** 2956 (1969).
11. Childs, C. W., *Inorg. Chem.*, **9,** 2465 (1970).
12. Chughtai, A., R. Marshall, and G. H. Nancollas, *J. Phys. Chem.*, **72,** 208 (1968).
13. Davies, C. W., *Ion Association*, Butterworth, London, 1962.
14. Davis, B. L., *Science*, **145,** 489 (1964).
15. Davis, B. L., and L. H. Adams, *J. Phys. Chem. Solids*, **24,** 787 (1963).
16. Deitz, V. R., H. M. Rootare, and F. G. Carpenter, *J. Colloid Sci.*, **19,** 87 (1964).
17. Dickens, B., Abstracts, American Crystallographic Association, Winter Meeting, Columbia, South Carolina, 1971, No. G-3.
18. Dickens, B., and W. E. Brown, *Inorg. Chem.*, **9,** 480 (1970).
19. Eanes, E. D., and A. S. Posner, *Trans. N.Y. Acad. Sci.*, **28,** 233 (1965).
20. Eanes, E. D., and A. S. Posner, *Calc. Tiss. Res.*, **2,** 38 (1968).
21. Eanes, E. D., I. H. Gillessen, and A. S. Posner, *Nature*, **208,** 365 (1965).
22. Elmore, K. L., and T. D. Farr, *Ind. Eng. Chem.*, **32,** 580 (1940).
23. Elmore, K. L., J. D. Hatfield, R. L. Dunn, and A. D. Jones, *J. Phys. Chem.*, **69,** 3520 (1965).
24. Frondel, C., *Am. Mineral.*, **26,** 145 (1941).
25. Frondel, C., *Am. Mineral.*, **28,** 215 (1943).
26. Garrels, R. M., and C. L. Christ, *Solutions, Minerals, and Equilibria*, Harper and Row, New York, 1965.
27. Gregory, T. M., E. C. Moreno, and W. E. Brown, *J. Res. Natl. Bur. Stand.*, **74A,** 461 (1970).
28. Guggenheim, E. A., *Thermodynamics: An Advanced Treatment for Chemists and Physicists*, Interscience, New York, 1949, p. 311.
29. Kay, M. I., R. A. Young, and A. S. Posner, *Nature*, **204,** 1050 (1964).
30. Lafon, G. M., *Geochim. Cosmochim. Acta*, **34,** 935 (1970).
31. LaMer, V. K., *J. Phys. Chem.*, **66,** 973 (1962).
32. Langmuir, D., *Geochim. Cosmochim. Acta*, **32,** 835 (1968).
33. Lippmann, F., *Naturwiss.*, **46,** 553 (1959).
34. MacGregor, J., and W. E. Brown, *Nature*, **205,** 359 (1965).
35. Mackay, A. L., "A Study of the Structure of Calcium Orthophosphate," Ph.D. Thesis, London University, London, England, 1952.
36. McDowell, H., "The Solubility of $CaHPO_4$ and Ion Pair Formation in the System $Ca(OH)_2$—H_3PO_4—H_2O as a Function of Temperature," Ph.D. Thesis, Howard University, Washington, D.C., 1968.

37. McDowell, H., W. E. Brown, and J. R. Sutter, *IADR Program and Abstracts of Papers*, 44th General Meeting, Miami, Florida, 1966, Abstract 24.
38. McDowell, H., B. M. Wallace, and W. E. Brown, *IADR Program and Abstracts of Papers*, 47th General Meeting, Houston, Texas, 1969, Abstract 340.
39. Marschner, H., *Science*, **165,** 1119 (1969).
40. Moreno, E. C., "Solubility and Thermodynamic Data for Calcium Phosphates," in *Structural Properties of Hydroxyapatite and Related Compounds*, W. E. Brown and R. A. Young (Eds.), Gordon and Breach, New York, Chapter 14, in preparation.
41. Moreno, E. C., W. E. Brown, and G. Osborn, *Soil Sci. Soc. Am. Proc.*, **24,** 99 (1960).
42. Moreno, E. C., T. M. Gregory, W. E. Brown, *J. Res. Natl. Bur. Stand.*, **70A,** 545 (1966).
43. Moreno, E. C., T. M. Gregory, and W. E. Brown, *J. Res. Natl. Bur. Stand.*, **72A,** 773 (1968).
44. Moreno, E. C., J. M. Patel, T. M. Gregory, and W. E. Brown, *IADR Program and Abstracts of Papers*, 48th General Meeting, New York, 1970, Abstract 183.
45. Newman, W. F., and M. W. Newman, *The Chemical Dynamics of Bone Mineral*, University of Chicago Press, Chicago, Illinois, 1958.
46. Patel, P. R., E. M. S. Miletta, and W. E. Brown, *IADR Program and Abstracts of Papers*, 49th General Meeting, Chicago, 1971, Abstract 807.
47. Patel, P. R., E. C. Moreno, and T. M. Gregory, *J. Res. Natl. Bur. Stand.*, **73A,** 43 (1969).
48. *Phosphates in Detergents and the Eutrophication of America's Waters*, 23rd Report, Committee on Government Operations, U.S. Government Printing Office, Washington, D.C., 1970, p. 27.
49. Rootare, H. M., V. R. Deitz, and F. G. Carpenter, *J. Colloid Sci.*, **17,** 179 (1962).
50. Sayce, I. G., *Talanta*, **15,** 1397 (1968).
51. Semenov, E. I., *Kristallogr.*, **9,** 109; *Am. Mineral*, **49,** 1151 (1964).
52. Smith, J. P., J. R. Lehr, and W. E. Brown, *Am. Mineral*, **40,** 893 (1955).
53. Stumm, W., *Equilibrium Concepts in Natural Water Systems*, American Chemical Society, Washington, D.C., 1967.
54. Vollenweider, R. A., *Water Management Research*, Organization for Economic Cooperation and Development, Directorate for Scientific Affairs, Paris, 1968.
55. Young, R. A., *Trans. N.Y. Acad. Sci.*, **29,** 949 (1967).
56. Young, R. A., and J. C. Elliott, *Arch. Oral Biol.*, **11,** 699 (1966).
57. Young, R. A., W. Van der Lugt, and J. C. Elliott, *Nature*, **223,** 729 (1969).

11

Hydrolysis of Phosphorus Compounds

C. Y. SHEN and F. W. MORGAN

Detergent and Fine Chemicals Division, Monsanto Industrial Chemicals Company, St. Louis, Missouri

The large amounts of phosphorus compounds used in detergent (13%), feeds (8%), water softener (2%), surface treatment (1%), plasticizers, gasoline additives, insecticides (1%), and other applications (3%) are often disposed of through surface water. The reactions of various phosphorus compounds with water plays the important role in the future fate of the phosphorus compounds. Most of the phosphorus compounds ultimately will convert to orthophosphates probably following the similar process to form the major phosphate minerals. The chemistry and the rate of the large-volume phosphorus compounds have been investigated for centuries, as their effects on the water quality have been recognized. This summary, however, is believed to be the first effort that discusses the hydrolysis of various phosphorus compounds.

SCOPE AND LIMITATIONS

The main purpose of this chapter is to present the chemical reactions involved and the rate determined to hydrolyze the type of phosphorus compounds that are commonly used in large-scale operations. It is not intended to cover all phosphorus compounds that can undergo chemical reactions with water. Certain phosphorus compounds, which readily react with air (compounds with —P—P— bonds as white phosphorus, and with —P—H bonds as PH_3) or water (compounds with —P—X bonds as PCl_3, and with —P—M bonds as Fe_2P) and have little chance to be found in large bodies of water, are not discussed here. The information on the hydrolysis of the —P—C— bond of phosphonates is very scarce. A recent publication indicated that the phosphonates will undergo photolytic reactions with sunlight to orthophosphates and will serve as a sole phosphorus source to aquatic plants (107). The rate of hydrolysis of the phosphonate esters is discussed in detail in this chapter.

Since the hydrolytic reaction is to break the phosphorus bond to a simple —P—O⁻ bond, some qualitative estimation is possible based on the known mechanism. For this reason, the various effects to change the reaction rates and the mechanism are described in length to facilitate estimations.

CHEMISTRY AND KINETICS

Inorganic Phosphorus Compounds

The major inorganic phosphorus compounds used on a large scale are polyphosphates as detergent builders and water conditioners, and calcium and ammonium orthophosphates and polyphosphates used in fertilizers. The quantities of phosphates from these applications are very large and usually water soluble. The importance of the reactions of various phosphates with water and the catalytic effects of the impurities present in water have been recognized and studied for more than a century (1). The large number of publications cited in a recent work (2) show the continuous interest and the complicated nature of the hydrolysis of polyphosphates and cyclic metaphosphates.

Hydrolysis Reactions of Orthophosphates. The phase equilibrium of most metal phosphates favors the hydrolytic disproportionation of the orthophosphates to the less soluble species and sometimes the undefined complexes which composition varies with pH, temperature, and the presence of other cations in the water solution (3). The classical case is the divalent metal orthophosphates that will undergo successful hydrolysis to form divalent metal hydroxyl phosphates according to the following types of reactions:

$$M(H_2PO_4)_2 + YH_2O \leftrightarrows MHPO_4 + H_3PO_4 + YH_2O \quad (1)$$

$$10MHPO_4 + ZH_2O \leftrightarrows M_{10}(PO_4)_6(OH)_2 + 4H_3PO_4 + (Z - 2)(H_2O) \quad (2)$$

The well-known examples are the calcium or strontium hydroxylapatite (4). This is probably one of the mechanisms related to bone formation (5), and the various naturally occurring apatites. Further sorption and substitution of atoms within the crystal lattice probably are the causes of the variable composition of the apatites (4).

In the presence of aluminum and ferric ions, the hydrolysis reactions of orthophosphates can form many complicated phosphates. Many soil chemists have attributed these reactions to be the cause of insolubilization of phosphates. The taranakites, $H_6K_3Al(PO_4)_8 \cdot 18H_2O$, which have been found in minerals and are relatively insoluble, can be easily formed from the hydrolyzed acidic phosphate components and aluminum minerals in the soil (6, 7).

Low Molecular Weight Polyphosphates. The low molecular weight polyphosphates are comprised of pyro-, tripoly-, and tetrapolyphosphates. They differ from the longer chain polyphosphates in that their hydrolysis reactions are primarily breaking the $\text{—}\overset{|}{\underset{|}{P}}\text{—O—}\overset{|}{\underset{|}{P}}\text{—}$ bond without the formation of cyclic metaphosphates. Pyro- and tripolyphosphate are the major ingredients used in built detergents, and are found in many biological compounds such as adenosine pyro- and tripolyphosphates. Considerable studies on the hydrolysis rates of pyro-, (2, 8–19), tripoly- (2, 8–11, 14, 15, 19–27), and tetrapolyphosphates (2, 9, 10, 28–31) have been reported.

In dilute water solutions, the reaction is first order at a given pH for most of the course of reaction. The hydrolysis involves mainly attack on one end phosphorus–oxygen bridge exclusively, as illustrated by the following diagram where the phosphorus–oxygen skeletons are used to represent the ions concerned.

$$\underset{\text{Tetra}}{\text{—P—O—P—O—P—O—P—}} \rightarrow \underset{\text{Tri}}{\text{P—O—P—O—P}} + \underset{\text{Ortho}}{\text{P}} \tag{3}$$

$$\downarrow$$

$$\underset{\text{Pyro}}{\text{P—O—P}} + \underset{\text{Ortho}}{\text{P}}$$

$$\downarrow$$

$$\underset{\text{Ortho}}{\text{P}}$$

Although the reaction follows the first-order relationship in more concentrated solutions (23), there are sufficient evidences based on the hydrolyzed product analysis of the large increase in the pyro species that the reaction probably proceeds by more than one route. Splitting of tetraphosphate to two pyrophosphates was suggested (28, 30). A tripolyphospate complex that could combine with itself to produce pyrophosphates was another possibility (23). This hypothesis was supported by much lower activation energies in concentrated solution. In dehydrating the solid sodium tripolyphosphate hexahydrate, the degradation reaction seems to follow second-order kinetics (20).

The apparent first-order reaction rate constants for hydrolytic degradation of various sodium polyphosphates are shown in Fig. 1 (2). Other factors that affect the hydrolysis rates are summarized below.

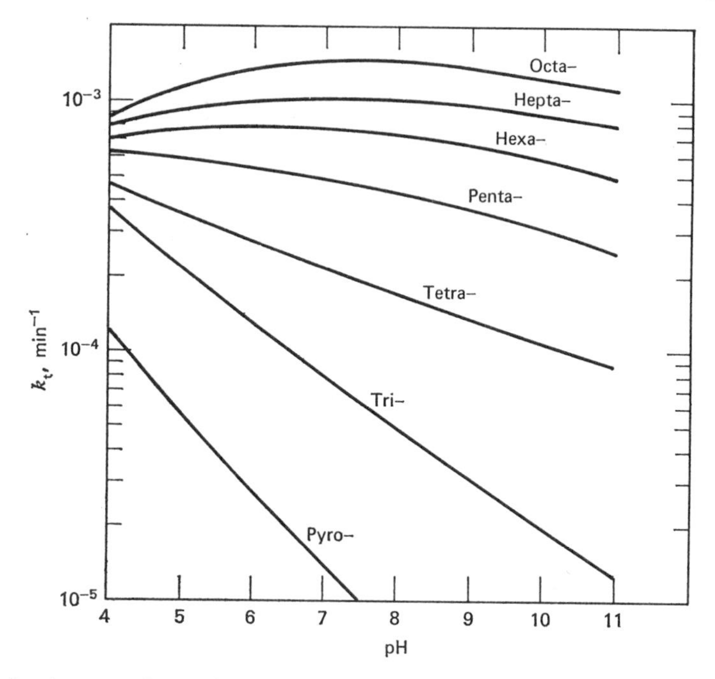

Figure 1. Apparent first-order hydrolytic degradation rate constants of sodium polyphosphate at 60°C (34).

Temperature. The hydrolysis rate increases with the increase of temperature, but the rate of increase is not uniformly dependent on temperature if the pH is not held constant (61). The activation energy appears to change from about 23 kcal/mole at pH = 10 to about 28 at pH = 4 for sodium pyro- and tripolyphosphates (8, 13). The activation energy for tetrapolyphosphate, however, changes much less with pH and remains nearly constant at 25 kcal/mole from pH = 11 to 7 (2). The range of activation energies of 20 to 25 kcal/mole is about the same as that required for breaking a —P—OP— bond. When the concentration of the phosphate solution increases, the activation energies are decreased by the amount of about 2 to 3 kcal/mole. It is reasonable to assume that this decrease is caused by the interaction of phosphate polyanions with the increased sodium ion concentration and hydration water molecules (23).

Foreign Ions. Metallic cations show a definite catalytic effect on the degradation of polyphosphates (8, 9, 19, 23, 27, 32, 33). This catalytic effect may be interpreted by the interaction of phosphate polyanions and complexed cations as detected by the P^{31} NMR shift (34). The exact effect, however, is difficult to measure or predict because this interaction is influenced by the presence of one or more hydration layers separating complexed cations and anions. Qualitatively, the observed catalytic effect is roughly proportional to the charges of the cations and to the logarithm of the cation concentrations (23). The anions were measured to have little effects on the degradation rates (23).

Enzymes. The enzymatic hydrolysis of chain and ring phosphates can be extremely fast, as much as 10^6 faster than the rate without enzyme (25, 35–37). The phosphatase catalytic action is specific and is greatly affected by metal ions and pH's. Many microorganisms, however, can convert simple orthophosphates to at least four different types of polyphosphates. Recent studies showed that these polyphosphates are playing important roles in the phosphorus metabolism in the algal cells (38).

High Molecular Weight Polyphosphates. Studies on the hydrolytic degradation of penta- (2, 39–41), hexa- (2, 10, 40, 41), hepta- (2, 40), octa- (2, 40), and short-chain polyphosphates are relatively scarce in comparison with polyphosphates of high chain lengths (2, 9–11, 15, 19, 27, 32, 33, 35, 37, 39–53). The mechanism involved in hydrolysis of long-chain polyphosphate is more complete and appears to proceed with the four following routes: (*a*) hydrolytic session in the interior of the chain to produce shorter chains; (*b*) clipping of the end group of the chain to form orthophosphate and a phosphate chain one phosphorus atom shorter; (*c*) splitting off a trimetaphosphate ring from an end of the chain; and (*d*) splitting out of trimetaphosphate from the interior of the chain. Since the formation of the trimetaphosphate by the routes (*c*) and (*d*) are not easily distinguishable, many investigators combined these two routes into one. Further simplification leads to three rate constants: k_t, overall disappearance of a phosphate species; k_e, clipping of the end group to form orthophosphate; and k_m, the formation of trimetaphosphate. All three rate constants for sodium polyphosphates appear to follow first-order kinetics (2). The formation of the orthophosphate could be zero order (27) if the orthophosphate is only produced from clipping the end of the chain.

Under first-order kinetics:

$$k_t = k_e + k_m \tag{4}$$

and

$$-dc/dt = k_t C = (k_e + k_m)C \tag{5}$$

where C is the concentration of the phosphate species being studied. The various factors affecting the rates are discussed in the following subsections.

pH. Examining the published data (47), it can be rationalized that polyphosphates undergo specific acid catalysis and are not subject to attack by nucleophiles to form the end groups from hydrolysis. The trimetaphosphates are formed from the rearrangement of the chain from specific site-binding and are usually proportional to the concentration of the polyphosphates. The trimetaphosphates are not formed from hydrolysis as shown by degradation carried in O^{18}-labeled water (27). The effect of the pH on the two reaction rates is different and causes the maximum rate at pH = 7 for longer chain polyphosphates (Fig. 1). The increase in the rate of the formation of orthophosphate at the chain end is higher from pH 11 to 7 than from pH 7 to 4 as the weakly acidic terminal groups of the chain are converted into undissociated —P(|)—OH groups. On the other hand, the increase in the rate of formation of trimetaphosphate is higher from pH 4 to 7 than from pH 7 to 11, because at pH $\simeq$ 4 the nonterminal strong acidic groups of the polyphosphates are converted to undissociated —P(|)—OH groups, which prevents the site-binding. Estimated rates of k_m and k_e consistent with the longest chain phosphate degradation rates are shown in Fig. 2. The

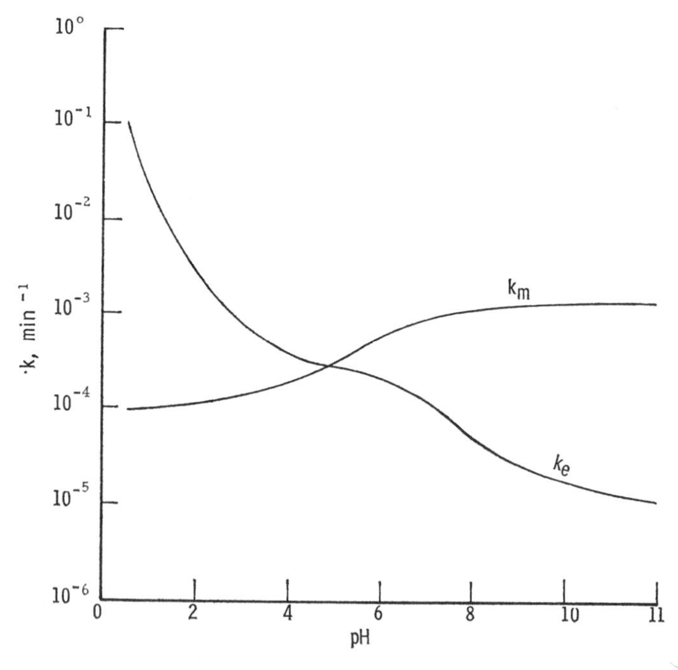

Figure 2. Effect of pH on the apparent rates of split-out trimetaphosphate (k_m) and end-group clipping (k_e) from long-chain polyphosphates at 60°C.

increase in the orthophosphate–trimetaphosphate ratio with a drop in pH is in good agreement with observed facts (47, 49).

Chain Length. The overall polyphosphate degradation rate increases with the increase of phosphate chain length and approaches a limiting value after the chain length is above about ten (2, 27). The increase between tripoly and tetrapoly is much greater than the increase of the rate constant between hepta- and octapolyphosphates under similar conditions as shown in Fig. 1.

Temperature. The activation energies of the three rate constants k_t, k_m, and k_e behave differently with pH. There seems a slight increase in the activation energies for k_t from about 20 at pH 4 to 7 to about 27 at pH 11. The activation energies for k_m increase with pH (from 16 at pH 4 to 25 at pH 11) and for k_e decrease with pH (from 20 at pH 4 to 12 at pH 11) for the sodium phosphate system (2). In other words, as the temperature increases, the proportion of trimetaphosphate to orthophosphate also increases.

Catalysis. Cations increases the degradation rate but the formation of trimetaphosphate is accelerated more than the clipping of the ends (27, 32, 33). Since the hydrogen ion inhibits the metal ion catalysis by forming undissociated —P—OH, it is suggested that the metal catalytic effect applies only to reaction occurring at the ends of the chain (32). A better explanation probably is from the chelate formation thus enhancing the reactivity with other phosphate groups (27). Certain ligands forming chelates with the metal ion can further increase the catalytic activity and selectivity (41). The cause of the change in selectivity with different ligands has not been determined. In enzymatic degradation of polyphosphate, a large influence of the metal salt was observed (35). Similarly, in soil, the adsorption of the polyphosphate on the clay particle to exhibit some type of complex increased the degradation rate (26).

Cyclic Metaphosphates. The most common cyclic metaphosphates are tri- and tetrametaphosphates which can be derived from directly heating some metal dihydrogen orthophosphates (54). Penta- and hexametaphosphates were isolated from the lithium or sodium phosphate glasses (55, 56). The hydrolysis of trimeta- and tetrametaphosphate will give corresponding tripoly- and tetrapolyphosphates (15). The polyphosphates from large cyclic metaphosphates are usually degraded to lower phosphate before all metaphosphate has reacted. The rates are catalyzed by both acid and base and are slowest at pH from 6 to 10. Metal ions catalyze the hydrolysis reaction (58) and are used as activators in enzymatic hydrolysis (60). The rates are somewhat slower than the corresponding polyphosphates in acidic solution but convert very rapidly in highly alkaline solution (15, 58). The large-ring metaphosphates are very resistant to hydrolysis in the pH range from 4 to

11 (56). The half-life for $(NaPO_3)_6$ in 0.1 *N* NaOH at 25° is about 1000 hr while that for $(NaPO_3)_3$ is about 4.5 hr (27). The activation energies for the trimeta and tetrameta are around 15 to 20 kcal/mole (46, 57, 58) but drops to 6 to 9 kcal/mole for the hexametaphosphate (56).

Pyrophosphites and Isohypophosphites. Pyrophosphites $\overset{3}{\text{—P}}\text{—O—}\overset{3}{\text{P}}\text{—}$ and isohypophosphites $\overset{3}{\text{—P}}\text{—O—}\overset{5}{\text{P}}\text{—}$ in aqueous solution are hydrolyzed to orthophosphites and orthophosphates and are catalyzed by both Brønsted acids and nucleophiles (2, 61, 62). The metal ion catalyzes the hydrolysis rate for isohypophosphite (61) more than the pyrophosphite (63). This is due to the fact that the pyrophosphite is a weak complexing agent. Although the isohypophosphite does not seem to form intermediates, the pyrophosphite often reacts with water to form an orange-yellow material that will gradually decompose and give dangerous phosphines (PH_3 and P_2H_4) (4). In general the pyrophosphites and isohypophosphites are several orders less stable than the pyrophosphate at a given temperature and pH (61, 63).

Cyclic Phosphimates and Imidophosphates. When phosphonitrilic chlorides are hydrolyzed, cyclic phosphimates are formed (4). The phosphimates then hydrolyze following a complex route (60–67) though the ultimate product from hydrolysis of these phosphorus–nitrogen compounds is orthophosphate. An example of this type of hydrolysis is shown in Fig. 3. The main sequence of hydrolysis follows the solid arrows. The reaction appears to hydrolyze the imide group to produce amide and phosphate groups that combine immediately to form —P—O—P linkage and ammonia

$$\text{—P—}\overset{\text{H}}{\text{N}}\text{—P—} + \text{HOH} \rightarrow \text{—P—NH}_2 + \text{HO—P—} \rightarrow \text{—P—O—P—} + \text{NH}_3 \tag{6}$$

The amide condensation can be observed by acidulation of amidotriphosphate (68, 69). The trimetaphosphate is formed very quickly at pH < 7 at room temperature.

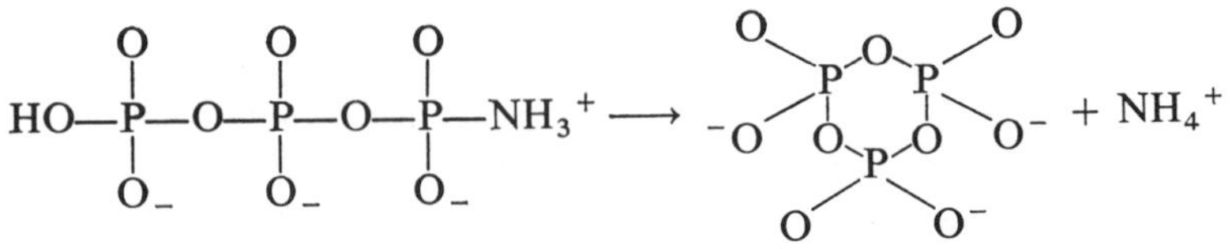

$$\text{HO—}\underset{\text{O}^-}{\overset{\text{O}}{\text{P}}}\text{—O—}\underset{\text{O}^-}{\overset{\text{O}}{\text{P}}}\text{—O—}\underset{\text{O}^-}{\overset{\text{O}}{\text{P}}}\text{—NH}_3^+ \longrightarrow (P_3O_9)^{3-} \text{ (cyclic trimetaphosphate)} + NH_4^+$$

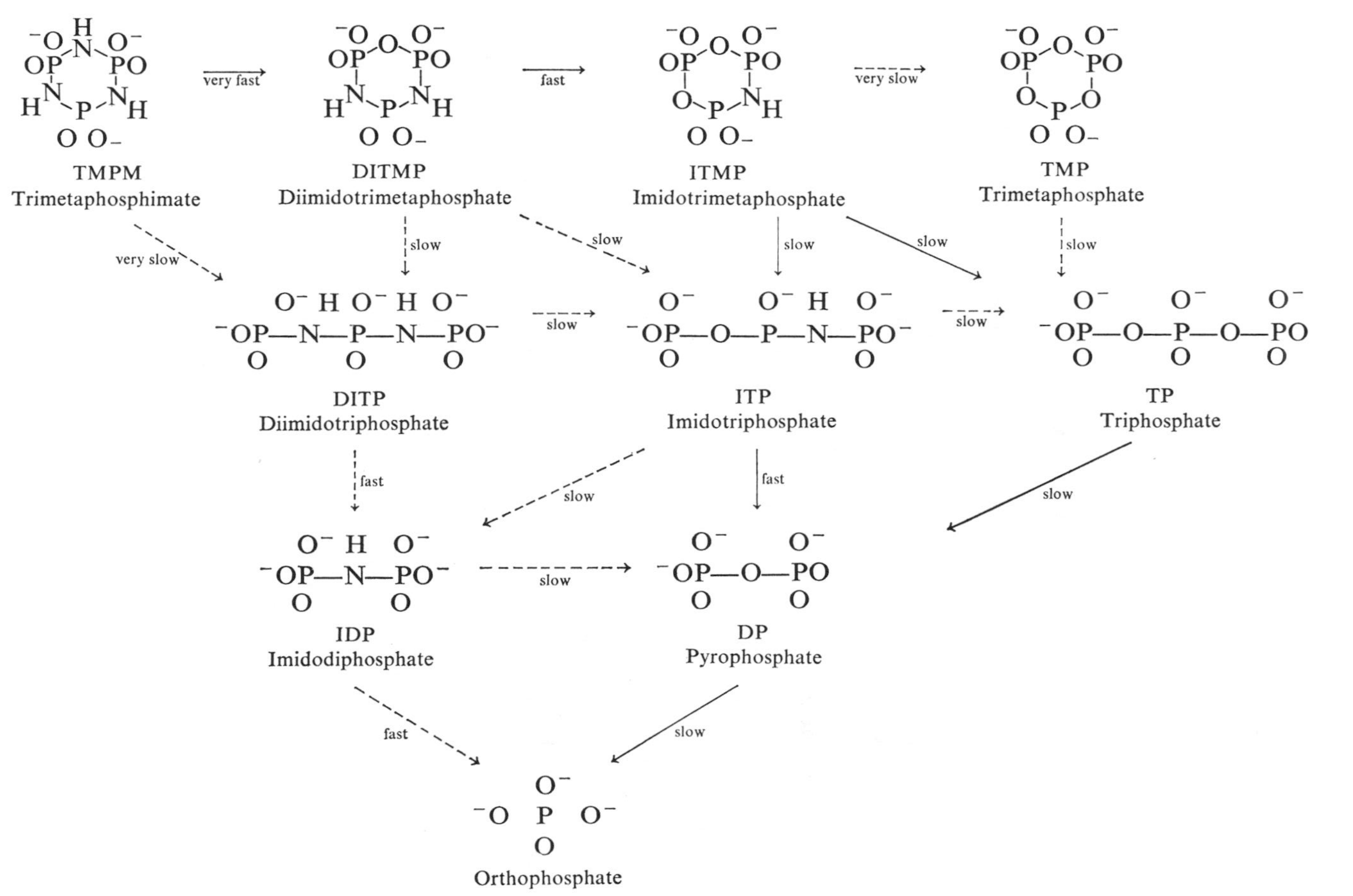

Figure 3. Hydrolysates from trimetaphosphimates (69).

The imidophosphate degrades rapidly in slightly acidic solution and is less stable than corresponding oxy links in pH 10 to 11 solution (68, 90). Imidotrimetaphosphate behaves exceptionally at pH 3.5 where the oxy links break slightly faster than the imide link.

Organic Phosphorus Compounds

Although organophosphorus compounds consume only about 4% of the elemental phosphorus produced annually in the United States, they are used widely as plasticizers, insecticides, gasoline and oil additives, flotation agents, stabilizers and antioxidants, and surfactants. A total of 350 million lb was produced in 1966 with volumes decreasing in the order above (71). Most of these compounds eventually contact plants, animals, soil, or groundwater. Having entered the natural environment, organophosphorus compounds are degraded by biological and/or chemical action to orthophosphate, the ultimate degradation product.

Chemical degradation proceeds primarily by hydrolysis although oxidation or isomerization of sulfur-containing phosphorus compounds may occur prior to or simultaneously with hydrolysis. This section is concerned chiefly with hydrolytic degradation of organophosphorus compounds that may have environmental application and with the modes of hydrolysis of intermediates created prior to the eventual production of orthophosphate.

Nomenclature and Types of Organic Phosphorus Compounds. The following types of compounds (as shown on p. 251) include most materials commonly used in agriculture and industry and encompass materials to be discussed. Substituents are included in the names of the compounds with an indication of the atom to which they are attached, for example, *N,N*-diethyl *O*-methyl *O*-*i*-propyl phosphoramidate. The atom of attachment may be eliminated,

$$(C_2H_5)_2N\text{—}P(=O)(OCH_3)(OC_3H_7)$$

however, when a compound can be unambiguously named. Anhydrides are

$$CH_3O\text{—}P(=S)(OC_3H_7)(OC_3H_7)$$

Methyl di-*n*-propyl phosphorothionate

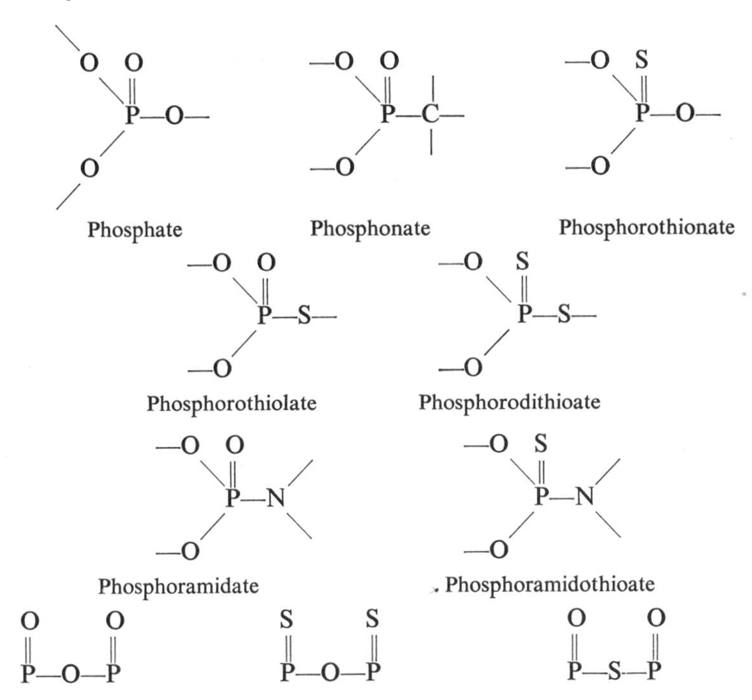

Phosphoric anhydride Thionophosphoric anhydride Thiolophosphoric anhydride

named as indicated above if they are symmetrical. Unsymmetrical anhydrides are named for the parent pyrophosphate (72).

Toxic Phosphate Esters. Certain phosphorus compounds containing one or more ester groups are being used as insecticides and fumigants. In addition, several war gases containing phosphorus have been developed. The toxic action of certain trisubstituted phosphate esters arises from their ability to inhibit transmission of nerve impulses. Inhibition results from phosphorylation of essential enzymes, called cholinesterases, according to the following equation:

$$EH + (RO)_2P(O)X \rightarrow (RO)_2POE + HX \tag{8}$$

That an actual chemical bond is formed to the enzyme (EH) has been established by several types of experimental investigations (73, 74). Phosphorylation can be considered to be an ordinary electrophilic attack, and groups that remove electron density from a phosphorus atom of an ester group would be expected to increase toxicity. In the series

$$(C_2H_5O)_2P(O)O\text{—}C_6H_4\text{—}y$$

where y is p-NH_2, p-Cl, o-Cl, o-NO_2, and p-NO_2, pI_{50} values of 4.0, 3.4, 4.2, 6.5, and 7.7, respectively, have been observed (75, 76). The pI_{50} (negative logarithm of the moles of compound needed in a given time to reduce the cholinesterase activity to 50% of its original value) roughly parallels electronegativity of the resulting phenol unit and parallels the acidity of the phenolic acid. The electronegativity considerations above also predict lower hydrolytic stability for toxic esters containing electron-withdrawing substituents. Straight-line relationships between hydrolysis constants and rate constants for reaction of inhibitors with enzymes in fact are observed (77).

Organophosphorus insecticides are the most widely used toxic esters and in general are readily hydrolyzed in basic solution. Most insecticides have the general structure:

$$(RO)_2P{=}O \text{ (or S)} \quad -O\text{(or S)}-X$$

where R is an alkyl group and X is an organic radical. Hydrolysis normally results in cleavage of the P—O—X linkage to form nontoxic products,

$$(RO)_2P(=O(S))-OH \quad \text{and} \quad HOX$$

The hydrolysis rates of insecticides approved for use in Great Britain were studied (78) and are listed in Table 1. Conditions close to expected field conditions were used (pH buffered at 6.0) except that a temperature of 70°C was used to allow measurable reaction within 4 days. Esters of the type

$$R'-O-\overset{\overset{\displaystyle S}{\|}}{P}-(OR)_2$$

are hydrolyzed largely to R'OH and

$$HO\overset{\overset{\displaystyle S}{\|}}{P}(OR)_2$$

Oxidizing agents and biological activity can convert $\gt P{=}S$ compounds to $\gt P{=}O$ compounds, which are usually more toxic and more readily hydrolyzed since sulfur is less electronegative than oxygen. Compounds containing

$$R-O-\overset{\overset{\displaystyle S}{\|}}{P}\lt$$

TABLE I Half-Lives of Insecticides at 70°C (Ethanol–Water, 20:80, Buffered at pH 6.0; Concentration 6 μg/ml)

Insecticide	Structure	Half-Life (hr)
1. Parathion methyl	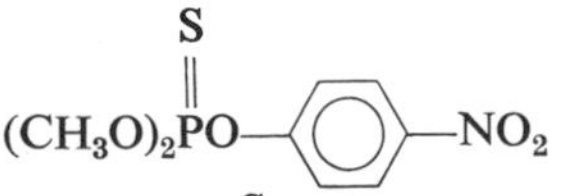 $(CH_3O)_2P(=S)O-C_6H_4-NO_2$	8.4
2. Parathion	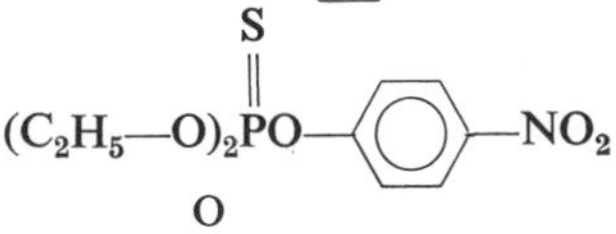 $(C_2H_5-O)_2P(=S)O-C_6H_4-NO_2$	43.0
3. Parathion-*O*-analog	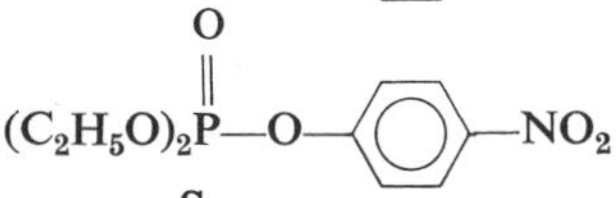 $(C_2H_5O)_2P(=O)-O-C_6H_4-NO_2$	28.0
4. Thiometon	$(CH_3O)_2P(=S)-SCH_2CH_2SC_2H_5$	17.0
5. Disulfoton	$(C_2H_5O)_2P(=S)-SCH_2CH_2SC_2H_5$	32.0
6. Demeton-*S*-methyl	$(CH_3O)_2P(=O)-SCH_2CH_2SC_2H_5$	7.6
7. Oxydemeton methyl	$(CH_3O)_2P(=O)-SCH_2CH_2S(\rightarrow O)C_2H_5$	17.1
8. Sulfone	$(CH_3O)_2P(=O)-SCH_2CH_2SO_2C_2H_5$	5.1
9. Demeton-*S*	$(C_2H_5O)_2P(=O)SCH_2SC_2H_5$	18.0
10. Phorate	$(C_2H_5O)_2P(=S)SCH_2SC_2H_5$	1.75
11. Phorate-*O*-analog	$(C_2H_5O)_2P(=O)SCH_2SC_2H_5$	0.5
12. Thiometon	$(CH_3O)_2P(=S)SCH_2CH_2SC_2H_5$	17.0
13. Demeton-*S*-methyl	$(CH_3O)_2P(=O)SCH_2CH_2SC_2H_5$	7.6

TABLE 1 (cont.)

Insecticide	Structure	Half-Life (hr)
14. Thionazin	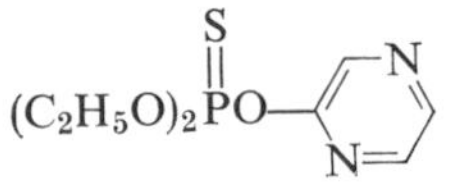 $(C_2H_5O)_2\overset{S}{\overset{\parallel}{P}}O$–(pyrazinyl)	29.2
15. Thionazin-*O*-analog	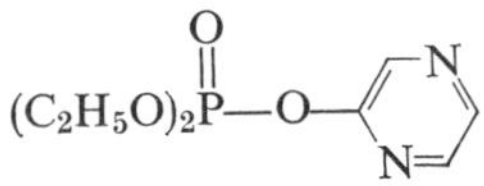 $(C_2H_5O)_2\overset{O}{\overset{\parallel}{P}}$—O–(pyrazinyl)	8.2
16. Malathion	$CH_3O\overset{S}{\overset{\parallel}{P}}$—$SCHCO_2C_2H_5$ with $CH_2CO_2C_2H_5$ on the CH	7.8
17. Malathion-*O*-analog	$(CH_3O)_2\overset{O}{\overset{\parallel}{P}}$—$SCHCO_2C_2H_5$ with $CH_2CO_2C_2H_5$ on the CH	7.0
Insecticide	Structure	Half-Life (hr)

may, in addition, undergo isomerization to more toxic materials containing

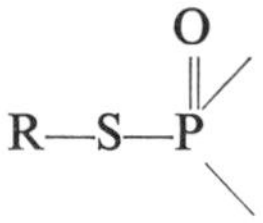

$$R\text{—}S\text{—}\overset{O}{\overset{\parallel}{P}}\!\!<$$

R—S— groups are more acidic and hence form better leaving groups than R—O— groups. It can also be seen from Table 1 that hydrolytic stability increases with complexity of alkyl substituents (parathion methyl versus parathion).

Among the toxic phosphates not listed in Table 1 are the pyrophosphates. Tetraethyl pyrophosphate is water soluble and hydrolyzes to nontoxic diethyl phosphate in aqueous solution according to the following reaction:

$$(C_2H_5O)_2\overset{O}{\overset{\parallel}{P}}\text{—}O\text{—}\overset{O}{\overset{\parallel}{P}}(OC_2H_5)_2 + H_2O \rightarrow 2(C_2H_5O)_2\overset{O}{\overset{\parallel}{P}}OH \qquad (9)$$

The reaction in water is first order in ester and has rate constants at 25 and 38°C of 1.70×10^{-3} and $3.50 \times 10^{-3}\ min^{-1}$, respectively (79). Alkyl halophosphates,

$$(RO)_2\overset{\displaystyle O}{\overset{\|}{P}}X$$

and alkylhalophosphonates,

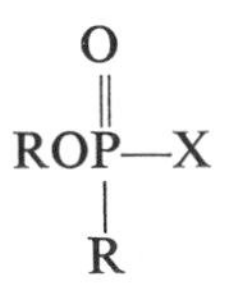

are highly toxic and members of these classes have been developed for war gases. Hydrolysis produces dialkylphosphates and monoalkyl phosphonates respectively, both of which are of low toxicity. Table 2 summarizes fluoro- and cyanophosphate hydrolyses.

Nontoxic Phosphate Esters. Nontoxic phosphate esters may be introduced into the natural environment by hydrolysis of insecticides or by application as plasticizers, surfactants, and so on. Esters of three degrees of substitution are possible, monoesters $ROP(O)(OH)_2$, diesters $(RO)_2P(O)OH$, and triesters $(RO)_3PO$. In each case hydrolysis is relatively slow compared to the hydrolysis of esters (93) considered in the previous section and varies in mechanism and position of bond cleavage from one class of esters to another. Table 3 compares the hydrolysis rates and position of bond cleavage for the methyl ester for all three classes.

Triesters. The most reactive of the orthophosphate esters are the triesters. Bond cleavage occurs predominantly in basic solution between phosphorus and oxygen in an overall second-order reaction (86). At low pH, however, water attacks the ester group with carbon–oxygen bond cleavage in a pseudo-first-order reaction (86). Two mechanisms have been postulated for the hydrolysis of triesters (92): (*a*) a nucleophilic attack similar to

TABLE 2 Second-Order Rate Constants for the Reaction of Alkyl Halophosphates and Alkyl Halophosphonates with Aqueous Base ($R_1R_2R_3P—O$)

Number	R_1	R_2	R_3	Temperature (°C)	k [l/(mole)(sec)]	E (kcal/mole)	Ref.
1	CH_3O	CH_3O	F	25	1.82×10	11.6	80
2	C_2H_5O	C_2H_5O	F	25	1.8	—	81
3	i-C_3H_7O	i-C_3H_7O	F	25	1	—	82
4	CH_3O	CH_3	F	25	1.06×10^2	10.5	83
5	C_2H_5O	CH_3	F	25	6.07×10	11.2	83
6	i-C_3H_7O	CH_3	F	25	2.58×10	9.1	83
7	$(CH_3)_2N$	C_2H_5O	CN	25	7.49	10.1	84
8	i-C_3H_7NH	i-C_3H_7NH	F	25.1	8.1×10^{-1}	11.2	85

TABLE 3 Rate Constants for Methyl Esters of Phosphoric Acid

Compound	Species	Cleavage	Temperature (°C)	k	Conditions	Ref.
H_3PO_4	$H_2PO_4^{\ominus}$	P—O	100	4.03×10^{-6} sec^{-1}	—	88
	H_3PO_4	P—O	100	1.28×10^{-6} sec^{-1}	—	88
	$H_4PO_4^{\oplus}$	P—O	100	5.45×10^{-6} l/(mole)(sec)	—	88
$CH_3OPO_3H_2$	$CH_3OPO_3H^{\ominus}$	P—O	100	8.23×10^{-6} sec^{-1}	$\mu = 0$	89
	$CH_3OPO_3H_2$	C—O	100	0.50×10^{-6} sec^{-1}	$\mu = 0$	89
	$CH_3OPO_3H_3^{\oplus}$	C—O	100	2.00×10^{-6} l/(mole)(sec)	$\mu = 0$	89
	$CH_3OPO_3H_3^{\oplus}$	P—O	100	1.08×10^{-6} l/(mole)(sec)	$\mu = 0$	89
$(CH_3O)_2PO_2H$	$[OH^{\ominus}][(CH_3O)_2PO_2^{\ominus}]$	P—O	125	$\sim 0.6 \times 10^{-6}$ l/(mole)(sec)	$\mu = 1.16$	90
	$[OH^{\ominus}][(CH_3O)_2PO_2^{\ominus}]$	C—O	125	$\sim 5.4 \times 10^{-6}$ l/(mole)(sec)	$\mu = 1.16$	90
	$(CH_3O)_2PO_2H$	C—O	100	3.3×10^{-6} sec^{-1}	$\mu = 0$	91
	$(CH_3O)_2PO_2H$	P—O	100	0.9×10^{-6} sec^{-1}	$\mu = 0$	91
	$(CH_3O)_2PO_2H_2^{\oplus}$	C—O	100	0.91×10^{-6} l/(mole)(sec)	$\mu = 0$	91
	$(CH_3O)_2PO_2H_2^{\oplus}$	P—O	100	0.11×10^{-6} l/(mole)(sec)	$\mu = 0$	91
$(CH_3O)_3PO$	$[OH^{\ominus}][(CH_3O)_3PO]$	P—O	35	3.3×10^{-4} l/(mole)(sec)	—	87

S_{N^2} reactions

$$OH^- + \overset{O\ \ OR}{\underset{OR\ \ \ OR}{P}} \rightarrow HO^-\cdots\overset{O}{\underset{OR\ \ OR}{P}}\cdots OR \rightarrow (RO)_2PO_2^- + HOR \quad (10)$$

and (*b*) an addition followed by elimination,

$$RO_3PO + OH^- \rightarrow RO\text{—}\overset{O^-}{\underset{OH\ OR}{P}}\text{—}OR \rightarrow RO\text{—}\overset{OH}{\underset{O^-\ OR}{P}}\text{—}OR \rightarrow (RO)_2PO_2^- + HOR \quad (11)$$

Most evidence supports mechanism (*a*); however, the two are indistinguishable if addition and elimination are extremely rapid.

Typical hydrolysis rates for nontoxic triesters are $k = 3.36 \times 10^{-4}$ l/(mole)(sec) at 35°C, E_{act} (activation energy) = 16.2 kcal/mole, for $(CH_3O)_3PO$ (87), and 3.73×10^{-5} l/(mole)(sec) at 37.5°C, E_{act} = 15.0 kcal/mole for $(C_2H_5O)_3PO$ (94).

Diesters. Diesters exist in solution as anions (p$K \sim 1.5$) except at low pH's (91, 92) and as a class are least reactive toward anionic nucleophiles (91). Cyclic five-membered phosphates, (for example, ethylene phosphate

$$\begin{array}{l} H_2C\text{—}O \\ \ \ | \qquad\quad \diagdown \\ \ \ | \qquad\qquad P(O)OH \\ \ \ | \qquad\quad \diagup \\ H_2C\text{—}O \end{array}$$

and certain diaryl phosphates (86) are exceptions, being considerably more reactive. The anions react slowly with hydroxide ion at 125°C and carbon–oxygen bond fission predominates. In acidic solution, the undissociated diester and its conjugate acid are attacked by water with carbon–oxygen bond cleavage. The monoanion of di-*p*-nitrophenyl phosphate reacts with hydroxide ion at 25°C with a second-order rate constant of 6×10^{-6} l/(mole)(sec) in 20% ethanol–80% water (94).

Cyclic five-membered esters of phosphoric acid are hydrolyzed in base or acid millions of times faster than their acyclic analogs (95). Ring strain has been proposed to account for acceleration of the opening of cyclic phosphates but does not satisfactorily explain similar enhancement of elimination of a methoxyl group from methyl ethylene phosphate (Fig. 5*a*) without ring opening. Westheimer (96) has proposed that hydrolysis external to the ring proceeds by a mechanism involving "pseudorotation" in which water adds

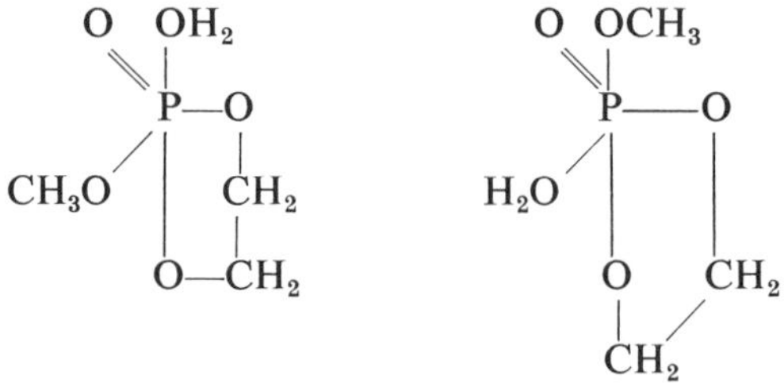

(a) (b)

Figure 4

to phosphorus to form a trigonal-bipyramidal intermediate with the water molecule and one oxygen atom from ethylene glycol in apical positions (Fig. 4*a*). The intermediate rearranges, by undergoing changes in bond angles, to a new configuration in which the methoxyl group is located at one apex (Fig. 4*b*). Methanol is subsequently eliminated with the destruction of the five-coordinate species. Ring opening can, however, proceed without "pseudorotation." Ring strain is reduced in the intermediate since a 90° O—P—O angle is possible if the ring occupies equatorial and apical positions. Methyl propyl phosphonate (Fig. 5*b*), on the other hand, is hydrolyzed rapidly and exclusively with ring opening to methyl hydroxypropyl phosphonate. The reaction product contained less than one part in 500 of methanol (97). Cyclic phosphinates (Fig. 5*c*) however, are hydrolyzed at rates very close to those of acyclic analogs (97). The following rules have been formulated to explain hydrolytic behavior of cyclic esters of phosphoric acid but apply in part to general phosphate ester hydrolyses. (*a*) Hydrolysis proceeds through a pentacoordinate intermediate of trigonal bipyramidal geometry. (*b*) Groups enter and leave the intermediate at axial positions only. (*c*) Electronegative groups preferentially occupy axial positions (alkyl groups therefore occupy equatorial positions). (*d*) Exchange between axial and equatorial positions is possible by means of "pseudorotation." Cyclic phosphonates therefore are expected to show accelerated ring opening but not accelerated hydrolysis external to the ring since "pseudorotation"

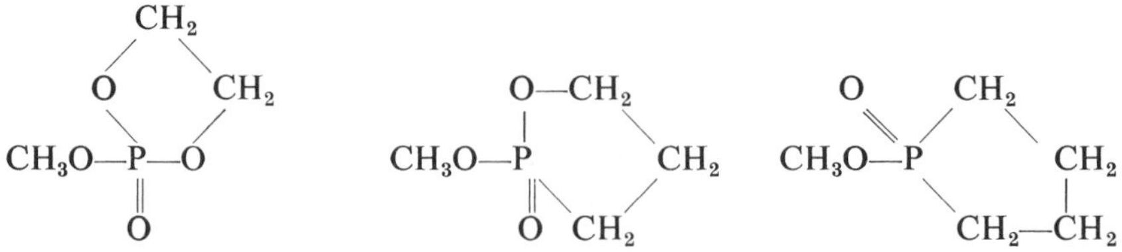

(a) (b) (c)

Figure 5

would either place a methylene unit in an apical position or place the five-membered ring in an unfavorable diequatorial configuration. Cyclic phosphinates for the same reasons are expected to show no enhancement at all.

Monoesters. Monosubstituted phosphate esters may exist in water as the undissociated ester, the monoanion, or the dianion (86, 92). Most monoalkyl phosphates show simple kinetic relationships between rate constants and pH. The rates of hydrolysis reach maximum at about pH 4, the region of maximum monoanion concentration; the rate is $k_1[ROPO_3H^-]$ where k_1 is the specific first-order constant for the monoanion and has the value of 8.23×10^{-6} sec^{-1} at 100.1°C (89). Phosphorus–oxygen bond fission occurs in all cases studied in the region of the rate maximum (89, 98). In strong acid solution two additional species are hydrolyzed, that is, the neutral molecule and its conjugate acid.

The mechanism by which the monoanion of monoalkyl phosphates is hydrolyzed is believed to involve loss of metaphosphate ion (86, 92, 99, 100)

$$ROPO_3H^- \rightarrow ROH + PO_3^- \tag{12}$$

Hydrolysis of the dianion becomes important only when breaking the P—O bond can produce a stable anion. The dianion of acetyl phosphate hydrolyzes readily at 39°C ($\mu = 0.6$), $k = 4.3 \times 10^{-2}$ min^{-1} versus a value of 13.0×10^{-2} min^{-1} for the monoanion (100). Both species eliminate metaphosphate ions. Substituted benzoyl phosphates show similar reactivity through the same mechanism (100).

Neutral monosubstituted phosphate esters are readily hydrolyzed if carbonium ions are easily formed from the alcohol groups, for example, *t*-butyl phosphate (101, 102).

Phosphonate Esters. Phosphonate diesters are similar in hydrolytic behavior to phosphate triesters and are readily hydrolyzed with the loss of one ester group as shown in Table 4. Phosphorus–oxygen bond fission is observed in basic solution and carbon–oxygen bond cleavage in acidic solution (103). Phosphonate monoesters are hydrolyzed in strongly acid solution, are very stable near pH 7, and are slowly hydrolyzed in alkaline solution.

The effect of electronegativity of alkyl and chloralkyl groups on diethyl phosphonates has been studied (104). For compounds of the type $RPO(OEt)_2$ where R = Et, Me, $ClCH_2$, and Cl_2CH, relative alkaline hydrolysis rates at 30°C based on Me = 1 are 0.13, 1, 15.6, and 108, respectively. Much greater differences have been observed for carbon esters.

Hydrolysis rates of phosphonates in alkaline solution are quite sensitive to the type of substituent attached to phosphorus; for example, for $(i\text{-}C_3H_7O)_2RPO$ relative rates based on R = Me = 1 are Et = 0.16, *n*-Pr = 0.062, *n*-Bu = 0.039, and *t*-Bu = 0.002. Under acidic conditions, however,

TABLE 4 Hydrolysis Rates for Some Phosphonates in Aqueous Base (106)

Number	Compound	Temperature (°C)	k^a [l/(mole)(sec)]	Activation Energy (kcal/mole)
1	$(CH_3O)_2(CH_3)PO$	100	2.35×10^{-1}	13.3
2	$(C_2H_5O)_2(CH_3)PO$	100	1.80×10^{-2}	11.6
3	$(i\text{-}C_3H_7O)_2(CH_3)PO$	100	4.82×10^{-4}	16.1
4	$(C_6H_5O)_2(CH_3)PO$	100	3.81	12.85
5	$(C_2H_5O)_2(C_2H_5)PO$	100	9.40×10^{-3}	13.9
6	$(i\text{-}C_3H_7O)_2(CH_3)PO$	100	4.82×10^{-4}	16.1
7	$(i\text{-}C_3H_7O)_2(n\text{-}C_4H_9)PO$	100	1.51×10^{-5}	22.7

[a] Calculated from data given.

little substituent effect is observed: Me = 1, Et = 0.5, *n*-Pr = 0.5, *n*-Bu = 0.33, and *t*-Bu = 0.33. Acid conditions are believed to involve attack on the C—O bond, whereas basic hydrolysis requires attack on phosphorus (105).

REFERENCES

1. Reynoso, A., *Ann.*, **83,** 98 (1852); *C. R.*, **34,** 795 (1852).
2. Griffith, E. J., and R. L. Buxton, *J. Am. Chem. Soc.*, **89,** 2884 (1967).
3. Shen, C. Y., and C. F. Callis, "Orthophosphoric Acids and Orthophosphates," in *Preparative Inorganic Reactions*, Vol. 2, W. Jolly (Ed.), Interscience, New York, 1965.
4. Van Wazer, J. R., *Phosphorus and Its Compounds*, Vol. I, Interscience, New York, 1958.
5. LaMer, V. K., *J. Phys. Chem.*, **66,** 973 (1962).
6. Taylor, A. W., E. L. Gurney, and A. W. Frazier, *Soil Sci. Soc. Am. Proc.*, **29,** 317 (1965).
7. Taylor, A. W., and E. L. Gurney, *J. Phys. Chem.*, **65,** 1613 (1961).
8. Van Wazer, J. R., E. J. Griffith, and J. F. McCullough, *J. Am. Chem. Soc.*, **77,** 287 (1955).
9. Thilo, E., in *Advances in Inorganic and Radiochemistry*, Vol. 4, H. J. Emeleus and A. G. Sharpe (Eds.), Academic Press, New York, 1962.
10. Strauss, U. P., and J. W. Day, *J. Polym. Sci., Part C*, **1967,** 2161.
11. Bell, R. N., *Ind. Eng. Chem.*, **39,** 136 (1947).
12. Bunton, C. A., and H. Cheirmovich, *Inorg. Chem.*, **4,** 1763 (1965).
13. Watanabe, M., T. Takahara, and T. Yamada, *Chubu Kogyo Daigaku Kiyo*, **4,** 141 (1968).

14. Friess, S. L., *J. Am. Chem. Soc.*, **74,** 4027 (1952).
15. Crowther, J. P., and A. E. R. Westman, *Can. J. Chem.*, **34,** 969 (1956).
16. Kiehl, S. J., and M. F. Moose, *J. Am. Chem. Soc.*, **60,** 47 and 257 (1938).
17. Munson, R. A., *J. Phys. Chem.*, **68,** 3374 (1965); **69,** 1761 (1965).
18. Campbell, D. O., and M. L. Kilpatrick, *J. Am. Chem. Soc.*, **76,** 893 (1954).
19. Topley, B., *Q. Rev.*, **3,** 345 (1949).
20. Buyers, A. G., *J. Phys. Chem.*, **66,** 939 (1962).
21. Zettlemoyer, A. C., C. H. Schneider, H. V. Anderson, and R. J. Fuchs, *J. Phys. Chem.*, **61,** 991 (1957).
22. Shen, C. Y., J. S. Metcalf, and E. V. O'Grady, *Ind. Eng. Chem.*, **51,** 717 (1959).
23. Shen, C. Y., and D. R. Dyroff, *Ind. Eng. Chem. Prod. Res. Dev.*, **5,** 97 (1966).
24. Griffith, E. J., *Ind. Eng. Chem.*, **51,** 240 (1959).
25. Karl-Kruppa, E., C. F. Callis, and E. Seifer, *Ind. Eng. Chem.*, **49,** 2061 (1957).
26. Lyons, J. W., *J. Colloid Sci.*, **19,** 399 (1964).
27. Thilo, E., *Angew. Chem.*, **4,** 1065 (1965).
28. Quimby, *J. Phys. Chem.*, **58,** 603 (1954).
29. Smith, M. J., *Can. J. Chem.*, **37,** 1115 (1959).
30. Thilo, E., and R. Ratz, *Z. Anorg. Allgem. Chem.*, **260,** 255 (1949).
31. Griffith, E. J., *J. Inorg. Nucl. Chem.*, **26,** 1381 (1964).
32. Miller, D. L., G. J. Krol, and U. P. Strauss, *J. Am. Chem. Soc.*, **91,** 6882 (1969).
33. Wicker, W., and E. Dido, *Z. Anorg. Allgem. Chem.*, **306,** 48 (1960).
34. Crutchfield, M. M., and R. R. Irani, *J. Am. Chem. Soc.*, **87,** 2815 (1965).
35. MacElroy, W., and B. Glass, *Phosphorus Metabolism*, Johns Hopkins Press, Baltimore, Maryland, 1951.
36. Roche, J., "Phosphatases," in *The Enzymes Chemistry and Mechanism of Action*, Vol. I, J. Sumner and K. Myrback (Eds.), Academic Press, New York, 1951.
37. Bamann, E., and E. Henmüller, *Naturwiss.* **28,** 535 (1940).
38. Aoki, S., and S. Miyachi, *Plant Cell Physiol.* (*Japan*), **5,** 241 (1964).
39. Miura, M., and Y. Moriguchi, *Bull. Chem. Soc. Japan*, **37,** 1522 (1965).
40. Otani, S., M. Miura, and T. Doi, *Kogyo Kagaku Zasshi*, **66,** 593 (1963).
41. Moriguchi, Y., and M. Miura, *Bull. Chem. Soc. Japan*, **38,** 678 (1965).
42. Wieker, W., A. R. Grimmer, and E. Thilo, *Z. Anorg. Allgem. Chem.*, **330,** 78 (1964).
43. Strauss, U. P., and T. L. Treithler, *J. Am. Chem. Soc.*, **78,** 3553 (1956); **77,** 1473 (1955).
44. Strauss, U. P., and G. J. Krol, *J. Polym. Sci.*, *Part C*, **1967,** 2171.
45. Thilo, E., G. Schulz, and E. M. Wichmann, *Z. Anorg. Allgem. Chem.*, **272,** 182 (1953).
46. Thilo, E., and W. Wieker, *Z. Anorg. Allgem. Chem.*, **291,** 164 (1957); *J. Polym. Sci.*, **291,** 164 (1963).
47. McCullough, J. F., J. R. Van Wazer, and E. J. Griffith, *J. Am. Chem. Soc.*, **78,** 4528 (1956).
48. Aiken, S. M., and J. B. Gill, *J. Inorg. Nucl. Chem.*, **28,** 2460 (1966).
49. Gill, J. B., and S. A. Riaz, *J. Chem. Soc.*, *A*, **1969,** 183.

50. Bamann, E., and E. Nowotny, *Ber.*, **81**, 463 (1948).
51. Piotrowska, M., and A. Swinarski, *Rocz. Chem.*, **35**, 432 (1961).
52. Shen, C. Y., N. E. Stahlheber, and D. R. Dyroff, *J. Am. Chem. Soc.*, **91**, 62 (1969).
53. Huffman, E. O., and J. D. Fleming, *J. Phys. Chem.*, **64**, 240 (1960).
54. Shen, C. Y., and D. R. Dyroff, "Condensed Phosphoric Acids and Condensed Phosphates," in *Preparative Inorganic Reactions*, Vol. 5, W. Jolly (Ed.), Interscience, New York, 1968.
55. Thilo, E., and U. Schuelke, *Z. Anorg. Allgem. Chem.*, **341**, 293 (1965).
56. Griffith, E. J., and R. L. Buxton, *Inorg. Chem.*, **4**, 549 (1965).
57. Shen, C. Y., *Ind. Eng. Chem. Prod. Res. Dev.*, **5**, 272 (1966).
58. Healy, R. M., and M. L. Kilpatrick, *J. Am. Chem. Soc.*, **77**, 5258 (1955); **79**, 6575 (1957).
59. Indelli, A., *Am. Chim. (Rome)*, **46**, 367 (1956).
60. Meyerhof, O., R. Shatos, and A. Kaplan, *Biochim. Biophys. Acta*, **12**, 121 (1953).
61. Carroll, R. L., and R. E. Mesmer, *Inorg. Chem.*, **6**, 1137 (1967).
62. Blaser, V. B., and K. H. Worms, *Z. Anorg. Allgem. Chem.*, **311**, 313 (1961).
63. Mesmer, R. E., and R. L. Carrol, *J. Am. Chem. Soc.*, **88**, 1381 (1966).
64. Riess, J. G., and J. R. Van Wazer, *Inorg. Chem.*, **5**, 178 (1966); and C. Y. Shen, unpublished information.
65. Quimby, O. T., A. Narath, and F. H. Lohmann, *J. Am. Chem. Soc.*, **78**, 4493 (1956); **82**, 1099 (1960).
66. Pollard, F. H., G. Nickless, and R. W. Warrender, *J. Chromatogr.*, **9**, 493, 513 (1962).
67. Pollard, F. H., G. Nickless, and A. M. Bigwood, *J. Chromatogr.*, **11**, 534 (1963); **12**, 527 (1963); **16**, 207 (1964).
68. Quimby, O. T., and T. J. Flautt, *Z. Anorg. Allgem. Chem.*, **296**, 220 (1958).
69. Feldmann, V. W., and E. Thilo, *Z. Anorg. Allgem. Chem.*, **327**, 159 (1964).
70. Nielson, M. L., R. R. Ferguson, and W. S. Coakley, *J. Am. Chem. Soc.*, **83**, 99 (1961).
71. Van Wazer, J. R., *Kirk-Othmer Encyclopedia Of Chemical Technology*, Vol. 15, Wiley, New York, 1968.
72. O'Brien, R. D., *Toxic Phosphorus Esters*, Academic Press, New York, 1960, Chapter 1.
73. Aldrige, W. N., *Biochem. J.*, **54**, 442 (1953).
74. Hartley, B. S., and Kilby, B. A., *Biochem. J.*, **50**, 672 (1952).
75. Ketelaar, J. A. A., *Trans. 9th Int. Congr. Entomol., Amsterdam, 1951* **2**, 318 (1953).
76. Ahmed, M. K., J. E. Casida, and R. E. Nichols, *J. Agric. Food Chem.*, **6**, 740 (1958).
77. Aldridge, W. N., and A. N. Davison, *Biochem. J.*, **52**, 663 (1952).
78. Ruzicka, J. H., J. Thomson, and B. B. Wheals, *J. Chromatogr.*, **31**, 37 (1967).
79. Coates, H., *Ann. Appl. Biol.*, **36**, 156 (1949).
80. Aksnes, G., *Acta Chem. Scand.*, **14**, 1515 (1960).
81. Hudson, R. F., and L. Keay, *J. Chem. Soc.*, **1960**, 1859.
82. Kilpatrick, M., and M. L. Kilpatrick, *J. Phys. Chem.*, **53**, 1371 (1949).
83. Larson, L., *Acta Chem. Scand.*, **11**, 1131 (1957).

84. Larson, L., *Acta Chem. Scand.*, **12,** 783 (1958).
85. Heath, D. F., *J. Chem. Soc.*, **1956,** 3796.
86. Bruice, T. C., and S. S. Benkovic, *Bio-organic Mechanisms*, Benjamin, New York, 1966, Chapters 5 and 6.
87. Barnard, P. W. C., C. A. Bunton, D. R. Llewellyn, C. A. Vernon, and W. A. Welch, *J. Chem. Soc.*, **1961,** 2670.
88. Bunton, C. A., D. R. Llewellyn, C. A. Vernon, and V. A. Welch, *J. Chem. Soc.*, **1961,** 1636.
89. Bunton, C. A., D. R. Llewellyn, K. G. Oldham, and C. A. Vernon, *J. Chem. Soc.*, **1958,** 3574.
90. Cox, J. R., Jr., Ph.D. Thesis, Harvard University, Cambridge, Massachusetts, 1959,
91. Bunton, C. A., M. M. Mhala, K. G. Oldham, and C. A. Vernon, *J. Chem. Soc.*, **1960,** 3293.
92. Cox, J. R., Jr., and O. B. Ramsay, *Chem. Rev.*, **64,** 317 (1961).
93. Ketelaar, J. A. A., H. R. Gersmann, and F. Hartog, *Rec. Trav. Chim.*, **71,** 1253 (1952).
94. Ketelaar, J. A. A., and H. R. Gersmann, *Rec. Trav. Chim.*, **77,** 973 (1958).
95. Kumamoto, J., J. R. Cox, Jr., and F. H. Westheimer, *J. Am. Chem. Soc.*, **78,** 4858 (1956).
96. Westheimer, F. H., *Acc. Chem. Res.*, **1,** 70 (1968).
97. Dennis, E. A., and F. H. Westheimer, *J. Am. Chem. Soc.*, **88,** 3431, 3432 (1966).
98. Butcher, W. W., and F. H. Westheimer, *J. Am. Chem. Soc.*, **77,** 2420 (1955).
99. Bunton, C. A., *Acc. Chem. Res.*, **3,** 257 (1970); *J. Chem. Educ.*, **45,** 21 (1968).
100. DiSabato, G., and W. P. Jenks, *J. Am. Chem. Soc.*, **83,** 4400 (1961).
101. Bunton, C. A., D. R. Llewellyn, K. G. Oldham, and C. A. Vernon, *J. Chem. Soc.*, **1958,** 3588.
102. Bunton, C. A., and E. Humeres, *J. Org. Chem.*, **34,** 572 (1969).
103. Keay, L., *Can. J. Chem.*, **43,** 2637 (1965).
104. Aksnes, G., and J. Songstad, *Acta. Chem. Scand.*, **19,** 893 (1965).
105. Kilpatrick, M., Jr., and M. L. Kilpatrick, *J. Phys. Colloid Chem.*, **53,** 1371, 1385 (1949).
106. Hudson, R. F., and L. Keay, *J. Chem. Soc.*, **1956,** 2463.
107. Libby, R. A., *Inorg. Chem.*, **10,** 386 (1971).

12

Dissociation Constants of Phosphorus—Containing Chelating Agents

A. FRED KERST

Director Research and Development, Michigan Chemical Corporation, St. Louis, Michigan

The ability of phosphates to form water-soluble complexes has been known for more than a century. However, it was only after utilization of this phenomenon to soften water that significant interest was generated in this technological area (1). The industrial interest in utilizing phosphates to form water-soluble metal complexes in water softening initiated activity that resulted in a concerted effort by a number of workers to better understand the practical and theoretical aspects of this technology. A number of publications review the properties of complexing agents that form soluble complexes with metal ions (2–5). This process is known as sequestration.

In determining stability constants for metal-ligand complexes the competition of hydrogen ions with metal ions for the ligand binding sites must be known. This competition is generally most pronounced in low pH, highly acidic media; however, even at a very high pH, hydrogen ions compete for ligand. A decrease in pH results in a shift of metal-ligand complex toward dissociation. Metal-ligand complexes will dissociate completely if the solution is sufficiently acidic. The proton dissociation properties of polyphosphates and polyphosphonates are important to understanding the metal complexing properties of these important classes of compounds. A number of workers have completed detailed studies of the acidity of these materials as well as other metal-chelating agents that find use in an array of industrial applications. It is hoped that a review of the proton dissociation constants of several of the various types of metal-chelating agents will assist the reader in better understanding the importance of this property as it affects metal complexing.

METHODS

A number of methods have been developed for determining the acidity of polyfunctional acids. The most frequently used methods include potentiometry (6–8), calorimetry (9), and spectrophotometry (10). Most of the potentiometric methods are based on the methods of Bjerrum (11).

The acid-base equilibria presented are discussed in terms of pK_n values ($-\log_{10} K_n$) and refer to the dissociation:

$$H_nL^{(a-n)} \rightleftharpoons H_{n-1}L^{(a+1-n)-} + H^+$$

$$K_n = \frac{[H^+][H_{n-1}L^{(a+1-n)-}]}{[H_nL^{(a-n)-}]}$$

$$pK_n = -\log_{10} K_n$$

where n is an integer from 0 to a, and a is an integer indicating the total number of protonated sites on ligand L. Experiments directed toward examining proton-ligand interactions must be conducted with careful consideration of ionic strength, temperature, and chemical purity of the materials being employed. It has long been recognized that polyphosphates (12, 13) and most other polyanions used in water treatment (14) form complexes with the alkali metals. The stability constants for these complexes are given in Table 1. Due to the formation of these complexes, salts of these metals should not be employed in studies of proton-ligand equilibria. If such salts are used as supporting electrolyte, the apparent acidity of the ligand will be increased due to complex formation. The tetramethylammonium salts are frequently used as supporting electrolytes since complex formation does not occur with the tetramethylammonium cation.

Table 2 presents dissociation constants for the most frequently encountered metal-complexing agents. A very thorough tabulation of acid dissociation

TABLE 1 Formation Constants K_{ML} for Alkali Metal Complexes with Pyrophosphate, Tri-, and Tetraphosphate Ions (μ = 1.0, 25°)

	Log K_{KL}	Log K_{NaL}	Log K_{LiL}	Ref.
$MP_2O_7^{3-}$	0.80 ± 0.06	1.00 ± 0.06	2.39 ± 0.06	15
$MHP_2O_7^{2-}$	Not detected	Not detected	1.03 ± 0.06	15
$MP_3O_{10}^{4-}$	1.39 ± 0.06	1.64 ± 0.06	2.87 ± 0.06	16
$MHP_3O_{10}^{3-}$	Not detected	0.77 ± 0.06	0.83 ± 0.06	16
$MP_4O_{13}^{5-}$	1.71	1.79	2.64	17
$MHP_4O_{13}^{4-}$	1.11	1.10	1.59	17

TABLE 2 pK Values for a Number of Phosphorus and Nonphosphorus Containing Chelating Agents

	pK_1	pK_2	pK_3	pK_4	pK_5	pK_6	Electrolyte[a]	Temperature (°C)	μ	Ref.
I. Phosphates										
Phosphoric acid	2.12	7.20	12.36				TMA^+Cl^-	25	0	9
			12.38					25	0	10
	2.36	6.61	11.1				TMA^+Br^-	25	1.0	22
Pyrophosphoric	−0.44	2.64	6.76	9.41			TMA^+Cl^-	25	0	9
		2.64	6.76	9.42			TMA^+Cl^-	25	0	15
			6.12	8.98			TMA^+Cl^-	25	0.1	27
	0.82	1.81	6.13	8.93			TMA^+Cl^-	25	1.0	23
	~1.7	1.75	5.98	8.74			TMA^+Br^-	25	1.0	22
Triphosphoric	−0.51	1.20	2.30	6.50	9.24		TMA^+Cl^-	25	0	9
			2.30	6.50	9.24		TMA^+Cl^-	25	0	15
				5.79	8.81		TMA^+Cl^-	25	0.1	27
	~0.51	1.15	2.04	5.69	8.56		TMA^+Br^-	25	1.0	22
		1 06	2.11	5.83	8.81		TMA^+Cl^-	25	1.0	24
Tetraphosphoric				2.59	6.75	8.50	$NaNO_3$	25	0	16
					7.38	9.11	TMA^+Cl^-	25	0	23
				2.20	5.95	7.54	$NaNO_3$	25	0.1	16
					5.91	8.88	TMA^+Cl^-	25	0.1	27
				1.78	5.52	7.07	$NaNO_3$	25	0.3	16
				1.30	4.96	6.48	$NaNO_3$	25	1.0	16
			1.36	2.23	6.63	8.34	TMA^+Cl^-	25	1.0	23
II. Phosphoramidates										
Phosphoramidic acid	2.74	8.05					KCl	26.5	0.5	26
	3.3	8.28					TMA^+Br^-	25	1.0	22
Imidodiphosphoric acid	~1.5	2.66	7.32	10.22			TMA^+Br^-	25	0.1	22
	~2	2.85	7.08	9.72			TMA^+Br^-	25	0.2	22
	~2	2.81	7.05	9.77			TMA^+Br^-	25	0.3	22
	~1.5	3.05	7.62	10.36			TMA^+Br^-	25	1.0	22
Diimidotriphosphoric acid	~1	~2	3.03	6.61	9.84		TMA^+Br^-	25	0.1	22
	~0	~2	3.83	7.02	9.92		TMA^+Br^-	25	0.2	22
	~0	~2	3.94	7.74	9.95		TMA^+Br^-	25	0.3	22
	~1	~2	3.36	6.68	10.00		TMA^+Br^-	25	1.0	22
III. Methylene phosphonates										
$H_2O_3PCH_2PO_3H_2$	2.2	2.87	7.45	10.69			TMA^+Br^-	25	0	35
		2.5	6.87	10.33			KCl	20	0.1	34
	1.7	2.75	7.33	10.42			TMA^+Br^-	25	0.1	35

$H_2O_3PCH_2CH_2PO_3H_2$	1.5	3.18	7.62	9.28			TMA^+Br^-	25–50	0	35
	1.5	2.86	7.50	9.08			TMA^+Br^-	25–50	0.1	35
$H_2O_3PCH_2CH_2CH_2PO_3H_2$	1.6	3.06	7.65	8.63			TMA^+Br^-	25–50	0	35
	1.6	2.81	7.50	8.43			TMA^+Br^-	25–50	0.1	35
		2.6	7.34	8.35			KCl	20	0.1	34
$H_2O_3P(CH_2)_nPO_3H_2$	1.7	3.19	7.78	8.58			TMA^+Br^-	25–50	0	35
$n = 4$	1.7	2.85	7.58	8.38			TMA^+Br^-	25–50	0.1	35
		2.7	7.54	8.38			KCl	20	0.1	34
$n = 6$	1.8	3.12	7.73	8.56			TMA^+Br^-	25–50	0	35
	1.8	3.07	7.65	8.34			TMA^+Br^-	25–50	0.1	35
$n = 10$	2.1	3.27	7.93	8.94			TMA^+Br^-	25–50	0	35
	2.1	3.15	7.74	8.83			TMA^+Br^-	25–50	0.1	35
IV. Aminomethyl phosphonates										
$N(CH_2PO_3H_2)_3$	<2	<2	4.30	5.46	6.66	12.34	KNO_3	25	1.0	25
$N(CH_2PO_3H_2)_2CH_2CO_2H$	1.73	2	5.01	6.37	10.80		KNO_3	25	0.1	29
$N(CH_2PO_3H_2)(CH_2CO_2H)_2$	2.0	2.25	5.57	10.76			KCl	20	0.1	30
			5.80	10.64			KCl	30	0.1	30
$ON(CH_2PO_3H_2)_2$	<2	<2	3.28	5.26	6.95	12.05	KNO_3	25	1.0	28
$CH_3CH_2N(CH_2PO_3H_3)_2$	<2	4.70	5.92	12.42			KNO_3	25	1.0	25
$(CH_3CH_2)_2NCH_2PO_3H_2$	5.79	12.32					KNO_3	25	1.0	25
$H_2O_3PCH_2CH_2N(CH_2CO_2H)_2$	1.9	2.45	6.54	10.46			KCl	20	0.1	29
$(H_2O_3PCH_2)_2NCH_2CH_2$— $N(CH_2PO_3H_2)_2$	1.46	2.72	5.05	6.18	6.63	7.43	KCl	25	0.1	31
						$pK_7 = 9.22$				
						$pK_8 = 10.60$				
$(HO_2CCH_2)_2NCH_2CH_2$— $N(CH_2PO_3H_2)_2$	2.83	3.52	5.46	7.72	10.6	11.2	KNO_3	25	1.0	32
V. Aminomethyl carboxylates										
$N(CH_2CO_2H)_3$	1.9	2.5	9.73				KNO_3	25	0.1	28
	1.8	2.49	9.73				KCl	20	0.1	29
$ON(CH_2CO_2H)_3$	<2	2.57	7.89				KNO_3	25	0.1	28
$CH_3N(CH_2CO_2H)_2$	2.1	9.65					KCl	20	0.1	29
$HO_2CCH_2CH_2N(CH_2CO_2H)_2$	2.0	3.69	9.66				KCl	20	0.1	29
$(HO_2CCH_2)_2NCH_2CH_2$— $N(CH_2CO_2H)_2$	2.0	2.67	6.16	10.26						33

[a] TMA^+ = tetramethylammonium ion.

constants and metal stability constants for many of these materials as well as other complexing groups has been published (18–20).

LINEAR CHELATING AGENTS CONTAINING PHOSPHORUS

The linear polyphosphates are the most frequently used complexing agents. These polymeric materials are composed of tetrahedral phosphate (PO_4^{3-}) groups. Titration curves of phosphates possess two inflection points, one near pH 4.5 and the other near pH 9.5. Each phosphate group contains a reasonably acidic proton that dissociates before reaching pH 4.5. Titration between pH 4.5 and 9.5 gives the number of terminal phosphate groups that possess a weakly acidic proton. Orthophosphoric acid (H_3PO_4) possesses a third proton which is not detected in most pH titrations. Figure 1 presents the titration curves obtained for phosphoric acid, pyrophosphoric acid, tripolyphosphoric acid, and tetraphosphoric acid. The ionization constants for removal of the final proton becomes more acidic as the phosphate chain increases. It has been predicted that the values of the dissociation constants for straight-chain polyphosphates would approach a value of 10^{-8} as the chain length increases (21).

Figure 2 plots the pK values corresponding to removal of the final two

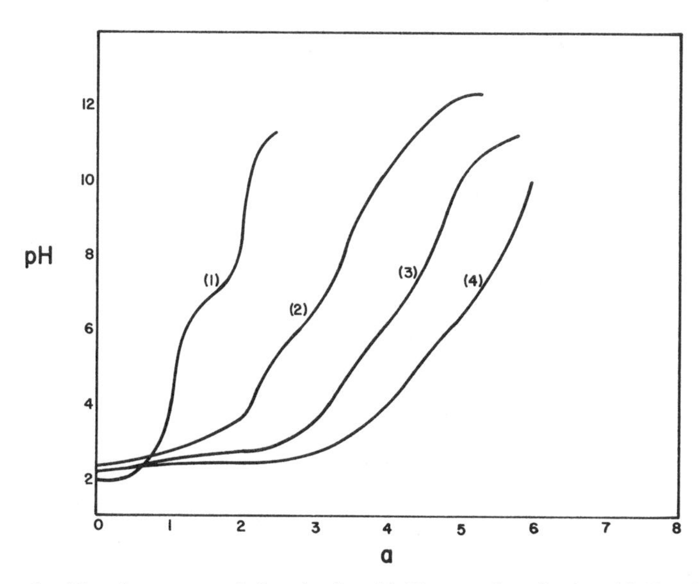

Figure 1. Titration curves of phosphoric acid (1), pyrophosphoric acid (2), triphosphoric acid (3), and tetraphosphoric acid (4) as a function of the equivalents (a) of tetramethylammonium hydroxide.

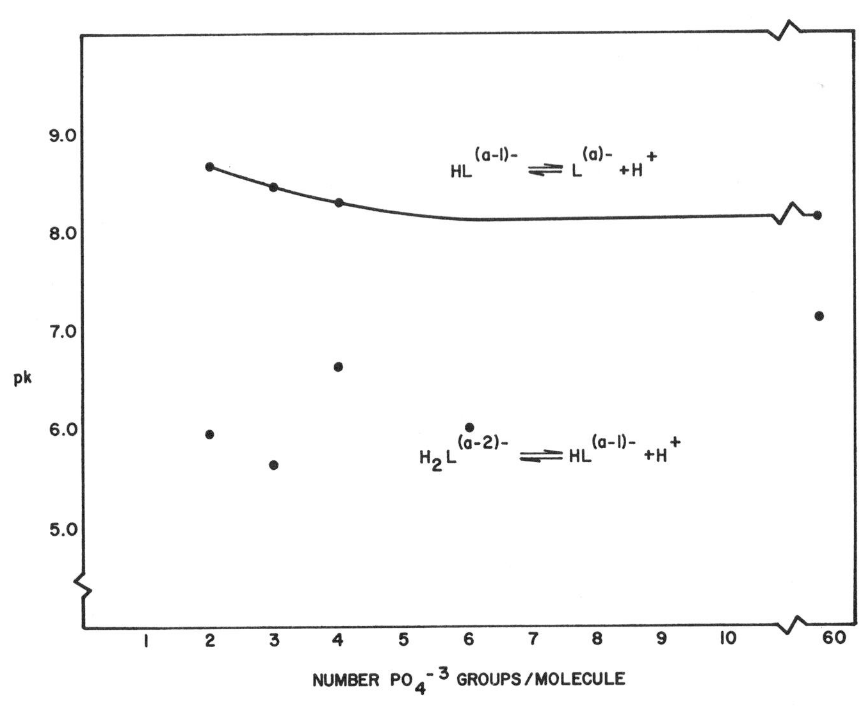

Figure 2. Plot of pK_n and pK_{n-1} versus number of phosphate groups contained in linear polyphosphates.

protons from polyphosphates as a function of the number of tetrahedral phosphate groups present. The data used in presenting this plot were obtained at 25° using 1.0 *M* tetramethylammonium salts as the electrolyte. Presented in this plot are data published by Irani and Callis (22) which includes the pK values for the final two dissociation constants of the linear polyphosphates,

$$\begin{array}{c} \text{O} \quad \text{O} \quad \text{O} \\ \text{HOP(OPO)}_n\text{POH} \\ \text{O} \quad \text{O} \quad \text{O} \\ \text{H} \quad \text{H} \quad \text{H} \end{array}$$

where $n = 4$ and 58. The pK values for the final two proton dissociations are 8.13 and 5.98 for $n = 4$ and 8.17 and 7.22 for $n = 58$. The values used for tetraphosphoric acid are those published by Watters and co-workers (23). Though more data are required to establish this trend with certainty, it is apparent that the acidity of the final proton increases with an increase in chain length.

The anomalously low value of pK_4 for $H_5P_3O_{10}$, which does not follow the general increase of pK_{n-1} values with increasing chain length, has been rationalized as being due to a coiling of the triphosphate ion around the first hydrogen atom to produce an exceptionally stable structure (23). This structure is disrupted when the second proton is associated, thus increasing the dissociation constant for $H_2P_3O_{10}^{3-}$. This behavior has been investigated using P^{31} nuclear magnetic resonance (24). The phosphoramidates are the class of compounds formed by linking phosphorus atoms by —NH— groups instead of oxygen atoms. The replacement of the oxygen with the more electronegative —NH— group decreases the acidity of the phosphoramidates relative to the phosphates. This is especially pronounced in the values of pK_n and pK_{n-1} for imidodiphosphoric acid and imidotriphosphoric acid in comparison to the same values for pyrophosphoric acid and triphosphoric acid. Methylenediphosphoric acid represents a third type of linear phosphorus containing metal-complexing agents where the phosphorus atoms are linked by a $—CH_2—$ group. The pK_n value for this material is greater than that of imidodiphosphoric acid which might be expected because of the greater electronegativity of the $—CH_2—$ group.

As the number of $—CH_2—$ groups joining the $—PO_3H_2$ groups is increased beyond 3, the pK_n and pK_{n-1} values approach constant values of 8.5 and 7.9, respectively. Statistical considerations show that $pK_n - pK_{n-1}$ should approach 0.6 as the separation between acid groups is increased to the point where intramolecular interaction is negligible. Evidently, this interaction is nearly nonexistent when more than three $—CH_2—$ groups separate the phosphonate groups (35).

NONLINEAR CHELATING AGENTS CONTAINING PHOSPHORUS

The exceptional metal-chelating properties of ethylenediaminetetraacetic acid **1** (EDTA) and nitrilotriacetic acid **2** (NTA) have generated considerable interest in this general class of chelating agents.

In addition to the carboxyl-containing aminomethyl polyanions, many aminomethyl phosphonates have been reported and investigated. Many

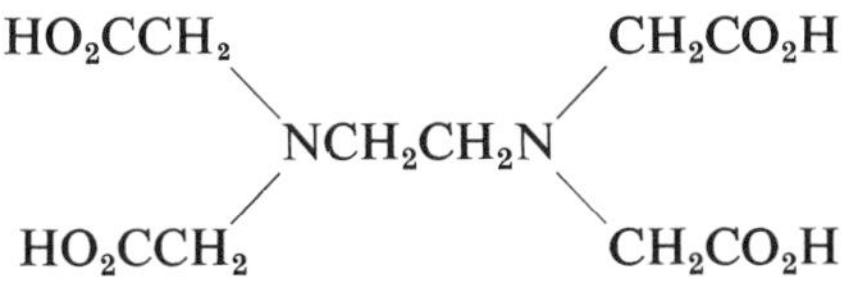

$$(HO_2CCH_2)_2NCH_2CH_2N(CH_2CO_2H)_2$$

1

N,N,N′,N′-Ethylenediaminetetracetic acid

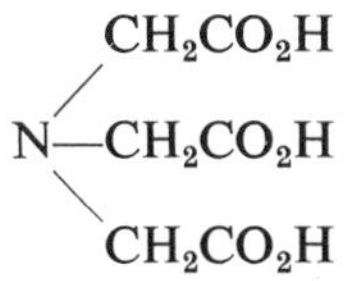

$$N(CH_2CO_2H)_3$$

2

Nitrilotriacetic acid

of the compounds that result by successive replacement of the acetate group of nitrilotriacetic acid and *N*,*N*,*N'*,*N'*-ethylenediaminetetracetic acid with a methyl phosphonate group have been synthesized and their metal-chelating properties investigated. The acid dissociation constants for a number of these compounds are presented in Table 2. The p*K* values corresponding to removal of the final proton from these materials are quite large and reflect attachment of this proton to nitrogen and not the phosphonate or carboxylate groups. Consideration of the acid equilibria in the case of EDTA (Scheme 1) and the ethylenediaminetetramethylphosphoric acid (Scheme 2) illustrates this point. Potassium salts have been used as supporting electrolytes in studying aminomethyl phosphonates because in the absence of potassium ions the dissociation constant of the least acidic proton is estimated to be less than 10^{-13} which is too small to be accurately measured from potentiometric titrations.

The similarity in the values of p*K* for the aminomethyl phosphonates and the similarity in the $\Delta pK = pK_n - pK_{n-1}$ for a number of these ligands support the existence of a dipolar ion. The results of P^{31} and H^1 nuclear magnetic resonance studies have been interpreted as supporting involvement of the nitrogen in the final deprotonation step of aminomethyl phosphonates (24). Although additional study is needed before a quantitative understanding of the importance of charge, the nitrogen atom, and structural effects can be developed, it is quite evident that the nitrogen atom is more important in

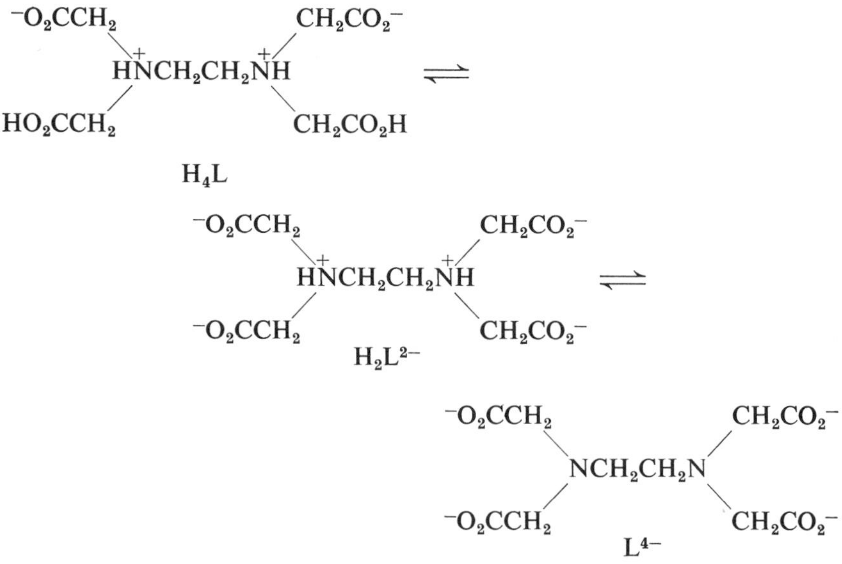

$$(^-O_2CCH_2)(HO_2CCH_2)\overset{+}{N}H-CH_2CH_2-\overset{+}{N}H(CH_2CO_2^-)(CH_2CO_2H) \rightleftharpoons$$

H_4L

$$(^-O_2CCH_2)_2\overset{+}{N}H-CH_2CH_2-\overset{+}{N}H(CH_2CO_2^-)_2 \rightleftharpoons$$

H_2L^{2-}

$$(^-O_2CCH_2)_2NCH_2CH_2N(CH_2CO_2^-)_2$$

L^{4-}

Scheme 1

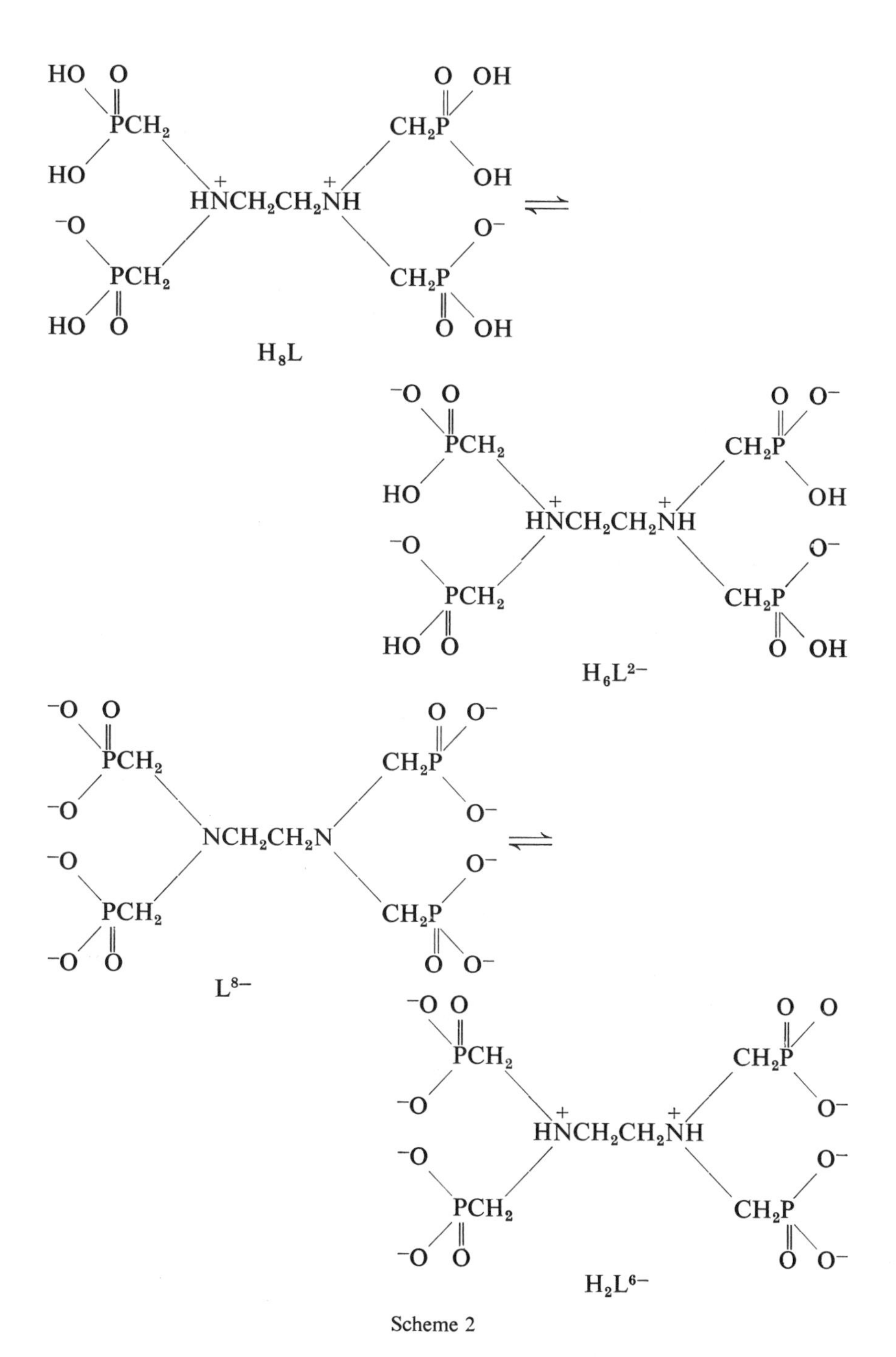

Scheme 2

metal complexing in the case of aminoacetates than in the case of the aminomethyl phosphonates. This suggestion is supported by comparing the stability constants for nitrilotriacetic acid and nitrilotri(methylenephosphoric) acid with that of their respective *N*-oxides. The stability constant for the calcium complex for $ON(CH_2CO_3H_2)_3$ is a factor of 10^4 less than for $N(CH_2CO_2H)_3$ while only a decrease by a factor of 10 is observed when going from $N(CH_2PO_3H_2)_3$ to $ON(CH_2PO_3H_2)_3$ (see Table 4). Evidently, in the case of $N(CH_2PO_3H_2)_3$, the six negative charges versus the three in the case of $N(CH_2CO_2H)_3$ provide the primary binding force.

THERMODYNAMIC PROPERTIES OF IONIZATION OF ORTHO-, PYRO-, AND TRIPOLYPHOSPHORIC ACID

The thermodynamic properties of proton-ligand and metal-ligand interactions offer great opportunity for further study. A thorough understanding of these properties would provide invaluable insight into the forces involved in complexation of protons and metals.

Irani and Taulli (9) completed a calorimetric analysis of the thermodynamics of ionization of ortho-, pyro-, and tripolyphosphoric acids. A potentiometric study of the ionization properties of triphosphoric acid at various temperatures has recently been reported (36). Data summarizing the results of both these groups are presented in Table 3.

ACIDITY AND METAL CHELATION

It has been generally accepted that anions of weaker acids form more stable complexes. The interaction between cationic size and charge versus anionic size, configuration, and charge must be important in considering metal-complexing properties. Table 4 presents the value of p*K* for dissociation of the final proton from a number of polyanions and the stability constant for calcium complexes of the completely ionized ligand. Except in a few instances, it is apparent that the concept that weaker acids yield more stable metal complexes must be applied cautiously.

Considerable work remains to be done before an understanding of the role of basicity as it affects the stability of metal complexes can be elucidated. Studies directed toward determining the thermodynamics of the ionization of complexing agents and the thermodynamics of the association of these same agents with metals would seem to offer the greatest potential in developing this understanding.

TABLE 3 Thermodynamic Parameters for Ionization of Ortho-, Pyro-, and Triphosphoric Acid at 25° (9). (Unit Molal Standard State)

	pK_0	$\Delta G°$ (kcal/mole)	$\Delta H°$ (kcal/mole)	$\Delta S°$ (eu)
$HPO_4^{2-} \rightleftharpoons H^+ + PO_4^{3-}$	12.36	16.86	2.6	−48
$H_2PO_4^- \rightleftharpoons H^+ + HPO_4^{2-}$	7.20	9.82	0.6	−31
$H_3PO_4 \rightleftharpoons H^+ + H_2PO_4^-$	2.12	2.89	−1.9	−16
$HP_2O_7^{3-} \rightleftharpoons H^+ + P_2O_7^{4-}$	9.41	12.84	0.4	−42
$H_2P_2O_7^{2-} \rightleftharpoons H^+ + HP_2O_7^{3-}$	6.76	9.22	0.1	−31
$H_3P_2O_7^- \rightleftharpoons H^+ + H_2P_2O_7^{2-}$	2.64	3.60	−3.0	−22
$H_4P_2O_7 \rightleftharpoons H^+ + H_3P_2O_7$	−0.44	−0.6	−3.0	−8
$HP_3O_{10}^{4-} \rightleftharpoons H^+ + P_3O_{10}^{5-}$	9.24 (9.4)[b]	12.60 (12.8)[b]	0.1 (−4.18)[a]	−42 (−57.13)[a]
$H_2P_3O_{10}^{3-} \rightleftharpoons H^+ + HP_3O_{10}^{4-}$	6.50 (6.1)[b]	8.86 (8.3)[b]	−1.5 (−4.48)[a]	−31 (−43.0)[a]
$H_3P_3O_{10}^{2-} \rightleftharpoons H^+ + H_2P_3O_{10}^{3-}$	2.30 (2.3)[b]	3.14 (3.2)[b]	−2.5 (−1.46)[a]	−19 (−15.51)[a]
$H_4P_3O_{10}^- \rightleftharpoons H^+ + H_3P_3O_{10}^{2-}$	1.20 (0.6)[b]	1.64 (0.8)[b]	−2.5 (−1.41)[a]	−14 (−7.29)[a]
$H_5P_3O_{10} \rightleftharpoons H^+ + H_4P_3O_{10}^-$	−0.51	−0.7	−2.5	−6

[a] Values are taken from Ref. 36.

[b] Calculated from $\Delta H°$ and $\Delta S°$ values given in Ref. 36.

TABLE 4 Values of pK_n and Stability Constants for Calcium Complexes of a Number of Polyanions

	pK_n	log K_{CaL}	Electrolyte	Temperature (°C)
I. Ligands bearing three negative charges				
$N(CH_2CO_2H)_3$	9.73 (29)[a]	6.41 (29)[b]	0.1 *N* KCl	20
$ON(CH_2CO_2H)_3$	7.89 (28)	2.46 (28)	0.1 *N* KNO_3	25
$HO_2CCH_2CH_2N(CH_2CO_2H)_2$	9.66 (29)	5.04 (29)	0.1 *N* KCl	20
$HO_3SCH_2CH_2N(CH_2CO_2H)_2$	8.16 (29)	5.44 (29)	0.1 *N* KCl	20
II. Ligands bearing four negative charges				
O O HOPOPOH O O H H	8.98 (27)	5.39 (7)	0.1 *N* TMA^+Cl^-	25
O O HOPNHPOH O O H H	10.22 (22)	5.36 (37)	0.1 *N* TMA^+Cl^-	25
$N(CH_2PO_3H_2)(CH_2CO_2H)_2$	10.76 (29)	7.18 (29)	0.1 *N* KCl	20
$CH_3CH_2N(CH_2PO_3H_2)_2$	12.42 (25)	3.36 (25)	1.0 *N* KNO_3	25
$H_2O_3PCH_2CH_2N(CH_2CO_2H)_2$	10.46 (29)	5.44 (29)	0.1 *N* KCl	20
$(HO_2CCH_2)_2NCH_2CH_2N(CH_2CO_2H)_2$	10.26 (33)	10.59 (33)		

III. Ligands bearing five negative charges				
O O O HOPOPOPOH O O O H H H	8.81 (27)	6.41 (7)	0.1 N TMA^+Br^-	25
O O O HOPNHPNHPOH O O O H H H	9.84 (22)	6.74 (37)	0.1 N TMA^+Br^-	25
$N(CH_2PO_3H_2)_2CH_2CO_2H$	10.80 (31)	6.17 (31)	0.1 KNO_3	25
IV. Ligands bearing six negative charges				
$N(CH_2PO_3H_2)_3$	12.34 (25)	6.68 (25)	1 N KNO_3	25
$ON(CH_2PO_3H_2)_3$	12.05 (28)	5.69 (28)	1 N KNO_3	25
V. Ligands bearing eight negative charges				
$(H_2O_3PCH_2)_2NCH_2CH_2N(CH_2PO_3H_2)_2$	10.60 (31)	5.74 (31)	0.1 KCl	25

[a] Reference from which pK_n values were taken.
[b] Reference from which $\log K_{CaL}$ values were taken.

REFERENCES

1. Hale, R. E., U.S. Patent 1,956,515 (1934).
2. Van Wazer, J. R., and C. F. Callis, *Chem. Rev.*, **58,** 1011 (1958).
3. Martell, A. E., "Complexing Agents," *Kirk-Othmer Encyclopedia of Chemical Technology*, 2nd ed., Vol. 6, Wiley, New York, 1965, p. 1.
4. Chaberek, S., and A. E. Martell, *Organic Sequestering Agents*, Wiley, New York, 1959.
5. Dwyer, F. P., and D. P. Mellor, Eds., *Chelating Agents and Metal Chelates*, Academic Press, New York, 1964.
6. Schwarzenbach, G., A. Willi, and R. O. Bach, *Helv. Chim. Acta*, **30,** 1303 (1947).
7. Irani, R. R., and C. F. Callis, *J. Phys. Chem.*, **64,** 1398 (1960).
8. Walters, J. I., E. Loughran, and S. M. Lambert, *J. Am. Chem. Soc.*, **78,** 4855 (1956).
9. Irani, R. R., and T. A. Taulli, *J. Inorg. Nucl. Chem.*, **28,** 1011 (1966).
10. Vanderzee, C. E., and A. S. Quist, *J. Phys. Chem.*, **65,** 118 (1961).
11. Bjerrum, J., *Metal Ammine Formation in Aqueous Solution*, P. Haase and Son, Copenhagen, 1941.
12. Mark, C. B., *J. Chem. Soc.*, **1949,** 427.
13. Van Wazer, J. R., and D. Campanella, *J. Am. Chem. Soc.*, **72,** 655 (1950).
14. Schwarzenbach, G., and H. Ackerman, *Helv. Chim. Acta*, **30,** 1798 (1947).
15. Lambert, S. M., and J. I. Watters, *J. Am. Chem. Soc.*, **79,** 4262 (1957).
16. Walters, J. I., S. M. Lambert, and E. Loughran, *J. Am. Chem. Soc.*, **79,** 3651 (1957).
17. Watters, J. J., and S. Matsumoto, *J. Inorg. Nucl. Chem.*, **29,** 2955 (1967).
18. Bjerrum, J., G. Schwarzenbach, and L. G. Sillen, *Stability Constants*, Part I, *Organic Ligands*, Special Publ. No. 6, The Chemical Society, London, 1957.
19. Bjerrum, J., G. Schwarzenbach, and L. G. Sillen, *Stability Constants*, Part II, *Inorganic Ligands*, Special Publ. No. 7, The Chemical Society, London, 1958.
20. Sillen, L. G., and A. E. Martell, Eds., *Stability Constants*, 2nd ed., The Chemical Society, London, 1964.
21. Van Wazer, J. R., and K. A. Holst, *J. Am. Chem. Soc.*, **72,** 639 (1950).
22. Irani, R. R., and C. F. Callis, *J. Phys. Chem.*, **65,** 934 (1961).
23. Watters, J. J., P. E. Sturrock, and R. E. Simonaitis, *Inorg. Chem.*, **2,** 765 (1963).
24. Crutchfield, M. M., C. F. Callis, R. R. Irani, and G. C. Roth, *Inorg. Chem.*, I, 813 (1962).
25. Carter, R. P., R. L. Carroll, and R. R. Irani, *Inorg. Chem.*, **6,** 939 (1967).
26. Peacock, C. J., and G. Nickless, *Z. Naturforsch.*, **24,** 245 (1969).
27. McNabb, W. M., J. F. Hazel, and R. A. Baxter, *J. Inorg. Nucl. Chem.*, **30,** 1585 (1968).
28. Carter, R. P., M. M. Crutchfield, and R. R. Irani, *Inorg. Chem.*, **6,** 943 (1967).
29. Schwarzenbach, G., H. Ackerman, and P. Ruckstuhl, *Helv. Chim. Acta*, **32,** 1175 (1949).
30. Ockerbloom, N., and A. E. Martell, *J. Am. Chem. Soc.*, **80,** 2351 (1958).
31. Westerback, S., K. S. Rajan, and A. E. Martell, *J. Am. Chem. Soc.*, **87,** 2567 (1965).

32. Rajan, K. S., J. Murase, and A. E. Martell, *J. Am. Chem. Soc.*, **91**, 4408 (1969).
33. Schwarzenbach, G., and E. Freitag, *Helv. Chim. Acta*, **34**, 1492 (1951).
34. Schwarzenbach, G., and J. Zurc, *Monatsh. Chem.*, **81**, 202 (1950).
35. Irani, R. R., and K. Moedritzer, *J. Phys. Chem.*, **66**, 1349 (1962).
36. Mitra, R. P., and B. R. Thukral, *Indian J. Chem.*, **8**, 347 (1970).
37. Irani, R. R., and C. F. Callis, *J. Phys. Chem.*, **65**, 296 (1961).

Distribution of Polyphosphate Deflocculants in Aqueous Suspensions of Inorganic Solids

JOHN W. LYONS

Commerical Development, Special Chemical Systems, Monsanto Company, St. Louis, Missouri

Deflocculation of inorganic solids suspended in water is required in a variety of industrial processes. The action needed is one of separating clusters of particles into the component, individual units, that is, of overcoming the attractive forces that normally produce aggregation. The result is a change in the behavior of the suspension, often marked. Viscosity may decrease by an order of magnitude, large particles will settle more readily, classification by sizes becomes easier, less water is required for a given level of viscosity producing subsequent savings in dewatering operations, and so on. Certain industrial processes would be economically unattractive or technically impossible without deflocculants. Among these are the processing and use of coating clays, production of latex paints, use of water-based drilling muds for oil wells, and a variety of applications for pigment-grade metal oxides. The application of kaolin coatings to paper, for example, is accomplished using aqueous suspensions at very high solids contents—60 to 70% by weight—made possible through the action of deflocculants. Without them the clay-water system would be an intractable semisolid. It would be workable only at, say, 30 to 40% solids; a tightly packed coating (after drying) would be more difficult to achieve. Similarly, paints without deflocculants would be too thick to handle at desired pigment-volume concentrations; more water would be required to produce the desired viscosity.

Deflocculation is used in a broad spectrum of applications where the kinds of results mentioned above are desired. This means that the deflocculants are encountered in a variety of industrial processes and find their way into a variety of waste-disposal systems. In what follows a description of the underlying technology is presented to enable the reader to judge for himself the potential impact of phosphorus-bearing deflocculants on the overall effluent question.

To achieve the deflocculated state a new balance of forces must be struck:

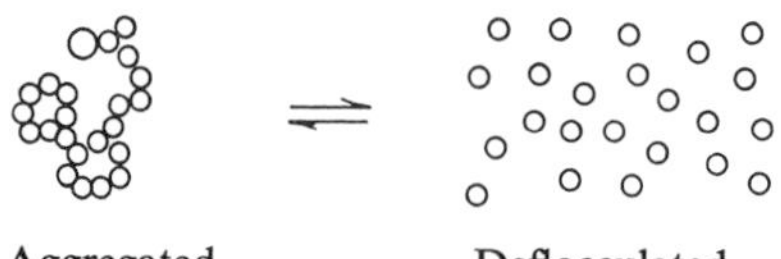

In the aggregated, or flocculated, condition the particles stick together; in the deflocculated state they are separated and act independently. The forces holding the particles together are generally long-range London-van der Waals attractions—like to like. The solvent (suspending medium) is not more strongly attracted to the surfaces than are other particles. There are at least two ways to alter the situation:

1. The particle-solvent interaction can be enhanced to the point where this interaction is stronger than the particle-particle force.
2. The particle-particle force can be reduced until it is less than the particle-solvent interaction.

This may be viewed in exactly the same way as the forces governing one versus two-phase systems. There is an important distinction in the present case arising because of the large size of the particles, namely, the long-range nature of the particle forces. Thus a single layer of solvent may not be enough to prevent aggregation. The particle-solvent attraction may have to be great enough to attract and orient many layers of solvent in order to cause sufficient separation to overcome the particle-particle attractions.

Chemically, the task is performed with (*a*) additives that sorb strongly on the particles and in turn cause strong solvent orientation or (*b*) additives that sorb strongly on the particles and build up a sufficient surface charge to produce electrostatic repulsion of neighboring, similarly charged particles. The first category includes the protective colloids—various gums that will adsorb and at the same time are very hydrophilic. The result is that many layers of water are strongly oriented about the particle and the interparticle distances become sufficiently great that the attractive forces are effectively screened. The second group, (*b*), includes the polyelectrolytes. Most inorganic suspensoids are somewhat negatively charged. If a polyanion will readily adsorb on the surface, the surface becomes strongly negative with accompanying repulsive effects. This adsorption is critical; not all polyelectrolytes are strongly adsorbed—those that are not are ineffective deflocculants. Occasionally, changing the pH will increase the surface charge by promoting surface ionization. In this sense, caustic soda or soda ash may sometimes be regarded as partial deflocculants. Sodium silicate works well in some systems, not in others. Polycarboxylates of intermediate molecular weight ($\approx$10 M or so) are sufficiently adsorbed to increase surface charge and to deflocculate. (Higher molecular weights lead to adsorption on two or more particles and hence to flocculation.)

The polyphosphates are strongly adsorbed, are efficient in terms of charge per unit weight, and are inexpensive. They are often the deflocculants of choice and find widespread use in industrial processing.

> It should be noted that in the use of polyphosphates as detergent builders these polyelectrolytes not only perform as sequestrants and as salts involved in the dynamics of micelles of surfactant molecules but they also serve as deflocculants. So, in a sense, the deflocculant use covers the application in detergency. All of what is said here for industrial deflocculants applies to the use in detergents as well.

The action of phosphates as deflocculants was reviewed in detail in 1961 (1). Most of what is known today was known at that time with some few exceptions. The reader is therefore referred to the earlier review for background. A more recent (1969) paper (2) gives results for a carefully defined kaolinite-water system and answers certain questions that arose in preparing the 1961 review. It, too, is recommended.

Figure 1 shows the most dramatic observable effect of adding polyphosphate to a clay suspension. The viscosity of this system is reduced tenfold

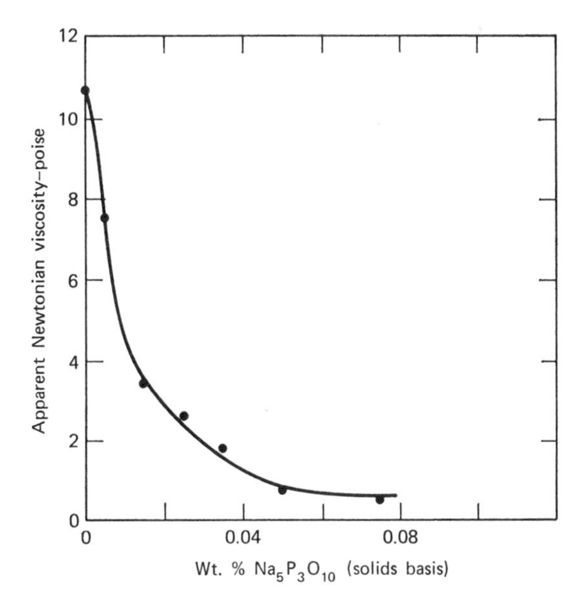

Figure 1. Viscosity change due to sodium triphosphate in 63% kaolinite–water suspension at pH 7.0 and 25°C, as measured by the Rotovisco (2).

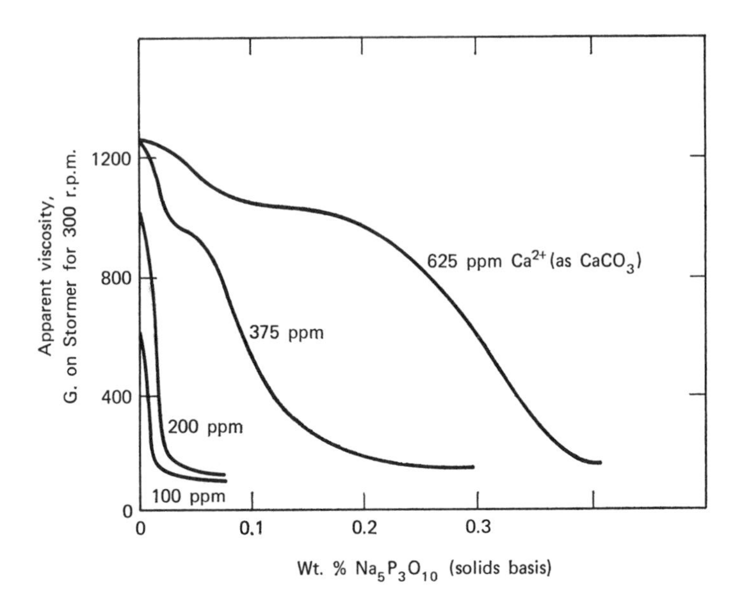

Figure 2. Effect of calcium ion on viscosity reduction in 60% kaolinite–water suspension with no pH adjustment as measured by Stormer viscometer (2).

by addition of 0.04% (solids basis) of sodium tri(poly)phosphate. Since the suspension was at 63% solids, this represents about 0.07% phosphate on the contained water. This clay had been washed free of multivalent cations that ordinarily interfere with polyphosphates and cause the requirements to be higher. Figure 2 shows the effect of calcium ion on the phosphate level; for this reason it is not unusual to have to add 0.1 to 0.2% (solids basis) to effect full deflocculation. For every ton of solids treated, some 2 to 4 lb of sodium polyphosphate may be used. The question now is as follows: What is the fate of that phosphate?

The easiest consideration is to ask what would happen should a treated suspension be centrifuged and the liquor sent to waste. Where is the phosphate—in the cake or in the liquor? First consider the adsorption isotherms in Fig. 3. These show (*a*) the polyphosphate is strongly adsorbed; (*b*) only about 0.02 mM/l are required in the solution phase to maintain full surface coverage on the clay particles; and (*c*) the point at which the surface is fully covered is the point at which full deflocculation occurs. The amount of tripolyphosphate in the centrifuge liquor for this example would be 7 mg/l. The P_2O_5 would be 4 mg/l. If the clay-water system be contaminated with multivalent cations, the phosphate required to deflocculate the system will be more. This extra phosphate is presumably in the solution phase in complex form with the cations. There will then be an increase in the P_2O_5

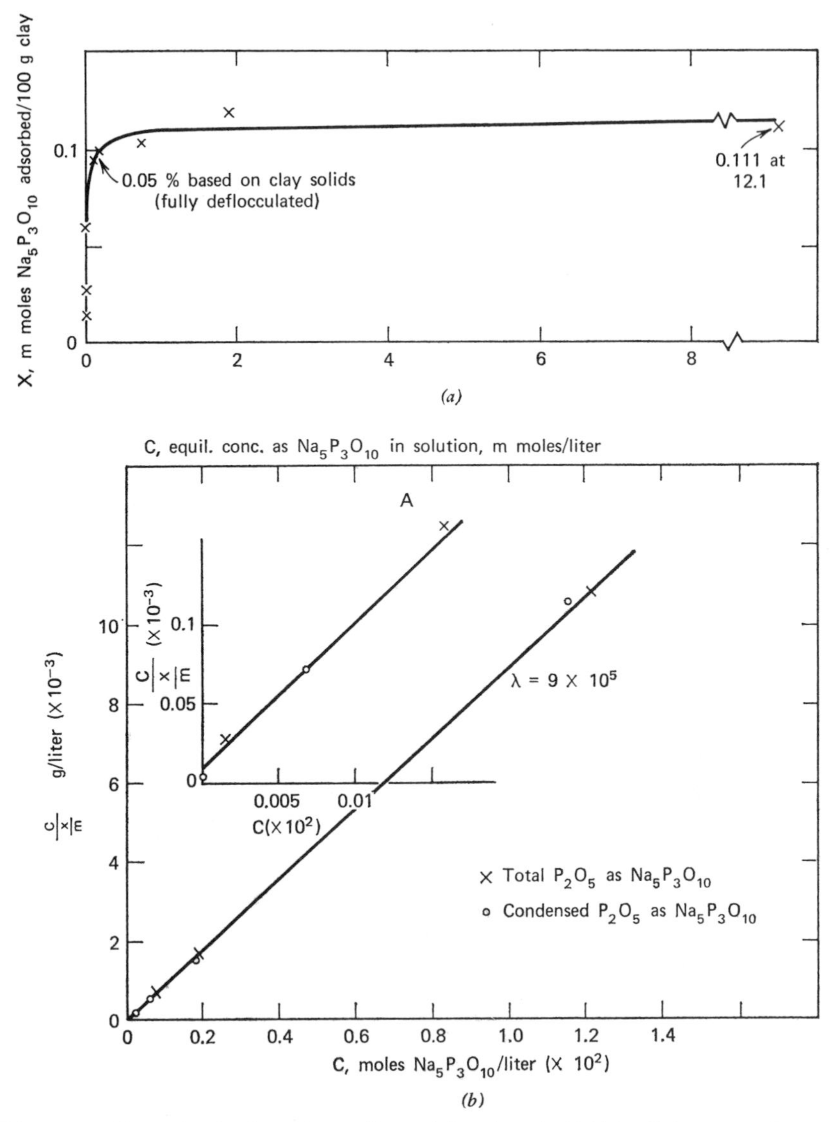

Figure 3. Absorption isotherm for sodium triphosphate in 33% kaplinite suspensions at pH 7.0: (*a*) linear plot, (*b*) Langmuir plot (2).

in the solution which is removed by centrifuging. The more the system is contaminated with extraneous soluble salts of calcium, iron, and so on, the more P_2O_5 will end up in the solution which is removed. Thus in a suspension requiring 0.2% phosphate, only 0.04% or so will be on the clay surface; the rest will be in solution.

If the amount of phosphate required to cover the surface is known, the excess in solution can be calculated and the contribution to the waste can be estimated rather closely. It is important to have characterized the particular system of concern, for each is different. Studies like that in Ref. 2 are rare in the literature; indeed, the work there was done because a close study of the literature showed that such work had not been published earlier. It will therefore be necessary for new studies to be made for many if not most such systems before statements can be made from basic thermodynamic information.

A word about the tenacity with which the deflocculant may be held on the particle surfaces. A calculation for the kaolinite–triphosphate equilibrium

$$\underset{\text{Clay phase}}{\text{site}} + \underset{\text{Ion in solution}}{P_3O_{10}} \leftrightharpoons \underset{\text{Sorbed ion on clay}}{\text{site} \cdot P_3O_{10} \text{ ion complex}}$$

showed that the binding constant is of the order of 10^5 l/mole (2). This means that the sorbed P_2O_5 will be difficult to leach from the clay with pure water; many washings would be needed with only small amounts desorbing in each. In one attempt to recover sorbed triphosphate ions for analytical purposes (3) it was found that only very strongly adsorbed anions would displace the triphosphate. Thus of competing ions studied only fluoride ion was found effective. The fluoride ion reacts with clay surfaces to form new aluminum and silicon compounds. Most monovalent anions would not be expected and indeed are found not to displace a polyanion from the surface except when used in large excess. The relationships are precisely those found useful in treating the equilibria in ion exchange. The polyphosphates are more tightly held on clay surfaces than orthophosphate. In one model study the dimer (pyrophosphate) was shown to compete for aluminum ion about twentyfold more than the monomer (ortho) (3). Thus the binding constant for orthophosphate may be of the order of 10^4 or so. (This indicates that even orthophosphate is held tightly at the clay surface.)

What is the meaning of all this in terms of P_2O_5 values in wastewaters? Much of the deflocculant used goes into systems where the water used is subsequently evaporated. The deflocculant employed remains 100% in the solid phase. Such is the case with paints, for example. In those cases where the deflocculated solid is separated from the water or solution phase mechanically, as in paper coating, clay beneficiating, and the like, the phosphate required to cover the mineral surface is tightly adsorbed and remains with the solid. This portion undoubtedly does not enter the wastewater. A variable amount of polyphosphate is required in the solution phase to maintain full surface coverage on the solid phase (maintain complete deflocculation). The amount of "free" polyphosphate is very small, of the order of 0.02 mM/l, but the total amount of polyphosphate in solution may be large because of complex ion formation with multivalent cations. Thus a system with much soluble calcium will have a good deal of polyphosphate in the solution phase at full deflocculation. Rarely does this exceed 0.2 to 0.4% on solids contained; however, a key conclusion of this brief look at P_2O_5 values arising from the deflocculation application is that the precise amount can be calculated from thermodynamic principles in advance provided that certain basic measurements have been made to characterize both the solid phase and the water to be used. Such measurements are warranted for many reasons; the question of water quality merely adds another.

REFERENCES

1. Lyons, J. W., in *Phosphorus and Its Compounds*, Vol. II, J. R. Van Wazer (Ed.), Interscience, New York, 1961, pp. 1655–1729.
2. Lyons, J. W., *J. Colloid Sci.*, **19**, 399 (1969).
3. Wilbanks, J. A., M. C. Nason, and W. C. Scott, *J. Agric. Food Chem.*, **9**, 174 (1961); unpublished data from G. D. Nelson, Research Department, Inorganic Chemicals Division, Monsanto Company, St. Louis, Missouri.

14

Colorimetric Methods of Analysis of Phosphorus at Low Concentrations in Water

J. E. HARWOOD and W. H. J. HATTINGH

National Institute for Water Research of the South African Council for Scientific and Industrial Research, Pretoria, Republic of South Africa

Phosphorus occurs widely in the environment, its importance being due to its nutrient role. In the aqueous environment, phosphorus is a prerequisite for microbial growth. Heavy algal growth (the so-called algal bloom) is undesirable in bodies of water (rivers, streams and lakes) since secondary pollution is created. The analysis of phosphorus is therefore of extreme importance to the limnologist and those concerned with water pollution. Normally, the phosphorus level in natural waters is low, so that high phosphorus levels are indicative of pollution (44). Eutrophication is becoming more and more serious in many parts of the world (68). In the sea, too, phosphorus plays the role of a micronutrient, its availability affecting the productivity of the sea.

Reliable methods exist for analyzing seawater samples (98) and are not discussed further in this chapter. Our attention here is mainly concerned with methods of phosphorus analysis for water and wastewaters because although standard procedures are available [see *Standard Methods* (94)], we feel that the methods chosen are not the best available.

Chemical Forms Present

Phosphorus may be present in aqueous samples in many different forms (24). In seawater the dissolved phosphorus may be phosphate or organic phosphorus compounds, while the suspended phosphorus may be insoluble or adsorbed phosphates or organically combined phosphorus in suspension (4). In addition, polyphosphates are used in detergents and water treatment (91) and therefore could be present in both freshwater and seawater samples (93). The sum of all the concentrations represents the total phosphorus content of the sample [*Standard Methods of Chemical Analysis* (95)] and is determined chemically rather than by addition.

A consideration of the chemical forms of phosphorus which might be present in any sample gives rise to the following broad classification:

1. orthophosphate;
2. condensed phosphorus;
3. organic phosphorus, that is, algal metabolites (107); and
4. total phosphorus (the sum of 1, 2, and 3 above)

Generally we are concerned mainly with 1 and 4 and phosphorus present in the "dissolved" fraction rather than the "suspended" fraction. Studies on the phosphorus cycle in lakes require analysis of the sediments in addition to analysis of the water phase. Sediments require their own methods (104).

Fractionation of Sample

The sample to be analyzed is split into dissolved and suspended fractions, usually by filtration. Generally a membrane filter (pore size 0.45 μ) is used (101, 98). This separation is rather arbitrary in that Rigler (80, 81) has found that the "dissolved" phosphate concentration was markedly dependent on the pore size of the membrane used in the filtration.

Analytical Procedures

Having separated the sample into dissolved and suspended fractions, various analyses can be performed on them. The most frequently performed analysis is that for orthophosphate. Analytical procedures have been improved upon (37) so that this analysis is now easily performed on a routine basis. The major problem is that, under the conditions of the analysis, phosphorus in other forms might be converted to orthophosphate. The trend is to rely on methods using the mildest possible conditions and the shortest reaction times. Methods using heating should be avoided to minimize the possibility of hydrolysis of nonphosphate phosphorus compounds to orthophosphate.

The second class of phosphorus compounds of interest in aqueous samples is the so-called condensed or acid-hydrolyzable phosphates (101). These are a heterogeneous group of compounds of which sodium polyphosphate glasses are an example.

No chemical methods exist for determining condensed phosphates as such. Use is made, instead, of their ease of hydrolysis in acidic solution, the condensed phosphates being converted to orthophosphate which is then determined. Boiling for 30 min in 1 N acid is generally adequate (95). The condensed phosphate is calculated from the difference in the orthophosphate concentration before and after acid hydrolysis. Since the concentration is based on differences between two other concentrations, the method is not very reliable at low concentrations. Furthermore, acid hydrolysis might convert organic phosphorus to orthophosphate, although this can be minimized by judicious choice of reagent strengths, boiling time, and so on (101).

The third category of interest is the level of organic phosphorus. The usual method for this group entails some sort of mineralization of the organic phosphorus to orthophosphate, with subsequent determination of the orthophosphate.

The analytical result is called "total phosphorus" since the procedure ensures conversion of condensed and organic phosphorus to orthophosphate. Organic phosphorus can be calculated by the difference between the total and the condensed phosphorus results. Because of doubts (above) as to the selectivity of the latter analysis, a value for organic (as opposed to total) phosphorus can be viewed with a certain amount of doubt.

In general, analyses for "ortho-" and "total" phosphorus are the most reliable. "Condensed" and "organic" phosphorus analyses do not have this reliability since they are often calculated as small differences between two readings, and are consequently of limited validity.

The following discussion is concerned mainly with the determination of orthophosphate and total phosphate.

ANALYTICAL METHODS

The phosphorus concentrations in the water samples are often at the trace (milligrams per liter) level in the absence of gross pollution. This naturally precludes the more insensitive analytical methods (i.e., gravimetric) from being applied directly to the samples. Consequently the methods most often applied are spectrophotometric methods, since they have adequate sensitivity.

Where large numbers of analyses are required, simple methods requiring a minimum of operator skill, and no critical time interval between adding reagents or measuring the color intensity, are distinct advantages.

Determination of Orthophosphate

The most commonly used methods are based on the reaction of phosphate in acidic medium with molybdate ions to form molybdophosphoric acid, which is subsequently measured (49). This type of reaction is not specific for phosphate since other ions (i.e., silicate, arsenate, germanate) also form heteropolyacids with molybdenum (63). Snell and Snell (92) give a detailed account of the earlier work. Orthophosphate is, however, the only form of phosphate that forms a complex with molybdate (95).

With relatively high phosphorus concentrations, the analysis depends on the formation of yellow vanadomolybdophosphoric acid in the presence of vanadium (71, 35, 1, 101). The method has been shown to be susceptible to silicate interference (64) although this is not serious in the case of water samples. Extraction of the molybdovanadophosphate helps reduce interferences from labile phosphorus compounds (76).

An alternative method for relatively high concentrations could be the use of lanthanum chloranilate (40). However, sulfate and fluoride interfere. The method covers the concentration range 3 to 300 mg P/l, and so is too insensitive to be suitable for water analyses.

With lower phosphate concentrations (i.e., in relatively unpolluted waters), increased sensitivity is obtained by reducing the phosphomolybdate to give a blue-colored complex, loosely known as a heteropolyblue. Heslop and Pearson (48) have, however, expressed reservations concerning methods based on measuring the reduced species. The formation of the 12-molybdophosphoric acid is very acid sensitive (52) so that careful control of the acidity is required. Methods using solvent extraction (77, 78) or reduction (89) shift the equilibrium of the reactions taking place, so that these methods do not require accurate control of the acidity.

In the past, aminonaphtholsulfonic acid and stannous chloride reductants have been recommended as standard methods of analysis (94). Neither of these satisfies the criteria suggested earlier for routine methods, requiring fixed time intervals before taking readings and also careful temperature control.

Stannous chloride in particular has gained wide popularity, and the reagent itself has been improved (90, 43). The serious shortcomings of this reductant have been fully discussed by Burton and Riley (13), and (82). The major shortcomings are as follows:

1. The absorbance is temperature dependent.
2. Color is relatively unstable, so that accurate time intervals for reading the color are necessary.
3. There is an appreciable salt error.
4. Arsenate interferes but can be removed chemically (58).
5. Copper interferes appreciably above 50 μg Cu/l.
6. The method is unsatisfactory at high concentration levels (69, 101).

The use of ascorbic acid as the reducing agent has eliminated most of these problems. The initial method (72) had a very stable color and no copper or arsenate interference. Color development took a long time (12 hr) or required boiling (27), with a probability of hydrolysis of organic phosphorus compounds. Catalysts for this reductant have been reported (54, 55), considerably reducing the period required for a stable color, although it now appears that bismuth is not a catalyst but takes part in the reaction (29). A subsequent modification, which uses antimony as catalyst (74), produced a method with a stable color, rapid color development, and a small salt error. Arsenate interference could be reduced by changes in the acid concentration (82, 50) or by using thiosulfate and bisulfite (102) or by solvent extraction (84).

The methods of Riley have been tested in comparative tests on seawater. The earlier method gave results (58) both higher and lower than the stannous chloride method. Jones (57) concluded that the improved technique was a good method due to its ease and color stability.

The method of Murphy and Riley (74) has also been applied to wastewaters (51, 23) and freshwaters (30).

The calibration graph was shown to be nonlinear at higher phosphorus concentrations. Edwards *et al.* (23) improved the linearity as well as the sensitivity of the method. Harwood *et al.* (37) extended linearity even further, making the method more suitable for samples of higher phosphorus content.

Strickland and Parsons (98) expressed the opinion that the use of an ascorbic acid–antimony reagent probably represented the ultimate in seawater analysis. Our experience is that it certainly represents a major advantage over stannous chloride and the other reductants. Vogler (103) has also pointed out that many errors occur with stannous chloride but that ascorbic acid gave constant readings.

A recent comparative test on phosphate methods for water samples (101) concluded that stannous chloride was the best method of those tested. However, no antimony–ascorbic acid method was among those being evaluated, although a few participants did use it. Hence, this study does not invalidate the claims made for the antimony–ascorbic acid technique.

Doubt exists over the specificity of the reaction, using antimony–ascorbic acid reductant. Jones and Spencer (58) considered that inorganic complex phosphates had been dissociated. Tri-, hexa-, and metaphosphates hydrolyze, while some of the other polyphosphates are stable (66). However, both Edwards *et al.* (23) and Harwood *et al.* (37) showed that calgon, a polyphosphate glass, gave minimal response in the orthophosphate procedure.

Murphy and Riley (73) observed no interference by a number of organic phosphorus compounds, while appreciable interference had been reported for sugar phosphates (37). Murphy and Riley (73) did mention, however, that very labile phosphorus compounds might give incorrect results. In biological materials, hydrolysis of labile organic phosphorus compounds is observed (95). These labile compounds are of two types, namely organic and inorganic phosphates that hydrolyze by splitting P—O—P links, and phosphate esters of aldehydes or ketones in the "enol" form. These hydrolyze by splitting the C—O—P links. Hence, adenosine triphosphate has two labile phosphorus atoms, and therefore interferes in phosphate analyses (37). Likewise, glucose-1-phosphate has a labile phosphorus atom which is linked to the "enol" oxygen. Glucose-6-phosphate has no such labile phosphorus atoms.

Doubts have also been expressed (80, 81) over the specificity of methods using molybdenum blue colors developed with stannous chloride reductants. The interference from labile phosphate esters can be overcome by an extraction prior to the reduction (31). This also reduces the salt effect, using stannous chloride reductant. Any chemical species capable of forming heteropoly blue compounds can interfere. These include arsenate, silicate, and germanate (63). Arsenate interference can be controlled by adjusting the acid strength (82), which also applies to silicate interference (45, 73) since by suitable adjustment of the acidity, it is possible to form the phosphomolybdate and not the silicomolybdate. Interferences by arsenate and silicate can also be readily controlled by means of solvent extraction (106). Isobutyl acetate is extremely selective for phosphorus under certain conditions (75). Arsenic(V) interference can be controlled by a coprecipitation technique (59) which is lengthy, or by reduction to arsenic(III) (22).

Of the other techniques used in trace analyses, atomic absorption spectrophotometry cannot be applied directly owing to the lack of sensitivity for phosphorus. Methods using a preconcentration of phosphorus by solvent extraction of the phosphomolybdic acid have nevertheless been developed (60, 79).

Amplification procedures (100, 10) in which the phosphorus is reacted with molybdate, which in turn is reacted with oxine (11), have also been reported. The molybdenum can also be determined spectrophotometrically (20), by atomic absorption (109, 60, 50), or by titration (61).

A recent development in analytical techniques has been the upsurge in the use of reaction-rate measurements to determine concentration, rather than in measuring the intensity of the final (stable) color. Use has been made of the fact that silicate reduction occurs much slower than with phosphate (88). Rate differentials also exist in the formation of the heteropoly blue from phosphate and from organophosphorus compounds (65). The advantage is that measurement can be made before side reactions have begun occurring (19), thus improving the selectivity of the method (53). The instrument could give a direct concentration readout from 0.1 mg P/l. However, while arsenate did not interfere, silica did, although the reduction of the silicomolybdate can be made considerably slower than that of phosphomolybdate (18).

It is also possible to use radiochemical methods (33). However, the method suffers from interferences similar to those of the chemical method, and does not represent any real advantage. Activation analysis is also possible (2).

The increased complexity of these other instrumental methods rather limits their usefulness. The simplicity of the spectrophotometric method using ascorbic acid–antimony reductants leads us to prefer these reagents.

Determination of Condensed Phosphate

Condensed phosphates are measured indirectly, after acid hydrolysis to the orthophosphate (94). Only inorganic polyphosphates and phosphorus bound as P—O—P, but not that bound as P—O—C or P—C, are hydrolyzed (7).

Since the possibility exists that organically bound phosphorus will also be hydrolyzed, this determination has rather limited validity. Vogler (105) has shown that this analysis is only reliable if no organic phosphorus is present. The concentration is also obtained by difference, and so is inaccurate.

Determination of Total Phosphorus

Organic phosphorus is converted to orthophosphate by oxidation by either wet or dry methods (94). Organic phosphorus can then be calculated by the difference between the total phosphorus and the condensed and orthophosphate phosphorus. Often this difference is small, so the result is not reliable. We prefer to report total phosphorus, and not organic phosphorus.

An alternative procedure for organic phosphorus is to use ultraviolet irradiation of the sample (6, 5) which releases organic phosphorus, while condensed phosphorus is unaffected (46). A difference calculation is required (in orthophosphate levels, before and after irradiation), so this method is also subject to inherent limitations in accuracy.

Wet Oxidation. Numerous mixtures have been used for the wet oxidation of organic phosphorus to orthophosphate. Perchloric acid (34, 49) has been widely used, and is recommended for seawater samples (97). Sulfuric acid with hydrogen peroxide (15) or with fusion of the residue (16) makes up two other methods. A mixture of sulfuric and perchloric acids (14) has also been used.

Gourley (32) has shown rather poor recoveries after refluxing with concentrated nitric acid and fuming with perchloric acid. Fuming with perchloric acid gave poor recoveries unless the samples were taken to dryness (13). Clearly, care is required to ensure complete mineralization with wet oxidations. Sanning (86) compared acid digestion with persulfate oxidation, and preferred the latter.

Dry Oxidation. Dry oxidation is the conventional ashing technique using a furnace. It is difficult to envisage problems of poor recovery of phosphorus due to nonbreakdown of organic phosphorus compounds if ashing is complete. Having oxidized all the carbonaceous material, only an inorganic residue should remain. Phosphorus might be present in forms other than orthophosphate (i.e., pyrophosphate) should the ashing temperature be too high, and so give inaccuracies in the subsequent reaction with molybdate.

Levine *et al.* (63) have ashed at 600°C but the method is not widely used, due possibly to formation of meta- and polyphosphates at this temperature (82). Redissolving the residue can have problems (62) even when the ashing itself is satisfactory. Saliman (85) ashed at 700°C and found low recoveries, due to the formation of pyrophosphates, which do not give the heteropoly blue reaction.

Harwood *et al.* (38), in a recent comparative study of five oxidation methods (three wet, two dry oxidations), tested some newer oxidants. These included the use of persulfate (68, 70, 28), a potassium nitrate–sodium nitrate eutectic melt (12), and 50% hydrogen peroxide (99, 21, 3). Conventional wet ashing with perchloric–sulfuric acids, and dry ashing with magnesium nitrate at 550° (8) were included for comparison purposes. Newer methods were tested since the older techniques have been investigated previously (83).

The methods were tested on both liquid and solid samples, mainly of biological origin. The magnesium nitrate method appeared very good, but the preliminary evaporation to dryness lengthened the procedure. The 50% hydrogen peroxide method was extremely quick, being even quicker than persulfate. Further experience has led us generally to prefer dry ashing, using magnesium nitrate. The analyses are performed on a semimicro scale, so that many samples can be processed in a conventional furnace at any one time.

For dilute samples, the small volume that can be placed in a crucible might limit the dry ashing. For such samples, a wet-ashing procedure with 50% hydrogen peroxide (38) is advantageous since larger samples can be used in this case. The orthophosphate calibration graph can be used for total phosphorus determination with 50% hydrogen peroxide oxidant (38). Normally a separate calibration graph is required (62).

Another oxidation procedure that could be applied to solid samples is the Schoniger (oxygen) flask technique (87). This technique has been applied to the determination of phosphorus in, among others, motor oils and additives (9) and in organic phosphorus compounds (26). The method would not appear to be readily adaptable to large numbers of liquid samples.

GENERAL

In the discussion above attention has been focused primarily on spectrophotometric methods. These have the advantage that changes in concentration can be readily accommodated by changes in the path length of the optical cell. For reasonable precision, the absorbance should lie between 0.3 and 1.0 (82). While commercially available equipment varies widely in complexity, a colorimeter is standard equipment in practically every analytical laboratory.

Furthermore, colorimetric methods are readily automated for analyses of ever-increasing sample numbers. Simple automation consists of using reagent dispensers set at the required volumes, and colorimeters with flow-through cells. Increasing complexity can be achieved by having direct concentration readout, with finally the whole method being performed automatically, including data output.

An important point to bear in mind when collecting samples for subsequent phosphorus analysis is the preservation of the phosphorus in the form in which it was present initially. Any changes with time can make a complete mockery of results obtained from careful analytical procedures. If total phosphate is the only analysis required, then preservation is not required (101). Loss of phosphate onto the walls of the sample container must be guarded against, however.

Phosphate has been shown (39) to be only slightly adsorbed by Pyrex glass, while polyethylene and polyvinyl chloride adsorbed phosphate strongly (72). A treatment with a solution of iodine in potassium iodide (47) enables polyethylene bottles to be used in sampling.

Generally, acidification of the sample helps to prevent phosphate adsorption onto container walls, while filtration prevents desorption of phosphate from the suspended matter present (82). Algae should be removed before preserving since phosphorus is released from them (25).

Studies have been conducted on the stability of samples to different types of preservation. The use of mercuric chloride has been shown to be generally satisfactory (41, 42, 101). Cooling, but preferably freezing, appears to be better (56, 46) if these facilities are available. Freezing should be used whenever a delay longer than a few hours is likely to occur between sampling and analyzing (97). Hydrolysis of polyphosphate can also occur in samples (101).

Naturally, as the concentration of the element to be analyzed decreases, so more and more stringent precautions are required to ensure that the samples do not become contaminated with phosphorus from extraneous sources. At trace levels, the glassware is kept separate, while phosphate-containing detergents should not be used (101). High-purity reagents are necessary and reagent blank analyses must be performed in addition to analyzing the samples. When membranes are used for filtering, they should be thoroughly washed and soaked before use. Samples themselves should be analyzed in duplicate, or even triplicate for more exact work, and differences in the optical cells themselves allowed for. The extent to which these precautions should be observed should be considered for each case by the analyst.

CONCLUSIONS

We have tried to give an indication of the methods and problems associated with analysis for the different forms of phosphorus in water samples. Of the many different methods available, molybdovanadate is a simple procedure for higher phosphate concentrations. At lower levels reduction to the heteropoly blue is required, preferably with ascorbic acid–antimony reductant.

For the mineralization of total phosphorus, a dry-ashing procedure can be readily adapted to large numbers of samples. Where large sample volumes have to be analyzed, either hydrogen peroxide (50%) or persulfate generally give satisfactory results.

The methods chosen are considered to be quick and easy, although with certain samples containing labile organophosphorus compounds the orthophosphate result will be inaccurate. Generally, the methods give satisfactory analyses without requiring tedious steps (i.e., solvent extraction) or expensive equipment.

REFERENCES

1. Abbot, D. C., G. E. Emsden, and J. R. Harris, *Anal., London*, **88**, 814 (1963).
2. Allen, H. E., and R. B. Hahn, *Environ. Sci. Technol.* **3**, 844 (1969).
3. Analytical Methods Committee, *Anal., London*, **92**, 403 (1967).
4. Armstrong, F. A. J., in *Chemical Oceanography*, Vol. 1, J. P. Riley and G. Skirrow (Eds.), Academic Press, New York, 1965, Chapter 8.
5. Armstrong, F. A. J., and S. Tibbits, *J. Mar. Biol. Assoc. U. K.*, **48**, 143 (1968).
6. Armstrong, F. A. J., P. M. Williams, and J. D. H. Strickland, *Nature*, **211**, 481 (1966).
7. Association of the American Soap and Glycerine Producers, Inc., *J. Am. Water Works Assoc.*, **50**, 1963 (1958).
8. Association of the Official Agricultural Chemists, *Official Methods of Analysis*, 7th ed., 1950, p. 196.

9. Barney, J. E., J. G. Bergman, and W. G. Tuskan, *Anal. Chem.*, **31**, 1394 (1959).
10. Belcher, R., *Talanta*, **15**, 357 (1968).
11. Belcher, R., and P. C. Uden, *Anal. Chim. Acta*, **42**, 180 (1968).
12. Bowen, H. J. M., *Anal. Chem.*, **40**, 969 (1968).
13. Burton, J. D., and J. P. Riley, *Mikrochim. Acta*, 1350 (1956).
14. Christopher, A. J., and T. R. F. W. Fennel, *Microchem. J.*, **12**, 593 (1967).
15. Cooper, L. H. N., *J. Mar. Biol. Assoc. U. K.*, **19**, 755 (1934).
16. Cooper, L. H. N., *J. Mar. Biol. Assoc. U. K.*, **21**, 673 (1937).
17. Crouch, S. R., and H. V. Malmstadt, *Anal. Chem.*, **40**, 1922 (1968).
18. Crouch, S. R., and H. V. Malmstadt, *Anal. Chem.*, **39**, 1090 (1967).
19. Crouch, S. R., and H. V. Malmstadt, *Anal. Chem.*, **40**, 1901 (1968).
20. Djurkin, V., G. F. Kirkbright, and T. S. West, *Anal., London*, **91**, 89 (1966).
21. Down, J. L., and T. T. Gorsuch, *Anal., London*, **92**, 398 (1967).
22. Duval, L., *Chim. Anal.*, **49**, 307 (1967).
23. Edwards, G. P., A. H. Molof, and R. S. Schoeman, *J. Am. Water Works Assoc.*, **57**, 917 (1965).
24. Eliassen, R., and G. Tchobanoglous, *J. Water Pollut. Control Fed.* **40**, May, R171, (1968).
25. Fitzgerald, G. P., and S. L. Faust, *Limnol. Oceanogr.*, **12**, 332 (1967).
26. Fleischer, K. D., B. C. Southworth, J. H. Hodecker, and M. M. Tuckerman, *Anal. Chem.*, **30**, 152 (1958).
27. Fogg, D. N., and N. T. Wilkinson, *Anal., London*, **83**, 406 (1958).
28. Gales, M. E., E. C. Julian, and R. C. Kroner, *J. Am. Water Works Assoc.*, **58**, 1363 (1966).
29. Goldman, H. D., and L. G. Hargis, *Anal. Chem.*, **41**, 490 (1969).
30. Golterman, H. L., *IBP Handbook No. 8*, Blackwell Scientific Publications, Oxford, 1969.
31. Golterman, H. L., and I. M. Wurtz, *Anal. Chim. Acta*, **25**, 295 (1961).
32. Gourley, D. R. H., *Nature*, **169**, 192 (1952).
33. Hahn, R. B., and T. M. Schmitt, *Anal. Chem.*, **41**, 359 (1969).
34. Hansen, A. L., and R. J. Robinson, *J. Mar. Res.*, **12**, 31 (1953).
35. Hanson, W. C., *J. Sci. Food Agric.*, **1**, 172 (1950).
36. Harvey, H. W., *J. Mar. Biol. Assoc. U. K.*, **27**, 337 (1948).
37. Harwood, J. E., R. A. van Steenderen, and A. L. Kühn, *Water Res.*, **3**, 417 (1969a).
38. Harwood, J. E., R. A. van Steenderen, and A. L. Kühn, *Water Res.*, **3**, 425 (1969b).
39. Hassentuefel, W., R. Jagitsch, and F. F. Koczy, *Limnol. Oceanogr.*, **8**, 152 (1963).
40. Hayashi, J., T. Danzuka, and K. Ueno, *Talanta*, **4**, 244 (1960).
41. Hellwig, D. H. R., *Int. J. Air Water Pollut.*, **8**, 215 (1963).
42. Hellwig, D. H. R., *Water Res.*, **1**, 79 (1966).
43. Henriksen, A., *Anal., London*, **88**, 898 (1963).
44. Henriksen, A., *Anal., London*, **90**, 29 (1965).
45. Henriksen, A., *Anal., London*, **91**, 290 (1966).
46. Henriksen, A., *Anal., London*, **95**, 601 (1970).

47. Heron, J., *Limnol. Oceanogr.*, **7,** 316 (1962).
48. Heslop, R. B., and E. F. Pearson, *Anal. Chim. Acta*, **39,** 209 (1967).
49. Hirata, A. A., and D. Appleman, *Anal. Chem.*, **31,** 2097 (1959).
50. Hurford, T. R., and D. F. Boltz, *Anal. Chem.*, **40,** 379 (1968).
51. Jankovic, S. G., D. T. Mitchell, and J. C. Buzzell, *Water Sewage Works*, **114,** 471 (1967).
52. Javier, A. C., S. R. Crouch, and H. V. Malmstadt, *Anal. Chem.*, **40,** 1922 (1968).
53. Javier, A. C., S. R. Crouch, and H. V. Malmstadt, *Anal. Chem.*, **41,** 239 (1969).
54. Jean, M., *Anal. Chim. Acta*, **14,** 172 (1956).
55. Jean, M., *Anal. Chim. Acta*, **31,** 24 (1964).
56. Jenkins, D., in "Trace Inorganics in Water," *Advances in Chemistry Series*, No. 73, American Chemical Society, Washington, D.C., 1968.
57. Jones, P. G. W., *J. Mar. Biol. Assoc. U. K.*, **46,** 19 (1966).
58. Jones, P. G. W., and C. P. Spencer, *J. Mar. Biol. Assoc. U. K.*, **43,** 251 (1963).
59. Kar, K. R., and G. Singh, *Mikrochim. Acta*, 560 (1968).
60. Kirkbright, G. F., A. M. Smith, and T. S. West, *Anal., London*, **92,** 411 (1967).
61. Kirkbright, G. F., A. M. Smith, and T. S. West, *Anal., London*, **93,** 224 (1968).
62. Lee, G. F., N. L. Clesceri, and G. P. Fitzgerald, *Int. J. Water Pollut.*, **9,** 715 (1965).
63. Levine, H., J. J. Rowe, and F. S. Grimaldi, *Anal. Chem.*, **27,** 258 (1955).
64. Lew, R. B., and F. Jakob, *Talanta*, **10,** 322 (1963).
65. Lorentz, K., *Z. Anal. Chem.*, **237,** 32 (1968).
66. Lucci, G. C., *Ann. Chim.*, **56,** 1343 (1966).
67. Mackenthun, K. M., *J. Am. Water Works Assoc.*, **60,** 1047 (1968).
68. May, R., *Anal. Chem.*, **31,** 308 (1959).
69. McAloren, J. T., and G. F. Reynolds, *Talanta*, **10,** 145 (1963).
70. Menzel, D. W., and N. Corwin, *Limnol. Oceanogr.*, **10,** 280 (1965).
71. Michelsen, O. B., *Anal. Chem.*, **29,** 60 (1957).
72. Murphy, J., and J. P. Riley, *Anal. Chim. Acta*, **14,** 318 (1956).
73. Murphy, J., and J. P. Riley, *J. Mar. Biol. Assoc. U. K.*, **37,** 9 (1958).
74. Murphy, J., and J. P. Riley, *Anal. Chim. Acta*, **27,** 31 (1962).
75. Pakalns, P., *Anal. Chim. Acta*, **40,** 1 (1968).
76. Parvin, R., and R. A. Smith, *Anal. Biochem.*, **27,** 65 (1969).
77. Paul, J., *Mikrochim. Acta*, 830 (1965).
78. Paul, J., *Anal. Chim. Acta*, **35,** 200 (1966).
79. Ramakrishna, T. V., J. W. Robinson, and P. W. West, *Anal. Chim. Acta*, **45,** 43 (1969).
80. Rigler, F. H., *Verh. Int. Verein. Limnol.*, **16,** 465 (1966).
81. Rigler, F. H., *Limnol. Oceanogr.*, **13,** 7 (1968).
82. Riley, J. P., in *Chemical Oceanography*, Vol. 2, J. P. Riley and G. Skirrow, (Eds.), Academic Press, New York, 1965, Chapter 21.
83. Robertson, R. S., *J. Am. Water Works Assoc.*, **52,** 483 (1960).
84. Ross, H. H., and R. B. Hahn, *Talanta*, **7,** 276 (1961).
85. Saliman, P. M., *Anal. Chem.*, **36,** 112 (1964).

86. Sanning, D. E., *Water Sewage Works*, **114,** 131 (1967).
87. Schoniger, W., *Mikrochim. Acta*, 123 (1955).
88. Shen, C. Y., and D. R. Dyroff, *Anal. Chem.*, **34,** 1367 (1962).
89. Sims, R. P. A., *Anal., London*, **86,** 584 (1961).
90. Sletten, O., and C. M. Bach, *J. Am. Water Works Assoc.*, **53,** 1031 (1961).
91. Smith, C. V., and S. J. Medlar, *J. Am. Water Works Assoc.*, **60,** 921 (1968).
92. Snell, F. D., and C. T. Snell, *Colorimetric Methods of Analysis*, 3rd ed. Vol. II, Van Nostrand, Princeton, N.J., 1949.
93. Solorzano, L., and J. D. H. Strickland, *Limnol. Oceanogr.*, **13,** 515 (1968).
94. *Standard Methods for the Examination of Water and Waste Water*, 12th ed., American Public Health Association, New York, 1965.
95. *Standard Methods of Chemical Analysis*, 6th ed., Vol. 1, N. H. Furman (Ed.), Van Nostrand, Princeton, N.J., 1966, Chapter 35.
96. Stephens, K., *Limnol. Oceanogr.*, **8,** 361 (1963)
97. Strickland, J. D. H., and T. R. Parsons, *A Manual of Sea-Water Analysis*, Fisheries Research Board of Canada, Bulletin No. 125, Ottawa, 1960.
98. Strickland, J. D. H., and T. R. Parsons, *A Manual of Sea-Water Analysis*, Fisheries Research Board of Canada, Bulletin No. 125 (2nd ed., revised) Ottawa, 1965.
99. Taubinger, R. P., and J. R. Wilson, *Anal., London*, **90,** 429 (1965).
100. Umland, D. F., and G. Wünsch, *Z. Anal. Chem.*, **213,** 186 (1965).
101. United States Public Health Service, Water Nutrient No. 2, Study Number 36, Conducted by the Analytical Reference Service, Environmental Health Service, Cincinnati, Ohio, 1970.
102. Van Schouwenburg, J. C., and I. Waunga, *Anal. Chim. Acta*, **37,** 271 (1967).
103. Vogler, P., *Fortschr. Wasserchem. Ihre Grenzgeb.*, **2,** 109 (1965a).
104. Vogler, P., *Int. Rev. Ges. Hydrobiol.*, **50,** 33 (1965b).
105. Vogler, P., *Fortschr. Wasserchem. Ihre Grenzgeb.*, **4,** 211 (1966).
106. Wadelin, C., and M. G. Mellon, *Anal. Chem.*, **25,** 1668 (1953).
107. Watt, W. D., and F. R. Hayes, *Limnol. Oceanogr.*, **8,** 276 (1963).
108. Wentz, D. A., and G. F. Lee, *Environ. Sci. Technol.*, **3,** 750 (1969).
109. Zaugg, W. S., and R. J. Knox, *Anal. Chem.*, **38,** 1759 (1966).

15

Chromatographic Analysis of Oxoacids of Phosphorus

Shigeru Ohashi

Kyushu University, Fukuoka, Japan

The most important form of phosphorus in natural waters is orthophosphate. However, some condensed phosphates have also been found in surface waters (1). Although these condensed phosphates are largely manmade for use in detergents, water treatment, and so forth, they are also generated by living organisms.

In this chapter chromatographic analysis of orthophosphate and condensed phosphates as well as oxoacid anions of phosphorus with oxidation numbers lower than 5 is briefly mentioned. The more detailed description on these problems can be seen in the volume *Analytical Chemistry of Phosphorus Compounds* edited by M. Halmann (2).

When paper or thin-layer chromatography or paper electrophoresis is adopted for analysis of oxoacid anions of phosphorus, 2 to 10 μl of a sample solution containing 1 to 5 mg P/ml for each oxoacid anion species is usually loaded at a starting line. On the other hand, when ion-exchange or gel chromatography is employed, 1 to 2 ml of a sample solution containing 0.1 to 1 mg P/ml for each oxoacid anion species is usually loaded at the top of a column. The above-mentioned concentrations and volumes of the sample solutions are deduced from the assumption that phosphorus is finally determined colorimetrically by the use of a blue color of reduced molybdophosphoric acid. If phosphorus is determined by activation analysis in the last step, a more diluted sample solution can be used for the chromatographic analysis.

When a sample concentration is too low for the chromatographic analysis, oxoacid anions of phosphorus in the sample solution must be concentrated to a desired level. Anion exchangers may be used for the collection of oxoacid anions of phosphorus, if concentrations of other anions are not high. A small amount of orthophosphate can be collected by coprecipitation with ferric hydroxide. A large amount of alkali metal chloride can be removed by dialysis from a small amount of polymeric oxoacid anions of phosphorus (3). Molecular-sieve gels such as Sephadex may be used for the same purpose.

Condensed phosphates and other polymeric oxoacid anions of phosphorus have a tendency to be hydrolyzed to monomers in an aqueous solution. These hydrolysis reactions proceed very rapidly in a strongly acidic solution at a high temperature and are accelerated by the presence of multivalent metallic ions. Therefore care must be taken for the hydrolytic degradation in the whole course of the analysis of polymeric oxoacid anions of phosphorus.

The abbreviated notations for the oxoacids of phosphorus listed in Table 1 are used in this chapter.

PAPER AND THIN-LAYER CHROMATOGRAPHY

Paper and thin-layer chromatographic techniques are the simplest methods for the identification of oxoacid anions of phosphorus. A large number of papers on the application of paper chromatography to the analysis of oxoacid anions of phosphorus have been published. In paper chromatography it is usually recommended to cut an edge of a sheet of filter paper into fishtail ends and to develop an unknown sample and a proper reference sample on a same sheet of paper as shown in Fig. 1.

Typical results for the R_f values of some oxoacid anions of phosphorus are shown in Tables 2 and 3 (4, 5). Table 2 indicates that for the identification of trimeta- and tetrametaphosphate in the presence of oligomeric linear phosphates, the basic solvent is better than the acidic one because both the cyclic phosphates move faster than any of the linear phosphates in the development with the basic solvent.

After the development, oxoacid anions of phosphorus on paper can be visualized as blue spots by spraying an acidic solution of ammonium molybdate and irradiating ultraviolet rays. The quantitative analysis of condensed phosphates by paper chromatography is usually composed of the following steps (6–8): (*a*) the phosphate spots or bands are cut out; (*b*) the phosphate is brought into an aqueous solution by extraction with aqueous ammonia or by ashing the paper followed by dissolving the ash; and (*c*) after the condensed phosphate is completely hydrolyzed to orthophosphate, the resulting orthophosphate is determined colorimetrically. Another method for the determination of a condensed phosphate separated by paper chromatography is based on the complete hydrolysis of the condensed phosphate to orthophosphate on the paper strip followed by the densitometric measurement of a blue color of reduced molybdophosphoric acid (9, 10). Oxoacid anions of phosphorus separated on paper are also detected or determined by the activation analysis based on the n-γ reaction of ^{31}P.

TABLE 1 Oxoacids of Phosphorus and Their Abbreviated Notations

Chemical Formula	Abbreviated Notation	Name of Salt or Anion
H_3PO_4	P_1	Orthophosphate
Condensed phosphoric acids (linear structure)		
$H_4P_2O_7$	P_2	Diphosphate (pyrophosphate)
$H_5P_3O_{10}$	P_3	Triphosphate
.	.	.
.	.	.
.	.	.
$H_{n+2}P_nO_{3n+1}$	P_n	Polyphosphate
Condensed phosphoric acids (cyclic structure)		
$(HPO_3)_3$	P_{3m}	Trimetaphosphate
$(HPO_3)_4$	P_{4m}	Tetrametaphosphate
.	.	.
.	.	.
.	.	.
$(HPO_3)_n$	P_{nm}	Metaphosphate
Lower oxoacids of phosphorus		
$HP^{I}H_2O_2$	$\overset{1}{P}$	Hypophosphite
$H_2P^{III}HO_3$	$\overset{3}{P}$	Phosphite
$H_3P^{II}P^{IV}HO_5$	$\overset{2}{P}-\overset{4}{P}$	
$H_4P_2{}^{IV}O_6$	$\overset{4}{P}-\overset{4}{P}$	Hypophosphate
$H_2P_2{}^{III}H_2O_5$	$\overset{3}{P}-O-\overset{3}{P}$	
$H_3P^{III}P^{V}HO_6$	$\overset{3}{P}-O-\overset{5}{P}$	
$H_5P^{III}P_2{}^{IV}O_8$	$\overset{4}{P}-\overset{3}{P}-\overset{4}{P}$	
$H_4P^{III}P_2{}^{IV}HO_8$	$\overset{3}{P}-O-\overset{4}{P}-\overset{4}{P}$	
$H_5P_2{}^{IV}P^{V}O_9$	$\overset{5}{P}-O-\overset{4}{P}-\overset{4}{P}$	
$H_6P_4{}^{IV}O_{11}$	$\overset{4}{P}-\overset{4}{P}-O-\overset{4}{P}-\overset{4}{P}$	
$(H_2P_2{}^{IV}O_5)_2$	$(-\overset{4}{P}-\overset{4}{P}-O-)_2$	
$(HP^{III}O_2)_6$	$(-\overset{3}{P}-)_6$	

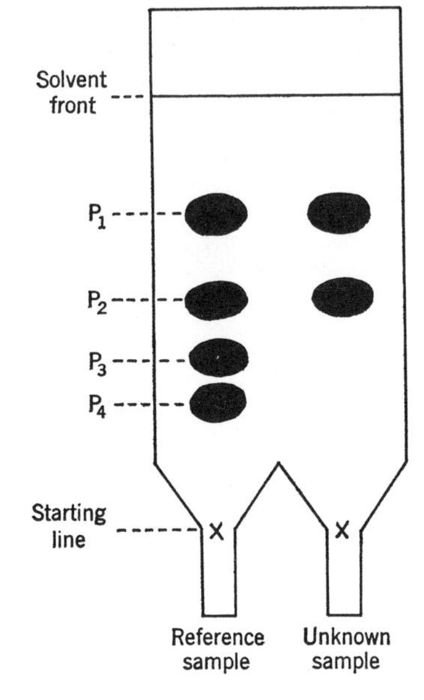

Figure 1. Schematic representation of paper chromatograms of a reference and an unknown sample.

TABLE 2 R_f Values of Some Condensed Phosphates Obtained by Ascending Paper Chromatography[a] (4)

Oxoacid Anion	R_f	
	Acidic Solvent[b]	Basic Solvent[c]
Linear phosphates		
P_1	0.73	0.33
P_2	0.53	0.24
P_3	0.39	0.24
P_4	0.29	(0.20)
P_5	0.22	—
P_6	0.16	—
P_7	0.11	—
P_8	0.08	—
Cyclic phosphates		
P_{3m}	0.32	0.53
P_{4m}	0.18	0.40

[a] Paper: Schleicher-Schüll No. 2040a. Temperature: 16°C.
[b] Acidic solvent (ml): isopropanol (70), water (10), 20% trichloroacetic acid (20), and 25% aqueous ammonia (0.3).
[c] Basic solvent (ml): isopropanol (40), isobutanol (20), water (39), and 25% aqueous ammonia (1).

TABLE 3 R_f Values of Some Lower Oxoacid Anions of Phosphorus Obtained by Ascending Paper Chromatography[a] (5)

Acidic Solvent[b]		Basic Solvent[c]	
Oxoacid Anion	R_f	Oxoacid Anion	R_f
$\overset{3}{P}$	0.79	$\overset{1}{P}$	0.58
P_1	0.77	$\overset{3}{P}—O—\overset{3}{P}$	0.50
$\overset{1}{P}$	0.66	$\overset{3}{P}$	0.34
$\overset{4}{P}—\overset{4}{P}$	0.39	$\overset{3}{P}—O—\overset{5}{P}$	0.32
$\overset{2}{P}—\overset{4}{P}$	0.36	$(—\overset{4}{P}—\overset{4}{P}—O—)_2$	0.29
$\overset{5}{P}—O—\overset{4}{P}—\overset{4}{P}$	0.27	P_1	0.28
$\overset{4}{P}—\overset{3}{P}—\overset{4}{P}$	0.18	$\overset{2}{P}—\overset{4}{P}$	0.28
$\overset{4}{P}—\overset{4}{P}—O—\overset{4}{P}—\overset{4}{P}$	0.13	$\overset{3}{P}—O—\overset{4}{P}—\overset{4}{P}$	0.26
$(—\overset{4}{P}—\overset{4}{P}—O—)_2$	0.04	$\overset{4}{P}—\overset{4}{P}$	0.16
$(—\overset{3}{P}—)_6$	0.00	$\overset{5}{P}—O—\overset{4}{P}—\overset{4}{P}$	0.13
		$\overset{4}{P}—\overset{3}{P}—\overset{4}{P}$	0.11
		$\overset{4}{P}—\overset{4}{P}—O—\overset{4}{P}—\overset{4}{P}$	0.10
		$(—\overset{3}{P}—)_6$	0.00

[a] Paper: Toyoroshi No. 51A. Temperature: 5 to 7°C.
[b] Acidic solvent (ml): acetone (325), 28% aqueous ammonia (1.75), water (150), and trichloroacetic acid (25 g).
[c] Basic solvent (ml): ethanol (190), isobutanol (150), 28% aqueous ammonia (2.7), and water (225.3).

Thin-layer chromatography has recently been rapidly developed for the separation of oxoacid anions of phosphorus on the basis of the knowledge concerning paper chromatographic behavior of condensed phosphates. Cellulose powders and silica gels are used as sorption materials.

In general thin-layer chromatography gives better resolution in less time in comparison with paper chromatography. For instance, a mixture of ortho-, di-, tri-, trimeta-, and tetrametaphosphate can be separated into single spots in only 12 min under the proper conditions (11). Linear and cyclic phosphates with degrees of polymerization higher than 4 are hardly separated from each other by the usual one-dimensional paper or thin-layer chromatography.

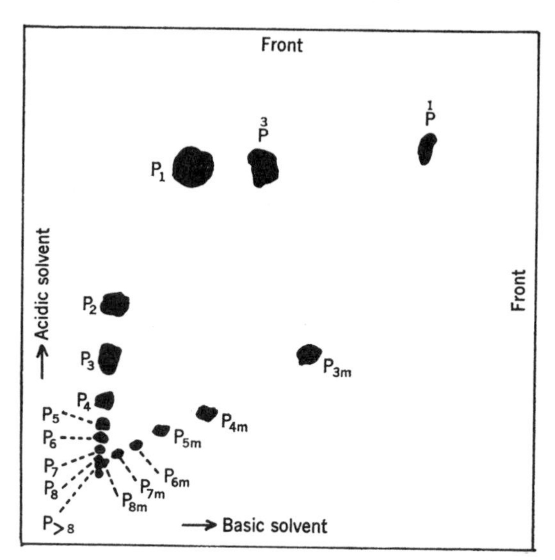

Figure 2. The two-dimensional thin-layer chromatogram of a complex mixture of hypophosphite, phosphite, and two series of linear and cyclic phosphates (12). Thin-layer: cellulose MN 300. Acidic solvent (vol. %): ethanol (10), *n*-propanol (50), water (40), and trichloroacetic acid (1 g/100 ml). Basic solvent (vol. %): ethanol (60), water (35), concentrated aqueous ammonia (5), and trichloroacetic acid (1.5 g/100 ml). Reproduced by permission of Springer-Verlag.

In order to overcome this defect, a two-dimensional technique has been developed. An example shown in Fig. 2 indicates the separation of a complex mixture of hypophosphite, phosphite, and both series of linear and cyclic phosphates (12).

PAPER ELECTROPHORESIS

Paper electrophoresis has also been employed for the analysis of oxoacid anions of phosphorus. The zone mobility of a given migrating ion in paper electrophoresis is expressed as a function of the charge and the molecular weight of the ion (13). Since the charge of a given oxoacid anion of phosphorus is dependent on the pH of a background electrolyte solution, its zone mobility varies with the pH (14). If a background electrolyte reacts with oxoacid anions of phosphorus to form complexes, these reactions must be taken into consideration.

Some investigators have recommended a mixed solution of zinc acetate and lactic acid as a background electrolyte solution for the separation of some oxoacid anions of phosphorus (15). The relative mobilities of ortho-, di-, and triphosphate and some other oxoacid anions of phosphorus with lower

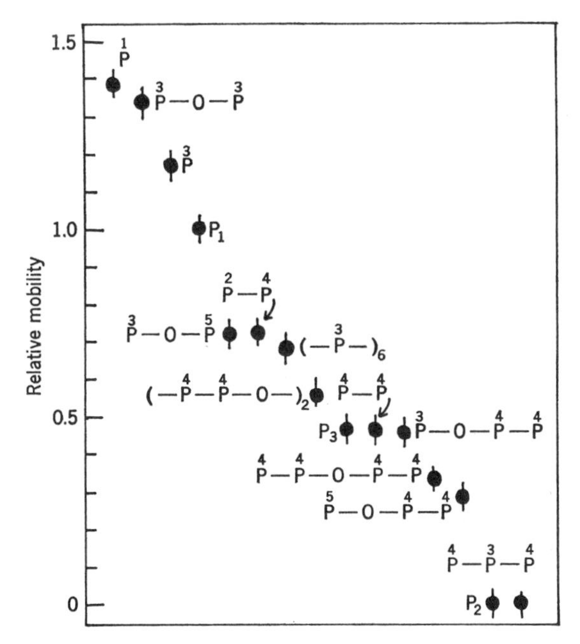

Figure 3. Relative mobilities of oxoacid anions of phosphorus in paper electrophoresis (16). Paper: Toyoroshi No. 51A, 2 × 60 cm. Background electrolyte solution: 0.03 *M* zinc acetate–0.04 *M* lactic acid, pH 4.2. Voltate: 28 V/cm. Temperature: 10°C. Reproduced by permission of the Japan Society for Analytical Chemistry.

oxidation numbers in this background electrolyte solution are shown in Fig. 3 (16). Diphosphate and $\overset{4}{P}—\overset{3}{P}—\overset{4}{P}$ anion remain at the starting line because of the formation of their insoluble zinc salts. A short running time is one of the most characteristic features of paper electrophoresis. The data given in Fig. 3 were obtained by migrating the sample ions for 2 hr.

ION-EXCHANGE CHROMATOGRAPHY

For the separation of oxoacid anions of phosphorus, resins such as Dowex 1 and Amberlite IRA-400 are usually employed. These resins are strong-base anion exchangers composed of cross-linked polystyrene. Sodium or potassium chloride is often used as an eluent.

The most important parameter in the ion-exchange chromatographic separation is the distribution ratio of a given ion between a resin phase and a solution phase. When a resin and an eluent are given, the distribution ratio of an oxoacid anion of phosphorus is determined by the pH and the concentration of an eluent solution.

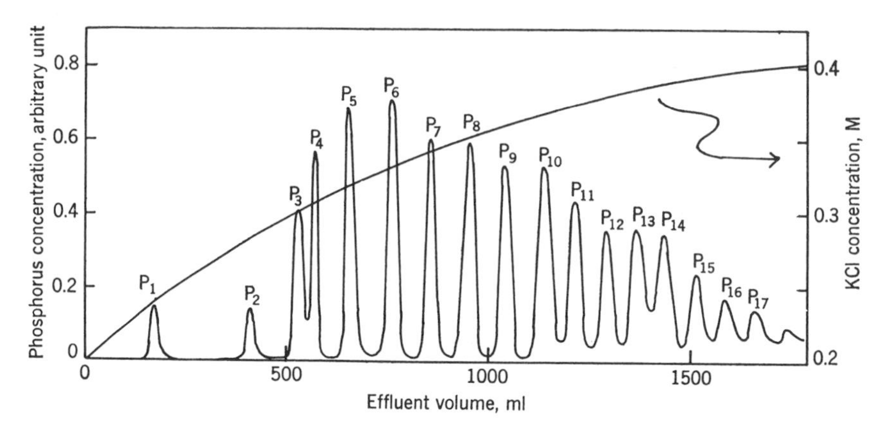

Figure 4. The elution curve of linear phosphates obtained by ion-exchange chromatography (18). Sample: a mixture of P_1, P_2, and glassy sodium polyphosphate with an average degree of polymerization of 10. Total amount of the phosphates loaded: 18 mg P. Resin: Dowex 1-X4 (100 to 200 mesh), Cl^- form. Bed size: 1.2 × 69 cm. Eluent: KCl solution, pH 8.0, buffered with borate. Initial concentration of KCl in a mixing bottle: 0.2 *M*. Concentration of KCl in a reservoir: 0.45 *M*. Volume of KCl solution in a mixing bottle: 1000 ml. Flow rate: 0.3 ml/min. Room temperature. Reproduced by permission of Elsevier Publishing Company.

Orthophosphate and a few condensed phosphates with relatively low degrees of polymerization such as di- and triphosphate can be separated on an ion-exchange resin column by the stepwise-elution technique which is based on the stepwise increase of an eluent concentration (17). However, a gradient-elution technique is usually more convenient for the separation of a complex mixture of condensed phosphates. As an example, the separation of a series of linear phosphates from ortho- to heptadecaphosphate is illustrated in Fig. 4, which also indicates the exponential increase of the concentration of potassium chloride as an eluent (18). For the separation of a series of linear phosphates, ion-exchange chromatography has so far given better resolution than any other chromatographic techniques.

Recently it was found that QAE-Sephadex A-25, which is an ion-exchange dextran gel, can be successfully employed for the separation of a series of cyclic phosphates from trimeta- to octametaphosphate (19).

GEL CHROMATOGRAPHY

Although the major factor for the separation mechanism in gel chromatography is a molecular-sieve effect, ion exclusion and adsorption must be considered as side effects in some cases. The investigation on the gel chromatographic behavior of a number of oxoacid anions of phosphorus indicates

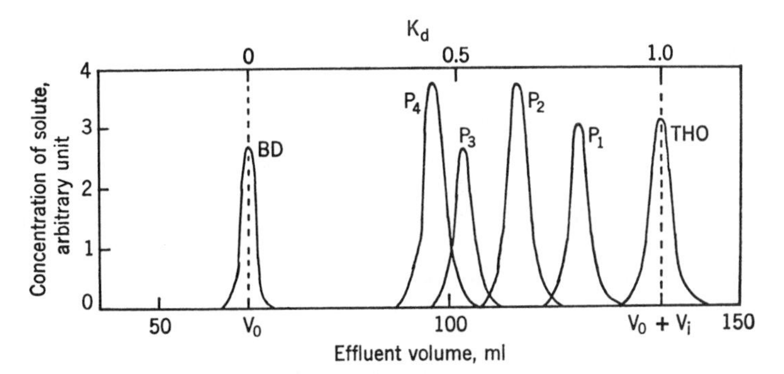

Figure 5. The elution curves of linear phosphates obtained by gel chromatography (20). Sample: 1 ml of each phosphate solution (2×10^{-3} to 5×10^{-3} M as P). Gel: Sephadex G-25 (20 to 80 μ). Bed size: 1.5×90 cm. Eluent: 0.1 M KCl. Flow rate: 25 to 35 ml/hr. Temperature: 20°C. Standard with $K_d = 0$: blue dextran 2000 (BD). Standard with $K_d = 1$: tritiated water (THO). Reproduced by permission of Elsevier Publishing Company.

that under the proper conditions the elution process is controlled by only the molecular-sieve effect (20). As an example, the elution curves of ortho-, di-, tri-, and tetraphosphate on a Sephadex G-25 (a cross-linked dextran gel) column are shown in Fig. 5 (20).

A number of oxoacid anions of phosphorus can be fractionated according to their effective sizes into three groups by means of gel chromatography as shown in Table 4 (20). The distribution coefficient, K_d, in Fig. 5 and Table 4 is defined as $K_d = (V_e - V_0)/V_i$, where V_e is the elution volume, V_0 the void volume outside the gel particles, and V_i the internal volume within the gel particles.

TABLE 4 Three Groups of Oxoacid Anions of Phosphorus Classified by the K_d Values in Gel Chromatography[a]

K_d	$0.46 \sim 0.52$	$0.59 \sim 0.68$	~ 0.78
Oxoacid anion	$\overset{4}{P}-\overset{4}{P}-O-\overset{4}{P}-\overset{4}{P}$, P_4 P_{4m}, $(-\overset{4}{P}-\overset{4}{P}-O-)_2$, $\overset{3}{P}-O-\overset{4}{P}-\overset{4}{P}$, $\overset{5}{P}-O-\overset{4}{P}-\overset{4}{P}$, P_3, P_{3m}	$\overset{3}{P}-O-\overset{3}{P}$, $\overset{3}{P}-O-\overset{5}{P}$, $\overset{4}{P}-\overset{3}{P}-\overset{4}{P}$, P_2, $\overset{2}{P}-\overset{4}{P}$, $\overset{4}{P}-\overset{4}{P}$	$\overset{1}{P}$, $\overset{3}{P}$, P_1

[a] Sample: 1 ml of a solution of each oxoacid anion of phosphorus (2×10^{-3} to 5×10^{-3} M as P). Gel: Sephadex G-25 (20 to 80 μ). Bed size: 1.5×90 cm. Eluent: 0.1 M KCl, pH 7.0. Flow rate: 25 to 35 ml/hr. Temperature: 20°C.

REFERENCES

1. McCarty, P. L., *et al.* (committee report), *J. Am. Water Works Assoc.*, **62,** 127 (1970).
2. Kiso, Y., M. Kobayashi, and Y. Kitaoka, "Chromatography in Solution," and S. Ohashi, "Phosphorus Oxides and Oxo Acids," in *Analytical Chemistry of Phosphorus Compounds*, M. Halmann (Ed.), Interscience, New York, 1972, pp. 93 and 409.
3. Griffith, E. J., and R. L. Buxton, *J. Am. Chem. Soc.*, **89,** 2884 (1967).
4. Grunze, H., and E. Thilo, *Sitz. Ber. Deut. Akad. Wiss. Berlin, Kl. Math. Allgem. Naturw.*, **1953,** No. 5, 26.
5. Ohashi, S., T. Nakamura, and S. Katabuchi, *J. Chromatogr.*, **43,** 291 (1969).
6. Karl-Kroupa, E., *Anal. Chem.*, **28,** 1091 (1956).
7. Smith, M. J., *Anal. Chem.*, **31,** 1023 (1959).
8. Rössel, T., *Z. Anal. Chem.*, **196,** 6 (1963).
9. Bernhart, D. N., and W. B. Chess, *Anal. Chem.*, **31,** 1026 (1956).
10. Kolloff, R. H., *Anal. Chem.*, **33,** 373 (1961).
11. Takai, N., T. Iida, and T. Yamabe, *Seisan Kenkyu*, **19,** 87 (1967).
12. van Ooij, W. J., and J. P. W. Houtman, *Z. Anal. Chem.*, **244,** 38 (1969).
13. Jokl, V., *J. Chromatogr.*, **13,** 451 (1964).
14. Kiso, Y., M. Kobayashi, Y. Kitaoka, K. Kawamoto, and J. Takada, *J. Chromatogr.*, **36,** 215 (1968).
15. Fenger, J., *Radiochim. Acta*, **10,** 138 (1968).
16. Deguchi, T., K. Ujimoto, and S. Ohashi, *Japan Anal.* (*Bunseki Kagaku*), **18,** 981 (1969).
17. Peters, T. V., Jr., and W. Rieman, III, *Anal. Chim. Acta*, **14,** 131 (1956).
18. Ohashi, S., N. Tsuji, Y. Ueno, M. Takeshita, and M. Muto, *J. Chromatogr.*, **50,** 349 (1970).
19. Kura, G., and S. Ohashi, *J. Chromatogr.*, **56,** 111 (1971).
20. Ueno, Y., N. Yoza, and S. Ohashi, *J. Chromatogr.*, **52,** 469 (1970).

16

Gravimetric Analysis of Phosphorus Compounds

THOMAS P. WHALEY and L. W. FERRARA.

International Minerals and Chemical Corporation, Libertyville, Illinois

One of the classical methods of chemical analysis is based on the weight of a precipitate and is therefore known as gravimetric analysis. In its usual form, it depends on the solubilization of the element in question and the subsequent formation of a precipitate of sufficiently low solubility in the solvent that the element is precipitated quantitatively from solution. Inasmuch as phosphorus forms the various phosphoxy anions so readily and the aqueous solubility of the resulting compounds can vary from very high to very low, depending on the cationic species, it is well suited to analysis by gravimetric methods.

ANALYSIS FOR TOTAL PHOSPHORUS

Although phosphorus forms many insoluble compounds that could serve as the basis for gravimetric analysis, the most generally acceptable methods involve the formation of insoluble phosphomolybdates, magnesium ammonium orthophosphate, or magnesium pyrophosphate. Precipitation as silver, barium, zirconium, iron, aluminum, or calcium phosphates can be used for gravimetric determination of phosphorus under certain conditions, but coprecipitation and occlusion often result in the precipitation of more than one type of insoluble phosphate species when several cations are present; uncertainties are thus introduced into the composition of the total precipitate. In addition, many phosphates are often poorly crystallized and precipitate slowly. For these reasons, a more specific phosphate precipitate is used as the basis for gravimetric procedures. In many environmental situations, however, a low phosphorus concentration and/or a need for continuous monitoring are better served by nongravimetric procedures.

In the analysis of materials for total phosphorus content, all of the phosphorus is first converted to the orthophosphate form, PO_4^{3-}. Organophosphorus compounds must be decomposed and condensed phosphates must be hydrolyzed. The organic matter is generally decomposed by oxidation with strong acids such as sulfuric acid, nitric–hydrochloric acid (1) or nitric–perchloric acid (2); it may also be removed by oxidation with $KClO_3$, as has been recommended for the analysis for total phosphorus in water (3). A continuous monitor for seawater utilizing ultraviolet light to convert organophosphorus compounds has also been described (4). In all of these procedures, the oxidation of organic matter is followed by analysis of the remaining solution for the total phosphorus content.

Use of strong oxidizing acids, such as nitric acid or perchloric acid, to destroy the organophosphorus compounds also serves to convert other forms of phosphorus to the orthophosphoric acid required for phosphomolybdate precipitation. Thus, reduced forms of phosphorus are oxidized to P^{5+} and any condensed forms of phosphorus such as tripolyphosphate $P_3O_{10}^{5-}$ or pyrophosphate $P_4O_7^{4-}$ are hydrolyzed to the orthophosphate PO_4^{3-} form.

Phosphomolybdate Precipitation

Phosphorus forms an acid-insoluble yellow complex with the molybdate anion that precipitates rapidly and is specific enough to use for gravimetric analyses; as is discussed later, the color of this complex is sufficiently intense that it can also serve as the basis for a colorimetric procedure for phosphorus analysis. In its most common form, the reaction leading to the formation of the phosphomolybdate can be represented as follows:

$$H_3PO_4 + 12(NH_4)_2MoO_4 + 21HNO_3 \rightarrow (NH_4)_3PO_4 \cdot 12MoO_3 + 21NH_4NO_3 + 12H_2O$$

Once the organic matter has been oxidized and all of the phosphorus has been converted to the orthophosphate, the resulting solution is then ready for further treatment to precipitate the phosphorus as phosphomolybdate. This is generally accomplished by addition of ammonium molybdate solution.

Ammonium molybdate solution is formed by first dissolving a known weight of reagent grade MoO_3 in a known volume of nitric acid to form molybdic acid, and then adding a known amount of ammonium hydroxide. Typically, 118 g of 85% molybdic acid is mixed with 400 ml of distilled water and then reacted with 80 ml of concentrated ammonium hydroxide. The solution is filtered and the filtrate treated with 60 ml of concentrated nitric acid. When cooled to room temperature, the solution is added with constant stirring to a cold solution of 400 ml of concentrated nitric acid in 600 ml of distilled water (5).

Gravimetric Analysis as Ammonium Phosphomolybdate. When the unknown sample is in solution, a volume equal to approximately 0.02 g phosphorus is diluted to 100 ml with distilled water and neutralized by addition of ammonium hydroxide solution to a methyl orange end point. The acidified solution is warmed to about 40°C and the ammonium molybdate solution added [60 ml of $(NH_4)_2MoO_4$ solution for every 0.1 g P_2O_5 present in the unknown—an excess of ten- to a hundred-fold]. The resulting solution is stirred until a cloudy yellow precipitate of ammonium phosphomolybdate appears, at which time the solution is set aside and the precipitation process allowed to continue. After about an hour, the precipitate is filtered and washed successively with 1% nitric acid and 1% KNO_3 or NH_4NO_3 and distilled water. The filtrate is then tested for complete removal of phosphorus by addition of more ammonium molybdate solution. The washed precipitate is dried for 2 hr in an oven at 110°C, then cooled in a desiccator and weighed. The weight of the precipitate multiplied by the factor 0.0165 gives the phosphorus content in grams.

Other Analyses Based on Phosphomolybdate Complex. The yellow acid-insoluble ammonium phosphomolybdate complex may also be reduced to a soluble blue complex that is well suited to colorimetric or spectrophotometric measurements. This is often the method of choice in water analysis (6) and may be based on any one of several reducing agents, the most common and most sensitive of which is stannous chloride. This is also known as Deniges' method and is preferred at concentrations of 0.05 to 3 mg PO_4/l. When aminonaphtholsulfonic acid is dissolved in a solution containing Na_2SO_3 and $Na_2S_2O_5$ (7) and used to reduce the phosphomolybdate complex to "molybdenum blue," the colorimetric method is most applicable at concentrations of 0.1 to 30 mg PO_4/l. The use of ascorbic acid as reducing agent for the phosphomolybdate complex has also been investigated and may be used in colorimetric analysis of phosphorus-containing waters.

The blue color is believed to be due to a partial reduction of the molybdenum rather than to any change in the central phosphorus atoms; in fact, the change from yellow to blue is believed to be due to reduction of only about 20 to 25% of the peripheral molybdenum atoms in the complex (8).

In a very similar vein, a yellow-colored complex is formed when vanadium is substituted for part of the molybdenum in the presence of phosphate and ammonium ion; both vanadium and molybdenum must be in the highest valence states (9). The resulting ammonium phosphomolybdovanadate is used as the basis for spectrophotometric measurements, rather than a truly gravimetric procedure. Similarly, ammonium molybdate and potassium antimonyltartrate react with dilute solutions of phosphate in an acid medium to form a phosphoantimonyl–molybdate complex that is reduced by ascorbic acid to an intense blue color. This may also be used as a colorimetric procedure for phosphorus (10).

Precipitation as Quinoline Phosphomolybdate

In the early 1950s, a method for precipitating phosphorus as quinoline phosphomolybdate was developed by Wilson (11). Initial attempts to use this method for gravimetric determination of phosphorus were thwarted by the difficulty in removing the last traces of water from the precipitate, so the precipitate was measured by an alkalimetric method, that is, by quantitative neutralization of the quinoline phosphomolybdate precipitate with standardized alkali solution, a volumetric method of analysis. It was later discovered that quinoline phosphomolybdate could be ignited at 550°C to phosphomolybdic anhydride, which could be used as the basis for gravimetric determination. The ignited sample contained only 3.947% phosphorus, a desirable feature for gravimetric determination. Unfortunately, the anhydride was somewhat hygroscopic and the ignition procedure at 550°C resulted in an erratic weight-loss pattern due apparently to transient reduction of molybdenum followed by air reoxidation. Ultimately, a compound corresponding to the formula $(CaH_7N)_3H_3(PO_4 \cdot 12MoO_3)$ was found to result from drying the quinoline phosphomolybdate precipitate at 200 to 250°C (12).

This modification of the Wilson method consists of first removing organic matter from the unknown by oxidation with hot nitric–perchloric acid (as discussed earlier) or nitric–hydrochloric acid, then precipitating the phosphorus from a known amount of the resulting solution with an excess of a citric–molybdate solution followed by addition of quinoline solution. The latter solutions are prepared (*a*) by dissolving a known weight of reagent-grade MoO_3 in heated sodium hydroxide solution and adding the resulting sodium molybdate solution to a solution of citric acid in moderately strong hydrochloric acid, and (*b*) by dissolving pure synthetic quinoline in dilute hydrochloric acid. The precipitation is carried out by adding the quinoline solution from a buret immediately after the citric–molybdate solution has been added to a known volume of the unknown, and constantly swirling the solution as the quinoline reagent is being added slowly in order to produce a precipitate with maximum particle size. The precipitate is then filtered, washed with distilled water, and dried to constant weight at 200 to 250°C. A blank is determined by carrying out the procedure with distilled water and is used in calculating the weight of the precipitate. After cooling the precipitate to room temperature in a desiccator, the P_2O_5 content is determined by multiplying the weight of the dried precipitate by 0.032074.

Magnesium Precipitation

Phosphorus may be determined gravimetrically by precipitation as magnesium pyrophosphate, $Mg_2P_2O_7$, or as magnesium ammonium orthophosphate hexahydrate, $MgNH_4PO_4 \cdot 6H_2O$ (13). These methods are more sensitive than phosphomolybdate precipitation to interference from heavy-metal impurities, however, and are most applicable to the analysis of systems that are free of metals whose phosphates are insoluble in ammoniacal solution. If heavy-metal contaminants are present, the interferences may be obviated by first precipitating the phosphorus as ammonium phosphomolybdate, filtering, and then converting the precipitate to magnesium ammonium orthophosphate followed by calcination to magnesium pyrophosphate (14). Following the procedure described earlier, the phosphomolybdate precipitate is filtered from solution and washed several times with 1% nitric acid. The precipitate is then dissolved in hot ammonium hydroxide solution and, after cooling to room temperature, the excess ammonia in the solution is neutralized with hydrochloric acid prior to addition of the magnesia reagent.

The magnesia reagent is prepared from reagent-grade $MgCl_2 \cdot 6H_2O$ or by first dissolving freshly ignited magnesia in dilute hydrochloric acid and then adding a small quantity of calcined magnesia in excess to precipitate any iron, aluminum, or phosphate (15). This solution is filtered and carefully neutralized again with hydrochloric acid. Reagent-grade ammonium chloride is added to the neutralized solution, followed by further addition of ammonium hydroxide and distilled water until the desired dilution is reached. The concentration of the solution is chosen so that 10 ml may be used for every 100 mg P_2O_5 present in the sample to be analyzed. When cool, this quantity of the magnesia reagent is added dropwise, with constant stirring, to the cooled solution containing the phosphomolybdate unknown.

When the solution becomes cloudy, the stirring is discontinued and the precipitate permitted to settle for several minutes. Concentrated ammonium hydroxide equal to about one-fourth the volume of the unknown solution is added, with stirring, and the precipitate allowed to settle for about 2 hr. The precipitate is then filtered, washed with dilute ammonium hydroxide, transferred to a porcelain crucible with addition of a little concentrated NH_4NO_3 solution to aid oxidation, and calcined to magnesium pyrophosphate at 1000 to 1100°C. Calcination temperature is important (16) because magnesium pyrophosphate slowly decomposes at 1200°C, yet constant weight is not achieved rapidly at temperatures of 1000°C or lower.

Direct Precipitation of $MgNH_4PO_4 \cdot 6H_2O$. If the phosphorus-containing unknown contains no heavy metals to interfere with the precipitation of

$MgNH_4PO_4$, a gravimetric analysis based on direct precipitation with the magnesia reagent may be employed, that is, without the initial precipitation of ammonium phosphomolybdate. The procedure involves a slow neutralization of an acid solution containing magnesium, ammonium, and acid phosphate ions (17). Salts of alkali metals are adsorbed during the initial precipitation, but these errors can usually be eliminated by redissolving the precipitate in acid and reprecipitating the magnesium ammonium phosphate. In the special case of potassium, several precipitations may be required to eliminate K^+ because the ionic radius of K^+ is so similar to the ionic radius of NH_4^+ that mixed crystals of $MgNH_4PO_4$—$MgKPO_4$ tend to form during the precipitation.

The third ionization constant of phosphoric acid becomes an important factor in solutions containing both phosphate and hydronium ions, and pH control becomes critical in the precipitation step in order to minimize contamination by $MgHPO_4$, $Mg_3(PO_4)_2$, or $Mg(OH)_2$. The formation of the latter is avoided by the presence of ammonium ion, which decreases the OH^- concentration, and the simple magnesium phosphates are minimized by a slow neutralization of the acid solution.

The direct precipitation of magnesium ammonium phosphate is carried out by acidifying a known amount of the phosphorus-containing unknown with hydrochloric acid and then adding the magnesia reagent in an amount equal to roughly 10 ml for each 100 mg of P_2O_5 believed to be present in the sample. Concentrated ammonium hydroxide is added until a precipitate forms and, after adding additional ammonium hydroxide to ensure an excess, the solution is permitted to stand for several hours. The precipitate is filtered, washed with more ammonium hydroxide solution, and redissolved with hot hydrochloric acid. The acidified solution is then diluted and treated again with the magnesia reagent and ammonium hydroxide to reprecipitate $MgNH_4PO_4 \cdot 6H_2O$.

The precipitate is filtered, washed with dilute ammonium hydroxide, and either calcined to magnesium pyrophosphate at 1000 to 1100°C or washed with alcohol and ether for weighing as the hexahydrate. The hexahydrate is stable enough at room temperature for use in gravimetric determinations and has the advantage of avoiding some of the problems associated with the calcination step, for example, decomposition of the pyrophosphate. However, the hexahydrate is unstable above about 40°C and tends to lose water with formation of the monohydrate. For this reason, the temperature must not exceed 40°C if the hexahydrate is to be used for direct determinations. If calcination to magnesium pyrophosphate is employed, the phosphorus content is determined by multiplying the weight of the $Mg_2P_2O_7$ by 0.2783.

Other Precipitation Methods

A phosphorus-containing compound of high molecular weight may be precipitated from a phosphomolybdic acid solution by the addition of a saturated solution of nitratopentammine cobaltic nitrate (18).

The solution containing the phosphate unknown is acidified to about 2 *N* with 6 *N* H_2SO_4, evaporated to 6 to 8 ml, and enough sodium molybdate solution added to provide about 0.2 g MoO_3 for each milligram of phosphorus expected to be present. Enough of a saturated solution (8.5 g/l) of nitratopentammine cobaltic nitrate is added to color the solution pink, and then an additional 3 to 5 ml is added. The solution is heated to 90°C and evaporated to about 18 to 20 ml. The system is then cooled and the precipitated $[Co(NH_3)_5NO_3]H_3PMo_{12}O_{41}$ is filtered through a weighed filter crucible, washed successively with dilute HNO_3, water, alcohol, and ether.

A similar reagent may be used to precipitate pyrophosphate and tripolyphosphate from a system containing several different phosphate species, including orthophosphate, trimetaphosphate, and tetrametaphosphate, in addition to the pyro- and tripolyphosphate species (19). This is discussed at greater length in a later section.

Several methods involve precipitation of phosphates, but the final determination does not depend on the weight of the precipitate. In precipitating phosphorus as ammonium phosphomolybdate, for example, the precipitate may either be isolated and weighed in a gravimetric procedure, or it may be titrated with standard sodium hydroxide according to the equation (20):

$$(NH_4)_3PO_4 \cdot 12MoO_3 + 23NaOH \rightarrow NaNH_4HPO_4 + (NH_4)_2MoO_4 + 11Na_2MoO_4 + 11H_2O$$

Procedurally, an aliquot of the sample containing roughly 0.05 g P_2O_5 is diluted and neutralized to a methyl orange end point with ammonium hydroxide. Ammonium nitrate is added, followed by ammonium molybdate solution to precipitate the ammonium phosphomolybdate. The precipitate is filtered from the system and washed with 1% KNO_3 solution. The yellow precipitate is dissolved in excess standard NaOH and the excess NaOH backtitrated with standard HNO_3. Phosphorus is calculated from the equation shown above.

Similarly, another phosphate precipitation method that uses titrimetry rather than gravimetry for the final determination involves precipitation as the silver salt. It may be used to determine the amount of orthophosphate in systems that may contain other phosphates or, by first converting all other forms to orthophosphate, it may be used to determine total phosphorus.

As applied to the analysis for total phosphorus, such as would be employed in most situations involving our environment, the aqueous sample is first acidified with an excess of nitric acid and the solution boiled gently to hydrolyze all condensed phosphates to the ortho form. The solution is then cooled to room temperature and the excess acidity neutralized with ammonium hydroxide until the pH of the solution is approximately 3 to 4. A volume of about 1 *N* $AgNO_3$ solution, in excess of the quantity required to precipitate all of the phosphate in the solution, is added and the yellow silver orthophosphate forms as an insoluble precipitate. This releases hydrogen ions that then can be determined by titration with standard sodium hydroxide solution.

ANALYSIS OF MIXTURES OF PHOSPHORUS COMPOUNDS

Zinc Precipitation

Procedures for determining total phosphorus must include a step for hydrolyzing condensed phosphates to orthophosphate, if the presence of the condensed species is suspected. On the other hand, methods for determining the amount of a condensed phosphate species in a mixture of several species must avoid such hydrolytic action. One such method involves the precipitation of zinc pyrophosphate and is known as the Britske-Dragunov titration; it has been modified (21) so that it can also be used to determine triphosphate as well as pyrophosphate in the presence of metaphosphates and orthophosphates. The Britske-Dragunov method is not, strictly speaking, a gravimetric procedure but it does involve precipitation. It is based on the addition of zinc sulfate to a pyrophosphate-containing solution at a pH of 3.8 to precipitate zinc pyrophosphate and liberate sulfuric acid according to the equations:

$$\text{(at pH 3.8)}\quad Na_4P_2O_7 + 2HCl \xrightarrow{25°C} Na_2H_2P_2O_7 + 2NaCl$$

$$Na_2H_2P_2O_7 + 2ZnSO_4 \rightarrow Zn_2P_2O_7 + H_2SO_4 + Na_2SO_4$$

The liberated sulfuric acid is then titrated with standardized 0.1 *N* NaOH solution at exactly 25 ± 1°C.

When only pyrophosphate is present, the method gives an accurate pyrophosphate determination, but any tripolyphosphate ion $P_3O_{10}{}^{5-}$ will also liberate H_2SO_4 by reaction with zinc sulfate. Consequently, the method must be modified to distinguish between pyrophosphate and tripolyphosphate. One such method is based on the insolubility of the zinc pyrophosphate compared with zinc tripolyphosphate and involves an initial adjustment of the unknown solution to exactly pH 3.8 by either dilute HCl or NaOH, followed by the addition of zinc sulfate and a titration of the liberated H_2SO_4 with 0.1 *N* NaOH to pH 3.8. This gives a total value for both the zinc pyrophosphate and zinc tripolyphosphate; however, only the pyrophosphate is insoluble and is precipitated as the titration proceeds. After filtering the precipitate, washing well with water, and igniting at 500 to 600°C, the precipitate is weighed as $Zn_2P_2O_7$. The gravimetric procedure gives the pyrophosphate concentration, which is subtracted from the total determined titrimetrically to give the tripolyphosphate concentration. This method can be used to determine pyrophosphate and tripolyphosphate in the presence of metaphosphate and orthophosphate, the latter being determined colorimetrically by formation of the orthophosphomolybdate complex and the metaphosphate being determined by precipitation as the barium salt (22).

Precipitation as Cobalt Complex Phosphates

Another method that will determine pyrophosphate and tripolyphosphate, alone or in the presence of orthophosphate and metaphosphate, depends on the precipitation of these species with tris(ethylenediamine)cobalt(III) chloride [or hexamminecobalt(III) chloride] (23). This procedure is based on the fact that the cobalt complex will precipitate pyrophosphate preferentially at a higher pH than the tripolyphosphate; for example, at pH 3.5, tris(ethylenediamine)cobalt(III) chloride precipitates tripolyphosphate but not pyrophosphate, and pyrophosphate but not tripolyphosphate at pH 6.5. The ethylenediamine complex does not precipitate orthophosphate, trimetaphosphate, or tetrametaphosphate under these conditions, so that this reagent can be used to determine pyrophosphate and tripolyphosphate in a mixture containing all five phosphates. By contrast, hexamminecobalt(III) chloride precipitates both pyrophosphate and tripolyphosphate at all pH values and also orthophosphate; consequently, its selectivity is not sufficient to permit its use in determining specific phosphate species in mixtures.

When the ethylenediaminecobalt complex is used to precipitate pyrophosphate and tripolyphosphate at pH 6.5 and 3.5, respectively, the precipitates contain acidic hydrogen. After drying at 110°C, the compositions of the precipitates are $Co(en)_3HP_2O_7$ and $Co(en)H_2P_3O_{10}\cdot 2H_2O$, respectively. By contrast, the precipitates formed by the hexamminecobalt(III) complex do not contain acidic hydrogen.

The ethylenediaminecobalt(III) complex precipitates about 93% of the pyrophosphate at pH 6.5 and 98 to 99.5% of the tripolyphosphate at pH 3.5 from solutions containing the pure salts. However, precipitation selectivity from solutions containing both species depends on the quantities of the two species that are present. McCune and Arquette (24) used radioactive tracer techniques with P^{32}-tagged sodium pyrophosphate and tripolyphosphate to determine the selectivity at both pH levels and found that coprecipitation caused gravimetric yields in excess of 100% when the tripoly-pyro ratio was between 3:1 and 1:1 (i.e., some of the pyrophosphate precipitated with the tripolyphosphate). When the pyrophosphate concentration was in excess, the precipitation was incomplete.

Although the precipitation with cobalt complexes could theoretically be used as the basis for a gravimetric procedure, the coprecipitation described above complicates such an approach. Isotope dilution methods using P^{32} and amperometric titrations (25) with solutions of the cobalt complexes have seemed more promising, and a colorimetric method for tripolyphosphate with a standard deviation of 0.8% absolute has been reported (26). In this procedure, an excess of tris(ethylenediamine)cobalt(III) solution is added to precipitate the phosphate, and the excess is measured colorimetrically. The method compensates for the error due to coprecipitation of the pyrophosphate by using a calibration curve and carefully controlled precipitation conditions.

Barium Precipitation

The methods just described for determing pyrophosphate and tripolyphosphate do not provide a way to measure the metaphosphate content of mixtures. A gravimetric procedure for determining metaphosphate content (27) involves precipitation as the barium salt. Trimetaphosphate is not precipitated by metallic ions such as silver or barium, but longer chain length polyphosphate is precipitated as the barium salt from an acid solution. Consequently, a solution containing both species can be analyzed by first acidifying and removing the long-chain phosphate as the barium salt, neutralizing with NaOH to precipitate the barium salts of other phosphates, and then determining the total phosphorus content of the filtrate to give the trimetaphosphate content.

Manganese Precipitation

Pyrophosphate may be separated from tripolyphosphate by precipitation as the manganous salt at pH 4.1 (28). In systems containing ortho-, meta-, pyro-, and tripolyphosphates, the metaphosphate may be removed as the barium salt from acid solution and the orthophosphate removed as the orthophosphomolybdate; following this, addition of manganous chloride solution, adjustment of pH to exactly 4.1, and addition of acetone permits manganous pyrophosphate to precipitate. The precipitation is slow when pyrophosphate concentration is very low, however, and must be allowed to continue for up to several hours to ensure completion; in solutions containing from 25 to 100 mg of pyrophosphate, the precipitate forms readily.

Precipitation of manganous pyrophosphate is retarded by the presence of tripolyphosphate and by lowering the temperature to 0°C; the precipitation rate is increased by raising the temperature somewhat. The most satisfactory conditions of temperature and time for the precipitation are found to be 20 to 30°C and 12 to 16 hr.

ENVIRONMENTAL ASPECTS OF PHOSPHORUS ANALYSIS

The selection of analytical methods for determining phosphorus concentrations in various environmental situations obviously varies with the type of environmental conditions. In measuring the phosphorus content of rivers, ponds, lakes, and so on, such as might be undertaken in a study of limnological eutrophication, the total phosphorus content would probably be the desired analysis. If the study involved many samples on a continuing basis, particularly if an autoanalyzer is used, the analytical scheme would probably be a colorimetric or spectrophotometric determination based on reduced phosphomolybdate, with appropriate hydrolysis and oxidation steps. Most of the concentrations in such studies range from about 7 μg P/l to over 500 μg/l (29) and thus fall nicely into the concentration range most suitable for the phosphomolybdate colorimetric procedure.

In some instances, environmental monitoring might attempt to distinguish between condensed phosphates and orthophosphates in water or sewage systems, such as in an attempt to differentiate between industrial and agricultural phosphates. Certainly one would expect at least partial hydrolysis of condensed phosphates that had spent several days in an aqueous environment, and possibly complete hydrolysis. Nevertheless, any attempts to make such a differentiation would probably be based on one of the schemes just described, such as the zinc precipitation, or formation of the ethylenediaminecobalt(III) complex, or the phosphomolybdate procedure with and without a hydrolysis step.

Organic Phosphorus

One of the most important forms in which phosphorus is found in the environment and which often complicates the analysis is organic phosphorus, that is, organic molecules containing one or more phosphorus atoms. This may range from trace quantities of organophosphorus insecticides such as Malathion and parathion to the algae that have taken up phosphorus from their environment (30). In all instances, the analytical procedure consists of an oxidation step that converts the organophosphorus molecule to a soluble inorganic phosphate followed by a step to analyze the system for total phosphorus. The oxidation procedure used is often a function of the organophosphorus form and may include oxidation with strong oxidizing acids, persulfate (31), and others. Comparative studies (32) have shown that a sample containing phosphorus as adenosine triphosphate (ATP) should be autoclaved with potassium persulfate for 60 min; this converts about 95% of the ATP to soluble inorganic phosphate.

Interferences (33)

Many metals interfere with the gravimetric determination of phosphorus, but the phosphomolybdate precipitation is specific enough to isolate the phosphorus from most other common cations and anions. This is a major factor in selecting the phosphomolybdate procedure for use in systems that might contain different contaminating species. Nevertheless, even this procedure is complicated somewhat by interferences from organic material, as well as several cationic impurities (such as bismuth, tin, zirconium, lead, silicon, and arsenic) and anionic impurities (such as selenium, tellurium, fluoride, chloride, and the sulfate ion). Any organic matter is removed as an interference by the oxidation step described earlier, however, and need not be discussed further.

Bismuth and tin form highly insoluble phosphates that coprecipitate with the phosphomolybdate, and lead forms an insoluble molybdate that also coprecipitates; these metal interferences can be removed by treatment with H_2S and subsequent removal as the sulfides. The H_2S treatment also removes arsenic, which interferes in pentavalent form by substituting for phosphorus in the phosphomolybdate structure.

Zirconium interferes by coprecipitation as an insoluble phosphate along with the phosphomolybdate. Tungsten, as well as vanadium, forms a heteropoly acid with phosphorus that is similar to phosphomolybdate. Silicon interferes by coprecipitation as gelatinous hydrated silica. All of these are removed as insoluble materials after treatment with an oxidizing acid and before the phosphomolybdate precipitation step; these residues usually contain traces of phosphorus.

Fluoride, chloride, selenium, tellurium, and sulfate interfere by delaying the phosphomolybdate precipitation. Fluorine interference can be eliminated, when present as HF, by addition of boric acid with which it forms a complex. Use of an excess of molybdate overcomes the interference due to chloride and sulfate.

A few other elements can also interfere; potassium interference, for example, has been discussed briefly in an earlier section. In most environmental situations, however, the concentrations of other interfering elements are not likely to contribute a significant error.

REFERENCES

1. Perrin, C. H., *J. Assoc. Off. Agric. Chem.*, **41**, No. 4, 760 (1958).
2. Proft, G., *Limnol.*, **2** (4), 407–409 (1964); *Chem. Abstr.*, **63**, 9654*f* (1965).
3. Datsko, V. G., and A. D. Semenov, *Sovrem. Metody Anal. Prir. Vod. Akad. Nauk SSSR*, **1962,** 112–114; *Chem. Abstr.*, **58,** 9957*f* (1963).
4. Grasshoff, K., *Z. Anal. Chem.* **220** (2), 89–95 (1966); *Chem. Abstr.*, **65,** 11976*g* (1966).
5. Furman, N. H., Ed., *Scott's Standard Methods of Chemical Analysis*, 6th ed. Vol. I, Van Nostrand, Princeton, N.J., 1962, p. 814.
6. Klein, L , *River Pollution, I. Chemical Analysis*, Academic Press, New York, 1959, p. 105; G. Deniges, *Bull. Soc. Pharm. Bord.*, **65,** 107 (1927); *C. R.*, **185,** 687 (1927).
7. *Standard Methods for the Examination of Water and Wastewater*, 12th ed., American Public Health Association, New York, 1965, pp. 230–238.
8. Van Wazer, J. R., *Phosphorus and Its Compounds*, Vol. 1, Interscience, New York, 1958, p. 568.
9. Wagner, W., C. J. Hull, and G. E. Markle, *Advanced Analytical Chemistry*, Reinhold, New York, 1956, p. 185.
10. *FWPCA Methods for Chemical Analysis of Water and Wastes*, U.S. Department of the Interior, Nov. 1969, pp. 239–243.
11. Wilson, H. N., *Anal.*, **76,** 65 (1951); **79,** 535 (1954).
12. Duval, C., *Anal. Chem.*, **23,** 1283 (1951); C. H. Perrin, Ref. 1.
13. Pierce, W. C., E. L. Hoenisch, and D. T. Sawyer, *Quantitative Analysis*, 4th ed., Wiley, New York, 1959, pp. 369–373.
14. Furman, N. H., Ed., *Scott's Standard Methods of Chemical Analysis*, 6th ed., Vol. 1, Van Nostrand, Princeton, N.J., 1962, pp. 812, 813.
15. Hoffman, J. I., and G. E. F. Lundell, *J. Res. Natl. Bur. Stand.*, **5,** 279 (1930).
16. Hoffman, J. I., and G. E. F. Lundell, *ibid.*
17. Pierce, W. C., E. L. Hoenisch, and D. T. Sawyer, Ref. 13.
18. Furman, N. H., Ed., *Scott's Standard Methods of Chemical Analysis*, 5th ed., Vol. I, Van Nostrand, Princeton, N.J., 1939, p. 696.
19. McCune, H. W., and G. J. Arquette, *Anal. Chem.*, **27,** No. 3, 401–405 (1955).
20. Furman, N. H., Ed., *Scott's Standard Methods of Chemical Analysis*, 6th ed., Vol. 1, Van Nostrand, Princeton, N.J., 1962, pp. 814–816; W. Wagner, C. J. Hull, and G. E. Markle, *Advanced Analytical Chemistry*, Reinhold, New York, 1956, pp. 184–186.
21. Bell, R. N., *Anal. Chem.*, **19,** No. 2, 97–100 (1947); R. N. Bell, A. R. Wreath, and W. T. Curless, *ibid.*, **24,** No. 12, 1997–1998 (1952).
22. Jones, L. T., *Ind. Eng. Chem. Anal.*, **14,** 536–542 (1942).
23. McCune, H. W., and G. J. Arquette, *Anal. Chem.*, **27,** No. 3, 401–405 (1955); H. J. Weiser, Jr., *ibid.*, **28,** No. 4, 477–481 (1956).
24. McCune, H. W., and G. J. Arquette, *Anal. Chem.*, **27,** No. 3, 401–405 (1955).
25. Laitenen, H. A., and L. W. Burdelt, *Anal. Chem.*, **22,** 833 (1950); **23,** 1265 (1951).
26. Weiser, H. J., Jr., *Anal. Chem.*, **28,** No. 4, 477–481 (1956).

27. Jones, L. T., Ref. 22.
28. Jones, L. T., *ibid.*
29. Brady, N. C., Ed., *Agriculture and the Quality of Our Environment*, A.A.A.S. Publ. 85, Washington, D.C., 1967, pp. 163–172.
30. Moore, H. G., Jr., and E. G. Fruh, "Surplus Phosphorus Uptake by Microorganisms—Algae," Technical Report to the Federal Water Pollution Control Administration, EHE-69-05, CRWR 39.
31. *FWPCA Methods for Chemical Analysis of Water and Wastes*, U.S. Department of the Interior, Nov. 1969, pp. 223–257.
32. Moore, H. G., Jr. and E. G. Fruh, in Ref. 30, pp. 53–56.
33. Furman, N. H., Ed., *Scott's Standard Methods of Chemical Analysis*, 6th ed., Vol. I, Van Nostrand, Princeton, N.J., 1962, p. 816.

pH Titration of Phosphorus Compounds

17

JOHN F. McCULLOUGH

Division of Chemical Development, Tennessee Valley Authority, Muscle Shoals, Alabama

One of the most useful tools in the analysis of phosphorus compounds is pH titration, and methods for analysis of phosphorus acids by colorimetric pH titration have been known for many years (1, 2). Most early analyses, however, were applicable only to individual phosphorus acids and their salts. More recently, the large-scale use of sodium polyphosphates in water treatment and in detergents has prompted the development of pH titration methods for differential analysis of phosphate mixtures. The development of pH meters for general laboratory use, and especially the development of automatic potentiometric titrators, has made pH titration methods both rapid and accurate. Elucidation of the molecular structures of the different classes of condensed phosphates has made possible the correct interpretation and maximum use of pH titration data, but general methods for differential analysis of mixtures of phosphorus compounds other than inorganic phosphates have not been developed. Many mixtures, however, can be analyzed by application of methods used for the inorganic phosphates.

INORGANIC PHOSPHATES

The validity of pH titration methods rests on the fact that orthophosphoric acid and condensed phosphoric acids contain only one strongly ionized hydrogen atom per phosphorus atom, and all other hydrogen atoms are weakly ionized (3). This experimentally demonstrable fact is in agreement with the antibranching rule (4, 5) which states that, of the four classes of phosphates (6)—orthophosphate, straight-chain phosphates or polyphosphates, cyclic phosphates or metaphosphates, and ultraphosphates (in which some of the phosphate units share three oxygen atoms with adjacent

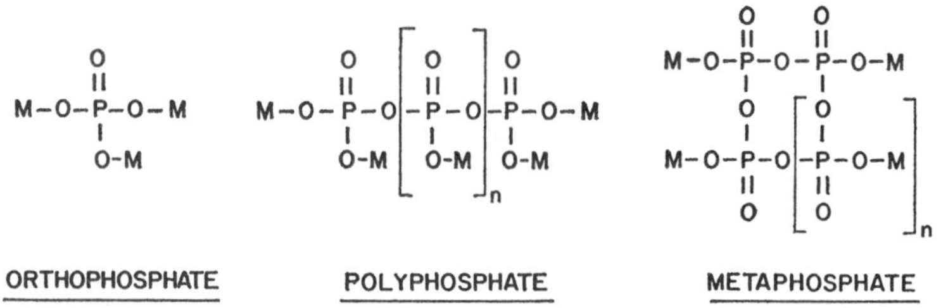

```
       O                O    ┌ O  ┐   O              O       O
       ||               ||   │ ||  │   ||             ||      ||
  M−O−P−O−M       M−O−P−O│−P−O│−P−O−M       M−O−P−O−P−O−M
       |                |    │ |   │   |              |       |
      O-M              O-M   │ O-M │  O-M             O     ┌ O  ┐
                             └     ┘n                 |     │ |  │
                                                 M−O−P−O│−P−O−M│
                                                      ||    │ || │
                                                      O     └ O  ┘n
```

ORTHOPHOSPHATE POLYPHOSPHATE METAPHOSPHATE

Figure 1. Phosphates stable in aqueous solution.

phosphate units)—only the first three are stable to any extent in aqueous solution. The structural formulas of the first three classes are shown in Fig. 1 in which M is the chemical equivalent of a metal or an ammonium or hydrogen ion, and *n* is any whole number from zero to infinity.

Until recently, phosphates were considered to be either orthophosphate (M_3PO_4), pyrophosphate ($M_4P_2O_7$), metaphosphate (MPO_3), or their mixtures. This obsolete classification of Graham (7) is used in most of the older literature and is found in even some recent publications.

The phosphate unit in all these phosphates consists of a phosphorus atom tetrahedrally bonded to four oxygen atoms as shown in Fig. 1. Orthophosphate is a single unit and contains one strong-acid function and two weak-acid functions; only one of the weak-acid functions is strong enough to be titrated. Each of the phosphate units at the ends of a polyphosphate chain—called end groups—shares one oxygen atom with the adjacent phosphate unit and contains one strong- and one weak-acid function. Each of the other units of polyphosphates—called middle groups—shares two oxygen atoms with adjacent phosphate units, and each such unit has one strong-acid function. Metaphosphates contain only middle groups and hence have only strong-acid functions, one for each phosphorus atom. The fourth class of phosphates—the ultraphosphates—decompose on dissolution into mixtures of the other three classes.

Titration curves for members of each class of stable phosphoric acids and for a mixture of all three classes are plotted in Fig. 2. These curves show that each phosphate species contains one equivalent of strong acid per mole of phosphorus and that all other acid functions are weak. The ionization constants of orthophosphoric acid and of all polyphosphoric acids and their mixtures are such that their pH titration curves have inflection points near pH 4.5 and 9. Metaphosphoric acid titrates entirely as a strong acid and has no weak-acid functions. The amount of alkali or acid required to titrate a phosphate solution from its original pH to the pH 9 end point (Fig. 3) is a measure of the hydrogen (or metal) ion associated with the phosphate. Titration from the end point at pH 4.5 to the end point at pH 9 measures

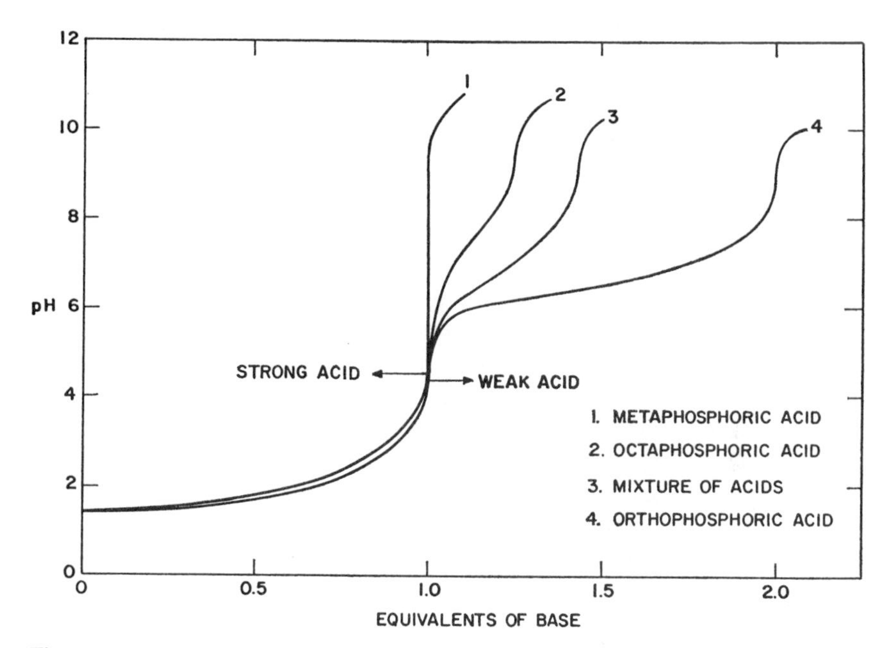

Figure 2. Titration of phosphoric acids containing 1 mole of phosphorus.

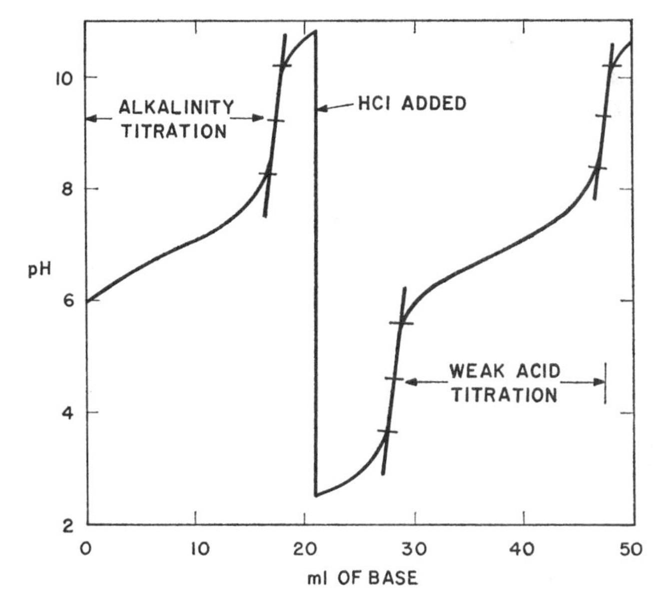

Figure 3. Alkalinity and weak-acid titration of phosphate solution with initial pH between 4.5 and 9.

the weak-acid function which is the sum of orthophosphate and end-group phosphate. The strong-acid function, which is equivalent to the total phosphorus, of all phosphates and their mixtures is then obtained by titration between pH 4.5 and 9 after hydrolysis of all phosphates to orthophosphate. Finally, orthophosphate is determined in the presence of other phosphates by titration of its third hydrogen after it is released by addition of a precipitating cation such as silver (8), as shown in Fig. 4. These four titrations yield the following information: total phosphate; distribution of phosphate between orthophosphate, end-group phosphate, and middle-group phosphate; and amount of metallic oxide and constituent water associated with the phosphate. (End-group phosphate and middle-group phosphate correspond, respectively, to pyrophosphate and metaphosphate in the notation of composition of condensed phosphates in the older literature.)

Since three components—total phosphate, weak-acid functions, and orthophosphate—of the phosphate mixtures are determined independently, any mixture of orthophosphate with two other known species, such as mixtures of the sodium salts of ortho-, pyro-, and tripolyphosphate, can be analyzed unequivocally (9). The number average chain length, which is used extensively in characterizing polyphosphates, can be calculated from pH titration data if metaphosphates are absent or if their amount is known.

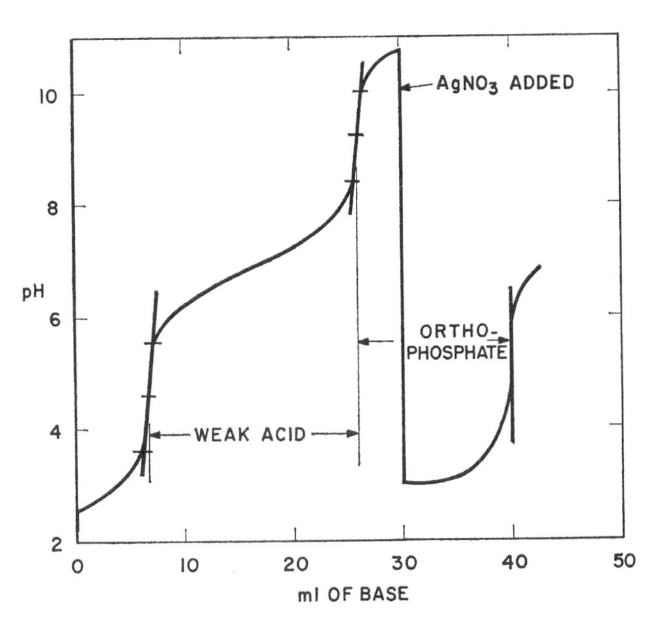

Figure 4. Titration of orthophosphate by silver nitrate method.

Many phosphate compositions obey the relation

$$P_2O_5 + M_2O + H_2O = \text{total weight} \quad (1)$$

in which the formulas represent the weight of the respective components, M is the chemical equivalent of a known metal or an ammonium ion that is associated with the phosphate, and H_2O represents the water of constitution (and the free acidity) of the phosphate. An example is a mixture of the anhydrous salts, NaH_2PO_4 and $Na_5P_3O_{10}$. In the analysis of these phosphates, the total phosphorus can be calculated from a material balance (10), thereby avoiding hydrolysis of the phosphates and an additional titration.

The literature contains many wet-chemical methods for the determination of individual species—or what was believed at the time to be individual species—in phosphate mixtures (11–13). Each of these methods depends on the selective precipitation of one or more of the phosphate species under closely controlled conditions, and in many of the methods the acid released by precipitation is titrated. In general, however, the selectivity of the precipitating ion is not high and each method amounts to a solubility fractionation. The methods generally give inaccurate results, as shown by Dewald (14) for mixtures of pyro- and tripolyphosphate, but they are satisfactory for many routine analyses of industrial products when the results can be standardized against known compositions. These methods have been largely superseded by paper chromatographic analyses (15) and are not considered further.

APPARATUS AND GENERAL TECHNIQUES

Apparatus

Although acid-base color indicators can be used, pH titrations of phosphates are most accurately carried out electrometrically. Manual titrations with an indicating pH meter equipped with suitable electrodes and titrating accessories give good results, but they are time-consuming, not only because of the time required for addition of the titrant but also because the titration curve must be plotted for graphical determinations of the end points. Equally reliable or even better results are obtained with automatic titrators which also save much time and labor.

In one type of automatic titrator the titration curve is plotted by a strip-chart recorder that is synchronized with an automatic device that delivers the titrant. As in manual titrations, the end points are determined graphically. While not entirely automatic, this type of titrator is accurate—within 0.05 ml—and is probably the most versatile type for nonroutine work. An example is the Precision-Dow Recording Titrometer of the Precision Scientific Company, Chicago, Illinois.

Another type of automatic titrator shuts off at a preselected pH. Although this type gives higher reproducibility than the other type—readings are reproducible to within 0.01 ml (16)—the exact pH at the end point must be known. This decreases the titrator's versatility because the pH at the end point varies with the species of phosphate and their concentrations, the concentration of neutral salt, and other factors. An example of this type is the Beckman Autotitrator of Beckman Instruments, Inc., South Pasadena, California.

Glass and calomel electrodes are recommended for pH titration of phosphates. Capillary-tip calomel electrodes behave erratically in the presence of large amounts of precipitated solids, as in the silver titration for orthophosphate. Sleeve-type calomel electrodes, however, give reliable results in both solutions and slurries of phosphates (9).

Determination of End Points

End points are determined graphically in manual electrometric titrations and in automatic titrations in which the titration curve is plotted. Any of the usual methods of locating end points are satisfactory—plots of $\Delta pH/\Delta V$ or $\Delta^2 pH/\Delta V^2$ against V (where V is volume of titrant), or simple bisection of the curve in the region of the end points as shown in Fig. 3. The bisection method appears to be as accurate as any and is considerably more rapid than the others. Considerable time is saved in manual titrations by first locating the approximate end points by a rough titration. End points can be established with color indicators also. Reasonably good results were obtained with bromocresol green for end points near pH 4.5, thymol blue or oleo red B for end points near pH 9, and methyl red for the silver titration (8). (Thymol blue and oleo red B are used in the presence of about 20 g of $NaNO_3$.) The most precise results are obtained by establishing the pH at the end points electrometrically so that exact color standards can be prepared. Since electrometric titrations of phosphates are more accurate than colorimetric titrations, and are also more rapid with automatic titrators, colorimetric titrations are not considered further.

Interferences

All cations that form insoluble phosphates under the conditions of the analysis interfere with the titrations. Cations of weak bases such as ammonia make the end points indistinct. Reliable titrations are made only with phosphoric acids, alkali phosphates, phosphates of strong organic bases, or phosphates that can be converted to these forms. Most interfering cations can be replaced with hydrogen or sodium by treatment with ion-exchange resins (17, 18), and potassium oxalate is also effective in removing most interfering cations from phosphate solutions (19).

Strong acids and their neutral salts do not interfere with the titrations. Anions of weak acids, such as carbonates and silicates, interfere by adding to the weak-acid titration. Carbonate interference is manifested by irregularities in the titration curve.

Concentration and Sample Size

About 0.005 *N* is the minimum concentration of weak-acid phosphate that shows a well-defined break in its titration curve at the end point near pH 9. The strength—degree of ionization—of a weak acid is increased significantly by the presence of a neutral salt such as sodium chloride (3). Sodium chloride is sometimes added to phosphate samples to sharpen the end points when the breaks in the titration curves are too vague because of low concentrations of weak acid. This practice is most advantageous in the titration of long-chain polyphosphates such as Graham's salt in which the ratio of end groups to total phosphorus is low.

Probably the largest systematic error in pH titrations is the uncertainty in location of the end points, and thus the relative error is decreased by taking as large a sample as is practical. Best results are obtained with samples that require 20 to 50 ml of titrant. Titrations of long-chain polyphosphates require large samples (as much as 50 g for the longest chain length that can be determined) to provide sufficient weak acid for accurate measurement. For example, 46 g of a sodium polyphosphate glass (Graham's salt) with a chain length of 300 is required for a 30-ml titration with 0.1 *N* NaOH.

Large samples of glassy phosphates dissolve slowly with normal stirring. The dissolution is accelerated markedly by stirring in a Waring blender or by suspending the sample so that the solution drains from the surface of the undissolved particles.

Determination of Alkalinity and Weak Acid

Most titrations are made with base, but acid titration is required when the initial pH is above 9. It is recommended that monopotassium phosphate be used as the primary standard for electrometric titrations. Carbonates should be removed from the NaOH used for preparing standard solutions by precipitation from a 50% solution (20) rather than by precipitation as barium carbonate.

Weigh out sufficient sample for a 20 to 50 ml titration and dissolve in 100 to 200 ml of distilled water. Titrate the solution from its original pH through the end point near pH 9; the exact pH at the end point depends on the phosphate species, the concentration of phosphate, and the concentration of neutral salts. When the original pH is far enough below the end point near pH 4.5 to allow that end point to be established, a single titration through it and on through the end point near pH 9 is sufficient for determination of both alkalinity and weak acid. Similarly, the same information is obtained when the original pH is above 9 by titrating with acid through the end point near pH 4.5. (When the pH is above the end point near pH 9, as with solutions of trisodium orthophosphate, the sample should be checked for carbonates.) If the original pH is between pH 4.5 and 9, the sample is titrated through the end point near pH 9 to obtain a measure of its alkalinity. The pH is then lowered to about 3 with 1 *N* HCl, and the sample is titrated through the end point near pH 4.5 and on through that near pH 9, as in Fig. 3, to obtain a measure of the weak acid. The sample should be titrated promptly after acidulation to minimize hydrolysis. Samples should never be heated in these titrations.

Determination of Orthophosphate

The determination of orthophosphate depends on the release of hydrogen ion as strong acid by precipitation of orthophosphate as a highly insoluble normal salt such as silver phosphate (8)

$$3AgNO_3 + Na_2HPO_4 = Ag_3PO_4 + 2NaNO_3 + HNO_3 \qquad (2)$$

Cations other than silver, such as calcium (21) and cerous (22), have been used, but they give no better results with mixed phosphates than does silver.

After the weak acid is titrated, adjust the pH to about 3 with 1 *N* HCl; if the weak-acid function was titrated with acid, however, no pH adjustment is necessary. Titrate with alkali to a point just beyond the end point near pH 4.5, stop the titration, and add 50 ml of 1 *N* silver nitrate solution. Resume the titration, but add the alkali titrant slowly—no faster than about 0.5 ml/min—until the moderately strong end point near pH 5.5 is reached. This titration—from pH 4.5 to 5.5—is a measure of the previously measured weak-acid function plus the orthophosphate.

An alternative and more rapid method that can be used if one-third or more of the total phosphate is present as orthophosphate and if the weak acid has been titrated with base is to add silver nitrate solution after titrating through the end point near pH 9. (The pH should fall to about 3; if it does not, the former method should be used.) The sample is then titrated through the end point near pH 5.5 as before. This titration—from the end point near pH 9 to the end point near pH 5.5—is a measure of orthophosphate only, as shown in Fig. 4.

A little less than the stoichiometric amount of base is always consumed in the silver titration, and it is recommended that the base be standardized under the conditions of the silver titration in addition to the regular standardization (9).

Determination of Total Phosphate

For determination of total phosphate, all condensed phosphates must be completely converted to orthophosphate. Condensed phosphates are stable in neutral solutions at ambient temperatures but are rapidly hydrolyzed at low pH and high temperature (23). Complete hydrolysis of any condensed phosphate is effected by boiling for 30 min in 1 *N* solutions of strong mineral acids.

The pH of the hydrolyzed sample is adjusted to about 3 with carbonate-free NaOH solution and the sample then is titrated through the end points near pH 4.5 and 9 as in the weak-acid titration. This titration is a measure of the total phosphate which is the same as the strong-acid function of the unhydrolyzed sample.

A possible source of error is contamination with carbonate during pH adjustment. This can be avoided by an alternate hydrolysis procedure in which the sample is evaporated to dryness in a concentrated HCl solution containing potassium chloride (24).

CALCULATIONS AND ACCURACY

The following calculations are based on the titrations outlined in the general procedures. Calculations are given only for titrations in which silver nitrate is added at pH above 9. When silver nitrate is added at pH below 9, the term $(J - E)$ can be substituted for G in all equations. All terms are defined after Eq. (24):

$$\text{Wt.\% } P_2O_5 = 7.1\ BK/W \tag{3}$$

$$\text{Wt.\% } P_2O_5 \text{ as orthophosphate} = 7.1\ BG/W \tag{4}$$

$$\text{percent of total } P_2O_5 \text{ as orthophosphate} = 100\ G/K \tag{5}$$

$$\text{percent of total } P_2O_5 \text{ as end groups} = 100\ (E - G)/K \tag{6}$$

$$\text{percent of total } P_2O_5 \text{ as middle groups} = 100\ (K - E)/K \tag{7}$$

$$\text{wt.\% } M_2O = 100\ RB(K + E - C)/W \text{ for original pH below 9} \tag{8}$$

$$\text{wt.\% } M_2O = 100\ R(BK + AD)/W \text{ for original pH above 9} \tag{9}$$

$$\text{wt.\% } H_2O \text{ of constitution} = 0.901\ B(C + G)/W \text{ for original pH below 9} \tag{10}$$

$$\text{wt.\% } H_2O \text{ of constitution} = 0.901\ (BJ - AD)/W \text{ for original pH above 9} \tag{11}$$

It is not necessary to hydrolyze samples whose compositions are described by Eq. (1) because the total phosphate can be obtained from the material balance equation (10).

$$W = \text{weight of middle groups} + \text{weight of end groups} + \text{weight of ortho} \quad (12)$$

End groups, ortho, and the unneutralized acid are determined from the titrations for alkalinity, weak acid, and orthophosphate. Thus, for samples having an original pH below the end point near pH 9,

$$S = \frac{W + B[(Q - 0.001)(C + G) - (2R + 0.071)(E - G) - (3R + 0.071)G]}{R + 0.071} \quad (13)$$

$$\text{wt.\% } P_2O_5 = 7.1(S + BE)/W \quad (14)$$

$$\text{percent of total } P_2O_5 \text{ as orthophosphate} = 100\, BG/(S + BE) \quad (15)$$

$$\text{percent of total } P_2O_5 \text{ as end groups} = 100\, B(E - G)/(S + BE) \quad (16)$$

$$\text{percent of total } P_2O_5 \text{ as middle groups} = 100\, S/(S + BE) \quad (17)$$

Similar equations can be derived for titrations in which the original pH is above 9.

Mixtures of orthophosphate with any two other species can be calculated from the simultaneous equations

$$XL_x + YL_y = B(K - G) \quad (18)$$

$$XU_x + YU_y = B(E - G) \quad (19)$$

The term K in Eq. (18) can be replaced by $(S + BE)/B$ when Eq. (1) describes the sample.

The average chain length ($\bar{n}$) of polyphosphates free of ortho and metaphosphates is equal to

$$\bar{n} = 2[\text{total } P_2O_5/\text{end group } P_2O_5] = 2K/E \quad (20)$$

When Eq. (1) is applicable, as for phosphate glasses,

$$\bar{n} = (2S/BE) + 2 \quad (21)$$

Polyphosphate glasses with $\bar{n}$ above about 75 may be considered to have the meta composition MPO_3 (10) and, thus,

$$\bar{n} = 2W/(M_2O + 0.071)BE \quad (22)$$

Correction must be made for the presence of ortho- and metaphosphates when calculating $\bar{n}$ of polyphosphates—long-chain polyphosphate glasses, for example, contain as much as 10% metaphosphates (25). Thus,

$$\bar{n} = 2\left[\frac{\text{total } P_2O_5 - \text{meta-}P_2O_5 - \text{ortho-}P_2O_5}{\text{end group } P_2O_5}\right] \quad (23)$$

If orthophosphate is included as part of the chain

$$\bar{n} = 2\left[\frac{\text{total } P_2O_5 - \text{meta-}P_2O_5}{\text{end group } P_2O_5 + 2 \text{ ortho-}P_2O_5}\right] \tag{24}$$

Definition of Terms

- A normality of acid.
- B normality of base.
- C ml of base for alkalinity titration (pH below 9).
- D ml of acid for alkalinity titration (pH above 9).
- E ml of base for weak-acid titration.
- F ml of acid for weak-acid titration.
- G ml of base for silver titration ($AgNO_3$ added at pH above 9).
- J ml of base for silver titration ($AgNO_3$ added at pH below 9).
- K ml of base for titration of hydrolyzed sample.
- L number of phosphorus atoms per molecule.
- M milliequivalents of monovalent metal, H, or NH_4.
- Q milliequivalent weight of M.
- R milliequivalent weight of M_2O.
- S milliequivalents of middle-group phosphate.
- U number of weak-acid groups per molecule.
- V volume of titrant (general).
- W weight of sample.
- X number of millimoles of species x.
- Y number of millimoles of species y.

Weak acid can be determined with an accuracy of better than 5 ppt by electrometric pH titration (16) and with a mean precision of 2 to 3 ppt with the Precision-Dow Titrometer (9).

The pH titration after addition of silver for determining orthophosphate in the presence of other phosphates is accurate to within about 5 pph (9). The method is very rapid and convenient, however, especially when weak-acid determinations are needed also, and it is sufficiently accurate for many purposes. When higher accuracy is required, either paper chromatography (15) or cold molybdate precipitation of orthophosphate (16) gives reliable results.

PHOSPHORUS COMPOUNDS OTHER THAN INORGANIC PHOSPHATES

The total acidity of any acid of phosphorus that is soluble in either water or suitable nonaqueous solvents (26–28) may be determined by the use of pH titration. The total phosphorus in any compound can, of course, be determined by converting a sample to orthophosphate (16, 29), removing interferences, and then titrating the weak acid.

The rule that the first acid group attached to a phosphorus atom is strongly ionized and that all other acid groups are weakly ionized appears to apply to all phosphorus acids; no exceptions have been reported. Furthermore, the ionization constants (5, 30) are such that pH titration curves for all acids of phosphorus are similar to those for inorganic phosphates—the end points are near pH 4.5 and 9 when weak acid is present, and near pH 7 when only strong acid is present. This makes possible the determination by pH titration of many mixtures of acids of phosphorus and their salts by applying the methods used for inorganic phosphates (28, 31, 32).

As an example of pH titration for determination of various mixtures of acids of phosphorus, consider a water-soluble mixture of mono-, di-, and trialkylphosphoric acids and orthophosphoric acid. Titration of the strong acid is a measure of the sum of orthophosphate and mono- and dialkyl phosphate. The weak-acid titration is a measure of the sum of orthophosphate and monoalkyl phosphate. Orthophosphate is determined directly by the silver titration. Thus, the difference between the strong- and weak-acid functions is a measure of the dialkyl phosphate, and the difference between the weak-acid and the silver titration is a measure of the monoalkyl phosphate. The trialkyl phosphate then is determined by difference after a sample is hydrolyzed to orthophosphate and titrated for total phosphorus.

REFERENCES

1. Kolthoff, I. M., I. H. Menzel, and N. H. Furman, *Volumetric Analysis*, Vol. II, Wiley, New York, 1929, pp. 143–146.
2. Smith, J. H., *J. Soc. Chem. Ind.*, **36,** 415 (1917).
3. Van Wazer, J. R., and K. A. Holst, *J. Am. Chem. Soc.*, **72** (2), 639 (1950).
4. Van Wazer, J. R., *J. Am. Chem. Soc.*, **78** (22), 5709 (1956).
5. Van Wazer, J. R., *Phosphorus and Its Compounds*, Vol. I, Interscience, New York, 1961, p. 437.
6. Thilo, E., *Angew. Chem. Int. Ed. Engl.*, **4** (12), 1061 (1965).
7. Graham, T., *Philos. Trans.*, **123,** 253 (1833); *Ann. Phys.*, **32,** 33 (1834).
8. Gerber, A. B., and F. T. Miles, *Ind. Eng. Chem., Anal. Ed.*, **13,** 406 (1941).
9. Van Wazer, J. R., E. J. Griffith, and J. F. McCullough, *Anal. Chem.*, **26,** 1755 (1954).
10. Griffith, E. J., *J. Am. Chem. Soc.*, **79** (3), 509 (1957).
11. Jones, L. T., *Ind. Eng. Chem., Anal. Ed.*, **14,** 536 (1942).
12. Bell, R. N., *Anal. Chem.*, **19,** 97 (1947); **24,** 1997 (1952).
13. Haywood, D., *Ind. Chem.*, **30,** 356 (1954).
14. Dewald, W., *Fette u Seifen*, **56,** 105 (1954).
15. Karl-Kroupa, E., *Anal. Chem.*, **28,** 1091 (1956).
16. *Scott's Standard Methods of Chemical Analysis*, 6th ed., Vol. 1, N. H. Furman (Ed.), Van Nostrand, Princeton, N.J., 1962.

17. Osborn, G. H., *Anal.*, **78,** 235 (1953).
18. Helrich, K., and W. Rieman, III, *Anal. Chem.*, **19,** 651 (1947).
19. Fischer, J., and G. Kraft, *Z. Anal. Chem.*, **149,** 29 (1956).
20. Kolthoff, I. M., and E. B. Sandell, *Textbook of Quantitative Inorganic Analysis,* Macmillan, New York, 1943, p. 551.
21. Greenfield, S., *Comprehensive Analytical Chemistry,* Vol. IC, C. L. Wilson and D. W. Wilson (Eds.), Elsevier, New York, 1962, p. 227.
22. Madsen, E. R., and T. Kjaergard, *Acta Chem. Scand.*, **7,** 735 (1953).
23. Van Wazer, J. R., E. J. Griffith, and J. F. McCullough, *J. Am. Chem. Soc.*, **77,** 287 (1955).
24. Griffith, E. J., *Anal. Chem.*, **28,** 525 (1956).
25. Thilo, E., and U. Schülke, *Z. Anorg. Allgem. Chem.*, **341,** 293 (1965).
26. Kreshkov, A. P., V. A. Drozdov, and N. A. Kolchina, *Zh. Anal. Khim.*, **19** (10), 1177 (1964).
27. Onoda, K., and K. Jamada, *Yukagaku,* **16** (7), 426 (1967).
28. Clark, F. B., and J. W. Lyons, *J. Am. Chem. Soc.*, **88,** 4401 (1966).
29. Kosolapoff, G. M., *Organophosphorus Compounds,* Wiley, New York, 1950, p. 7.
30. Makitie, O., and V. Konttinen, *Acta Chem. Scand.*, **23** (4), 1459 (1969).
31. Manevich, I. Ya., Yu. A. Buslaev, K. A. Andrianov, and E. I. Matrosov, *Izv. Akad. Nauk SSSR, Neorg. Mater.*, **1** (8), 1241 (1965).
32. Gordievskii, A. V., and V. A. Moskinov, *Tr. Mosk. Khim.-Tekhnol. Inst.*, **43,** 29 (1963).

18

Spectrometric Methods for Determining Phosphorus and Its Compounds

EMMETT F. KAELBLE

Monsanto Company, St. Louis, Missouri

This chapter covers important spectrometric techniques for determining phosphorus and its compounds which are not discussed elsewhere in the book. The widely used colorimetric methods (e.g., molybdenum blue) are covered in a separate chapter.

For each technique discussed, the principles are stated briefly and selected applications are given. Sensitivities, limitations, and types of samples which can be analyzed are included to provide guidance as to when a given technique might be the method of choice.

X-RAY SPECTROSCOPY

X-ray spectroscopy has been used to determine phosphorus in a variety of environmental materials—geological, biological, organic, inorganic, air pollutants, and lake sediments. It has been applied to phosphorus by itself and in combination with separation or concentration techniques, to solids and liquids, from microgram levels to high percentages. Like most spectrometric techniques, it has great advantages of speed and ease over chemical methods of analysis. And since the sample is not consumed in the analysis, it can be saved for other analyses or for future reference.

In principle, x-ray spectroscopy involves causing the sample to emit characteristic x-rays that are then measured and related to the concentration of the element of interest. The excitation source is most commonly a high-intensity x-ray tube, although radioisotopes and electron and proton beams are also used. The x-ray wavelength of interest (e.g., that of the phosphorus $K\alpha$ line) is separated from other wavelengths present by a crystal monochromator, by electronic discrimination, or by a combination of both. Detection and measurement of the x-rays are accomplished with a radiation detector (most commonly a gas-flow proportional counter for phosphorus) plus suitable amplification and readout devices. For a more thorough discussion of the principles of x-ray spectroscopy, see Ref. 18.

In geological samples, phosphorus usually is found in combination with calcium. This presents a problem to the x-ray spectroscopist in that the second-order calcium $K\beta_1$ and $K\beta_5$ lines interfere spectrally with the first-order phosphorus $K\alpha$ line. Fabbi (11) eliminated this interference by combining maximum pulse height discrimination (electronic) with a mathematical correction for peak overlap. Various chemically analyzed rock standards were used for calibration purposes, and the method was then used routinely to determine P_2O_5 in sedimentary rocks: clay shales, silty shales, siltstones, phosphatic siltstones, dolomitic limestones, and limestones. Agreement between x-ray results and chemical analysis values was excellent. Deviations of no more than 0.04% were observed in the concentration range 0 to 16% P_2O_5.

The spectral interference of calcium can also be eliminated by use of a germanium analyzing crystal, which gives almost no second-order reflection.

Interelement effects can present serious problems in x-ray spectroscopy. The x-ray intensity measured for an element can be affected (either positively or negatively) by the other elements present. The effect in phosphate rock analysis is particularly severe because rock from different sources varies widely in composition and because four of the elements normally determined (Mg, Al, Si, and P) are adjacent to each other in the periodic table.

A number of techniques have been developed to minimize interelement effects. McKinney and Rosenberg (22) developed a mathematical model to correct for matrix variations in each sample. The model utilized the x-ray intensity from each of the six elements considered and the mass absorption coefficients of the elements to calculate concentrations. The x-ray spectrometer was interfaced with a small computer which performed the calculations on each sample and printed the results.

Accurate analyses were obtained for P, Si, Ca, Fe, Al, and Mg in less than 10 min. Samples were analyzed in the range 27.4 to 33.6% P_2O_5 with a standard deviation of 0.18, a precision equal to that found for chemical analysis. The samples were dried, ground, and pressed into pellets prior to analysis.

Phosphorus was among 41 elements determined quantitatively by x-ray spectroscopy in an extensive study of the composition of lake sediments (9). Levels found ranged from 0.07 to 0.25%, expressed as P_2O_5.

As conventionally applied, the detection limit of x-ray spectroscopy for phosphorus is about 0.01% (as P_2O_5). Combined with separation and concentration techniques, such as thin-layer chromatography, paper chromatography, ion exchange, or precipitation, x-ray spectroscopy is capable of detecting phosphorus compounds in microgram amounts. Libby (19), for example, used x-ray spectroscopy and thin-layer chromatography in combination to measure quantitatively microgram amounts of phosphates and phospholipids. The solute zones were cut from the chromatograms and placed directly into the spectrograph.

Luke (20) determined 1 to 40 μg of phosphorus using a coprecipitation technique. All phosphorus was converted to orthophosphate and precipitated by the addition of excess beryllium, with arsenate added as coprecipitant. The precipitate was collected on a filter disk which was then analyzed in the x-ray spectrograph.

The limit of detection of x-ray spectroscopy for phosphorus and many other elements has been extended even further by the use of heavy particles for x-ray excitation and solid-state detectors. Johansson *et al.* (15) have described a system in which the material to be analyzed is deposited on a very thin carbon foil and placed in an irradiation chamber connected to the beam pipe of a Van de Graaff generator. A well-collimated proton beam traverses the sample, and the x-rays generated are recorded by a silicon detector and a 4096 channel analyzer. The energy resolution of the system is sufficient to permit separation and identification of the $K\alpha$ lines of elements heavier than silicon. A single drop of sample solution can be evaporated on the carbon foil and analyzed. The method was tested for measuring air contaminants by analyzing a foil which had been placed outdoors for a day. The resulting spectrum showed 13 elements including 10^{-9} g phosphorus.

Electron-probe microanalysis is an important special form of x-ray spectroscopy. A focused beam of electrons is used to excite characteristic x-rays in an area as small as 1 μ in diameter on the sample surface. Thus the importance of this technique is that it can measure local variations in composition as opposed to the average composition of the whole specimen (4). An example of a phosphorus application of the electron probe is the work of Walitsky and Hopkins (32). They found that calcium fluorapatite crystals grown from nonstoichiometric melts contained defects of either CaF_2 or $Ca_3(PO_4)_2$ depending on the melt composition. These defects affected the performance of the crystals as laser materials.

FLAME SPECTROSCOPY

The term "flame spectroscopy" is used broadly here to include the techniques of flame emission, atomic absorption, and atomic fluorescence spectroscopy. In flame-emission spectroscopy, thermal energy from a flame excites some of the sample atoms to higher energy levels. When these atoms return to their lower energy levels, the energy acquired from the flame is released, partly in the form of light. This emitted light can be used analytically because its wavelength is characteristic of the emitting element, and its intensity is proportional to the concentration of that element in the sample.

When a solution is aspirated into a flame, only a relatively small number of its atoms are normally excited to a state from which emission of light occurs. Most of the atoms are reduced to the ground state, unionized and unexcited. In this state, they are capable of absorbing light of characteristic wavelengths. Use is made of this capability in atomic absorption spectroscopy.

In absorbing light, atoms gain energy. They eventually lose this energy and return to their lower energy levels, emitting radiation in some cases in the process. Use of this process for analytical purposes is called atomic fluorescence spectroscopy.

The flame spectrometric methods are among the most sensitive analytical techniques available today, capable of detecting low parts per billion of many elements. Phosphorus is rarely determined directly by flame spectroscopy because its principal atomic lines lie in the vacuum ultraviolet portion of the spectrum and are, therefore, outside the range of conventional spectrometers. Nevertheless, the flame techniques are important in phosphorus analysis because of the indirect ways in which they are applied.

The indirect applications are of two general types. One makes use of the depressive effect phosphorus has on the response of other elements. The other involves measurement of an element, for example, molybdenum, which has been stoichiometrically reacted with the phosphorus in the sample.

The depressive effect of phosphorus on flame emission from the alkaline earth elements has been known for a long time. While this has been a major interference in the determination of the alkaline earths, it is used to advantage in the determination of phosphorus, since the effect is linearly proportional to orthophosphate concentration. Cause of the effect is thought to be formation of refractory alkaline earth phosphates in the flame which do not dissociate completely. After the phosphorus in the sample has been converted to orthophosphate and cations have been removed by passage through a cation-exchange resin, a known amount of an alkaline earth is added, and the intensity of a suitable emission line of the alkaline earth is measured. The concentration of phosphorus is then determined from a calibration curve prepared using identical conditions. Less than 10 ppm phosphorus can be detected in this way. In a recent application, Sanyal *et al.* (26) used the depressive effect of phosphate on calcium emission to determine phosphorus in phosphate rock. Phosphate has a similar depressive effect on the response of alkaline earths in atomic absorption spectroscopy. Singhal *et al.* (27) developed a quick and accurate atomic absorption procedure for determining phosphorus in phosphate rock using the depression of strontium absorption. The calibration curve was linear between 0 and 5 ppm phosphorus.

Modifications of the phosphomolybdate method have been widely used to determine phosphorus indirectly by atomic absorption spectroscopy. In these procedures, the phosphomolybdic acid complex is formed and isolated either by precipitation or by solvent extraction. The molybdenum in the complex is then measured by atomic absorption, and the corresponding phosphorus concentration determined. Phosphorus concentrations as low as 0.01 ppm have been measured in solutions of such substances as fertilizers, feeds, steel, rock, and water (25, 28, 33, 34).

The flame photometric detector for gas chromatography provides a sensitive and selective way of detecting phosphorus and its compounds as they are eluted from a chromatographic column. This detector, which depends on phosphorus spectral emissions in a fuel-rich hydrogen-air flame, is described in Chapter 19.

EMISSION SPECTROSCOPY

When used by itself, the term "emission spectroscopy" is generally understood to refer to the spectrochemical technique that employs an electrical discharge (arc or spark) to excite the sample. The visible or ultraviolet radiation generated is dispersed by a prism or grating and detected photographically or by means of photomultiplier tubes. The radiation emitted is characteristic in wavelength of the elements in the sample, and its intensity is proportional to the concentration of these elements.

Emission spectroscopy is a very sensitive technique that is applicable to all metals and to many nonmetals. Phosphorus is among the nonmetals that can be determined spectrographically. The lines at 2535.7 and 2553.6 Å are the most commonly used for phosphorus determination.

Bedrosian *et al.* (2) were able to detect as little as 2 ppm of phosphorus in various biological materials including animal tissue, blood serum, feces, bone, and plant leaves. Phosphorus was one of more than 60 elements included in this study, pointing up one of the chief advantages of emission spectroscopy: the ability to determine many elements simultaneously. Great speed is another advantage. Relatively little sample preparation was involved in this work. Twenty-five milligrams of the dried biological material was mixed with internal standard (lutetium and yttrium as oxides) and graphite, pelletized, and inserted into the cup of a graphite electrode. Maximum sensitivity was obtained with a 20-sec exposure and no preburn. Spectra were recorded on photographic plates. A computer program was used to process the data. Precision of results was better than $\pm 15\%$ of the amount present.

Almost any type of sample can be analyzed by emission spectroscopy.

In alloy analysis, for example, the sample itself is commonly used as one or both electrodes. A number of procedures have been devised for analyzing liquids and solutions. A rotating disk is frequently used to carry the sample solution into the spark discharge. This was the technique used by Jones and Warner (16) to determine phosphorus and 15 other elements in plant tissue. The fresh plant material was dried at 80°C, ground to pass a 20-mesh screen, ashed at 500°C, and dissolved in acid. Lithium was added as internal standard and the solutions analyzed with a direct-reading spectrometer.

SPARK SOURCE MASS SPECTROMETRY

Perhaps the best spectrometric technique for broad spectrum surveys of trace impurities is spark source mass spectrometry. This technique can simultaneously detect practically every element in the periodic table with an almost uniform sensitivity limit of about 10 ppb. (The exceptions are hydrogen, carbon, nitrogen, and oxygen, for which the detection limit is in the low parts-per-million region.) Because of its primary use as a survey technique, one finds few specific references to phosphorus determination by spark source mass spectrometry in the literature. It is safe to say, however, that laboratories equipped with a spectrometer find the technique valuable for phosphorus analysis as for most other elements. Because of the high cost of the equipment (currently about $120,000), relatively few laboratories are so equipped (an estimated 50 to 100) (12).

A spark source mass spectrometer consists essentially of the three components common to all mass spectrometers: an ionization source, an analyzer to separate the ions formed according to their mass-charge ratio, and a detector to record the position and intensity of the ion beams in the spectrum. In the spark source mass spectrometer, the components generally are, respectively, a pulsed radiofrequency spark source, a double-focusing analyzer, and a photographic detection system. Currently there is a trend toward electrical detection, particularly when the objective is to determine a few elements rather than make a broad survey. Electrical detection offers advantages in precision, sensitivity, and speed while sacrificing versatility.

Techniques have been devised for analyzing a wide variety of samples—solids, liquids, and powders. Because sample consumption is small (about a hundredfold less than in emission spectroscopy), the spark source technique is sometimes used for relatively high concentrations when the amount of sample available is too small for other techniques. The spark source technique has also become useful as a microprobe for studying small-scale inhomogeneities. As such it supplements the electron probe microanalyzer described earlier in this chapter under X-Ray Spectroscopy. Ahearn has edited a valuable general reference (1) for readers interested in a detailed discussion of spark source mass spectrometry.

FLUOROMETRY

If a metal ion reacts both with an organic ligand to form a fluorescent complex and with an anion of interest, the concentration of the anion can be measured fluorometrically. By complexing the metal, the anion prevents it from reacting with the organic ligand. As a result, the fluorescence intensity is diminished by an amount that is proportional to the concentration of the anion. Measurements are made with a spectrofluorometer or filter fluorometer.

This approach has been applied by Guyon and Shults (13) to the determination of phosphate in well and pond waters. They developed rapid and simple procedures capable of measuring phosphate levels varying from a few parts per billion to several hundred parts per million. Interaction of phosphate with the aluminum–morin system was used for measuring phosphate concentrations in the parts-per-billion range. The tin–flavonol system was used for higher concentrations. Interferences were few and could be easily removed. An average relative error of 1 to 5% was observed for the tin–flavonol system and 2% for the aluminum–morin system.

X-RAY POWDER DIFFRACTION

X-ray powder diffraction is a powerful analytical technique that can be applied to the analysis of crystalline compounds of phosphorus. Every crystalline powder produces a characteristic x-ray diffraction pattern. Qualitative analysis is accomplished by comparing the pattern of an unknown with data in a file of standards. Quantitative analysis is possible because peak intensities are proportional to concentration. X-ray powder diffraction can be thought of as the reverse of x-ray emission spectroscopy discussed earlier in this chapter. In the latter, a crystal of known interplanar spacing is used to disperse and measure unknown wavelengths. In diffraction, x-rays of known wavelength are used to measure unknown interplanar spacings. X-ray emission spectroscopy is a method for determining the *elements* present in a sample; diffraction identifies *crystalline compounds* present. It is often the only satisfactory method of distinguishing among polymorphs or detecting a compound in the presence of others containing the same elements. Patterns may be recorded electronically or photographically. Details of theory and principles can be found in Ref. 17 or other volumes on x-ray analysis.

A very comprehensive catalog of standard data, the *Powder Diffraction File*, is published by the Joint Committee on Powder Diffraction Standards.* The 1971 edition of this file contains over 21,500 numeric diffraction patterns of crystalline materials. Included in the file are data for approximately 1300 inorganic phosphorus compounds and 270 organophosphorus compounds. The data are available in a number of forms that allow easy systematic searching, including magnetic tape with search and match computer programs. The latter is especially useful for identifying the components in complex mixtures.

In x-ray powder diffraction, the sample must be crystalline. Amorphous phosphates which might be present would be overlooked. This and other limitations have restricted the quantitative use of the technique. Particle size can affect peak intensities, and sample particles must be randomly oriented. The method of Mabis and Quimby (21) is a classic example of quantitative x-ray powder diffraction. The three common forms of sodium tripolyphosphate (anhydrous forms I and II and the hexahydrate) and tetrasodium pyrophosphate can be determined. Magnesium oxide is added as an internal standard, and steps are taken to control particle size and to avoid preferred orientation. The average deviation of the results is of the order of 3% absolute. Intensity measurements are made with a spectrometer. The method has been widely used by manufacturers and users of sodium tripolyphosphate.

INFRARED SPECTROSCOPY

Infrared absorption spectroscopy, an exceedingly powerful analytical method for organic analysis and structure determination, is far less useful for the determination of phosphorus compounds. Inorganic phosphates in general produce rather ill-defined spectra. In some cases, however, the spectra are sufficiently intense and characteristic to be useful for qualitative purposes, particularly when used in conjunction with other techniques, for example, emission spectroscopy and x-ray diffraction. Miller and Wilkins' (23) tabulation of the infrared spectra of 159 inorganic compounds includes data for 20 phosphates. Powdered samples are generally prepared by mulling with heavy paraffin oil or by incorporation into a pressed pellet of potassium bromide.

Corbridge and Lowe (8), using the KBr pellet technique, developed successful quantitative methods for commercial mixtures of condensed phosphates containing forms I and II of sodium tripolyphosphate, tetrasodium pyrophosphate, and sodium trimetaphosphate. Accuracy within ±5%

* 1601 Park Lane, Swarthmore, Pennsylvania 19081.

was reported, and results agreed favorably with the x-ray diffraction method described in the preceding section of this chapter. In general, the importance of infrared spectroscopy as a quantitative technique has decreased partly because of inherent difficulties and partly because of the emergence of more convenient methods of quantitation, particularly gas chromatography (10).

The characteristic infrared absorption frequencies of organophosphorus compounds have been studied very comprehensively by Thomas and Chittenden (5–7, 29–31).

ELECTRON SPECTROSCOPY

The newest member of the spectroscopic team is electron spectroscopy, a technique that has great potential in both structural and analytical chemistry and is applicable to almost every element in the periodic table. In this technique, binding energies of electrons in molecules are measured by determining the energies of electrons ejected by the interactions of a molecule with a monoenergetic beam of x-rays or photons. Actually, the term electron spectroscopy includes two closely related techniques: "electron spectroscopy for chemical analysis (ESCA)," in which the ionizing source is a high-energy x-ray beam; and "photoelectron spectroscopy (PES)," in which the source is a beam of photons, for example, from a helium discharge. The terms are commonly used interchangeably as they basically measure the same things, although the instrumentation, information obtained, and interpretive procedures differ somewhat.

The energy of the ionizing source largely determines the kind of information obtained. High-energy radiation, for example, x-ray, ejects core electrons and allows their binding energies to be determined. With low-energy excitation, for example, vacuum ultraviolet, valence electrons are ejected, permitting determination of ionization potentials. Qualitatively, electron spectroscopy yields structural information similar to infrared or NMR spectroscopy. Quantitatively, results are comparable to x-ray fluorescence.

Solid, liquid, and gaseous samples have been studied by electron spectroscopy. Liquid samples must be frozen on a cold finger to reduce vapor pressure, however, a requirement which has rendered the technique ineffective to date for studying solutions. Another requirement is that the sample be stable under high vacuum. Since photoejected electrons do not emerge from depths greater than 100 Å, electron spectroscopy is basically a surface technique. Spectra have been reported from samples as small as 10^{-8} g.

Electron spectroscopy has been the subject of two excellent reviews (3, 14). The technique has been used to measure the phosphorus 2*p* electron binding energies of 53 phosphorus compounds (24).

REFERENCES

1. Ahearn, A. J., Ed., *Mass Spectrometric Analysis of Solids*, Elsevier, Amsterdam, 1966.
2. Bedrosian, A. J., R. K. Skogerboe, and G. H. Morrison, *Anal. Chem.*, **40**, 854 (1968).
3. Betteridge, D., and A. D. Baker, *Anal. Chem.*, **42** (1), 43A (1970).
4. Birks, L. S., *X-Ray Spectrochemical Analysis*, 2nd ed., Interscience, New York, 1969, p. 111.
5. Chittenden, R. A., and L. C. Thomas, *Spectrochim. Acta*, **20**, 1679 (1964).
6. Chittenden, R. A., and L. C. Thomas, *Spectrochim. Acta*, **21**, 861 (1965).
7. Chittenden, R. A., and L. C. Thomas, *Spectrochim. Acta*, **22**, 1449 (1966).
8. Corbridge, D. E. C., and E. J. Lowe, *Anal. Chem.*, **27**, 1383 (1955).
9. Cowgill, U. M., in *Developments in Applied Spectroscopy*, Vol. 5, L. R. Pearson and E. L. Grove (Eds.), Plenum Press, New York, 1966, p. 3.
10. Ewing, G. W., *Instrumental Methods of Chemical Analysis*, 3rd ed., McGraw-Hill, New York, 1969, Chapter 5.
11. Fabbi, B. P., *Appl. Spectrosc.*, **25**, 41 (1971).
12. Fergason, L. A., Monsanto Company, personal communication, 1971.
13. Guyon, J. C., and W. D. Shults, *J. Am. Water Works Assoc.*, **61**, 403 (1969).
14. Hercules, D. M., *Anal. Chem.*, **42** (1), 20A (1970).
15. Johansson, T. B., R. Akelsson, and S. A. E. Johansson, *Nucl. Instrum. Methods*, **84**, 141 (1970).
16. Jones, J. B., Jr., and M. H. Warner, in *Developments in Applied Spectroscopy*, Vol. 7A, E. L. Grove and A. J. Perkins (Eds.), Plenum Press, New York, 1969, p. 152.
17. Kaelble, E. F., Ed., *Handbook of X-Rays*, McGraw-Hill, New York, 1967, Part 2.
18. Kaelble, E. F., Ed., *Handbook of X-Rays*, McGraw-Hill, New York, 1967, Chapter 30.
19. Libby, R. A., *Anal. Chem.*, **40**, 1507 (1968).
20. Luke, C. L., *Anal. Chim. Acta*, **45**, 365 (1969).
21. Mabis, A. J., and O. T. Quimby, *Anal. Chem.*, **25**, 1814 (1953).
22. McKinney, C. N., and A. S. Rosenberg, in *Advances in X-Ray Analysis*, Vol. 13, B. L. Henke *et al.* (Eds.), Plenum Press, New York, 1970, p. 125.
23. Miller, F. A., and C. H. Wilkins, *Anal. Chem.*, **24**, 1253 (1952).
24. Pelavin, M., *et al.*, *J. Phys. Chem.*, **74**, 1116 (1970).
25. Ramakrishna, T. V., J. W. Robinson, and P. W. West, *Anal. Chim. Acta*, **45**, 43 (1969).
26. Sanyal, R. M., K. C. Singhal, and B. K. Banerjee, *Technology*, **5**, 316 (1968).
27. Singhal, K. C., A. C. Banerji, and B. K. Banerjee, *Technology*, **5**, 117 (1968).
28. Slavin, W., *Atomic Absorption Spectroscopy*, Interscience, New York, 1968, p. 140.
29. Thomas, L. C., and R. A. Chittenden, *Spectrochim. Acta*, **20**, 467 (1964).
30. Thomas, L. C., and R. A. Chittenden, *Spectrochim. Acta*, **20**, 489 (1964).
31. Thomas, L. C., and R. A. Chittenden, *Spectrochim. Acta*, **21**, 1905 (1965).
32. Walitsky, P. J., and R. H. Hopkins, *Appl. Spectrosc.*, **23**, 490 (1969).
33. Zaugg, W. S., *At. Absorpt. Newslett.*, **6**, 63 (1967).
34. Zaugg, W. S., and R. J. Knox, *Anal. Chem.*, **38**, 1759 (1966).

Gas Chromatography Detectors for Phosphorus and Its Compounds

EMMETT F. KAELBLE

Monsanto Company, St. Louis, Missouri

Gas chromatography plays an important role in the determination of phosphorus compounds (particularly organophosphorus pesticides) in the environment. To be sensitive to the low concentrations encountered, special detectors have been developed which, along with great sensitivity, provide a high degree of selectivity. These detectors are described briefly in this chapter.

MICROCOULOMETRIC DETECTOR

In this detector (4), the phosphorus compounds eluting from the chromatographic column are converted to phosphine by reduction with molecular hydrogen at 950°C. The reduction products are led into a coulometric titration cell equipped with silver generating and sensing electrodes, and silver phosphides of undetermined composition are precipitated. The electrical imbalance created by this is sensed by a microcoulometer, and silver equivalent to the amount precipitated is regenerated from the generator anode. The current required to regenerate the silver ion is measured in the form of a gas chromatographic peak.

The detection system per se is not specific for phosphorus because reduction products of compounds containing chlorine or sulfur also precipitate silver ion. Specificity for phosphorus compounds can be achieved, however, by passing the reduced vapors through a short column containing aluminum oxide before they reach the titration cell. This material binds hydrogen chloride and hydrogen sulfide while permitting free passage of phosphine. This detector is capable of detecting less than microgram amounts of phosphine.

THERMIONIC EMISSION DETECTOR

A modification of the well-known flame ionization detector, this device is sensitive to subnanogram amounts of organophosphorus compounds. While the detector responds also to compounds containing halogens or nitrogen, selectivity for phosphorus compounds can be achieved by proper design and choice of operating parameters. It operates on the principle that phosphorus increases the volatility of alkali metal heated in a hydrogen flame and also increases the ionization of the alkali metal.

Karmen (5) has described a detector system consisting of two flames, one mounted above the other, between which is placed a screen coated with an alkali metal salt. The lower flame combusts organic compounds and delivers their combustion products to the heated area of the screen. The upper flame detects alkali metal released from the screen by flame ionization. Hydrocarbons passing through the lower flame are converted to carbon dioxide and water, which are not detected by the upper flame. Small amounts of phosphorus (or halogens or nitrogen) in the lower flame cause an increase in ionization current in the upper flame, thus providing specific, sensitive detection of these elements.

FLAME PHOTOMETRIC DETECTOR (3)

This detector consists basically of a fuel-rich hydrogen-air flame, a narrow bandpass interference filter, and a photomultiplier tube. Sample components, after separation on a chromatographic column, are swept into the flame. Phosphorus compounds emit light at 526 nm which is measured and related to the concentration of the compound being eluted.

Detection limits as low as 10^{-12} g have been reported for phosphorus compounds with this detector. Hydrocarbons and compounds containing halogens, nitrogen, and oxygen are essentially undetected in the flame photometric mode. The detector can also be operated as a sulfur detector by substitution of a filter which transmits light at 394 nm, or as a conventional flame ionization detector. Accessories are commercially available that allow simultaneous observation of phosphorus and sulfur emissions plus ionization response. Thus, this detector possesses considerable versatility in addition to high sensitivity and selectivity.

All volatile phosphorus compounds respond, including the free element. Traces of elemental phosphorus in water, mud, biological samples (1), and air (2) can be determined in a few minutes by collection of the phosphorus in a suitable solvent followed by injection into a gas chromatograph equipped with a flame photometric detector.

A unique application is the determination of traces of inorganic phosphate in aqueous media (6). Following isolation and concentration of the phosphate by extraction with an organic solution of a quaternary ammonium salt, the phosphate is silylated directly in the organic extract and gas chromatographed. As little as 100 ppb of phosphate can be determined.

REFERENCES

1. Addison, R. F., and R. G. Ackman, *J. Chromatogr.*, **47**, 421 (1970).
2. Bohl, C. D., and E. F. Kaelble, *Am. Ind. Hyg. Assos. J.*, to be published, 1973.
3. Brody, S. S., and J. E. Chaney, *J. Gas Chromatogr.*, **4**, 42 (1966).
4. Burchfield, H. P., *et al.*, *J. Gas Chromatogr.*, **3**, 28 (1965).
5. Karmen, A., *J. Chromatogr. Sci.*, **7**, 541 (1969).
6. Matthews, D. R., *Anal. Chem.*, **43**, 1582 (1971).

Phosphate Measurements in Natural Waters—A Critique*

20

WILLIAM CHAMBERLAIN

Indiana State University, Terre Haute, Indiana

and

JOSEPH SHAPIRO

University of Minnesota, Minneapolis, Minnesota

The molybdenum blue procedure for assaying the inorganic phosphate concentration of solutions is well established in scientific circles, and the current literature on this method is voluminous (24). Numerous modifications of the method, first proposed by Osmond in 1887, have appeared, and the reaction is the basis of almost all routinely employed techniques for assaying phosphate in natural waters.

At least when applied to inorganic phosphate standards, the method appears satisfactory. Even without concentration procedures, by using scrupulous techniques and a long light path, for example, 15 cm, it is possible to determine accurately the phosphate concentration of solutions containing as little as 1 μg $PO_4 \cdot P/l$. Even greater sensitivities can be obtained if one concentrates phosphomolybdenum blue, the reaction product, by extracting it into a small volume of suitable organic solvent.

In view of the simplicity, sensitivity, and reproducibility of the method, initially there was little reason to suspect that the molybdenum blue method was not providing valid estimates of the dissolved inorganic phosphate concentration of natural waters, and hence of the amount of phosphorus readily available to organisms. However, some authors (26, 34) did suggest that under the acid conditions necessary for assay, some hydrolysis of natural organic phosphorus compounds might occur and consequently that the molybdenum blue method might overestimate the dissolved inorganic phosphate concentration of natural waters. For this reason Strickland and Parsons (34) introduced the term "soluble reactive phosphorus" [defined

* Contribution No. 103 from the Limnological Research Center, University of Minnesota.

as that phosphorus in a sample of Millipore (HA) filtrate which reacts in 5 min with acidified molybdate] to distinguish what was being measured by current techniques from what was actually present in the water sample.

More recently the suspicion that "soluble reactive phosphorus" and dissolved inorganic phosphate are not equivalent has received considerable experimental support, and the results suggest that the discrepancy is frequently not insignificant. For example, Jones and Spencer (18) attempted to use columns of Amberlite IRA-400 ion-exchange resin to obtain phosphate-free seawater for use as blanks in phosphate determinations. They found that this resin would remove 98% of the added dissolved inorganic phosphate from a 0.55 *M* NaCl solution containing 31 μg PO_4·P/l, but with seawater containing 15.5 μg PO_4·P/l (as natural soluble reactive phosphate), only 30% of the phosphate was removed. These experiments and similar experiments by Rigler (29) with the same type of resin and with ^{32}P added as dissolved inorganic phosphate to a sample of lake-water filtrate suggest that a significant fraction (25 to 70%) of the "soluble reactive phosphorus" is not identical with dissolved inorganic phosphate.

At this point it might be argued that the inability of an ion-exchange resin to remove a portion of the molybdate-reactive material is no indication that this portion is not as readily available to organisms as is dissolved inorganic phosphate. However, such a conclusion is suggested by the results of bioassay experiments. For instance, Kuenzler and Ketchum (19) cultured *Phaeodactylum tricornutum* in supplemented seawater containing varying amounts of phosphate. They then compared those estimates of dissolved inorganic phosphate obtained by chemical analysis with the values based on measurements of the apparent equilibrium distribution of ^{32}P between suspended particles and the solution. These calculated values of dissolved inorganic phosphate should have been higher than the measured values. However, contrary to expectation, in all experiments but one the calculated value was between one and two orders of magnitude lower than the measured value. From this the authors concluded that only a small fraction of the molybdate-reactive material was present as "free phosphate in solution."

Essentially similar results have been obtained in short- (29) and long-term (4) studies with lake water. ^{32}P added as inorganic phosphate to samples of freshly collected lake water is rapidly removed from solution, and within a few minutes approximately 90% is incorporated by the seston. Over the next 2 to 24 hr no detectable change occurs in the distribution of tracer between seston and solution. This apparent equilibrium distribution of tracer between seston and solution has been regarded as an indication of phosphorus exchange between inorganic phosphate in solution and at least part of the sestonic phosphorus (27). Although the size of the exchanging compartment in the seston is not known, it is possible, following reasoning similar to that employed by Kuenzler and Ketchum, to calculate from the following relationship a minimum estimate of the amount of sestonic phosphorus involved in the exchange:

$$\text{"exchanging sestonic phosphorus"} = \text{soluble reactive phosphorus} \times (^{32}\text{P seston}/^{32}\text{P solution})$$

When such a calculation is performed one routinely obtains estimates for "exchanging sestonic phosphorus" which exceed not only the measured amount of phosphorus in the seston but often the estimated total phosphorus concentration of the water sample. This discrepancy can most easily be resolved by assuming, as did Kuenzler and Ketchum, that the amount of molybdate-reacting material is much greater than the amount of true dissolved inorganic phosphate in the sample.

What is most interesting biologically is that these results do indicate that the ability of material to react with molybdate is no indication that the material is readily available to organisms. Indeed Rigler (28) has concluded from radiobiological assays that the concentration of dissolved inorganic phosphate in the surface waters of some lakes is from 10 to 100 times lower than that indicated by chemical analysis, in close agreement with the findings of Kuenzler and Ketchum for seawater.

Thus it would appear that, on the basis of both biological and chemical criteria, there is good reason to suspect that not all of the molybdate-reactive material in natural waters is free phosphate ion, but the extent of the error is still open to some doubt. Whereas the ion-exchange studies indicate that some 30 to 75% of the molybdate-reactive material may be free orthophosphate ion, the bioassay results suggest that free phosphate ion is present only in trace amounts and that essentially all of the color obtained in the molybdenum blue procedure is attributable to other substances.

As correct interpretation of the soluble reactive phosphorus values ultimately determines what information such measurements can contribute to an understanding of the role of phosphorus in aquatic ecosystems, possible reasons for the discrepancy are considered before we attempt to evaluate what, specifically, is being measured by the molybdenum blue procedure.

HIDDEN BLANK

One factor that might account for the apparent failure of the molybdenum blue method would be the existence of a hidden-blank error. By this we mean that in the absence of specific substances capable of complexing with molybdate to form reducible compounds, due to differences in the ionic milieu the reagents form color in natural waters, whereas in distilled water they do not. As Shatkay (32) has illustrated, the magnitude of an error such as this is difficult to estimate, and neither the standard addition method (6) nor the successive dilution technique (12), methods frequently used to obtain working curves for colorimetric analysis will eliminate such an error. However, for the following reasons it appears unlikely that a hidden-blank error can account for the failure of the molybdenum blue method, except where inappropriate methods are used [see Jones and Spencer (18) concerning the Wooster and Rakestraw technique].

In our studies of Minnesota lakes (5) three different procedures were used to assay for dissolved phosphate. In two of the techniques (Harvey and 6 sec) both the complexing reaction and the reduction step were carried out in the aqueous sample, whereas in the third technique (extraction method) the complex was extracted into an organic solvent, the extract washed with dilute hydrochloric acid, and the complex then reduced. In seven of the lakes where arsenic concentrations were not high enough to invalidate a comparison of the results obtained by these three methods, the values obtained by the extraction procedure averaged 0.7% greater than the values obtained by the Harvey method and 7% greater than the 6-sec method. When one considers that the combination of extraction and washing virtually eliminates from the organic extract both unreacted molybdate and molybdoarsenic acid, a compound very similar to 12-molybdophosphoric acid in reactivity, one would expect interference to be minimal in the extraction procedure. Hence if a hidden blank accounted for a substantial fraction of the color which developed in the lake-water samples, one would expect the extraction results to be lower than those obtained by the direct analyses, but this was not the case. One might argue in this regard that the efficiency of extraction of 12-molybdophosphoric acid from lake water was more efficient than from distilled water standards, but we did not find this. Furthermore, despite wide variations in chemical composition of the various lake waters, all three methods showed the same sensitivity in both lake and distilled water samples supplemented with known amounts of phosphate.

Although these observations do not conclusively prove there is no hidden-blank error, they suggest it is highly unlikely that differences in ionic milieu can account for the discrepancies of the molybdenum blue method.

ARSENATE

Until recently, at least, arsenate would have been regarded as an unlikely source of interference in freshwaters, but this assumption no longer seems valid. In six of the Minnesota lakes which we studied, the arsenic concentration, as measured by the silver diethyldithiocarbamate procedure [American Public Health Association, 1965, but see Whitnack and Martens (37) for an evaluation of the technique] averaged 144 μg AsO_4 As/l (64 to 224), and high values have been reported by Angino *et al.* (2) for Kansas waters. In the case of the Minnesota waters, judging by the strong positive interference that was detected when waters were analyzed by the arsenate-sensitive Harvey and 6-sec methods, the arsenic was present in solution as arsenate.

Undoubtedly these values are much above normal levels and can largely be attributed to man's activities (31), but in view of the widespread uses of arsenic as a pesticide and herbicide, checks should be made for this substance; and in instances where high arsenate is present, special methods for assaying phosphate must be employed.

Fortunately arsenate interference may be eliminated by resorting to one of the many arsenate-insensitive extraction techniques that have been reported in the literature (23, 33, 35, 5), but the choice of method will to some extent depend on the circumstances. In the first two and the last-mentioned methods the same organic solvent, isobutyl alcohol, is used, but whereas we recommend reduction after extraction, the other workers suggest extraction after reduction. This, together with the absence of an acid wash step in the other methods, may account for the lesser arsenate interference in our method.

Unfortunately no comparison was made between our method and that of Sugawara and Kanamori (35) because our experience indicated that the *n*-butanol–chloroform mixture, which these workers employ, forms a rather stable emulsion with many lake waters, and a rapid, clean separation of the aqueous and organic phases is not achieved.

HYDROLYSIS

Hydrolysis of organic phosphorus compounds has long been suspected as a possible source of error in the molybdenum blue method, and various approaches have been used in an attempt to circumvent the problem. As early as 1925, for example, Fiske and Subbarow attempted to avoid hydrolysis by working rapidly. This approach has been followed by many, including recently Crouch and Malmstadt (8, 9), Javier *et al.* (17), and Chamberlain and Shapiro (5). In the case of Javier *et al.*, the ultimate was achieved, it being possible to perform quantitative phosphate analyses in the millisecond time range. However, this method requires expensive equipment, whereas our somewhat longer 6-sec procedure can be done in any laboratory equipped with a photometer or spectrophotometer. A somewhat different approach was used by Lowry and Lopez (1946) who attempted to avoid hydrolysis by performing the determination at pH 4. Interestingly enough this pH is, according to Weil-Malherbe and Green (36), optimum for the molybdate-catalyzed hydrolysis of the terminal phosphate of ATP. A third approach has been that originally devised by Berenblum and Chain (3), in which the phosphomolybdate is extracted from the solution into an organic solvent that is then quickly separated from the rest of the solution. Golterman and Wurtz (13) applied this procedure to lake water and described a special separatory funnel for its application. Our own extraction procedure is similar and allows complete reaction and extraction of the phosphomolybdate with as little as 10 sec shaking time. A similar procedure has been adapted for an autoanalyzer by Hillman and Henry (14).

However, despite the suspicion of hydrolysis as a factor and despite the attempts made to obviate it, evidence showing that hydrolysis is important as a source of error is scant and indirect. Nothing is known about the precise chemical composition of the so-called soluble organic phosphorus fraction of natural waters, so the lability of the material is impossible to assess. However, the results of Weil-Malherbe and Green (36) can be used to determine a probable upper limit for the magnitude of error attributable to hydrolysis.

In tests with eight organic phosphate esters these workers detected major differences in their rates of hydrolysis. At pH 1.0, a molybdate concentration of 2.5%, a temperature of 70°C, and an exposure time of 30 min, the most labile compound was glucose-1-phosphate, which was 60% hydrolyzed, but the average for all eight compounds was only 11%.

They also determined that with increasing molybdate concentration, temperature, and length of exposure time, the amount of hydrolysis increased, but the effect of these three factors differed greatly. Whereas the relationship between the percentage of hydrolysis and either molybdate concentration or exposure time was roughly linear, the curve relating the percentage of hydrolysis to temperature was exponential. Their results suggest that the hydrolysis of, for example, glucose-1-phosphate, under the conditions normally employed in assaying for phosphate in natural waters (pH 1.0, molybdate concentration 1.0%, temperature 20°C, exposure time 10 min), would be only 1% or some 60 times lower than that determined under the above-mentioned conditions. Hence with a lake water containing 25 μg P/l (29), if all this material were as labile as glucose-1-phosphate, then the error attributable to hydrolysis would be 0.25 $\mu gPO_4 \cdot P/l$.

Although an error of this magnitude is not great enough to account for the approximately 11 μg $PO_4 \cdot P/l$ discrepancy reported by Spencer and Jones for ion-exchanged seawater, or the discrepancy between radiobiological and chemical estimates (19, 28), it could account for the discrepancy between chemical estimates of soluble reactive phosphate in lake waters treated and untreated by ion exchange by Rigler (29). However, it should be stressed that the estimated error attributable in this way to hydrolysis is most probably a maximum because it is based on the assumption that all of the soluble organic phosphate esters that occur in natural waters are of more than average lability. Were this the case one would conclude, on the basis of published values for the hydrolytic sensitivity constants of organic phosphate esters (7), that progressive hydrolysis would be easily measurable, but in our study of Minnesota lakes such a phenomenon was not detected. A comparison of the results of analyses on replicate samples from 11 lakes that were exposed to acid and molybdate for 30 sec or 15 min revealed no consistent differences between the estimates of soluble reactive phosphate obtained under the two regimens. In only two instances were higher values obtained with the samples exposed for 15 min, and in each case the absolute difference between the 30-sec and 15-min values was within the limits of error of the procedure.

On the basis of this evidence we previously concluded (1969) that there was no evidence that progressive hydrolysis was occurring and indicated that if any acid-labile material had been present in the sample, it must have completely hydrolyzed in the first 30 sec of exposure to the acid. After a more thorough review of the information on the hydrolytic sensitivity of organic phosphate esters, we would suggest that it is highly unlikely that the waters we analyzed did contain easily acid-labile molybdate-reactive material. Further we suggest that with assay procedures employing normal temperatures and short acid exposure times, it is improbable that the hydrolysis of organic phosphate esters, in HA Millipore filtrate, if it occurs at all, will represent a significant error in the assay of inorganic phosphate in natural surface lake waters.

ACID-LABILE INORGANIC PHOSPHATE

In 1941 Hutchinson (15) reported that part of the reactive phosphorus in samples of lake water could be removed by filtering through a membrane with pores approximately 0.5 μ in diameter. He suggested that this so-called acid-soluble sestonic phosphorus was suspended ferric phosphate and concluded that such material was present in significant amounts only during, and for a short time after, the autumnal circulation. It now appears likely that a substantial fraction of even the material in 0.5-μ lake-water filtrate might be particulate. Chamberlain (4), for example, reported that approximately 90% of the ^{32}P in HA Millipore filtrate prepared from lake water to which labeled inorganic phosphate had been added could be removed by refiltering the water through a 0.01-μ membrane filter. To establish that this removal was not due exclusively to adsorption by the filter, other aliquots of HA filtrate were subjected to ultracentrifugation, and under this treatment 80% of the ^{32}P was precipitated.

More recently Rigler (29) has presented similar observations suggesting that when ^{32}P is added as inorganic phosphate to samples of HA filtrate prepared from some lake waters, part of the tracer does not remain in solution as free phosphate ion. Such observations, he suggests, are consistent with the hypothesis that HA filtrate contains a phosphorus compartment that exchanges with soluble orthophosphate and, although he does not specifically suggest it, the presence of a particulate component capable of adsorbing phosphate would account for his results.

Although the observations above concerning added ^{32}P do not demonstrate conclusively that any of the soluble reactive phosphorus is particulate, other results obtained by Chamberlain (unpublished data) indicate this may be the case. He found that when HA filtrate of lake water was refiltered through a 0.10 μ membrane filter, the soluble reactive phosphorus concentration of the water was reduced from 3.9 to 0.8 μg $PO_4 \cdot P/l$ and the total soluble phosphorus decreased from 10.4 to 1.5 μg $PO_4 \cdot P/l$. Thus, at least in this particular lake, approximately 80% of the soluble reactive phosphorus and 90% of the soluble unreactive phosphorus were associated with particles less than 0.45 μ but greater than 0.10 μ in diameter.

As yet the chemical composition of the particulate "soluble" reactive phosphorus fraction has not been determined, but various observations suggest that two types of inorganic phosphate complexes may occur in natural waters. The first and probably the simplest component would be an insoluble metal phosphate, and in view of the work by Einsele (10) and the observations of Hutchinson, ferric phosphate is a prime suspect. Recently, however, Menar and Jenkins (22) have presented evidence indicating that in hard-water sewage, when the pH rises above 8, a calcium phosphate precipitate is formed and settles out with the flocculating sludge. Whether such a precipitate can form in alkaline eutrophic lakes has not been established, but if it does, then in all probability significant amounts of acid-labile colloidal calcium phosphate could remain in suspension in the lake water.

The second type of particulate component would be an organometallic phosphate complex such as has been reported by Levesque and Schnitzer (21). These workers have found that water-soluble fulvic acid extracted from soils has the ability to bind metal ions such as aluminum and iron and that the bound metal ions can form complexes with phosphate. Although the bound phosphorus was not dialyzable and showed limited extractability with Dowex A-1 ion-exchange resin, it reacts as does inorganic orthophosphate ion in assay procedures for inorganic phosphate. As fulvic acid has been isolated from lake sediments by Ishiwatari (16), and Shapiro (30) has isolated, from lake water, organic matter with the ability to bind iron, it seems highly probable that lake water contains such complexed phosphate. The existence of such material could easily account for the discrepancies between chemical estimates of soluble reactive phosphate and radiobiological assay or ion-exchange estimates of free orthophosphate ion (see below).

BIOLOGICALLY AVAILABLE PHOSPHORUS

Thus, in our estimation, it seems unlikely that either a hidden-blank error or the hydrolysis of organic phosphate esters will seriously bias the results obtained by the molybdenum blue procedure. For this reason we suggest that in all probability, except in instances where arsenate is a problem, the molybdate-reactive material present in natural waters may be regarded as inorganic phosphate, although not necessarily as free phosphate ion. As to the possibility of distinguishing between the two it seems unlikely that currently employed analytical techniques will be effective. Even with our 6-sec method, approximately 10% of the phosphorus in a prepared ferric phosphate suspension of 56 ppb was hydrolyzed in this brief exposure to acid, and when calcium phosphate solutions were analyzed we were unable to distinguish between bound and free orthophosphate ion. Exactly how much of the phosphorus in an organometallic phosphate complex would be hydrolyzed under such conditions is not known. Our observations did show that in arsenate-free filtered water the 6-sec method gives slightly lower results than either our extraction technique or the Harvey method, and this provides some indication that at least part of any acid-labile inorganic phosphate in the water does not react during a brief exposure to acid. If this is so, it may be possible to reduce further this hydrolytic interference in lake waters by adapting the reaction-rate method of Javier *et al.* (17) as the required reaction time by this method is only a few milliseconds. However, to achieve this a 100- to 1000-fold increase in sensitivity will be necessary and sophisticated instrumentation is required.

Exactly what would be achieved by such refinement is at present moot, for there is as yet little information on either the quantitative importance or the biological availability of this bound inorganic phosphate. As Rigler (29) has pointed out, "one is not yet forced to reject the hypothesis that soluble reactive phosphorus is equivalent to available phosphorus," and within limits this is supported by our results on the availability of phosphorus in Minnesota lakes. We found that in waters whose soluble reactive phosphate concentration ranged from 0.50 to 30 $\mu g\ PO_4 \cdot P/l$, as measured by the 30-sec extraction procedure, virtually all of this phosphorus was incorporated by phosphorus-starved *Microcystis aeruginosa* within a 1-hr period. It is certainly possible that the availability of the soluble reactive phosphorus differed from that of free orthophosphate ion in that the latter may be absorbed more rapidly. We have no information on this, but if one is prepared to accept that phosphorus sufficiently mobile to be absorbed by organisms within a 1-hr period is readily available biologically, then our results suggest that in arsenate-free waters "biologically available phosphorus" can be measured by one of the short-time (e.g., 10 sec) methods for measuring soluble reactive phosphate. This is particularly true if unfiltered samples are used. Shapiro (unpublished) found, in an unfiltered sample of slightly turbid Cedar Creek water in Minnesota, that the amount of dissolved phosphate increased with exposure to the acid conditions. In this case filtration helped somewhat, but not completely (Table 1). A further experiment was done in which an HA Millipore-filtered aliquot of the water sample was exposed to phosphorus-starved *Microcystis* for 3 hr. The 6-sec phosphate concentration fell to 0, but the chemically measured increase from 6 to 60 sec remained at 9.4 μg $PO_4 \cdot P/l$, which is about 75% of the 12.6 $\mu g/l$ found in the control with no algae. Thus the 6-sec phosphate represented the "biologically available" phosphorus—or at least immediately available phosphorus, with a much slower uptake of the 60-sec phosphorus taking place.

Clearly then the question of time must be involved when speaking of availability. The recent studies of Fitzgerald and Nelson (11) and Kuenzler (20) on phosphatase and on the ability of algae to utilize organic phosphorus compounds indicate that any attempt to differentiate or define phosphorus availability without considering the time element is futile.

Nonetheless we insist that if investigators continue to measure concentrations of dissolved inorganic phosphorus and refer to them as such that they be aware of the problems we have discussed, and attempt to circumvent them. To this end we urge adoption of a technique for surface waters such as our 6-sec procedure where arsenic is definitely shown to be absent, or a short-period (e.g., 10 sec) extraction technique where arsenic is, or may be, present, either method to be done on 0.45-μ filtered water. Only in this way will "soluble reactive phosphorus" and "dissolved inorganic phosphorus" begin to approach each other and only then perhaps will phosphate concentration estimates become meaningful.

TABLE 1 Dissolved Orthophosphate as μg $PO_4 \cdot P/l$ in Cedar Creek Water Exposed to Acid Conditions for Varying Lengths of Time[a]

	Time					
Pretreatment	6 sec	12 sec	30 sec	1 min	8 min	16 min
None	13.0	16.2	28.4	>45	>45	>45
Paper filtration	10.8	14.8	26.4	38.0	>45	>45
Millipore (HA) filtration	12.0	14.6	21.0	24.6	26.5	26.5

[a] The procedure, except for the timing, was that used in the 6-sec method.

ACKNOWLEDGMENTS

Financial assistance for various aspects of this study was received from the Minnesota Resources Commission, the Federal Office of Water Resources Research, the Federal Water Pollution Control Administration, and the National Science Foundation.

REFERENCES

1. American Public Health Association, *Standard Methods for the Examination of Water and Waste Water*, 12th ed., American Public Health Association, New York, 1965, 769 pages.
2. Angino, E. E., L. M. Magnuson, T. C. Waugh, O. K. Galle, and J. Bredfeldt, "Arsenic in Detergents: Possible Danger and Pollution Hazard," *Science*, **168,** 389–390 (1970).
3. Berenblum, J., and E. Chain, "Studies on the Colorimetric Determination of Phosphate," *Biochem. J.*, **32,** 286 (1938).
4. Chamberlain, W. M., "A Preliminary Study of the Nature and Importance of Soluble Organic Phosphorus in the Phosphorus Cycle of Lakes," Ph.D. Thesis, University of Toronto, Canada, 1968, p. 231.
5. Chamberlain, W. M., and J. Shapiro, "On the Biological Significance of Phosphate Analysis; Comparison of Standard and New Methods with a Bioassay," *Limnol. Oceanogr.*, **14,** 921–927 (1968).
6. Chow, T. J., and T. G. Thompson, "Flame Photometric Determination of Calcium in Sea Water and Marine Organisms," *Anal. Chem.*, **27,** 910–913 (1955).
7. Colowick, S. P., and N. O. Kaplan, *Methods in Enzymology*, Vol. 3, Academic Press, New York, 1957, p. 1145.
8. Crouch, S. R., and H. V. Malmstadt, "A Mechanistic Investigation of Molybdenum Blue Method for Determination of Phosphate," *Anal. Chem.*, **39,** 1084–1093 (1967).
9. Crouch, S. R., and H. V. Malmstadt, "Determination of Inorganic Phosphate in the Presence of Adenosine Triphosphate by the Automatic Reaction Rate Method," *Anal. Chem.*, **40,** 1901–1902 (1968).
10. Einsele, W., "Uber chemische und kolloidchemische Vorgange in Eisen-Phosphat-Systemen unter limnochemischen und limnogeologischen Gesichtspunkten," *Arch. Hydrobiol.*, **33,** 361–367 (1938).
11. Fitzgerald, G. P., and T. C. Nelson, "Extractive and Enzymatic Analysis for Limiting or Surplus Phosphorus in Algae," *J. Phycol.*, **2,** 32–37 (1966).
12. Gilbert, P. T., Jr., "Determination of Cadmium by Flame Photometry," *Anal. Chem.*, **31,** 110–114 (1959).
13. Golterman, H. L., and I. M. Wurtz, "A Sensitive Rapid Determination of Inorganic Phosphate in Presence of Labile Phosphate Esters," *Anal. Chim. Acta*, **25,** 295–297 (1961).
14. Hillman, H., and S. Henry, "The Measurement of Low Concentrations of Inorganic Phosphate, Technicon Symposium on Automation in Analytical Chemistry," 1967, pp. 183–185.
15. Hutchinson, G. E., "Limnological Studies in Connecticut: IV. Mechanisms of Intermediary Metabolism in Stratified Lakes," *Ecol. Monogr.*, **11,** 21–60 (1941).
16. Ishiwatari, R., "An Estimation of the Aromaticity of a Lake Sediment Humic Acid by Air Oxidation and Evaluation of It," *Soil Sci.*, **107,** 53–57 (1969).

17. Javier, A. C., S. R. Crouch, and H. V. Malmstadt, "An Automated Fast Reaction-Rate System for Quantitative Phosphate Determinations in the Millisecond Range," *Anal. Chem.*, **41,** 239–243 (1969).
18. Jones, P. G. W., and C. P. Spencer, "Comparison of Several Methods of Determining Inorganic Phosphate in Sea Water," *J. Mar. Biol. Assoc. U.K.*, **43,** 251–273 (1963).
19. Kuenzler, E. J., and B. H. Ketchum, "Rate of Phosphorus Uptake by *Phaeodactylum tricornutum*," *Biol. Bull.*, **123,** 134–145 (1962).
20. Kuenzler, E. J., "Dissolved Organic Phosphorus Excretion by Marine Phytoplankton," *J. Phycol.*, **6,** 7–13 (1970).
21. Levesque, M., and M. Schnitzer, "Organo-Metallic Interactions in Soils: 6. Preparation and Properties of Fulvic Acid–Metal Phosphates," *Soil Sci.*, **103,** 183–190 (1967).
22. Menar, A. B., and D. Jenkins, "Fate of Phosphorus in Waste Treatment Process: Enhanced Removal of Phosphate by Activated Sludge," *Environ. Sci. Technol.*, **4,** 1115–1121 (1970).
23. Murphy, J., and J. P. Riley, "A Single-Solution Method for the Determination of Phosphate in Natural Waters," *Anal. Chem. Acta*, **27,** 31–36 (1962).
24. Olson, S., "Recent Trends in the Determination of Orthophosphate in Water," in *Chemical Environment in the Aquatic Habitat*, H. L. Golterman and R. S. Clymo (Eds.), N. V. Noord-Hollandsche Uitgevers Maatschappij, Amsterdam, 1967, pp. 63–105.
25. Osmond, F., "Sur une reaction pouvant servir on dosage colorimetrique du phosphore dans les fontes, les aciers," *Bull. Soc. Chem. Paris*, **47,** 745–748 (1887).
26. Rigler, F. H., "The Circulation of Phosphorus in Lake Water," Ph.D. Thesis, University of Toronto, Canada, 1954, 117 pages.
27. Rigler, F. H., "A Tracer Study of the Phosphorus Cycle in Lakewater," *Ecol.*, **37,** 550–562 (1956).
28. Rigler, F. H., "Radiobiological Analysis of Inorganic Phosphorus in Lakewater," *Verh. Int. Verein. Limnol.*, **16,** 465–470 (1966).
29. Rigler, F. H., "Further Observations Inconsistent with the Hypothesis that the Molybdenum Blue Method Measures Orthophosphate in Lake Water," *Limnol. Oceanogr.*, **13,** 7–13 (1968).
30. Shapiro, J., "Iron in Natural Waters: Its Characteristics and Biological Availability as Determined with the Ferrigram," *Int. Verein. Theor. Angew. Limnol.*, **17,** 456–466 (1969).
31. Shapiro, J., "Arsenic and Phosphate: Measured by Various Techniques," *Science*, **171,** 234 (1971).
32. Shatkay, A., "Photometric Determination of Substances in Presence of Strongly Interfering Unknown Media," *Anal. Chem.*, **40,** 2097–2106 (1968).
33. Stephens, K., "Determination of Low Phosphate Concentrations in Lake and Marine Waters," *Limnol. Oceanogr.*, **8,** 361–362 (1963).
34. Strickland, J. D. H., and T. R. Parsons, "A Manual of Seawater Analysis," *Bull. Fish. Res. Board Can.*, **125,** 1–185 (1960).
35. Sugawara, K., and S. Kanamori, "The Spectrophotometric Determination of Trace Amounts of Arsenate and Arsenite in Natural Waters with Special Reference to Phosphate Determination," *Bull. Chem. Soc. Japan*, **37,** 1358–1363 (1964).
36. Weil-Malherbe, H., and R. H. Green, "The Catalytic Effect of Molybdate on the Hydrolysis of Organic Phosphate Bonds," *Biochem. J.*, **49,** 286–292 (1951).
37. Whitnack, G. C., and H. H. Martens, "Arsenic in Potable Desert Groundwater: An Analysis Problem," *Science*, **171,** 383–385 (1971).

Low-Level Phosphorus Detection Methods

21

RICHARD A. KIMERLE

Monsanto Company, St. Louis, Missouri

and

WILLIAM RORIE

Industrial Testing Laboratories, Inc., St. Louis, Missouri

In attempting to define man's chemical environment qualitatively and quantitatively, one has no difficulty justifying the search for better methods of chemical analysis. The need for simple and accurate methods for determining phosphorus at the low parts-per-billion range is particularly relevant to current eutrophication and water-pollution problems. Phosphorus is one algal nutrient the significance of which at very low levels is currently being extensively studied. It is therefore the purpose of this paper to review some of the pertinent literature on low-level phosphorus-detection methods as related to (*a*) the established methodology found in limnology, oceanography, and sanitary engineering; (*b*) newer methods of analysis from other disciplines; and (*c*) the problems of what forms of phosphorus are actually being measured, irrespective of the method used. It is hoped that the reader will be able to correct some of the inherent problems associated with the "accepted" methods or be able to substitute an alternative better suited to his needs.

LITERATURE REVIEW

Rieman and Beukenkamp (44) reviewed the variety of methods that are used in many disciplines to analyze for phosphorus. These include gravimetric, titrimetric, photometric, spectrophotometric, chromatographic, and radioactivation methods of analysis. When the literature is surveyed for papers specifically related to low parts-per-billion (ppb) phosphorus-detection methods, it quickly becomes obvious that the spectrophotometric molybdenum blue method dominates all other methods in the literature, especially prior to the late 1960s. Therefore any discussion of this topic must necessarily include a review of the molybdenum blue method of phosphorus analysis.

The molybdenum blue method was first published by Osmond in 1887 (42). Since that time, the method has been more or less adopted as the "standard method" of analysis for phosphorus in every branch of science where phosphorus is routinely measured. This diversity of use has led to hundreds of papers through the years where the authors report changes in the basic method to improve its performance for their particular use. This fact makes a review of the method a horrendous task. Therefore the discussion here is limited to those papers dealing specifically with low-level phosphorus-detection methods. This segment of the literature happens to appear mostly in the area of water chemistry.

The molybdenum blue method depends on the reaction of an acid ammonium molybdate solution with orthophosphate ions to form a molybdophosphoric acid complex. The molybdophosphoric acid can be reduced to molybdenum blue in the aqueous phase (2–5, 11, 13, 14, 30, 40) or be reduced after extraction with an appropriate solvent (2, 3, 6, 18, 25, 26, 36, 43, 48, 49, 51, 54). Many of the papers above discuss the topics of acidity, reagent concentration, and various types of reducing agents.

One of the most widely used methods of phosphate analysis is that developed by Murphy and Riley (40) for use in seawaters. They claim the use of ascorbic acid as a reducing agent, and the addition of antimony rapidly produces a molybdenum blue color in 10 min that is stable for 24 hr. This method also has the advantage in ocean-water analysis of being less sensitive to the salt error associated with stannous chloride reducing agent. Arsenate interferes with the method but only at the levels considered greater than would normally be encountered in seawater. They considered the method to be free of interferences from hydrolysis of organic phosphorus compounds, but admitted that very labile phosphorus compounds might lead to high orthophosphate results. Silicate interferences were not investigated.

Edwards *et al.* (13) modified the Murphy-Riley method to extend the detectable concentration of phosphate. Interferences from three hydrolyzable condensed phosphates were also investigated. After 10 min of color-development time, the error from sodium tripolyphosphate, tetrasodium pyrophosphate, and sodium hexametaphosphate in concentrations of 1.00 mg/l phosphorus was 2.5, 1.7, and 1.9%, respectively. Orthophosphate was also determined in a mixture of 1.00 mg/l orthophosphate and 1.00 mg/l pyrophosphate. The resulting orthophosphate was 1.03 mg/l phosphorus instead of 1.00 mg/l phosphorus added as orthophosphate—an error of 30 μg/l phosphorus. Arsenate concentrations of 0.10 mg/l interfered with phosphate determinations. The effect of silicates was not studied.

Strickland and Parsons (50) recommended the use of the Murphy-Riley (40) method. They stated that the method "is so superior to the other methods in terms of rapidity and ease of analysis that it probably represents the ultimate in sea-going techniques." They also recommended a low-level detection method for oligotrophic waters incorporating the Murphy-Riley method with the extraction procedures of Proctor and Hood (43) and Stephens (49).

Harwood *et al.* (23) also reviewed methods of phosphate analysis, emphasizing a method of extending the linear range of the methods of Murphy-Riley (40) and Edwards *et al.* (13). The Murphy-Riley method was preferred and tested for interferences. Sodium hexametaphosphate gave an orthophosphate response equivalent to 2% of the total added, similar to the 1.9% found by Edwards *et al.* (13). Tests with several organic phosphorus compounds resulted in an error of 10% for the most labile compounds and 0.1% with the least labile compounds. Harwood *et al.* (23) concluded that the importance of this interference would depend on the relative concentrations of labile phosphorus compounds to orthophosphate concentration, thus limiting the applicability of the method for trace orthophosphate analysis with high concentrations of labile phosphorus compounds. Harwood *et al.* (23) also lengthened the reaction time from 10 to 30 min during winter months because of marked temperature effect on color development. Arsenate and silicate interferences were not investigated.

Jones and Spencer (29) also reached the conclusion that hydrolysis of dissolved organic phosphorus compounds would occur in orthophosphate analysis of waters high in biological activity.

Chamberlain and Shapiro (8) in an attempt to determine the biological significance of phosphate analysis studied new methods to circumvent both the problems of hydrolysis of labile phosphorus compounds and arsenate interferences. They concluded that their 6-sec method may reduce the interference from labile phosphorus compounds because of the shorter time period in which hydrolysis could take place. They also suggested that the 6-sec method was less sensitive to arsenate interference because the rate of formation or reduction, or both, of phosphomolybdic acid was more rapid than for molybdoarsenic acid. Chamberlain and Shapiro (8) did not evaluate the effect of silicate on their 6-sec method.

Initial studies on the extraction of heteropoly acids were performed by chemists and biochemists (6, 12, 18, 36, 38, 54) and later by scientists concerned with water chemistry (3, 8, 43, 48, 49, 51). Procedures using isobutyl alcohol for the extraction of phosphomolybdic acid to analyze for phosphate were originally developed by Berenblum and Chain (6). They claimed the method to be completely free of interfering substances and insensitive to slight changes in acidity and reagent concentration. Martin and Doty (36) improved the performance of the solvent extraction by using a 1:1 mixture of benzene and isobutyl alcohol. This reduced the number of washings to one and the shaking time to 15 sec. Ging (18), in studying extraction procedures, evaluated several reducing agents for color stability and found 1-amino-2-naphthol-4-sulfonic acid to be the best. He pointed out that the labile phosphorus compounds tested did not interfere because the molybdate or molybdic acid was removed after a short period of exposure to the sample. The extraction also had the advantage of preventing interference from substances that inhibit color development. The extraction step separated the interfering ions before the reducing agent was added. The American Soap and Glycerine Producers (3), after studying methods of phosphate analysis in surface waters, recommended the extraction procedure of Martin and Doty (36). They concluded that the reduced contact time of 15 sec of the acid molybdate and sample avoided the hydrolysis of condensed phosphates. The extraction method was also reported to eliminate interferences of other ions. Shapiro *et al.* (47) developed a variation of the extraction procedure of Stephens (49). This new method was reported to eliminate arsenate interference with a second wash of 100 ml of 0.5 *N* HCl (saturated with isobutyl alcohol), thus removing excess molybdate. They also suggested that their extraction procedure would not be as sensitive to interfering labile phosphorus compounds as Stephens' method because the exposure time of sample to reagents was much less. Chamberlain and Shapiro (8) established that phosphate measured by extraction procedures was very similar to biologically available phosphate as determined by their bioassay technique.

In a search for a spectrophotometric method of determining submicron quantities of orthophosphate in natural waters, Sugawara and Kanamori (51) used *n*-butanol–chloroform as the extraction solvent. The extracted phosphomolybdate was then decomposed and the liberated molybdenum spectrophotometrically determined as thiocyanate complex, thus enabling the limit of detection to be extended to 0.02 μg/250 ml sample, with an error of 5%. Using the *n*-butanol—chloroform solvent, no significant interferences were found from the 32 ions tested.

Numerous other authors have recognized the problem of silicon, arsenic, and germanium interferences with the molybdenum blue method of phosphate analysis and developed nonextraction techniques to circumvent the problem (35, 53, 7, 28). Levine *et al.* (35) described a technique in which phosphate was precipitated with aluminum hydroxide, centrifuged, and treated with a combination of acids to volatilize any coprecipitated silicon, arsenic, or germanium. The accuracy of their method was shown by the excellent recovery of added phosphorus to seawater containing 1.7 γ P_2O_5, even in the presence of 0.01 to 1.00 mg As_2O_5, SiO_2, and GeO_2. Van Schouwenburg and Walinga (53) used a reducing agent of sodium metabisulfite and sodium thiosulfate in the Murphy-Riley method to prevent interference from arsenic. A similar method was presented by Johnson (28) in which the arsenate was determined as the difference in absorbance between a reduced and unreduced sample. Campbell and Thomas (7) calculated the silicate interference in phosphate analysis by measuring both silica and phosphate in the same sample. The phosphomolybdate complex was then completely destroyed with oxalic acid (50 g/l) and the silicomolybdate complex reduced and read at 815 mμ. Apparent P due to silica was then subtracted from the apparent soluble P determination, thus giving a corrected soluble P value. The limit of phosphorus detection was given in the low microgram-per-liter range.

Only a few alternate methods to the molybdenum blue method of parts per billion phosphorus analysis can be found in the literature. The use of gas chromatography for determination of trace quantities of phosphates in aqueous media was reported by Matthews and Dean (37). They utilized some of the methods developed in the gas chromatographic analysis of nucleic acids and organophosphorus pesticides. Phosphates were concentrated fiftyfold by extraction with quaternary ammonium salts in tolueneoctanol. Derivatization in organic phase was achieved with *N*,*O*-bis(TMS)-trifluoroacetamide—1% trimethylchlorosilane. The organic solution was injected into a gas chromatograph equipped with a phosphorus-selective flame photometric detector. Quantitative results were obtained from 20 to 100 ppb phosphorus. Neutron activation analysis of phosphorus in water was reported by Wayman (55). The level of detection was 0.01 to 10.0 ppm. A rapid and simple fluorometric method of phosphate analysis was developed by Guyon and Shults (19). The method depends on the fluorescence of a complex of an organic ligand and a metal ion. However, in the presence of a particular anion (phosphate) a metal–phosphate complex is formed. This results in diminished fluorescence proportional to the concentration of the anion. Methods of avoiding interferences were discussed. The level of sensitivity was good below 100 ppb.

A number of methods exist that still partially depend on the reaction of phosphate and molybdate ions. Atomic absorption spectroscopy (56) was used to measure the molybdenum from the reduction of phosphomolybdic acid. Allen and Hahn (1) and Hahn and Schmitt (20) used activation analysis by counting the activity of tungsten-185 that was incorporated into the phosphomolybdic acid complex.

Phosphorus occurs in the environment in various forms. Numerous authors have presented separation schemes based on chemical types and physical states to facilitate analysis of each of the phosphate components of the total phosphorus in a sample (2, 41, 50). Most methods of determining trace quantities of phosphorus depend on the phosphorus being an orthophosphate with respect to chemical type. Condensed and organic phosphates are determined after subjecting the sample to various degrees of hydrolization or oxidation (2, 10, 16, 21, 22, 27, 34, 46). If hydrogen peroxide is used in the digestion procedure, care should be taken to use phosphorus-free H_2O_2. Some H_2O_2 producers use phosphorus compounds as stabilizers (33).

Olsen (41) discussed the problems of separating and measuring phosphorus in the various physical states of liquid, colloidal, and solid. *Standard Methods* (2) recommended the use of 0.45 μ membrane filters to divide the sample into "soluble" and particulate portions. However, Olsen (41) felt that membrane filtration alone resulted in a liquid and solid fraction with undefined portions of the colloidal phosphorus. He stated that hardly any of the published data on dissolved orthophosphate really fall into this category. The only satisfactory method to achieve absolute separation of dissolved and colloidal fractions was with ultrafiltration of the water sample. Olsen further indicates that total phosphates are the only analyses that can be and have been satisfactorily performed ever since the beginnings of phosphate determination in waters. All the other forms of phosphorus—ortho, poly, and residual—occurring in the dissolved, colloidal, and solid phases, cannot be accurately measured because of poor separation techniques. Those forms that can be determined are of little general interest. Olsen suggested that the status of limnological phosphate research would not be improved until ultrafiltration or centrifugation techniques become universally adopted to better define the colloidal fraction of water samples.

Chemical types of phosphorus can be effectively separated using thin-layer and paper chromatographic techniques (9, 31, 52). Karl-Kroupa (31) demonstrated that paper chromatographic techniques can be used successfully to separate the various forms of condensed phosphates. After separating the phosphate components, the spots were cut out of the paper, extracted, and quantitatively determined using the method of Martin and Doty (36). The level of detection was reported to be 0.01 ppm. Clesceri and Lee (9) reported on the use of thin-layer chromatography to separate ortho- and pyrophosphates. No quantitative estimates were made of the phosphates. Anion-exchange thin-layer chromatography was used by Tanzer *et al.* (52) to separate several linear and cyclical phosphates. Approximate chain lengths were also determined.

The problems of how samples are handled prior to analysis and the changes that can occur in phosphate levels have received considerable attention. Karl-Kroupa *et al.* (32) and Clesceri and Lee (10) have shown that condensed and organic phosphorus compounds can undergo hydrolytic and enzymatic degradation. The best technique to avoid the problem is to perform the analysis immediately after sampling (39). If samples must be stored, they should be filtered first (15) and then treated with chloroform (17). Polyethylene bottles should not be used because they adsorb phosphate ions (39).

DISCUSSION

Considering the numerous alternative procedures in the basic molybdenum blue method, the most logical approach to accurate and universally accepted phosphorus data is to adopt the "best" method of analysis. The method used most is probably that of Murphy and Riley (40) because the analysis is rapid and easy to perform. However, as is evident in the literature, this method is subject to interferences from labile phosphorus compounds and arsenate. We feel therefore that the method is not universally applicable for all waters in the United States. This is particularly true when the phosphorus content is low and the possibility exists that labile phosphorus compounds and/or arsenate could be present at significant levels in the water sample. Shapiro (48) has recently called attention to the fact that some lakes in Minnesota contain as much as 224 μg/l, thus suggesting that much of the accumulated phosphorus data could be erroneously high. Our experience with the Murphy and Riley (40) method, and similar methods in which the molybdenum blue is developed in the aqueous phase without extraction, has led us to believe that these methods should be avoided unless interferences are known to be absent or if the orthophosphate concentration is high relative to the concentration of interfering ions.

From the literature it seems evident that extraction techniques offer us the best method of phosphorus analysis in the presence of ubiquitous interfering ions. Although extraction methods may be more time-consuming and require more glassware than the direct methods, the accuracy of the results obtained more than justify the effort. Shapiro *et al.* (47) clearly demonstrated in comparative studies that solvent extraction of phosphomolybdic acid, and subsequent reduction to blue color, yields results of significantly lower phosphorus levels than methods employing color development in the aqueous phase. Differences in their data were roughly attributed to arsenate interference.

The nonextraction, 6-sec method of Shapiro, Chamberlain, and Barrett (8, 47) seems to offer only a partial solution to the problem of interferences. They stated that the 6-sec method was less sensitive than the Harvey method to arsenate interference because the formation of molybdoarsenic acid proceeds at a slower rate than does phosphomolybdic acid. The interferences from the hydrolysis of labile phosphorus compounds would also be minimized because of the reduced time of the reaction. However, the advantages of the new 6-sec method must be weighed against the new problem of critical timing for reproducibility. Shapiro *et al.* (47) showed that 75% of the molybdenum blue color was developed after 6-sec. The rate of color development up to 15 sec is quite rapid; thus, a small error in timing could result in a disproportionately larger error in color intensity. It would be essential therefore that persons performing the analysis be made well aware of the importance of timing.

For routine analysis of phosphorus in surface waters of undefined chemical purity, it is our opinion that the extraction method of Martin and Doty (36) as outlined by the American Association of Soap and Glycerine Producers (3) and American Public Health Association (2) is the most accurate, precise, and broadly applicable method available for low microgram per liter phosphorus analysis. We have tested the method for interference by silicate at 100 mg/l and found none. Arsenate begins to interfere with a 20-ppb phosphorus sample at a concentration of about 500 μg/l as As, a concentration far greater than would be found in surface waters. Condensed phosphates have not been found to be significantly hydrolyzed during the 15-sec shaking period. Table 1 shows the results of our tests of the 1:1

TABLE 1 Absorbances for Low μg/l Phosphorus Analysis Using the 1:1 Benzene-Isobutanol Extraction Method (1, 2)[a]

P Added (μg/l)	Absorbance			X	Standard Error
A. Distilled water					
0.0	0.002	0.006	—	0.004	0.002
0.1	0.002	0.006	—	0.004	0.002
0.5	0.006	0.012	—	0.009	0.003
1.0	0.018	0.015	0.016	0.016	0.001
5.0	0.090	0.082	0.085	0.086	0.002
10.0	0.178	0.184	0.175	0.179	0.003
B. Lake water					
0.0	0.036	0.034	0.037	0.036	0.001
1.0	0.051	0.058	0.053	0.054	0.002
Difference	0.015	0.024	0.016	0.018	—
C. Lake water spiked with: 0.2 mg/l As 2.0 mg/l SiO_2	0.035	0.038	0.039	0.037	0.001

[a] Sample size 250 ml, 10-cm cells, Coleman 124D Spectrophotometer, 725 mμ.

benzene–isobutanol extraction technique. The purpose of the study was to determine the absolute low limit of detection and demonstrate its insensitivity to arsenate and silicate during parts per billion phosphorus analysis.

The data in Table 1 were obtained under scrupulously clean conditions, an absolute must for low microgram per liter phosphorus analysis. Data in part A, using distilled water, show that 1 μg/l of phosphorus can be determined with an error of less than 5%. At 0.5 μg/l the error was increased to approximately 30%, and 0.1 μg/l P cannot be detected. We therefore suggest that under optimum conditions, using 10-cm cells, the best anyone can expect to do with any degree of accuracy is 1 μg/l. Part B of Table 1 demonstrates that even 1 μg/l of phosphorus can be recovered from natural lake water with an initial concentration of approximately 2 μg/l. Part C shows that, after spiking the lake water with 0.2 mg/l As and 2 mg/l SiO_2, the absorbance was identical to the unspiked water in part B. The method was clearly free of arsenate and silicate interferences, even when measuring 1 μg/l phosphorus.

Special attention should be given to the method of Sugawara and Kanamori (51) by those persons needing analysis at the submicrogram concentration. The minimum level of detection, 0.02 μg/250 ml, with 5% error, is significantly lower than that possible with normal extraction and reduction techniques. They also established that the method was free of interferences from the 32 ions tested.

The new, rapid, and simple fluorimetric method of phosphorus analysis of Guyon and Shults (19) is interesting because it appears to be quite capable of low parts per billion level of analysis. Also, many limnological laboratories are now using fluorescence for routine analysis of organic dyes and chlorophyll-*a*. Thus, fluorimetric analysis of phosphorus would not necessarily require purchasing new equipment. However, before the method can be widely used, further studies need to be performed on all aspects of the analysis.

In addition to the need for developing new approaches of phosphorus analysis, a greater need exists for better methods of characterizing the forms of phosphorus present in a sample. Olsen (41) takes the position that many of the limnologically important phosphorus fractions cannot be measured because adequate separation techniques are not available. Total phosphorus in a sample is probably the only analysis that can be performed with any degree of accuracy on samples that contain the numerous forms of phosphorus in the dissolved, colloidal, and particulate phases. The paper chromatographic techniques presented by Karl-Kroupa (31) seem well suited to help solve the problem of separation of chemical states of phosphorus at low concentrations, but the procedure has not been widely used.

SUMMARY AND CONCLUSIONS

Numerous methods are available to analyze for phosphorus in the environment. However, at the present time alternate methods to the molybdenum blue method of parts per billion phosphorus analysis are very few. We therefore support the use of the 1:1 benzene-isobutyl alcohol extraction procedure presented by the American Association of Soap and Glycerine Producers (3) and the American Public Health Association (2). Some of the advantages of this method over the many others include the following:

1. Interferences from labile phosphorus compounds, arsenate, and silicate are minimized or eliminated by the 15-sec mixing time of the acid molybdate and sample.
2. Interference from complexed arsenate and/or silicate, at reasonable concentrations, is eliminated because of the apparent partial selectivity of the solvent for phosphomolybdic acid.
3. The limit of sensitivity at 1 μg/l, with less than 5% error, is considered adequate for environmental studies. Sensitivity can be increased by increasing the size of the sample and the length of the light-path cell.
4. The method is amenable to the capabilities of the personnel in laboratories where phosphorus is routinely measured, because it is a relatively easy analysis requiring equipment normally found in most laboratories.

Accepting the premise that orthophosphate can be accurately measured, using the techniques above, we are still confronted with the problems of inadequate sampling of the environment, effect of sample containers, hydrolysis of phosphorus compounds between the time of sampling and analysis, and separation and quantification of each phosphorus component of the total phosphorus. Then, after all factors have been considered, we must still determine the biological or ecological significance of the data, a formidable task.

REFERENCES

1. Allen, H. E., and R. B. Hahn, "Determination of Phosphate in Natural Waters by Activation Analysis of Tungstophosphoric Acid," *Environ. Sci. Technol.*, **3** (9), 844–848 (1969).
2. American Public Health Association, Inc., *Standard Methods for the Examination of Water and Wastewater*, 13th ed., published jointly by American Public Health Association, American Water Works Association, and Water Pollution Control Federation, 1971.

3. Association of American Soap and Glycerine Producers Committee Report, "Determination of Orthophosphate, Hydrolyzable Phosphate, and Total Phosphate in Surface Waters," *J. Am. Water Works Assoc.*, **50** (12), 1563–1574 (1958).
4. Atkins, W. R. G., "The Phosphate Content of Fresh and Salt Water in Its Relationship to the Growth of the Algal Plankton," *J. Mar. Biol. Assoc., U.K.*, **13,** 119–150 (1923).
5. Bell, R. D., and E. A. Doisy, "Rapid Colorimetric Methods for the Determination of Phosphorus in Urine and Blood," *J. Biol. Chem.*, **44,** 55–67 (1920).
6. Berenblum, I., and E. Chain, "An Improved Method for the Colorimetric Determination of Phosphate," *Biochem. J.*, **32,** 295–298 (1938).
7. Campbell, F. R., and R. L. Thomas, "Automated Method for Determining and Removing Silica Interference in Determination of Soluble Phosphorus in Lake and Stream Waters," *Environ. Sci. Technol.*, **4** (7), 602–604 (1970).
8. Chamberlain, W., and J. Shapiro, "On the Biological Significance of Phosphate Analysis; Comparison of Standard and New with a Bioassay," *Limnol. Oceanogr.*, **14** (6), 921–927 (1969).
9. Clesceri, N. L., and G. F. Lee, "Thin Layer Chromatographic Separation of Orthophosphate and Pyrophosphate," *Anal. Chem.*, **36** (11), 2207–2208 (1964).
10. Clesceri, N. L., and G. F. Lee, "Hydrolysis of Condensed Phosphates—I: Non-Sterile Environment," *Int. J. Air Water Pollut.*, **9,** 723–742 (1965).
11. Deniges, G., "Reaction de Coloration Extrêment Sensible des Phosphates et des Arseniates. Ses Applications," *C. R. Acad. Sci.*, **171,** 802–804 (1920).
12. DeSesa, M. A., and L. B. Rogers, "Spectrophotometric Determination of Arsenic, Phosphorus, and Silicon in the Presence of Each Other," *Anal. Chem.*, **26** (8), 1381–1383 (1954).
13. Edwards, G. P., A. H. Molof, and R. W. Schneeman, "Determination of Orthophosphate in Fresh and Saline Waters," *J. Am. Water Works Assoc.*, **57,** 917–925 (1965).
14. Fiske, C. H., and Y. Subarow, "The Colorimetric Determination of Phosphorus," *J. Biol. Chem.*, **66,** 375–400 (1925).
15. Fitzgerald, G. P., and S. L. Faust, "Effect of Water Sample Preservation Methods on the Release of Phosphorus from Algae," *Limnol. Oceanogr.*, **12** (2), 332–334 (1967).
16. Gales, M. E., Jr., E. C. Julian, and R. C. Kroner, "Method for Quantitative Determination of Total Phosphorus in Water," *J. Am. Water Works Assoc.*, **58,** 1363–1368 (1966).
17. Gilmartin, M., "Changes in Inorganic Phosphate Concentration Occurring during Seawater Sample Storage," *Limnol. Oceanogr.*, **12** (2), 325–328 (1967).
18. Ging, N. S., "Extraction Method for Colorimetric Determination of Phosphorus in Microgram Quantities," *Anal. Chem.* **28,** 1330–1334 (1956).
19. Guyon, J. C., and W. D. Shults, "Rapid Phosphate Determination by Fluorimetry," *J. Am. Water Works Assoc.*, **61,** 403–404 (1969).
20. Hahn, R. B., and T. M. Schmitt, "Determination of Phosphate in Water with Tungsten-185," *Anal. Chem.*, **41** (2), 359–360 (1969).
21. Hansen, A. L., and R. J. Robinson, "The Determination of Organic Phosphorus in Sea Water and Perchloric Acid Oxidation," *J. Mar. Res.*, **12** (1), 31–42 (1950).
22. Harwood, J. E., R. A. van Steenderen, and A. L. Kuhn, "A Comparison of Some Methods for Total Phosphate Analysis," *Water Res.*, **3,** 425–432 (1969).
23. Harwood, J. E., R. A. van Steenderen, and A. L. Kuhn, "A Rapid Method for Orthophosphate Analysis at High Concentrations in Water," *Water Res.*, **3,** 417–432 (1969).
24. Henriksen, A., "Application of a Modified Stannous Chloride Reagent for Determining Orthophosphate," *Anal.*, **88,** 898 (1963).

25. Henriksen, A., "An Automated Method for Determining Low-Level Concentrations of Phosphate in Fresh and Saline Waters," *Anal.*, **90,** 29–34 (1965).
26. Isaeva, A. B., "Determination of Small Amounts of Phosphates in Sea Water after Their Preliminary Extraction as Phosphomolybdic Acid, and Determination of Phosphates in Turbid Water," *J. Anal. Chem. USSR*, **24,** 1505–1508 (1969).
27. Jankovic, S. G., D. T. Mitchell, and J. C. Buzzell, Jr., "Measurement of Phosphorus in Wastewater," *Water and Sewage Works*, **114,** 471–474 (1967).
28. Johnson, D. L., "Simultaneous Determination of Arsenate and Phosphate in Natural Waters," *Environ. Sci. Technol.*, **5** (5) 411–414 (1971).
29. Jones, P. G. W., and C. P. Spencer, "Comparison of Several Methods of Determining Inorganic Phosphate in Sea Water," *J. Mar. Biol. Assoc., U.K.*, **43,** 251–273 (1963).
30. Juday, C., E. A. Birge, G. I. Kemmerer, and R. J. Robinson, "Phosphorus Content of Lake Waters of Northeastern Wisconsin," *Trans. Wis. Acad. Sci.*, **23,** 233–248 (1928).
31. Karl-Kroupa, E., "Use of Paper Chromatography for Differential Analysis of Phosphate Mixtures," *Anal. Chem.*, **28** (7), 1091–1097 (1956).
32. Karl-Kroupa, E., C. F. Callis, and E. Siefter, "Stability of Condensed Phosphates in Very Dilute Solutions," *Ind. Eng. Chem.*, **49** (12), 2061–2062 (1957).
33. Kirk, R. E., and D. F. Othmer, *Encyclopedia of Chemical Technology*, 2nd ed., Vol. 11, Interscience, New York 1966, p. 395.
34. Lee, G. F., "Analytical Chemistry of Plant Nutrients," *Water and Sewage Works*, **116,** R38–R43 (1969).
35. Levine, H., J. J. Rowe, and F. S. Grimaldi, "Molybdenum Blue Reaction and Determination of Phosphorus in Waters Containing Arsenic, Silicon, and Germanium," *Anal. Chem.*, **27,** 258–262 (1955).
36. Martin, J. B., and D. M. Doty, "Determination of Inorganic Phosphate," *Anal. Chem.*, **21** (8), 965–967 (1949).
37. Matthews, D. R., and J. A. Dean, "Extraction-Derivitization-Gas Chromatographic Determination of Trace Phosphates in Aqueous Media," *Anal. Chem.*, **43** (12), 1582–1585 (1971).
38. Morrison, G. H., and H. Freise, *Solvent Extraction in Analytical Chemistry*, Wiley, New York, 1957, p. 269.
39. Murphy, J., and J. P. Riley, "The Storage of Sea-Water Samples for the Determination of Dissolved Inorganic Phosphate," *Anal. Chim. Acta*, **14,** 318–319 (1956).
40. Murphy, J., and J. P. Riley, "A Modified Single Solution Method for the Determination of Phosphate in Natural Waters," *Anal. Chim. Acta*, **27,** 31–36 (1962).
41. Olsen, S., "Recent Trends in the Determination of Orthophosphate in Water," in *Chemical Environment in the Aquatic Habitat*, Proceedings of an I.B.P. symposium held in Amsterdam and Nieuwersluis, October 10–16, 1966, 1967, pp. 63–105.
42. Osmond, F., "Sur une Réaction Pouvant Servir en Dosage Colorimetrique du Phosphore dans les Fontes, les Aciers," *Bull. Soc. Chem., Paris*, **47,** 745–748 (1887).
43. Proctor, C. M., and D. W. Hood, "Determination of Inorganic Phosphate in Sea Water by an Isobutanol Extraction Procedure," *J. Mar. Res.*, **13,** 122–131 (1954).
44. Rieman, W., III, and J. Beukenkamp, "Phosphorus," *Treatise on Analytical Chemistry*, Vol. 5, I. M. Kolthoff, P. J. Elving, and E. B. Sandell (Eds.), Interscience, New York, 1961, Part 2, Section A, pp. 317–394.
45. Rigler, F. H., "Further Observations Inconsistent with the Hypothesis that the Molybdenum Method Measures Orthophosphate in Lake Water," *Limnol. Oceanogr.*, **13** (1), 7–13 (1968).
46. Sanning, D. E., "Phosphorus Determination—A Method Evaluation," *Water and Sewage Works*, **114,** 131–133 (1967).

47. Shapiro, J., W. Chamberlain, and J. Barrett, "Factors Influencing Phosphate Use by Algae," *Adv. Water Pollut. Res.*, 149–167 (1969).
48. Shapiro, J., "Arsenic and Phosphate: Measured by Various Techniques," *Science*, **171,** 234 (1971).
49. Stephens, K., "Determination of Low Phosphate Concentrations in Lake and Marine Waters," *Limnol. Oceanogr.*, **8,** 361–362 (1963).
50. Strickland, J. D. H., and T. R. Parsons, *A Practical Handbook of Seawater Analysis*, Bulletin 167, Fisheries Research Board of Canada, Ottawa, 1968, p. 311.
51. Sugawara, K., and S. Kanamori, "Spectrophotometric Determination of Submicrogram Quantities of Orthophosphate in Natural Waters," *Bull. Chem. Soc. Japan*, **34** (2), 258–261 (1961).
52. Tanzer, J. M., M. I. Krichevsky, and B. Chassy, "Separation of Polyphosphates by Anion Exchange Thin Layer Chromatography," *J. Chromatogr.*, **38,** 526–531 (1968).
53. Van Schouwenburg, J. C., and I. Walinga, "The Rapid Determination of Phosphorus in the Presence of Arsenic, Silicon, and Germanium," *Anal. Chim. Acta*, **37,** 271–274 (1967).
54. Wadelin, C., and M. G. Mellon, "Extraction of Heteropoly Acids—Application of Determination of Phosphorus," *Anal. Chem.*, **25** (11), 1668–1673 (1953).
55. Wayman, C. H., "Simultaneous Determination of Sulfur and Phosphorus in Water by Neutron Activation Analysis," *Anal. Chem.*, **36** (3), 665–666 (1964).
56. Zaugg, W. S., and R. J. Knox, "Indirect Determination of Inorganic Phosphate by Atomic Absorption Spectrophotometric Determination of Molybdenum," *Anal. Chem.*, **38** (12), 1759–1760 (1966).

22

The Role of Phosphate in Photosynthesis

ANDRE T. JAGENDORF

Division of Biological Sciences, Cornell University, Ithaca, New York

Photosynthesis, best known as the transformation of CO_2 and water into sugar and oxygen by green plants, is placed in a deeper perspective when thought of as a magnificent natural machine for the conversion of energy from the inanimate to the living milieu. Overall, the energy input is electromagnetic radiation—visible light—and the final energy output lies in the burnability (i.e., energy-releasing potential) of free O_2 and fixed carbon—sugars, wood, and so on. Between the input and the output lie a number of stages with the energy found in several transient forms. Many specific biochemical steps occur, driven by intermediate products of light energy. It is in these intermediate stages that compounds of phosphate play an unquestioned essential role. A sure indication is the failure of plants to photosynthesize or to grow when they are deficient in phosphate.

Phosphates are crucial in two or three sometimes overlapping ways. In the first place the covalent ester bond between two phosphorus atoms is at a higher "energy level" than are covalent bonds between many other sorts of atoms. That is, it takes more energy for these to be synthesized, and conversely they release more energy when they are either hydrolyzed or participate in alternative reactions (such as phosphate addition to other molecules, etc.) (1). The usual example of this sort of "high-energy" bond is found between the first and second and between the second and third phosphate atoms of adenosine triphosphate (Fig. 1). The enzymatic phosphorylation of adenosine diphosphate:

$$\mathrm{ADP} + \mathrm{P}_i \rightleftarrows \mathrm{ATP} + \mathrm{H_2O}$$

requires energy input, and correspondingly such a phosphorylation is one way of storing chemical energy until it is needed later. Note that formation of ATP is a dehydration, and its breakdown requires hydrolysis. The standard free energy of this hydrolysis is approximately 7000 kcal/mole (but this varies with pH, etc.) (2). This compares with the energy of other phosphate bonds and some other covalent bonds as shown in Table 1. Notice that

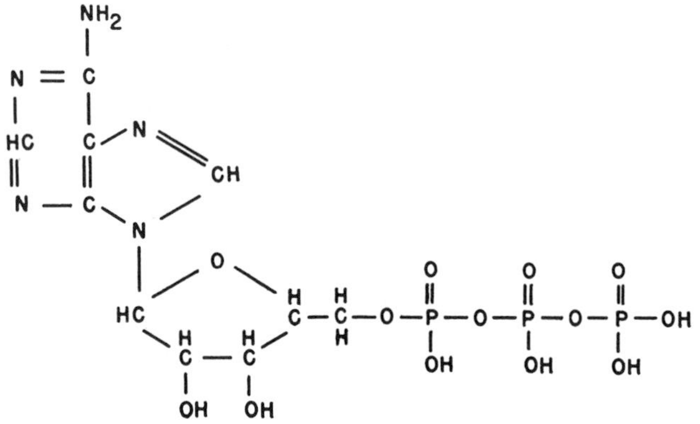

ADENOSINE TRIPHOSPHATE

Figure 1. Structure of adenosine triphosphate.

inorganic pyrophosphate is another high-energy compound, while sugar phosphates are at a lower energy level.

Utilization of the energy in the phosphates of ATP occurs by "driving" other chemical reactions. The simplest case might be that of direct phosphorylation of some other oxygen-containing compound, as in the phosphorylation of glucose by the enzyme hexokinase:

$$\text{ATP} + \text{glucose} \rightarrow \text{glucose-1-P} + \text{ADP}$$

A second type of use is in changing the energy of the phosphate–phosphate bond to that of a phosphate–other molecule by a group-transfer reaction. An example is the formation of ADP–glucose:

$$\text{glucose-1-P} + \text{ATP} \rightleftarrows \text{ADP–glucose} + \text{PP}_i$$

The product ADP–glucose has a similar amount of energy in the bond between glucose and phosphate. Accordingly ADP–glucose is an energetic glucose donor, and indeed is the substrate for synthesis of starch, the most common ultimate photosynthesis product:

$$\text{ADP–glucose} + (\text{glucose})_n \rightarrow (\text{glucose})_{n+1} + \text{ADP}$$
$$\text{(Starch)}$$

One can say that the glucose has been activated, and the ADP becomes a good "leaving group" as the new bond is formed between glucose and the growing end of the starch molecule.

TABLE 1 Energy Levels of Some Phosphate Bonds[a]

Compound	Bond	$\Delta G'_o$ for Hydrolysis (cal/mole)
Phosphenol pyruvate	(Pyruvate)—C(=)(—)—O—P	−13,300
1,3-Diphosphoglyceric acid	(P-Glyceric)—C(=O)—O—P	−13,600
Acetyl phosphate	(Acetate)—C(=O)—O—P	−10,500
Uridinediphosphoglucose	(Glucose)—C(—O)(H)—O—P—O—P	−7,600
Adenosine triphosphate (Mg+ complex)	(Adenylic)—O—P—O—P—O—P	−7,000
Inorganic pyrophosphate	P—O—P	−8,900
Glucose-1-phosphate	(Glucose)—C(—O)(—H)—O—P	−5,000
Fructose-6-phosphate	(Fructose)—C(H)(H)—O—P	−3,800
Glucose-6-phosphate	(Glucose)—C(H)(H)—O—P	−3,001
Glycerol-1-phosphate	(Glycerol)—C(H)(H)—O—P	−2,200

[a] Taken in large part from Table 3, p. 94, of *Biochemists' Handbook*, C. Long, (Ed.), Spon, London, 1961.

Reactions similar in principle involve formation and use of AMP–R or of PP–R, where because of its association with phosphate esters "R" becomes activated and can form a new covalent bond with other molecules. One example is activation of amino acids, the first step in protein synthesis:

$$\text{amino acid} + \text{ATP} \rightleftarrows \text{AMP–amino acid} + \text{PP}_i$$

$$\text{AMP–amino acid} + t\text{-RNA} \rightleftarrows t\text{-RNA–amino acid} + \text{AMP}$$

where PP_i stands for inorganic pyropyrophosphate.

Activated pyrophosphate intermediates are illustrated in the use of isopentenyl pyrophosphate as an intermediate in the synthesis of terpenoids, including carotenes and phytol, some of the pigments used in photosynthesis:

$$\text{isopentyl-PP} + \text{dimethylallyl-PP} \rightarrow \text{geranyl-PP} + \text{PP}_i$$

(Geranyl pyrophosphate can be further converted to photosynthetic pigments.)

Thus formation of a phosphate ester is an essential prerequisite for many sorts of group-transfer reactions, leading to new covalent bonds. Thanks to its use in numerous reactions of this type, ATP (or more properly the two terminal phosphate atoms of the ATP molecule) has been called the "energy coinage of the cell."

A second role for phosphorus is that of a component part of some electron carriers, mediating biological oxidation-reduction reactions. This is true, for instance, of both NADP (nicotinic–adenine dinucleotide phosphate) and of FAD (flavin–adenine dinucleotide). Both of these are active in photosynthetic electron transport (see Fig. 5); FAD is the active part of the ferredoxin–NADP reductase enzyme.

Finally, even apart from considerations of energy levels or oxidation-reduction phosphorus plays a very important role in most metabolism. In the course of evolution enzymes have been developed to metabolize the phosphorylated versions of innumerable compounds rather than the molecules themselves. This is certainly true of the intermediary metabolism in photosynthesis, as can be seen in the diagrams of the carbon pathway shown in Fig. 2.

Figure 2 shows the major route of carbon atoms, from carbon dioxide to sugars (3). The overall pattern (or "reductive pentose phosphate pathway") is sometimes called the "Calvin cycle" in honor of the leader of the laboratory where it was postulated and demonstrated. It is present in essentially all photosynthetic cells. Ribulose-5-P, a 5-carbon sugar, is phosphorylated by ATP to form ribulose diphosphate, the actual acceptor of the first carbon dioxide molecule to be fixed. The product, at least in vitro, is 2 moles of the 3-carbon sugar acid, phosphoglyceric acid. In the cell kinetic evidence suggests that only one free phosphoglyceric acid may be formed, and the second 3-carbon piece might be reduced while on an enzyme surface to the level of

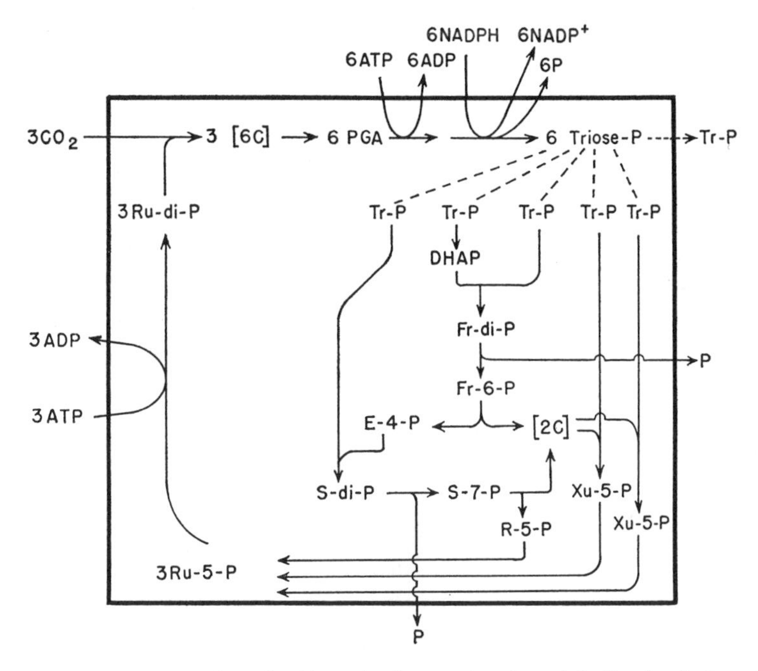

Figure 2. Fixation of carbon dioxide and subsequent early metabolism in photosynthesis: the reductive pentose phosphate pathway. Abbreviations include: Ru-di-P for ribulose diphosphate (5 carbons); PGA for 3-phosphoglyceric acid (3 carbons); Tr-P for triose phosphate (i.e., 3-phosphoglyceraldehyde, 3 carbons); DHAP for dihydroxyacetone phosphate (3 carbons); Fr-di-P for fructose diphosphate; Fr-6-P for fructose-6-phosphate (6 carbons); Xu-5-P for xylulose-5-P (5 carbons); E-4-P for erythrose-4-phosphate (4 carbons); S-di-P for sedoheptulose diphosphate (7 carbons); S-7-P for sedoheptulose-7-phosphate; R-5-P for ribose-5-phosphate; and Ru-5-P for ribulose-5-phosphate. Note that another pathway for early fixation of carbon dioxide, by attachment to phosphoenolypyruvic acid to form oxaloacetic acid first, then other 4-carbon dicarboxylic acids, is prevalent in tropical grasses. This "Hatch-Slack" pathway (15) is not shown here.

triose phosphate. The observable metabolism, however, continues with first the phosphorylation of phosphoglyceric acid by another ATP, then reduction of the diphosphoglyceric acid to triose phosphate by NADPH, together with the release of one inorganic phosphate per mole.

In an energetic sense, this is all there is to it. Triose phosphate is at the oxidation level of sugars (i.e., it *is* a 3-carbon sugar), and it contains the newly fixed molecule of CO_2. The entire process is driven by ATP and by NADPH, in *only* those four reactions discussed so far. The energetic chemical products (ATP and NADPH) formed from the light steps in photosynthesis are used in these reactions only. The rest of the cycle is a series of enzymatic steps, an integrated device for taking most of the triose phosphate molecules and reconverting them into ribulose diphosphate, the acceptor molecule for carbon dioxide.

The first step of the return to ribulose is the formation of dihydroxyacetone phosphate from one of the triose phosphate (3-phosphoglyceraldehyde) moieties, then condensation with another triose phosphate to form fructose diphosphate. The fructose diphosphate loses one phosphate by action of a phosphatase, to become fructose-6-phosphate. Fructose-6-phosphate can now be attacked by a "transketolase" enzyme, which lifts the top two carbon atoms from the 6-carbon sugar, reattaching them to a third triose phosphate to make the 5-carbon sugar, xylulose-5-phosphate. The leftover 4-carbon sugar, erythrose-4-P, is combined with a fourth triose phosphate to form the 7-carbon sedoheptulose diphosphate. The sedoheptulose loses another inorganic phosphate, to become sedoheptulose-7-P. This is attacked by a transketolase just as fructose-6-P was, with the top two carbon moieties removed (leaving behind the 5-carbon sugar, ribulose-5-phosphate) and added on to the top end of a fifth triose-P, to form another molecule of xylulose-5-P. Thus three 5-carbon sugars are re-formed, starting with the input of five triose phosphate molecules. Epimerase and isomerase steps can convert the pentose sugars into their isomer, ribulose-5-P, and the cycle is ready to start again.

The cycle described above includes within it three steps that are, in a metabolic sense, largely irreversible. These are the ribulose diphosphate carboxylase, and the two phosphatase steps—hydrolysis of fructose-1,6-diphosphate and of sedoheptulose-1,7-diphosphate. Since the equilibrium values for these reactions are so far over toward the products, their inclusion within a larger sequence provides "directionality" to net carbon flow within the pathway. Other things being equal, the phosphatase reactions will "pull" the steps prior to fructose diphosphate and sedoheptulose diphosphate, and will "push" the succeeding steps. These enzymes then are ideal places in which to exert metabolic control; conditions which inhibit or accelerate these steps in particular will have the largest effects on the overall process, and indeed natural selection has led to significant controls over most phosphatase and equivalent steps in metabolism. In the case of photosynthesis the exact nature of the controls is not known yet; however, it is clear from tracer data with whole cells (4) that these three enzymes are active in the light, but rapidly become inactive in the dark. Examples of metabolic controls over the more energetic biochemical reactions, including phosphorylation or phosphatase reactions in many instances, abound in biochemistry.

To drive the carbon metabolism of photosynthesis, light energy has to be changed into the energy of NADPH in an atmosphere of 20% O_2, and into the energy of the terminal phosphate bonds of ATP. Whereas the carbon metabolism including phosphorus functions outlined above went on in the

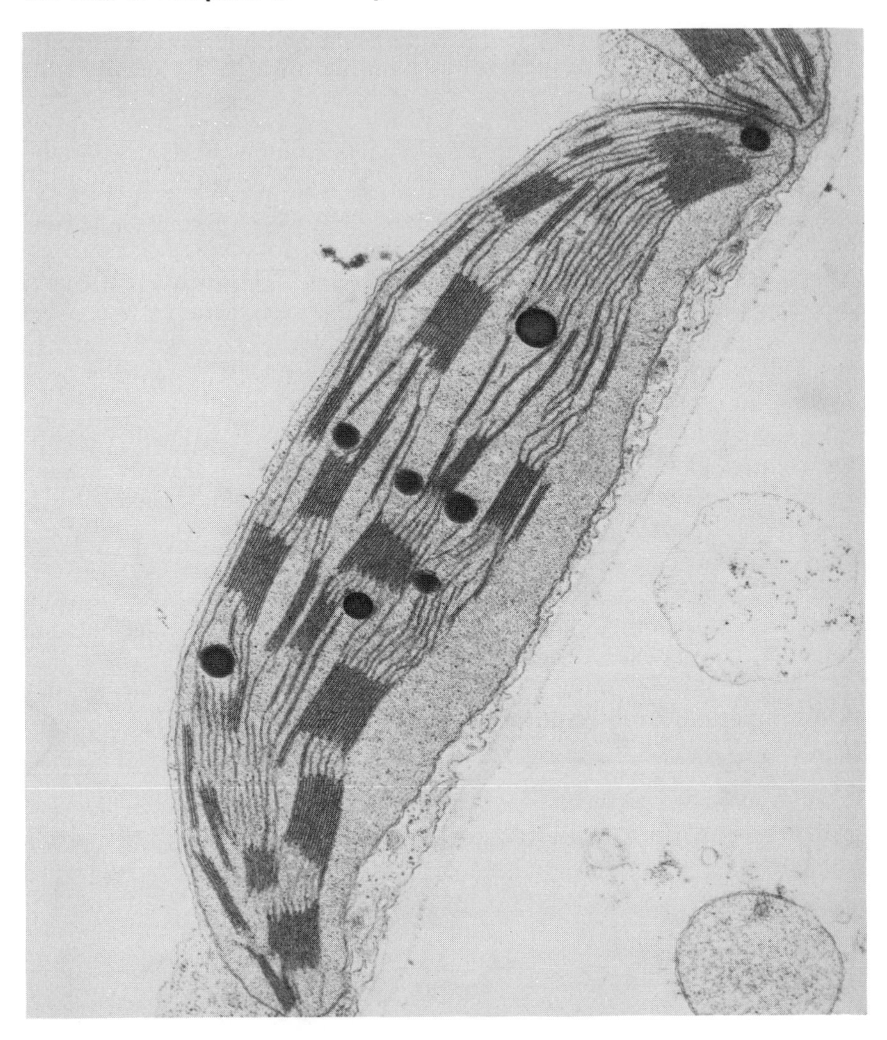

Figure 3. Ultrathin section of cell, showing fine structure of oat coleoptile chloroplast. Notice extensive series of inner membranes, which bear both pigments and the electron transport and photosynthetic phosphorylation enzymes. Photograph is published through the courtesy of Dr. M. Parthasarathy, Cornell University, Ithaca, New York, Magnification, ×21,000.

"soluble," unstructured stroma region of chloroplasts (Fig. 3), or in the cytoplasm of blue-green algae or of photosynthetic bacteria, the basic reactions of conversion of light energy into ATP and NADPH plus O_2 occur on the semisolid lipoprotein internal membrane system ("lamellae" and "grana" of chloroplasts, "chromatophores" of bacteria). These membranes contain chlorophylls, accessory pigments, electron-transferring enzymes, and the machinery for both ion transport and phosphorylation of ADP, all bound together. Sequential steps in the process can be defined fairly clearly by now (5, 6).

1. Absorption of a photon by a molecule of chlorophyll. Time constant, approximately 10^{-15} sec, as the chlorophyll goes into its "excited state."
2. Migration of the quantum of energy, either from accessory pigment to chlorophyll, or from one chlorophyll to another, by means of a physical process called "induced resonance energy transfer." The quantum of energy ends up eventually at the Reaction Center of this agglomeration of pigment molecules. The Reaction Center consists of a chlorophyll molecule situated next to both an electron donor and an electron acceptor. The nature of these donors and acceptors is currently still being sought.
3. Conversion of excited-state energy to oxidation-reduction energy. The chlorophyll at the active center, in its excited state, is a better electron donor (at a more reducing potential) than is the ground-state molecule (see Fig. 4). It passes an electron on to the acceptor (A), a redox carrier at a very reducing potential. The chlorophyll itself is now in its ground-state oxidized form, a very powerful oxidant. Accordingly it can take electrons away from a donor (D^-) which may have a strongly oxidizing midpoint potential. The net effect is movement of electrons away from D^- where they do not want to leave, through a "pump" or light step, and on to an acceptor A where they ordinarily do not want to go. This process is usually abbreviated

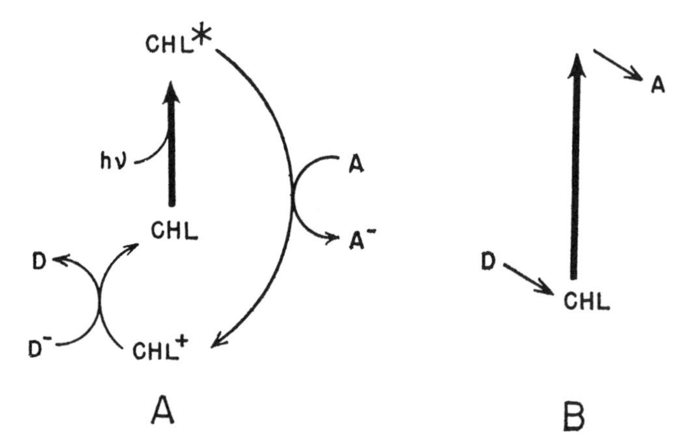

Figure 4. (*a*) Nature of the primary light reaction in photosynthetic systems. The excited state of the pigment serves as a powerful electron donor; the resulting oxidized ground-state form is a powerful electron acceptor. (*b*) The usual symbolic abbreviation of this reaction; the upward-pointing arrow representing input of energy via the excited state of the pigment, and horizontal arrows showing actual electron-transfer pathway.

in the form shown in Fig. 4*b*, where the vertical arrow represents the change in energy level of the electron on chlorophyll due to light activation, and the horizontal arrows are an abbreviation for electron transfer from one molecule to the next.

4. The electron transport chain. In higher plants and algae, two light reactions of the sort discussed above are embedded in an electron transport sequence (7, 8) shown in Fig. 5. A number of the components are inferred but not identified (especially those labeled Q, A, M, and X). There is some controversy still over the exact sequence of a few of the carriers, and the existence of others as yet neither identified nor inferred cannot be ruled out. However, there is general agreement on the outlines: There are two light

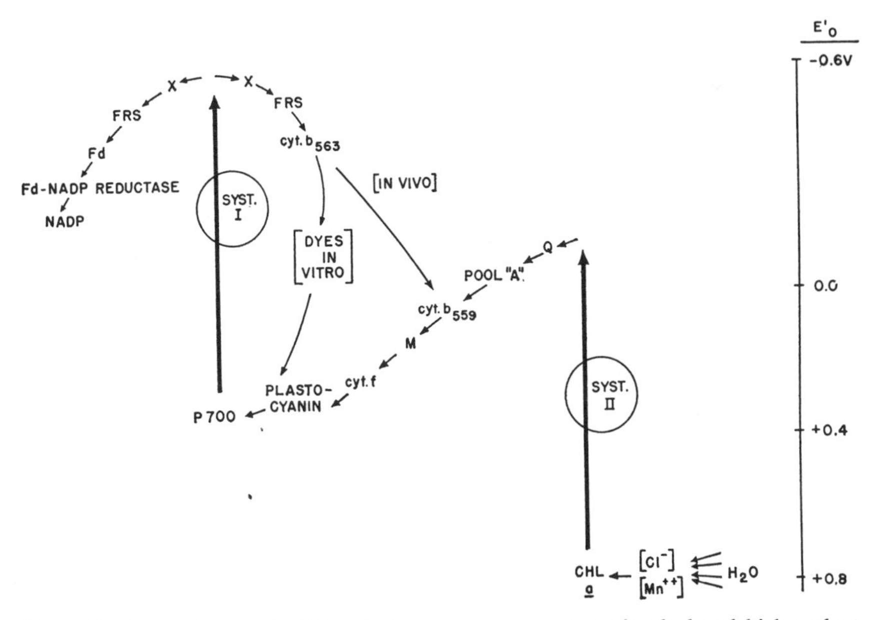

Figure 5. Current concept of the electron transport sequence in algal and higher plant chloroplasts. Notice two light reactions in series (system I and system II), each one driven by a somewhat separate set of bulk light-gathering pigment molecules. Phosphorylation of ADP accompanies downhill electron transport, probably at one point between the two light reactions, on second place in the cyclic electron flow, and possibly at other locations as well. Abbreviations include Q for quencher (an unknown compound); cyt. b_{559} for cytochrome b with an absorption peak at 559 nm; cyt. f for cytochrome f (absorption peak at 552 nm); P700 for a photooxidizable minor chlorophyll component with peak absorption of the reduced form at 700 nm; FRS for a recently discovered "ferredoxin reducing substance"; cyt. b_{563} for a b-type cytochrome with absorption peak at 563 nm; Fd for ferredoxin; Fd–NADP reductase for a flavoprotein enzyme mediating electron flow from ferredoxin to NADP; and NADP for nicotine–adenine dinucleotide phosphate.

reactions, almost certainly in sequence; electron transport from the reducing end of each light reaction is thermodynamically downhill; the initial electron donor is water, yielding free O_2 as a byproduct, and the ultimate electron acceptor in a physiological sense is NADPH, to be used for reducing phosphoglyceric acid in the stroma. Also, the downhill electron-transfer steps, most probably those involving the cytochromes as one of the carriers, do lead to ATP formation from ADP and phosphate, and can also lead to ion transport as an alternate use and/or storage form of energy.

5. Ion transport. All of the intact photosynthetic organelles studied so far have the property of causing the medium to go alkaline when electron transport is proceeding. This apparently represents an uptake of protons due to an oriented "proton pump" which must be a part of the membrane system. These reactions are observed only with topological spheres—whole chloroplasts or chromatophores, or parts of membranes rounded up to form vesicles (bubbles). Counterions move at the same time (either cations in the opposite direction or anions in the same direction, depending on circumstances) and although most evidence suggests the proton pump is the primary driving force, some possibility remains that other ion pumps do the work and protons move to preserve electrical neutrality. In any event the resulting ion gradients (as for instance from acid inside to alkaline outside) represent a form of stored energy. Peter Mitchell has advanced (9) and championed the idea that this store of energy represents the most immediate form of coupling between oxidation-reduction reactions and ATP formation, that the pH gradient is the high-energy intermediate connecting electron flow with the dehydration reactions forming chemical high-energy intermediates. While this concept is still controversial, with chloroplasts it was possible to demonstrate the conversion of an artificially imposed acidity gradient into ATP synthesis, in the dark, and without the intervention of electron transport reactions (10).

6. ATP synthesis. As indicated above, the energy of at least some of the electron-transfer steps is conserved by the simultaneous formation of ATP from ADP and inorganic phosphate (11). The way in which the oxidation-reduction reaction is coupled to formation of the anhydro bond in ATP is still not known, although the existence of the similar process was discovered first in 1937 in mitochondria, then in bacterial chromatophores (by Frenkel, Geller, and Lipmann), and in chloroplasts (by Arnon, Whatley, and Allen) in 1954. Neither the number of phosphorylation sites in the chloroplast electron transport chain nor the specific redox reactions supporting phosphorylation are known with certainty yet. Current evidence indicates the existence of at least one site between the two light steps, and probably a second distinct site in the cyclic electron flow running around photoreaction

I (Fig. 5). Nevertheless indirect evidence at times has pointed to the existence of one additional site in the linear path between water and NADPH, and even possibly one additional site in the cyclic electron pathway around system I.

While the detailed chemical mechanism for phosphorylation is completely unknown, at least one of the enzymes concerned has been identified (12). This is a protein of molecular weight about 300,000, containing perhaps 12 protein subunits. It has been visualized in the electron microscope, especially by negative staining procedures, and is seen to sit on the outside of the bag-like vesicles derived from isolated chloroplasts' green internal membranes. It can be detached, and then the chloroplast fragments lose the ability to do phosphorylation. Either on the membrane, or when detached, it can be activated by special procedures to demonstrate ATP hydrolyzing ability (apparently water does not easily enter the active center of this enzyme when it is in its native state in the chloroplast). Similar enzymes are found in bacterial chromatophores, which also show electron transport and phosphorylating ability. The key to the mechanism for this crucial reaction of photophosphorylation of ADP will certainly involve the action of this terminal enzyme.

A remarkable recent discovery (13) is the existence in some photosynthetic bacteria of a mechanism for converting inorganic phosphate into inorganic pyrophosphate rather than ATP. The phosphorylation of phosphate itself is stimulated by reagents that prevent ATP formation, and hence it must occur at an earlier step, closer to the electron transport pathway itself. This reaction may very well represent an evolutionary throwback—suggesting that living cells (precursors of bacteria) learned to convert oxidation-reduction energy into pyrophosphate bonds as such perhaps even before ADP and ATP were invented.

Finally it should be mentioned that the entire machinery of photosynthesis—either chloroplasts or chromatophores—includes phosphorus compounds in both structural and developmental roles (14). Phosphorproteins and phospholipids are constituents of chloroplast and chromatophore membranes. The chloroplast in higher plants in some ways seems to be a remnant of an invading, independent organism. Chloroplasts contain enzymes for synthesizing amino acids, long-chain fatty acids, lipids, carotenes, chlorophylls, proteins, and nucleic acids. They contain their own DNA, RNA, and ribosomes, and while many or most of the genes for chloroplast proteins are localized in the cell nucleus, at least some must be present in the chloroplast itself. On all accounts, the replication of chloroplasts and development of the photosynthetic machinery would not be possible without nucleic acids, in which phosphorus plays a major role.

REFERENCES

(For a general view of photosynthesis see the article by R. P. Levine in the November 1969 issue of *Scientific American*. A little more complete is the short monograph by G. E. Fogg, *Photosynthesis*, published by American Elsevier Press, New York 1968.)

1. Lipmann, F., *Adv. Enzymol.*, **1**, 99 (1941).
2. Benzinger, T., R. Hems, K. Burton, and C. Kitzinger, *Biochem. J.*, **71**, 400 (1959).
3. Bassham, J. A., and M. Calvin, *The Path of Carbon in Photosynthesis*, Prentice-Hall, Englewood Cliffs, N.J., 1957.
4. Bassham, J. A., and G. H. Krause, *Biochim. Biophys. Acta*, **189**, 207 and 221 (1969).
5. Clayton, R. K., *Molecular Physics in Photosynthesis*, Random House (Blaisdell), New York, 1965.
6. Kamen, M., *Primary Processes in Photosynthesis*, Academic Press, New York, 1963.
7. Levine, R. P., *Ann. Rev. Plant Physiol.*, **20**, 523 (1969).
8. Boardman, N. K., *Adv. Enzymol.*, **30**, 1 (1968).
9. Mitchell, P., *Fed. Proc.*, **26**, 1370 (1967).
10. Jagendorf, A. T., *Fed. Proc.*, **26**, 1370 (1967).
11. Arnon, D. I., *Ann. Rev. Plant Physiol.*, **7**, 325 (1956).
12. McCarty, R. E., and E. Racker, *J. Biol. Chem.*, **243**, 129 (1968).
13. Baltscheffsky, M., H. Baltscheffsky, and L.-V. von Stedingk, *Brookhaven Symp. Biol.*, **19**, 246 (1966).
14. Kirk, J. T. O., and R. A. E. Tilney-Bassett, *The Plastids*, Freeman, San Francisco, California, 1967.
15. Hatch, M. D., and C. R. Slack, *Ann. Rev. Plant Physiol.*, **21**, 141 (1970).

Transformations of Naturally Occurring Organophosphorus Compounds in the Environment

LEWIS G. SCHARPF, JR.

Monsanto Company, St. Louis, Missouri

CLASSIFICATION OF NATURALLY OCCURRING ORGANOPHOSPHORUS COMPOUNDS

Phosphorus-containing biochemical compounds are ubiquitous in nature and as dissolved forms represent 30 to 60% (3) of the total phosphorus content of natural waters. While much progress has been made in elucidating the structure and role of these compounds in biological systems, little information is available on their transformations outside the living cell.

The past 50 years of biochemical research have uncovered a vast assortment of organic phosphate esters in living systems. In the cell, these compounds are involved in nearly all phases of energy transfer and the resultant transport and metabolism of nutrients. Structurally, the phosphate-containing nucleic acids hold and duplicate the genetic code, and organophosphates such as the techoic acids give strength and rigidity to cell walls. The biochemistry and physiology of these compounds have been treated in numerous texts and review articles (34, 54, 75). Although the biological esters of phosphoric acid cover a wide range of type and complexity, they may be conveniently categorized on the basis of substitution on the ortho or condensed phosphoric acid molecules (Table 1). In these compounds, the carbon-containing portion of the organophosphorus molecule is attached to oxygen or nitrogen on the phosphorus atom. Several reviews (43, 63) cover the chemistry and role of the organophosphates in biological systems.

In 1959, a class of organophosphorus compounds previously unrecognized to be of biological origin was discovered by Horiguchi and Kandatsu (36) who announced the isolation from protozoa of a compound having a carbon-to-phosphorus bond. Since that time, a number of phosphonic acid analogs of naturally occurring biochemicals have been reported (65). The presence of compounds with the C—P bond in lipids has been demonstrated, and they also seem to be an integral part of certain soluble and insoluble structural

TABLE 1 Classification of Naturally Occurring Organophosphorus Compounds

Substitution on Ortho or Condensed Phosphoric Acid	Nature of Substituent	Customary Nomenclature of Compounds	Examples
I. X—P(=O)(OH)—OH	1. X = RO—	Phosphoric acid monoesters	Sugar phosphate Glycerol phosphate Mononucleotides Phosphoserine Phytic acid Phosphoenol pyruvic acid
	2. X = R—C(=Z)—O—	Enol phosphates	Acetyl phosphate
	3. X = RNH—	Phosphoramidates	Creatine phosphate Arginine phosphate
	4. X = RC—	Phosphonates	2-Aminoethylphosphonic acid Phosphonoalanine
II. X—P(=O)(OH)—O—P(=O)(OH)—OH	X = Nucleoside-5′		Nucleoside 5′-diphosphate Thiamine pyrophosphate
III. X—P(=O)(OH)—O—P(=O)(OH)—O—P(=O)(OH)—OH	X = Nucleoside-5′		Nucleoside 5′-triphosphate
IV. X—P(=O)(OH)—O—P(=O)(OH)—Y	X = Nucleoside-5′ Y = RO		Nucleotide coenzymes
V. X—P(=O)(OH)—Y	1. X = Nucleoside-5′ Y = Acyloxy group	Phosphoric acid diesters	Acyl adenylates
	2. X = RO— Y = Sulfate group	Phosphosulfates	3′-Phosphoadenosine 5′-Phosphosulfate
	3. X = RO— Y = R′O—		Nucleic acids Vitamin B_{12} Phospholipids Techoic acids Cyclic nucleotides
	4. X = RO— Y = RC—		Phosphonolipids

proteins (41). The mechanisms of synthesis and the reason for nature's inclusion of these compounds in biological systems remain obscure. The chemistry, biochemistry, and distribution of organophosphonates in the animal and plant kingdom have recently been reviewed by Quin (65) and Cassaigne (18).

RELEASE OF ORGANOPHOSPHORUS BY MEMBERS OF THE ECOSYSTEM

At any given time, organophosphorus deposition by plants, animals, and microorganisms comprises a major portion of available phosphate nutrients in the environment. The compounds may be excreted by living systems or may be the result of decomposition of dying or dead cells (see Fig. 1).

Bacteria, molds and fungi, zooplankton, insects, higher plants, and animals continually excrete measurable amounts of organophosphate. Bacteria such as *Bacillus* (60) and *E. coli* (37) strains secrete substantial amounts of nucleotides and ribonucleic acids into their culture media. Certain species of amphipods excrete an amount of organophosphorus equal to one-third of the total soluble phosphorus released (39). Animal excretions contribute considerable amounts of organophosphates to the soil, particularly in heavily grazed soil (14, 20). The largest portions of soil organophosphorus, however, are probably of microbial origin (14) and decayed vegetation (2).

The types and quantities of free organophosphorus compounds occurring in nature are summarized in other chapters. The nature and extent of these compounds in soil and water have been reviewed in detail by Anderson (4), Larsen (48), and the American Water Works Association (3), respectively. In contrast to inorganic phosphorus, very little is known about the organic phosphorus fraction of soils and natural bodies of water.

PHYSICOCHEMICAL TRANSFORMATIONS OF NATURALLY OCCURRING ORGANOPHOSPHORUS COMPOUNDS

Interaction with Metal Ions

Organic phosphates, like inorganic phosphates, have the ability to form complexes, chelates, and insoluble salts with many metal ions. The extent of complex and chelate formation will depend on such parameters as concentrations of the phosphate and the metal ions, pH, and the presence of other ligands (81). The acidity constants of many of the naturally occurring organic phosphates are such that they exist as anionic species in neutral to

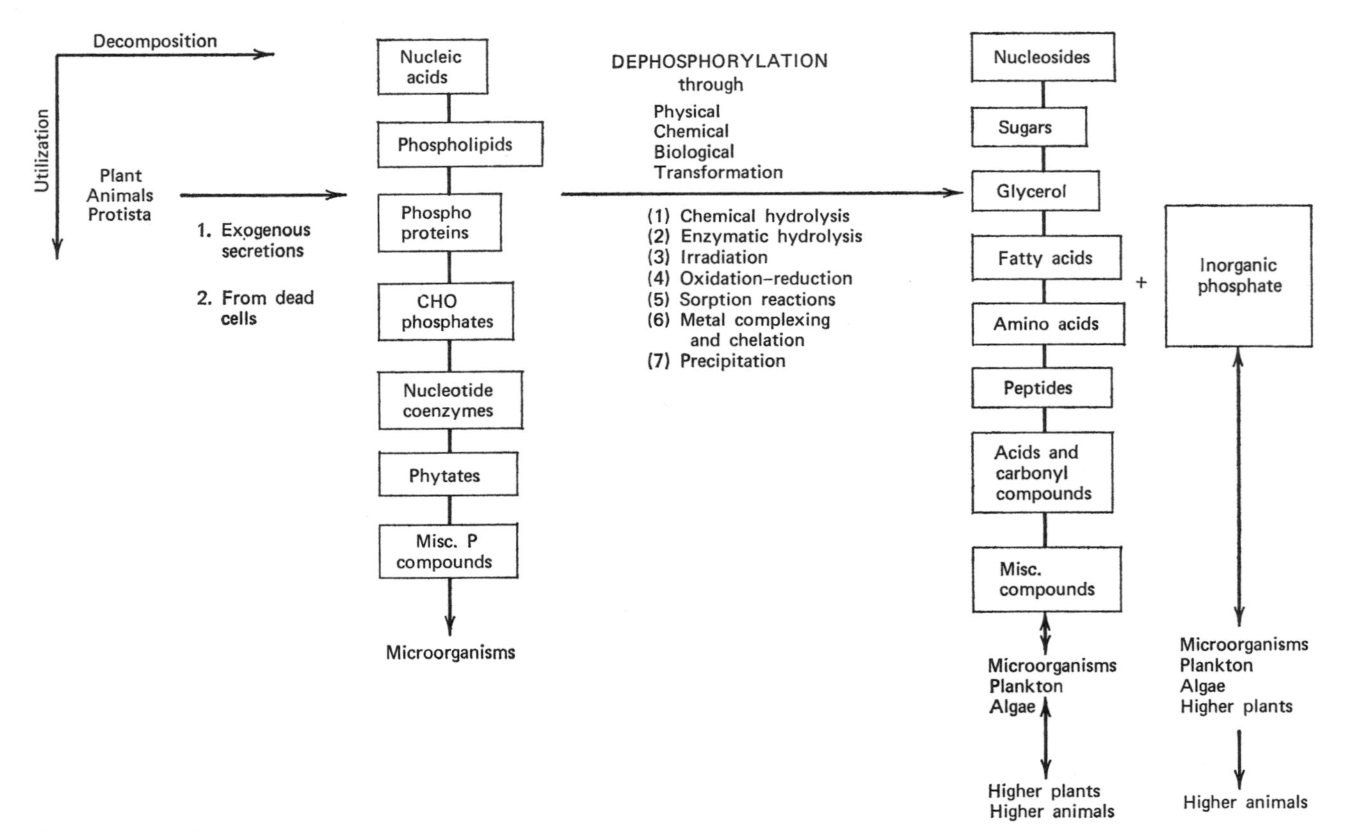

Figure 1. Transformations and utilization of biological phosphates in the environment.

alkaline solutions (3). Equilibrium constants for complex formation reactions between selected organophosphorus compounds and metals ions are given in Table 2 and range from 10 to 10^4. In a solution containing 10^{-5} *M* orthophosphate at pH 7.0, 60% of the phosphate is bound to calcium when the latter is present at 300 mg/ml (3). While these data cannot be directly extrapolated to heterogeneous solution systems of organic phosphates, it nevertheless indicates that complex formation can occur in the environment. Since both magnesium and calcium are present in water and soil in large amounts relative to the organophosphates (3), they may greatly influence the distribution and properties of the latter.

TABLE 2 Stability Constants for Selected Naturally Occurring Organophosphates

Compound	Temperature (°C)	Media	Log Equilibrium Constant		
			$(H^+)^a$	$(Ca^{2+})^b$	$Mg^{(2+)b}$
Adenosine triphosphate	20	0.1 *M* KCl	6.50	3,60	4.00
Creatine	25	0.16 Ionic strength	—	0	—
Flavin adenine dinucleotide	23	0.1 *M* NaCl	—	2.02	2.02
Fructose dihydrogen phosphate	20	0.1 *M* KCl	5.84	1.47	1.59
Glucose-1-phosphate	20	—	6.50	2.46	—
Glycerol-1-phosphate	20	Extrapolated to zero ionic strength	6.64	1.66	1.80
Guanosine triphosphate	23	0.1 *M* NaCl	—	3.58	4.02
Inosine triphosphate	23	0.1 *M* NaCl	—	3.76	4.04
Phosphoserine	25	0.15 Ionic strength	9.74	2.30	2.40
Uridine triphosphate	23	0.1 *M* NaCl	—	3.71	4.02

[a] For protonation of the ligand, $H^+ + L \rightleftarrows HL$, where $K_1 = [HL]/[H^+][L]$.
[b] For combination of the ligand with metal ion, $ML_{n-1} + L \rightleftarrows ML_n$, where $K_n = [ML_n]/[ML_{n-1}][L]$. Compiled from Sillen and Martell (70).

Note. *K* values measured in Na^+ or K^+ solutions will be low due to weaker but significant complexing of alkali metal ion with $K = 10^1$ to 10^2.

Bivalent metal ions greatly influence the decomposition of biologically important organophosphates. To explain the role of the metal ion, Koshland (44) emphasized the importance of chelate structure such as the formation of a cyclic intermediate by the metal ion and the phosphate. Kosower (45), on the other hand, suggests that chelate formation is not the predominant effect but rather reduction of the ionic charge of the phosphate anion by interaction with a metal cation. Recent NMR and IR results (59) on D_2O solutions of acetyl phosphates favor the view of charge neutralization which facilitates solvolysis.

The hydrolysis of certain biological polyphosphates is accelerated by calcium and magnesium. For example, adenosine triphosphate, which is moderately stable in biological systems (45), shows an increase of 54% in hydrolysis rate at pH 8.5 to 8.8 in the presence of calcium (57). Accelerations of hydrolysis rates have been noted for cerium and lanthanum ions and the mercuric ion effects almost instantaneous hydrolysis of naturally occurring vinyl phosphates such as phosphenol pyruvic acid (45). In contrast, Mn^{2+}, Co^{2+}, Ni^{2+}, Cu^{2+}, and Zn^{2+} have essentially no effect on the rate of α-D-glucose-1-phosphate hydrolysis (56).

Ions of group IV elements of the periodic table can dephosphorylate sugar phosphates and nucleotides and are just as efficient catalysts for dephosphorylation as certain enzymes (79). Thorium and zirconium act catalytically over the pH range 2 to 10 and have pH optima in alkaline and acid media (79).

Polynucleotides reportedly are degraded in the presence of divalent and trivalent inner transition-metal ions (27). For example, polyribonucleotides such as polyadenylic acid are degraded into mononucleotides and low molecular weight oligonucleotides in a 20-hr period at 64° by the action of Mn^{2+}, Ni^{2+}, and Cu^{2+} (17) and in a 2-hr period by Zn^{2+} and La^{3+} (27). With the trivalent metal ions, cleavage of the polynucleotide chains occurs at the 5′-phosphate linkages (27). The 3′ mononucleotides produced, however, required over 100 hr of further heating at 64° to produce nucleoside and orthophosphate (27). The strength of specific binding of monovalent metal cations for the polynucleotides is in the order $Li^+ > Na^+ > K^+ > Rb^+ > Cs^+$ (7).

The lack of data for the various phases of biological phosphates precludes quantitative evaluation of their solubilities. General observations on solubilities, however, can be cited. Certain metal ions may form difficultly soluble or insoluble precipitates of some naturally occurring organophosphates. The iron and aluminum salts of myoinositol hexaphosphate are insoluble in acid media and the calcium, magnesium, and barium salts are insoluble in alkaline media (4). Barium forms difficultly soluble salts of 3-phosphoglyceric acid and hexosediphosphates at pH 8.2 (61).

Tervalent cations (73) and cupric salts (88) yield insoluble complexes of DNA apparently by binding DNA molecules together in lateral aggregates by chelation with phosphate groups of adjacent chains. Thermally denatured DNA is precipitated by lead ions (74).

Phospholipids are quite insoluble in soil because they usually exist as metal and other inorganic complexes (5).

Sorption Reactions

The sorption of phosphates and polyphosphates onto solid phase surfaces, particularly minerals, is well known (81). Phosphates are sorbed onto clays in soil by two mechanisms—chemical bonding of the anions to positively charged edges of clays and substitution of phosphates for silica in the clay (3). Nucleic acids (13), nucleotides (13), nucleoproteins (13), and inositol phosphates (31) are strongly adsorbed by bentonites and kaolins. Functional groups such as —OH, COOH, —CHO, —$CONH_2$, —NH_2, SH, and —COOM are likely to be involved in the sorption of organophosphates onto soil clays (87). These sorption phenomena are likely to play important roles in the biodegradation of organophosphates in the environment.

Conflicting opinions exist on the effect of soil clay on the decomposition of organic matter. Waksman (85) noticed enhanced degradation of certain organic matter in the presence of clay, which suggests that certain adsorption sites on the clay minerals catalyze biological degradation (87). Proteins, on the other hand, are thought to degrade more slowly in the presence of clays (53), apparently because sorption binds certain hydrolytically active sites on the enzyme. Moreover, Bower (13) found that dephosphorylation of nucleic acids and guanine nucleotides by nuclease enzymes decreased considerably when clays are present.

The inhibition of the enyzmatic hydrolysis of glycerophosphate is probably caused by the adsorption of the enzyme rather than the substrate on clay (55), although the reverse is probably true with nucleotide and phytate enzymatic hydrolysis (52).

Chemical Hydrolysis

Sugar Phosphates. In light of their biological significance, it is surprising that only few reports are available on the hydrolytic mechanisms of sugar phosphates in aqueous solution. Careful studies have been carried out on α-D-glucose-1-phosphate (25) and it is clear that the rate of hydrolysis increases slowly from pH 8.0 to 4.0 and then more rapidly as the pH decreases further. The rates of hydrolysis of this molecule have been interpreted in terms of two reactions: (*a*) with the monoanion and proceeding with phosphorus–oxygen bond fission, and (*b*) with the neutral molecule and proceeding with carbon–oxygen bond fission (16). Since the phosphate group of glucose-1-phosphate is attached to the hemiacetal hydroxyl group and in glucose-6-phosphate to the primary alcoholic group, differences in hydrolytic rates and mechanism of the latter may be expected. In contrast to the alkali stability of glucose-1-phosphate, glucose-6-phosphate undergoes a rather rapid alkaline hydrolysis probably with a β-elimination mechanism (15). In general, the aldose-1-phosphates are stable in alkaline media, whereas sugar phosphates having a free aldehyde group are unstable (22).

Attention has been given to the mechanism of the alkali-catalyzed degradation of D-glucose-6-phosphate. Among the hydrolysis products are D-glucometasaccharinic acid 6-phosphate, D-fructose-6-phosphate, and D-glyceraldehyde-3-phosphate (23). The D-fructose-6-phosphate then undergoes a reverse aldol reaction to yield glyceraldehyde-3-phosphate and dihydroxyacetone. These products finally yield lactic acid and orthophosphate (28). Degani and Halmann (24) compared this reaction with the anaerobic enzymic metabolism of the ester, which was strikingly similar. Out of this comparison came the interesting suggestion that metabolic pathways in chemical evolution were available before the enzymes required for their metabolism (24).

The hydrolyses of D-fructose-1-phosphate and -1,6-diphosphate have been studied at various pH values. In these compounds fragmentation of the carbon chain occurs concurrently with cleavage of the phosphate ester bonds (28).

Other phosphoric acid esters of carbohydrate are chemically unstable, particularly the lower sugar phosphates. For example, deoxy sugar phosphates are very labile in acidic environments, particularly when the sugar is in the furanose form (76). To illustrate this point, the biochemically important 2-deoxy-α-D-erythropentafuranosyl phosphate has a half-life of only 10 to 15 min at pH 4 and 23°C (30).

The considerable biochemical interest of several of the ketose-1-phosphates has led to studies of their acid hydrolysis. Normally the ketose phosphates hydrolyze more rapidly in acid than their corresponding aldose phosphates (76). The 3- and 6-phosphates of 2-deoxy-D-galactose, for example, are hydrolyzed considerably more rapidly than are the corresponding phosphates of D-galactose (29). In alkaline media, the ketose-3-phosphates are far less stable than the 5-phosphates presumably because of elimination of the phosphate group β to the carbonyl (45).

Several of the deoxyribose and -hexose sugars demonstrate instability in the form of phosphate migration, which has been reviewed by Szabo (76). These transformations may be represented by the accompanying reaction. It is thought that the driving force for such migrations in these sugars is the

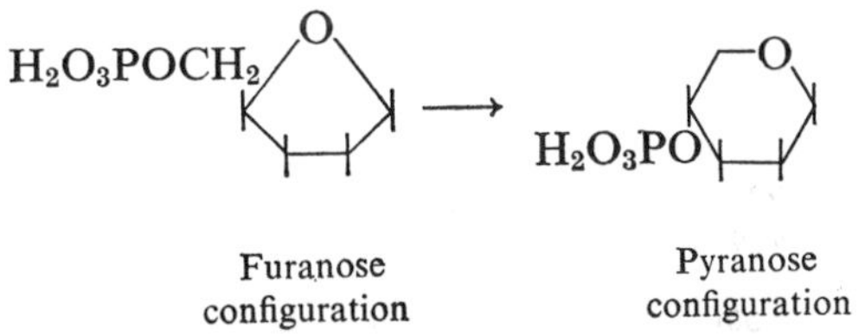

$H_2O_3POCH_2$ (Furanose configuration) → H_2O_3PO (Pyranose configuration)

Furanose configuration

Pyranose configuration

furanose to pyranose transformation, which always occurs (76). In solution, the pyranose configuration of the sugar phosphate is the thermodynamically preferred form. (See Table 3).

Nucleic Acids and Nucleotides. The secondary structure of nucleic acids in solution is the first to be degraded in the series of chemical breakdowns ultimately resulting in the liberation of orthophosphate. X-Ray analyses reveal that the most important stabilizers of secondary structure are the hydrogen bonds between base pairs (34, 54). These hydrogen bonds may be disrupted by increasing the temperature, by treatment with acid or alkali, or by dilution of the ionic concentration generally below 10^{-4} *M* (54). In the DNA structure, these disruptions result in an unwinding of the double helical structure of DNA and ultimate separation of the two complementary

TABLE 3 Hydrolysis Rates for Selected Organophosphate Compounds[a]

Compound	Constant[b] $k \times 10^3$	Hydrolysis Temperature (°C)	Medium	Ester Group Form
2-Amino-2 deoxy-D-glucose-6-phosphate	86	26	N H_2SO_4	Disodium *N*-acetyl
2-Deoxy-D-ribose-5-phosphate	50	100	*N* HCl	Free acid
D-Ribofuranose-5-phosphate-1-pyrophosphate[c]	30	65	Acetate buffer pH 4.0	—
D-Glyceric acid 1,3-diphosphate	26	38	Water	Free acid
β-Lactosyl phosphate	6	37	NHCl	Barium salt
α-D-Ribofuranosyl phosphate	1.25	20	0.01 *N* HCl	Free acid
D-Mannitol-1-phosphate	<0.5	100	*N* HCl	Free acid

[a] Compiled from Maher and Wolfrom (50).
[b] $k = 0.30/\text{time}$ in minutes for 50% hydrolysis of the ester linkage.
[c] For the pyrophosphate group.

strands. The secondary structure of RNA is generally that of a single strand of polynucleotides (54) and considerably less is known about the effects of the parameters above on its structure.

Deoxyribonucleic acids are stable to mild alkaline conditions, as are a number of the ribonucleic acids such as di- and triadenylic acids (47). More vigorous treatment with hot alkali, however, yields mixtures of oligodeoxynucleotides from DNA (54) and 2′,3′ cyclic phosphate intermediates from RNA via transphosphorylation (54). Mild acidic hydrolysis of the nucleic acid yields mononucleotides by mechanisms similar to those involved in alkaline degradation. Vigorous acid hydrolysis of RNA, however, yields primarily the purine bases and pyrimidine nucleotides (82). Ribonucleic acids are also hydrolyzed to varying amounts of nucleosides, nucleotides, and oligonucleotide materials by metal hydroxides and such nucleophiles as hydrazines and hydroxyl amines. These reactions have been reviewed in detail by Michelson (54).

The mononucleotides formed by the hydrolysis of nucleic acids are further degraded in alkaline media to give the corresponding nucleoside and inorganic phosphate (54). This reaction is thought to proceed via glycosyl fission to give the sugar phosphate which is then degraded. The presence of the phosphate group contributes to the stability of the glycosyl linkage in nucleotides toward acidic hydrolysis (54). In acidic media, the pyrimidine nucleotides are considerably more stable than the purine nucleotides, and liberation of inorganic phosphate is much faster in the case of the latter (54). In general, the hydrolytic conditions must be rigorous for measurable hydrolysis to occur. Most of the 5′ ribonucleotides, for example, are stable at 100° at pH 3, 5, 7, and 9 for days (46).

The kinetic parameters for the hydrolysis of nucleoside 2′,3′-phosphates have recently been elucidated. These compounds are subject to hydrolysis by second-order hydrogen catalysis, first-order hydroxide ion catalysis, and at elevated temperatures (>82°C) in the pH range of 4 to 9 (1). At temperatures above 100°C and pH values below 7, the phosphomonoester products are rapidly hydrolyzed to the nucleosides and inorganic phosphate (1).

The decomposition of nicotinamide adenine dinucleotide (NAD) in various media has been investigated. Under alkaline conditions nicotinamide is formed along with ADP and inorganic phosphate (35). NAD is totally destroyed after boiling for 10 to 15 min in the presence of 0.01 *M* sodium carbonate. The rate of hydrolysis of NAD at 100°C under neutral conditions varies greatly with the type of buffer used to maintain constant pH. The rate constants obtained at neutral pH in nitrogen base buffers were considerably lower than those obtained with such buffers as phosphate, β-glycerol phosphate, and pyrophosphate (4).

Although acid anhydrides, like acetic anhydride and phosphoryl chloride, are readily hydrolyzed by water at physiological pH, the biological anhydrides such as adenosine triphosphate and 3′-phosphoadenosine-5′-phosphosulfate (PAPS) are relatively stable (9). At that pH, they probably exist in the stable dianionic form (9). Upon activation by enzymes, either protonation of the phosphoryl oxygen or complexation with metal ion should produce active sulfate (9) or phosphate capable of carrying out the steps of sulfation or phosphorylation, respectively.

Phospholipids. Certain phospholipids such as the phosphatidic acids are relatively unstable, particularly in acid media. Hydrolytic decomposition of these compounds occurs through intramolecular attack by the phosphoric acid group and may be quite rapid on the acyl ester. Distearoylphosphatidic acid completely decomposes to free fatty acid and glycerophosphoric acid after 5 month of storage (33). Alkaline hydrolysis of the phosphatidic acids, in contrast to acid hydrolysis, proceeds without migration of the phosphate group (80).

The myoinositol phosphates derived from soil are chemically very stable, particularly in strong alkaline media (5). The stability of the esters differ in weakly alkaline media, depending on the number and position of phosphate groups in the molecule. Inositol monophosphate is quite stable at pH 7.5, whereas myoinositol is slowly hydrolyzed (26).

Enzymatic Transformations

The rate of decomposition and transformation of naturally occurring organophosphate is greatly accelerated in the presence of specific enzymes. A large number of enzymes have now been extracted from plant and animal sources capable of performing an astounding variety of catalytic operations on these compounds. While some enzymes are strictly degradative, others catalyze the synthesis or transformation of biologically important molecules. Biosynthetic reactions in nature frequently occur at the expense of organophosphates—that is, during a biosynthetic step, these enzymes concomitantly catalyze the degradation of a phosphate intermediate. Examples of these enzymes, as well as the purely degradative ones, are summarized in Table 4.

Oxidation-Reduction Reactions

Of the redox transformations occurring at the P center of biological phosphates, oxidative phosphorylation is perhaps the most significant. Biochemically, this term describes the transfer of energy derived from transport of electrons or oxidation-reduction processes to production of "high-energy

TABLE 4 Classification of Enzymes Capable of Degrading and Transforming Biological Organophosphates[a]

Enzyme Class	Function	Substrates Acted upon	Example
I. Oxidoreductases	Catalyze oxidation-reduction reactions	A. CH—OH group of donors	D-Glucose-6-phosphate: NADP oxidoreductase
		B. Carbonyl group of donors	D-Glyceraldehyde-3-phosphate: NAD oxidoreductase (phosphorylating)
II. Transfearses	Catalyze group-transfer reactions	A. One-carbon groups	Carbamoyl phosphate: L-aspartate carbamoyl transferase
		B. Carbonyl groups	D-Sedoheptulose-7-phosphate: D-glyceraldehyde-3-phosphate glycoaldehyde transferase
		C. P-containing groups	ATP: glucose-6-phospho-transferase
		D. Acyl groups	Acetyl–CoA orthophosphate acetyl transferase
		E. Glycosyl groups	Maltose orthophosphate glucosyl transferase
		F. Pentosyl groups	Purine nucleoside: orthophosphate ribosyl transferase

III. Hydrolases	Catalyze hydrolysis of organophosphorus compounds	A. Orthophosphate monoesters	Orthophosphoric monoester phosphohydrolase
		B. Orthophosphate diesters	Orthophosphoric diester phosphohydrolase
		C. Pyrophosphates	Pyrophosphate phosphohydrolase
		D. Polyphosphates	Adenosine triphosphate phosphohydrolase
IV. Lyases	Catalyze additions of groups to double bonds or vice versa	A. Carbon–carbon bonds	Orotidine-5-phosphate carboxylyase
		B. Aldehyde	Ketose-1-phosphate aldehyde lyase
V. Isomerases	Catalyze isomerization	A. Carbohydrates and derivatives	D-Ribulose-5-phosphate-3-epimerase
VI. Ligases	Catalyze condensation of two molecules	A. Forming C—O bonds	L-Tyrosine: SRNA ligase (AMP)
		B. Forming C—S bonds	Acetate–CoA ligase (AMP)
		C. Forming C—N bonds	Xanthosine-5′-phosphammonia ligase (AMP)
		D. Forming C—C bonds	Pyruvate: carbon dioxide ligase (ADP)

[a] Based on the classification system of the Commission on Enzymes of the International Union of Biochemistry (51).

phosphates" (11). Mechanistic schemes for oxidative phosphorylation in biological systems have been proposed as follows (66):

$$AH_2 + B + X \rightleftarrows A \sim X + BH_2 \quad (OX)$$
$$A \sim X + Pi \rightleftarrows A + X \sim P$$
$$X \sim P + ADP \rightleftarrows X + ATP$$

where the energy of electron transport ($AH_2 \rightarrow BH_2$) is transferred to an intermediate, $A \sim X$, containing a high-energy bond.

Although not cited in the literature, chemical phosphoryl transfer reactions involving oxidative steps must certainly be involved in transformation of organophosphorus compounds of the environment. Examples of known reactions of this type are as follows (58):

$$P - \text{transfer: } S + ATP \rightarrow S - P + ADP$$
$$P \sim \text{transfer: } S + ATP \rightarrow S \sim P + ADP$$
$$P \sim P \text{ transfer: } S + ATP \rightarrow S - P \sim P + AMP$$

The oxidation of biochemical compounds is facilitated by the high-energy phosphate bond. For example, the redox potential for the oxidation of acetaldehyde to acetate has a large negative value, but phosphorylation raises the E_0' by 200 to 400 mu. This elevation E brings the reaction within the range of physiological enzymes (58).

Radiation-Induced Reactions

Visible Light. The field of radiobiochemistry is rapidly expanding to include the effects of visible and ultraviolet light and ionizing radiation on biological macromolecules. Of the phosphorus-containing macromolecules, nucleic acids are the most labile in the presence of radiation. In fact, the most profound biological genetic effects of radiation result from its action on nucleic acids. The photochemical and radiation-induced reactions of phosphorus compounds have been reviewed in detail by Halmann (32), and only the important biochemical species are considered herein.

Under the influence of visible light in the presence of Fe^{3+}, viral ribonucleic acid degrades and loses infectivity at 0°C and neutral pH. Under these conditions the bases of the RNA are destroyed and released from the glycosidic linkages and breakage of the phosphodiester bond apparently occurs (71).

Ultraviolet Light and Other Radiation. The biological effects of ultraviolet light have been reviewed by Shugar (69) and Wacker (84). The exposure of DNA to ultraviolet light results in dimer formation by two adjacent thymine bases (34). In RNA, ultraviolet irradiation results in dimerization of both uracil and uridine (83). Irradiation of aqueous solutions of DNA causes irreversible cleavage of the hydrogen bonds and also results in the appearance of heat-stable linkages between the two complementary polynucleotide strands (54).

Large doses of ultraviolet radiation are required to damage DNA to an extent sufficient for detection by chemical means. Only the sensitive biological tests such as enzyme induction are required for detection of effects of lower doses. Irradiation of high molecular weight DNA results in depolymerization and disruption of the hydrogen bonds. High dosages of x-ray irradiation (200 kV) induce scission of the internucleotide phosphate bonds after attack by radicals produced by action of the x-rays on the water solvent (21).

A complex series of reactions occur when solutions of nucleotides are irradiated with x-rays. Free radicals formed by initial splitting of the water molecules attack the nucleotides (68) and ammonia and unstable phosphate esters are liberated (21). ^{60}Cobalt γ-irradiation degrades oxygenated solutions of polynucleotides with the formation of 5′-phosphate and groups and liberation of inorganic phosphates (86). γ-Irradiation of ribomononucleotides in aqueous solution results in polymer formation, apparently through formation of phosphodiester bonds (78). A review on the radiation chemistry of nucleic acid constituents has been written by Ponnamperuma (64).

Upon irradiation with 200 kV x-rays, α- and β-glycerophosphates release inorganic phosphate. Additionally, the α form yields dihydroxyacetone phosphate and the β forms give a phosphate ester labile to acid at elevated temperatures (68).

Irradiation of hydrolecithin in the absence of air results in splitting of the carboxylic ester linkage producing fatty acids and choline phosphate derivatives (32).

DEGRADATION OF ORGANIC PHOSPHATES IN SOIL

Tracer experiments have shown that the solid phase of soil is in dynamic equilibrium with the soil solution, and the rate of ion exchange between the two is fast (12). Therefore the concentration of phosphorus in the soil remains constant even when being exchanged among microorganisms and plants. The biodegradation of organic phosphorus compounds in soil is largely due to the combined activities of the soil microflora and free enzymes present in the soil.

Free enzymes in soil are derived from the living organisms present in the soil. Some of the enzymes are exoenzymes and are secreted into the soil by living cells. Other enzymes are intracellular, and it is only after the death and rupture of cell walls of the producing organism that they are released into the soil. Since enzymes are proteins, they may on excretion enter the food chain and again be metabolized by the soil organism. It must be assumed, however, that many enzymes remain active in the soil and persist for at least some period of time (72).

Because of the technical difficulty of selectively differentiating the activity of free enzymes from enzymes bound to or integrally a part of the organisms, the origin of reported enzyme activities of soil is unknown. Nevertheless, the presence of enzymatic activities in soil capable of degrading organophosphorus compounds has been reported (72). The sod-podzolic soils possess a much greater number of hydrolytic enzymes than oxidative ones (6).

Microorganisms capable of extensively decomposing organic insoluble phosphorus compounds have been isolated from soil. Bacteria such as *Bacillus* and *Pseudomonas* species produce phosphatases that degrade glycerophosphates, nucleic acids, and phytin (42). Up to 100% of the phosphorus in phytin and 50% in nucleic acids and certain phospholipids are ultimately dissolved by many bacterial strains (42). Carbohydrate is apparently required as an energy source for degradation of organic phosphates in soil by microorganisms. Although it is thought that the ability to solubilize organophosphorus is a common property of soil microorganisms (42), no evidence has been found to suggest the existence of a morphological or physiological group especially well adapted to this function (77).

A number of phosphorus-containing compounds including nucleic acids, lecithin, and phytin are dephosphorylated when added to soil (62). Ribonucleic acid has been reported to be more stable than deoxyribonucleic acid (8). Microbial populations of soil rapidly degrade added nucleotides (20) and nucleic acids (Table 5).

Organophosphorus compounds in soil are gradually mineralized with time (20). Birch (10), however, found that no pronounced mineralization of plant organic phosphorus takes place during the first 3 months of decomposition. Mineralization of several organic phosphates in soil has been reviewed by Cosgrove (20).

TABLE 5 Decomposition of Organophosphates in Soil[a]

Soil Amendment	Organic P (%)	Acid-Soluble Inorganic P (ppm)			
		0 Days	5 Days	45 Days	90 Days
Lecithin (soybean meal)	0.526	28.5	32.0	41.5	45.5
RNA	7.80	25.5	79.0	56.0	62.5
Phytin	19.10	25.5	24.0	40.0	65.0

[a] From Pearson *et al.* (62).

DEGRADATION OF ORGANOPHOSPHATES IN WATER

Organic phosphates in natural waters may be present in soluble, colloidal, and suspended forms (38). Approximately 30% of the phosphorus of algae and associated organisms is resistant to degradation and forms refractory material in the bottom sediments of natural waters (3). This material is probably further degraded and utilized at rates of no more than a few percent per year (3).

Quantitative and even qualitative data are unavailable on the nature and kinetics of organophosphorus decomposition in natural waters.

UTILIZATION OF ORGANOPHOSPHORUS BY MEMBERS OF THE ECOSYSTEM

Plants and animals normally do not directly utilize the naturally occurring organophosphates of soil (2) but depend on microorganisms to degrade these compounds first to inorganic phosphate. It is well known that bacteria and fungi are capable of hydrolyzing condensed phosphates to orthophosphate under a wide variety of environmental conditions (19, 40, 67). Mono- and diesters of phosphoric acids such as sugar phosphates and nucleic acids are readily dephosphorylated by microorganisms (Table 2). The inorganic orthophosphate produced from these reactions may then be assimilated by higher plants and animals, respectively, or reutilized by lower forms. In marine ecosystems, marine bacteria are capable of utilizing up to 80% of the dissolved inorganic phosphate while growing diatoms can absorb 40% of the same material (39).

REFERENCES

1. Abrash, H. I., C. S. Cheung, and I. C. Davis, *Biochem.*, **6**, 1298 (1967).
2. Alexander, M., *Introduction to Soil Microbiology*, Wiley, New York, 1964.
3. American Water Works Association, "Final Report, Water Quality Division Committee on Nutrients in Water," *J. Am. Water Works Assoc.*, **62**, 127 (1970).
4. Anderson, B. M., and C. D. Anderson, *J. Biol. Chem.*, **238**, 14–75 (1963).
5. Anderson, G., "Nucleic Acids, Derivatives, and Organic Phosphates," in *Soil Biochemistry*, A. D. McLaren and G. H. Peterson (Eds.), New York, 1967, pp. 67–90.
6. Aseeva, I. V., and V. A. Vanyarkho, *Biol. Nauk.*, **11**, 128 (1969); *Chem. Abstr.*, **72**, 77991*t* (1970).
7. Barber, R., and M. Noble, *Biochem. Biophys. Acta*, **123**, 205 (1966).

8. Bartholomew, M. V., and C. A. I. Goring, *Soil Sci. Soc. Am. Proc.*, **13**, 238 (1948).
9. Benkovic, S. J., and R. C. Hevey, *J. Am. Chem. Soc.*, **92,** 4971 (1970).
10. Birch, H. F., *Plant Soil Sci.*, **15,** 347 (1961).
11. Blackburn, G. M., and J. S. Cohen, "Chemical Oxidative Phosphorylation," in *Topics in Phosphorus Chemistry*, Vol. 6, M. Grayson and E. J. Griffith (Eds.), Wiley, New York, 1968.
12. Bowen, H. J. M., *Trace Elements in Biochemistry*, Academic Press, London, 1966.
13. Bower, C. A., Iowa State College Engineering Experimental Station Bulletin No. 362, 1949.
14. Bromfield, S. M., *Aust. J. Agric. Res.*, **12,** 111 (1961).
15. Brown, D. M., F. Hayes, and A. R. Todd, *Chem. Ber.*, **90,** 936 (1957).
16. Bunton, C. A., D. R. Llewellyn, K. G. Oldham, and C. A. Vernon, *J. Chem. Soc.*, **1958,** 3588.
17. Butzow, J. J., and G. L. Eichorn, *Biopolym.*, **3,** 95 (1965).
18. Cassaigne, A., *Recherches sur les Acides Aminoalkylphosphoniques*, Imprimerie E. Drouillard, Bordeaux, 1968.
19. Clesceri, N. L., and G. F. Lee, *Air Water Pollut.*, **9,** 723 (1965).
20. Cosgrove, D. J., "Metabolism of Organic Phosphates in Soil," in *Soil Biochemistry*, A. D. McLaren and G. H. Peterson (Eds.), Dekker, New York, 1967, pp. 216–228.
21. Daniels, M., G. Scholes, J. Weiss, and C. M. Wheeler, *J. Chem. Soc.*, **1957,** 226.
22. Degani, C., and M. Halmann, *J. Am. Chem. Soc.*, **88,** 4075 (1966).
23. Degani, C., and M. Halmann, *Israel J. Chem.*, **5,** 59P (1967).
24. Degani, C., and M. Halmann, *Nature*, **216,** 1207 (1967).
25. Desjobert, A., *Bull. Soc. Chim. Biol.*, **33,** 42 (1951).
26. Desjobert, A., *Bull. Soc. Chim. Biol.*, **36,** 1293 (1954).
27. Eichhorn, G. L., and J. J. Butzow, *Biopolym.*, **3,** 79 (1965).
28. Ferrier, R. J., and R. D. Guthrie, *Carbohydrate Chemistry*, Part I, *Mono-, Di-, and Trisaccharides and Their Derivatives*, The Chemical Society, London, 1969, pp. 55, 56.
29. Foster, A. B., W. G. Overend, and M. Stacey, *J. Chem. Soc.*, **1951,** 980.
30. Friedkin, M., and D. Roberts, *J. Biol. Chem.*, **207,** 257 (1954).
31. Goring, C. A. I., and W. U. Bartholomew, *Soil Sci.*, **74,** 149 (1952).
32. Halmann, M., "Photochemical and Radiation-Induced Reactions of Organophosphorus Compounds," in *Topics in Phosphorus Chemistry*, Vol. 4, M. Grayson and E. J. Griffith (Eds.), Wiley, New York, 1967.
33. Hanahan, D. J., *Lipide Chemistry*, Wiley, New York, 1960.
34. Harbers, E., G. F. Domagk, and W. Muller, *Introduction to Nucleic Acids—Chemistry, Biochemistry and Functions*, Reinhold, New York, 1968.
35. Hilvers, A. G., and K. van Dam, *Biochem. Biophys. Acta*, **81,** 394 (1964).
36. Horiguchi, M., and M. Kandatsu, *Agric. Biol. Chem.*, **28,** 408 (1964).
37. Hurwitz, C., C. L. Rosano, and R. A. Peabody, *Biochem. Biophys. Acta*, **72,** 80 (1963).
38. Hutchinson, G. E., *A Treatise on Limnology*, Vol. I, Wiley, New York, 1957.
39. Johannes, R. E., *Limnol. Oceanogr.*, **9,** 224 (1964).

40. Karl-Kroupa, E., C. F. Callis, and E. Seifter, *Ind. Eng. Chem.*, **49,** 2061 (1957).
41. Kittredge, J. S., and E. Roberts, *Science*, **164,** 37 (1969).
42. Kobus, J., *Rocz. Nauk Roln., Ser. D91*, **5** (1961); *Chem. Abstr.*, **66,** 6417 (1962).
43. Kohrana, H. G., *Some Recent Developments in the Chemistry of Phosphate Esters of Biological Interest*, Wiley, New York, 1961.
44. Koshland, D. E., *J. Am. Chem. Soc.*, **74,** 2286 (1952).
45. Kosower, E. M., *Molecular Biochemistry*, McGraw-Hill, New York, 1962.
46. Kuriyama, C., M. Fushizaki, and K. Murata, *Eiyo To Shokuryo*, **17,** 337 (1965); *Chem. Abstr.*, **64,** 9988 (1966).
47. Lane, B. G., and G. C. Butler, *Biochim. Biophys. Acta*, **33,** 281 (1959).
48. Larsen, S., "Soil Phosphorus," in *Advances in Agronomy*, A. G. Norman (Ed.), Vol. 19, Academic Press, New York, 1967, pp. 179–182.
49. Lewak, S., and L. Szabo, *J. Chem. Soc.*, **1963,** 3795.
50. Maher, G. C., and M. L. Wolfrom, "Carbohydrate Phosphate Esters," in *Handbook of Biochemistry*, H. A. Sober (Ed.), Chemical Rubber Publishing Company, Cleveland, Ohio, 1969, pp. D34–D38.
51. Mahler, H. R., and E. H. Cordes, *Biological Chemistry*, Harper and Row, New York, 1966.
52. Mattingly, G. E. G., and O. Talibudeen, "Progress in the Chemistry of Fertilizer and Soil Phosphorus," in *Topics in Phosphorus Chemistry*, Vol. 4, M. Grayson and E. J. Griffith (Eds.), Wiley, New York, 1967, pp. 157–290.
53. Mattson, S., *Soil Sci.*, **23,** 41 (1932).
54. Michelson, A. M., *The Chemistry of Nucleosides and Nucleotides*, Academic Press, London, 1963.
55. Mortland, M. M., and J. E. Gieseking, *Soil Sci. Soc. Am. Proc.*, **16,** 10 (1952).
56. Murakami, Y., and M. Tanaka, *Bull. Chem. Soc. Japan*, **39,** 122 (1966).
57. Nanninga, L. B., *J. Phys. Chem.*, **61,** 1144 (1957).
58. Needham, A. E., *The Uniqueness of Biological Materials*, Pergamon Press, New York, 1965.
59. Oestreich, C. H., and M. M. Jones, *Biochem.*, **6,** 1515 (1967).
60. Ogata, K., A. Imada, and Y. Nakao, *Agric. Biol. Chem.*, **26,** 586 (1962).
61. Omura, T., *J. Agric. Chem. Soc. Japan*, **28,** 240 (1954).
62. Pearson, R. W., A. G. Norman, and C. Ho, *Soil Sci. Soc. Am. Proc.*, **6,** 168 (1941).
63. Picard, J., *Trait. Biochem. Gen.*, **3,** 451 (1967).
64. Ponnamperuma, A., U.S. Energy Commission UCRL 10053, 1962.
65. Quin, L. D., "The Carbon–Phosphorus Bond," in *Topics in Phosphorus Chemistry*, Vol. 4, M. Grayson and E. J. Griffith (Eds.), Wiley, New York, 1967, pp. 23–48.
66. Racker, E., *Mechanisms in Bioenergetics*, Academic Press, New York, 1965.
67. Scharpf, L. G., Jr., and T. P. Kichline, *J. Agric. Food Chem.*, **15,** 787 (1967).
68. Scholes, G., W. Taylor, and J. Weiss, *J. Chem. Soc.*, **1957,** 235.
69. Shugar, D., *The Nucleic Acids*, Vol. III, Academic Press, New York, 1960.
70. Sillen, L. G., and A. E. Martell, *Stability Constants of Metal Ion Complexes*, Special Publ. No. 17, The Chemical Society, London, 1964.
71. Singer, B., and H. Fraenkel-Conrat, *Biochem.*, **4,** 226 (1965).

72. Skujins, J. J., "Enzymes in Soil," in *Soil Biochemistry*, A. D. McLaren and G. H. Peterson (Eds.), Dekker, New York, 1967.
73. Stern, K. G., and M. A. Steinberg, *Biochim. Biophys. Acta*, **11,** 553 (1953).
74. Stevens, V. L., and E. L. Duggan, *J. Am. Chem. Soc.*, **79,** 5703 (1957).
75. Strominger, J. L., "The Biosynthesis of Bacterial Cell Walls," in *The Bacteria*, Vol. III, I. C. Gunsalus and R. Y. Stanier (Eds.), Academic Press, New York, 1962.
76. Szabo, L., "The Synthesis and Reactions of Phosphorylated Deoxy Sugars," in *Deoxy Sugars, Advances in Chemistry*, R. F. Gould (Ed.), American Chemical Society, Washington, D.C., 1968.
77. Szember, A., *Ann. Univ. Mariae Curie-Sklodowska, Lublin-Polonia*, **15,** 133 (1960); *Chem. Abstr.*, **57,** 17168 (1961).
78. Torres, R. E., *Anal. Fac. Quim. Farm., Univ. Chile*, **14,** 41 (1962); *Chem. Abstr.*, **59,** 5858 (1963).
79. Trapmann, H., and M. Devani, *Enzymol.*, **28,** 65 (1964).
80. Uhlenbrock, J. H., and P. E. Verkade, *Rec. Trav. Chim.*, **72,** 558 (1953).
81. Van Wazer, J. R., Ed. *Phosphorus and its Compounds*, Vol. II, Interscience, New York, 1961.
82. Vischer, E., and E. Chargaff, *J. Biol. Chem.*, **176,** 715 (1948).
83. Wacker, A., D. Weinblum, L. Träger, and Z. H. Moustafa, *J. Mol. Biol.*, **3,** 790 (1961).
84. Wacker, A., *Prog. Nucleic Acid Res.*, **1,** 369 (1963).
85. Waksman, S. A., *Humus, Origin, Chemical Composition and Importance in Nature*, Williams and Wilkins, Baltimore, Maryland, 1936.
86. Ward, J. F., and M. M. Urist, *Int. J. Radiat. Biol.*, **12,** 209 (1967).
87. Wayman, C. H., "Adsorption on Clay Mineral Surfaces," in *Principles and Applications of Water Chemistry*, S. D. Faust and J. V. Hunter (Eds.), Wiley, New York, 1967, pp. 127–168.
88. Zubay, G., and P. Doty, *Biochim. Biophys. Acta*, **29,** 47 (1958).

24

Phosphorus in Human Nutrition

LAWRENCE J. MACHLIN

Life Science Department, New Enterprise Division, Monsanto Company, St. Louis, Missouri

FUNCTION

The adult body contains about 650 g of phosphorus. About 86% of this occurs as hydroxyapatite [$3Ca_3(PO_4)_2Ca(OH)_2$], an extremely hard, difficultly soluble calcium phosphate salt which occurs in bones and teeth. Thus a primary function of phosphorus is for the growth and maintenance of skeletal tissue and teeth (1–3). The remaining phosphorus occurs in the soft tissues where it is

1. essential for anabolic and catabolic reactions as exemplified by the role of phosphorus in high-energy bond formation (e.g., adenosine triphosphate), as phosphorylated intermediates of carbohydrate metabolism, and as a constituent of RNA;
2. a precursor for phospholipids which are important in the formation of cell membranes and in cellular permeability;
3. a precursor in the synthesis of genetically significant substances, particularly DNA;
4. a contributor to the buffering capacity of body fluids and cells.

A discussion of phosphorus function in soft tissues would touch almost all aspects of modern biochemistry and is obviously beyond the scope of this review.

ABSORPTION AND METABOLISM

It is likely that most, if not all, phosphorus is absorbed as free phosphate. The various inorganic esters must thus be hydrolyzed, probably by various phosphatases prior to absorption. Usually about 70% of the dietary phosphorus is absorbed. Polyphosphates are not absorbed although intestinal bacteria do hydrolyze 10 to 20% of the dietary polyphosphates to free phosphate. The amount absorbed varies inversely with the amount of calcium and other elements that form insoluble salts. However, even with a high calcium intake (over 2 g/day), children absorbed over 60% of the dietary phosphorus (4). The mechanism of phosphorus absorption by the gastrointestinal tract is unknown. Increasing levels of phosphate ions in the intestinal lumen linearly enhanced net absorption (5), a response typical of a diffusion-like process. In animals, vitamin D has no effect on absorption of inorganic phosphorus although it has enhanced the availability of phytate phosphorus (6).

When casein is digested there is some evidence (7) that phosphopeptides are released which can effectively chelate calcium and other metal ions. The calcium phosphopeptide is highly soluble and is readily absorbed from the gastrointestinal tract of animals (8) and infants (9). Calcium bound in this form will allow bone calcification to occur in rachitic children in some cases without simultaneous administration of vitamin D. The high availability of calcium in milk as compared with inorganic calcium may be due in part to the formation of such a complex.

Most of the phosphorus in plasma exists in an ionized form (10) (Table 1). In contrast to calcium, the concentration of plasma phosphorus varies with age, diet, and hormonal status. In the human adult, the range of values is 2.5 to 4.3 mg/100 ml. In younger children, the normal range is 5.0 to 6.0 mg/100 ml (11).

The major excretory route for phosphate is the kidney. A 70-kg man on a diet containing 900 mg of phosphate per day will excrete 600 mg in the urine.

Urinary phosphorus is largely inorganic phosphate. Oxidation of sugar requires phosphorus, and there is a temporary lowering of serum phosphates and of phosphate excretion after ingestion of carbohydrates. Catabolism of

TABLE 1 Forms of Phosphate in Normal Plasma[a]

Phosphate	mM/Liter	Percent of Total
Free HPO_4^{2-}	0.50	43.0
Free H_2PO_4	0.11	10.0
Protein bound	0.14	12.0
$NaHPO_4^-$	0.33	29.0
$CaHPO_4^-$	0.04	3.0
$MgHPO_4^-$	0.03	3.0
Total	1.15	100.0

[a] From Walser, M., *J. Clin. Invest.*, **40,** (10) 723 (1961).

body tissues, as occurs in starvation and acidosis, releases considerable phosphorus which is excreted by the kidney.

Urinary excretion is reduced by a dietary phosphate deficiency, lactation, and hypoparathyroidism. Its excretion is increased by vitamin D excess, hyperparathyroidism, and renal tubular disorders.

DIETARY SOURCES

Meat, fish, poultry, eggs, and cereal products are the primary sources of phosphorus in the average United States diet (Tables 2 and 3). Foods rich in protein and calcium are also rich in phosphorus.

TABLE 2 Estimates of Average Daily Intake of Calcium and Phosphorus[a]

Food Category	Calcium (mg/day)	Phosphorus (mg/day)
Bakery products	101	205
Whole-grain products	27	137
Eggs	25	99
Fresh vegetables	41	140
Root vegetables	17	16
Milk	642	548
Poultry	25	384
Fresh fish	25	356
Flour	30	274
Macaroni	24	14
Rice	2	16
Meat	3	548
Shellfish	30	3
Dried beans	2	27
Fresh fruit	8	54
Potatoes	37	82
Canned fruits	16	5
Fruit juices	4	5
Canned vegetables	5	27
Total	1051	2940

[a] Summarized by Henrikson (47) based on U.S.D.A. survey on food consumption (51) and calcium and phosphorus values of Watt and Merrill (12).

TABLE 3 Phosphorus and Calcium Content of Representative Foods[a]

Food Type	Examples	Calcium (mg/100g)	Phosphorus (mg/100g)	Ca/P Ratio
Meats	Beef, round, broiled	12	250	0.05
	Pork, ham, roasted	10	236	0.04
Poultry	Broilers	9	201	0.04
Fish	Halibut, broiled	16	248	0.06
	Cod, broiled	31	274	0.11
	Tuna, canned	16	190	0.08
Cereals and cereal products	Oatmeal	9	57	0.16
	Rice, white, cooked	10	28	0.36
	Bread, whole wheat	84	254	0.33
	Bread, white	70	87	0.80
Eggs	Egg	54	205	0.26
Dairy products	Milk, whole	118	93	1.26
	Cheddar cheese	750	478	1.57
	Cottage cheese	94	152	0.62
Root vegetables	Potatoes	7	53	0.13
	Beets	14	23	0.61
Green vegetables	Lettuce	20	22	0.90
	Peas, cooked	23	99	0.23
Legume seeds	Lima beans, cooked	47	121	0.39
	Beans, dry, canned	68	121	0.56
	Beans, snap, cooked	50	37	1.35
Fruit	Apples	7	10	0.70
	Peaches	9	19	0.47

[a] Based on data of Watt and Merrill (12).

Phosphorus occurs in plants as phytin (the calcium–magnesium salt of inositol hexaphosphate or phytic acid). The percentage of total phosphorus occurring as phytin varies considerably in different varieties and strains of edible plants: cereals (40 to 90%), legumes (5 to 72%), fruits (0 to 16%), nuts (12 to 50%), and root vegetables (5 to 23%) (13).

Most of the phytic acid in cereals is present in the outer bran layers. When different cereals are milled, varying proportions of phytic acid are removed. Thus when wheat is milled, nearly 90% of the phytin phosphorus is removed. On the other hand, polished rice contains considerable phytin phosphorus. Phytin phosphorus is not as readily absorbed as inorganic phosphorus. Phytase, an enzyme that hydrolyzes phytic acid to inositol and phosphate, is believed to be absent from the human gastrointestinal tract, but may be present in foods and be active during food preparations

(14). The availability of phytin phosphorus in human diets is not accurately known. However, phytin cannot be disregarded as a source of phosphorus in view of a number of studies (14–16) suggesting that at least 50% of the phytin phosphorus is available. For countries such as India, where cereals constitute the largest part of the diet, as much as 60% of the total dietary phosphorus may be phytin phosphorus. The normal Indian diet is high in both total and phytic acid phosphorus, containing as much as 2 g of total phosphorus with about 1 g of phytin phosphorus.

DIETARY REQUIREMENTS

Animal experiments have demonstrated quite clearly that dietary phosphorus is necessary for growth (17) and prevention of renal calculi (18) and that all organic forms of biological importance can be synthesized in the animal body from the simple orthophosphates in food (19). Long-term human studies with phosphorus-low diets have not been carried out to our knowledge. Therefore, no dietary requirements have been established experimentally for man and few national or international agencies have established minimum requirements for phosphorus. The Food and Nutrition Board of the U.S. National Research Council (20) has set a dietary allowance for phosphorus using the following logic:

> Consideration of phosphorus along with calcium is appropriate since they are the major mineral constituents of bone and teeth. These tissues place major demands on the total requirement for these elements, particularly during growth. The ratio of calcium to phosphorus in bones is about 2:1. In addition to the phosphorus present in bones and teeth, phosphorus is present in blood and cells as soluble phosphate ion, as well as in lipids, proteins, carbohydrates, and energy-transfer enzymes. The 2:1 ratio of calcium to phosphorus in bones would suggest this to be the ideal dietary ratio, particularly during growth. However, the much higher ratio of phosphorus to calcium in soft tissues and the practical consideration that with ordinary diets the phosphorus intake invariably equals or exceeds the calcium intake (21–23) lead to a recommended allowance for phosphorus equal to that for calcium for all age groups except the young infant (Table 4).
>
> In considering a phosphorus allowance for the infant, it is recognized that a liberal phosphorus intake, such as is provided by cow's milk, may contribute to the occurrence of hypcalcoemic tetany during the first week of life (24, 25). The ratio of calcium to phosphorus in cow's milk is 1.2:1, compared with a ratio of 2:1 in human milk. Current evidence supports the recommendation that in early infancy the calcium:phosphorus ratio in the diet be 1.5:1. For older infants the phosphorus allowance is raised to about 80% of the calcium allowance or to a ratio to calcium similar to cow's milk. This is consistent with the practical consideration that infants of this age generally receive cow's milk as the dietary staple.

TABLE 4 Food and Nutrition Board, National Academy of Sciences, National Research Council Recommended Daily Dietary Allowances for Calcium and Phosphorus[a] (17)

	Age (years)	Calcium (g)	Phosphorus (g)
Infants	0–$\frac{1}{6}$	0.4	0.2
	$\frac{1}{6}$–$\frac{1}{2}$	0.5	0.4
	$\frac{1}{2}$–1	0.6	0.5
Children	1–2	0.7	0.7
	2–6	0.8	0.8
	6–8	0.9	0.9
	8–10	1.0	1.0
Males	10–12	1.2	1.2
	12–18	1.4	1.4
	18–75+	0.8	0.8
Females	10–12	1.2	1.2
	12–18	1.3	1.3
	18–75+	0.8	0.8
Pregnancy		+0.4	+0.4
Lactation		+0.5	+0.5

[a] Seventh edition, 1968. These recommendations are designed to be adequate for practically all of the populations and allow a margin of safety for individual variation. Individuals whose diets do not meet the RDA are not necessarily deficient.

Evidence for a phosphorus-depletion syndrome in man has been reported in patients receiving prolonged treatment with nonabsorbable antacids such as magnesium–aluminum hydroxides (26). Such antacids limit gastrointestinal absorption of phosphorus resulting in a syndrome characterized by hypophosphatemia, increased gastrointestinal absorption of calcium, hypercalcemia, increased resorption of skeletal calcium and phosphorus, and debility, with anorexia, bone pain, and malaise.

MINERAL INTERRELATIONSHIPS

Calcium

The significance of the dietary calcium–phosphorus ratio has been discussed for many years (23, 27, 28). In the past it has been stressed that the ideal calcium–phosphorus ratio for human beings ranges between 2:3 and 1:1 and that any significant departure is dangerous. It was suggested that high phosphorus relative to calcium mostly interferes with calcium absorption, presumably by causing formation of insoluble calcium phosphate in the gastrointestinal tract. This dogma was based primarily on animal experiments. Experiments with human subjects demonstrated that addition of a large amount of phosphorus to the diet has almost no effect on calcium absorption (29). Based on the study of Malm (29), some nutritionists (27, 30) assume that the variations in the calcium–phosphorus ratio in most diets are of no practical significance in human nutrition.

In newborn infants, however, excessive phosphorus relative to calcium may be hazardous. A tetany of the newborn which can be cured by administration of calcium salts has been observed for many years. This tetany is associated with the feeding of cow's milk, decreased serum calcium, and increased serum phosphorus levels (24, 25, 31, 32). Cow's milk has a much lower calcium–phosphorus ratio than breast milk (Table 5). Moreover, addition of phosphorus is known to reduce blood calcium levels (31). Evidently the newborn is particularly sensitive to the high phosphorus load presented by ingestion of cow's milk and can easily develop a hypocalcemic tetany.

On the other hand, the phosphorus level of breast milk is so low that for the newborn (first week), calcification and growth of soft tissue may be limited by this nutrient. At this age phosphate addition to breast-fed babies promoted absorption of calcium and increased retention of calcium (33).

When phosphorus was added to white bread as sodium phytate, a decreased absorption of calcium was observed (34). Ca^{45} was absorbed more readily

TABLE 5 Calcium and Phosphorus Content of Cow's Milk and of Human Milk in Neonatal Period

	Calcium (mg/100 ml)	Phosphorus (mg/100 ml)	Ca/P Ratio
Human milk (days 1 to 5 of lactation)[a]	19.7	8.7	2.26
Human milk[b]	33.0	14.0	2.35
Cow's milk (fresh, skimmed, undiluted)[a]	122.0	96.0	1.27
Cow's milk product with normal dilution[a]	52.0	53.0	1.00

[a] From Gardner *et al.* (24).
[b] From Watt and Merrill (12).

when a phytate-free cereal (farina) was consumed than with a phytate-rich cereal (oatmeal) (35, 36). However, sodium phytate has an even greater effect on inhibiting calcium absorption than the phytin from oatmeal. Phytates as the acid or in the form of a soluble salt apparently interfere with absorption of calcium in man while phytates in food substances are not as available for interference. The effect of phytates is related to the level of calcium in the diet. If the diet is high in phytates and low in calcium, absorption of calcium will be reduced (37). It is generally agreed that in this country there is no significant problem from phytin (38) although in other areas such as India with calcium deprivation and high phytate intakes, the nutritional role of phytates may assume considerable importance.

Magnesium

In experimental animals, high phosphorus level interferes with absorption of magnesium (39), thus increasing the dietary requirements for magnesium for the prevention of calcification of the kidney, aorta, and other soft tissues (39–43). Calcium also interferes with the absorption of magnesium. The effect of calcium and phosphorus in increasing the dietary requirement for magnesium appears to be additive (44).

None of these effects has been studied in humans. It is not possible to relate these findings directly to the problem of renal calculus in man and related clinical phenomena at the present time.

Iron

Absorption of iron, calcium, and phosphorus is dependent on the relative amounts of each of these substances in the diet (45). Iron absorption may fall below 3% in diets with a low calcium, high phosphorus, and high phytic acid content. The limiting effect of phosphates and phytic acid on iron absorption may be reduced by excess of iron and/or calcium. Excess dietary calcium and a low phosphorus intake may decrease iron absorption, whereas increased absorption may result from a high intake of both calcium and phosphate.

EFFECT OF DIETARY PHOSPHATE ON DENTAL CARIES AND PERIODONTAL DISEASE

Caries

Many animal experiments have demonstrated that various phosphates mixed in food, baked in bread, or supplied in water can give a substantial (up to more than 90%) caries reduction (46). Supplying phosphate in pills or by stomach tube is ineffective, indicating that the effect of phosphate is local. Apparently phosphate takes the place of carbonate in the superficial enamel layers resulting in greater acid resistance. In man the effect is less easily demonstrated. Dalderup (46) states that "although at this moment it does not seem justified to supplement various foodstuffs for human consumption with phosphates, it seems certainly justified to choose a diet with an ample supply of phosphate while poor in sugar."

Periodontal Disease

Periodontal disease is a disease of the supporting tissues of the teeth which may lead to detachment of teeth. It is generally accepted that bacterial plaque and calculus are the most common direct causes of periodontal disease. Henrikson (47) and Krook *et al.* (48), however, have found that in studies of dietary calcium deficiency and/or phosphorus excess, there is a generalized bone loss. The bones of the jaw exhibit the earliest signs of excess resorption and also show the most severe bone loss with time. These observations led these workers to postulate that "the primary lesion in periodontal disease is excessive resorption of alveolar bone as a manifestation of dietary calcium deficiency."

A U.S.D.A. survey in 1955 showed that the production of calcium-containing food items provides 1.05 g per capita per day in the United States (Table 2). This is more than the recommended allowance of the Food and Nutrition Board (Table 4). However since there is considerable food wastage and very unequal food distribution, a large proportion of the United States population is probably receiving less than the recommended allowance. Equally significant is the considerable excess of phosphorus in our diet (the average calcium-to-phosphorus ratio is 1:2.8) (Table 2). It is known from animal experimentation that even with optimal calcium levels in the diet, excessive phosphorus results in a lowering of blood calcium leading to a secondary hyperparathyroidism and skeletal disease. The 1:2.8 ratio which was found in the United States diet is about the same ratio as that used to induce severe osteopenia with periodontal disease in experimental animals (49). Henrikson (47) states that "a diet in which the calcium content meets or even exceeds standard recommendations, but in which phosphorus exceeds calcium, is as detrimental to the skeleton as a simple calcium deficiency."

Whether the high levels of phosphorus present in United States diets do present a significant hazard in regard to the etiology of periodontal disease remains to be established. A U.S.D.A. survey in 1965 (50) showed that during the period of 1955 to 1965, milk production decreased 10% in the

United States and the meat production increased by 10%. Milk is the major source of calcium in the diet and meat the richest source of phosphorus so that the excess of phosphorus compared to calcium in the diet is growing worse. It is therefore quite relevant to provide more definitive evidence for the hypothesis that an excessive phosphorus relative to calcium can lead to periodontal disease.

Lang (52), however, considers the maximum harmless intake to be 6.6 g of phosphorus daily. Since the daily intake of phosphorus rarely exceeds 4 g, this work suggests that there is little likelihood of hazards arising from excess phosphorus.

SUMMARY

Phosphorus is present in relatively large amounts in many foods such as meat, fish, milk, eggs, beans, and cereals. As a consequence a deficiency of this element in humans is unusual. Of more concern are the possible problems arising from excessive phosphorus relative to other elements. Specifically, an excess relative to calcium can result in a hypocalcemic tetany in the newborn. In the adult an excess may interfere with iron absorption and may be involved in the etiology of periodontal disease. In animals an excess of phosphorus relative to magnesium increases the dietary requirement for magnesium, thus exacerbating the calcification of soft tissues (e.g., renal calculus) resulting from a magnesium deficiency. It should be emphasized that the only unequivocally demonstrated consequence of phosphorus excess in humans is the hypocalcemic tetany of the newborn. However, since animal experiments suggest a possible role of phosphorus excess in the etiology of renal calculus and periodontal disease, more extensive investigations with humans are certainly warranted.

REFERENCES

1. Neuman, W. F., and M. W. Neuman, *The Chemical Dynamics of Bone Mineral*, University of Chicago Press, Chicago, Illinois, 1958.
2. McClean, F. C., and M. R. Urist, *Bone, An Introduction to the Physiology of the Skeleton*, University of Chicago Press, Chicago, Illinois, 1958.
3. Irving, J. T., "Dynamics and Function of Phosphorus," in *Mineral Metabolism II*, C. L. Comar and F. Bronner (Eds.), Academic Press, New York, 1964, Part A, Chapter 18, p. 249.
4. Lutwak, L., L. Laster, M. Fox, and G. D. Whedon, *Am. J. Clin. Nutr.*, **14,** 76 (1964).
5. McHardy, G. J. R., and D. R. Parsons, *Q. J. Exp. Physiol.*, **41,** 398 (1956).

6. Pileggi, V. J., H. F. DeLuca, and H. Steenbock, *Arch. Biochem. Biophys.*, **58,** 194 (1955).
7. Mellander, O., *Nutr. Rev.*, **13,** 161 (1955).
8. Mellander, O., and B. Isaksson, *Acta Soc. Med. Ups.*, **55,** 239 (1950).
9. Mellander, O., *Acta Soc. Med. Ups.*, **55,** 247 (1950).
10. Walser, M., *J. Clin. Invest.*, **40,** 723 (1961).
11. Rasmussen, H., *Textbook of Endocrinology*, R. H. Williams (Ed.), Saunders, Philadelphia, Pennsylvania, 1968, p. 847.
12. Watt, B. K., and A. L. Merrill, *Compositions of Foods*, Handbook No. VIII, U.S. Department of Agriculture, Washington, D.C., 1963.
13. Harris, R. S., *Nutr. Rev.*, **13,** 257 (1955).
14. Anonymous, *Nutr. Rev.*, **25,** 218 (1967).
15. McCance, R. A., and E. M. Widdowson, *J. Physiol.*, **101,** 44 (1943).
16. Kurier, P., M. Swaminathan, and V. Subrahmanyan, *Ann. Biochem. Exp. Med.*, **21,** 41 (1961).
17. Osborne, J. B., and L. B. Mendel, *J. Biol. Chem.*, **34,** 131 (1918).
18. Schneider, H., and H. A. Steenbock, *J. Biol. Chem.*, **128,** 159 (1939).
19. McCollum, E. V., *Am. J. Physiol.*, **25,** 120 (1909).
20. *Recommended Dietary Allowances*, Seventh Edition Publication 1694, National Academy of Sciences, Washington, D.C., 1968.
21. Patton, M. B., E. D. Wilson, J. M. Leichserring, L. M. Norris, and C. M. Dienhart, *J. Nutr.*, **50,** 373 (1953).
22. Schofield, F. A., D. E. Williams, E. Monell, B. B. McDonald, E. Brown, and E. L. McLeod, *J. Nutr.*, **59,** 561 (1956).
23. Sherman, H. C., *Calcium and Phosphorus in Foods and Nutrition*, Columbia University Press, New York, 1947.
24. Gardner, L. I., E. A. MacLachlen, W. Pick, M. L. Terry, and A. M. Butler, *Pediatrics*, **5,** 228 (1950).
25. Gardner, L. I., *Pediatrics*, **9,** 534 (1952).
26. Lotz, M., E. Zisman, and F. C. Barter, *New England J. Med.*, **278,** 409 (1968).
27. Harris, R. S., and A. E. Nizel, *J. Dent. Res.*, **43,** 1090 (1964).
28. Wasserman, R. H., *Fed. Proc.*, **19,** 636 (1960).
29. Malm, O. J., *Scand. J. Clin. Lab. Invest.*, **5,** 75 (1963).
30. FAO/WHO Expert Group, *Calcium Requirements*, WHO Technical Report Series No. 230, Rome, 1962, pp. 16–17.
31. Bakwin, H., *Am. J. of Diseases of Children*, **54,** 1211 (1937).
32. Oppé, T. E., and D. Redstone, *Lancet*, **1,** 1045 (1968).
33. Widdowson, E. M., R. A. McCance, G. E. Harrison, and A. Sutton, *Lancet*, 1250 (1963).
34. McCance, R. A., and F. M. Widdowson, *J. Physiol.*, **101,** 44 (1942).
35. Bronner, F., R. S. Harris, C. J. Maletskos, and C. E. Benda, *J. Nutr.*, **59,** 393 (1956).
36. Bronner, F., R. S. Harris, C. J. Maletskos, and C. E. Benda, *J. Nutr.*, **54,** 523 (1954).
37. Brine, C. L., and F. A. Johnston, *J. Am. Diet. Assoc.*, **31,** 883 (1955).
38. Harris, R. S., *Nutr. Rev.*, **13,** 257 (1955).

39. Meyer, D. L., and R. M. Forbes, *J. Nutr.*, **93,** 361 (1967).
40. MacKay, E. M., and J. Oliver, *J. Med.*, **61,** 319 (1935).
41. Hamuro, Y., *J. Nutr.*, **101,** 635 (1971).
42. Bunce, S., P. G. Reeves, and T. S. Oba, *J. Nutr.*, **88,** 406 (1965).
43. Bunce, G. E., K. J. Jenkins, and P. H. Phillips, *J. Nutr.*, **76,** 17, 23 (1962).
44. Morris, E. R., and B. L. O'Dell, *J. Nutr.*, **81,** 175 (1963).
45. Anonymous, *Nutr. Rev.*, **25,** 218 (1967).
46. Dalderup, L. M., *World Rev. Nutr. Diet.*, **7,** 72 (1967).
47. Henrikson, P., *Acta Odontol. Scand. Suppl. 50*, **26** (1968).
48. Krook, L., L. Lutwak, P. Henrikson, and J. Whalen, Cornell Nutrition Conference, Buffalo, N.Y., November 1970, p. 10.
49. Krook, L., and J. E. Lowe, *Path. Vet., Suppl. 1*, **1** (1964).
50. *U.S. Department of Agriculture*, Agricultural Research Service, 62-17, U.S. Government Printing Office, Washington, D.C., 1968.
51. *Dietary Levels of Households in the United States*, U.S. Department of Agriculture, Household Food Consumption Survey 1955, Report No. 6, U.S. Government Printing Office, Washington, D.C., 1957, pp. 1–68.
52. Lang, K., *Médizin Und Ernáehrung*, **2,** 49 (1961).

Biomineralogy of Phosphates and Physiological Mineralization

DUNCAN McCONNELL

Ohio State University, Columbus, Ohio

Biomineralogy has not achieved the recognition that it deserves from either the biological or the physical scientists until comparatively recently. Despite the verve that has arisen during the past decade, one might compare its status with that of clay mineralogy about half a century ago.

To be sure, there had been compendia such as those by Clarke and Wheeler (12) and Vinogradov (95), but these works merely tabulated the inorganic oxides present in ashed organisms or in their hard parts (endo- or exoskeletons). While this information might give important clues to the types of minerals present, polymorphism precludes the possibility of surmising anything about the crystallography from the composition, even when it can be assumed that the particular composition is that of a single solid phase. For example, Clarke (11) shows analyses of crabs and shrimps which range from about 15 to 50% of tricalcium phosphate, but the minerals remain "undefined" according to Lowenstam (54).

Added to this difficulty was an equally important one concerning the range in composition that might be expected for a single solid phase through isomorphic substitution. Thus it must be appreciated that mineralogical knowledge comprises sound information on both the chemical composition and the crystal structure.

The crystal chemistry (structure and composition) of biominerals, despite the existence of many sophisticated analytical methods, has not progressed rapidly for two reasons: (*a*) the inherent complexities of the substances and (*b*) the inexperience or lack of fundamental knowledge of many investigators. The latter statement acquires major importance in conjunction with the apatitic mineral of teeth and bones.

Up to this point the discussion has been concerned with biomineralogy in general and it should be mentioned that the carbonates are somewhat exceptional insofar as the two principal minerals (aragonite and calcite) have been investigated in some detail with respect to minor constituents such as magnesia and strontia. The resulting data may soon lead to interpretations of importance with respect to paleoecology, marine geochemistry, and related topics, particularly when combined with a knowledge of isotopic compositions.

However, even when working with such nice samples as shells of pelecypods, difficulties arise, and there remain knotty problems involving vaterite and several hydrated carbonates—some of which contain cations other than calcium and magnesium. The carbonate biominerals are mentioned because some scrutiny of this area focuses attention on the vastness of our ignorance on the biomineralogy of the phosphates, but also because the biochemistry of phosphate precipitation is related to the ever-present carbonate ion in natural environments.

Indeed, one may visualize the connecting link between biological chemistry and inorganic chemistry as carbon dioxide in one or another of its diverse forms of combination. Consequently any synthetic system that fails to consider the presence of carbonate (or bicarbonate) ions is naturally unrealistic at the outset, and although carbonate groups may or may not enter the minerals' structure, the possible influence of carbonate ions must be recognized because carbon dioxide is an end product of animal metabolism.

Biomineralogy is not a new topic, but it might not be amiss to indicate what is meant. A *mineral*, according to most definitions, is a solid phase that occurs naturally and has fairly well-defined physical and chemical characteristics. It stands in contradistinction with the term *rock* which is heterogeneous (composed of more than one mineral, in general) and which comprises a portion of the earth's solid mass. One immediately may ask: Why, then, is there merit to adding the prefix *bio* insofar as *mineral* already includes natural inorganic solids?

Traditionally, earth scientists have been concerned with minerals that are not of biogenic origin and some definitions of *mineral* quite deliberately exclude substances of obviously organic origin. Therefore solely for the purpose of emphasizing the relation to living things, several authors have assumed that *biomineralogy* is a distinctive category. Most earth scientists, unfortunately, do not devote attention to this topic, possibly because they are fully occupied with the petrography and petrology of igneous, metamorphic, and sedimentary rocks. On the other hand, physical scientists cannot immediately feel at home among the myriads of living species that produce the biominerals. These species range from bacteria to man, and in the final analysis will require extensive knowledge of physiological chemistry.

Far more courageous has been the group best characterized as "health scientists." While their writings are voluminous, much of this material is based on inadequate experience, and consequently it has not brought about significant progress.

PHOSPHATE BIOMINERALS

Aside from those produced by certain bacteria, phosphatic precipitations are associated solely with the animal kingdom. Insofar as is known, there are no true plants that form inorganic phosphates which are analogous to the carbonate depositions of calcareous algae or the siliceous tests of diatoms. Although phosphorus is essential to plant life, the physiological chemistry of plants does not seem to involve storage of phosphates as inorganic compounds of comparatively low solubility.

Some marine invertebrates, however, are phosphate accumulators. The living brachiopod genus *Lingula* has a francolite shell (65), and Lowenstam (54) has indicated that certain mollusks and arthropods have the ability to form skeletal tissues of calcium phosphate. In addition to the fossil inarticulate brachiopods, essentially all conodont remains are phosphatic, as discovered in the sedimentary strata, and it seems highly probable that this was their composition during life, although there may have been fluorine enrichment subsequently. In other situations, although fossil remains may be phosphatic, there is virtually no evidence that they were not originally organic or calcareous in view of the compositions of related recent genera.

McConnell (57) tabulated the minerals "likely to occur in rock phosphates," that is, in sedimentary rocks that have been subjected merely to low-temperature metasomatism. Such minerals—capable of forming at atmospheric or physiologic temperatures and pressures—are tabulated in Table 1, with revisions and extensions that are currently required. The present list is longer, although confined to phosphates of cations of the commoner elements. No minerals are included that contain significant amounts of beryllium (uralolite), boron (seamanite), copper (sampleite), lithium (tavorite), manganese (strunzite), zinc (hopeite), and the like. Also excluded are those that contain uranium, vanadium, rare earths, and other elements unlikely to be accumulated beyond traces by living organisms.

Only those minerals shown boldface in Table 1 have been recognized as resulting directly from biochemical influences. Others are secondary minerals that might form under conditions of biochemical alteration of pegmatitic phosphates, for example. Biogeochemical processes involving the interaction of vertebrate excrement (primarily that of birds) with various types of rocks account for some of the aluminum and iron phosphates. Hutchinson (37) published an extensive survey of such deposits.

The work of Haseman *et al.* (32) on aluminum and iron phosphates containing potassium and ammonium should be mentioned despite the fact that their synthetic preparations do not necessarily show a close coincidence with mineral species; another useful source of data on fertilizer compounds

TABLE 1 Phosphatic Minerals Likely to Form under Biochemical Influences

Mineral[a]	Chemical Composition	Ref.
Brushite*	$CaHPO_4 \cdot 2H_2O$	89, 74
Monetite*	$CaHPO_4$	80
Whitlockite*	β-$Ca_3(PO_4)_2$	89, 5
Dahllite*	Carbonate hydroxyapatite	60, 6
Francolite	Carbonate fluorapatite	65
Dehrnite	Na-containing carbonate apatite	55
Lewistonite	K-containing carbonate apatite	55
Bobierrite*	$Mg_3(PO_4)_2 \cdot 8H_2O$	25
Vivianite*	$Fe_3(PO_4)_2 \cdot 8H_2O$	98, 21
Newberyite*	$MgHPO_4 \cdot 3H_2O$	89, 50
Struvite*	$NH_4MgPO_4 \cdot 6H_2O$	89, 50, 5a
Stercorite*	$NH_4NaHPO_4 \cdot 4H_2O$	Reported from guano deposits; see *Dana's System*[b]
Hannayite*	$Mg_3(NH_4)_2H_4(PO_4)_4 \cdot 8H_2O$	50, 25
Dittmarite	$(NH_4)MgPO_4 \cdot H_2O$	73
Phosphorrösslerite	$MgHPO_4 \cdot 7H_2O$	30, 26
Schertelite*	$Mg(NH_4)_2H_2(PO_4)_2 \cdot 4H_2O$	25
Collinsite*	$Ca_2(Mg, Fe)(PO_4)_2 \cdot 2H_2O$	98
Messelite*[c]	$Ca_2Fe(PO_4)_2 \cdot 2H_2O$	8
Crandallite*	$CaAl_3(PO_4)_2(OH)_5 \cdot H_2O$	78
Millisite*	$(Na, K)CaAl_6(PO_4)_4(OH)_9 \cdot 3H_2O$	78
Wardite*	Similar to millisite	17, 78, 35
Taranakite*	Highly hydrated phosphate of (Al, Fe) with (K, Na, NH_4, Ca)	74, 50
Montgomeryite*	$Ca_2Al_2(PO_4)_3(OH) \cdot 7H_2O$	71
Overite*	$Ca_3Al_8(PO_4)_8(OH)_6 \cdot 15H_2O$	44
Brazilianite*	$NaAl_3(PO_4)_2(OH)_4$	94
Minyulite*	$KAl_2(PO_4)_2(OH, F) \cdot 4H_2O$	32
Gordonite*	$MgAl_2(PO_4)_2(OH)_2 \cdot 8H_2O$	45, 77
Davisonite[d]	$Ca_3Al(PO_4)_2(OH)_3 \cdot H_2O$ (?)	43
Englishite*	Ca, K, Al hydrated phosphate	43
Lehiite	Highly hydrated phosphate of Al, Ca, and (Na, K)	43
Coeruleolactite	$CaAl_6(PO_4)_4(OH)_8 \cdot 4H_2O$	20
Kingite*	$Al_3(PO_4)_2(OH, F)_3 \cdot 9H_2O$	40, 75
Wavellite*	$Al_3(PO_4)_2(OH)_3 \cdot 5H_2O$	29, 21
Variscite*	$AlPO_4 \cdot 2H_2O$ (orthorh.)	56, 9
Metavariscite*	$AlPO_4 \cdot 2H_2O$ (monocl.)	56, 9
Barrandite*	$(Al, Fe)PO_4 \cdot 2H_2O$ (orthorh.)	56, 9, 1
Clinobarrandite	$(Al, Fe)PO_4 \cdot 2H_2O$ (monocl.)	56, 9, 1
Strengite*	$FePO_4 \cdot 2H_2O$ (orthorh.)	56, 9, 1, 3, 39
Phosphosiderite*	$FePO_4 \cdot 2H_2O$ (monocl.)	56, 9, 1, 3, 39

TABLE 1 (continued)

Mineral[a]	Chemical Composition	Ref.
Vauxite*	Hydrated ferrophosphate of Al	See *Dana's System*[b]
Metavauxite*	Hydrated ferrophosphate of Al	See *Dana's System*[b]
Paravauxite*	Hydrated ferrophosphate of Al	See *Dana's System*,[b] 77
Beraunite*	$Fe_3(PO_4)_2(OH)_3 \cdot 2\frac{1}{2}H_2O$ (?)	19, 21, 72
Cacoxenite*	$Fe_4(PO_4)_3(OH)_3 \cdot 12H_2O$ (?)	39, 92
Tinticite*	$Fe_3(PO_4)_2(OH)_3 \cdot 3\frac{1}{2}H_2O$	87
Sigloite*	$(Fe'', Fe''')Al_2(PO_4)_2(O, OH) \cdot 8H_2O$	36
Barbosalite*	$FeFe_2(PO_4)_2(OH)_2$	49
Laubmannite*	Ferric hydroxyphosphate	72
Dufrenite*	Basic ferroferriphosphate	27, 72
Rockbridgeite*	Basic ferroferriphosphate	39, 47, 72
Cyrilovite*[e]	$NaFe_3(PO_4)_2(OH)_4 \cdot 2H_2O$	76, 48, 88
Leucophosphite*	Hydrated ferriphosphate of K	3, 85
Anapaite*	$Ca_2Fe(PO_4)_2 \cdot 4H_2O$	98, 21
Mitridatite*	Hydrated ferriphosphate of Ca	10
Calcioferrite	$Ca_2Fe_2(PO_4)_3(OH) \cdot 7H_2O$	71
Efremovite	Hydrated ferriphosphate of Ca	14
Ardealite	$CaHPO_4 \cdot CaSO_4 \cdot 4H_2O$	83
Bradleyite*	$Na_3Mg(PO_4)(CO_3)$	16
Viséite*	Hydrated aluminum phosphate silicate of Na and Ca	58
Kribergite	Hydrated phosphate sulfate of Al	13
Destinezite*[f]	Hydrated ferric phosphate sulfate	4
Diadochite*[f]	Similar	4
Egueïite	Hydrated ferriphosphate of Ca	See *Dana's System*
Borickyite	Similar (amor. ?)	See *Dana's System*[b]
Foucherite	Similar (amor.)	See *Dana's System*[b]
Azovskite	Hydrated ferriphosphate	See *Dana's System*[b]
Bolivarite	Hydrated phosphate of Al (amor.)	93, 94a
Vashegyite*	Hydrated phosphate of Al	41, 21
Evansite	Hydrated phosphate of Al (amor.)	41
Delvauxite	Hydrated ferriphosphate (amor.)	4
Richellite*[g]	Hydrated ferriphosphate of Ca	66, 92

[a] Starred minerals (*) are included in *Inorganic Index to the Powder Diffraction File*, American Society for Testing and Materials, Philadelphia, 1969, 812 pages.
[b] "See *Dana's System*" refers to C. Palache, H. Berman, and C. Frondel, *The System of Mineralogy*, 7th ed., Vol. 2, Wiley, New York, 1951, 1124 pages. However, another important reference is H. Strunz, *Mineralogische Tabellen*, 5th ed., Akademische Verlagsgesellschaft, Geest and Portig, Leipzig, 1970, 621 pages.
[c] Powder pattern indexed as neomesselite.
[d] Originally known as dennisonite.
[e] Powder pattern indexed as avelinoite.
[f] Powder patterns are given for destinezite (diadochite) and for diadochite.
[g] Powder pattern was obtained after heating virtually amorphous material.

is by Lehr *et al.* (46). There may be synonyms in Table 1; descriptions of some of these substances are quite inadequate by present-day standards. Several "amorphous" (amor.) aluminum and iron phosphates, such as bolivarite (93), are considered, although they have not been recognized as directly connected with biochemical processes.

Pathological Mineralizations

Although hopeite [$Zn_3(PO_4)_2 \cdot 4H_2O$] has been omitted (Table 1), it has been reported (79) as a component of a urolith. Nevertheless, even under pathological conditions one must suspect that ingestion of excessive amounts of zinc may have occurred in conjunction with a particular patient because of the unusual etiology.

Besides the oxalate "stones," human renal "stones" may be composed of phosphates, either in major or minor amounts. Lonsdale (53) and her collaborators have found these stones to be mainly calcium and magnesium–ammonium phosphates. Some of these phosphatic minerals seem to have precursors of lower stability. Octacalcium phosphate [$Ca_8H_2(PO_4)_6 \cdot 5H_2O$], for example, does not occur naturally except as calculi, and is said to hydrolyze readily to produce an apatite.

Lagergren (42) and Prien and Frondel (80) do not report the occurrence of newberyite ($MgHPO_4 \cdot 3H_2O$) in renal calculi, and Lonsdale (53) suggests that it may be a secondary, decomposition product of struvite ($MgNH_4PO_4 \cdot 6H_2O$). Other recognized phosphates that occur are apatite, whitlockite [β-$Ca_3(PO_4)_2$], and brushite ($CaHPO_4 \cdot 2H_2O$). Although apatite varieties have been reported as being free from carbonate, all such samples have indicated a positive test when examined by the writer's methods. The principal precaution in performing this test consists of keeping the pH sufficiently low to prevent the carbon dioxide from passing from the solid substance with immediate dissolution in the liquid. It is highly probable that carbonates per se do not occur in urinary calculi and this has given rise to a general comment (42): "Moreover, carbonate was never a component of calculi which did not contain apatite."

The conditions which contribute to precipitation of urinary calculi in vivo are poorly understood, but Lonsdale (53) tends to relate their formation essentially to inorganic chemical factors, such as relative solubility, pH, and temperature, at the same time pointing out that the solubility of calcium oxalate is affected comparatively little by such changes. Therefore, Lonsdale (53) suggests that either weddellite or whewellite serves as a nucleus, and precipitation of phosphates takes place through epitaxic relations.

All of these relations surely must be recognized, but the physiologic environment deserves equal consideration despite its extreme complexity.

As will be mentioned in conjunction with in vitro production of dental calculus, certain biochemical substances can act as catalysts or inhibitors, and certain bacteria can induce precipitation of apatite, according to Ennever (15) and Rizzo *et al.* (81). Indeed, the presence of pyrophosphate ions in urine is believed by Fleisch *et al.* (22) to be a significant inhibitory factor with respect to urolithiasis.

According to Herman *et al.* (33) fluorine can occur in uroliths, presumably in the apatitic phase which has a well-known affinity for this element.

"Stones" that occur in animals include the aforementioned phosphate minerals, and in addition an enterolith was reported (38) to contain a substance with a composition between bobierrite [$Mg_3(PO_4)_2 \cdot 8H_2O$] and vivianite [$Fe_3(PO_4)_2 \cdot 8H_2O$]. Apatitic corneal calcification has been induced in rabbits according to Fine *et al.* (18), and some types of deposition in human cornea are supposedly similar.

Hard substances may be deposited in various tissues and organs of humans, and the major component is frequently a carbonate apatite. Capen *et al.* (7) found that experimental hypervitaminosis D produced cardiovascular mineralization in cattle which was extensive after 3 weeks' feeding of 30 million U.S.P. units of vitamin D daily. The inorganic deposition of the arteries was carbonate hydroxyapatite. Some of the inorganic depositions within the lungs—associated with histoplasmosis, for example—are apatitic also. Indeed, with the exception of statoliths (or otaconia), which are usually calcareous, and the oxalate uroliths, the various inorganic hard tissues that form (normally or pathologically) within vertebrates are predominately phosphatic, and apatite is the most common mineral. In connection with humans the report by Frondel and Prien (28) is of interest.

Dental calculus has been reported to contain inorganic substances other than apatite, including brushite, whitlockite, and possibly monetite (96), but Schroeder and Bambauer (84) indicate that brushite and octacalcium phosphate occur most frequently in younger specimens. This observation is consistent with the supposition that whitlockite and apatite are more insoluble and more stable in physiologic environments that are normal or approximately normal.

The question of the mineralogy of dental calculus is somewhat complicated, however, by the fact that some of the widely used dentifrices contain solid calcium phosphates that could readily become incorporated within the subgingival plaque. Stones from salivary glands and ducts, while less likely to be contaminated by the constituents of toothpastes, are not nearly as common as depositions on the surfaces of teeth and have not been extensively investigated. The few that have been examined by the writer (using x-ray diffraction) have proved to be essentially apatite. Such a stone is illustrated in vivo as Fig. 1.

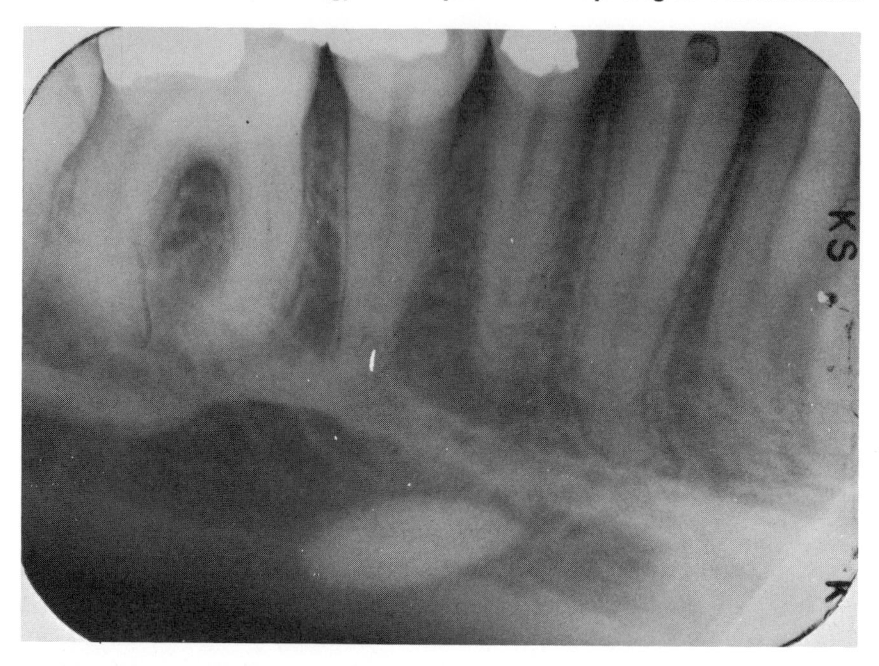

Figure 1. Concretionary body (sialolith) within the submandibular gland (photo by Gus C. Pappas, D.D.S.).

Numerous attempts to produce in vitro oral calculus have been successful in yielding deposits from saliva on various types of surfaces. One of the most interesting outgrowths of such experimentation was the discovery that a common enzyme (carbonic anhydrase) had a pronounced catalytic effect not only for boiled saliva but also for synthetic solutions simulating the composition with respect to salinity and calcium and phosphate contents. McConnell and co-workers (69) also found that sulfanilamide inhibited the reaction and no deposition took place when the enzymatic activity was thus counteracted (Fig. 2). These experiments lead to the conclusion that the solubility of the carbonate hydroxyapatite is closely related to the availability of carbonate (or bicarbonate) ions and is not solely a function of the calcium and phosphate ionic concentrations.

From the theoretical viewpoint these experiments represented a significant advance because a connecting link between the inorganic chemistry and the physiological chemistry of the mineralization process was demonstrated.

Normal Mineralizations

With the exception of statoliths, the normal hard tissues of vertebrates are now recognized as containing, in addition to organic matter, only one

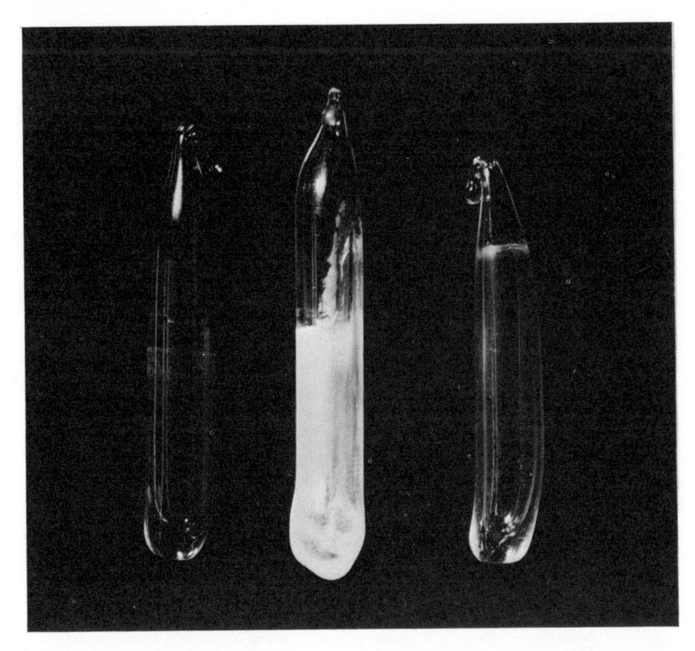

Figure 2. In vitro calculus experiments with glass plummets: control (left); deposit obtained with enzymatic catalyst (carbonic anhydrase) added (center); and with both catalyst and inhibitory agent (sulfanilamide) added (right). The deposit is crystallochemically virtually identical with bone mineral (69). Reprinted from McConnell *et al.* (69) by permission. Copyright 1961 by the American Association for the Advancement of Science.

inorganic phosphatic mineral, dahllite (carbonate hydroxyapatite; see Fig. 3). For dental enamel, which contains comparatively small amounts of organic matter, fairly good x-ray and electron-diffraction patterns can be obtained. Bone and dentin produce somewhat more diffuse x-ray diffraction patterns (Fig. 4), and one of the principal reasons is undoubtedly related to the smaller sizes of the individual crystallites. Taking into account the size of the crystallites and the presence of larger amounts of organic matter permits one to conclude that the inorganic components are qualitatively the same and that all three substances are dahllite.

Any authoritative discourse on bone [such as the work of McLean and Urist (70)] will reveal that the cytology or histochemistry of bone and tooth formation is not within our present knowledge. Attempts to calculate solubility product constants are misleading generally because they do not take into account complexing of the several ions present and do not involve an accurate expression of the composition of the solid phase (bone mineral). In other words, such calculations disregard carbonate ions that are present in both the solid and the solution, as well as the presence of organic catalysts

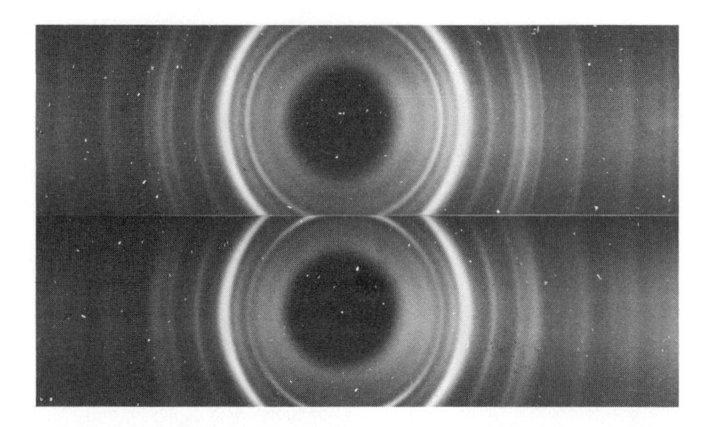

Figure 3. X-Ray diffraction patterns of: (*a*) deposit obtained on center plummet (Fig. 2), and (*b*) crystalline substance within cartilage of a shark. The crystallochemical nature of the substances is virtually identical; both are carbonate apatites (69). Reprinted from McConnell *et al.* (69) by permission. Copyright 1961 by the American Association for the Advancement of Science.

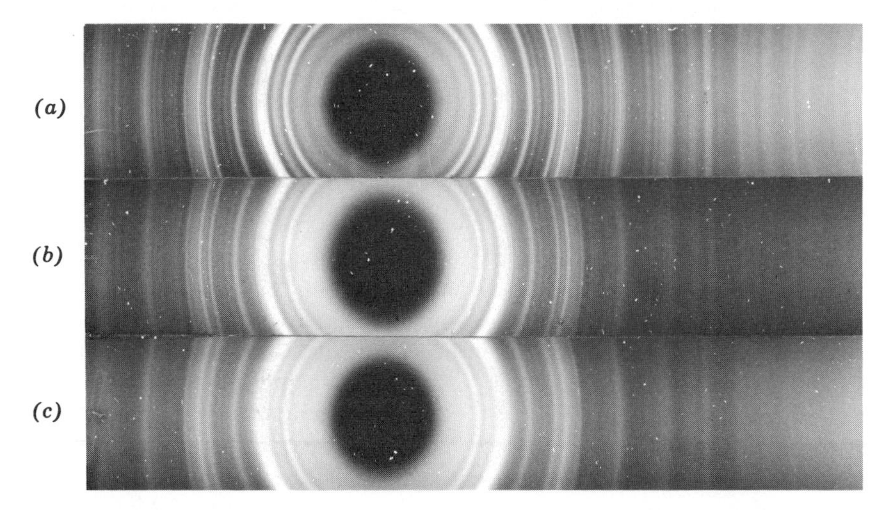

Figure 4. Comparison of x-ray diffraction patterns of (*a*) synthetic carbonate hydroxy-apatite (prepared by R. Klement), (*b*) marginal portion of the shell of *Lingula*, and (*c*) human dentin. The resolution of the diffraction decreases from top to bottom. Reprinted by permission of the Geological Society of America (65).

and inhibitors, and are self-defeating to the extent that they attempt to eliminate those physiochemical factors wherein the secret of bone formation must be sought.

It should be recalled that the work of McConnell *et al.* (69) demonstrated that a synthetic solution (simulating a physiologic solution with respect to inorganic ions) would produce a precipitate upon addition of a few parts per million of crystalline carbonic anhydrase and that this precipitate was crystallochemically bone mineral. That a phospholipid may be involved in the mineralization process is implied by Urist (91) and further supported by the work of Takazoe *et al.* (90) in connection with *Bacterionema matruchotii.*

That carbonic anhydrase may be related to the production of the francolite shell of *Lingula* is suggested by the work of Hammen (31), but an unknown factor, of course, is the reason for the production of a phosphatic shell rather than a calcareous one. Besides being an accumulator of phosphorus, *Lingula* is an accumulator of fluorine (65), and by inference this is true of some inarticulate brachiopods since Cambrian time.

Besides the endoskeleton, the scales of bony fish are apatitic. Another deposition of apatitic character occurs in the human pineal body; it is called brain sand (*corpora arenacea*) and its quantity seems to increase with age. The function of the "sand" is unknown, so the question arises whether its formation is normal or a gerontological manifestation.

BIOLOGICAL APATITES

Probably the analyses of bone by Armstrong and Singer (2) can be accepted as standards; for dry, fat-free, bovine, cortical bone they obtained:

Ca	Mg	Na	K	Sr	P	F	Cl	CO_2	Sum
26.70	0.44	0.73	0.06	0.04	12.47	0.07	0.08	3.48	44.06

When citric acid and nitrogen are added the summation is slightly below 50%. These percentages are for the elements (except CO_2 and citric acid), and can be converted to weight percentages of oxides, thus:

CaO	MgO	Na_2O	K_2O	SrO	P_2O_5	CO_2	F	Cl	Sum*
37.56	0.72	0.99	0.07	0.04	28.58	3.48	0.07	0.08	71.54

Adding the nitrogen and citric acid yields 77.32% of the material as accounted for. The remainder (>22%) is undoubtedly organic carbon, chemically combined water, and such moisture as would remain present in the "dry" bone. There seems to be little probability of ascertaining the amount of chemically combined water directly from bone in view of these uncertainties.

* The sum has been corrected for F and Cl by −0.05.

However, it does seem feasible to measure the amounts of chemically combined water directly in samples of dahllite, francolite, and some well-crystallized synthetic carbonate apatites. Particularly for francolites, the sum of (OH + F) usually exceeds the two atoms called for at $(0, 0, \frac{1}{4})$ within the apatite structure—sometimes by a most significant factor. A very simple explanation would be that the substitution for phosphate groups by carbonate groups is accompanied by the substitution of a fluorine or hydroxyl ion for an oxygen; that is,

$$PO_4 \rightarrow CO_3F \quad (\text{or } CO_3OH)$$

The persistent fact, however, is that *there is no correlation whatever between carbon and* (*F* + *OH*), as is clearly indicated in Table 2, and as was pointed out several years ago (60).

Before continuing with a more elaborate discussion of crystal-chemical relationships of carbonate apatites, another analytical difficulty should be mentioned in connection with bone. Besides nitrogen (4.92%), organic carbon, and moisture, it must be assumed that the organic components also contain phosphate and possibly a lesser amount of calcium ions in chemical combination. During the analytical procedures these constituents become dissociated and are recorded as though they were inorganic in

TABLE 2 Significant Ratios of Hydrated Francolites Compared with Minor Constituents

Example	Minor Constituents (percent by weight)					Ratios[a]		
	H_2O	CO_2	F	Cl	SO_3	Ca′/P′	Ca/(P′ + C)	(F + OH + Cl − 2)/C
F-1[b]	1.32	1.89	4.22	0.10	0.83	1.79	1.69	3.5
F-2	0.91	1.32	4.15	N.D.	0.18	1.73	1.64	4.0
F-3	1.29	2.58	4.13	0.02	0.54	1.82	1.64	2.5
F-4	2.97	3.20	4.12	N.D.	1.57	1.87	1.64	3.4
F-5	0.39	2.05	3.86	N.D.	0.01	1.78	1.64	1.6
F-6	3.57	0.52	2.82	N.D.	0.05	1.71	1.67	21.2
F-7	1.76	1.52	3.60	Trace	0.28	1.75	1.65	3.2
F-8[c]	0.63	2.70	3.89	Trace	—	1.78	1.61	0.9
F-9[d]	0.47	3.40	3.71	None	None	1.80	1.57	0.6
F-10[e]	0.82	3.00	1.71	0.19	Excl.	1.87	1.67	−0.1

[a] Ca′ = Ca + Na + Mg + ⋯ (on the basis of atoms per unit cell); P′ = P + S. For $Ca_{10}(PO_4)_6F_2$, Ca′/P = $\frac{10}{6}$ = 1.67.

[b] F-1 to F-7 are from Whippo and Murowchick (97), their numbers 1 to 7 of Table 4, Recalculated to Impurity Free Basis (including leaching with triammonium citrate to remove $CaCO_3$).

[c] F-8 is from Brophy and Nash (6); francolite from Staffel, Germany.

[d] F-9 is from Sandell *et al.* (82); francolite from Tavistock, Devon.

[e] F-10 is from Hoffman and Trdlička (34); francolite from Kutná Hora. Sulfur was reported as 0.22%, but was excluded from the carbonate apatite as an impurity, as also was 9.85% SiO_2, by the authors. (Thus, the constituents shown above should be increased by about 10% in order to obtain a basis of 100% for the apatite.)

nature, thus altering the true Ca/P ratio of the bone mineral. Whereas all Ca/P ratios for francolites (Table 2) exceed the theoretical value (1.67), probably the best value for bone is about 1.65 to 1.66. Armstrong and Singer (2) found values between 1.58 and 1.70, depending on the method used to remove organic matter. In addition they found differences in the carbonate ion to range from 1.40 to 2.08 meq/g.

The advantages to be gained by analyzing dental enamel are appreciable because of the enamel's significantly lower organic content. Little (51) found human dental enamel to contain

Ca	Mg	Na	K	P	CO_2	N
36.41	0.21	0.70	0.04	17.48	2.24	0.06

(average for two determinations). Little's two Ca/P ratios are 1.692 and 1.67, and she found weight losses to be 2.8 and 2.5% at 500°C, a temperature too low to remove either all of the carbon dioxide or all of the chemically combined water.

Little and Casciani (52) follow the suggestion of McConnell (62) and claim to have recognized three different increments of chemically combined water in dental enamel, the last increment of which is removable only between 900 and 1300°. McConnell (60) suggested that the reason for failure to find any correlation between (OH + F) for francolites and dahllites arises from the possible entry of water into the structure through more than one mechanism. In addition to the usual site at $(0, 0, \frac{1}{4})$ he suggested that H_3O^+ might occur on the simple three-fold axes at $(\frac{1}{3}, \frac{2}{3}, 0)$, thereby replacing a calcium ion. Two other possibilities exist also: (*a*) a phosphorus vacancy to yield a "tetrahedral group of hydroxyls," as is well known (24) for silicon in the hydrogarnets, and (*b*) replacement of the hydroxyl at $(0, 0, \frac{1}{4})$ by either H_2O or H_3O^+ (61, 67). The works of Simpson (86) and Whippo and Murowchick (97) offer tentative confirmation of McConnell's proposals for synthetic apatite and for phosphorite, respectively.

In view of the accumulations of data which qualitatively indicate an excess of (OH + F) in many natural carbonate apatites, McConnell (68) has proposed a semiquantitative relation for predicting the a dimension of the unit cell based on the composition (with respect to H_2O, CO_2, F, and Cl) and has referred to such substances as "hydrated carbonate apatites." The present limitation to the use of his method for predicting the amount of chemically combined water in bone mineral (from $a = 9.44$ Å) stems from the fact that only the total number of protons can be considered, without regard for their mode of entry into the complex structure which McConnell (59, 63) has proposed. Nevertheless, he described models for hydrated apatites (with and without carbonate) which have a range for a up to at least 9.49 A, and with Ca/P values from 1.50 to 2.25.

Although numerous persons have advocated the use of the fluorine content to estimate the age of fossil bones, this method is not recommended for several reasons even when the fluorine content is determined by chemical analysis. Indirect methods, such as attempts to estimate the fluorine content from the *a* periodicity, are unrealistic (64) because the *a* dimension obtained by x-ray diffraction is a function of at least three important variables in addition to fluorine, and by far the most significant component is chlorine with respect to its effect on both *a* and *c* dimensions (68).

In summary, then, it can be said with reasonable certainty that the knowledge of the composition and structure of biological apatites will be resolved in terms of application of sound crystallochemical theory, and not through mystical assumptions about the presence of additional phases, the existence of which has not been proved. In the history of mineralogical chemistry, there probably has never been a topic that has been so energetically attacked by persons so completely unqualified. In this respect "bone mineral" falls in a category that is unique.

REFERENCES

1. Arlidge, E. Z., V. C. Farmer, B. D. Mitchell, and W. A. Mitchell, "Infra-red, X-Ray and Thermal Analysis of Some Aluminium and Ferric Phosphates," *J. Appl. Chem.*, **13,** 17–27 (1963).
2. Armstrong, W. D., and L. Singer, "Composition and Constitution of the Mineral Phase of Bone," *Clin. Orthop.*, **38,** 179–190 (1965).
3. Axelrod, J. M., M. K. Carron, C. Milton, and T. P. Thayer, "Phosphate Mineralization at Bomi Hill and Bambuta, Liberia, West Africa," *Am. Mineral.*, **37,** 883–909 (1952).
4. Bouška, V., E. K. Lazarenko, J. M. Melnik, and E. Slánský, (A contribution to the knowledge of destinezite), *Acta Univ. Carolinae, Geol., Prague*, **1960,** 127–152 (1960); see *Mineral. Abstr.*, **16,** 64 (1963).
5. Braithwaite, C. J. R., "Diagenesis of Phosphatic Carbonate Rocks on Remire, Amirantes, Indian Ocean," *J. Sed. Petrol.*, **38,** 1194–1212 (1968).

5a. Bridge, P. J., "Analyses of Struvite from Skipton, Victoria," *Mineral. Mag.*, **38,** 381–382 (1971).

6. Brophy, G. P., and T. J. Nash, "Compositional, Infrared, and X-Ray Analysis of Fossil Bone," *Am. Mineral.*, **53,** 445–454 (1968).
7. Capen, C. C., C. R. Cole, and J. W. Hibbs, "The Pathology of Hypervitaminosis D in Cattle," *Pathol. Vet.*, **3,** 350–378 (1966).
8. Cech, F., and K. Padera, "Messelit aus den Phosphatnestern im Granit bei Pribyslavice (Bohmen) und das Messelitproblem," *Chem. Erde*, **19,** 436–449 (1958); see *Am. Mineral.*, **44,** 469 (1959).
9. Čech, F., and E. Slánský, "X-Ray Powder Study and Thermal Investigation of $AlPO_4{\cdot}2H_2O$ Minerals," *Acta Univ. Carolinae, Geol., Prague*, **1,** 1–30 (1965).

10. Chukhrov, F. V., V. A. Moleva, and L. P. Ermilova (New data on mitridatite), *Bull. Acad. Sci. USSR, Sér. Géol.*, **8,** 16–26 (1958); see *Mineral. Abstr.*, **14,** 138 (1959).
11. Clarke, F. W., *The Data of Geochemistry*, U.S. Geological Survey Bulletin **770,** 1924, 841 pages.
12. Clarke, F. W., and W. C. Wheeler, *The Inorganic Constituents of Marine Invertebrates*, U.S. Geological Survey Prof. Paper **124,** 1922, 62 pages.
13. Du Rietz, T., "Kribergite, a New Mineral from Kristineberg Mine, Västerbotten," *Geol. För. Förh.*, **67,** 78, 79 (1945); see Fleischer (23).
14. Efremov, N. F., "Calcium Ferriphosphate, a New Mineral of the Phosphate Class," *Mem. Soc. Russe Minéral. Sér. 2*, **65,** 225–232 (1936); see Fleischer (23).
15. Ennever, J., "Microbiologic Calcification," *New York Acad. Sci., Ann.*, **109,** 4–13 (1963).
16. Fahey, J. J., and G. Tunell, "Bradleyite, a New Mineral, Sodium Phosphate–Magnesium Carbonate," *Am. Mineral.*, **26,** 646–650 (1941).
17. Fanfani, L., A. Nunzi, and P. F. Zanazzi, "The Crystal Structure of Wardite," *Mineral. Mag.*, **37,** 598–605 (1970).
18. Fine, B. S., J. W. Berkow, and S. Fine, "Corneal Calcification," *Science*, **162,** 129, 130 (1968).
19. Fischer, E., "Über Beraunit-Eleonorit," *Heidelb. Beitr. Min. Petr.*, **5,** 204–209 (1956).
20. Fischer, E., "Über die Beziehungen zwischen Coeruleolactit, Planerit, Türkis, Alumochalkosiderit, und Chalkosiderit," *Beitr. Mineral. Petr.*, **6,** 182–189 (1958); for planerite see also *Am. Mineral.*, **47,** 1221, 1222 (1962).
21. Fisher, D. J., "Pegmatite Phosphates and Their Problems," *Am. Mineral.*, **43,** 181–207 and (addendum) 609, 610 (1958).
22. Fleisch, H., S. Bisaz, and A. D. Care, "Effect of Orthophosphate on Urinary Pyrophosphate Excretion and the Prevention of Urolithiasis," *Lancet*, **1964,** 1065–1067 (1964).
23. Fleischer, M., "Index of New Mineral Names, Discredited Minerals, and Changes of Mineralogical Nomenclature in Volumes 1–50 of *The American Mineralogist*," *Am. Mineral.*, **51,** 1247–1357 (1966).
24. Foreman, D. W., Jr., "Neutron and X-Ray Diffraction Study of $Ca_3Al_2(O_4D_4)_3$, a garnetoid," *J. Chem. Phys.*, **48,** 3037–3041 (1968).
25. Frazier, A. W., J. R. Lehr, and J. P. Smith, "The Magnesium Phosphates Hannayite, Schertelite, and Bobierrite," *Am. Mineral.*, **48,** 635–641 (1963).
26. Friedrich, O. M., and J. Robitsch, "Phosphorrösslerit ($MgHPO_4 \cdot 7H_2O$) als Mineral aus dem Stüblbau zu Schellgaden," *Zentrlb. Mineral. Abt. A*, **1939,** 142–155 (1939).
27. Frondel, C., "The Dufrenite Problem," *Am. Mineral.*, **34,** 513–540 (1949).
28. Frondel, C., and E. L. Prien, "Deposition of Calcium Phosphates Accompanying Senile Degeneration and Disease," *Science*, **103,** 326 (1946).
29. Gordon, S. G., "Crystallographic Data on Wavellite from Llallagua, Bolivia, and on Cacoxenite from Hellertown, Pennsylvania" (Abstract), *Am. Mineral.*, **35,** 132 (1950).
30. Hägele, G., and F. Machatschki, "Röntgenographische Untersuchungen an $MgHPO_4 \cdot 7H_2O$," *Zentrlb. Mineral. Abt. A*, **1939,** 297–300 (1939).
31. Hammen, C. S., personal communication, 1965; see also *Comp. Biochem. Physiol.*, **5,** 185–191 (1962).

32. Haseman, J. F., J. R. Lehr, and J. P. Smith, "Mineralogical Character of Some Iron and Aluminum Phosphates Containing Potassium and Ammonium," *Proc. Soil Sci. Soc. Am.*, **15**, 76–84 (1951).

33. Herman, J. R., B. Mason, and I. Light, "Fluorine in Urinary Tract Calculi," *J. Urol.*, **80**, 263–268 (1958).

34. Hoffman, V., and Z. Trdlička, "Über ein Carbonat-Apatit (Francolith) von Kutná Hora," *Acta Univ. Carolinae, Geol., Prague*, No. 3, 195–202 (1967).

35. Hurlbut, C. S., Jr., "Wardite from Beryl Mountain, New Hampshire," *Am. Mineral.*, **37**, 849–852 (1952).

36. Hurlbut, C. S., Jr., and R. Honea, "Sigloite, a New Mineral from Llallagua, Bolivia," *Am. Mineral.*, **47**, 1–8 (1962).

37. Hutchinson, G. E., "Survey of Contemporary Knowledge of Biogeochemistry. 3. The Biogeochemistry of Vertebrate Excretion," *Bull. Am. Mus. Nat. Hist.*, **96**, 554 p. (1950).

38. Hutton, C. O., "The Nature of an Enterolith," *N.Z. J. Sci. Technol.*, **26B**, 304–307 (1945).

39. Kahler, E., "Sekundäre Phosphate von der Koralpe Steiermark," *N. Jahrb. Mineral. Abh.*, **98**, 1–13 (1962).

40. Kato, T., "Cell Dimensions of the Hydrated Phosphate, Kingite," *Am. Mineral.*, **55**, 515–517 (1970).

41. Koch, S., and I. Sarudi, "The Hydrated Basic Aluminum Phosphates of Zeleznik (Vashegy), Slovakia (ČSSR)," *Acta Min. Petr. Univ. Szeged.*, **16**, 3–10 (1963); see *Mineral. Abstr.*, **18**, 203 (1967).

42. Lagergren, C., "Biophysical Investigation of Urinary Calculi," *Acta Radiol., Suppl.* **133**, 71 p. (1956).

43. Larsen, E. S., and E. V. Shannon, "The Minerals of the Phosphate Nodules from near Fairfield, Utah," *Am. Mineral.*, **15**, 307–337 (1930).

44. Larsen, E. S., III, "Overite and Montgomeryite: Two New Minerals from Fairfield, Utah," *Am. Mineral.*, **25**, 315–326 (1940).

45. Larsen, E. S., III, "The Mineralogy and Paragenesis of the Variscite Nodules from near Fairfield, Utah," *Am. Mineral.*, **27**, 281–300, 350–372, 441–451 (1942).

46. Lehr, J. R., E. H. Brown, A. W. Frazier, J. P. Smith, and R. D. Thrasher, "Crystallographic Properties of Fertilizer Compounds," *Chem. Eng. Bull.*, **6**, T.V.A., 166 p. (1967).

47. Lindberg, M. L., "Frondelite and the Frondelite-Rockbridgeite Series," *Am. Mineral.*, **34**, 541–549 (1949).

48. Lindberg, M. L., "Relationship of the Minerals Avelinoite, Cyrilovite, and Wardite," *Am. Mineral.*, **42**, 204–213 (1957).

49. Lindberg, M. L., and W. T. Pecora, "Tavorite and Barbosalite, Two New Phosphate Minerals from Minas Gerais, Brazil," *Am. Mineral.*, **40**, 952–966 (1955).

50. Lindsay, W. L., A. W. Frazier, and H. F. Stephenson, "Identification of Reaction Products from Phosphate Fertilizers in Soils," *Proc. Soil Sci. Soc. Am.*, **26**, 446–452 (1962).

51. Little, M. F., "Studies on the Inorganic Carbon Dioxide Component of Human Enamel," *J. Dent. Res.*, **40**, 903–914 (1961).

52. Little, M. F., and F. S. Casciani, "The Nature of Water in Sound Human Enamel. A Preliminary Study," *Arch. Oral Biol.*, **11,** 565–571 (1966).

53. Lonsdale, K., "Human Stones," *Science*, **159,** 1199–1207 (1968).

54. Lowenstam, H. A., "Biologic Problems Relating to the Composition and Diagenesis of Sediments," in *The Earth Sciences*, T. W. Donnelly (Ed.), Rice University, Semicentennial Publications, 1963, pp. 137–195.

55. McConnell, D., "A Structural Investigation of the Isomorphism of the Apatite Group," *Am. Mineral.*, **23,** 1–19 (1938).

56. McConnell, D., "Clinobarrandite and the Isodimorphous Series, Variscite-Metavariscite," *Am. Mineral.*, **25,** 719–725 (1940).

57. McConnell, D., "The Petrography of Rock Phosphates," *J. Geol.*, **58,** 16–23 (1950).

58. McConnell, D., "Viséite, a Zeolite with the Analcime Structure and Containing Linked SiO_4, PO_4 and H_xO_4 Groups," *Am. Mineral.*, **37,** 609–617 (1952).

59. McConnell, D., "The Problem of the Carbonate Apatites. IV. Structural Substitutions Involving CO_3 and OH," *Bull. Soc. Fr. Minéral.*, **75,** 428–445 (1952).

60. McConnell, D., "The Crystalchemistry of Dahllite," *Am. Mineral.*, **45,** 209–216 (1960).

61. McConnell, D., "The Stoichiometry of Hydroxyapatite," *Naturwiss.*, **47,** 227 (1960).

62. McConnell, D., "Recent Advances in the Investigation of the Crystal Chemistry of Dental Enamel," *Arch. Oral Biol.*, **3,** 28–34 (1960).

63. McConnell, D., "The Crystal Structure of Bone," *Clin. Orthop.*, **23,** 253–268 (1962).

64. McConnell, D., "Dating of Fossil Bones by the Fluorine Method," *Science*, **136,** 241–244 (1962).

65. McConnell, D., "Inorganic Constituents in the Shell of the Living Brachiopod *Lingula*," *Geol. Soc. Am. Bull.*, **74,** 363–364 (1963).

66. McConnell, D., "Thermocrystallization of Richellite to Produce a Lazulite Structure (Calcium Lipscombite)," *Am. Mineral.*, **48,** 300–307 (1963).

67. McConnell, D., "Deficiency of Phosphate Ions in Apatite," *Naturwiss.*, **52,** 183 (1965).

68. McConnell, D., "Crystal Chemistry of Bone Mineral: Hydrated Carbonate Apatites," *Am. Mineral.*, **55,** 1659–1669 (1970).

69. McConnell, D., W. J. Frajola, and D. W. Deamer, "Relation between the Inorganic Chemistry and Biochemistry of Bone Mineralization," *Science*, **133,** 281–282 (1961).

70. McLean, F. C., and M. R. Urist, *Bone: An Introduction to the Physiology of Skeletal Tissue*, 2nd ed., University of Chicago Press, Chicago, Illinois, 1961, 261 pages.

71. Mead, C. W., and M. E. Mrose, *Solving Problems in Phosphate Mineralogy with the Electron Probe*, U.S. Geological Survey Prof. Paper **600-D**, 1968, pp. D204–D206.

72. Moore, P. B., "Crystal Chemistry of the Basic Iron Phosphates," *Am. Mineral.*, **55,** 135–169 (1970).

73. Mrose, M. E., "New Mineral Data for Hydrated Phosphates and Sulfates," *U.S. Geol. Surv., Prof. Paper* **750**-A, A115 (1971).

74. Murray, J. W., and R. V. Dietrich, "Brushite and Taranakite from Pig Hole Cave, Giles County, Virginia," *Am. Mineral.*, **41,** 616–626 (1956).

75. Norrish, K., L. E. R. Rogers, and R. E. Shapter, "Kingite, a New Hydrated Aluminum Phosphate Mineral from Robertstown, South Australia," *Mineral. Mag.*, **31,** 351–357 (1957).

76. Novotný, M., and J. Stanek, "Novy mineral fosforecnan cyrilovit," *Acta Acad. Sci. Moravo-Silesiacae*, **25**, 325–336 (1953); see *Mineral Abstr.*, **12**, 512 (1952).
77. Nuffield, E. W., "Relation of Paravauxite and Gordonite" (Abstract), *Am. Mineral.*, **32**, 205 (1947).
78. Owens, J. P., Z. S. Altschuler, and R. Berman, "Millisite in Phosphorite from Homeland, Florida," *Am. Mineral.*, **45**, 547–561 (1960).
79. Parsons, J., "Zinc Phosphate Identified as a Constituent of Urinary Calculi," *Science*, **118**, 217 (1953).
80. Prien, E. L., and C. Frondel, "Studies in Urolithiasis: I. The Composition of Urinary Calculi," *J. Urol.*, **57**, 949–991 (1947).
81. Rizzo, A. A., G. R. Marlin, D. B. Scott, and S. E. Mergenhagen, "Mineralization of Bacteria," *Science*, **135**, 439–441 (1962).
82. Sandell, E. B., M. H. Hey, and D. McConnell, "The Composition of Francolite," *Mineral. Mag.*, **25**, 395–401 (1939).
83. Schadler, J., "Ardealit, ein neues Mineral, $CaHPO_4 \cdot CaSO_4 \cdot 4H_2O$," *Zentrb. Mineral. Abt. A*, **1931**, 40–41 (1931).
84. Schroeder, H. E., and H. U. Bambauer, "Stages of Calcium Phosphate Crystallization during Calculus Formation," *Arch. Oral Biol.*, **11**, 1–14 (1966).
85. Simmons, G. C., "Leucophosphorite, a New Occurrence in the Quadrilátero Ferrífero, Minas Gerais, Brazil," *Am. Mineral.*, **49**, 377–386 (1964).
86. Simpson, D. R., "Substitution in Apatite: I. Potassium-Bearing Apatite," *Am. Mineral.*, **53**, 432–444 (1968).
87. Stringham, B., "Tinticite, a New Mineral from Utah," *Am. Mineral.*, **31**, 395–400 (1946).
88. Strunz, H., "Identität von Avelinoit und Cyrilovit," *N. Jahrb. Mineral. Monatsh.*, **1956**, 187–189 (1956).
89. Sutor, D. J., and S. Scheidt, "Identification Standards for Human Urinary Calculus Components, Using Crystallographic Methods," *Br. J. Urol.*, **40**, 22–28 (1968).
90. Takazoe, I., J. Vogel, and J. Ennever, "Calcium Hydroxyapatite Nucleation by Lipid Extract of *Bacterionema matruchotii*," *J. Dent. Res.*, **49**, 395–398 (1970).
91. Urist, M. R., "Origins of Current Ideas about Calcification," *Clin. Orthop.*, **44**, 13-39 (1966).
92. Van Tassel, R., "Strengite, phosphosidérite, cacoxénite et apatite fibroradiée de Richelle," *Bull. Soc. Belge Géol.*, **68**, 360–368 (1959).
93. Van Tassel, R., "Bolivarite Restudied," *Mineral. Mag.*, **32**, 419–420 (1960).
94. Van Tassel, R., "La brasilianite de Buranga, Rwanda occidental," *Bull. Acad. Roy. Sci. d'Outre-Mer*, **7**, 404–409 (1961).
94a. Van Hambeke, L., "The Uranium-Bearing Mineral Bolivarite: New Data and a Second Occurrence," *Mineral. Mag.*, **38**, 418–423 (1971).
95. Vinogradov, A. P., *The Elementary Chemical Composition of Marine Organisms*, Sears Foundation Marine Research, Yale University Memoir 2, New Haven, Connecticut, 1953, 647 pages.
96. Westerden, E. M., and K. Little, "Some Observations on the Composition of Calculus" (Abstract), *J. Dent. Res.*, **37**, 749–750 (1958).
97. Whippo, R. E., and B. L. Murowchick, "The Crystal Chemistry of Some Sedimentary Apatites," *Trans. Am. Inst. Min. Eng.*, **238**, 257–263 (1967).
98. Wolfe, C. W., "Classification of Minerals of the Type $A_3(XO_4)_2 \cdot nH_2O$," *Am. Mineral.*, **25**, 738–753, 787–809 (1940).

26

Phosphorus Nutrition of Algae

JOSEPH C. O'KELLEY

University of Alabama,
Biology Department, University, Alabama

Phosphorus has been known for many years to be a required nutrient element for algae (1–5). The extensive literature dealing specifically with the uptake and metabolism of phosphorus by algae was discussed comprehensively by Kuhl (6) in 1962. As in all other living organisms, in the algae compounds containing phosphorus play important roles in many phases of metabolism and particularly in energy-transfer reactions. In the physiology of photosynthesizing organisms, including the algae, this element receives additional attention because phosphorylated compounds participate in photosynthesis.

METABOLISM OF PHOSPHORUS

Phosphorus metabolism in the algae shows one feature, however, not generally seen in the other green plants. This is the capacity of a number of commonly cultured algal species to produce and store large quantities of polyphosphates (7). This polyphosphate in *Chlorella pyrenoidosa* has been separated into an acid-soluble fraction, designated polyphosphate A, and three acid-insoluble polyphosphates, B, C, and D (8). Fraction A is highly condensed and lacks tri- or tetrametaphosphate, pyrophosphate, or tripolyphosphate (9). In *Scenedesmus quadricaudata* the polyphosphate that accumulates is mostly acid insoluble; this and other phosphate fractions follow a diurnal course (10, 11). In synchronous cultures polyphosphate C appeared to be a reservoir of phosphorus for polyphosphate A and for nucleotide-labile phosphate. Phosphate for DNA synthesis in light originated from polyphosphate A; in darkness polyphosphate B specifically decreased with a concomitant increase of phosphate in RNA (12, 13). Voletin granules in algae contain mainly acid-soluble polyphosphate and RNA, and chloroplasts also contain two forms of acid-soluble polyphosphate as well as DNA (14). Polyphosphate–RNA complexes have also been isolated from synchronously grown *Chlorella* and from blue-green algae (15, 16).

NUTRIENT PHOSPHORUS SOURCES

The sources of phosphate that can be used by algae are varied. The primary source in nature is recognized to be inorganic orthophosphate, but other sources can also be of importance. Galloway and Krauss (17) determined that external polyphosphate with chain lengths up to 53 will support growth of *Chlorella* at the same rate as will potassium phosphate. An adaptive enzyme associated with the cell wall hydrolyzes pyrophosphate bonds, and the orthophosphate released appears to be absorbed as such. *Chlorella* cells can also deacylate phospholipids on their surface and absorb the released phosphate (18). *Chara* can use the phosphate of glycerophosphate (19). Kuenzler and Perras (20, 21) have shown that some species of marine algae, but not others, have phosphate-repressible surface alkaline phosphatases that will split glucose-6-phosphate. The glucose released remains outside the cell, but the phosphate is assimilated. Marine algae can also release and absorb inorganic orthophosphate from AMP (adenosine monophosphate) and from α-glycerophosphate.

PHOSPHATE UPTAKE MECHANISMS

The mechanism of phosphate uptake is at least in part active. Incorporation by *Ankistrodesmus braunii* has been shown to be dependent on temperature both in light and in darkness (22). Amytal, DNP (2,4-dinitrophenol), and DCMU [3-(3,4-dichlorophenyl)-1,1-dimethyl urea] affected light-dependent phosphorylation by intact *Ankistrodesmus* cells in a different way than they affected dark phosphorylation; while oxidative phosphorylation appeared to be inhibited by light, preillumination stimulated subsequent dark phosphorylation. Light increased phosphate uptake in phosphorus-deficient *Scenedesmus* in the absence of CO_2, and CO_2 inhibited phosphate absorption either in darkness or in light (23). Uptake of ^{32}P by *Anaebaena cylindrica* increased in light in a nitrogen atmosphere (24). Phosphate-starved *Euglena* cells show an uptake that is saturable, competitively inhibited by arsenate, sensitive to DNP, and appears to result in accumulation against both concentration and electrical gradients (25). There is an induced acid phosphatase in *Euglena gracilis*, located in specific helical regions of the cell surface, that may be an agent of phosphate uptake

(26). While phosphate uptake in *Nitella* has been shown to be light enhanced and ATP dependent, characteristic of active transport systems (6), a study of its close relative *Chara* failed to produce any evidence for polar transport within the organism (19). Phosphate and other ions interact in ways that affect uptake. In *Scenedesmus*, phosphate is necessary for Na uptake (27) and affects the uptake of sulfate (28). Conversely, sodium stimulated phosphate absorption in *Ankistrodesmus* in light or darkness when small concentrations of phosphate were supplied; with Na present there was a marked increase in TCA(trichloroacetic acid)-soluble organophosphate (29). Chloride absorption, in contrast, is inhibited by adding phosphate to deficient *Scenedesmus* cells (30).

PHOSPHORUS LEVELS REQUIRED FOR GROWTH

The level of phosphate required for optimal growth differs among species of algae. For example, preliminary culture experiments with *Pteromonas varians* indicated that a high ratio of phosphate to other nutrients is required (31). Growth of *Cyclotella nana* and *Thallasosira fluviatilis* in the laboratory with artificial seawater, supplemented with nutrients and vitamin B_{12}, was limited by concentrations of 100 μg, or less, of phosphate per liter of medium (32). In similar laboratory experiments the phosphate requirement on a cellular basis for *Cylindrotheca closterium* was 0.0135, and for *Cyclotella nana* was 0.0129×10^{-15} g atoms P/μ^3 of cell volume; the algae for these determinations were cultured at 15.5°C and exposed to 5380 lux on a light-dark cycle of 19 to 5 hr (33). At the other extreme, *Chara* spp. have been shown to grow best in low concentrations of phosphate, below 20 μg/l (34, 35).

Azad and Borchardt (36) have studied phosphate uptake in relation to growth and have identified several phosphate fractions in cells of *Chlorella* and *Scenedesmus*. One was designated "CCP," the critical concentration of phosphorus, or that concentration required for maximal growth under the prevailing cultural conditions. Any amount of phosphorus absorbed by these cells above the "CCP" implied a storage of unused nutrient; this excess was designated "LUP," the luxury uptake phosphorus. Azad and Borchardt found that *Scenedesmus* and *Chlorella* cells could accumulate as much as 10% phosphate by dry weight, while the "CCP" was only about 3% of the dry weight. This excess phosphate could be utilized for growth when the medium was made phosphate deficient. One thing has been made clear from these and related studies; that is, that phosphate can be accumulated in algae and used for growth when external phosphate becomes scarce. Marine algae, and possibly freshwater species also, can excrete soluble forms of organophosphorus, or DOP (dissolved organophosphorus).

It has been shown in laboratory cultures of *Cyclotella cryptica*, *Thallassosira fluviatilis*, *Dunaliella tertiolecta*, and *Synechococcus* sp. that more than 20% of the total phosphorus in a system may be excreted in this form (37). Furthermore, this DOP can be readily reassimilated either by the species that produced it or by certain other species. In these species, the ability to utilize the DOP is related to the presence of an alkaline phosphatase.

PHOSPHORUS AND ALGAL BLOOMS

A great amount of attention is currently being directed to the role of phosphorus in eutrophication and the development of algal blooms. Determining the amount of phosphate available for growth on the basis of soluble phosphate alone does not take into account the presence of phosphorus in particulate detritus or absorbed on particles of silt or clay, and both are available for algal growth. Fitzgerald (38) has shown that a variety of relatively insoluble forms of phosphorus, including animal teeth, can serve as phosphate sources for algal growth.

According to Talling (39) it was widely accepted for a number of years that the growth of phytoplankton in many bodies of freshwater tends to be limited by the supply of inorganic phosphate. However Rodhe (40) has estimated that as little as 20 μg P/l in natural lake waters may sustain maximum phytoplankton growth, and at least one algal bloom has been reported in waters having no more than 10 μg/l of soluble phosphorus in the upper 30 ft of a lake during the entire preceding year (41). In unpolluted bodies of freshwater, concentrations of phosphate phosphorus are often below 10 μg/l and in some lakes may be reduced to 1 μg/l or less by uptake during a period of algal growth (42). In seawater, also, phosphorus (as well as nitrogen) is normally present in small amounts and has often been indicated as a factor limiting the growth of marine phytoplankton (43).

It probably should be stated that it is not possible to describe the importance of phosphorus as a limiting factor for algal growth in general terms and have it hold for specific situations; a reconsideration of Blackman's principle of limiting factors (44) should make this clear. Working from this principle, the level of phosphate in water will regulate growth only when phosphorus is limiting. If phosphorus is originally limiting and the phosphate level is increased, growth will increase proportionately, but only until a different factor becomes limiting. At this point extra phosphate will have no appreciable stimulatory effect. Thus, consideration has to be given to all possible limiting factors, including nitrogen and carbon. In waters where everything else that is required for growth is present in abundant supply, algal blooms will result from an influx of phosphorus; on the other hand, eutrophication should be preventable if the phosphorus level can be kept at a limiting concentration, regardless of the quantity or concentration of other factors or nutrients.

Recently, papers have appeared which indicate that other factors than phosphate frequently limit algal growth in lakes and streams as well as in oceans. In 1967 Lange (45) published accounts of experiments which showed that the addition of carbohydrates to cultures of blue-green algae also containing bacteria stimulated algal growth. Additional experiments in 1970 (46) with phosphate-rich culture medium diluted to simulate nutrient levels of Lake Erie water also showed that additions of carbohydrate enhanced blue-green algal growth. Lange's conclusions were that the added carbohydrate stimulated bacterial growth and that bacterial respiration produced extra CO_2 which, in turn, stimulated photosynthesis in and growth of the blue-green algae. These conclusions imply that the limiting factor in these studies was not phosphate, but CO_2. Kuentzel (47, 48) and King (49), reviewing literature on eutrophication, have also concluded that the supply of carbon, rather than phosphorus, commonly limits algal growth in freshwater situations. Kerr and co-workers (50) have presented the results of current studies showing that the carbon dioxide supply of lakes of the southeastern United States is a more common limiting factor than is phosphate. Rhyther and Dunstan (51) have studied the potential role of phosphorus in eutrophication of coastal and marine waters and have indicated that about twice the limiting amount of phosphate is normally present. This phosphate surplus is a consequence of the low nitrogen-to-phosphorus ratio in terrigenous contributions including human waste, and of the fact that phosphorus regenerates more quickly than ammonia from decomposing organic matter.

Conversely, in freshwater bodies in northern regions where the dissolved carbonate concentration is higher than in the southeastern United States, high phosphate may contribute significantly to eutrophication (52). This is partly because the content of nitrogen, as well as phosphorus, is high in domestic sewage as well as in many industrial wastes. Of the elements that may be involved in eutrophication, phosphorus can be most readily removed from wastewater before it is released to streams or lakes (53). Finally, it appears that the efforts to remove phosphorus from influents to lake ecosystems are not a waste of time and money if phosphorus can be removed to a level that becomes growth limiting.

ACKNOWLEDGMENT

The author would like to thank Professor Louis G. Williams for assistance in surveying literature for this paper.

REFERENCES

1. Myers, J., *Ann. Rev. Microbiol.*, **5**, 157 (1951).
2. Ketchum, B. H., *Ann. Rev. Plant Physiol.*, **5**, 55 (1954).
3. Krauss, R. W., *Ann. Rev. Plant Physiol.*, **9**, 207 (1958).
4. Provasoli, L., *Ann. Rev. Microbiol.*, **12**, 279 (1958).
5. O'Kelley, J. C., *Ann. Rev. Plant Physiol.*, **19**, 89 (1968).
6. Kuhl, A., "Inorganic Phosphorus Uptake and Metabolism," in *Physiology and Biochemistry of Algae*, R. A. Lewin (Ed.), Academic Press, New York, 1962, p. 211.
7. Kuhl, A., in "Beitrage zur Physiologie und Morphologie der Algen," *Vortr. Gesamtgeb. Bot. Deut. Bot. Ges.*, **1**, 157 (1962).
8. Kanai, R., S. Aoki, and S. Miyachi, *Plant Cell Physiol.*, **6**, 467 (1965).
9. Aoki, S., and S. Miyachi, *Plant Cell Physiol.*, **5**, 241 (1964).
10. Overbeck, J., *Arch. Mikrobiol.*, **41**, 11 (1962).
11. Overbeck, J., *Ber. Deut. Bot. Ges.*, **76**, 276 (1963).
12. Miyachi, S., and S. Miyachi, *Plant Cell Physiol.*, **2**, 415 (1961).
13. Miyachi, S., *Plant Cell Physiol.*, **3**, 1 (1962).
14. Hase, E., S. Miyachi, and S. Mihara, "A Preliminary Note on the Phosphorus Compounds in Chloroplasts and Volutin Granules Isolated from *Chlorella* Cells," in *Studies on Microalgae and Photosynthetic Bacteria*, Japanese Society of Plant Physiologists (Ed.), The University of Tokyo Press, Tokyo, Japan, 1963, p. 619.
15. Correll, D. L., and N. E. Tolbert, *Plant Cell Physiol.*, **5**, 171 (1964).
16. Correll, D. L., *Plant Cell Physiol.*, **6**, 661 (1965).
17. Galloway, R. A., and R. W. Krauss, "Utilization of Phosphorus Sources by *Chlorella*," in *Studies on Microalgae and Photosynthetic Bacteria*, Japanese Society of Plant Physiologists (Ed.), The University of Tokyo Press, Tokyo, Japan, 1963, p. 569.
18. Miyachi, S., S. Miyachi, and A. A. Benson, *Plant Cell Physiol.*, **6**, 789 (1965).
19. Littlefield, L., and C. Forsberg, *Physiol. Plant.*, **18**, 291 (1965).
20. Kuenzler, E. J., *J. Phycol.*, **1**, 156 (1966).
21. Kuenzler, E. J., and J. P. Perras, *Biol. Bull.*, **128**, 271 (1965).
22. Simonis, W., and W. Urbach, "Untersuchungen zur lichtabhängigen Phosphorylierung bei *Ankistrodesmus braunii* IX. Beinflussung durch Phosphatkonzentration, Temperature, Hemmstoffe, Na^+-Ionen und Vorbelichtung," in *Studies on Microalgae and Photosynthetic Bacteria*, Japanese Society of Plant Physiologists (Ed.), The University of Tokyo Press, Tokyo, Japan, 1963, p. 597.
23. Kylin, A., *Physiol. Plant.*, **19**, 644 (1966).
24. Talpasayi, E. R. S., *Plant Cell Physiol.*, **3**, 189 (1962).
25. Sommer, J. R., and J. J. Blum, *J. Cell. Biol.*, **24**, 235 (1965).
26. Blum, J. J., *J. Cell. Biol.*, **24**, 223 (1965).
27. Kylin, A., *Physiol. Plant.*, **17**, 422 (1964).
28. Kylin, A., *Physiol. Plant.*, **17**, 384 (1964).
29. Simonis, W., and W. Urbach, *Arch. Mikrobiol.*, **46**, 265 (1963).

30. Kylin, A., *Z. Pflanzenphysiol.*, **56,** 70 (1967).
31. Evans, J. H., *Br. Phycol. Bull.*, **2,** 317 (1964).
32. Fuhs, G. W., *J. Phycol.*, **5,** 312 (1969).
33. Carpenter, E. J., *J. Phycol.*, **6,** 28 (1970).
34. Forsberg, C., *Nature*, **201,** 517 (1964).
35. Forsberg, C., *Physiol. Plant.*, **18,** 275 (1965).
36. Azad, H. S., and J. A. Borchardt, *Environ. Sci. Technol.*, **4,** 737 (1970).
37. Kuenzler, E. J., *J. Phycol.*, **6,** 7 (1970).
38. Fitzgerald, G. P., *J. Phycol.*, **6,** 239 (1970).
39. Talling, J. F., "Freshwater Algae," in *Physiology and Biochemistry of Algae*, R. A. Lewin (Ed.), Academic Press, New York, 1962, p. 743.
40. Rodhe, W., *Symb. Bot. Ups.*, **10,** 1 (1948).
41. Mackenthun, K. M., L. E. Keup, and R. K. Stewart, *J. Water Pollut. Control Fed.*, **42,** 2035 (1970).
42. Hutchinson, G. E., *A Treatise on Limnology*, Vol. I, *Geography, Physics and Chemistry*, Wiley, New York, 1957.
43. Yentsch, C. S., "Marine Plankton," in *Physiology and Biochemistry of Algae*, R. A. Lewin (Ed.), Academic Press, New York, 1962, p. 771.
44. Blackman, F. F., *Ann. Bot.*, **19,** 281 (1905).
45. Lange, W., *Nature*, **215,** 1277 (1967).
46. Lange, W., *J. Phycol.*, **6,** 230 (1970).
47. Kuentzel, L. E., *J. Water Pollut. Control Fed.*, **41,** 1737 (1969).
48. Kuentzel, L. E., *Proc. of the Natl. Symp. Hydrobiol., Am. Water Res. Assoc.*, **8,** 321 (1970).
49. King, D. F., *J. Water Pollut. Control Fed.*, **42,** 2035 (1970).
50. Kerr, P. C., D. F. Paris, and D. L. Brockway, Water Pollution Control Research Series 16050 FGS 07/70, Department of the Interior, Federal Water Quality Administration, 1970.
51. Rhyther, J. H., and W. M. Dunstan, *Science*, **171,** 1008 (1971).
52. Likens, G. E., A. F. Bartsch, G. H. Lauff, and J. E. Hobbie, *Science*, **172,** 873 (1971).
53. Liebmann, H., "The Regeneration of Some Large Lakes in Bavaria," in *Algae, Man, and the Environment*, Daniel F. Jackson (Ed.), Syracuse University Press, Syracuse, N.Y., 1968, p. 403.

27

The Role of Phosphorus in Bacteria and Viruses

P. H. HODSON*

Monsanto Company, St. Louis, Missouri

There are few places on the surface of our planet which are devoid of microbial life. This biosphere of microbial activity plays an integral role in the cyclic nature of the elements that are so necessary to support life. One of these elements is phosphorus, and it is ubiquitous in all protoplasm whether it be protista, plant, or animal. Phosphorus is a key element in nearly all facets of metabolism such as the biosynthesis of protein, nucleic acids, complex carbohydrates, lipids, and other cellular constituents.

Considering the important role phosphorus plays in so many aspects of life it is surprising how recently even sketchy details of the role of phosphates in life processes have been investigated. We know that life cannot exist in the absence of phosphorus.

Water and soil phosphorus is found only as phosphates. These phosphates occur in many forms, but they are generally found as orthophosphates. In living systems phosphorus is present not only as orthophosphates but also as polyphosphates of various molecular weights, phosphoproteins, phosphonates, various ortho- and polyphosphate esters, and complex nucleoproteins.

Because of the necessary space limitations it is impossible to cover this subject in any great depth. However, it is hoped that sufficient key references are given so that, if desired, one can obtain further information on the subject being discussed.

SOLUBILIZATION OF PHOSPHORUS IN THE BIOSPHERE

Many types of microorganisms participate in the solubilization of soil phosphates (116, 85, 3, 118, 108, 68). Rose (103) and Louw and Webley (76) found several microbial isolates which dissolve mineral phosphates. These isolates produce organic acids such as oxalic, citric, and lactic. Other

* *Current address:* Eli Lilly and Company, Clinton, Indiana.

organic acids may participate as well in the solubilization of phosphates. Tardieux-Roche (117) demonstrated that when the medium was acidified by the organism relatively large amounts of phosphorus were liberated, whereas if the medium was alkalinized, only small amounts of phosphorus were released. Although his data favor the theory of solubilization through acidification they also support the theory of solubilization through enzyme action. Berman (12) has shown that alkaline phosphatase may play a role in phosphorus nutrition in aquatic environments. The important role of phosphorus in symbiotic nitrogen fixation has been investigated by several workers (84, 65). By inoculation of soil with phosphate-solubilizing bacteria, increased growth of oats (36) and corn (97) has been obtained.

Phosphate-solubilizing fungi are frequently isolated from seeds and root zones of plants (64). Subba-Rao and Bajpai (115) isolated a *Penicillium* from the surface of legume root nodules which released nearly four times as much of phosphate as the phosphate-dissolving bacteria.

Organic phosphates are also solubilized by microorganisms (108). Phosphorus is liberated from bone meal (35), calcium glycerophosphate, and nucleic acids (68). Kobus (67) concluded that organic phosphorus compounds are more rapidly decomposed than mineral ones. After studying almost 1000 strains of soil microorganisms he concluded that the ability to solubilize phosphorus is a very common property. The degree of solubilization is influenced by the presence of utilizable carbon sources such as carbohydrates and amino acids (67, 102). Recently inorganic phosphorus regeneration from decomposition of diatoms in seawater has been found to be dependent on the dissolved oxygen (88). An exponential relation was found between inorganic phosphorus regeneration and oxygen consumption. Watt and Hayes (127) using radioactive phosphorus demonstrated the change of phosphorus phases through particulate phosphorus, dissolved organic phosphorus, and inorganic phosphate, and inferred the role of bacteria in the process. Lear (70) made a study of the ability of oceanic marine bacteria to regenerate inorganic phosphate from organic phosphate substrate. He found evidence that marine bacteria can regenerate inorganic phosphate from organophosphorus, compounds of marine organisms. Isolates of the genera *Pseudomonas*, *Vibrio*, and *Achromobacter* were capable of regenerating phosphates from zooplankton, phytoplankton, and fish meal. It is possible that all oceanic bacteria are capable of regenerating phosphate. The mechanism of phosphate regeneration was associated with autolysis of bacterial cells and seemed to be regulated by the amount of phosphate in solution.

Organic phosphorus and insoluble phosphates cannot be utilized by plants. By their actions, the fungi, bacteria, and actinomycetes make the bound phosphorus available to plants. Therefore microorganisms play a vital role in maintaining plant life.

ADSORPTION OF PHOSPHATES

Radioactive phosphorus has been employed by many investigators in following its adsorption into living systems. It is generally believed that the penetration of phosphorus into bacteria is not entirely an osmotic process but is regulated by cellular anabolism. A number of compounds such as selenite (32), vitamin K (66), nitrate (62), magnesium (14, 15), sodium azide, methylene blue, 2,6-dichloroindophenol (80), arsenate (48), and gramicidin (81) can affect phosphate transport into cells. Several workers have shown in yeast that PO_4^{3-} is actively transported into the cell by a glycolysis mechanism even against a hundredfold concentration gradient (13, 38). Leggett and Olsen (71) concluded that phosphate entered the yeast cell by passive diffusion into a space constituting over 80% of the cell volume. The phosphate was further accumulated by a metabolically dependent uptake from this space. In view of this, it is not surprising that respiratory poisons can affect phosphorus transport. Varma (120) and Luoma (77) have shown that pH can influence absorption of phosphate.

Nelson and Kornberg (90) studied phosphate metabolism during germination of bacterial spores. They report that within 5 to 10 min phosphate was incorporated into acid-insoluble material (RNA) and acid-soluble materials such as glucose-6-phosphate, fructose diphosphate, ATP, and L-α-glycerophosphate. They concluded that energy-yielding metabolism begins very early in germination. The paths of transfer for phosphate groups during uptake of inorganic orthophosphate by *Micrococcus pyogenes* were investigated by Mitchell and Moyle (86). These investigators followed the incorporation of ^{32}P-labeled phosphate groups into cellular phospholipid, acid-soluble inorganic, and organic, deoxyribonucleic acid, ribonucleic acid, and glycerophosphoprotein complex fractions. In resting cells, reciprocal exchange of inorganic phosphate across the osmotic barrier was one-half complete in 70 min. During respiration or growth, phosphates moved inward through the osmotic barrier at the same rate as with resting cells, but outward movement was stopped. During respiration phosphate accumulated mainly in the acid-soluble inorganic, acid-soluble organic, and phospholipid fractions, whereas accumulation was observed in all fractions during growth. Most of the phosphate of the organic fractions is drawn through the acid-soluble inorganic fraction.

It can be concluded that generally inorganic phosphate can rapidly enter a cell's metabolism and incorporate into a variety of cellular components. Phosphate entrance into a cell may be by diffusion or by an active transport system. Depending on the stage of cellular growth, it is possible that a combination of the two systems may operate.

PHOSPHORUS AS A GROWTH REGULATOR

Because phosphate is an essential nutrient for all living systems it is only reasonable that this element can limit microbial growth. Arsenate competitively inhibits the growth of *Streptococcus faecalis* by competing with phosphate for a common transport system. Mutants of this organism have been found with a defective phosphate system which therefore limits growth (47). Growing yeast in 99% D_2O O'Brien (94) found that phosphate uptake was inhibited. This was due to a direct isotope effect on the transport process. In a continuous culture of *Escherichia coli B* the cell division can be synchronized by periodic phosphate feeding (39). Postgate and Hunter (98) found that accelerated death of *Aerobacter aerogenes* was caused by starvation in the presence of growth-limiting phosphate. A biological assay using *Aspergillus niger* for phosphorus has been reported (106). Employing *Penicillium chrysogenum*, Defrance (28) also found a growth response to phosphate, and Nordheim and Rieche (92) showed that *Candida utilis* has a lower oxygen utilization when phosphate was limiting growth. This indicates an obligatory coupling of respiration to cellular phosphate. This same effect has been observed in *Escherichia coli* (56).

The production of extracellular metabolites can be influenced by phosphate concentrations. Erythromycin, for example, is produced in its highest concentration when phosphate is low (10 to 7 μg/ml), whereas higher concentrations linearly decreased the antibiotic yield (1).

Heinen (53, 54) has shown a relationship between silicate and phosphate metabolism of bacteria. There can be a partial exchange of silicate for phosphate. The total silicate of the cells can be completely exchanged with phosphate when the cells are cultured in silicate-free phosphate medium, whereas if silicate is present in the phosphate medium, the cellular silicate will not extrude.

There are numerous references in the literature which demonstrate that microbial growth requires the presence of available forms of phosphorus. In environments where phosphorus is limiting, its addition will stimulate microbiological activities. Deficiencies are not usually found in nature unless an abundant carbohydrate supply is present as the biosphere inhabitants appear to be highly efficient in mobilizing the large, natural reservoir of the element.

DETOXIFICATION OF ORGANOPHOSPHATE PESTICIDES

Degradation of pesticides by soil microorganisms is an important factor affecting the persistence of pesticides in soil. In addition to killing or reducing numbers of specific parasites, these chemicals directly or indirectly applied to the soil may alter the microbial soil population by killing, reducing numbers, or stimulating growth of certain saprophytic microorganisms. Most organic pesticides are decomposed at varying rates by one or more soil microorganisms (82). Persistence in soil depends on ease of decomposition, solubility, adsorption, volatility, absorption by plants, soil pH, organic matter, and other factors.

Lichtenstein and Schultz (74) found that soils containing low numbers of microorganisms or low in moisture (low microbial activity) parathion persisted for a relatively long time. However, in moist loam soil 95% of the applied methyl parathion and 30% of parathion were lost in 12 days. Yeast was primarily responsible for the reduction of parathion to aminoparathion. More recently Lichtenstein *et al.* (75) studied the effect of sterilizing agents on persistence of parathion and diazinon in soils and water. They concluded that sterilizing agents, such as sodium azide, or autoclaving reduced the number of bacteria in soils resulting in increased persistence of parathion.

Chemical hydrolysis of malathion, systox, and phosdrin appears to be the primary mechanism of detoxification in soils. However, Matsumura and Boush (83) have isolated two organisms (*Trichoderma viride* and *Pseudomonas* sp.) capable of metabolizing melathion. Although chemical hydrolysis plays a significant role in the hydrolysis of diazinon (88) Bro-Rasmussen *et al.* (17) concluded that soil microorganisms also play an important role in diazinon disappearance.

Even though the biosphere is capable of metabolizing organophosphate pesticides there is ample work to show that certain of these compounds will control some crop fungal diseases. An excellent review in this area was prepared by Scheinpflug and Jung (107). This review contains many chemical structures of the most important commercial and experimental products of organophosphorus compounds which have a fungitoxic action. Bent (11) also recently reviewed fungicides. The alkylation, dealkylation, condensation, hydrolysis, oxidation, and reduction of fungicides were reviewed by Owens (95).

Little is known about the specific metabolic sequences by which microorganisms detoxify and assimilate organophosphate pesticides. Each new molecule presented to the biosphere represents a new challenge for the microbial population. We are indeed fortunate that living microorganisms have an inherent capability to mineralize foreign substances and return the elements to their respective cycles in nature.

PHOSPHORUS METABOLISM

Coenzymes

Many coenzymes are organophosphorus compounds and act in cooperation with enzymes to carry out specific catalytic transformations. A coenzyme can generally be separated from the enzyme with which they function. Also, the same coenzyme can participate with different types of enzymes to bring about numerous reactions. Unlike enzymes, coenzymes are relatively small molecular weight substances, dialyzable, nonproteinaceous, and are heat stable. Table 1 lists the most prominent coenzymes that contain phosphorus. Others can be found in *The Enzymes*, Volumes 2 and 3 (16). These coenzymes function in all aspects of intermediary metabolism.

The most important coenzyme is probably ATP. This is not to say, however, that a cell could survive without one or more of the other coenzymes. Each coenzyme has its necessary cellular functions but the number of individual enzymes known to require ATP as substrate or energy source is very large (30). If the pure moiety of ATP is inosine, uridine, cytidine, or guanosine, then the coenzymes are called inosine triphosphate, uridine triphosphate, and so on. If one orthophosphate group is removed by hydrolysis, the compounds become diphosphate, that is, ADP, IDP, and so on. If two phosphate groups are removed, they then become monophosphates.

From a chemical viewpoint, the marriage broker and common coin of the biochemical universe is the ATP molecule (63). This is because potentially endergonic reactions are converted into exergonic ones in many of the enzymatic reactions so necessary to sustain life. In a broad sense, the utilization of carbohydrates is simply a process for making ATP. The exact chemical reactions are known and can be found in any biochemistry textbook. The transformations are essentially the same for all forms of life. It is known that the enzyme isolated from muscle and yeast requires diphosphopyridine nucleotide to bring about the transformation from glyceraldehyde-3-phosphate to 1,3-diphosphoglyceric acid. However, the enzyme system from plant tissue contains two dehydrogenases, one specific for diphosphopyridine nucleotide and the other for triphosphopyridine nucleotide (121). Nevertheless, the end result is the same.

Nucleic Acids

Miescher in 1869 isolated a cellular component from salmon sperm which he called nuclein (22). This material, in a more purified form, is now known as desoxyribonucleic acid (DNA). The important role DNA plays in genetics was not realized until after the work of Avery and co-workers was published in 1944 (8). These investigators isolated an agent that was responsible for the transformation of pneumococcal rough strains to smooth strains. They presented evidence which supported the belief that a nucleic acid containing desoxyribose was the fundamental agent that brought about the observed transformation. This transformation was not transient and cells now had the smooth encapsulated properties of the original strain from which the DNA

TABLE 1 Phosphorus-Containing Coenzymes

Name	Structure
Adenosine triphosphate (ATP)	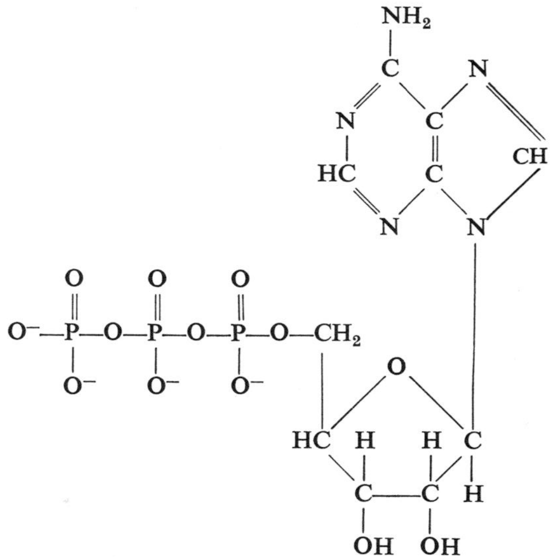 NH_2 C N N C HC C CH N N O O O O^-—P—O—P—O—P—O—CH_2 O^- O^- O^- O HC H H C C—C H OH OH

TABLE 1 (continued)

Name	Structure
Nicotinamide adenine dinucleotide in which R = H(NAD). If R = $PO(OH)_2$, then coenzyme is called nicotinamide adenine dinucleotide phosphate (NADP)	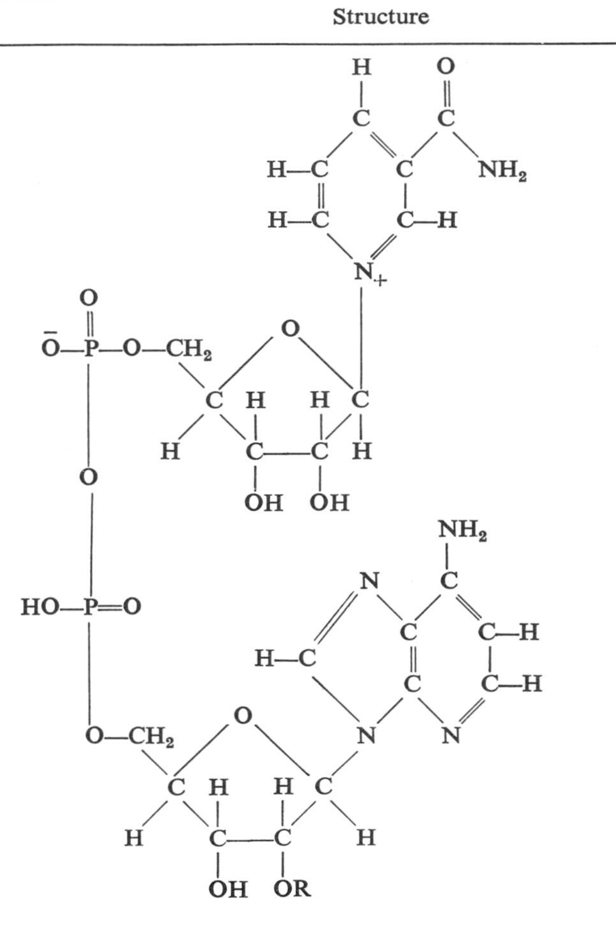

Riboflavin phosphate (FMN). If a molecule of adenine ribose phosphate and FMN are joined at the phosphate group, then the structure is known as flavin adenine dinucleotide (FAD)

Pyridoxal phosphate

Thiamine pyrophosphate

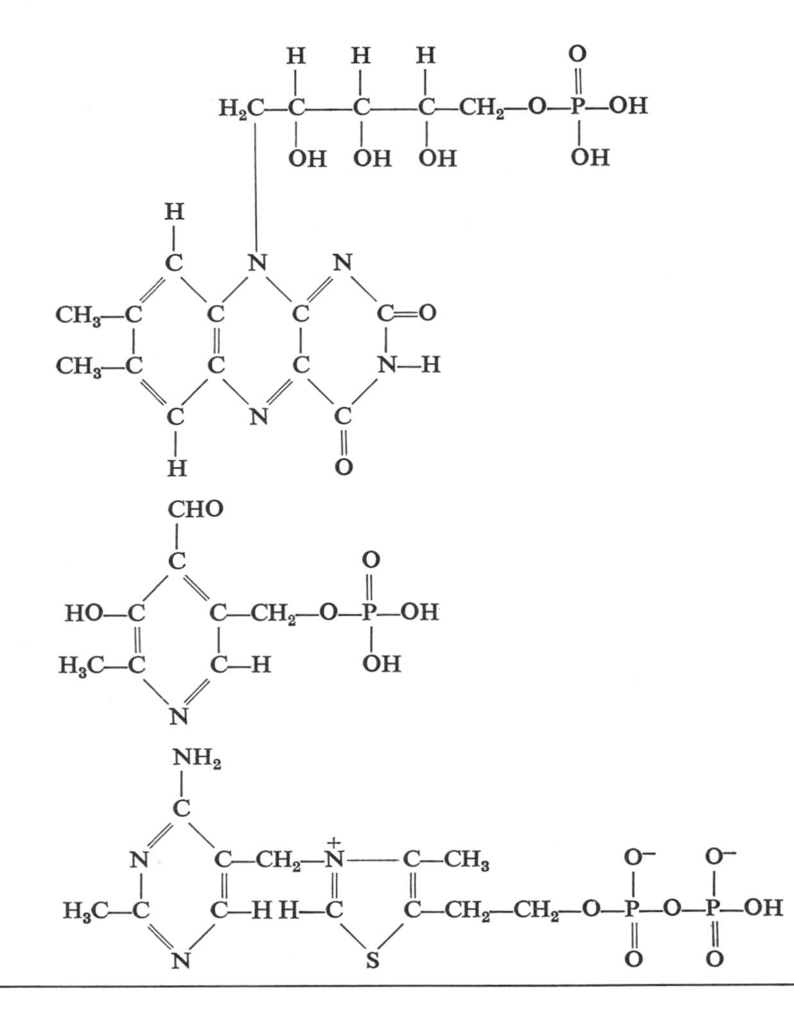

TABLE 1 (continued)

Name	Structure
Coenzyme A	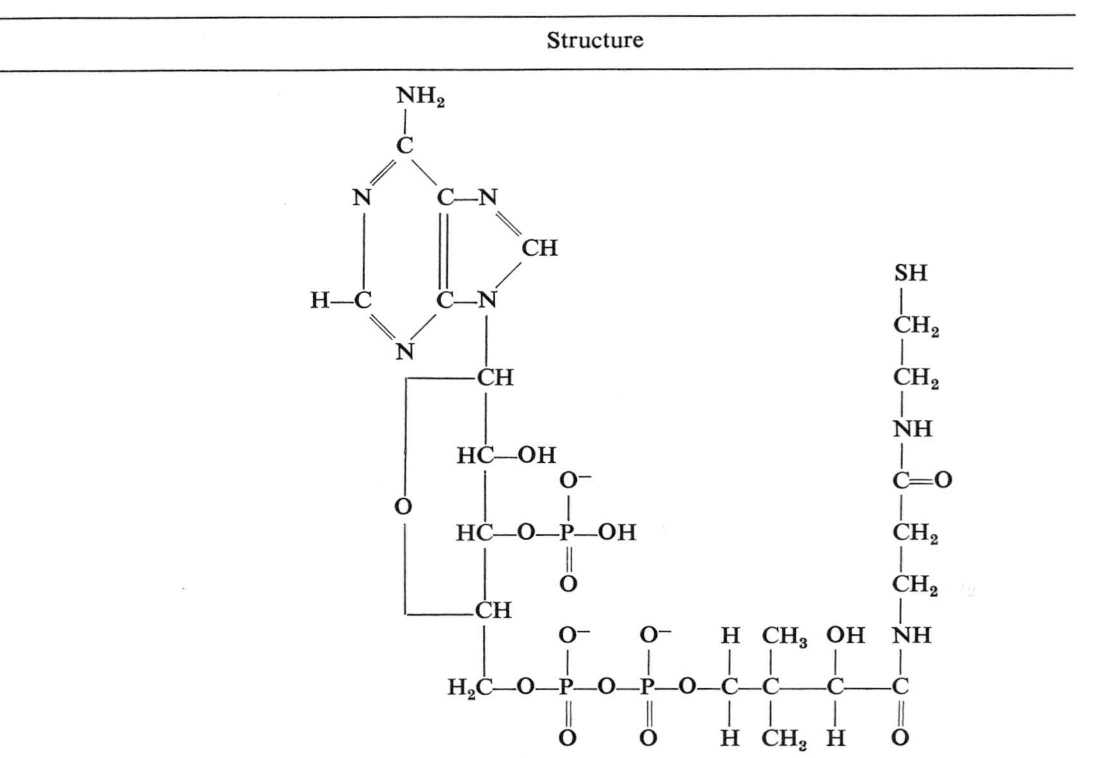

Uridine diphosphate sugars

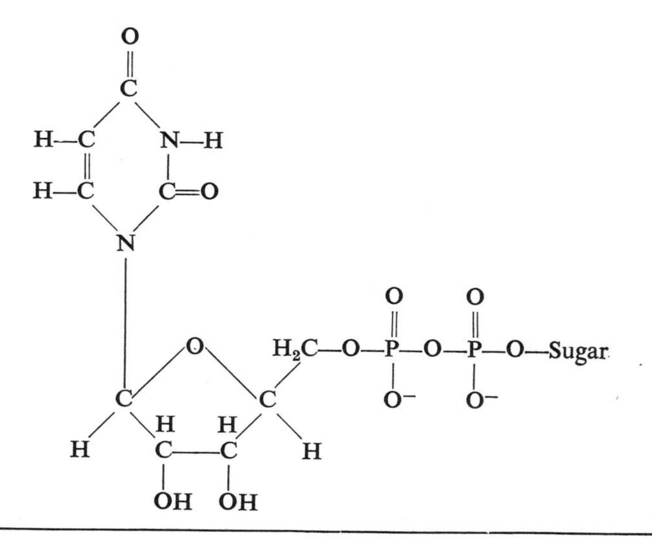

was derived. This observation brought together the two sciences of chemistry and genetics.

Miescher recognized the presence of phosphorus in his nuclein preparation. His best preparations contained on the average of 9.59% phosphorus and 13.09% nitrogen. Chargaff (22) felt the higher phosphorus values obtained by Miescher were due to the use of mineral acid for his nuclein isolation. The acid extraction most likely cleaved some of the purine constituents from the DNA polymer. We now know that the DNA is a double-stranded helical polymeric molecule that contains 10 sugar–orthophosphate units per turn of the helix and has a complementary arrangement of purines and pyrimidines (123, 124). This structure was formulated by the aid of x-ray diffraction patterns (131, 35). Information leading to the conclusion that DNA is a constituent of genes has been summarized by Sinsheimer (112). Much attention has been given to the science of DNA and the subject has been amply reviewed and documented (22, 21, 27).

In addition to DNA there is in cells a second nucleic acid. The compound is ribonucleic acid (RNA) and varies in a number of ways from DNA and yet functions very closely with DNA in cellular synthesis. A phosphate–sugar backbone is common to both nucleic acids. The pentose sugar is bonded at the 3′- and 5′- positions to the linking orthophosphate groups. The pentose sugar moieties in the two molecules differ only in the substitution of a hydrogen atom in DNA for a hydroxyl in the 2′ position of the RNA. Adenine, guanine, and cytosine occur in both types, but the uracil of RNA is replaced by a methylated uracil (thymine) in DNA. With few exceptions the nitrogenous, pentose, and phosphate constituents mentioned above compose the nucleic acids of plants, animals, and protista. The bulk of phosphorus in the bacterial cell is in RNA. This nucleic acid usually accounts for one-third to one-half of all the phosphorus. DNA contributes from 5 to 10% of the total cellular phosphorus.

There has been reported, however, nitrogenous moieties in DNA of plants (20), bacteria (10, 31), and animal (19) which differ from those usually found. The biological significance of the occasional occurrence in DNA of methylated purines or pyrimidines is by no means clear.

Another baffling property of DNA researchers were slow to realize is its lability to shear (6, 25, 26). It is believed that the depolymerization of DNA by mechanical means involves the rupture of phosphodiester bonds because the appearance of new end groups has been demonstrated (100, 101).

Mitchell and Moyle (86) found in *Micrococcus pyogenes* that RNA is formed from the nucleotides of the acid-soluble organic fractions, while the DNA phosphate is derived either directly from the acid-soluble inorganic or from intermediates in the acid-soluble organic phosphate fractions. They concluded there was little or no turnover of RNA or DNA phosphate during growth and that phosphate is not transferred directly or indirectly between the two nucleic acids.

Enzymes have been isolated that will polymerize nucleoside triphosphate to RNA or DNA. The first enzyme that would convert deoxynucleoside polyphosphates into polymeric material was isolated from extracts of *E. coli* by Lehman and his associates (72). The DNA polymerase substrates were the deoxynucleoside triphosphates and not the diphosphates. DNA primer molecule was necessary to initiate the polymerization reaction with liberation of pyrophosphate. RNA will not substitute for the DNA primer.

Systems have been found which will polymerize nucleoside polyphosphates to RNA. Grunberg-Mango and Ochoa isolated such a system from *Azotobacter vinelandii* (40). This enzyme catalyzes the formation of RNA from the nucleoside diphosphates and is specific for the diphosphates and for ribose. The reaction is stimulated by the presence of RNA primers although there is no relationship between structure and product structure. A review of this interesting topic of enzymatic synthesis of nucleic acids is given by Grunberg-Mango (41).

Even though we are able to synthesize polymeric nucleic acids we do not have sufficient control on the sequencing during synthesis. Coupled with this problem a detailed determination of the nucleotide sequence of a DNA molecule is beyond our present means, nor is it likely to occur in the near future. The molecular weight of the largest DNA molecule lies in the range of 2×10^8 to 2×10^9; this means that a single strand would be composed of 3×10^5 to 3×10^6 nucleotides. Even the smallest functional phages DNA (119) must contain about 5000 nucleotides. Chargaff (22) concludes that we may leave the task of reading the complete nucleotide sequence of a DNA molecule to the twenty-first century.

Carbohydrates

No attempt is made here to detail the many metabolic pathways involving carbohydrates but rather to show, in general, the involvement of phosphate in carbohydrate chemistry. A good review of carbohydrate metabolism can be found in Mahler and Cordes (79, Chapter 10).

Glycolysis and fermentation are essentially the same process but are carried out in different systems. In glycolysis, glycogen (mammalian starch) must be converted to glucose phosphate; whereas in fermentation, glucose is converted to glucose phosphate to start the reaction sequence. The requirement of phosphate for microbial fermentation dates back to the time of Pasteur. But it was Harden and Young (43, 44) who demonstrated that orthophosphate disappeared and could be rate limiting during fermentation. These workers isolated a phosphorylated hexose sugar (fructose-1,6-diphosphate) which became known as the Harden-Young-ester. The consecutive events that occur in fermentation and glycolysis were elucidated one by one as the enzymes, intermediates, and cofactors were discovered. As the compounds were discovered they were named after the workers who found them. Glucose-1-phosphate was known as the Cori-ester, glucose-6-phosphate the Robison-ester, fructose-6-phosphate the Neuberg-ester, glyceraldehyde-3-phosphate the Fischer-ester, 1,3-diphosphoglyceric acid the Negelein-ester, and 3-phosphoglyceric acid the Nilsson-ester (33).

Microbial fermentation of glucose to ethanol, lactate, glycerol, glycols, and a wide variety of other products all make use of what is now known as the Embden-Meyerhof-Parnas (EMP) scheme of glycolysis. For each molecule of free glucose traversing the EMP scheme two molecules of ATP are consumed and four molecules are produced (Fig. 1).

The EMP scheme is an anaerobic process, that is, no molecular oxygen is required. For further breakdown and release of energy of the products the aerobic citric acid cycle is usually brought into play. The citric acid cycle in addition to supplying additional cellular ATP also supplies critical intermediates for biosynthetic anabolic processes. The intermediates in this cycle are not phosphorylated. However, phosphate-containing coenzymes are involved in the reactions.

In microbial metabolism a second major pathway for hexose catabolism has been demonstrated. This pathway is known by several names, such as the hexose monophosphate shunt, pentose cycle, phosphogluconate pathway, the oxidative pathway of hexose metabolism, and the hexose monophosphate pentose pathway. The intermediates are phosphate esters, some of which are common to the EMP scheme. The new compounds include D-glucono-∂-lactone-6-phosphate, 6-phospho-D-gluconate, D-ribulose-5-phosphate, D-ribose-5-phosphate, D-sedoheptulose-7-phosphate, and D-xylulose-5-phosphate. For a detailed discussion of the enzymes, coenzymes, and reactions the reader is referred to Mahler and Cordes (79, p. 448). The net result of this pathway is

$$6 \text{ glucose-6-P} + 6O_2 \rightarrow 5 \text{ glucose-6-P} + 6CO_2 + 6H_2O + \text{Pi}$$

In cellular metabolism there are numerous nonoxidative interconversions of hexoses and pentoses. These interconversions also involve the phosphorylated compounds. Because a large number of microorganisms are able to interconvert hexoses, pentoses, and trioses the cells are able to grow on a variety of carbon sources. A discussion of generalized sugar interconversion can be found in Mahler and Cordes (79, p. 470). It is known that some enzymatic reactions oxidize unsubstituted sugars rather than their phosphate esters. Many of these reactions have been found in microbial cells.

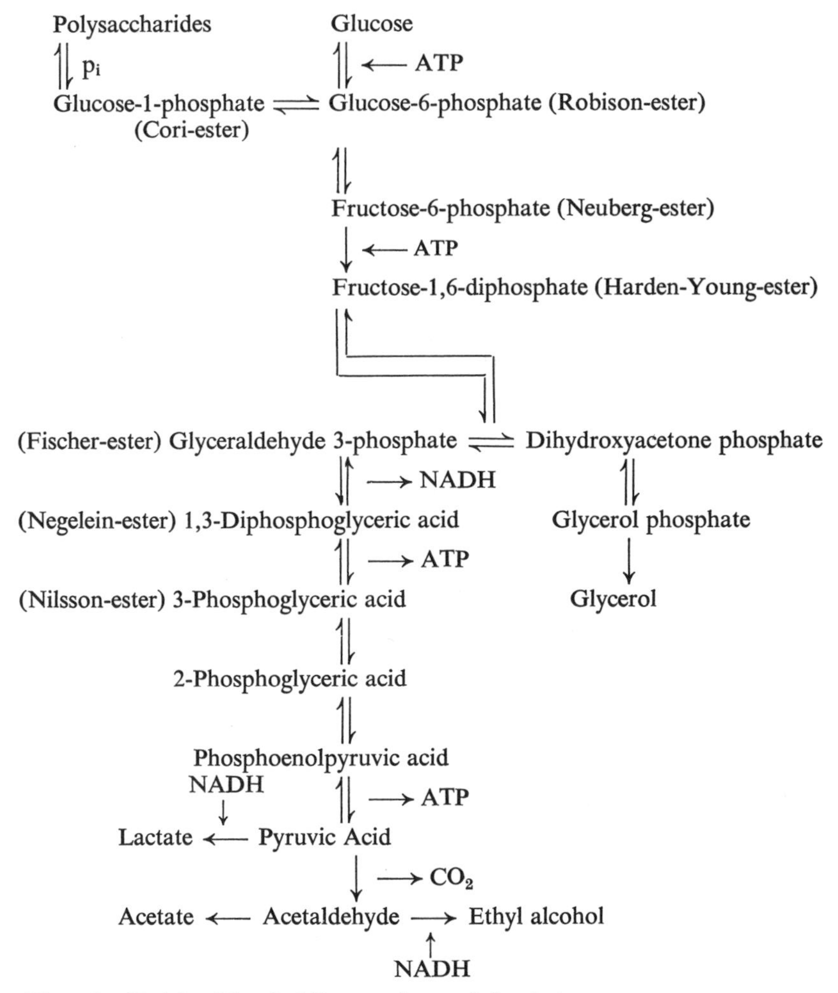

Figure 1. Embden-Meyerhof-Parnas scheme of glycolysis.

There are other minor pathways that are unique to certain bacterial species such as the Entner-Doudoroff pathway in *Pseudomonas* (49), but generally the pathways above serve as the major ones for the assimilation of carbohydrates. Needless to say phosphorus is a key element in carbohydrate metabolism.

Phospholipids

Phospholipids are compounds in which phosphate is combined with a lipid. Until the mid-1950s our knowledge of bacterial phospholipids (phosphatides) was somewhat limited due to the difficulties in separation of these compounds. At this time chromatographic techniques in lipid research became widespread. There are numerous lipids and phosphatides found in microbial systems. Kates in 1964 reviewed bacterial lipids (64) and in 1967 Ikawa again reviewed this subject (59). Even though the information on phosphatides is incomplete and diversified some general conclusions can be made. As might be expected members of a single bacterial family generally contain similar phosphatides (63). Gram-negative organisms contain more complex phosphorus-containing lipopolysaccharides than the gram-positive bacteria. The latter group of bacteria contain larger quantities of the nonnitrogenous phosphatidylglycerol and di- and polyphosphatidylglycerols. Phosphatidylethanolamine seems to be a common consitituent of the gram-negative bacteria. However, phosphatidylcholine occurs in relatively few of the eubacteria studied (65).

Virtually all the lipids in gram-positive bacteria are associated with the cytoplasmic membrane, whereas in gram-negative bacteria the cell wall as well as the membrane contains lipids. Asseleneau and Lederer (7) in reviewing the chemistry and metabolism of bacterial lipids concluded that bacteria differ radically from other organisms in the absence of sterols.

Macfarlane in reviewing lipids of bacterial membranes (78) noted that the membrane in various gram-positive bacteria, at least, is known to be analogous to the mitochondria of animal tissues. Both structures contain the cytochrome oxidases and various dehydrogenase activities and further resemble each other by their remarkable high content of phosphatidylglycerols.

Bacterial cells also contain lipoamino acids, the most common one appearing to be phosphatidylserine. These substituted amino acids have not been implicated as precursors or active intermediates in protein synthesis (68). Their exact function in cellular metabolism remains unexplained.

Phospholipids containing glucose are rare in nature; however, several bacteria have been reported containing phosphatidylglucose (111). Glucosyl diglycerides are characteristic of the lipids of gram-positive bacteria (109). The function of these compounds in cellular metabolism is not clear at this time but it is felt that they may play an important part of membrane metabolism (111). Morman and White (87) in their work on phospholipid metabolism in *Bacillus licheniformis* felt that the involvement of phospholipids in membrane functions has proved confusing. Perhaps a better understanding of the membrane could be obtained if the membrane is fractionated into functional subgroups before the involvement of the lipid is analyzed. It can be concluded, however, that phosphatides play an important role in bacterial cycloplasmic membranes. This is evident because lipid and protein make up at least 70% and in some cases almost all of the weight of the membrane (65).

VIRUSES

Viruses are submicroscopic particles that can penetrate and multiply in suitable host cells. Viruses can infect most, if not all, plants and animals. The viruses that infect bacterial cells are known as bacteriophages (phage). The mechanism of viral infection has been most intensively studied in bacterial-phage systems.

Since the initial biochemical studies by Cohen (23) we now recognize that the production of bacteriophage occurs by specific metabolic sequences within the bacteria they infect. Much of the bacterial cellular metabolism is involved in the production of phage because phages do not have the ability to reproduce outside of a host cell. It is known that phages are composed of essentially a protein coat and a nucleic acid core.

Phosphorus has played an essential role in elucidating the infection process. Phages were labeled with radioactivity phosphorus, that is the nucleic acid was labeled, then it was determined that the radioactivity was, as it were, injected into the host cell. If the phage is labeled with radioactive sulfur, that is, the proteins are labeled, the label remains outside of the bacterial cell after infection (55).

All viruses, when extracellular, are completely inert. We now realize that viruses are small pieces of genetic material and that the progeny of viral particles resemble their parents because they contain identical chromosomes. For this reason viruses are convenient systems for quickly studying the consequences of the sudden introduction of new genetic material into a cell.

There are DNA and RNA viruses; however, no virus is known to contain both nucleic acids. The nucleic acid (both RNA and DNA) may be either single or double stranded (125). The smallest well-characterized viruses are the RNA bacterial phages of the F2 group that reproduce in *E. coli*. The F2 RNA chain contains only about 3000 nucleotides which would allow coding for 1000 amino acids. The smallest known DNA virus is the mouse polyoma virus. The DNA in this virus is a double helix with a molecular weight of 3×10^6 (126).

Stahl (113) reviewed the inactivation of phage by the decay of incorporated radioactive phosphorus. Phages so labeled are called suicide phage because as the radioactive phosphorus decays there can be sufficient nucleic acid damage such that the mutations become lethal. These types of experiments show the importance of phosphorus in maintaining viral integrity.

For an excellent review on the physiology and genetics of bacteriophage the reader is referred to Hayes' book, *The Genetics of Bacteria and Their Viruses* (52, Part 7).

Not only does phosphorus play a vital role in the viral nucleic acids but its presence or absence can influence viral attachment to the host cell. Hargrove *et al.* (46) investigated the effects of various phosphates on the proliferation of bacteriophages in milk-souring cultures. They found that 2% orthophosphate had a strong inhibiting effect. It was believed that the phosphates tie up the calcium required by the phages. In fact, 13 strains of lactic acid bacteria, strongly contaminated with phages, were freed of the phages by passing the cultures through milk in which the temperature, pH, and phosphate concentration were controlled. The presence of high phosphate concentrations has been shown to inhibit phage production in *Bacillus megaterium* (93) as well as in *Shigella paradysenteriae* (122). The effect of phosphate has also been investigated for numerous plant viruses (61).

Not only does phosphorus play an essential role in the cellular metabolism of the viral host cell, it is also essential for the nucleic acids of the virus. Also, extracellular phosphate can influence phage attachment. If the right phosphate concentrations are employed, the propagation of certain phages can be prevented because the virus is unable to attach to the host cell. If attachment is prevented, then viral propagation is terminated.

AUTOLYSIS

If the integrity of the bacterial membrane is not maintained, there will be a "leakage" of cellular constituents into the cell's environment. Alteration of permeability for the release of metabolites from the microbial cell was reviewed by Demain and Birnbaum (29). It has been reported that phosphates and pyrophosphates promote the autolysis of numerous species of enteric bacteria (104). These phosphates appear to activate a cellular autolytic enzyme system. Phosphorus compounds are involved in the autolysis of *Staphylococcus aureus* (130). Low concentration of K_2HPO_4, $[NaPO_3]_n$, $Na_4P_2O_7$, and sodium glycerophosphate promotes autolysis. However, sodium hypophosphite inhibited autolysis, whereas sodium phosphite had no effect. Phosphate has no direct effect on the development of lytic activity but it aids in the liberation of the lytic enzyme system from the bacterial cells. Phosphate, inorganic or organic, is necessary for the formation of *Staphylococcus aureus* autolysin (129).

Inorganic phosphates prevent the lethal action of streptomycin on *Escherichia coli* and promote excretion of nucleotides by exponentially growing bacteria exposed to the antibiotic (58). It has been found that some bacterial cells are osmotically fragile and begin to lyse if not stabilized by a hypertonic medium or other means after growth ceased. It was determined that glucose-1-phosphate prevented lysis of this culture, *Streptococcus diacetilactis* (89).

There are numerous ways by which cells can be lysed and it is evident that in some bacteria phosphorus does play a significant role in the lytic process. Shockman (110) reviewed unbalanced growth and autolysis of bacterial cells.

MOVEMENT OF PHOSPHORUS IN SOILS

Jones and Bromfield (60) found that when microbial growth was inhibited with formalin the inorganic phosphate was readily leached from soil, and that microbial activity largely prevented this loss.

A soil may be expected to contain almost any naturally occurring organic compound because there are present large numbers of living and dead microorganisms. However, very little nucleic acids exist in soil (1 ppm), perhaps no more than the microbes themselves. Adams *et al.* (2) found that the total organic phosphorus ranged from 300 to 500 ppm. The concentration of inositol hexaphosphate has been found to range from 100 to 300 ppm (5). Nucleic acids, lecithin, and phytin are all dephosphorylated when added to soil (96). The metabolism of organic phosphates in soil was reviewed by Cosgrove (24).

CONCLUSIONS

Since Harden (45) demonstrated the dependence of alcoholic fermentation on the presence of inorganic phosphate, biochemical research has uncovered an amazing variety of manifestations and functions of phosphate esters in cellular metabolism. There is hardly anything that goes on in the cell in which esters of phosphoric acid in one form or another are not involved.

Essentially all of our biochemical information is based on in vitro studies of isolated and often purified systems. We then attempt to translate our findings back to cellular metabolism which is, indeed, a complex puzzle. We lack the technology of crowded systems that exist in living cells. It is difficult to visualize the numerous events that simultaneously take place in a living cell. Everything happens on top of each other, precursors, intermediates, and end products are formed apparently without getting into each other's way. Nucleic acids, phospholipids, proteins, polysaccharides, and other cellular constituents are all assembled and deposited in their proper place. All of this metabolism is presumably under the control of the genome which at the same time is busily reproducing itself.

Even though a high degree of biochemical unity does exist among the various forms of life, it has become apparent that a great deal of biochemical diversity is also present. This diversity in the protesta is so common that the lower forms of life comprise a heterogeneous group. Yet, all life is linked together with nucleic acids and the components of nucleic acids are linked together with phosphorus.

Microorganisms bring about many transformations of phosphorus. These include (*a*) altering the solubility of inorganic compounds of phosphorus; (*b*) mineralizing organic compounds with the release of orthophosphates; (*c*) converting the inorganic available anion into cell protoplasm, an immobilization process analogous to nitrogen fixation; and (*d*) bringing about an oxidation or reduction of inorganic phosphorus compounds. Particularly important to the phosphorus cycle in nature are the microbial mineralization and immobilization reactions (4).

Therefore, if it were not for the work of the biosphere, phosphorus would rapidly become unavailable to plant life, and life as we know it would cease.

REFERENCES

1. Abou-Zeid, A. A., and A. A. El-Gamal, *Indian J. Exp. Biol.*, **7**, 276 (1969) (English).
2. Adams, A. P., W. V. Bartholomew, and F. E. Clark, *Soil Sci. Soc. Am. Proc.*, **18**, 40 (1954).
3. Albertson, O. E., U.S. Patent No. 3, 485, 750 (1967).
4. Alexander, M., *Soil Microbiology*, Wiley, New York, 1964.
5. Anderson, G. J., *Soil*, **12**, 276 (1961).
6. Anderson, N. G., *Nature*, **172**, 807 (1953).
7. Asseleneau, J., and E. Lederer, in *Lipide Metabolism*, K. Bloch (Ed.), Wiley, New York, 1960, p. 337.
8. Avery, O. T., C. M. MacLeod, and M. McCarty, *J. Exp. Med.*, **79**, 137 (1944).
9. Barth, E. F., and M. B. Ettinger, U.S. Patent No. 3, 480, 144 (1968).
10. Belozersky, A. N., and A. S. Spirin, *The Nucleic Acids*, E. Chargaff and J. N. Davidson (Eds.), Vol. III, Academic Press, New York, 1960, p. 147.
11. Bent, K. J., *Endeavour*, **28**, 129 (1969).
12. Berman, T., *Nature*, **224**, 1231 (1969).
13. Borst-Pauwels, G. W. F. H., *Biochim. Biophys. Acta*, **65**, 403 (1962) (English).
14. Borst-Pauwels, G. W. F. H., *Naturwiss.*, **49**, 454 (1962) (English).
15. Borst-Pauwels, G. W. F. H., *Acta Bot. Neer.*, **16**, 125 (1967) (English).
16. Boyer, P. D., H. Lardy, and K. Myrback, *The Enzymes*, Vols. 2 and 3, Academic Press, New York, 1960.
17. Bro-Rasmussen, F., E. Noddegaard, and K. Voldum-Clausen, *J. Sci. Food Agric.*, **19**, 278 (1968).
18. Cecil, L. K., *Chem. Eng. Prog. Symp., Ser.* **63**, 159 (1967).

19. Chargaff, E., C. F. Compton, and R. Lipshitz, *Nature*, **172,** 298 (1953).
20. Chargaff, E., *The Nucleic Acids*, E. Chargaff and J. N. Davidson (Eds.), Vol. I, Academic Press, New York, 1955, p. 307.
21. Chargaff, E., and J. N. Davidson, Eds., *The Nucleic Acids*, Vol. I (1955), Vol. II (1955), Vol. III (1960), Academic Press, New York.
22. Chargaff, E., *Prog. Nucleic Acid Res. Mol. Biol.*, **8,** 297 (1968).
23. Cohen, S. S., *Fed. Proc.*, **20,** 641 (1963).
24. Cosgrove, D. J., *Soil Biochemistry*, A. D. McLarsen and G. H. Peterson (Eds.), Dekker, New York, 1967, p. 216.
25. Davidson, P. F., *Proc. Natl. Acad. Sci. U.S.*, **45,** 1560 (1959).
26. Davidson, P. F., and D. Freifelder, *J. Mol. Biol.*, **16,** 490 (1966).
27. Davidson, P. F., *Cold Spring Harbor Symp. Quant. Biol.*, **33,** 403 (1968).
28. Defrance, A., *C. R.*, **259,** 864 (1964).
29. Demain, A., and J. Birnbaum, *Curr. Top. Microbiol. Immunol.*, **46,** 1 (1968).
30. Dixon, M., and E. C. Webb, *Enzymes*, 2nd ed., Academic Press, New York, 1964.
31. Doskoeil, J., and Z. Sormova, *Biochim. Biophys. Acta*, **95,** 513 (1965).
32. Falcone, G., *Giorn. Microbiol.*, **8,** 129 (1960) (English).
33. Farber, F., *History of Phosphorus*, National Museum Bulletin No. 240, 1965, p. 177.
34. Feng, T. H., *Water and Sewage Works*, **109,** 431 (1962).
35. Franklin, R. E., and R. G. Gosling, *Acta Crystallogr.*, **6,** 678 (1953).
36. Gerretsen, F. C., *Plant and Soil*, **1,** 51 (1948).
37. Goodman, B. L., and K. A. Mikkelson, *Chem. Eng.*, **77,** 75 (1969).
38. Goodman, J., and A. Rothstein, *J. Gen. Physiol.*, **40,** 915 (1957).
39. Goodwin, B. C., *Eur. J. Biochem.*, **10,** 515 (1969) (English).
40. Grunberg-Mango, M., and S. Ochoa, *J. Am. Chem. Soc.*, **77,** 3165 (1955).
41. Grunberg-Mango, M., *Ann. Rev. Biochem.*, **31,** 301 (1962).
42. Hall, M. W., and R. S. Engelbrecht, *Water Wastes Eng.*, **6,** 50 (1969).
43. Harden, A., and W. Young, *J. Chem. Soc.*, **21,** 189 (1905).
44. Harden, A., and W. Young, *Proc. Roy. Soc. (London)*, **B129,** 174 (1906).
45. Harden, A., *Alcoholic Fermentations*, 3rd ed., Longmans, Green, and Company, New York, 1923.
46. Hargrove, R. E., F. E. McDonald, and R. P. Tittsler, *J. Dairy Sci.*, **44,** 1799 (1961).
47. Harold, F. M., R. L. Harold, and A. Abrams, *J. Biol. Chem.*, **240,** 3145 (1965).
48. Harold, F. M., and J. R. Baarda, *J. Bacteriol.*, **91,** 2257 (1966).
49. Harris, I., B. P. Meriwether, and J. H. Park, *Nature*, **198,** 154 (1963).
50. Harwood, J. E., R. A. Steenderen, and A. L. Kuhn, *Water Res.*, **3,** 417 (1969).
51. Harwood, J. E., R. A. Steenderen, and A. L. Kuhn, *Water Res.*, **3,** 425 (1969).
52. Hayes, W., *The Genetics of Bacteria and Their Viruses*, 2nd ed., Wiley, New York, 1968.
53. Heinen, W., *Arch. Microb.*, **41,** 229 (1962).
54. Heinen, W., *Arch. Microb.*, **45,** 145 (1963).
55. Hershey, A. D., and M. Chase, *J. Gen. Physiol.*, **36,** 39 (1952).
56. Holzer, H., and H. Grunicke, *Biochim. Biophys. Acta*, **53,** 591 (1961).

57. Horn, J. A., B. L. DePrater, and J. L. Witherow, *Water Wastes Eng.*, **6**, 40 (1969).
58. Hurwitz, C., C. L. Rasano, and R. A. Peabody, *Biochim. Biophys. Acta*, **72**, 80 (1963).
59. Ikawa, M., *Bacteriol. Rev.*, **31**, 54 (1967).
60. Jones, O. L., and S. M. Bromfield, *Aust. J. Agric. Res.*, **20**, 653 (1969).
61. Kado, C. I., *Nature*, **197**, 925 (1963).
62. Kaltwasser, H., G. Vogt, and H. G. Schlegel, *Arch. Mikrobiol.*, **44**, 259 (1962).
63. Katchman, B. J., *Phosphorus and Its Compounds*, Vol. II, J. R. Van Wazer (Ed.), Interscience, New York, 1961, p. 1337.
64. Kates, M., "Bacterial Lipids," in *Advances in Lipid Research*, Vol. 2, R. Paoletti and D. Kritchevsky (Eds.), Academic Press, New York, 1964, p. 17.
65. Katznelson, H., E. A. Peterson, and J. W. Rouatt, *Can. J. Bot.*, **40**, 1181 (1962).
66. Kitamikado, M., *Hakko Zasshi*, **37**, 242 (1959); *Chem. Abstr.*, **54**, 16534*C* (1959).
67. Kobus, J., *Rocz. Nauk Roln., Ser. D*, **91**, 5 (1961); *Chem. Abstr.*, **66**, 6417*a* (1961).
68. Kobus, J., *Acta Microbiol. Pol.*, **11**, 265 (1962) (English).
69. Kroner, R. C., and M. E. Gales, *U.S. Fed. Water Pollut. Control Adm.*, 28 pp. (1966).
70. Lear, D. W., Jr., "Bacterial Participation in Phosphate Regeneration in Marine Ecosystems," *Diss. Abstr.*, **B27** (2), 359 (1966), Order No. 66-4407.
71. Leggett, J. E., and R. Olsen, *Plant Physiol.*, **39**, 387 (1964).
72. Lehman, I. R., M. J. Bessman, E. S. Simms, and A. Kornberg, *J. Biol. Chem.*, **233**, 163 (1958).
73. Levin, G. V., and J. Shapiro, *J. Water Pollut. Control Fed.*, **37**, 800 (1965).
74. Lichtenstein, E. P., and K. R. Schultz, *J. Econ. Entomol.*, **57**, 618 (1964).
75. Lichtenstein, E. P., T. W. Fuhremann, and K. R. Schultz, *J. Agric. Food Chem.*, **16**, 870 (1968).
76. Louw, H. A., and D. M. Webley, *J. Appl. Bacteriol.*, **22**, 227 (1959).
77. Luoma, H., *Arch. Oral Biol.*, **13**, 1331 (1968) (English).
78. Macfarlane, M. G., *Metabolism and Physiological Significance of Lipids*, R. M. C. Dawson and D. N. Rhodes (Eds.), Wiley, New York, 1964, p. 399.
79. Mahler, H. R., and E. H. Cordes, *Biological Chemistry*, Harper and Row, New York, 1966.
80. Malkov, A. M., and V. E. Deeva, *Mikrobiol.*, **30**, 229 (1961); *Chem. Abstr.*, **56**, 3908*b* (1961).
81. Mallin, M. L., and N. O. Kaplan, *J. Bacteriol.*, **77**, 125 (1959).
82. Martin, J. P., "Pesticides, Their Effects on Soils and Water," Symposium Papers, Columbia, Ohio, 1965, pp. 95–108.
83. Matsumura, F., and G. M. Boush, *Science*, **153**, 1278 (1966).
84. McLachlan, K. D., and B. W. Norman, *J. Aust. Inst. Agric. Sci.*, **27**, 244 (1961).
85. Merzari, A. H., S. F. Suarez, and A. Godoy, *Arg. Rep., Com. Nacl. Energ. At., Inf.*, **93**, 8 (1963).
86. Mitchell, P., and J. M. Moyle, *J. Gen. Microbiol.*, **9**, 257 (1953).
87. Morman, M. R., and D. C. White, *J. Bacteriol.*, **104**, 247 (1970).
88. Motohashi, K., C. Matsudaria, and J. Tohoku, *Agric. Res.*, **20**, 73 (1969).
89. Moustafa, H. H., and E. B. Collins, *J. Bacteriol.*, **95**, 592 (1968).

90. Nelson, D., and A. Kornberg, *J. Biol. Chem.*, **245,** 1146 (1970).
91. Nesbitt, J. B., *J. Water Pollut. Control Fed.*, **41,** 701 (1969).
92. Nordheim, W., and A. Rieche, *Naturwiss.*, **53,** 441 (1966).
93. Northrop, J. H., *J. Gen. Physiol.*, **34,** 715 (1957).
94. O'Brein, R. T., *Proc. Soc. Exp. Biol. Med.*, **117,** 555 (1964).
95. Owens, R. G., *Ann. N.Y. Acad. Sci.*, **160,** 114 (1969).
96. Pearson, R. W., A. G. Norman, and C. Ho, *Soil Sci. Soc. Am. Proc.*, **6,** 168 (1941).
97. Pikovskaya, R. I., *Soobshch. Akad. Nauk. Gruz. SSR*, **56,** 413 (1969); *Chem. Abstr.*, **72,** 97561*f* (1969).
98. Postgate, J. R., and J. R. Hunter, *J. Gen. Microbiol.*, **34,** 459 (1964).
99. Rice, A. H., U.S. Patent No. 3,171,804 (1965).
100. Richards, O. C., and P. D. Boyer, *J. Mol. Biol.*, **11,** 327 (1965).
101. Richardson, C. C., *J. Mol. Biol.*, **15,** 49 (1966).
102. Roche, A., and H. deBarjac, *Ann. Inst. Pasteur*, **96,** 781 (1959).
103. Rose, R. E., *N.Z. J. Sci. Technol.*, **38B,** 773 (1957).
104. Salmon, J., *C. R. Soc. Biol.*, **145,** 1413 (1951).
105. Scalf, M. R., F. M. Pfeffer, L. D. Lively, J. L. Witherow, and C. P. Priesing, *J. Sanit. Eng. Div., Am. Soc. Civil Eng.*, **95,** 817 (1969).
106. Schalscha, B. E., and O. Bentjerodt, *Agric. Technol.*, **29,** 24 (1969); *Chem. Abstr.*, **71,** 80220*a* (1969).
107. Scheinpflug, H., and H. F. Jung, *Pflangenschultz-Nachr. Bayer.*, **21,** 79 (1968).
108. Sen, A., and N. B. Paul, *Curr. Sci. (India)*, **26,** 222 (1957).
109. Shaw, N., and J. Baddiley, *Nature*, **217,** 142 (1968).
110. Shockman, G. D., *Bacteriol. Rev.*, **29,** 345 (1965).
111. Short, S. A., and D. C. White, *J. Bacteriol.*, **104,** 126 (1970).
112. Sinsheimer, R. L., *Science*, **125,** 1123 (1957).
113. Stahl, F. W., *The Viruses*, Vol. 2, F. M. Burnet and W. M. Stanley (Eds.), Academic Press, New York, 1959.
114. Stethers, H., *Department of Research and Control, Research Report*, 1968, 25 pages.
115. Subba-Rao, N. S., and P. D. Bajpai, *Experientia*, **21,** 386 (1965).
116. Sundara-Rao, W. V. B., and M. K. Sinha, *Indian J. Agric. Sci.*, **33,** 272 (1963).
117. Tardieux-Roche, A., *Ann. Inst. Pasteur*, **103,** 314 (1962).
118. Tardieux-Roche, A., *Ann. Inst. Pasteur*, **103,** 565 (1964).
119. Thomas, C. A., and L. A. MacHattie, *Ann. Rev. Biochem.*, **36,** 485 (1967).
120. Varma, M. M., *Water and Sewage Works*, **110,** 364 (1963).
121. Velick, S. F., *Methods in Enzymology*, Vol. 1, S. P. Colowick and N. O. Kaplan (Eds.), Academic Press, New York, 1955, p. 401; E. G. Krebs, *ibid.*, Vol. I, p. 407; M. Gibbs, *ibid.*, Vol. I, p. 411.
122. Wahl, R., and L. Blum-Emerique, *Ann. Inst. Pasteur*, **80,** 155 (1951).
123. Watson, J. D., and F. H. C. Crick, *Cold Spring Harbor Symp. Quant. Biol.*, **18,** 123 (1953).
124. Watson, J. D., and F. H. C. Crick, *Nature*, **171,** 737 (1953).
125. Watson, J. D., *Molecular Biology of the Gene*, Benjamin, New York, 1965, p. 342.

126. Watson, J. D., *Molecular Biology of the Gene*, Benjamin, New York, 1965, p. 359.
127. Watt. W. D., and F. R. Hayes, *Limnol. Oceanogr.*, **8,** 276 (1963).
128. Weaver, P. J., *J. Water Pollut. Control Fed.*, **41,** 1647 (1969).
129. Welsch, M., J. Salmon, and C. Heusghem, *C. R. Soc. Biol.*, **143,** 1152 (1949).
130. Welsch, M., *C. R. Soc. Biol.*, **141,** 1253 (1951).
131. Wilkins, M. F. H., A. R. Stokes, and H. R. Wilson, *Nature*, **171,** 738 (1953).

28

The Phosphatases

JOSEPH FEDER

New Enterprise Division, Monsanto Company, St. Louis, Missouri

The general title of phosphatase has been used to describe a broad group of enzymes which catalyze the hydrolysis of both esters and anhydrides of phosphoric acid (1). The Commission on Enzymes of the International Union of Biochemistry has classified all of these enzymes into five major groups (2). These include the phosphoric monoester hydrolases (EC 3.1.3), phosphoric diester hydrolases (EC 3.1.4), triphosphoric monoester hydrolases (EC 3.1.5), enzymes acting on phosphoryl-containing anhydrides (EC 3.6.1), and enzymes acting on P–N bonds (EC 3.9) such as the phosphoamidases (EC 3.9.1.1). This chapter presents a limited review of the distribution and physical, chemical, and enzymatic properties of these enzymes, with emphasis on the phosphomonoesterases and only cursory examination of the others. Derivatives of phosphoric acid play a major role in all aspects of cellular metabolism. All living systems, from the primitive monocellular organism to the most complicated mammalian structure, are equipped with various enzymes that can catalyze the hydrolysis of these compounds. The recent growth of interest and concern over environmental problems has focused attention on the effect of phosphate on our ecology. A general review of the phosphatases would be of interest in trying to understand the role of biological systems such as various microorganisms in maintaining a normal ecology after stresses by polyphosphate and other phosphate derivatives.

PHOSPHOMONOESTERASES (PHOSPHORIC MONOESTER HYDROLASE, EC 3.1.3)

The phosphomonoesterases catalyze the hydrolysis of monoesters of orthophosphoric acid. Among the enzymes that have been subclassified under this heading are the alkaline phosphatases and acid phosphatases with rather broad range of substrate specificities and the more specific enzymes such as the phosphoserine phosphatases, nucleotidases, hexose mono- and diphosphatases, and glycerol phosphatases. The phosphomonoesterases have been shown to catalyze P—O bond cleavage (1, 3). These enzymes (as well as phosphodiesterases) also have transferase activity (1, 4–9). The stereochemical specificity of the transferase activity appears to be much more selective than the hydrolase activity (10, 11). The reader is referred to the review edited by W. H. Fishman (11a).

Alkaline Phosphatases (EC 3.1.3.1)

The rather nonspecific alkaline phosphatases have been isolated from a wide range of sources including microorganisms to mammalian tissues. They share characteristic alkaline pH optima and act on a wide range of phosphate substrates. The most extensively studied phosphomonoesterase is the alkaline phosphatase from *Escherichia coli* (12–36). Horiuchi *et al.* (15) and Torriani (16) showed that various *E. coli* strains produced high levels of intracellular alkaline phosphatase in growth media depleted of phosphate, suggesting a negative feedback mechanism with phosphate as the repressor. This was in marked contrast to the intracellular acid phosphatase which was present at all times at a constant level. The enzyme has been isolated by a variety of procedures from broken cell extracts (17–20).

Malamy and Horecker (21) isolated crystalline enzyme from the supernatant fraction obtained by converting *E. coli* cells into spheroplasts with lysozyme and EDTA (22, 23). Evidence has been presented by Brockman and Heppel (24) that the *E. coli* alkaline phosphatase together with the acid phosphatase and cyclic phosphodiesterase is localized near the surface of the cell, external to the protoplasmic membrane. The enzyme is isolated from the cell as a dimer of about 86,000 molecular weight (25) containing two identical subunits (26).

Reynolds and Schlesinger (27–29) have demonstrated that the dimer can be reversibly dissociated in the presence of 6 *M* guanidine hydrochloride. The alkaline phosphatases have been shown to be metalloenzymes containing tightly bound zinc which was functionally required (30–32). The *E. coli* alkaline phosphatase has been shown to contain 4 g atoms of zinc per dimer of protein of molecular weight of 89,000 (20, 33, 34). Simpson and Vallee (35) have presented evidence that two of these zinc atoms have a catalytic function and two of them have some structural role. The alkaline phosphatases catalyze the hydrolysis of a broad range of phosphate esters including *p*-nitrophenyl phosphate (17–20), nucleoside polyphosphates, deoxynucleoside polyphosphate, inorganic pyrophosphate, tripolyphosphate, and short-chain metapolyphosphate (36). None of these enzymes exhibits phosphodiesterase activity (13, 14). Garen and Levinthal (17, 18) reported that the *E. coli* enzyme catalyzed the hydrolysis of *p*-nitrophenyl phosphate, glucose-1-phosphate, adenosine-3′- or 5′-phosphate, riboflavin-5′-phosphate, L-histidinol phosphate and a few pyrimidine mononucleotides with equal facility. Anderson and Nordlie (37) demonstrated that highly purified alkaline phosphatase from *E. coli* also catalyzed the transfer of a phosphoryl group from pyrophosphate to glucose. This supported a mechanism involving the formation of phosphoryl–enzyme intermediate. The formation of a phosphoprotein upon incubation of alkaline phosphatases including the calf intestinal enzyme with labeled inorganic phosphate was reported by Engstrom and Agren (38–41) and others (42–44), and the phosphate was shown to be covalently linked to serine by Schwartz and Lipmann (19). Several workers have presented kinetic evidence to support the thesis that a phosphoryl–enzyme intermediate is formed during hydrolysis of phosphate esters by alkaline phosphates (45–51).

The presence of an alkaline phosphatase in extracts of *Neurospora crassa* has been reported (52–54). Kuo and Blumenthal (55) purified the enzyme from the mold mycelium and reported a maximum activity at about pH 9.0 with β-glycerol phosphate as substrate. The enzyme was active toward a number of phosphate derivatives including α- and β-glycerol phosphate, glucose-6-phosphate, glucosamine-6-phosphate, ribose-5-phosphate, adenosine-5′-phosphate, L-histidinol phosphate, *p*-nitrophenyl phosphate and phospho-L-homoserine but inactive toward acetyl phosphate, potassium phosphoramidate, potassium pyrophosphate, thiamine pyrophosphate, and phospho-DL-serine and phospho-DL-threonine. The formation of this alkaline phosphatase, unlike the *E. coli* enzyme and similar to the acid phosphatase, was not repressed by the presence of phosphate in the growth medium.

An orthophosphate-repressible alkaline phosphatase has been shown to exist in a wild-type strain of *N. crassa* which is harvested from media that have limiting concentrations of inorganic phosphate (56). The enzyme was shown to be a dimer of molecular weight of 154,000 which could be dissociated into 77,000 molecular weight subunits in the presence of 6 *M* guanidine–mercaptoethanol (57). The enzyme, which was a glycoprotein (11.5% carbohydrate), represented 1% of the total protein in derepressed cultures of the mold (57). Kadner and Nyc (58) also found that unlike the *E. coli* and other alkaline phosphatases, EDTA stimulated the activity. The pH dependence of the activity toward *p*-nitrophenyl phosphate in the absence of EDTA exhibited a plateau between pH 6.5 and 8.0 with increasing activity above pH 8.0 (58). Gorman and Hu (59) isolated a nonspecific alkaline phosphatase from *Saccharomyces cerevisiae* together with a histidinol phosphatase. The alkaline phosphomonoesterase hydrolyzed a variety of phosphomonoesters including L-histidinol phosphate, *p*-nitrophenyl phosphate, 5′-AMP, α-glycerophosphate, glucose-6-phosphate, and *O*-phosphoserine. The enzyme had optimum activity toward both *p*-nitrophenyl phosphate and L-histidinol phosphate between pH 8.5 and 8.8 with *Km* values of 5.2×10^{-4} and 1.2×10^{-4} for histidinol phosphate and *p*-nitrophenyl phosphate, respectively. The separation of two alkaline phosphatases from baker's yeast extracts has been reported by Attias and co-workers (60). One fraction with broad substrate specificity exhibited a pH optimum of 10 with *p*-nitrophenyl phosphate and was activated by Mg^{2+} and inhibited by Be^{2+} and Zn^{2+}. The other fraction was specific for nitrophenyl phosphate with an optimum activity at pH 8.4. Wood and Tristram (61) reported the localization of alkaline phosphatase of *B. subtilis* in the protoplasmic membrane. Nonspecific alkaline phosphatases have also been isolated from a variety of mammalian tissue (13) including kidney, bone marrow (62, 63), and intestinal mucosa (32, 39, 64–67). The calf intestinal alkaline phosphatase has been studied in great detail by Engstrom (32, 39) and has been shown to incorporate inorganic phosphate-forming phosphorylserine. Rosenblum *et al.* (68) purified an alkaline phosphate from guinea pig bone marrow which contained 3 to 5 g atoms Zn/400,000 g protein. Alkaline phosphatases have been prepared also from amniotic fluid (69) and placenta (70–72).

Unlike alkaline phosphatases of other tissues, the human placental enzyme is heat stable (72) and is the major source of serum alkaline phosphatase during pregnancy (73, 74). The presence of nonspecific alkaline monophosphatases in several snake venoms has been reported (75–77).

Acid Phosphatases (Orthophosphoric Monoester Phosphohydrolases, EC 3.1.3.2)

The nonspecific acid phosphatases like the alkaline phosphatases display broad specificity toward a number of phosphomonoesters but with optimum activity in acidic solution. The enzymes are widely distributed (78) having been isolated from yeast (79–82), *Escherichia coli* (83–86), *Escherichia freundii* (87), *Neurospora crassa* (54, 55, 88–90), potatoes (91–96), wheat leaf (97–99), wheat germ (100), tobacco leaf (101), lupine seedlings (102), soybean (103, 104), and a number of animal tissues including liver (105–110), brain (111–113), prostate (114–119), spleen (120–122), placenta (123–126), milk (127), erythrocytes (128–132), and bone (133–137).

Van Hofsten and Porath (83) reported the presence of several acid phosphatases in *Escherichia coli* and purified one of these by chromatography, zone electrophoresis, and gel filtration. The enzyme was very stable in 1 *M* acetic acid in the absence of salts. The enzyme exhibited optimum activity toward *p*-nitrophenyl phosphate between pH 4 and 5 with a *Km* of 3×10^{-3} *M*. Unlike the alkaline phosphatase from *E. coli*, the formation of acid phosphatase was insensitive to phosphate repression (15, 16, 83, 83). The minimal molecular weight has been estimated to be 13,000 to 14,000 (83).

The *E. coli* enzyme catalyzed the hydrolysis of glucose-6-phosphate, glucose-1-phosphate, and *p*-nitrophenyl phosphate, but not α-glycerophosphate and adenosine-5-phosphate (83). A crystalline acid phosphatase was prepared from extracts of *Escherichia freundii* by Tani and workers (87). The enzyme had optimum activity at pH 6.0 with *p*-nitrophenyl phosphate and also hydrolyzed sugar phosphates, nucleosides, phenyl phosphate, and acetyl phosphate. The enzyme also possessed transphosphorylation activity toward pyridoxine concomitant with the hydrolase.

The studies of Rautanen and Kaerkkaeinen (81) with *Torulopsis utilis* and those of Schmidt *et al.* (82) with baker's yeast revealed the presence of intracellular phosphate-repressible acid phosphatases. Schmidt *et al.* (80) suggested that the acid phosphatase of baker's yeast was localized in a cell compartment near the cell surface. The enzyme had an optimum activity near pH 3.6 toward phenyl phosphate. A number of workers have shown that the acid phosphatase of baker's yeast with a pH optimum between 3 and 4 was in the cell wall (80, 138–141).

Baer and Steyn-Parve (142–144) isolated and studied the mechanism of action of this acid phosphatase. They were not able to isolate a phosphoryl–enzyme intermediate and consequently proposed a noncovalent single displacement mechanism for the yeast acid phosphatase analogous to the nonenzymatic hydrolysis of phosphomonoesters. This is quite different from the alkaline phosphatases and the human prostatic acid phosphatase which Greenberg and Nachmansohn (145) have suggested form phosphoryl–enzyme intermediates.

Kuo and Blumenthal purified an acid phosphomonoesterase from *Neurospora crassa* (54, 88). The enzyme had a pH optimum of 5.6 with β-glycerol phosphate and glucosamine-6-phosphate. The substrate specificity for phosphomonoester bond was quite broad, including phosphomonoesters of both primary and secondary alcohols, but no catalytic activity was observed toward phosphodiesters, inorganic or organic pyrophosphate, or phosphoprotein. The addition of various metallic ions did not affect the activity and likewise no inhibition was observed with various chelating agents such as EDTA, potassium cyanide, and diethyl dithiocarbonate (54).

Nyce (90) described the presence of a phosphate-repressible alkaline phosphatase in the culture media and mycelia of a normal strain of *N. crassa* grown on limiting phosphate. The enzyme had an optimal activity at pH 5.6. This enzyme was more substrate specific than the other phosphomonoesterases of *N. crassa* exhibiting phosphodiesterase activity and is also active toward phosphoenolpyruvate and phenyl phosphate but inactive toward phosphocholine, phosphoserine, sugar phosphates, β-glycerophosphate, DPN, AMP, pyrophosphate, and trimetaphosphate (90).

Alvarez (94) studied the kinetics of the potato acid phosphatase-catalyzed hydrolysis of β-glycerophosphate and nitrophenyl phosphate and proposed a mechanism for the hydrolysis of phosphoric acid monoesters involving a phosphorylated enzyme intermediate. A nucleophilic attack of the tertiary nitrogen of imidazole on the phosphorus atom of the substrate was suggested for the formation of the phosphoryl–enzyme intermediate.

Hsu and his associates (96) have reported a rapid method for preparation of a potato nonspecific phosphomonoesterase and a study of its mechanism of action with *p*-nitrophenyl acetate and β-glycerophosphate. Studying the product inhibition of these reactions they concluded that the mechanism must have an ordered release of products, first the alcohol and then the inorganic phosphate. The data also suggested the involvement of two enzyme–phosphate complexes.

The isolation and purification of a nonspecific acid phosphatase from tobacco leaves have been reported by Shaw (101). The enzyme was shown to catalyze both the hydrolysis of a number of phosphomonoesters including *p*-nitrophenyl and phenyl phosphate, hexose phosphates and diphosphates, α- and β-glycerophosphate, and mono- and diphosphoglyceric acid mononucleosides, and also pyrophosphates such as ATP, ADP sodium pyrophosphate, thiamine pyrophosphate, and NADP. The enzyme exhibited an optimum activity at pH 5.5 to 5.7 with several substrates and did not require divalent cations for activity.

Among the acid phosphatases from animal tissue, the enzyme from prostate has been known for some time (78, 146). Ostrowski and co-workers (114, 116) have purified the enzyme from human prostate gland by a method involving ammonium sulfate fractionation, DEAE cellulose chromatography, and gel filtration. This enzyme had a pH optimum of 4.8 with *p*-nitrophenyl phosphate.

Nigam *et al.* (119, 147) reported optimum activity for the human acid phosphatase at pH 4.9 to 5.0 with phenyl phosphate, pH 4.9 to 5.5 for nitrophenyl phosphate, and pH 5.5 to 6.0 with β-glycerophosphate, depending on the buffer employed. All three substrates were markedly inhibited by L-tartrate.

A molecular weight of 95,000 has been reported for this enzyme (114). The presence of two unreactive sulfhydryl groups per molecule has been reported which when reacted with *p*-chloromercuribenzoic acid or Ag^+ causes only partial loss of activity (117).

Greenberg and Nachmansohn (145) found that diisopropyl fluorophosphate inhibited the acid phosphomonoesterases but that, unlike the inhibition observed with acyl esterases or proteases, the inhibition was removed by removal of the inhibitor by dialysis or dilution. No stable phosphorylated enzyme intermediate could be isolated from these experiments.

Incubation of the acid phosphatase with labeled inorganic phosphate resulted in the binding of about 0.3 moles of phosphate per mole of enzyme, but only traces of *O*-phosphorylserine were isolated after acid hydrolysis of the enzyme. Although a phosphoryl–enzyme intermediate has been postulated for the acid phosphomonoesterases, it was suggested that the phosphate was attached to a nitrogen atom such as a histidine (145).

Heinrikson (105) has reported the presence of multiple forms of acid phosphatases in various mammalian tissues using gel filtration. He isolated three acid phosphatases from bovine kidney, three from bovine liver, and three from bovine pancreas as well as two or three from bovine spleen, porcine kidney and liver, and rat kidney, liver, and spleen. The bovine liver acid phosphatase **3** was isolated and purified and shown to be a low molecular weight enzyme of 16,590. The enzyme has a pH optimum of 5.5 toward *p*-nitrophenyl phosphate, with a Km of 7.5×10^{-4} M. Although the enzyme was most active toward *p*-nitrophenyl phosphate, good activity was also observed toward flavin mononucleotide and galactose-6-phosphate. The enzyme was stabilized by phosphate but was inactivated by Mg^{2+} or mercaptoethanol in the absence of EDTA (105) unlike the spleen (bovine) and yeast enzymes (79) which require mercaptoethanol or Mg^{2+}, respectively, for optimum activity.

Igarashi *et al.* (107) reported the role of a single histidine for the enzymatic activity of an acid phosphatase from rat liver. A^{32}P-labeled phosphohistidine was isolated from alkaline hydrolyzates of the enzyme which had been incubated in ^{32}P–phosphate. Multiple acid phosphatases have been found in liver extracts from mouse (109, 110, 148–150) and other species (151, 152). The presence of acid phosphatase in brain tissue has been reported (112, 113, 153–155).

Chaimovich and Nome (111) purified and characterized a low molecular weight acid phosphatase from bovine brain. The enzyme was active toward *p*-nitrophenyl phosphate, α-glycerophosphate, *o*-phosphorylethanolamine, and UMP but had no pyrophosphatase activity, glucose or fructose phosphatase, or ATPase or ADPase activity. A broad pH optimum was observed between pH 4.8 and 5.8 with *p*-nitrophenyl phosphate ($Km = 7 \times 10^{-5}$ M). The brain enzyme was completely inhibited by Hg^{2+}, *p*-chloromercuribenzoate, and *N*-ethylmaleimide suggesting a role for an SH group in the activity. Nonspecific acid phosphomonoesterases have been isolated from red blood cells (129–132, 156). Georgatsos (128) separated two acid phosphatases from human erythrocytes by gel filtration. The two enzymes differ with respect to activation or inhibition, pH optima (pH 6.0 and 5.2), and molecular weight. The presence of two or more acid phosphatases in placental extracts has been reported by a few investigators (123–126). DiPietro and Zengerle (123) separated three distinct acid phosphatases from human placental millus tissue by gel filtration. These enzymes had molecular weights of 35,000, 105,000, and greater than 200,000 with pH optima of 5.5, 4.0, and 5.5, respectively. Acid phosphatases also have been isolated and characterized from bone (133–137), spleen (120–122), and milk (127).

Phosphoserine Phosphohydrolase (EC 3.1.3.3)

The phosphoserine phosphatase are specific phosphomonoesterases which catalyze the hydrolysis of phosphoserine (157). The enzyme has been found in microorganisms including *E. coli* (158, 159), baker's yeast (160, 161), and in animal tissues including brain (162) and chicken (11, 163, 164) and rat liver (165).

Pizer (158) isolated the *E. coli* phosphoserine phosphatase by DEAE cellulose chromatography. The enzyme was activated by divalent metal ions such as Mg^{2+}, Co^{2+}, Ni^{2+}, and Mn^{2+} and inhibited by fluoride. The phosphatase exhibited phosphotransferase activity, transferring a phosphoryl group from phosphoserine to another serine. The enzyme was inhibited by serine (158) and was very specific for phosphoserine. No phosphatase activity was found with *p*-nitrophenyl phosphate, DL-phosphothreonine, α- and β-glycerophosphate, and phosphoethanolamine.

A very specific phosphoserine phosphatase also was isolated from dried baker's yeast by Schramm (160, 161). The enzyme required a divalent cation for activity with maximal rates in the presence of magnesium ions. A pH optimum of 6.5 to 7.0 was obtained with L-phosphoserine with a *Km* of 4×10^{-4} *M*. Fluoride, manganese, and calcium ions were inhibitors. The enzyme did not dephosphorylate 3-phosphoglyceric acid, various phosphosugars, phosphothreonine, and adenosine mono-, di- and triphosphates. Neuhaus and Byrne (166) purified a phosphatase specific for phosphoserine from aqueous extracts of chicken liver acetone powder. The enzyme catalyzed transphosphorylation reactions between phosphoserine and serine. The enzyme had optimum phosphatase activity at pH 5.9 to 6.6 and optimum transphosphorylase activity at pH 6.9 to 7.3. A divalent cation, preferably Mg^{2+}, was required. A double displacement mechanism including a phosphoryl–enzyme intermediate was proposed.

Bridgers (162) partially purified a phosphoserine phosphatase from mouse brain. Both phosphatase activity, pH optimum 6.2, and phosphotransferase activity, pH optimum 7.5, were exhibited by the enzyme. The two activities can be inhibited separately; urea inhibits hydrolytic activity and sucrose inhibits transferase activity, suggesting a different mechanism from that proposed by Neuhaus and Byrne (166).

Phosphatidate Phosphatase (L-α-Phosphatidate Phosphohydrolase, EC 3.1.3.4)

The phosphatidate phosphatase catalyzes the hydrolysis of L-α-phosphatidate to a diglyceride and orthophosphate. The presence of this enzyme in chicken liver and various mammalian tissues has been reported (167–174). The phosphatidic acid phosphatase from chicken liver was specific for phosphatidic acid, did not catalyze the hydrolysis of lecithin, and was only very slightly active toward α- and β-glycerophosphate and glucose-6-phosphate (168). The enzyme was active over a pH range between 5.0 and 8.0 with maximum activity at pH 6.0 to 6.5. The enzyme was inhibited by small amounts of divalent cations including magnesium, calcium, manganese, and barium. Hubscher and Coleman (169) purified a phosphatidic acid phosphatase from microsomal fractions of pig kidney. The enzyme exhibited a pH optimum at 6.0 for phosphatidic acid ($Km = 2.2 \times 10^{-4}\ M$) and was inhibited by fluoride ions. Inhibitor studies suggested that thiol groups were essential for activity.

5′-Nucleotidase and 3′-Nucleotidase (5′-Ribonucleotide Phosphohydrolase, EC 3.1.3.5; 3′-Ribonucleotide Phosphohydrolase, EC 3.1.3.6)

The 5′- and 3′-nucleotidases catalyze the hydrolysis of 5′- and 3′-ribonucleotides, respectively, to yield orthophosphate and a ribonucleoside. The enzymes are very specific for the 5′- or 3′-ribonucleotides. The presence of the 5′-nucleotidase has been reported in numerous tissues (175–194).

The earliest studies of 5′-nucleotidase were reported by Reis (175, 191–193) who studied various mammalian tissues. Heppel and Hilmoe (177) purified a 5′-nucleotidase from bull seminal plasma and showed that it was active with adenosine-5-phosphate, inosine-5-phosphate, nicotinamide ribose-5′-phosphate, uridine-5-phosphate, and cytidine-5-phosphate. The activity toward adenosine-3-′-phosphate was removed during purification.

Segal and Brenner (185) reported studies on the 5′-nucleotidase from rat-liver microsomes. The liver enzyme had maximal activity at neutral pH and exhibited no requirement for added magnesium ions in contrast to the seminal plasma enzyme (177). The enzyme was inhibited by inorganic phosphate and various nucleosides including uridine, adenosine, inosine, and cytidine. A Km of about $1.0 \times 10^{-5}\ M$ was obtained for adenosine-5′-phosphate substrate. The substrate specificity studies indicated 5′-AMP to be the best substrate, 5′-deoxy-AMP quite poor, while 2′-AMP, 3′-AMP, and ribose-5-phosphate were not substrates. Song and Bodansky (189) showed that the 5′-nucleotidase of rat liver was localized in the microsomal membranes. In the absence of added metal ions, the enzyme exhibited optimum activity with AMP at pH 7.5 but in the presence of 0.01 *M* Mg^{2+} a second pH optimum was reported at pH 9.3. The pH dependence and double pH optimum were similar to what was reported for the 5′-nucleotidases from bull seminal plasma and for the enzyme from human liver (195). A mechanism for this double pH optimum and the effect of magnesium has been proposed by Levin and Bodansky (188) involving four sites on the enzyme for substrate interaction.

Itoh *et al.* (196) partially purified a 5′-nucleotidase from chicken liver and found it to be kinetically distinct from 5′-nucleotidases from other sources. The enzyme had an optimum activity at pH 6.5 with 5′-IMP, 5′-GMP, or 5′-AMP. The enzyme was inactive in the absence of divalent metal ions. Maximum activity was obtained with Mg^{2+}, with Co^{2+}, Mn^{2+}, and Ni^{2+} giving less activity, and no effect observed with Ca^{2+}. The substrate specificity of the chicken enzyme was quite different from other 5′-nucleotidases with 5′-IMP, 5′-GMP, and 5′-XMP being hydrolyzed considerably faster than 5′-AMP, 5′-UMP, and 5′-CMP. Inosine, guanosine, *p*-chloromercuribenzoate, and NaF inhibited the enzyme. Ipata (187) reported the presence in sheep brain of a 5′-nucleotidase with a molecular weight of about 140,000. The enzyme was inhibited by ATP, UTP, and CTP characterized as a mixed competitive and noncompetitive type which was increased at elevated temperatures. The inhibition by ATP was reversed by inorganic phosphate which is not an inhibitor. Treatment with *p*-mercuribenzoate desensitized the enzyme to ATP and UTP inhibition but not to CTP inhibition. Gulland and Jackson (176) first showed the presence of 5′-nucleotidase activity in venoms of various species of snake, and methods have been reported for their separation from other venom enzymes (197, 198).

Sulkowski *et al.* (181) purified 5′-nucleotidase from the venom of *Bothrops atrox.* approximately 1000-fold. The enzyme, which had a pH optimum around 9.0, was most active toward 5′-AMP, but also catalyzed the hydrolysis of the phosphate from 5′-IMP, 5′-UMP, 5′-CMP, 5′-GMP, and the corresponding deoxynucleotides. No activity, however, was observed with 3′-AMP, ATP, ribose-5-phosphate, inorganic pyrophosphate, or *p*-nitrophenyl phosphate. EDTA inactivated the enzyme.

Burger and Lowenstein (190) prepared a 5′-nucleotidase from the smooth muscle from porcine small intestine. The enzyme was strongly inhibited by ADP and the α,β-methylene phosphonate analog of ADP. Three pH optima were observed, the lowest in the absence of inhibitory buffer anions and the highest only in the presence of Mg^{2+} as was reported for other 5′-nucleotidases. Neu and Heppel (183) reported the release of a Co^{2+} stimulated 5′-nucleotidases from *E. coli* cells when they were converted to spheroplasts.

The enzyme, which was quite specific for 5′-ribonucleotides and 5′-deoxyribonucleotides, was inactive in the absence of divalent metal ions. The most effective stimulation was obtained with 0.01 *M* Co^{2+} and 0.01 *M* Ca^{2+}. The pH optimum was 5.8.

Neu and Chou (199) showed that the *Enterobacteriaceae* contains 5′-nucleotidases which can be released from the cells by osmotic shock. The purification and characterization of some of these enzymes have been reported (185). Shuster and Kaplan (200, 201) first described a β′-nucleotidase in a number of plant materials and purified the enzyme from germinating barley and rye grass. It had a pH optimum of 7.5 and was competitively inhibited by the 5′-nucleotides that were not substrates.

Cunningham (202) reported that the rye-grass enzyme did not phosphorylate deoxyribonucleoside 3′-phosphates. This was further defined by Walwick and co-workers (203) who proposed that the 3′-nucleotidase of grass requires a free cis hydroxyl at C-2 of the carbohydrate.

Barker and Lund (204) showed that the enzyme was specific for ribonucleoside 3′-phosphate and were not able to catalyze the hydrolysis of an arabinonucleoside 3′-phosphate.

Glucose-6-Phosphatase (D-Glucose-6-Phosphate Phosphohydrolase, EC 3.1.3.9)

Glucose-6-phosphatase is a microsomal enzyme isolated from liver (205–208, 210), small intestine (212), and kidney (211) which not only catalyzes the hydrolysis of glucose-6-phosphate but which also catalyzes the transfer of a phosphoryl group to the 6 position of glucose from pyrophosphate nucleoside di- or triphosphate or other sugar phosphate esters (157, 205–212). The enzyme also catalyzes the hydrolysis of inorganic pyrophosphate and various di- and triphosphonucleosides (206, 207, 210). A reaction mechanism suggesting a phosphoryl–enzyme intermediate has been proposed (213). This work has been reviewed by Nordlie (205). The pH dependence of the enzyme is changed as a result of deoxycholate treatment. The untreated enzyme exhibited a pH optimum of 6.0, 5.0, and 4.5 for glucose-6-phosphatase, pyrophosphatase, and phosphotransferase activity, respectively, as compared to pH 6.5, 6.0 and 6.0 for the respective activities with the treated enzyme (205). Orthophosphate was shown to inhibit both the phosphohydrolase and phosphotransferase activity of the liver enzyme (214–216). Bicarbonate also has been reported to inhibit the enzyme (217). The inhibition of the rat-liver microsomal glucose-6-phosphatase by 1,10-phenanthroline, sodium azide, diethyl dithiocarbonate, and 8-hydroxyquinoline has suggested that this enzyme is a metalloenzyme (218).

Glucose-1-Phosphatase (D-Glucose-1-Phosphate Phosphohydrolase, 3.1.3.10)

A specific glucose-1-phosphatase from silkworm blood was reported by Faulkner (219). Only glucose-1-phosphate, galactose-1-phosphate, and *p*-nitrophenyl phosphate were hydrolyzed by the enzyme. A pH optimum of 4.0 to 4.5 was exhibited with glucose-1-phosphate substrate. The enzyme was inhibited by fluoride, phosphate, and arsenate. Turner and Turner (220, 221) reported a phosphatase from pea seeds that catalyzed the hydrolysis of both glucose-1-phosphate and glucose-6-phosphate. Maximum activity toward both the glucose-1-phosphate and glucose-6-phosphate was observed at pH 5.4 to 5.7. Both activities were competitively inhibited by orthophosphate.

Hexosediphosphatase (D-Fructose-1,6-Diphosphate 1-Phosphohydrolase, EC 3.1.3.11) (227–261)

Hexosediphosphatase, a monoester phosphohydrolase, has been isolated from a variety of sources including rabbit liver (222–232), rat liver (233), the yeast *Candida utilis* (234–236), spinach leaves (237), pig kidney (238), *Euglene gracilis* (239), and *Pseudomonas saccharophilia* (240).

The rabbit-liver fructose diphosphatase was isolated and characterized by Pontremoli *et al.* (227). A molecular weight of 130,000 was obtained for the enzyme. Incubation with ^{32}P orthophosphate did not label the enzyme unlike the alkaline phosphatases (227). A pH optimum of 9 to 9.5 was obtained with fructose-1,6-diphosphate substrate. Treatment with dinitrofluorobenzene modified the catalytic activity of the enzyme (228). The modified enzyme exhibited an altered pH optimum with constant activity between pH 7.5 to 9.5, and an increase in catalytic activity with lower concentrations of the reagent, while high concentration of dinitrofluorobenzyne resulted in activity loss. The reaction of fructose diphosphatase with fluorodinitrobenzene results in dinitrophenylation of two sulfhydryl groups and one lysine ϵ-amino group. Presence of the substrate fructose-1,6-diphosphate prevents the modification (229, 230) while sedoheptulose-1,7-diphosphate, also a substrate, does not affect the dinitrophenylation. Treatment of the rabbit-liver enzyme with acetylimidazole resulted in acetylation of four tyrosyl groups and loss of activity (231). This can be prevented by presence of the substrate or reversed by deacetylation with hydroxylamine (231).

Rosen and Rosen (234–236) reported the iodination of fructose-1,6-diphosphatase from *Candida utilis* which resulted in loss of activity associated with iodination of tyrosine residues. Fructose diphosphatases are inhibited by adenosine monophosphate. Pontremoli *et al.* (231) reported that the rabbit-liver fructose diphosphatase bound four equivalents of AMP at pH 7.5.

Hanson *et al.* (247) have reported observations which indicate that the adenosine-5′-triphosphate–glucose and phosphoenolpyruvate–glucose phosphotransferase activities of rat-liver mitochondria are due to the microsomal glucose-6-phosphatase. Rabbit-liver fructose-1,6-diphosphatase requires Mg^{2+} or Mn^{2+} for activity (222). The enzyme is allosterically inhibited by AMP (241, 243, 246). The enzyme possesses four binding sites for the substrate and an equal number for the allosteric inhibitor AMP (231, 248, 249). It binds four equivalents of Mn^{2+} at neutral pH and four additional equivalents above pH 8.5 (250). The enzyme is dissociated in sodium dodecyl sulfate or by treatment with maleic anhydride into two kinds of subunits of molecular weight 29,000 to 31,000 and 35,000 to 39,000 (251).

The binding of substrate and inhibitor has been shown to induce changes in the state of ionization of a limited number of tyrosyl residues (230) and the reactivity of the thiol groups (252). Tamburro *et al.* (253) have reported circular dichroism studies with the rabbit-liver fructose-1,6-diphosphatase which suggest a strong rigidity of the overall conformation of the enzyme. The enzyme did not exhibit any significant conformational changes in response to substrate, inhibitor, or pH.

Kolb and Grodsky (254) have described a quantitative displacement radioimmunoassay for fructose-1,6-diphosphatase. Benkovic *et al.* (255) studied the effect of various inhibitors on the fructose-1,6-diphosphatase and concluded that the furanose configuration of fructose-1,6-diphosphate is acted upon by the enzyme. The activation of several vertebrate fructose-1,6-diphosphatases by certain monovalent cations, particularly K^+ or NH_4^+, has been reported (256).

Diphosphoglycerate Phosphatase (2,3-Diphospho-D-Glycerate-2-Phosphohydrolase, EC 3.1.3.13)

Sutherland *et al.* (257) first reported the existence of a phosphatase in rabbit-muscle extracts which attacked 2,3-diphosphoglyceric acid. Rapoport and Luebering (258) isolated the enzyme from rat and rabbit skeletal muscle and demonstrated that it catalyzed the hydrolysis of 2,3-diphosphoglyceric acid to the 3-phosphoglyceric acid. The enzyme activity was stimulated by 2-phosphoglycerate, Hg^{2+}, and Ag^+, and inhibited by 3-phosphoglycerate and Cu^{2+}. The preparation of diphosphoglycerate phosphatase from human and porcine erythrocytes which was activated by bisulfite was reported by Manyai and Varady (259). Joyce and Grisolia (260, 261) purified a glycerate-2,3-diphosphatase from baker's yeast and chicken breast muscle. Similar enzymatic activity was reported in homogenates of beef heart, brain, muscle, liver, and kidney (260), as well as acetone powders of rabbit brain, heart, muscle, liver, and kidney. The yeast enzyme exhibited a broad pH optimum between 5 to 7 while the muscle enzyme had a sharper pH optimum of about pH 7.0. The purest muscle preparation was inactive with ATP, β-glycerophosphate, 3-phosphoglyceric acid, 2-phosphoglyceric acid, and fructose-1-6-diphosphate. The yeast enzyme had traces of activity toward 3-phosphoglyceric acid, 2-phosphoglyceric acid, and fructose-1,-diphosphate DeVerdier (262) isolated the phosphatase from human erythrocytes. The enzyme was active in the presence of EDTA and inactivated by Hg^{2+}. The enzyme had a pH optimum of about 6.5. Inhibition was observed with 3-phosphoglycerate and 2-phosphoglycerate but not with inorganic phosphate or phosphoenolpyruvate. Harkness and Roth (263) reported the isolation and separation of two diphosphoglycerate phosphatases from human erythrocytes. One was stimulated by bisulfite, apparently similar to the enzyme described by Manyai and Varady (259). The enzyme had a pH optimum at 7.5 and optimal activity in the presence of 20 to 30 mM bisulfite. Rose and Liebowitz (264) found that the human erythrocyte enzyme was activated by chloride, phosphate, fluoride, bromide, arsenate, fluorophosphate, and sulfite. The most potent activator was glycolate-2-phosphate which gave a 1600-fold stimulation.

Histidinol Phosphatase (L-Histidinol Phosphate Phosphohydrolase, EC 2.1.3.15)

Ames (265) first isolated histidinol phosphatase from *Neurospora crassa* and demonstrated that it catalyzed the hydrolysis of L-histidinol phosphate to L-histidinol and inorganic phosphate, an important reaction in the biosynthesis of histidine. It did not hydrolyze ribose phosphate or ethanolamine phosphate and was relatively insensitive to beryllium ions or EDTA which inhibit the nonspecific alkaline phosphatases from *Neurospora*. The enzyme had a pH optimum at 9.0 and yielded a Km of 4.2×10^{-3} M for histidinol phosphate at that pH. Gorman and Hu (59) isolated a specific histidinol phosphatase from *Saccharomyces cerevisiae*. The enzyme, unlike some nonspecific alkaline phosphatases that also catalyze this hydrolysis (17, 55), was specific for histidinol phosphatase. It was insensitive to beryllium ions or EDTA, similar to the *Neurospora* enzyme.

PHOSPHORIC DIESTER HYDROLASES (EC 3.1.4)

These enzymes catalyze the hydrolysis of a phosphoric diester to yield a phosphoric monoester and an alcohol. This group is subdivided further based on the specific substrate attached. The least specific of these enzymes are the phosphodiesterases (orthophosphoric diester phosphohydrolase, EC 3.1.4.1). The spleen enzyme (phosphodiesterase **2**) with a pH optimum near neutrality catalyzes the hydrolysis of synthetic monoalkyl esters of ribo- and deoxyribonucleoside 3′-phosphates (266, 267). Higher molecular weight polyribonucleotides, nucleic acids, and certain oligodeoxyribonucleotides and oligoribonucleotides also are hydrolyzed to yield nucleoside 3′-phosphates (266, 268–274). The enzyme acts stepwise as an exonuclease from the 5′-hydroxyl end of the molecule. The presence of a phosphodiesterase in snake venom (phosphodiesterase **2**) was first demonstrated by Uzawa (275). This enzyme together with closely related enzymes from kidney and other tissues catalyzes the hydrolysis of *p*-nitrophenyl thymidine-5′-phosphate and oligoribo- and deoxyribonucleotides with free 3′-hydroxyl end groups to form mononucleoside 5′-phosphates (276–278).

Glycerophosphorylcholine Diesterase (EC 3.1.4.2) (L-3-Glycerophosphorylcholine Glycerophosphohydrolase)

Hayaishi and Kornberg (279) demonstrated the presence of an enzyme in extracts of *Serratia plymuthica* which catalyzed the hydrolysis of L-3-glycerophosphorylcholine to glycerophosphate and choline. Glycerophosphorylethanolamine also was a substrate. Similar enzymes have been shown to be present in rat liver and various nerve tissues of a number of animals which are inhibited by EDTA and reactivated by Mn^{2+}, Mg^{2+}, and Ca^{2+} (280, 281).

Phospholipase C (Phosphatidylcholine Cholinephosphohydrolase, EC 3.1.4.3) (Lipophosphodiesterase 1) (14)

McFarlane and Knight (282) first discovered that the hemolytic factor of *Clostridium welchii* α-toxin converted lecithins into phosphorylcholine and diglycerides. This enzyme hydrolyzes all major types of phosphatides including the lecithins, sphingomyelins, and kephalins present in red cell membranes (283). Similar enzymes have been demonstrated in *Clostridium perfringens* α-toxin (284) and *Clostridium oedematiens* culture filtrate (285). The enzymes are activated by Ca^{2+} and inhibited by NaF (284, 286).

Phospholipase D (Phosphatidylcholine Phosphatidohydrolase, EC 3.1.4.4) (Lipophosphodiesterase 2) (14)

Phospholipase D catalyzes the hydrolysis of lecithin to yield choline and a phosphatidic acid. The enzyme is widely distributed in plants including carrots (287, 288), cabbage leaves (289–293), and cottonseed (23). The leaf enzymes from spinach and cabbage have been shown to be confined to the chloroplast fraction (294, 295). The cabbage enzyme was shown to require Ca^{2+} for hydrolysis and exhibited a sharp pH optimum at pH 5.4 (290). The hydrolysis of lecithin was inhibited by protamine sulfate, choline, and ethanolamine. Cationic amphipathic substances inhibited the hydrolysis, while anionic amphipathic substances such as dodecyl sulfate, phosphatidic acid, triphosphoinositide, and monoacetylphosphoric acid were activators (290). Transphosphatidylation activity was obtained with purified cabbage phospholipase D (293).

ENZYMES CATALYZING THE HYDROLYSIS OF NUCLEIC ACIDS

A number of enzymes have been described that catalyze the hydrolysis of nucleic acids, but that differ both with respect to the mechanism of action and specificity. Some of these enzymes, the exonucleases, are phosphodiesterases and have been described above. The ribonucleases which catalyze the hydrolysis of ribonucleic acid (RNA) are not hydrolases but rather transferases which result in an overall splitting of RNA. The deoxyribonucleases and the various microbial nucleases are true phosphoric diester hydrolases with varying specificities. Only a few comments are made concerning certain of these enzymes without attempt to review this vast literature. The reader is referred to some specific reviews in this area (296–306, 319, 320).

Deoxyribonuclease (Deoxyribonucleate Oligonucleotidohydrolase, EC 3.1.4.5)

Deoxyribonuclease catalyzes the hydrolysis of DNA and synthetic oligonucleotides to yield a mixture of mono- and oligonucleotides, each terminating in a 5′-phosphate. The enzyme has been isolated from various tissues including bovine pancreas, spleen, and microorganisms (302, 307–314). The pancreatic deoxyribonuclease has a molecular weight of about 31,000 (314).

Deoxyribonuclease 2 (Deoxyribonucleate 3′-Nucleotidohydrolase, EC 3.1.4.6)

Deoxyribonuclease 2, isolated from porcine spleen and present in other tissues, cleaves DNA to yield mononucleotides and oligonucleotides terminating in a 3′-phosphate (302, 304, 305, 315–325). The spleen enzyme has been shown to consist of two subunits (323) and its specificity has been further clarified (324).

Micrococcal Nuclease (EC 3.1.4.7) [Ribonucleate (Deoxyribonucleate) 3′-Nucleotidohydrolase]

The nucleases have been isolated from *Staphylococcus* and *Streptococcus* and they hydrolyze both DNA and RNA to yield nucleotides with 3′-phosphates (305, 326–335). The *Staphylococcal* nuclease catalyzes the hydrolysis of a number of *p*-nitrophenyl ester derivatives of deoxythymidine-5′-phosphate (331). The enzyme has been shown to consist of a single polypeptide chain of 149 amino acids without any disulfide bridges (336, 337).

Ribonuclease (EC 2.7.7.16) [Ribonucleotide-2-Oligonucleotide Transferase (Cyclizing)]

The ribonucleases are transferases rather than hydrolases, the action of which brings about the cleavage of RNA. The enzyme transfers the 3′-phosphate of a pyrimidine nucleotide residue of a polynucleotide from the 5′ position of the adjoining nucleotide to the 2′ position of the pyrimidine nucleotide itself, thus forming a cyclic nucleotide. In the subsequent reaction the cyclic phosphates are hydrolyzed to yield terminal 3′-phosphates. The enzyme is widely distributed, having been isolated from pancreatic and other tissues (303, 304, 339, 340, 347), various microorganisms (266, 296, 297, 299–303, 338, 341–344, 346), and snake venom (345). The pancreatic ribonuclease has been studied extensively (348–371). The primary sequence (372–376) and the three-dimensional structure (377, 378) of ribonuclease have been determined.

Recently the total synthesis of ribonuclease A, the first enzyme synthesized, was achieved by Gutte and Merrifield (379). At the same time ribonuclease S was synthesized by the Merck group led by Denkewalter and Hirschman (380–384) which in the presence of the S peptide exhibited ribonuclease activity.

TRIPHOSPHORIC MONOESTER HYDROLASE (EC 3.1.5)

Deoxy-GTP-Triphosphohydrolase (EC 3.1.5.1)

Extracts of *Escherichia coli* contain an enzyme, deoxy-GTPase, which catalyzes the hydrolysis of deoxyguanosine triphosphate to deoxyguanosine and triphosphate (385). Guanosine triphosphate is also split at a slower rate. The enzyme does not act upon guanosine diphosphate or the triphosphate of adenosine, uridine, cytidine, or thymidine.

HYDROLASES ACTING ON PHOSPHORIC ACID ANHYDRIDES (EC 3.6.1)

The enzymes that catalyze the hydrolysis of phosphoryl-containing anhydrides include the inorganic pyrophosphatases, trimetaphosphatases, various ATPases, acyl phosphatases, nucleoside dephosphatases, nucleotide pyrophosphatase, and the polyphosphatases (14, 386). Most of these enzymes are activated by monovalent or divalent cations (14).

Inorganic Pyrophosphatase (EC 3.6.1.1) (Pyrophosphate Phosphohydrolase)

The enzyme that catalyzes the hydrolysis of inorganic pyrophosphate to orthophosphate is widespread (387, 388), its presence having been reported in yeast (389–396), bacteria, including *E. coli* (397, 398), *B. megaterium* (399), *Streptococcus faecalis* (400), *Ferrobacillus ferroxidans* (401), insects (402, 403), various plant sources (404–408), the chrysophycean flagellate *Prymnesium parvum* Carter (409), and a number of mammalian tissues (410–417).

Josse (397, 398) reported the isolation of an intracellular highly specific constitutive enzyme from *E. coli.* The enzyme exhibited a pH optimum of 9.1 and catalyzed the hydrolysis of inorganic pyrophosphate, tripolyphosphate, and tetrapolyphosphate with decreasing velocities, respectively (398). No activity was observed toward phosphomonoesters such as glucose-6-phosphate, ribo- or deoxyribopolynucleotides, ribo- or deoxyribonucleoside, mono-, di-, or triphosphates, cyclic tri- or tetrametaphosphates, or phosphofluoridates (398). Divalent cation was required for activity and Mg^{2+}, Mn^{2+}, Zn^{2+}, and Co^{2+} gave active enzyme. The enzyme was very stable, particularly in the presence of 0.01 *M* Mg^{2+}. It was suggested that the active site of the *E. coli* pyrophosphatase carries one or more positive charges at pH 9.1. The role of the magnesium seems to be that of forming a complex with the substrate. It was proposed that the divalent cation closed a stable, six-membered ring in the complex restricting rotation about the pyrophosphate–phosphoanhydride bond (398). Ricketts (409) described a magnesium requiring inorganic pyrophosphatase from the chrysophycean flagellate *Prymnesium parvum* Carter. The enzyme, which had a pH optimum of 8.9, was inactive toward ATP, ADP, AMP, thiamine pyrophosphate, and β-glycerophosphate (409). Calcium inhibited both the magnesium- and manganese-activated activity. Inhibition was observed with fluoride and alloxan, but not iodoacetate (409).

The yeast enzyme was isolated by Kunitz (395) by a combination of ammonium sulfate fractionation of the plasmolyzed yeast, calcium phosphate gel treatment, and crystallization with ethyl alcohol (270). A molecular weight of 63,000 was reported for this enzyme by Schachman (418). The enzyme requires the presence of Mg^{2+}, Co^{2+}, or Mn^{2+} for activity (270). In the presence of Mg^{2+} the enzyme is specific for inorganic pyrophosphate and inactive toward ATP, ADP, or thiamine pyrophosphate (387), which is somewhat changed in the presence of Zn^{2+}. The inorganic pyrophosphatase from *Ferrobacillus ferroxidans* (*Thiobacillus ferroxidans*) was shown by Howard and Lundgren (401) to require Mg^{2+} for maximum activity, with only 10% of the activity obtained by substitution by Mn^{2+}, Zn^{2+}, and Co^{2+}. The enzyme had a pH optimum between 7.5 and 8.5 with a magnesium-to-pyrophosphate ratio of 1. The enzyme was quite thermally stable in the presence of Mg^{2+}.

Trimetaphosphatase (EC 3.6.1.2) (Trimetaphosphate Phosphohydrolase)

The hydrolysis of trimetaphosphate (cyclic) by various cellular extracts has been reported (419–424). Kornberg (424) separated and purified a trimetaphosphatase from the tripolyphosphatase from extracts of baker's yeast. He demonstrated that tripolyphosphate was the product from the enzyme-catalyzed hydrolysis of the trimetaphosphate.

ATPases (EC 3.6.1.3, 3.6.1.4, 3.6.1.5, 3.6.1.8)

The various ATPases catalyze the hydrolysis of ATP and other nucleoside triphosphates to the di- or mononucleotides and orthophosphate or pyrophosphate. These enzymes are mentioned here only briefly. The ATP phosphohydrolases (EC 3.6.1.3) include myosin ATPase (386, 425, 428), mitochondrial ATPase (430–432), and membrane-associated ATPase (429, 434–437).

The Na^{+}- and K^{2+}-dependent ATPases that require Mg^{2+} (EC 3.6.1.4) are found in the microsomal fraction of various tissues (442–444) and are considered to be part of the mechanism of active transport of Na^{+} and K^{+} across biological membranes (437–440). These enzymes appear to hydrolyze ATP via a mechanism involving phosphorylation of the enzyme, which requires Na^{+} and a subsequent hydrolysis of the phosphorylated intermediate which is K^{+} dependent (438, 440, 441).

***Apyrase** (**ATP Diphsphohydrolase, EC 3.6.1.5**).* Apyrase (449) has been isolated primarily from plants such as potato and has been shown to hydrolyze both ATP and ADP (445–451). The enzyme requires divalent cation such as Ca^{2+}, Mn^{2+}, or Mg^{2+} for maximal activity (441, 445, 450, 451), and there appears to be more than one apyrase enzyme in the potato (446, 450, 451).

***ATP Pyrophosphohydrolase** (**EC 3.6.1.8**).* This ATPase yields pyrophosphate and the nucleoside monophosphate upon hydrolysis of ATP (452–456). It has been isolated from rabbit erythrocytes and reticulocytes (452), cobra venom (453), bovine seminal fluid (454), liver plasma membranes (455), and phage-infected *Escherichia coli* (456). These enzymes have fairly alkaline pH optimum with different substrate specificities and all require Mg^{2+} for activity (452).

Polyphosphatase (Polyphosphate Polyphosphohydrolase, EC 3.6.1.10)

A polyphosphatase has been isolated from *A. niger* which catalyzes the hydrolysis of polymetaphosphate of high molecular weight to tetra- or pentaphosphates (457, 458). The enzyme exhibited a pH optimum at pH 5.7 (457) and was activated by Mn^{2+}, Mg^{2+}, and Ca^{2+} (459). The enzyme also has been reported in *Proteus vulgaris* (459).

Acyl Phosphatase (Acyl Phosphate Phosphohydrolase, EC 3.6.1.7)

Acyl phosphatase, which cleaves acyl phosphates to the acid and orthophosphate, has been isolated in pure form from bovine brain (460, 461). The enzyme catalyzes the hydrolysis of acetyl phosphate, carbomoyl phosphate (462, 463), and 1,3-diphosphoglycerate (463). It has been suggested that acyl phosphatase is involved in the active transport of sodium and potassium across cell membrane (464, 465). The enzyme has been reported to have a very low molecular weight (8732) and is unusually stable to heat and other physical treatment (460).

PHOSPHOAMIDE HYDROLASE (EC 3.9.1.1)

The phosphoamidases (phosphoramidases) catalyze the hydrolysis of N—P linkage of phosphoramidates. The enzyme has been found in both mammalian tissues (466–469) and bacteria (470). The bovine spleen enzyme hydrolyzed the N—P bond of phosphoramidate, *N*-(*p*-chlorophenyl)-phosphoramidate,

N-(*p*-tolyl)-phosphoramidate, creatine phosphate, *N*-phosphorylglycine, and α-*N*-phosphoryl-DL-tryptophan methyl ester (466). The enzyme activity, which was maximal around pH 6.0, was enhanced by sulfhydryl compounds (466).

REFERENCES

1. Schmidt, G., and M. Laskowski, Sr., in *The Enzymes*, P. D. Boyer, H. Lardy, and K. Myrback (Eds.), Academic Press, New York, 1961, p. 3.
2. Florkin, M., and E. H. Stotz, *Comprehensive Biochemistry*, Vol. 13, Elsevier, Amsterdam, 1964.
3. Stein, S. S., and D. E. Koshland, Jr., *Arch. Biochem. Biophys.*, **39**, 229 (1952).
4. Morton, R. K., *Biochem. J.*, **70**, 139, 150 (1958).
5. Nigam, V. N., and W. H. Fishman, *J. Biol. Chem.*, **234**, 2394 (1959).
6. Meyerhof, O., and N. Green, *J. Biol. Chem.*, **183**, 377 (1950).
7. Tsuboi, K. K., and P. B. Hudson, *Arch. Biochem. Biophys.*, **43**, 339 (1953).
8. Jeffries, G. M., *Biochim. Biophys. Acta*, **23**, 155 (1957).
9. Tunis, M., and E. Chargaff, *Biochim. Biophys. Acta*, **21**, 204 (1956).
10. Borkenhagen, L. F., and E. P. Kennedy, *J. Biol. Chem.*, **234**, 849 (1959).
11. Neuhaus, F. C., and W. L. Byrne, *Biochim. Biophys. Acta*, **28**, 223 (1958).

11a. Fishman, W. H., Ed., *Ann. N.Y. Acad. Sci.*, **166**, Art. 2, 365 (1969).

12. Malamy, M., and B. L. Horecker, in *Methods in Enzymology*, Vol. 9, S. P. Colowick and N. O. Kaplan (Eds.), Academic Press, New York, 1966, p. 639.
13. Stadtman, T. C., in *The Enzymes*, Vol. 5, P. D. Boyer, H. Lardy, and K. Myrback (Eds.), Academic Press, New York, 1961, p. 55.
14. Morton, R. K., in *Comprehensive Biochemistry*, Vol. XVI, M. Florkin and E. H. Stotz (Eds.), Elsevier, New York, 1965, p. 55.
15. Horiuchi, T., S. Horiuchi, and D. Mizuno, *Nature*, **183**, 1529 (1959).
16. Torriani, A., *Biochim. Biophys. Acta*, **38**, 460 (1960).
17. Garen, A., and C. Levinthal, *Biochim. Biophys. Acta*, **38**, 470 (1960).
18. Garen, A., and H. Echols, *J. Bacteriol.*, **83**, 297 (1962).
19. Schwartz, J. H., and F. Lipmann, *Proc. Natl. Acad. Sci. U.S.*, **47**, 1996 (1961).
20. Plocke, D. J., C. Levinthal, and B. L. Vallee, *Biochem.*, **1**, 373 (1962).
21. Malamy, M. H., and B. L. Horecker, *Biochem.*, **3**, 1893 (1964).
22. Malamy, M. H., and B. L. Horecker, *Biochem. Biophys. Res. Commun.*, **5**, 104 (1961).
23. Malamy, M. H., and B. L. Horecker, *Biochem.*, **3**, 1889 (1964).
24. Brockman, R. W., and L. A. Heppel, *Biochem.*, **7**, 2554 (1968).
25. Schlesinger, M. J., and K. Barrett, *J. Biol. Chem.*, **240**, 4284 (1965).
26. Rothman, F., and R. Byrne, *J. Mol. Biol.*, **6**, 330 (1963).
27. Reynolds, J. A., and M. J. Schlesinger, *Biochem.*, **6**, 3552 (1967).
28. Reynolds, J. A., and M. J. Schlesinger, *Biochem.*, **7**, 2080 (1968).

29. Reynolds, J. A., and M. J. Schlesinger, *Biochem.*, **8,** 588 (1969).
30. Trubowitz, S., D. Feldman, S. W. Morgenstern, and V. M. Hunt, *Biochem. J.*, **80,** 369 (1961).
31. Mathies, J. C., *J. Biol. Chem.*, **233,** 1121 (1958).
32. Engstrom, L., *Biochim. Biophys. Acta*, **52,** 36 (1961).
33. Plocke, D. J., and B. L. Vallee, *Biochem.*, **1,** 1039 (1962).
34. Simpson, R. T., B. L. Vallee, and G. H. Tait, *Biochem.*, **12,** 4336 (1968).
35. Simpson, R. T., and B. L. Vallee, *Biochem.*, **7,** 4343 (1968).
36. Heppel, L. A., D. R. Harkness, and R. J. Hilmoe, *J. Biol. Chem.*, **237,** 841 (1962).
37. Anderson, W. B., and R. C. Nordlie, *J. Biol. Chem.*, **242,** 114 (1967).
38. Engstrom, L., and G. Agren, *Acta Chem. Scand.*, **12,** 357 (1958).
39. Engstrom, L., *Biochim. Biophys. Acta*, **52,** 49 (1961).
40. Engstrom, L., *Biochim. Biophys. Acta*, **56,** 606 (1962).
41. Engstrom, L., *Biochim. Biophys. Acta*, **92,** 71, 79 (1964).
42. Schwartz, J. H., *Proc. Natl. Acad. Sci. U.S.*, **49,** 871 (1963).
43. Milstein, C., *Biochem. J.*, **92,** 410 (1964).
44. Pigretti, M. M., and C. Milstein, *Biochem. J.*, **94,** 106 (1965).
45. Aldridge, W. N., T. E. Barman, and H. Gutfreund, *Biochem. J.*, **92,** 23C (1964).
46. Fernley, H. N., and P. G. Walker, *Nature*, **212,** 1435 (1966).
47. Fife, W. K., *Biochem. Biophys. Res. Commun.*, **28,** 309 (1967).
48. Williams, A., *Chem. Commun.*, **1966,** 676.
49. Ko, S. H. D., and F. J. Kezdy, *J. Am. Chem. Soc.*, **89,** 7139 (1967).
50. Barrett, H., R. Butler, and J. B. Wilson, *Biochem.*, **8,** 1042 (1969).
51. Reid, T. W., M. Pavlic, D. J. Sullivan, and I. B. Wilson, *Biochem.*, **8,** 3184 (1969).
52. Ames, B. N., *J. Biol. Chem.*, **226,** 583 (1957).
53. Nicholas, D. J. D., and K. Commisiong, *J. Gen. Microb.*, **17,** 699 (1957).
54. Kuo, M. H., and H. J. Blumenthal, *Biochim. Biophys. Acta*, **52,** 13 (1961).
55. Kuo, M. H., and H. J. Blumenthal, *Biochim. Biophys. Acta*, **54,** 101 (1961).
56. Nyce, J. F., R. J. Kadner, and B. J. Cracken, *J. Biol. Chem.*, **241,** 1468 (1966).
57. Kadner, R. J., J. F. Nyc, and D. M. Brown, *J. Biol. Chem.*, **243,** 3076 (1968).
58. Kadner, R. J., and J. F. Nyc, *J. Biol. Chem.*, **244,** 5125 (1969).
59. Gorman, J. A., and A. S. L. Hu, *J. Biol. Chem.*, **244,** 1645 (1969).
60. Attias, J., J. L. Bonnet, and J. C. Sauvagnargues, *Biochim. Biophys. Acta*, **212,** 315 (1970).
61. Wood, D. A. W., and H. Tristram, *J. Bacteriol.*, **104,** 1045 (1970).
62. Robison, R., *Biochem. J.*, **17,** 286 (1923).
63. Robison, R., and J. Soames, *Biochem. J.*, **18,** 740 (1924).
64. Levene, P. A., and R. T. Dillon, *J. Biol. Chem.*, **88,** 753 (1930).
65. Portmann, P., *Z. Physiol. Chem. Hoppe-Seylers*, **309,** 87 (1957).
66. Saini, P. K., and J. Done, *Fed. Eur. Biochem. Soc. Lett.*, **7,** 86 (1970)
67 Oide, M., *Comp. Biochem. Physiol.*, **36,** 241 (1970).
68. Rosenblum, D., S. R. Hemmelhack, E. A. Peterson, W. H. Evons, and M. G. Mage, *Arch. Biochem. Biophys.*, **141,** 303 (1970).

69. Kasper, D., and K. K. Y. Leung, *Environ. Sci. Technol.*, **11,** 429 (1970).
70. Beratis, N. G., W. Seegers, and K. Hirschhorn, *Biochem. Genet.*, **4,** 689 (1970).
71. Moss, D. W., and E. J. King, *Biochem. J.*, **84,** 192 (1962).
72. Neale, F. C., J. S. Clubb, D. Hotchkiss, and S. Posen, *J. Clin. Pathol.*, **18,** 359 (1965).
73. Ghosh, N. K., and W. H. Fishman, *Can. J. Biochem.*, **47,** 147 (1969).
74. Boyer, S. H., *Science*, **134,** 1004 (1961).
75. Richards, G. M., G. du Vair, and M. Laskowski, Sr., *Biochem.*, **4,** 501 (1965).
76. Georgatsos, J. G., and M. Laskowski, Sr., *Biochem.*, **1,** 288 (1962).
77. Sulkowski, E., W. Bjork, and M. Laskowski, Sr., *J. Biol. Chem.*, **238,** 2477 (1963).
78. Schmidt, G., in *The Enzymes*, Vol. V, P. D. Boyer, H. Lardy, and K. Myrback (Eds.), Academic Press, New York, 1961, p. 37.
79. Tsuboi, K. K., G. Weiner, and P. B. Hudson, *J. Biol. Chem.*, **224,** 621 (1957).
80. Schmidt, G., G. Bartsch, M. C. Laumont, T. Herman, and M. Liss, *Biochem.*, **2,** 126 (1963).
81. Rautanen, N., and V. Kaerkkaeinen, *Acta Chem. Scand.*, **5,** 1216 (1951).
82. Schmidt, G., K. Seraidarian, L. M. Greenbaum, M. D. Hickey, and S. J. Thannhauser, *Biochim. Biophys. Acta*, **20,** 135 (1956).
83. Von Hofsten, B., and J. Porath, *Biochim. Biophys. Acta*, **64,** 1 (1962).
84. Von Hofsten, B., *Biochim. Biophys. Acta*, **48,** 171 (1961).
85. Von Hofsten, B., *Acta Chem. Scand.*, **15,** 1791 (1961).
86. Roger, D., and F. J. Reithel, *Arch. Biochem. Biophys.*, **89,** 97 (1960).
87. Tani, Y., T. Tochikura, H. Yamada, and K. Ogata, *Biochem. Biophys. Res. Commun.*, **28,** 769 (1967).
88. Blumenthal, H. J., A. Hemerline, and A. Roseman, *Bact. Proc.*, **1956,** 109.
89. Kilsheimer, G. S., and B. Axelrod, *Nature*, **182,** 1733 (1958).
90. Nyce, J. F., *Biochem. Biophys. Res. Commun.*, **27,** 183 (1967).
91. Helferick, B., and G. V. Bruck, *Z. Physiol. Chem.*, **295,** 114 (1953).
92. Andreu, M., E. F. Alvarez, and M. Lora-Tamayo, *An. Real Soc. Espan. Fis. Quim (Madrid), Ser. B56*, **1060,** 67.
93. Jorgensen, O., *Acta Chem. Scand.*, **13,** 900 (1959).
94. Alvarez, E. F., *Biochim. Biophys. Acta*, **59,** 663 (1962).
95. Lora-Tamayo, M., E. F. Alvarez, and M. Andreu, *Bull. Soc. Chem. Biol.*, **44,** 501 (1962).
96. Hsu, R. Y., W. W. Cleland, and L. Anderson, *Biochem.*, **5,** 799 (1966).
97. Roberts, D. W. A., *Can. J. Biochem.*, **45,** 401 (1967).
98. Roberts, D. W. A., *Can. J. Biochem. Physiol.*, **41,** 113, 1275, 1727 (1963).
99. Roberts, D. W. A., *Enzymol.*, **39,** 151 (1970).
100. Joyce, B. K., and R. Grisolia, *J. Biol. Chem.*, **235,** 2278 (1960).
101. Shaw, J. G., *Arch. Biochem. Biophys.*, **117,** 1 (1966).
102. Newmark, M. Z., and B. S. Wenger, *Arch. Biochem. Biophys.*, **89,** 110 (1960).
103. Mayer, F. C., R. E. Campbell, A. K. Smith, and L. L. McKinney, *Arch. Biochem. Biophys.*, **94,** 301 (1961).
104. Giri, K. V., *Z. Physiol. Chem.*, **245,** 185 (1937).

105. Heinrikson, R. L., *J. Biol. Chem* , **244,** 299 (1969).
106. Shibko, S., and A. L. Tappel, *Biochem. Biophys. Acta*, **73,** 76 (1963).
107. Igarashi, M., H. Takahashi, and N. Tsuyama, *Biochim. Biophys. Acta*, **220,** 85 (1970).
108. Igarashi, M., and V. P. Hollander, *J. Biol. Chem.*, **243,** 6084 (1968).
109. Neil, M. W., and M. W. Horner, *Biochem. J.*, **92,** 217 (1964).
110. Neil, M. W., and M. W. Horner, *Biochem. J.*, **93,** 220 (1964).
111. Chaimovich, H., and F. Nome, *Arch. Biochem. Biophys.*, **139,** 9 (1970).
112. Koenig, H., D. Gaines, T. McDonald, R. Gray, and J. Scott, *J. Neurochem.*, **11,** 729 (1964).
113. Sellinger, C. Z., D. L. Rucker, and F. de B. Verster, *J. Neurochem.*, **11,** 271 (1964).
114. Ostrowski, W., and J. Rybarska, *Biochim. Biophys. Acta*, **105,** 196 (1965).
115. Boman, H. G., *Ark. Kemi*, **12,** 453 (1958).
116. Ostrowski, W., and A. Tsugita, *Arch. Biochem. Biophys.*, **94,** 68 (1961).
117. Domonski, J., L. Konieczay, and W. Ostrowski, *Biochim. Biophys. Acta*, **92,** 405 (1964).
118. Kilsheimer, G. S., and B. Axelrod, *J. Biol. Chem.*, **227,** 879 (1957).
119. Davidson, H. M., and W. H. Fishman, *J. Biol. Chem.*, **234,** 526 (1959).
120. Glomset, J. A., and J. Poroth, *Biochim. Biophys. Acta*, **39,** 1 (1960).
121. Glomset, J. A., *Biochim. Biophys. Acta*, **32,** 349 (1959).
122. Revel, H. R., *Methods Enzymol.*, **6,** 211 (1963).
123. DiPietro, D. L., and F. S. Zengerle, *J. Biol. Chem.*, **242,** 3391 (1967).
124. Ahmed, Z., and E. J. King, *Biochim. Biophys. Acta*, **34,** 313 (1959).
125. Ma, L., and M. C. C. Chen, *Clin. Chem. Acta*, **12,** 153 (1965).
126. DiPietro, D. L., and F. S. Zengerle, *Fed. Proc.*, **26,** 853 (1967).
127. Bingham, E. W., and C. A. Zittle, *Arch. Biochem. Biophys.*, **101,** 471 (1963).
128. Georgatsos, J. G., *Arch. Biochem. Biophys.*, **110,** 354 (1965).
129. Abul-Fadl, M. A. M., and E. J. King, *Biochem. J.*, **45,** 51 (1949).
130. Tsuboi, K. K., and P. B. Hudson, *Arch. Biochem. Biophys.*, **55,** 206 (1955).
131. Iio, M., T. Hashimoto, and H. Yoshikawa, *J. Biochem.* (*Tokyo*), **55,** 321 (1964).
132. Tsuboi, K. K., and P. B. Hudson, *Arch. Biochem. Biophys.*, **53,** 341 (1954).
133. Wergedal, J. E., *Proc. Soc. Exp. Biol. Med.*, **134,** 244 (1970).
134. Wergedal, J. E., and D. Boylink, *J. Histochem. Cytochem.*, **17,** 799 (1964).
135. Vaes, G., *Biochem. J.*, **97,** 393 (1965).
136. Vaes, G., and P. Jacques, *Biochem. J.*, **97,** 380 (1965).
137. Wergedal, J. E., *Calc. Tiss. Res.*, **3,** 55 (1969).
138. Tonino, G. J. M., and E. P. Steyn-Parve, *Biochim. Biophys. Acta*, **67,** 453 (1963).
139. Schaffner, A., and F. Krumey, *Z. Physiol. Chem.*, **255,** 145 (1938).
140. Suomalainen, H., M. Linko, and E. Oura, *Biochim. Biophys. Acta*, **37,** 482 (1960).
141. McLellan, W. L., and J. O. Lompen, *Biochim. Biophys. Acta*, **67,** 324 (1963).
142. Baer, P., and E. P. Steyn-Parve, *Biochim. Biophys. Acta*, **128,** 400 (1966).
143. Baer, P., and E. P. Steyn-Parve, *Biochim. Biophys. Acta*, **171,** 360 (1969).
144. Baer, P., and E. P. Steyn-Parve, *Biochim. Biophys. Acta*, **206,** 281 (1970).

145. Greenberg, H., and D. Nachmansohn, *J. Biol. Chem.*, **240,** 1639 (1965).
146. Kutscher, W., and H. Wolberg, *Z. Physiol. Chem.*, **236,** 237 (1935).
147. Nigam, V. N., H. M. Davidson, and W. H. Fishman, *J. Biol. Chem.*, **234,** 1550 (1959).
148. Neil, M. W., *Biochem. J.*, **81,** 418 (1961).
149. Goldberg, L., L. E. Martin, and J. Leigh, *Biochem. J.*, **85,** 56 (1962).
150. Neil, M. W., and M. W. Horner, *Biochem. J.*, **84,** 32p (1962).
151. Shibko, S., and A. L. Toppel, *Biochim. Biophys. Acta*, **73,** 76 (1963).
152. Moore, B. W., and P. U. Angeletti, *Ann. N.Y. Acad. Sci.*, **94,** 659 (1961).
153. Choy, A. E., and H. Croviato, *J. Histochem. Cytochem.*, **16,** 582 (1968).
154. Hunter, G. D., and G. C. Millson, *J. Neurochem.*, **13,** 375 (1966).
155. Rosenszajn, L., and B. Epstein, *J. Lab. Clin. Med.*, **72,** 786 (1968).
156. Roche, J., *Biochem. J.*, **25,** 1724 (1931).
157. Byrne, W. L., in *The Enzymes*, Vol. 5, P. D. Boyer, H. Lardy, and K. Myrback (Eds.), Academic Press, New York, 1961, p. 73.
158. Pizer, L. I., *J. Biol. Chem.*, **238,** 3934 (1963).
159. Smith, R. A., C. W. Shuster, S. Zimmerman, and I. C. Gunsalus, *Bacteriological Proceedings*, Society of American Bacteriologists, Baltimore, Maryland, 1956, p. 107.
160. Schramm, M., *J. Biol. Chem.*, **233,** 1169 (1958).
161. Schramm, M., in *Methods in Enzymology*, Vol. 6, S. P. Colowick and N. O. Kaplan (Eds.), Academic Press, New York, 1963, p. 215.
162. Bridgers, W. F., *J. Biol. Chem.*, **242,** 2080 (1967).
163. Neuhaus, F. C., and W. L. Byrne, *J. Biol. Chem.*, **233,** 109 (1958).
164. Neuhaus, F. C., and W. L. Byrne, *Fed. Proc.*, **17,** 282 (1958).
165. Borkenhagen, L. F., and E. P. Kennedy, *Biochim. Biophys. Acta*, **28,** 222 (1958).
166. Neuhaus, F. C., and W. L. Byrne, *J. Biol. Chem.*, **234,** 113 (1959).
167. Weiss, S. B., S. W. Smith, and E. P. Kennedy, *Nature*, **178,** 594 (1956).
168. Smith, S. W., S. B. Weiss, and E. P. Kennedy, *J. Biol. Chem.*, **228,** 915 (1957).
169. Hubscher, G., and R. Coleman, *Biochim. Biophys. Acta*, **56,** 479 (1962).
170. Coleman, R., and G. Hubscher, *Biochem. J.*, **80,** 11P (1961).
171. Hokin, M. R., and L. E. Hokin, *J. Biol. Chem.*, **234,** 1381 (1959).
172. Rossiter, R. J., and K. P. Strickland, *Ann. N.Y. Acad. Sci.*, **72,** 70 (1959).
173. Hokin, L. E., and M. R. Hokin, *Nature*, **189,** 836 (1961).
174. Hokin, L. E., and M. R. Hokin, *J. Gen. Physiol.*, **44,** 61 (1960).
175. Reis, J. L., *Bull. Soc. Chim. Biol.*, **16,** 385 (1934).
176. Gulland, J. M., and E. M. Jackson, *Biochem. J.*, **32,** 597 (1938).
177. Heppel, L. A., and R. J. Hilmoe, *J. Biol. Chem.*, **188,** 665 (1951).
178. Wang, T. P., *J. Bacteriol.*, **68,** 128 (1954).
179. Hermon, E. C., Jr., and B. E. Wright, *J. Biol. Chem.*, **234,** 122 (1959).
180. Brown, E. B., Jr., and E. R. Stadtman, *J. Biol. Chem.*, **235,** 2928 (1960).
181. Sulkowski, E., W. Bjock, and M. J. Laskowski, Sr., *J. Biol. Chem.*, **238,** 2477 (1963).
182. Neu, H. E., and L. A. Heppel, *Biochem. Biophys. Res. Commun.*, **17,** 215 (1964).

183. Neu, H. C., and L. A. Heppel, *J. Biol. Chem.*, **240,** 3685 (1965).
184. Stoh, R., A. Mitsui, and K. Tsushima, *Biochim. Biophys. Acta,* **146,** 151 (1967).
185. Segal, H. L., and B. M. Brenner, *J. Biol. Chem.*, **235,** 471 (1960).
186. Neu, H. C., *Biochem.*, **7,** 3766 (1968).
187. Ipata, P. L., *Biochem.*, **7,** 507 (1968).
188. Levin, S. J., and O. Bodansky, *J. Biol. Chem.*, **241,** 51 (1966).
189. Song, C. S., and O. Bodansky, *J. Biol. Chem.*, **242,** 694 (1967).
190. Burger, R. M., and J. M. Lowenstein, *J. Biol. Chem.*, **245,** 6274 (1970).
191. Reis, J., *Enzymol.*, **2,** 183 (1937).
192. Reis, J., *Enzymol.*, **5,** 251 (1938).
193. Reis, J., *Bull. Soc. Chim. Biol.*, **22,** 36 (1940).
194. Bodansky, O., and M. K. Schwartz, *J. Biol. Chem.*, **238,** 3420 (1963).
195. Song, C. S., and O. Bodansky, *Biochem. J.*, **101,** 5C (1966).
196. Itoh, R., A. Mitsui, and K. Tsushima, *Biochim. Biophys. Acta,* **146,** 151 (1967).
197. Koerner, J. F., and R. L. Sinsheimer, *J. Biol. Chem.*, **228,** 1049 (1957).
198. Bjork, W., and H. G. Boman, *Biochim. Biophys. Acta,* **34,** 503 (1959).
199. Neu, H. C., and J. Chou, *J. Bacterial.*, **94,** 1934 (1967).
200. Shuster, L., and N. O. Kaplan, *J. Biol. Chem.*, **201,** 535 (1953).
201. Shuster, L., and N. O. Kaplan, in *Methods in Enzymology*, Vol. 2, S. P. Colowick and N. O. Kaplan (Eds.), Academic Press, New York, 1955, p. 551.
202. Cunningham, L., *J. Am. Chem. Soc.*, **80,** 2546 (1958).
203. Walwick, E. R., W. K. Roberts, and C. A. Dekker, *Proc. Chem. Soc.*, **84** (1959).
204. Barker, G. R., and G. Lund, *Biochim. Biophys. Acta,* **55,** 987 (1962).
205. Nordlie, R. C., in W. H. Fishman (Ed.), *Ann. N.Y. Acad. Sci.*, **166,** 365 (1969).
206. Nordlie, R. C., and W. J. Arion, *J. Biol. Chem.*, **239,** 1680 (1964).
207. Stetter, M. R., *J. Biol. Chem.*, **239,** 3576 (1964).
208. Nordlie, R. C., and W. J. Arion, *J. Biol. Chem.*, **240,** 2155 (1965).
209. Arion, W. J., and R. C. Nordlie, *J. Biol. Chem.*, **239,** 2752 (1964).
210. Rafter, G. W., *J. Biol. Chem.*, **235,** 2475 (1960).
211. Nordlie, R. C., and J. F. Soodsma, *J. Biol. Chem.*, **241,** 1719 (1966).
212. Lygre, D. G., and R. C. Nordlie, *Biochem.*, **7,** 3219 (1968).
213. Parvin, R., and R. A. Smith, *Biochem.*, **8,** 1749 (1969).
214. Vianna, A. L., and R. C. Nordlie, *J. Biol. Chem.*, **244,** 4027 (1969).
215. Beaufay, H., H. G. Hers, J. Berthet, and C. DeDune, *Bull. Soc. Chim. Biol.*, **36,** 1539 (1954).
216. Hass, L. F., and W. L. Byrne, *J. Am. Chem. Soc.*, **82,** 947 (1960).
217. Dyson, J. E. D., W. B. Anderson, and R. C. Nordlie, *J. Biol. Chem.*, **244,** 560 (1969).
218. Nordlie, R. C., and P. T. Johns, *Biochem.*, **7,** 1473 (1968).
219. Faulkner, P., *Biochem. J.*, **60,** 590 (1955).
220. Turner, D. H., and J. F. Turner, *Biochem. J.*, **74,** 486 (1960).
221. Turner, D. H., and J. F. Turner, *Aust. J. Biol. Sci.*, **10,** 302 (1957).
222. Gomari, G., *J. Biol. Chem.*, **148,** 139 (1943).

223. Pogell, B. M., and R. W. McGilvery, *J. Biol. Chem.*, **197**, 293 (1952).

224. Pogell, B. M., and R. W. McGilvery, *J. Biol. Chem.*, **208**, 149 (1954).

225. Mokrasch, L. C., and R. W. McGilvery, *J. Biol. Chem.*, **221**, 909 (1956).

226. Pogell, B. M., in *Fructose 1,6-Diphosphatase and Its Role in Gluconeogenesis*, R. W. McGilvery and B. M. Pogell (Eds.), American Institute of Biological Services, Washington, D.C., 1961, p. 20.

227. Pontremoli, S., S. Traniella, B. Luppis, and W. A. Wood, *J. Biol. Chem.*, **240**, 3459 (1965).

228. Pontremoli, S., B. Luppis, W. A. Wood, S. Traniello, and B. L. Horecker, *J. Biol. Chem.*, **240**, 3464 (1965).

229. Pontremoli, S., B. Luppis, S. Travillo, W. G. Wood, and B. L. Horecker, *J. Biol. Chem.*, **240**, 3469 (1965).

230. Pontremoli, S., E. Grazi, and A. Accorsi, *J. Biol. Chem.*, **244**, 6177 (1969).

231. Pontremoli, S., E. Grazi, and A. Accorsi, *Biochem.*, **7**, 1655 (1968).

232. Pogell, B. M., *Biochem. Biophys. Res. Commun.*, **7**, 225 (1962).

233. Bonsignore, A., G. Mangiarotti, M. A. Mangiorotti, A. DeFlora, and S. Pontremoli, *J. Biol. Chem.*, **238**, 315 (1963).

234. Rosen, O. M., and S. M. Rosen, *Proc. Natl. Acad. Sci. U.S.*, **55**, 1156 (1966).

235. Rosen, S. M., and O. M. Rosen, *Biochem.*, **7**, 2094 (1967).

236. Rosen, O. M., S. M. Rosen, and B. L. Horecker, *Arch. Biochem. Biophys.*, **112**, 411 (1965).

237. Racker, E., and E. A. R. Schraeder, *Arch. Biochem. Biophys.*, **74**, 326 (1958).

238. Mendicino, J., and F. Vasarhely, *J. Biol. Chem.*, **238**, 3528 (1963).

239. App, A. A., and A. T. Jagendorf, *Biochim. Biophys. Acta*, **85**, 427 (1964).

240. Fossitt, D. D., and I. A. Bernstein, *J. Bacteriol.*, **86**, 598 (1963).

241. Takata, K., and B. M. Pogell, *Biochem. Biophys. Res. Commun.*, **12**, 229 (1963).

242. Newsholme, E. A., *Biochem. J.*, **89**, 388 (1963).

243. Goncedo, C. M., C. Salas, A. Giner, and A. Sols, *Biochem. Biophys. Res. Commun.*, **20**, 15 (1965).

244. Krebs, N. A., and M. Woodford, *Biochem. J.*, **94**, 436 (1965).

245. Fraenkel, D. G., S. Pontremoli, and B. L. Horecker, *Arch. Biochem. Biophys.*, **114**, 4 (1966).

246. Opie, L. H., and E. A. Newsholme, *Biochem. J.*, **104**, 353 (1967).

247. Hanson, T. L., J. D. Lueck, R. N. Horne, and R. C. Nordlie, *J. Biol. Chem.*, **245**, 6078 (1970).

248. Pontremoli, S., C. Grazi, and A. Accorsi, *Biochem.*, **7**, 3628 (1968).

249. Pontremoli, S., E. Grazi, and A. Accorsi, *Biochem. Biophys. Res. Commun.*, **33**, 335 (1968).

250. Pontremoli, S., E. Grazi, and A. Accorsi, *Biochem. Biophys. Res. Commun.*, **37**, 597 (1969).

251. Sia, C. L., S. Traniella, S. Pontremoli, and B. L. Horecker, *Arch. Biochem. Biophys.*, **132**, 325 (1969).

252. Rao, G. J. S., S. M. Rosen, and O. M. Rosen, *Biochem.*, **8**, 4904 (1969).

253. Tamburro, A. M., A. Scatturin, E. Grazi, and S. Pontremoli, *J. Biol. Chem.*, **245,** 6624 (1970).
254. Kolb, H. J., and G. M. Grodsky, *Biochem.*, **9,** 4900 (1970).
255. Benkovic, S. J., M. M. DeMaine, and J. J. Kleinschuster, *Arch. Biochem. Biophys.*, **139,** 248 (1970).
256. Hubert, E., J. Villanueva, A. M. Gonzales, and F. Marcus, *Arch. Biochem. Biophys.*, **138,** 590 (1970).
257. Sutherland, E. W., T. Pasternak, and C. F. Cori, *J. Biol. Chem.*, **181,** 683 (1949).
258. Rapoport, S., and J. Luebering, *J. Biol. Chem.*, **189,** 683 (1951).
259. Manyai, S., and Z. Varady, *Biochim. Biophys. Acta*, **20,** 594 (1956).
260. Joyce, B. K., and S. Grisolia, *J. Biol. Chem.*, **233,** 350 (1958).
261. Grisolia, S., in *Methods in Enzymology*, Vol. 5, S. P. Colowick and N. O. Kaplan (Eds.), Academic Press, New York, 1962, p. 243.
262. DeVerdier, C. H., *Biochem. J.*, **92,** 38P (1964).
263. Harkness, D. R., and S. Roth, *Biochem. Biophys. Res. Commun.*, **34,** 849 (1969).
264. Rose, Z. B., and J. Liebowitz, *J. Biol. Chem.*, **245,** 3232 (1970).
265. Ames, B. N., *J. Biol. Chem.*, **226,** 583 (1957).
266. Khorana, H. G., in *The Enzymes*, Vol. 5, P. D. Boyer, H. Lardy, and K. Myrback (Eds.), Academic Press, New York, 1961, p. 79.
267. Razzell, W. E., in *Methods in Enzymology*, Vol. 6, S. P. Colowick and N. O. Kaplan (Eds.), Academic Press, New York, 1963, p. 236.
268. Heppel, L. A., P. J. Ortiz, and S. Ochoa, *J. Biol. Chem.*, **229,** 679 (1957).
269. Brown, D. M., L. A. Hippel, and R. J. Hilmoe, *J. Chem. Soc.*, **1959,** 40.
270. Heppel, L. A., and R. J. Hilmoe, in *Methods in Enzymology*, Vol. II, S. P. Colowick and N. O. Kaplan (Eds.), Academic Press, New York, 1955, p. 565.
271. Heppel, L. A., and J. C. Rabinowitz, in *Annual Review of Biochemistry*, Vol. 27, J. M. Luck (Ed.), Annual Reviews, Palo Alto, California, 1958, p. 613.
272. Heppel, L. A., P. R. Whitfield, and R. Markham, *Nature*, **171,** 1152 (1953).
273. Koerner, J. F., and S. Sinsheimer, *J. Biol. Chem.*, **228,** 1049 (1957).
274. Turner, A. F., and H. G. Khorana, *J. Am. Chem. Soc.*, **81,** 4651 (1959).
275. Uzawa, T., *J. Biochem. (Tokyo)*, **15,** 19 (1932).
276. Razzell, W. E., *J. Biol. Chem.*, **236,** 3028, 3031 (1961).
277. Razzell, W. E., and H. G. Khorana, *J. Biol. Chem.*, **234,** 2105 (1959).
278. Razzell, W. E., and H. G. Khorana, *J. Biol. Chem.*, **236,** 1144 (1961).
279. Hayaishi, O., and A. Kornberg, *J. Biol. Chem.*, **206,** 647 (1954).
280. Webster, G. R., E. A. Marples, and R. H. S. Thompson, *Biochem. J.*, **65,** 374 (1957).
281. Dawson, R. M. C., *Biochem. J.*, **62,** 689 (1956).
282. Macfarlane, M. G., and B. C. J. G. Knight, *Biochem. J.*, **35,** 884 (1941).
283. DeGier, J., G. H. DeHaas, and L. L. M. VanDeenen, *Biochem. J.*, **81,** 33P (1961).
284. Ikezawa, H., and R. Murata, *J. Biochem. (Japan)*, **55,** 217 (1964).
285. Macfarlane, M. G., *Biochem. J.*, **42,** 590 (1948).
286. Macfarlane, M. G., *Biochem. J.*, **42,** 587 (1948).
287. Hanahan, D. J., and I. L. Chaikoff, *J. Biol. Chem.*, **168,** 233 (1947).
288. Hanahan, D. J., and I. L. Chaikoff, *J. Biol. Chem.*, **169,** 699 (1947).

289. Hanahan, D. J., and I. L. Chaikoff, *J. Biol. Chem.*, **172,** 191 (1948).
290. Dawson, R. M. C., and N. Hemington, *Biochem. J.*, **102,** 76 (1967).
291. Einset, E., and W. L. Clark, *J. Biol. Chem.*, **231,** 703 (1958).
292. Tookey, H. L., and A. K. Balls, *J. Biol. Chem.*, **218,** 213 (1956).
293. Yang, S. F., S. Freer, and A. G. Benson, *J. Biol. Chem.*, **242,** 477 (1967).
294. Kates, M., *Nature*, **172,** 814 (1953).
295. Kates, M., *Can. J. Biochem. Physiol.*, **32,** 571 (1954).
296. Egami, F., and K. Nakamura, *Microbial Ribonucleases*, Springer-Verlag, New York, 1969.
297. Egami, F., K. Takahashi, and T. Uchida, in *Progress in Nucleic Acid Research and Molecular Biology*, Vol. 3, J. N. Davidson and W. E. Cohn (Eds.), Academic Press, New York, 1964, p. 59.
298. Lehman, I. R., in *Progress in Nucleic Acid Research*, Vol. 2, J. N. Davidson and W. E. Cohn (Eds.), Academic Press, New York, 1963, p. 83.
299. Witzel, H., in *Progress in Nucleic Acid Research*, Vol. 2, J. N. Davidson and W. E. Cohn (Eds.), Academic Press, New York, 1963, p. 221.
300. Elson, D., and T. Kivity-Vogel, in *Regulation of Nucleic Acid and Protein Biosynthesis*, V. V. Koningsberger and L. Bosch (Eds.), Elsevier, Amsterdam, 1967, p. 196.
301. Barnard, E. A., in *Annual Reviews of Biochemistry*, Vol. 38, E. E. Snell (Ed.), Annual Reviews, Palo Alto, California, 1969, p. 677.
302. Lehman, I. R., in *Annual Reviews of Biochemistry*, Vol. 36, E. E. Snell (Ed.), Annual Reviews, Palo Alto, California, 1967, p. 645.
303. Anfinsen, C. B., and F. H. White, Jr., in *The Enzymes*, 2nd ed., Vol. 5, P. D. Boyer, H. Lardy, and K. Myrback (Eds.), Academic Press, New York, 1961, p. 95.
304. McDonald, M. R., in *Methods in Enzymology*, Vol. II, S. P. Colowick and N. O. Kaplan (Eds.), Academic Press, New York, 1955, p. 437.
305. Laskowski, M., Sr., in *The Enzymes*, 2nd ed., Vol. 5, P. D. Boyer, H. Lardy, and K. Myrback (Eds.), Academic Press, New York, 1961, p. 123.
306. Scheraga, H. A., and J. A. Rupley, in *Advances in Enzymology*, Vol. 24, F. F. Nord (Ed.), Interscience, New York, 1962, p. 161.
307. Gehrmann, G., and S. Okada, *Biochim. Biophys. Acta*, **23,** 621 (1957).
308. Polson, A., *Biochim. Biophys, Acta*, **22,** 61 (1956).
309. Weissbach, A., and D. Korn, *J. Biol. Chem.*, **238,** 3383 (1963).
310. Lee, C. Y., and S. H. Zbarsky, *Can. J. Biochem.*, **45,** 39 (1967).
311. Price, P. A., T. Y. Liu, W. H. Stein, and S. Moore, *J. Biol. Chem.*, **244,** 917 (1969).
312. Price, P. A., S. Moore, and W. H. Stein, *J. Biol. Chem.*, **244,** 924, 929, 933 (1969).
313. Ralph, R. K., R. A. Smith, and H. G. Khorana, *Biochem.* **1,** 131 (1962).
314. Lindberg, U., *Biochem.*, **6,** 335, 343 (1967).
315. Koerner, J. F., and R. L. Sinsheimer, *J. Biol. Chem.*, **228,** 1039, 1047 (1957).
316. Bernardi, G., and M. Griffe, *Biochem.* **3,** 1419 (1964).
317. Bernardi, G., E. Appella, and R. Zito, *Biochem.* **4,** 1725 (1965).
318. Cardonnier, C., and G. Bernardi, *Can. J. Biochem.*, **46,** 989 (1968).
319. Bernardi, G., *Adv. Enzymol.* **31,** 1 (1968).

320. Laskawski, M., *Adv. Enzymol.*, **30,** 165 (1967).
321. Bernardi, G., and C. Sadron, *Nature*, **191,** 809 (1961).
322. Bernardi, G., and C. Sadron, *Biochem.*, **3,** 1411 (1964).
323. Bernardi, G., *J. Mol. Biol.*, **13,** 603 (1965).
324. Carrara, M., and G. Bernardi, *Biochem.*, **7,** 1121 (1968).
325. Lindahl, T., J. A. Gally, and G. M. Edelman, *J. Biol. Chem.*, **244,** 5014 (1969).
326. Anfinsen, C. B., M. T. Rumley, and H. Tanuichi, *Acta Chem. Scand.*, **17,** S270 (1963).
327. de Meuron-Landholdt, M., and M. P. de Carille, *Biochim. Biophys. Acta*, **91,** 433 (1964).
328. Tanuichi, H., C. B. Anfinsen, and A. Lodja, *J. Biol. Chem.*, **242,** 4752 (1967).
329. Cuatrecascus, P., S. Fuchs, and C. B. Anfinsen, *J. Biol. Chem.*, **242,** 3063, 4759 (1967).
330. Anfinsen, C. B., and L. G. Corley, *J. Biol. Chem.*, **244,** 5149 (1969).
331. Cutarecasas, P., M. Wilchek, and C. B. Anfinsen, *Biochem.*, **8,** 2277 (1969).
332. Omenn, G. S., D. A. Ontjes, and C. B. Anfinsen, *Biochem.*, **9,** 304, 313 (1970).
333. Yasmineh, W. G., E. D. Gray, and L. W. Wannamaker, *Biochem.*, **7,** 91 (1968).
334. Gray, E. D., and W. G. Yasmineh, *Biochem.*, **7,** 98, 105 (1968).
335. Nestle, M., and W. K. Roberts, *J. Biol. Chem.*, **244,** 5213, 5219 (1969).
336. Taniuchi, H., and C. B. Anfinsen, *J. Biol. Chem.*, **243,** 4778 (1968).
337. Cuatrecasas, P., H. Taniuchi, and C. B. Anfinsen, *Brookhaven Symp. Biol.*, **21,** 172 (1968).
338. Lees, C. W., and R. W. Hortley, Jr., *Biochem.*, **5,** 3951 (1966).
339. Hummel, J. P., and G. Kalnitsky, *Ann. Rev. Biochem.*, **33,** 25 (1964).
340. Smyth, D. G., W. H. Stein, and S. Moore, *J. Biol. Chem.*, **238,** 227 (1963).
341. Rushizky, G. W., A. E. Greco, R. W. Hartley, Jr., and H. A. Sober, *Biochem.*, **2,** 787 (1963).
342. Hartley, R. W., G. W. Rushizky, A. E. Greco, and H. A. Sober, *Biochem.*, **2,** 794 (1963).
343. Glitz, D. G., and C. A. Dekker, *Biochem.*, **3,** 1391, 1399 (1964).
344. Rushizky, G. W., J. H. Mazejko, D. L. Rogerson, Jr., and H. A. Sober, *Biochem.*, **9,** 4966 (1970).
345. McLennon, B. D., and B. G. Lane, *Can. J. Biochem.*, **46,** 93 (1968).
346. Anderson, J. H., and C. E. Carter, *Biochem.*, **4,** 1102 (1965).
347. Delaney, R., *Biochem.*, **2,** 438 (1963).
348. Fujioku, H., and H. A. Scheraga, *Biochem.*, **4,** 2197, 2206 (1965).
349. Simpson, R. T., and B. L. Vallee, *Biochem.* **5,** 2531 (1966).
350. Ginsburg, A., and W. R. Carroll, *Biochem.*, **4,** 2159 (1965).
351. Bello, J., and E. F. Nowoswiot, *Biochem.*, **8,** 628 (1969).
352. del Rosario, E. J., and G. G. Hammes, *Biochem.*, **8,** 1884 (1969).
353. Roberts, C. K., D. H. Meadows, and O. Jardetzky, *Biochem.*, **8,** 2053 (1969).
354. Bello, J., *Biochem.*, **8,** 4535, 4542, 4550 (1969).
355. Ettinger, M. J., and C. H. W. Hirs, *Biochem.*, **7,** 3374 (1968).

356. Anderson, D. G., G. G. Hammes, and F. G. Walz, Jr., *Biochem.*, **7,** 1637 (1968).
357. Gross, E., and B. Witkop, *Biochem.*, **6,** 745 (1967).
358. Richards, F. M., *C. R. Trov. Lab. Carlsberg*, **29,** 329 (1955).
359. Richards, F. M., and P. J. Vithayathil, *J. Biol. Chem.*, **234,** 1459 (1959).
360. Gross, E., and B. Witkop, *J. Biol. Chem.*, **237,** 1856 (1962).
361. Smyth, D. G., *et al.*, *J. Biol. Chem.*, **237,** 1845 (1962).
362. Potts, J. T., A. Berger, F. Cooke, and C. B. Anfinsen, *J. Biol. Chem.*, **237,** 1851 (1962).
363. Ho, N. W. Y., and P. T. Gilham, *Biochem.*, **6,** 3632 (1967).
364. Hammes, G. G., and H. A. Schleraga, *Biochem.*, **5,** 3690 (1966).
365. Cathou, R. E., G. G. Hammes, and P. R. Schimmel, *Biochem.*, **4,** 2687 (1965).
366. Rhiem, J. P., and H. A. Scheroga, *Biochem.*, **4,** 772 (1965).
367. Broomfield, C. A., J. P. Rhiem, and H. A. Schewga, *Biochem.*, **4,** 751 (1965).
368. Rhiem, J. B., C. A. Broomfield, and H. A. Schewga, *Biochem.*, **4,** 760 (1965).
369. Goldberger, R. F., and C. B. Anfinsen, *Biochem.*, **1,** 401 (1962).
370. Allendi, J. E., and F. M. Richards, *Biochem.*, **1,** 295 (1962).
371. Neumann, N. P., S. Moore, and W. H. Stein, *Biochem.*, **1,** 68 (1962).
372. Smyth, D. G., W. H. Stein, and S. Moore, *J. Biol. Chem.*, **238,** 227 (1963).
373. Hirs, C. H. W., S. Moore, and W. H. Stein, *J. Biol. Chem.*, **235,** 633 (1960).
374. Smyth, D. G., W. H. Stein, and S. Moore, *J. Biol. Chem.*, **237,** 1845 (1962).
375. Potts, J. T., G. Berger, J. Cooke, and C. B. Anfinsen, *J. Biol. Chem.*, **237,** 1851 (1962).
376. Gross, E., and B. Witkop, *J. Biol. Chem.*, **237,** 1856 (1962).
377. Wyckoff, H. W., K. D. Hordmon, N. M. Allewell, T. Inagami, L. N. Johnson, and F. M. Richards, *J. Biol. Chem.*, **242,** 3984 (1967).
378. Kartha, G., J. Bello, and D. Harker, *Nature*, **213,** 862 (1967).
379. Gutte, B., and R. B. Merrifield, *J. Am. Chem. Soc.*, **91,** 501 (1969).
380. Denkewalter, R. G., D. F. Veber, F. W. Holly, and R. Hirschman, *J. Am. Chem. Soc.*, **91,** 503 (1969).
381. Strachan, R. G., W. J. Paleveda, R. F. Nutt, R. A. Vitali, D. F. Veber, M. J. Dickinson, V. Garsky, J. E. Deak, E. Walton, S. R. Jenkins, F. W. Holly, and R. Hirschman, *J. Am. Chem. Soc.*, **91,** 503 (1969).
382. Jenkins, S. R., R. F. Nutt, R. S. Dewey, D. F. Veber, F. W. Holly, W. J. Paleveda, Jr., T. Lanza, Jr., R. G. Strachan, E. F. Schoenewaldt, H. Barkemeyer, M. J. Dickinson, J. Sondery, R. Hirschman, and E. Walton, *J. Am. Chem. Soc.*, **91,** 505 (1969).
383. Veber, D. F., S. L. Varga, J. D. Milkawski, H. Joshua, J. B. Conn, R. Hirschman, and R. G. Denkewalter, *J. Am. Chem. Soc.*, **91,** 506 (1969).
384. Hirschman, R., R. F. Nutt, D. F. Veber, R. A. Vitali, S. L. Varga, T. A. Jacob, F. W. Holly, and R. G. Denkewatter, *J. Am. Chem. Soc.*, **91,** 507 (1969).
385. Kornberg, S. R., I. R. Lehman, M. J. Bessman, E. S. Simms, and A. Kornberg, *J. Biol. Chem.*, **233,** 159 (1958).
386. Kielley, W. W., in *The Enzymes*, Vol. 5, P. D. Boyer, H. Lardy, and K. Myrback (Eds.), Academic Press, New York, 1961, p. 149.

387. Kunitz, M., and P. W. Robbins, in *The Enzymes*, Vol. 5, P. D. Boyer, H. Lardy, and K. Myrback (Eds.), Academic Press, New York, 1961, p. 169.
388. Schmidt, G., in *Phosphorus Metabolism*, Vol. II, W. D. McElroy and B. Glass (Eds.), Johns Hopkins Press, Baltimore, Maryland, 1951, p. 443.
389. Boyland, C., *Biochem. J.*, **24,** 341 (1930).
390. Bauer, E., *Naturwiss.*, **23,** 866 (1935).
391. Bauer, E., *Z. Physiol. Chem.*, **239,** 195 (1936).
392. Bauer, E., *Z. Physiol. Chem.*, **248,** 213 (1937).
393. Bailey, K., and E. C. Webb, *Biochem. J.*, **38,** 394 (1944).
394. Kunitz, M., *J. Am. Chem. Soc.*, **73,** 1387 (1951).
395. Kunitz, M., *J. Gen. Physiol.*, **35,** 423 (1952).
396. Heppel, L. A., and R. J. Hilmoe, *J. Biol. Chem.*, **192,** 87 (1951).
397. Josse, J., *Fed. Proc.*, **24,** 410 (1965).
398. Josse, J., *J. Biol. Chem.*, **241,** 1938, 1948 (1966).
399. Levinson, H. S., J. D. Sloan, Jr., and M. T. Hyatt, *J. Bacteriol.*, **75,** 291 (1958).
400. Oginsky, E. L., and H. L. Rumbaugh, *J. Bacteriol.*, **70,** 92 (1955).
401. Howard, A., and D. G. Lundgren, *Can. J. Biochem.*, **48,** 1302 (1970).
402. Gilmour, D., and J. H. Calaby, *Enzymol.*, **16,** 34 (1953).
403. McElroy, W. D., J. Coulombre, and R. Hays, *Arch. Biochem. Biophys.*, **32,** 207 (1951).
404. Naganna, B., B. Venugopol, and C. E. Sripatti, *Biochem. J.*, **60,** 224 (1955).
405. Naganna, B., A. Ramon, B. Venugopol, and C. E. Scripathi, *Biochem. J.*, **60,** 215 (1955).
406. Moyer, K., and M. Klinga-Moyer, *Hoppe Seylers Z. Physiol. Chem.*, **267,** 115 (1940).
407. Fleury, P., and J. Courtois, *Enzymol.*, **1,** 377 (1937).
408. Fleurry, P., and J. Courtois, *Enzymol.*, **5,** 254 (1938).
409. Ricketts, T. R., *Arch. Biochem. Biophys.*, **110,** 184 (1965).
410. Rafter, S. W., *J. Biol. Chem.*, **230,** 643 (1958).
411. Norberg, B., *Acta Chem. Scand.*, **4,** 601 (1950).
412. Elliot, W. H., *Biochem. J.*, **65,** 315 (1957).
413. Lohmann, K., *Biochem. Z.*, **262,** 137 (1933).
414. Naganna, B., and V. K. N. Menon, *J. Biol. Chem.*, **174,** 501 (1948).
415. Gordon, J. J., *Biochem. J.*, **46,** 96 (1950).
416. Robbins, E. A., M. P. Stalberg, and P. D. Boyer, *Arch. Biochem. Biophys.*, **54,** 215 (1955).
417. Kay, H. D., *Biochem. J.*, **22,** 1446 (1928).
418. Schachman, H. K., *J. Gen. Physiol.*, **35,** 451 (1952).
419. Kitosato, T., *Biochem. Z.*, **197,** 257 (1928).
420. Kitosato, T., *Biochem. Z.*, **201,** 206 (1928).
421. Meyerhof, O., R. Shatas, and A. Kaplan, *Biochim. Biophys. Acta*, **12,** 121 (1953).
422. Man, T., *Biochem. J.*, **38,** 339, 345 (1944).
423. Mattenheimer, H., *Biochem. Z.*, **322,** 36 (1951).

424. Kornberg, S. R., *J. Biol. Chem.*, **218,** 23 (1956).
425. Brahms, J., and C. M. Kay, *J. Biol. Chem.*, **238,** 198 (1963).
426. Kielley, W. W., and W. F. Harrington, *Biochim. Biophys. Acta*, **41,** 401 (1960).
427. Ellenbogen, E., R. Iyengar, H. Stern, and R. E. Olsen, *J. Biol. Chem.*, **235,** 2642 (1960).
428. Gergely, J., in *Annual Review of Biochemistry*, Vol. 35, P. D. Boyer (Ed.), Annual Reviews, Palo Alto, California, 1966, p. 628.
429. Albers, R. W., in *Annual Review of Biochemistry*, P. O. Boyer (Ed.), Annual Reviews, Palo Alto, California, 1967, p. 727.
430. Pullman, M. E., H. S. Penefsky, A. Datta, and E. Rachen, *J. Biol. Chem.*, **235,** 3322 (1960).
431. Penefsky, H. S., M. E. Pullman, A. Dotta, and E. Racker, *J. Biol. Chem.*, **235,** 3330 (1960).
432. Penefsky, H. S., and R. C. Warner, *J. Biol. Chem.*, **240,** 4694 (1965).
433. Ulrich, F., *Biochim. Biophys. Acta*, **105,** 460 (1965).
434. Judah, J. D., K. Ahmed, and A. E. M. McLeon, *Biochim. Biophys. Acta*, **65,** 472 (1962).
435. Skou, J. C., *Biochim. Biophys. Acta*, **42,** 6 (1960).
436. Auditore, J. V., and L. Murray, *Arch. Biochem. Biophys.*, **99,** 372 (1962).
437. Borgman, R. J., J. F. Manery, and L. Pinteric, *Biochim. Biophys. Acta*, **203,** 506 (1970).
438. Neufeld, G. H., and H. M. Levy, *J. Biol. Chem.*, **244,** 6493 (1969).
439. Skou, J. C., *Physiol. Rev.*, **45,** 596 (1965).
440. Albers, R. W., *Ann. Rev. Biochem.*, **36,** 727 (1967).
441. Siegel, G. J., G. J. Koval, and R. Wayne Albers, *J. Biol. Chem.*, **244,** 3264 (1969).
442. Kielley, W. W., and O. Meyerhof, *J. Biol. Chem.*, **176,** 591 (1948).
443. Goldberg, M., and D. Gelman, *Arch. Biochem. Biophys.*, **51,** 411 (1954).
444. Caffrey, R. W., R. Trembloy, B. W. Gabrio, and F. M. Huennekins, *J. Biol. Chem.*, **223,** 1 (1956).
445. Kalckar, H. M., *J. Biol. Chem.*, **153,** 355 (1944).
446. Krishnan, P. S., *Arch. Biochem.*, **20,** 261, 272 (1949).
447. Traverso-Cori, A., H. Charmovich, and D. Cori, *Arch. Biochem. Biophys.*, **109,** 173 (1965).
448. Molnar, J., and L. Lorand, *Arch. Biochem. Biophys.*, **93,** 353 (1961).
449. Libecq, C., A. Lallemond, and M. J. Degueldre-Guillaume, *Arch. Biochem. Biophys.*, **97,** 609 (1962).
450. Liebecq, C., A. Lallemond, and M. J. Degueldre-Guillaume, *Bull. Soc. Chim. Biol.*, **45,** 573 (1963).
451. Molnar, J., and L. Lorand, *Arch. Biochem. Biophys.*, **93,** 353 (1961).
452. Chern, C. J., A. B. MacDonald, and G. J. Morris, *J. Biol. Chem.*, **244,** 5489 (1969).
453. Johnson, M., M. A. G. Kaye, R. Hems, and H. A. Krebs, *Biochem. J.*, **54,** 625 (1953).
454. Heppel, L. A., and R. J. Hlimae, *J. Biol. Chem.*, **202,** 217 (1953).
455. Lieberman, I., A. I. Lansing, and W. E. Lynch, *J. Biol. Chem.*, **242,** 736 (1967).

456. Zimmerman, S. B., and A. Kornberg, *J. Biol. Chem.*, **236,** 1480 (1961).
457. Malmgren, H., *Acta Chem. Scand.*, **6,** 16 (1952).
458. Lindeberg, G., and H. Malmgren, *Acta Chem. Scand.*, **6,** 27 (1952).
459. Ingelman, B., and H. Malmgren, *Acta Chem. Scand.*, **3,** 157 (1949).
460. Diederich, D. A., and S. Grisolia, *J. Biol. Chem.*, **244,** 2412 (1969).
461. Raijman, L., S. Grisolia, and H. Edelhoch, *J. Biol. Chem.*, **235,** 2340 (1960).
462. Grisolia, S., J. Caravaco, and B. K. Joyce, *Biochim. Biophys. Acta*, **29,** 432 (1958).
463. Harary, I., *Biochim. Biophys. Acta*, **26,** 434 (1957).
464. Bader, H., and A. K. Sen, *Biochim. Biophys. Acta*, **118,** 116 (1966).
465. Sachs, G., J. D. Rose, and B. T. Hirschowitz, *Arch. Biochem. Biophys.*, **119,** 277 (1967).
466. Singer, M. F., and J. S. Fruton, *J. Biol. Chem.*, **229,** 111 (1957).
467. Holzer, M. E., D. J. Burrow, and R. A. Smith, *Biochim. Biophys. Acta*, **56,** 491 (1962).
468. Holzer, M. E., K. D. Johnson, and R. A. Smith, *Biochim. Biophys. Acta*, **122,** 232 (1966).
469. Parvin, R., and R. A. Smith, *Biochem.*, **8,** 1748 (1969).
470. Fujimoto, A., and R. A. Smith, *Biochim. Biophys. Acta*, **56,** 501 (1962).

Vertical Movement of Phosphate in Freshwater

29

H. L. GOLTERMAN

Limnological Institute, Nieuwersluis, The Netherlands

The distribution of phosphate in a water column of a lake is generally not homogeneous. The dissolved phosphates, of course, tend to reach uniform concentration throughout the column but this tendency is counteracted by the various particulate phosphate compounds. Such compounds can be either organic (biochemical) or inorganic. The organic types are formed by the phosphate being taken up metabolically by the phytoplankton, so that these cellular phosphate molecules must inevitably follow the active migration processes of the phytoplankton and the sinking rate of the dead cells. Both processes will lead to local differences in the particulate phosphate concentration. But there are other features that work against this. Due to the normally short lifetime and rapid mineralization of most organic matter of algal cells in lakes (see p. 516) and also due to the geochemically loose binding of the phosphate molecule to the organic matter of the algal cells (P—O—C bonds, which hydrolyze easily), the phosphate molecules escape easily from sedimentation with dead phytoplankton. Both facts smooth out the pattern of phosphate sedimentation much more than, for example, that of silicates.

Primary production in deep lakes is mainly limited to their upper layers (e.g., epilimnion), and consequently this metabolic activity will enhance the vertical nonhomogeneity in thermally stratified lakes. The seasonal periodicity of both primary production and thermal stratification means that the vertical exchange is strongly influenced by the seasons.

Inorganic compounds, which lead to a nonhomogeneous distribution of phosphate, may be released into the water from the sediments of lakes. These undissolved substances are mineral phosphates of, for instance, calcium, aluminum, and iron, among which is to be considered also the humic bound phosphate, because the phosphate in these complexes is more likely bound to the iron than to the organic part of these molecules. Due to changes in the chemical properties of the water, such as pH and redox potential, part of this material sometimes comes into solution. The best known example is the dissolving of phosphates from $FePO_4$ after reduction of the ferric iron to ferrous iron in the anaerobic hypolimnion of stratified lakes. The occurrence of $AlPO_4$ has been remarked by some authors, but this has been mostly in the pure chemical analysis literature (16, 17, 24, 56). It is certainly not of importance in all lakes.

Insoluble phosphates like those of calcium and iron are not only formed in the mud but also in the water column. When the concentrations are sufficiently high, and depending on pH and redox values, they will then sink to the bottom of the lake. It seems likely that they may also be formed in the mud after the release of orthophosphate from mineralizing algal cells, and its conversion to calcium or iron phosphate due to relatively high Ca and Fe concentrations in the interstitial water. At the present moment it is not possible to distinguish between precipitation of insoluble phosphates formed in the lake and the phosphates supplied to the lake in insoluble forms, for instance, as fine rock particles, or as hydroxyapatite, or phosphates enclosed in the lattice of mineral clays.

For the sake of convenience we can make a distinction between an "internal or metabolic phosphate cycle":

$$(PO_4\text{—}P_{\text{water}} \xrightarrow[\text{productivity}]{\text{primary}} \text{cell } PO_4 \xrightarrow{\text{mineralization}} PO_4\text{—}P_{\text{water}} + \text{organic } P_{\text{water}})$$

and an "external phosphate cycle":

$$(PO_4\text{—}P_{\text{water}} \rightarrow \text{sediments} \rightarrow PO_4\text{—}P_{\text{water}} + \text{organic } P_{\text{water}}\ (?))$$

The first cycle summarizes metabolic aspects, the second is a geochemical cycle (see Fig. 1).

Processes in the first cycle usually occur quickly (in a few days), and their nature is normally biochemically well known. The second cycle may be very slow, especially the solubilization part. Very little is known about the processes involved and practically nothing about processes in the sediments. Hynes and Greib (33) showed that phosphate moves readily through undisturbed mud and that the movement is not a biotic process. It seems likely that the part of the mineralization of dead algal cells which takes place in the sediments does so in the uppermost layer only.

IDENTIFICATION AND OCCURRENCE OF SEDIMENT PHOSPHATES

For the characterization of insoluble mud phosphates the classical extraction procedure of Chang and Jackson for soils is usually used [see Jackson, (34)], sometimes after a modification (56). With the following extractions different

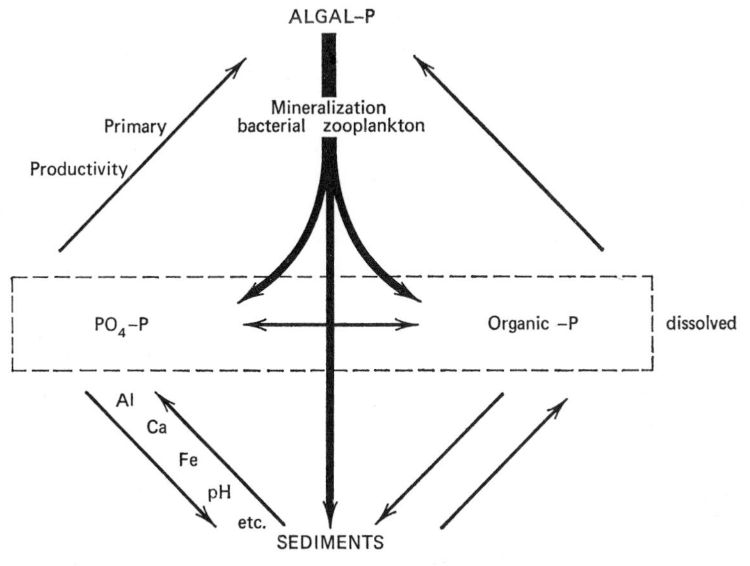

Figure 1. Metabolical ("internal") and geochemical ("external") phosphate cycles (respectively above and below the dissolved compounds).

inorganic phosphates are supposed to come into solution:

	Extraction Step	Extracted Inorganic Phosphates
I.	0.5 *N* NH_4F (1 hr, pH = 8.2)	Aluminum phosphate
II.	0.1 *N* NaOH (17 hr)	Iron phosphate
III.	0.3 *M* $Na_3C_6H_5O_7{\cdot}2H_2O$ + $Na_2S_2O_4$ (75–80°C) 5 min	Reductant soluble (iron) phosphate
IV.	0.5 *N* H_2SO_4 (1 hr)	Calcium phosphate

One of the major problems in this kind of analysis is that the insoluble phosphates are rendered soluble during the analysis, and so are no longer in their original forms. Thus there is no chemical proof that the phosphates of the different fractions really are the compounds named. For the determination of the organic phosphates the situation is even worse as these are measured as the difference between the total and the sum of the inorganic phosphate (sum of fractions I, II, III, IV). Yet there is no doubt that in the supposed inorganic fractions, organic phosphates are also found, and also it seems likely that parts of the inorganic phosphates are not extracted quantitatively in the extractions I, II, III, and IV. For instance, iron phosphate ($FePO_4$) added to mud is not quantitatively recovered in fraction II, illustrating that the situation is probably more complex than has previously been suggested (21). But for the sake of convenience the inorganic fractions are referred to as iron phosphates, and so on, in this review.

A different extraction method has been proposed by Mehta *et al.* (45). They determined inorganic and organic phosphates in combined supernatants of cold and hot HCl and cold and hot NaOH extractions. They suggested that, in the Harpster clay loam they used, hydrolysis was not a serious source of error, but in a more organic mud this may not be so. The real proof of their statement is very difficult. Frink (16) does find differences between the inorganic phosphates in the extracts as described in the table above and in those made according to Mehta *et al.* and believes—probably correctly—that the latter procedure is too drastic. A further improvement on Frink's method would be the determination of organic phosphate in the cold NaOH extract. Williams *et al.* (75) have shown that the phosphates released with the NH_4F and the NaOH extractants may be readsorbed onto calcareous materials presumably due to the formation of a CaF_2–orthophosphate complex or hydroxyapatite. This process would then underestimate the aluminum and iron phosphates and overestimate the calcium and iron occluded phosphates. The authors showed furthermore that in different lake samples the reductant soluble phosphate (fraction III) is also HCl extractable, as may be expected in view of the reducing conditions in most underwater soils. Their paper gives an interesting discussion of the many existing extraction schemes.

The readsorption of NaOH-extractable phosphate on calcareous material may be avoided by a previous HCl extraction, which renders it impossible to distinguish between iron- and calcium-bound phosphates, while in iron-rich sediments a resorption on the iron may occur.

Golterman (unpublished) found that more phosphate was extracted with NaOH in the presence of a strong anion exchanger of which only a larger fraction was used. The anion exchanger was separated from the mud through a suitable sieve and eluted with stronger NaOH. This phenomenon could be explained by the resorption as discussed above and could be used to avoid the resorption. When not all phosphate was removed from the NaOH extract (a smaller fraction sometimes remained in solution) the extract was treated a second time with the anion exchanger and both quantities were mixed and eluted.

Not all fractions will always be present in all muds. For example, in gyttja types of sediment we found that fractions I and III were negligibly low, while iron and calcium phosphates were high. Mackereth (42) does not consider calcium phosphate to be important in the sediments of Windermere because it is a soft-water lake. From the theoretical point of view the existence of $AlPO_4$ in lake sediments seems doubtful, due to the expected hydrolysis of a salt of a weak acid and a weak base. Not many quantitative data are available. Frink (17) measured quantities up to 300 μg $AlPO_4$/g dry weight, but remarks that in view of the experiments of Harter (24) (see also p. 528) the "Al–P fraction in these sediments is not clearly defined." Mackereth (42) has studied mud cores from three lakes that differ in their degree of productivity: Esthwaite water, Windermere, and Ennerdale. He analyzed not only the phosphate content of the cores but many other elements as well, such as carbon, iron, and so on, and he was therefore able to show that the sediments of Ennerdale contain a precipitate of organic phosphate, due to the sinking of phytoplankton, and a precipitate of particle-bound phosphate from material eroded from leachable phosphates from the catchment area. (In the following section we return to Mackereth's calculations concerning C/P ratios in the sediments and algae.)

Apart from this mechanism as found for Ennerdale, precipitation of iron phosphate also plays a role in Windermere and Esthwaite. Mackereth showed that in Windermere and Ennerdale release of phosphate from the sediments has never been important, although in Esthwaite a loss of phosphate has occurred combined with a loss of iron due to the mechanism explained in the subsection on the Fe—PO_4—S system. He argued reasonably that the same phenomenon must have also occurred in the sediments of Linsley Pond (41, 31).

Serruya (65) showed a high linear correlation between P and Fe in the subsurface sediments of Lake Kinneret. In two deeper cores Serruya found a negative correlation between Ca and P, and a positive correlation between Fe and P, except below 30 cm in the deepest core where the Ca sharply decreased. The Fe/P ratio, however, dropped also, although less so.

Besides Serruya's explanation of these effects (a diminished primary production or a higher rate of mineralization) an increased precipitation of a compound containing no organic material, Ca, Fe, or P seems possible. The correlations were calculated by us and were found to be:

	Ca/P	Fe/P
Core A	$r = -0.62$ (significant to 5%)	$r = 0.34$ (not significant)
Core B	$r = -0.84$ (significant to 0.5%)	$r = 0.73$ (significant to 1%)

It is evident that in the deeper core (B) the correlation between Fe and P is strongly significant, unlike in core A where the correlation is insignificant. It seems possible that the stronger the correlation between Fe and P, the stronger is the negative correlation between Ca and P. The fact that iron is correlated with P in the five subsurface samples in the whole lake, but not in a deeper core, suggests a continuous change even in the greater depths in the core. It seems likely that these facts are due to a natural grading process of iron phosphate particles to the deeper parts of the lake as has been suggested for Lake Mendota by Delfino *et al.* (9). The withdrawal of phosphate with the iron to these deeper parts would decrease the possible precipitation of $Ca_5(PO_4)_3 \cdot OH$. The same has been described by Frink (17, pp. 369–372) for clay, ignition loss, and total phosphorus in Bantam Lake, Connecticut. Frink, who also found a negative correlation of calcium phosphate with depth and a positive one for iron phosphate, suggests that the calcium phosphate is correlated with the sand fraction. As the sand itself appears to be a function of depth, the withdrawal of phosphate with iron to the deeper parts would cause the same correlations.

In the future more insight into the phenomena associated with the sediments can be expected from studies like these of Mackereth and Serruya, especially when combined with a differential extraction of the phosphates, which would give a better chemical proof of the presence of certain compounds than do the calculated correlations.

Little information is provided by the P content of the sediments itself, as this content is the result of P sedimentation, P release from the sediments, and sedimentation of P-free silt. A sudden decrease in the P profile may be explained by an increased release or an increased sedimentation of silt. Indications from other sources may sometimes yield strong arguments, but real proof is difficult to obtain. An example of this kind of argument is found in the discussion about the sediment from Linsley Pond by Hutchinson and Wollack (31) who, assuming a constant sedimentation of organic phosphates, suggested an increase of silting with a factor of 2.6 in order to explain the difference in the ash content of the surface mud and that of a 3-ft-deep sample.

Stangenberg-Oporowska (67) found very high organic phosphate contents in the upper 5 cm of bottom sediments of fishponds, while aluminum phosphates amounted to 12 to 32% of the mineral phosphate. It seems not unlikely that in these very fresh sediments the NH_4F-extractable phosphate is loosely bound (organic?) phosphate. Furthermore, Stangenberg-Oporowska (67) suggested that they found protein phosphates with Krause's fractionation procedure. It should be pointed out, however, that Krause's method was developed for organisms, and that it certainly will extract humic phosphates also. This could also account for the extracted amino acids, which might have been bound in the humic material.

VERTICAL EXCHANGE AND INTERNAL PHOSPHATE CYCLE

Phytoplankton is extremely efficient in removing phosphate from the water, although not in all lakes will values as low as 0.1 μg/l of P be reached as described by Heron (30).

Thomas (69) has shown the importance of the removal of phosphate by the phytoplankton. In the Zürichsee he has found a mean annual decrease in springtime from 90 to 10 tons of PO_4—P for the years 1960 to 1964 (see the section concerning the influence of water removal on phosphate sedimentation). See Fig. 2.

Gächter (18) has made a very detailed study of the phosphate kinetics in the Horwer Bucht in the Vierwaldstättersee. His orthophosphate–phosphorus isopleths varied from less than 1 up to 46 μg/l in the anaerobic hypolimnion during the summer period, whereas during the winter the phosphate concentration was homogenous throughout the whole water column. The other soluble phosphates showed a remarkable homogeneity over the whole year. The particulate phosphate concentrations did show a few maxima in the deeper layers, each being associated with a decrease of phosphate in the trophogenic layer, indicating a sinking of particulate phosphates. As the differences were small and the particulate P is measured as difference between

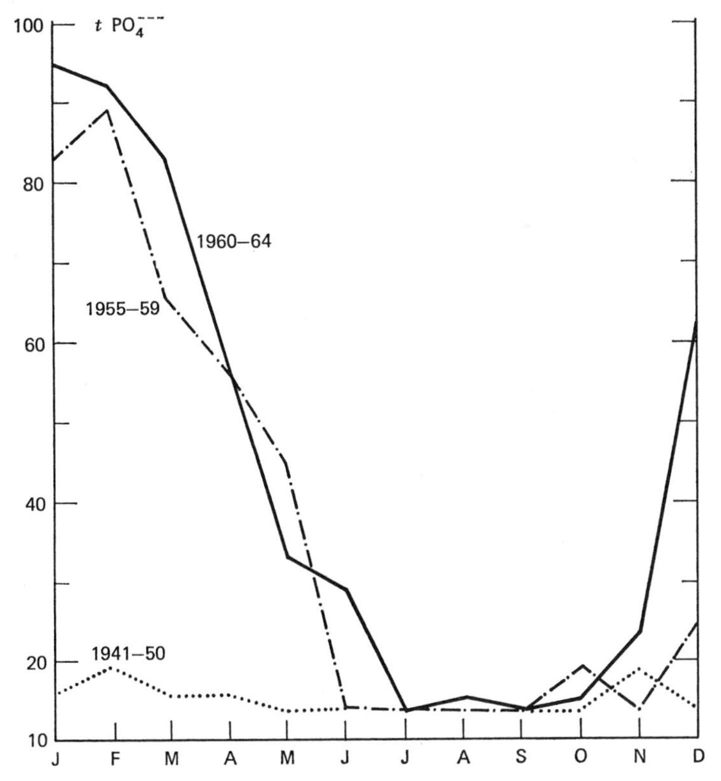

Figure 2. Phosphate concentration in tons in the 0 to 10 m layer in Lake Zürich; mean values for different year classes [see Thomas (69)].

total P minus soluble P, the sinking rate cannot be calculated from these figures.

There is a good deal of evidence indicating that mineralization of the dead plankton cells probably takes place mainly in the epilimnion (37, 52, 65), and that a high turnover rate of phosphate exists. As a direct demonstration of this phenomenon is extremely difficult, it can be demonstrated by comparing the phosphate concentrations in the water with phosphate uptake connected with primary production. The daily phosphate uptake is easily estimated. For example, indirect calculations may be based on the measurements of primary production assuming a given C/P ratio or a phosphorus content of the dry weight of, for example, 1%. From Rodhe's figures for primary production in the Zürichsee, 500 mg/m^2 of C, on May 9, 1957 (62), a daily phosphorus uptake of about 10 mg/m^2 may be assumed and with an actual phosphate concentration of about 7 mg/m^3 of PO_4—P (64) a turnover time* for the internal phosphate cycle may be estimated as 7 to 14 days, indicating rapid turnover of the phosphate in the upper layer.

Gächter has made similar calculations from his data. Assuming a P/C ratio of 1:40, and assuming that the primary production in the observation period (10:00 to 14:00 p.m.) was 50% of the daily primary production, he found a maximum turnover time of 10 days for particulate phosphate during a summer period, increasing to 30 days during the winter. He has also calculated the turnover rates for PO_4—P and Tot—P_{diss} (total dissolved phosphate), which are, however, maximal turnover times because he did not take the former phosphates into account in the calculation for the latter, and vice versa.

It must be pointed out, that the P/C ratio may be as low as 1:10, which would increase the daily assimilated amount of phosphate by a factor of 4. These high turnover times agree with the increasing evidence that a greater part of the material formed by photosynthesis is mineralized in the epilimnion and it seems therefore probable that sedimentation of dead phytoplankton only plays a minor part in the sedimentation of phosphates.

The high turnover time can only be explained with a rapid liberation of phosphate from dying cells by autolysis. Before dying cells are metabolically inactive, autolysis may cause a liberation of phosphates into the water. Golterman (19) found that during autolysis of a few hours, 50%, and after a few days even 70 to 80%, of the particulate phosphate was returned into solution. These processes reduce the sedimentation of the particulate phosphate. As the cells after autolysis contained sufficient phosphate for breakdown by bacteria the phosphate taken up by the bacteria or released to the

* This "biological" turnover time should not be confused with Rigler's (59) "physical" turnover time, which is mostly due to exchange processes between P^{31} and P^{32}. See footnote p. 526.

water is also not lost to the sediments. As the mineralization of P takes place in a few hours, but the mineralization of C and N in a few days, it is to be expected that the ratio of C (or N) over P changes during the sedimentation of the phytoplankton. It seems likely therefore that the fact that Mackereth (see p. 513) found the same ratio in the sediments as in the phytoplankton is due to other processes, accidentally giving the same ratios. Although the release of phosphate from dying algae has not yet been proved in nature, a sinking of particles containing an unusually high proportion of chlorophyll compared to phosphorus as found by Ketchum and Corwin (36) indicates such a release.

Not much is known about the kind of organic compounds that remain undecomposed, but it seems reasonable to suppose that the final products will be humic iron phosphate compounds or other products that are biochemically not very active. Humic phosphates are always found in the NaOH extracts of sediments, but very little is known about the activity of these compounds.

Ketchum (35) showed that regeneration of nutrients from the phytoplankton by the zooplankton is important, in addition to bacterial breakdown. No figures for phosphate can yet be given, but it seems likely that it may be as high as mentioned for the nitrogen. The amount of nitrogen excreted by zooplankton may cover 40 to 70% of the daily phytoplankton requirement (35). (See Chapter 34.)

Studies which combine particulate P analysis with phytoplankton population counts are especially wanted. If the algae would have been counted in Ambühl's (1) otherwise detailed study, he might then have been able to establish whether the peak in the particulate P which is observed at 20 m in Lake Lucerne followed a decline in the phytoplankton.

Gächter in the same publication as mentioned above quotes the experiments of Blöch who, with a sediment trap, found that 2 to 3% of the phosphate from the 0 to 7½ m layer sedimented into deeper layers. The difficulties in the technique of measuring sedimentation are outlined below (see p. 526). Gächter himself points out that the outflow was not taken into account, and that it is not certain whether the measured phosphate concentrations were of the same body of water during the whole experiment. In Gächter's figures no decrease of total P of the upper layer can be found, which Gächter explains as being due to the inflow of allochthonous phosphates. He showed that the daily supply of phosphate into the 0 to 10 m layer due to sewage (estimated at 1 mg/m^3 of P on the assumption of a production of 2 g of PO_4—P per person, which seems low) is sufficient to work against the sedimentation to keep the concentration constant. If we assume, however, an input of 4 g of PO_4—P per person, the supply would be double, and so the sedimentation should be higher. This could mean that the sediment collection was not as efficient as hoped.

Working with this idea of the effect of the high turnover time acting strongly against sedimentation, we could not find any significant difference between the summer and winter values for the particulate P during observations over 5 years, 1966–1970, in the sandpit Vechten, which stratifies during the summer (2). In the anaerobic hypolimnion of the sandpit maximal values between 500 and 1000 mg/m³ of PO_4–P were found near the bottom in the same period. As the sandpit has no water outlet or inlet it seems likely that its phosphate metabolism is autochthonic, that sedimentation plays no part in it, and that the phosphate in the hypolimnion is regulated by the external cycle only. The estimated turnover time of PO_4—P in this sandpit is in the same order of magnitude as mentioned above for the Zürichsee (p. 516) (7 to 14 days). The results of the different concentrations of phosphate in 1966 are given in Table 1.

From the figures in Table 1 for the particulate P it appears that there is no great loss of phosphate from the epilimnion. In this lake the primary production does not show a summer stagnation. Both facts indicate a rapid turnover time of the phosphate in the "internal" phosphate cycle. The estimated turnover time of PO_4—P from the primary production and phosphate concentrations in this sandpit is 7 to 14 days. The increase of particulate P

TABLE 1 Concentrations of Phosphate Phosphorus and Organic Phosphate Phosphorus in the Epilimnion, and of Particulate Phosphorus in Epi- and Hypolimnion in the Sandpit "Vechten" in 1966

			Particulate P (μg/l)	
Date	PO_4—P (μg/l)	Org-P_{diss} (μg/l)	Epilimnion	Hypolimnion (±50 cm above mud)
January 17	12	8	7	8
February 14	5	12	5	7
February 28	5	2	5	6
March 14	8	6	4	5
March 28	14	11	6	7
April 12	5	5	6	5
May 9	11	0	6	11
June 6	64	—	6	34
July 4	1	22	3	159
August 15	8	23	5	59
September 12	6	7	4	79
October 24	9	0	5	110
November 21	58	0	9	7
December 19	44	—	5	7

in the hypolimnion in October may be due to precipitation of $FePO_4$. The high value on July 4 cannot be explained.

VERTICAL EXCHANGE IN THE EXTERNAL PHOSPHATE CYCLE*

The Fe–PO_4–S System

When a lake stratifies and the hypolimnion becomes anaerobic (and the redox potential therefore falls) (55), the iron in the mud which is otherwise present in the ferric form is reduced to the ferrous form. The reducing agent may be H_2S, formed by reduction of sulfate by anaerobic bacteria such as *Desulfovibrio desulfuricans*:

$$H_2SO_4 + \text{H donor} \rightarrow H_2S + H_2O \tag{1}$$

According to Einsele (11) the reduction of the Fe^{3+} follows the reaction

$$2Fe^{3+} + S^{2-} \rightarrow 2Fe^{2+} + S \tag{2}$$

It is probably better to write

$$\begin{array}{c} Fe(OH)_3 \\ \text{or} \\ 2FePO_4 \end{array} + S^{2-} + 2H^+ \rightarrow 2Fe^{2+} + \begin{array}{c} 4OH^- + 2H_2O \\ \\ 2HPO_4^{2-} \end{array} + S \tag{3}$$

The Fe^{2+} appears in the water and is balanced by HCO_3^- ions. The course of this process is unknown; the equilibrium

$$FeS + 2CO_2 + 2H_2O \rightleftharpoons Fe(HCO_3)_2 + H_2S \tag{4}$$

as proposed by Ohle (50) and Einsele (11) seems to be to the left, as can be calculated from the pH's of H_2CO_3 and H_2S.

The reduction system has been studied by Einsele (11), who found that the reaction is not stoichiometric. In his experiments an excess of H_2S was necessary to cause all the iron to appear into solution. It is not clear why the first-formed Fe^{2+} ions are not directly precipitated by the remaining H_2S.

FeS is a very stable insoluble compound even under aerobic conditions, so that the back oxidation seems unlikely. It is, however, possible that FeS dissolves more easily when in contact with mud (Hogendijk, unpublished) than in an inorganic chemical system.

When lake stratification is destroyed, the dissolved Fe^{2+} is oxidized by O_2.

* Parts of this paragraph are taken from Golterman (20).

As the pH is usually greater than 6.0, the Fe^{3+} ions that are formed precipitate as $Fe(OH)_3$ which is the starting point of the reaction (2). When phosphates are present these coprecipitate, either as $FePO_4$ or adsorbed onto $Fe(OH)_3$. The latter is not a purely academic distinction, as adsorbed phosphorus is much more readily exchangeable than the phosphorus of $FePO_4$. (Whether algae can use $FePO_4$, which is practically insoluble as a phosphate source, is discussed the last section of this chapter.)

Much work on the iron phosphate system has already been done by Einsele (11–13), Ohle (50, 51), and Mortimer (46, 47). Ohle (50, 51) has demonstrated that phosphorus bound to a $Fe(OH)_3^-$ precipitate can be brought into solution by H_2S or H_2CO_3 and $Ca(HCO_3)_2$ more easily than the phosphate of $FePO_4$. Einsele (13) described the adsorption of the phosphate onto $Fe(OH)_3$ by the adsorption isotherm $C_d = KC_a^n$, in which C_d is the phosphorus concentration in the water, C_a the amount of phosphate adsorbed to 1 mg of $Fe(OH)_3$, and n and K are constants. The amount of $Fe(OH)_3$ in the mud is usually unknown, so the formula cannot be applied in practice.

The relationship between iron and phosphate of these processes was investigated in detail within natural situations by Mortimer (46, 47). He found that a redox potential below 0.2 V (associated with an oxygen concentration of 0.1 mg/l) correlated with the occurrence of iron in the ferrous form and that this redox potential is present in "oxidized" mud only a few centimeters below the mud-water surface. The thickness of the oxidized microzone is affected by turbulent displacement of the uppermost sediments into the overlying aerated water, as well as by the reducing power of the sediment itself (and therefore probably the productivity of the whole lake or pond).

Gorham (22) and Edwards (10) showed that the depth of the surface layer of sludge, in which the redox potential was high, was increased by the introduction of larvae of *Chironomus reparius*. Hayes and his associates (26) presented some evidence that the real oxidized layer is only 1 mm thick or less by placing iron and copper wires vertically in the mud. They thought that the break in the electrode potential curve at about 1 cm is caused by such factors as wind-induced water currents and presence of sulfide, or by depression of the redox layer by the electrode. Most of the phosphorus sorbed by the mud appeared to be in a layer of less than 1 mm when determined by radioautography.

For scientific and practical purposes we ought to know what happens to this sorbed phosphate when the oxidized crust is reduced, either due to the overlying water becoming anaerobic or because a new deposit has fallen on the surface and the older mud below has become anaerobic because of the O_2 demand in the fresh deposit. Furthermore, we should like to know how far the oxidized crust inhibits all chemical and physical movement of molecules and ions upwards and downwards. Some information regarding this comes from the work of Fair *et al.* (14), who described a loss of iron from deposits, probably by reduction of the iron followed by its removal from the mud by gas bubble (methane or hydrogen?) and the flow of the water. When there was no movement in the water the ferrous compounds were oxidized and formed the layer of $Fe(OH)_3$ as described by Mortimer (48).

A strong argument against the reduction of iron leading to phosphate release into the water can be found in the work of Thomas (68). He proved that $FePO_4$ in activated sludge does not go into solution during anaerobic fermentation, and he suggested the low solubility of $Fe(PO_4)_2$ (vivianite) as an explanation. If this is true, the reason for the increase of phosphate concentration in the deeper layers of hypolimnia may be a breakdown of organic phosphate compounds. During the period that a lake is unstratified the released phosphates from these sources are then washed away by mixing.

The Ca–CO_3–PO_4 System

It is likely that in natural hard waters the Ca content of the water influences the phosphorus concentration. There are two possibilities as to how this may happen: (*a*) precipitation of $Ca_5(PO_4)_3{\cdot}OH$ [or $3Ca_3(PO_4)_2{\cdot}Ca(OH)_2$] due to the very low solubility product of this compound, or (*b*) coprecipitation of phosphate molecules with a $CaCO_3$ precipitate. Hepher (29) has given considerable attention to the problem of the precipitation of calcium phosphate and has given a useful diagram for the calculation of the amounts of calcium and phosphate which can exist together in solutions of different pH values. His experimental results sometimes agree with the calculated values; the differences (see Table 6 of his paper) are too large to be neglected, however, but may be explained by assuming the precipitation of hydroxyapatite

$$3Ca_3(PO_4)_2{\cdot}Ca(OH)_2 \quad \text{or} \quad Ca_5(PO_4)_3(OH)$$

as under conditions leading to the formation of phosphorite in nature, the product is not $Ca_3(PO_4)_2$ but hydroxyapatite. This compound is also formed when tertiary calcium phosphate is suspended in water. This compound cannot therefore be prepared in the pure state by precipitation from a solution. Even $CaHPO_4$ is decomposed by water in the reaction

$$7CaHPO_4 + H_2O \rightarrow Ca_5(PO_4)_3OH + 2Ca(H_2PO_4)_2$$

For other information see Remy (58, pp. 624 and 638).

It would have been interesting to know the chemical composition of the precipitates occurring in Hepher's experiments, as it can be concluded that calcium was precipitated in quantities much greater than those of the phosphate. This suggests that very complicated precipitates were produced. Hepher used 1×10^{-25} as the solubility product of $Ca_3(PO_4)_2$ in his calculation, and mentioned still lower values. Even 1×10^{-25} is, however, much lower than can be calculated from the solubility of $Ca_3(PO_4)_2$, which is 20 to 30 mg/l (*Handbook of Chemistry and Physics*). This works out to a solubility product of 10^{-20} to 10^{-21}.

According to Van Wazer (70) discrepancies of 10^{11} are found in experiments to find the solubility product. The enormous difference between these values may be caused by the fact that $Ca_3(PO_4)_2$ in water forms compounds such as hydroxyapatites or even the chloride or fluoride analogs (58, p. 624 and 638). Larsen (39) presented the hypothesis that the solubility of hydroxyapatite depends on its colloid-chemical properties as found by Mattson *et al.* (44). According to Mattson the hydroxyapatite would behave as an amphoteric electrolyte with an electrical double layer that is regulated by carbonate ions or carbon dioxide. We have measured the solubility product in an "ecological manner" by mixing phosphate with calcium solutions and subsequently increasing the pH. From the measured calcium and phosphate concentrations after the precipitate was settled, we calculated the solubility product, assuming that the calcium and phosphate that were left in solution were in a "true" solution. The results are given in Table 2, in which values for two lakes are listed also for comparison. It is clear that the Tjeukemeer water is supersaturated with calcium phosphate.

Hepher's paper certainly shows a correlation between the $CaCO_3$ content of the mud and $Ca_3(PO_4)_2$ precipitation, but no proof of his proposed reaction between $CaCO_3$ and PO_4^{3-} ions. Barett (4) also supported the fact that the disappearance of added phosphorus from epilimnion water is related to alkalinity. Furthermore, Hepher's idea that no adsorption onto heavy

TABLE 2 Solubility of Calcium Phosphate in Inorganic Solutions and in Two Lakes

	mg Ca/l	mg P/l	pH	P_1	P_2[a]
Experiment					
1	20.0	11.2	6.95	1.2×10^{-51}	6.8×10^{-29}
2	20.4	21.2	6.95	8.4×10^{-51}	2.6×10^{-28}
3	24.3	20.6	6.81	4.9×10^{-51}	2.1×10^{-28}
4	30.0	27.4	6.73	8.9×10^{-51}	3.3×10^{-28}
Loosdrechtse plassen	50	0.015	8.0	1.25×10^{-53}	5.0×10^{-32}
Tjeukemeer	45	0.400	8.9	7.3×10^{-46}	2.0×10^{-26}

$P_2 = [Ca^{2+}]^3 \times [PO_4^{3-}]^2$

$P_1 = [Ca^{2+}]^5 \times [PO_4^{3-}]^3 \times [OH^-]$

[a] Although P_1 is the solution product that should be used, the P_2 values are also given for comparison with the literature.

metals could take place, because of the high pH of his ponds, is not necessarily true. Ohle (50, 51) showed indeed that PO_4^{3-} sorbed onto $Fe(OH)_3$ could be washed out with $Ca(HCO_3)_2$ solutions, as mentioned by Hepher as an explanation for no adsorption, but in ion-exchange processes in a flowing or turbulent system, however, the continuous renewal of the solution shifts the equilibrium

$$Fe(OH)_3 \sim P \rightleftharpoons Fe(OH)_3 + H_3PO_4$$

to the right by permanently removing one of the reaction products. Ohle's experiment only proved that sorption of phosphate to $Fe(OH)_3$ is less at pH 8.6 than at pH 6.0, but not that there is no sorption at pH 8.6.

We have demonstrated (Golterman, unpublished) that precipitates of $CaCO_3$ can adsorb phosphate ions. We brought 500 mg of $CaCO_3$ into solution by bubbling CO_2 through the suspension. After addition of 1 mg/l of PO_4–P we allowed the CO_2 to escape, which increased the pH to 8.4. After the precipitate settled the phosphate was found to have disappeared from the solution also. In a control with the same concentration of $CaCl_2$ instead of $Ca(HCO_3)_2$ no phosphate precipitation was found at the same pH. It seems likely that these kind of precipitations could play an important part in the sedimentation of phosphate in deeper lakes, for example, the Zürichsee.

INFLUENCE OF WATER RENEWAL ON PHOSPHATE SEDIMENTATION

Thomas (69) studied phosphate concentrations* in Zürichsee in three layers: 0 to 10, 10 to 20, and 20 to 136 m. He described (see Fig. 2) a decrease of PO_4–P quantity in the epilimnion from about 100 tons in January to 10 tons in the summer. From the decrease, and a decrease in the 10 to 20 m layer and an increase in the 20 to 136 m layer, Thomas calculated that 30 to 40 tons of PO_4 is supplied from allochthonous sources to the deeper water, so that about two-thirds of the increase of the hypolimnion comes from the upper layers. The decrease of the inorganic phosphate in the upper layer is due to a conversion into cell phosphates. As these also decrease with depth† (see Table 3) Thomas believes that the decrease of the phosphate in the 0 to 10 m layer is caused by sedimentation of algal cells.

* As Thomas expresses his concentrations in milligrams per liter of PO_4 the author follows him in this part for an easier comparison with Thomas' original work.

† The author wishes to express his appreciation for the use of Professor Thomas' collection of unpublished observations. The author hopes that Professor Thomas will be able to publish all these data so that this vast amount of useful scientific information is available to all limnologists.

TABLE 3 Phosphate and Oxygen Concentrations in Zürichsee near Thalwil on August 8, 1968. From Thomas (unpublished).[a]

Depth (m)	PO_4^{3-}	$(PO_4)_{tot}$	O_2
0.3	<0.01	0.05	11
10	0.01	0.05	7
20	0.21	0.26	6
40	0.27	0.28	8
60	0.29	0.33	8.8
80	0.31	0.34	8.5
100	0.33	0.35	8.1
120	0.42	0.45	6
130	0.45	0.5	3.5
136	0.75	0.95	0.7

[a] Concentrations of inorganic and total phosphate expressed as mg/l of PO_4; concentration of oxygen in mg/l.

Besides the sinking of the organically bound phosphates we must consider also the sedimentation of inorganic compounds. In the case of Zürichsee this is the coprecipitation of phosphate on $CaCO_3$ or even the precipitation of hydroxyapatite $Ca_5(PO_4)_3(OH)$ (see the preceding subsection). From Thomas' (69) figures it is clear that $CaCO_3$ does precipitate so that coprecipitation (see p. 523) is a definite possibility. But under the circumstances given (i.e., $Ca^{2+} = 40$ mg/l; $PO_4 = 0.3$ mg/l; pH $= 7.6$) (Thomas, personal communication) the ionic product $(Ca)^5 \times (PO_4^{3-})^3 \times (OH) = 2.7 \times 10^{-53}$, is near the solubility product of hydroxyapatite (see p. 522). If the pH should furthermore increase to 8 (e.g., due to photosynthesis during the day), the ionic product will become 10^{-51}, so that the lake is saturated with hydroxyapatite, and the possible precipitation of this compound might even be considered. Because not many experimental field data are available we have to leave open the question of the importance of this sedimentation of phosphates.

Besides sedimentation the outwash of phosphate should also be taken into account. Vollenweider (71) gives a formula for the computation of a substance budget for lakes in which sedimentation and outwash are occurring. He assumed the sedimentation to be the result of a monomolecular reaction. As the phosphate may sediment due to sinking of algal cells and to iron or calcium phosphates, Vollenweider's simplification cannot be used here. If sinking of algal cells were the main cause of the sedimentation, one would expect a phosphate maximum, for example, at 20 or 40 m. This could not be found in Table 3, nor in any of Thomas' unpublished profiles. Therefore we calculated the time in which the concentration could have been reduced to 50% ($T_{1/2}$) with Biffi's formula for a substance budget (71).

$$-\frac{dx}{dt} = \frac{a}{b} \cdot x - c \tag{1a}$$

Substituting $t = T_{1/2}$ for $x = \frac{1}{2}x_0$ we may write

$$\ln \frac{ax_0 - bc}{\frac{1}{2}ax_0 - bc} = \frac{a}{b} \cdot T_{1/2} \tag{1b}$$

in which a is the throughflow per day (6.3×10^6 m^3); b is the lake volume (3.4×10^9 m^3) or the epilimnion volume (0.7×10^9 m^3); c is the daily supply (120 kg/day or 420 kg/day); x_0 is the amount of PO_4 on January 1 (epilimnion: 10^5 kg; total lake: 7×10^5 kg), see Thomas (69); and $T_{1/2}$ is the time to reduce the phosphate concentration to 50%.

Vollenweider (71) used in his calculation the replenishment coefficient (R_pC) for the whole lake. As we want to calculate the $T_{1/2}$ from winter until summer, we have used—besides the R_pC of the whole lake—also a partial R_pC, as the incoming water of the Ober Zürichsee only flows through the epilimnion. Furthermore we have calculated the $T_{1/2}$ for an annual phosphate supply of 120 kg/day (inflow from Ober Zürichsee only) and for a supply of 420 kg/day (120 kg as above plus 300 kg from sewage). This value is 40% of 750 kg PO_4, the daily sewage supply according to Thomas (69).

Forty percent of the total sewage input is used because this is the highest percentage, where $\frac{1}{2}ax_0 - bc$ (see Eq. (1b)) is still larger than 10% of $\frac{1}{2}ax_0$. If bc becomes still larger and equals $\frac{1}{2}ax_0$, the formula no longer applies and a phosphate increase instead of an outwash ought to be found. This shows that a considerable part of the sewage phosphate gets into the hypolimnion without being used by the algae.

We have found the following $T_{1/2}$:

PO_4 Supply per Day	120 kg/day	420 kg/day
Inflow through whole lake	420 days	800 days
Inflow through epilimnion	90 days	300 days

With the assumption that the throughflow takes place through the epilimnion only, and with the daily supply of 120 kg/day, a $T_{1/2} = 90$ days was found, which is not far from the value to be found in Thomas (69, Fig. 2). The conclusion must be that the decrease of the phosphate in the 0 to 10 m layer was partly due to an outwash of phosphate. But as long as no exact measurements are obtainable as to whether or not the incoming water goes through the epilimnion only, no exact calculation for outwash and sedimentation can be given. We feel therefore that Vollenweider in his calculations has underestimated these effects in his rationalization of the Zürichsee budget. Several investigators have tried to solve the problem of sedimentation by collecting sinking material under field conditions. The main difficulty is, however, that the traps so far used increased the sedimentation rate by decreasing the turbulence of the water in their vicinity. Furthermore living material caught in the trap may die and may be digested by bacteria. These may give out phosphate to the water or may take it up from the water, depending on the C/N/P ratio in their substrate. Even dead material may undergo such alterations. If bactericides and the like are added to the trap, they may kill off living algae, and increase the mineralization.

These difficulties are demonstrated, for example, by Kleerekoper (38), who collected sinking material with his sedimentation apparatus in Lake Lauzon (37). Although the collected material did show a high percentage of mineralization, the phosphate content was quite high. It is not unlikely that this is due to bacterial growth in the sedimentation trap, or to a precipitation of $FePO_4$, which seems possible looking at the very high Fe content of the collected material and the high content of iron in the sediments of Lake Lauzon.

Analyses of total phosphates from these traps are therefore unreliable. So far no attempts have been made to fractionate the phosphates in the traps as described on page 523 of this chapter. This kind of analysis would certainly produce useful information.

PHOSPHATE ADSORPTION AND DESCRIPTION BY SEDIMENTS AND EXCHANGE PROCESSES*

Most sediments of lakes are capable of adsorbing large amounts of phosphate from the water, while desorption takes place when the phosphate concentration in the water falls.

The importance of $Fe(OH)_3$ as adsorbant has already been described under the Fe–PO_4–S system. The adsorbance is influenced by the pH, the rH, and the concentration of ions as HCO_3^- in the system. Another system for adsorption and desorption involves the clay minerals such as kaolinite and, more often, montmorillonite. This system is again pH dependent, and Table 4 gives the relation between pH and the binding capacity of different anions to these two types of clay.

* By exchange processes is meant here the exchange of two isotopes (for instance, ^{31}P with ^{32}P) with each other.

TABLE 4 The Dependence of the Adsorption of Chloride, Sulfate, and Phosphate ions in mval/100 g on the pH through the Clay Fractions of Two Soils (63)

Nipe Clay (Predominantly Kaolinite)				Sharkey Clay (Predominantly Montmorillonite)		
pH	Cl	SO_4	PO_4	pH	Cl	PO_4
7.2	0.0	0.0	31.2	6.8	0.0	22
6.7	0.3	2.0	41.2	5.6	0.0	36.5
6.1	1.1	5.5	46.5	4.0	0.05	47.4
5.8	2.4	7.1	50.8	3.2	0.1	64.0
5.0	4.4	10.5	66.1	3.0	0.1	73.5
4.0	6.0	—	88.2	2.8	0.4	100

$SiO_4 < PO_4 \ll SO_4 < Cl \sim NO_3$

It is clear from this table that the acidification of an anaerobic hypolimnion would lead to an increased adsorbance of PO_4 ions onto montmorillonite, which counteracts the normal findings in hypolimnia. Although SO_4 ions are much less adsorbed than PO_4 ions, due to the latter normally being present in natural waters at concentrations about a thousand times higher than that of the SO_4 ions, the influence of the SO_4 ions should not be neglected. Information about the sorption mechanism of phosphate on the clay has been recently summarized in F. E. Bear (1964). Unlike Cl^- and NO_3^-, which are not bound, PO_4^{3-} can be bound on clay at all pH's occurring in normal lakes. But the adsorption is strongly influenced by the type of clay minerals, contents of hydrous oxides, and organic matter. The mechanism of the sorption to the clay may be an exchange with other ions, preferably the OH^- ions, for example, $ROH_2^+ \cdot OH^- + H_2PO_4^- \rightleftharpoons ROH_2^+ \cdot H_2PO_4^- + OH^-$. The $H_2PO_4^-$ absorbed in this way is called saloid-bound. It occurs essentially below the equionic point. Quantitatively probably more important may be the binding of phosphate ions as a structural nondiffusible unit by the displacement of lattice OH. This type of bonding may explain the strong displacement of the phosphate ions exerted by alkali. This type of bonding is independent of the equionic point so that $H_2PO_4^-$, HPO_4^{2-}, and PO_4^{3-} may be bound. A third possibility may be an extension of the clay lattice by sorption of phosphate on the edges of the silica tetrahedron. Radioactivity was shown on the edges, when P^{32} was adsorbed (Hendriks, 1944). Finally it is not unlikely that sorption of phosphate onto clay may occur through clay–humic acids–phosphate complexes.

Salt effects on soil clay are intensively studied in soil science. It should be kept in mind, however, that clay in constant contact with lake water forms a rather constant system in which changes of the salt content may occur but will remain relatively small. Furthermore it should be realized that clay in soils is submitted to weathering and aging processes, which in the clay-lake water system will probably come to an end, so that less changes will occur in the lattice structure of an underwater soil. Larsen and Widdowson (40) have shown that the ratio of the solubility of phosphate—in contact with a clay loam— over the amount of isotopically exchangeable phosphate decreased during 5 months at least when the soil was suspended in water. As long as these aging processes are not well understood the results of studies of phosphate sorption on soil clays should be used with great care.

Müller and Tütz (49) showed that in Lake Constance the phosphorus content of the sediments depends primarily on the particle size and increases with clay content. In the tributaries where sewage-water inflow takes place a higher phosphate content is found due to sorption of the phosphate on the clay mineral. Although findings may vary from lake to lake, these results may be of great help toward a clearer understanding of the phosphorus balance.

Quantitative studies on whole sediments have been made by Harter (24) and Williams *et al.* (75). Harter found that between half and all of the phosphate adsorbed by mud was recovered in the NH_4F ("$AlPO_4$") and NaOH ($FePO_4$) extracts, the recovery decreasing with increased phosphate added. Since none of the adsorbed phosphate was found in a form extractable with H_2SO_4 (calcium phosphate) the remainder was assumed to have been incorporated into organisms.

Whereas all phosphate extractable with NaOH appeared to occur only as an iron phosphate, NH_4F apparently extracted phosphate bound by two different mechanisms for which Harter assumed that, besides $AlPO_4$, a more loosely bound phosphate was involved as well, which was also extractable by successive water treatments. Harter gave no suggestion as to what this loosely bound phosphate was, although he discussed its biological meaning in eutrophication.

Williams *et al.* (75) compared noncalcareous and calcareous sediments from eight Wisconsin lakes. They showed that soils with $CaCO_3$ had a lower phosphate-binding capacity than soils without $CaCO_3$. This probably means that $CaCO_3$ may coprecipitate phosphate ions during its formation (as discussed earlier), but has no exchange capacity when formed. Williams *et al.* found a close relationship between the capacity of the sediments to adsorb and retain added inorganic phosphate and the levels of total and inorganic phosphate which had accumulated in the sediments. They suggested that the P-retention capacity of a sediment is an important factor determining the levels of phosphate which accumulate in lake sediments under natural conditions. This statement seems in contrast to the adsorption mechanism that can be described by the adsorption isotherm. As they did not fractionate the phosphate in the sediments nor the adsorbed phosphate this principle should not be applied too easily to other sediments.

The same results as found by Williams *et al.* were obtained by Shukla *et al.* (66) who also found less phosphate sorption on calcareous material. Furthermore they showed that the amount of oxalate-extractable iron was the best single criterion explaining phosphate sorption on both calcareous and noncalcareous sediments, but they did not study other mechanisms as calcium oxalate formation. They suggested that hydrated iron oxide along with smaller amounts of organic matter, Al_2O_3, and associated $Si(OH)_4$ appears to be the major contributor to the sorption of inorganic phosphates by both soil types.

Livingstone and Boykin (41) studied the phosphorus in the sediment of Linsley Pond and suggested that a large quantity of the phosphate was bound by sorption to the mineral material below the gyttja. The mineral part was not present as apatite because it was insoluble in acids. In their interesting discussion they remark that in lake mud high phosphorus-binding capacity is correlated with high mineral content. A comparison of the sorption capacity of the gyttja and the mineral material as a function of the phosphorus concentration of the overlying water would have been very interesting.

Macpherson and co-workers (43) investigated the effect of the pH on phosphate adsorption onto dried mud samples that had been taken from eight lakes of different productivity grades. The results were compared to parallel experiments using the ash of these muds. The technique was to shake up the sample of mud or ash with distilled water, with the pH adjusted, and to measure the amount of phosphate released into the water. Both mud and ash showed maximum adsorption in the pH range 5 to 7, the ash releasing more phosphate than the dried mud. Thus the organic part of the mud appeared to moderate the phosphate release at the ends of the pH range. This was confirmed by shaking the mud with a solution containing 1.0 mg/l of phosphate. However, it is not possible to tell from their figures whether they measured changes in the sorption reaction caused by a pH effect or an effect caused by changing the ionic strength. The activity of the ash compared with the mud as a whole may also be an activation effect owed to heating. They also discussed the similarity between the pH effect on lake mud ash which they had studied and the effect on bentonite, Fuller's earth, and ferric hydroxide reported by other workers. It is not certain that drying of mud has no effect on its exchange capacities. For instance, we have found differences in phosphate-binding capacities between dried and undried ferric hydroxyde.

More detailed insight into the adsorption mechanism may be obtained by the use of radioactive phosphorus, but in such experiments, however, a rapid exchange between P^{31} and P^{32} can always be found. This exchange, which must be distinguished from sorption processes, can give some information about the nature of the phosphorus compounds present, but not about the quantity of the phosphorus "stored" in the mud. In the literature a clear distinction between the exchange process

$$XP^{31}O_4 + P^{32}O_4^{3-} \rightleftharpoons XP^{32}O_4 + P^{31}O_4^{3-}$$

and the sorption process

$$X(Y)_3 + PO_4^{3-} \rightleftharpoons XPO_4 + 3Y^-$$

is not always made.

An example of radiophosphorus work is that of Hayes *et al.* (25), who put $KH_2{}^{32}PO_4$ in a whole lake, and could demonstrate a turnover of phosphate between water and sediments.

The distribution of P^{32} between mud and water was studied by Hayes and Phillips (27), who found first that in natural Jenkin sampler cores, in artificial cores, and in bottle in which dredged surface mud was packed by centrifugation, the phosphorus equilibration pattern and rate of equilibration were the same, which showed that layering was relatively unimportant. Secondly they found that in the absence of added antibiotic the amount of P^{32} remaining in the water at equilibrium was greater than where antibiotic had been added. Besides the two explanations Hayes and Phillips have given, it is also possible that the antibiotics themselves influenced the sorption system.

They found turnover times to be about 5 min for bacteria or phytoplankton [confirming the results of Coffin *et al.* (8)], 1 week for the water of the whole lake, half a week for mud in a bottle (2 weeks without bacteria), and 1 month for lake sediments in nature (including rooted aquatics). Zooplankton were unable to use phosphorus until bacteria had combined it in organic form; their turnover time was then 1 day.

Pomeroy *et al.* (57) studied the phosphorus exchange of P^{32} with cores and suspended sediments that contained kaolinite and montmorillonite as principal clay minerals. They found a two-step ion exchange between the clay minerals and the phosphate of the water, plus an exchange between microorganisms and water. The latter appeared to be trivial in undisturbed sediments, but to equal inorganic processes in the suspended sediments. They do not mention whether the sediments contained precipitates of calcium carbonate or calcium phosphate.

Serruya (65) suggested that in Lake Kinneret sediments a certain amount of phosphate is adsorbed on montmorillonite, and that this is *Eh* (extent of reduction) indifferent. It is not clear whether this phosphate is exchangeable. If it is indeed not dependent on the *Eh*, only a change of pH could cause a release or a binding of phosphate.

Radiophosphorus experiments in lake sediments were carried out by Olsen (53, 54), who made the distinction between exchange and sorption, but whose paper did not receive the attention it deserved. He used a coarse sediment from water 3 m deep and a fine-grained deep-water sediment, both in the reduced and oxidized state. He found that, the relations between phosphorus in the water and in the sediment in both oxidized or reduced conditions could be described by the same mathematical formula, although with reservation in the case of phosphate released from the reduced sediments. The relation is a hyperbolic function of the form

$$b = K_b \cdot C^{-V_b}$$

in which b is the amount of exchangeable phosphorus per gram of dry matter, K_b and V_b are constants, and C is the concentration of phosphorus in the solution. Addition of this exchange quantity to the net adsorbed quantity (a) gives the gross adsorbed quantity (A), which Olsen described by the Freundlich adsorption isotherm where

$$A = K \cdot C^V$$

The full mathematical description of the direct measurable net adsorbed quantity was calculated by Olsen as

$$a = A - b = K \cdot C^V - K_b \cdot C^{-V_b}$$

It was shown by Olsen that the exchange of phosphorus is a very rapid process and that the uptake of phosphorus from the water by algae will be followed by a release of phosphorus from the sediments, with different constants for oxidized or reduced sediments. In the oxidized state, sediment exchanges and adsorbs more phosphorus than in the reduced state. This is another reason for investigating the (Eh) of shallow lake muds. In the reduced state the sediment from deeper water showed no sorption of phosphorus (and therefore only release) when the concentration of the phosphorus in the water was below the relatively high value of 2 mg/l.

It would be interesting to know whether or not the different behavior of the phosphorus in reduced "deep" sediments is related to redox potentials and iron metabolism. It may be that the reduced "deep" sediment has lost all its active iron and that the sorption of phosphorus from solutions containing more than 2 mg/l of phosphorus takes place by a different process, for example, precipitation with calcium.

Very few experiments with radiophosphorus have been done so far in whole lakes. Hutchinson and Bowen (32) showed a transport of phosphates from the epilimnion to hypolimnion and mud. The latter was rather rapidly regenerated, passing again into the free water. Only in the oxidized zone were phosphates precipitated, probably as $FePO_4$. Hutchinson and Bowen suggest that sinking phytoplankton is the transport mechanism. No information has been given about the possible role of zooplankton. From the data a possible turnover time of the plankton phosphates is calculated as to be greater than 3 weeks.

THE EXCHANGE IN SHALLOW LAKES; AVAILABILITY OF MUD PHOSPHATES

It seems likely that the same precipitation processes and exchange reactions as discussed for deeper lakes exist also in the shallow lakes, but in shallow lakes there is usually a more vigorous mixing of the water column compared to a deeper lake such that insoluble phosphates are mechanically transported to the trophogenic layer from the sediments, and the availability of these phosphates for metabolism should be considered. Armstrong and Harvey (3) mentioned that a phosphorus-depleted diatom grew in water containing $FePO_4$, but this growth probably followed the hydrolysis of the $FePO_4$ which takes place in seawater, as they demonstrated a concurrent fall in pH from 8 to 7, at which pH the process stopped. Golterman and his associates (21) showed that $FePO_4$ could be utilized completely as a source of phosphate in a medium containing sufficient $NaHCO_3$ to keep the pH around 8. The growth rate was less than with an equal amount of KH_2PO_4 (1000 μg/l of PO_4—P). The observed growth rate on $FePO_4$ was equal to that of a lower concentration of KH_2PO_4 (100 μg/l), which was therefore called the "apparent" PO_4—P concentration. Furthermore they showed that hydroxyapatite could only be used for one-third, and the growth rate was equivalent to 35 μg/l of PO_4—P of KH_2PO_4, and thus lower than for $FePO_4$.

Mud could be used also as a source for phosphates (see Fig. 3) and from the curves obtained the percentage of mud phosphate used and the "apparent" phosphate concentration of the mud phosphate could be found. Although the different forms of the growth curves for the different mud samples show interesting points for further work and insight in the mud-phosphate chemistry the results are too preliminary to be discussed here, so that we refer to the original paper for the discussion of these aspects.

It is worth mentioning here that the increase of the cell phosphate was equal to the decrease of the sum of iron phosphate and calcium phosphate, while the organic phosphates were not used. As the iron phosphates were not completely used there seems to be, besides the already mentioned chemical differences, a biological difference of the mud iron phosphate with "chemical inorganic" iron phosphate.

Fitzgerald (15), working with mud from Lakes Wingra and Mendota, could not detect growth of *Selenastrum capricornutum* on mud phosphates:

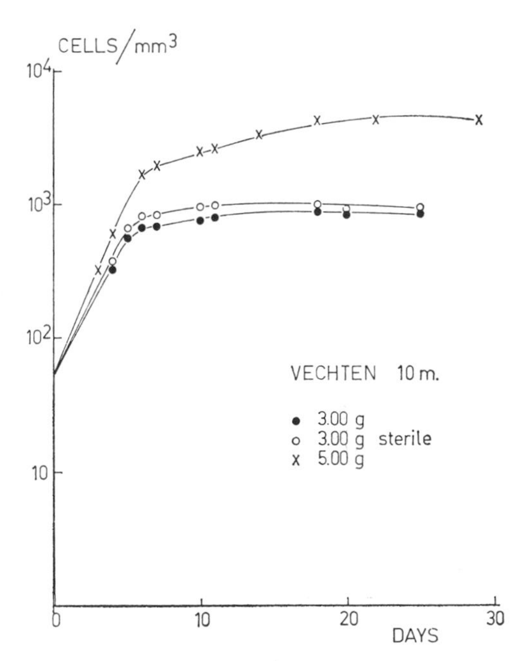

Figure 3. Growth of *Scenedesmus* cells on culture solutions with lake sediments as only source of phosphate [see Golterman *et al.* (21)].

with *Cladophora* a small growth was just measurable on mud from 18-m-deep Lake Mendota. The other two muds tested did not yield any growth. Fitzgerald suggests that the availability of phosphate from the muds tested with *Scenedesmus* by Golterman *et al.* (21) must have been due either to differences in the testing procedures or to special characteristics of the mud samples. As Fitzgerald did not fractionate his mud phosphates nothing can be said about the latter possibility, but the first hypothesis is certainly true. Golterman *et al.* (21) could also not detect growth with *Scenedesmus* if the mud sample was placed in a dialysis tube instead of being mixed through the culture solution. It seems possible that contact between algal cell and phosphate particle is necessary. It is also possible that the diffusion through the dialysis tube is so slow that the phosphates are sorbed on the mud before they can diffuse into the tube. Fitzgerald showed a sorption of 0.05 mg of PO_4—P on 0.4 g mud. Probably this lowers the free PO_4—P concentration so much that diffusion becomes zero. This did not happen in the control experiment in which phosphate was placed in the tube without mud, so that diffusion and thus growth could take place in these flasks.

Wentz and Lee (72, 73), working with sedimentary phosphates in lake cores, defined "available" phosphate as phosphate extractable with dilute

HCl—H_2SO_4, according to the agricultural definitions. With this extraction procedure $FePO_4$ is only dissolved for 4.4%, which means that they use the word "available" not in the sense as "available for algae." It may be that the rooted vegetations use different phosphates than algae. The procedure removes sorbed phosphates and apatites from the sediments as well. They found, furthermore, that the rate of available phosphate deposition in Lake Mendota (73) was constant in the marl and increased concomitantly with the change from marl to sludge, reaching a maximum at about 30 cm. The available phosphate was associated with the $CaCO_3$ portion of the sediments. The correlation with the clastic material and iron seemed equal if not better than the correlation with $CaCO_3$ in certain layers of the sediments. The authors discuss the possible interrelations of the observed changes with the history of the eutrophication of Lake Mendota.

Very little is known about conversions between the different phosphate compounds in sediments over long time periods. If phosphate is released in deeper layers into soluble inorganic phosphate, Hynes and Greib (33) showed that it may move upward in a reasonable time and is not necessarily trapped in the sediments. Mud samples kept in our laboratory at 4°C showed no significant change in phosphate composition in one year.

NOTES ADDED IN PROOF: Recently O'Melia (76) published a paper with a model for phosphate flux in stratifying lakes with formulae for a phosphate balance and for diffusion from hypolimnion into the epilimnion. For phosphate sedimentation a single (yearly) coefficient is used, which seems unlikely to be justified. For instance in Lake of Geneva a sudden decrease of the phosphate concentration occurred after the concentration reached a value above 180 μg/l in May 1970 (77). The fact that the precipitate occurred after the solubility product of hydroxy apatite was exceeded indicates that hydroxy apatite precipitated and not phosphate adsorbed onto $CaCO_3$ (see p. 523).

In a recent paper Golterman (78) more intensively discussed clay as an absorbent for phosphate and the availability of clay phosphates for algal growth. He found evidence that available phosphate can be extracted from sediment with a 0.01 *M* NTA solution.

REFERENCES

1. Ambühl, H., "Die neueste Entwicklung des Vierwaldstättersees (Lake of Lucerne)," *Verh. Int. Verein. Limnol.*, **17**, 219–230 (1969).
2. Annual Report of the Limnological Institute, Nieuwersluis, over the year 1969, *Verh. Kon. Nedl. Akad. Wet., Afd. Natuurkd., 2e Reeks*, dl **59,** (3), 66–78 (1970).
3. Armstrong, F. A. J., and H. W. Harvey, "The Cycle of Phosphorus in the Waters of the English Channel," *J. Mar. Biol. Assoc., N. S.*, **29**, 145–162 (1951).

4. Barett, P. H., "Relationship between Alkalinity and Adsorption and Regeneration of added Phosphorus in Fertilized Trout Lakes," *Trans. Am. Fish. Soc.*, **82,** 78–90 (1953).

5. Blösch, unpublished, cited in Gächter.

6. Brunskill, G. J., "Fayetteville Green Lake, New York, II. Precipitation and Sedimentation of Calcite in a Meromectic Lake with Laminated Sediments," *Limnol. Oceanogr.*, **14** (6), 830–847 (1969).

7. *Chemistry of the Soil*, 2nd ed., F. E. Bear (Ed.), Reinhold, New York, 1964.

8. Coffin, C. C., F. R. Hayes, L. H. Jodrey, and S. G. Whiteway, "Exchange of Material in a Lake as Studied by the Addition of Radioactive Phosphorus," *Can. J. Res., Sect. D.*, **27**, 207–222 (1949).

9. Delfino, J. J., G. C. Bortleson, and G. F. Lee, "Distribution of Mn, Fe, P, Mg, K, Na, and Ca in the Surface Sediments of Lake Mendota, Wisconsin," *Environ. Sci. Technol.*, **3,** 1189–1192 (1969).

10. Edwards, R. W., "The Effect of Larvae of *Chironomus riparius* Meigen on the Redox Potentials of Settled Activated Sludge," *Ann. Appl. Biol.*, **46** (3), 457–464 (1958).

11. Einsele, W., "Über die Beziehungen des Eisenkreislaufs zum Phosphatkreislauf im eutrophen See," *Arch. Hydrobiol.*, **29,** 664–686 (1936).

12. Einsele, W., "Physikalisch-chemische Betrachtungen einiger Probleme des limnischen Mangan- und Eisenkreislaufs," *Verh. Int. Verein. Limnol.*, **5** (3), 69–84 (1937).

13. Einsele, W., "Über chemische und kolloidchemische Vorgänge in Eisen-Phosphat Systemen unter limnochemischen und limnogeologischen Gesichtspunkten," *Arch. Hydrobiol.*, **33,** 361–387 (1938).

14. Fair, G. M., E. W. Moore, and H. A. Thomas, "The Natural Purification of River Muds and Pollutional Sediments," *Sewage Works J.*, **13** (2), 270–307 (1941).

15. Fitzgerald, G. P., "Aerobic Lake Muds for the Removal of phosphorus from Lake Waters, *Limnol. Oceanogr.*, **15** (4), 550–555 (1970).

16. Frink, C. R., "Fractionation of Phosphorus in Lake Sediments: Analytical Evaluation," *Soil Sci. Soc. Am. Proc.*, **33,** 326–328 (1969).

17. Frink, C. R., "Chemical and Mineralogical Characteristics of Eutrophic Sediments," *Soil Sci. Soc. Am. Proc.*, **33** (3), 369–372 (1969).

18. Gächter, R., "Phosphorhaushalt und planktische Primärproduktion im Vierwaldstättersee (Horwer Bucht)," *Schweiz. Z. Hydrol.*, **30** (1), 1–66 (1968).

19. Golterman, H. L. "Studies on the Cycle of Elements in Freshwater, *Acta Bot. Neerl.*, **9,** 1–58 (1960).

20. Golterman, H. L., "Influence of the Mud on the Chemistry of Water in Relation to Productivity," in H. L. Golterman and R. S. Clymo, *Proceedings of an I.B.P. Symposium Held in Amsterdam and Nieuwersluis*, 10–16 *October*, 1966, "Chemical Environment in the Aquatic Habitat," 1967, pp. 297–313.

21. Golterman, H. L., C. C. Bakels, and J. Jakobs-Möglin, "Availability of Mud Phosphates for the Growth of Algae, *Verh. Int. Verein. Limnol.*, **17,** 467–479 (1969).

22. Gorham, E., "Observations on the Formation and Breakdown of the Oxidized Microzone at the Mud Surface in Lakes," *Limnol. Oceanogr.*, **3** (3), 291–298 (1958).

23. *Handbook of Chemistry and Physics*, a ready reference book of chemical and physical data, 45th ed., Chemical Rubber Publishing Co., Cleveland, Ohio, 1964–1965.

24. Harter, R. D., "Adsorption of Phosphorus by Lake Sediment," *Soil Sci. Soc. Am. Proc.*, **32,** 514–518 (1968).

25. Hayes, F. R., J. A. McCarter, M. L. Cameron, and D. A. Livingstone, "On the Kinetics of Phosphorus Exchange in Lakes," *J. Ecol.*, **40** (1), 202–216 (1952).

26. Hayes, F. R., B. L. Reid, and M. L. Cameron, "Lake Water and Sediment, II. Oxidation-Reduction Relations at the Mud-Water Interface," *Limnol. Oceanogr.*, **3** (3), 308–317 (1958).

27. Hayes, F. R., and J. E. Phillips, "Lake Water and Sediment, IV. Radiophosphorus Equilibrium with Mud, Plants, and Bacteria under Oxidized and Reduced Conditions,"*Limnol. Oceanogr.*, **3** (4), 459–475 (1958).

28. Hendricks, S. B., Base Exchange of the Clay Mineral Montmorillonite for Organic Cations and Its Dependence upon Adsorption Due to van der Waals Forces, *J. Phys. Chem.* **45,** 65 (1941).

29. Hepher, B., "On the Dynamics of Phosphorus Added to Fish Ponds in Israel," *Limnol. Oceanogr.*, **3,** 84–100 (1958).

30. Heron, J., "The Seasonal Variation of Phosphate, Silicate, and Nitrate in Waters of the English Lake District," *Limnol. Oceanogr.*, **6,** 338–346 (1961).

31. Hutchinson, G. E., and A. Wollack, "Studies on Connecticut Lake Sediments, II. Chemical Analyses of a Core from Linsley Pond, North Branford," *Am. J. Sci.*, **238** (7), 493–517 (1940).

32. Hutchinson, G. E., and V. T. Bowen, "Limnological Studies in Connecticut, IX. A Quantitative Radiochemical Study of the Phosphorus Cycle in Linsley Pond," *Ecol.*, **31,** 194–203 (1950).

33. Hynes, H. B. N., and B. J. Greib, "Movement of Phosphate and Other Ions from and through Lake Muds," *J. Fish. Res. Board Can.*, **27** (4), 653–668 (1970).

34. Jackson, M. L., *Soil Chemical Analysis*, Prentice Hall, Englewood Cliffs, N.J. 1958, 498 pages.

35. Ketchum, B. H., "Regeneration of Nutrients by Zooplankton," *Rapport P.-V. Réun. Cons. Perm. Int. Explor. Mer*, **153,** 142–147 (1962).

36. Ketchum, B. H., and N. Corwin, "The Cycle of Phosphorus in a Plankton Bloom in the Gulf of Maine," *Limnol. Oceanogr., Suppl.*, **10,** R148–R161 (1965).

37. Kleerekoper, H., "A New Apparatus for the Study of Sedimentation in Lakes," *Can. J. Zool.*, **30,** 185–190 (1952).

38. Kleerekoper, H., "The Mineralization of Plankton," *J. Fish. Res. Board Can.*, **10** (5), 283–291 (1953).

39. Larsen, S., "Solubility of Hydroxyapatite," *Nature*, **212** (5062), 605 (1966).

40. Larsen, S., and A. E. Widdowson, "Aging of Phosphate Added to Soil," *J. Soil Sci.*, **22** (1), 5–7 (1971).

41. Livingstone, D. A., and J. C. Boykin, "Vertical Distribution of Phosphorus in Linsley Pond Mud," *Limnol. Oceanogr.*, **7** (1), 57–62 (1962).

42. Mackereth, F. J. H., "Chemical Investigation of Lake Sediments and Their Interpretation," *Proc. Roy. Soc., B*, **161,** 295–309 (1965).

43. Macpherson, L. B., N. R. Sinclair, and F. R. Hayes, "Lake Water Sediment, III. The Effect of pH on the Partition of Inorganic Phosphate between Water and Oxidized Mud or Its Ash," *Limnol. Oceanogr.*, **3** (3), 318–326 (1958).

44. Mattson, S., E. G. Williams, E. Koutler-Anderssond, and E. Barkhoff, "Phosphate Relationships of Soil and Plant, V. Forms of P in the Lanna Soil," *Ann. Roy. Agric. Coll. Sweden*, **17,** 130–140 (1950).

45. Mehta, N. C., J. O. Legg, C. A. I. Goring, and C. A. Black, "Determination of Organic Phosphorus in Soils, I. Extraction Method," *Soil Sci. Soc. Proc.*, **18** (4), 443–449 (1954).

46. Mortimer, C. H., "The Exchange of Dissolved Substances between Mud and Water in Lakes," *J. Ecol.*, **29,** 280–329 (1941).

47. Mortimer, C. H., "The Exchange of Dissolved Substances between Mud and Water in Lakes," *J. Ecol.*, **30,** 147–201 (1942).

48. Mortimer, C. H., "Underwater "Soils"; A review of Lake Sediments," *J. Soil Sci.*, **1,** 63–73 (1949).

49. Müller, K., and G. Tietz, "Der Phosphatgehalt der Bodenseesedimente," *Neues Jahrb. Mineral., Abh.*, **105,** 41–62 (1966).

50. Ohle, W., "Kolloidgele als Nährstoffregulatoren der Gewässer," *Naturwiss.*, **25** (29), 471–474 (1937).

51. Ohle, W., "Die Bedeutung der Austauschvorgänge zwischen Schlamm und Wasser für den Stoffkreislauf der Gewässer," *Jahrb. Wasser*, **13,** 87–97 (1938).

52. Ohle, W., "Primärproduktion des Phytoplanktons und Bioaktivität der Seen, Methoden und Ergebnisse," Limnologisymposion, Helsinki, Finland, 1964, pp. 24–43.

53. Olsen, S., "Phosphate Adsorption and Isotopic Exchange in Lake Muds, Experiments with P^{32}," *Verh. Int. Verein. Limnol.*, **13,** 915–922 (1958).

54. Olsen, S., "Phosphate Equilibrium between Reduced Sediments and Water, Laboratory Experiments with Radioactive Phosphorus," *Verh. Int. Verein. Limnol.*, **15,** 333–341 (1964).

55. Persall, W. H., and C. H. Mortimer, "Oxidation-Reduction Potentials in Waterlogged Soils, Natural Waters, and Mud," *J. Ecol.*, **27,** 483–501 (1939).

56. Petersen, G. W., and R. B. Corey, "A Modified Chang and Jackson Procedure for Routine Fractionation of Inorganic Soil Phosphates," *Soil Sci. Soc. Am. Proc.*, **30,** 563–565 (1966).

57. Pomeroy, L. R., E. E. Smith, and C. M. Grant, "The Exchange of Phosphate between Estuarine Water and Sediments," *Limnol. Oceanogr.*, **10** (2), 167–172 (1965).

58. Remy, A., *Treatise on Inorganic Chemistry*, Vol. 1, Elsevier, Amsterdam, 1956, pp. 624 and 638.

59. Rigler, F. H., "A Tracer Study of the Phosphorus Cycle in Lake Water," *Ecol.*, **37** (3), 550–562 (1956).

60. Rigler, F. H., "The Phosphorus Fractions and the Turnover Time of Inorganic Phosphorus in Different Types of Lakes," *Limnol. Oceanogr.*, **9,** 511–518 (1964).

61. Rigler, F. H., "Further Observations Inconsistent with the Hypothesis that the Molybdenum Blue Method Measures Orthophosphate in Lake Water," *Limnol. Oceanogr.*, **13** (1), 7–13 (1968).

62. Rodhe, W., "Standard Correlations between Pelagic Photosynthesis and Light," *Mem. Ist. Ital. Idrobiol., Suppl.*, **18,** 365–381.

63. Scheffer, F., and P. Schachachtschabel, *Lehrbuch der Bodenkunde; siebente, völlig neubearb. Aufl.*, Enke, Stuttgart, 1970, pp. 142, Tabel 40.

64. Schürmann, J., "Untersuchungen über organische Stoffe im Wasser des Zürichsees," *Viertel jahrsschr. Naturforsch. Ges. Zürich*, **109,** 409–460 (1964).

65. Serruya, C., "Lake Kinneret," *Limnol. Oceanogr.* **16** (3); 510–521 (1971).

66. Shukla, S. S., J. K. Syers, J. D. H. Williams, D. E. Armstrong, and R. F. Harris, "Sorption of Inorganic Phosphate by Lake Sediments," *Soil Sci. Soc. Am. Proc.*, **35**, 244–249 (1971).
67. Stangenberg-Oporowska, K., "Forms of Phosphorus in the Bottom of Carp Ponds," *Acta Hydrobiol.* (*Kraków*), **12**, 125–142 (1970).
68. Thomas, E. A., "Phosphat-Elimination in der Belebtschlammanlage von Männedorf und Phosphat-Fixation in See- und Klärschlamm," *Vierteljahrsschr. Naturforsch. Ges. Zürich*, **110**, 419–434 (1965).
69. Thomas, E. A., "Die Phosphattrophierung des Zürichsees und andere Schweizer Seen," *Mitt. Int. Verein. Limnol.*, **14**, 231–242 (1968).
70. Van Wazer, J. R., *Phosphorus and Its Compounds*, Vol. 1, *Chemistry*, Interscience, New York, 1958, 954 pages.
71. Vollenweider, R. A., "Möglichkeiten und Grenzen elementarer Modelle der Stoffbilanz von Seen," *Arch. Hydrobiol.*, **66** (1), 1–36 (1969).
72. Wentz, E. A., and G. F. Lee, "Sedimentary Phosphorus in Lake Cores—Analytical Procedure," *Environ. Sci. Technol.*, **3**, 750–754 (1964).
73. Wentz, E. A., and G. F. Lee, "Sedimentary Phosphorus in Lake Cores—Observations on Depositional Pattern in Lake Mendota," *Environ. Sci. Technol.*, **3**, 754–759 (1969).
74. Williams, J. D. H., J. K. Syers, and R. F. Harris, "Adsorption and Desorption of Inorganic Phosphorus by Lake Sediments in a 0.1 *M* NaCl System," *Environ. Sci. Technol.*, **4**, 517–519 (1970).
75. Williams, J. D. H., J. K. Syers, R. F. Harris, and D. E. Armstrong, "Fractionation of Inorganic Phosphate in Calcarous Lake Sediments, *Soil Sci. Soc. Am. Proc.*, **35**, 250–255 (1971).
76. O'Melia, Ch. R., "Approach to the Modeling of Lakes," *Schw. Z. für Hydrologie*, **34**(1), 1–34 (1972).
77. Revelly, P., "Rapports sur les Études Entreprises sur le Lac Léman et ses Affluents en 1969 et 1970. I," *Rapport sur l'Évolution Physico-Chimique du Léman* (1970).
78. Golterman, H. L., "Natural Phosphate Sources in Relation to Phosphate Budgets: A contribution to the Understanding of Eutrophication," *Water Res.*, **7** (in press).

A Dynamic View of the Phosphorus Cycle in Lakes

30

F. H. RIGLER

Department of Zoology, University of Toronto

As Hutchinson showed (29) in 1941, phosphorus in lakes is not a conservative element. That is, the movements and cycling of phosphorus within lakes may be influenced more by biogenic processes than by simple physical processes such as vertical eddy diffusion. Also, phosphorus differs from other essential elements such as nitrogen or carbon in that it is almost entirely associated with living and nonliving particulate matter. These and many other observations have fostered the widespread belief that the rapid eutrophication of lakes throughout the world is largely being caused by increased input of phosphorus resulting from human activities. This attitude is exemplified by the U.S. Academy of Sciences volume *Eutrophication, Causes, Consequences, Correctives* (1969), and has led to many intensive investigations of phosphorus in aquatic systems.

Many different approaches have been taken in the study of phosphorus in lakes and each offers certain advantages. For example, Vollenweider's (66) elegantly developed models of phosphorus budgets that predict phosphorus concentration from simple parameters such as input and output rates, flushing time, and lake morphometry, yet ignore all biological and temporal complexities, have immediate predictive value. However, to provide insight into the role and economy of phosphorus in aquatic systems, detailed studies of phosphorus cycling must also be pursued. Information about phosphorus cycling can be obtained in two ways. One can take advantage of non-steady-state conditions within a system or perturb the system by massive introductions of P, as was done by Einsele (17), to measure *net* transfer of P from one organism to another or one part of the system to another; alternatively, one can use radioactive phosphorus (^{32}P or ^{33}P) to measure gross transfer of P. The latter technique, introduced to limnology by Hutchinson and Bowen (31), is potentially the most informative, and has provided the basis for most of the following presentation.

When the results of experiments with radioisotopes are to be used to give a quantitative measure of the cycle of an element, they are usually analyzed by the method of compartmental analysis as outlined by Riggs (52) and Sólomon (61). Without chemical analyses one does not know if steady-state conditions exist during the experiment, and that it can be conceptually divided into a number of chemically or spatially recognizable compartments. In natural ecosystems, such as lakes, the number of recognizable compartments is so large that one must be content with a partial subdivision in which many recognizable compartments are lumped in one operationally defined compartment. For example, the open water of lakes contains a multitude of different living and nonliving particles containing phosphorus. The nonliving particles (tripton) may be autochthonous or allochthonous and the living particles (plankton) are represented by hundreds of species of monerans, protistans, plants, and animals. In theory, each species could be considered as a separate compartment, and furthermore each could be divided into subcompartments. This is not done in practice because even if it were possible to separate individuals of each species, mathematical analysis of the results would be intractible. In the early studies of phosphorus cycling within the open water of lakes, only two compartments were used, phosphorus that passed through a membrane or paper filter and phosphorus in seston (tripton and plankton) removed by the filter. There are two consequences of this gross but necessary oversimplification. First, different workers may use different schemes for compartmentalizing the same system, and secondly the schemes first used are likely to become progressively more complex as experimentation reveals discrepancies between predicted and observed results.

Although this chapter is concerned with the kinetics of phosphorus cycling, the results of tracer experiments cannot be fully interpreted without a knowledge of the phosphorus content of the compartments under investigation. Without chemical analyses, the assumption that steady-state conditions exist cannot be tested. Also the rate of loss of tracer from a compartment tells only the proportion of phosphorus leaving that compartment per unit time. To calculate the amount of phosphorus leaving, one must also know the amount of phosphorus in the compartment. Since much of the model of phosphorus cycling to be presented is built on premises about the interpretation of analytical results which differ considerably from the generally accepted premises [see Hutchinson (30) for the classical approach], I discuss first the chemical analysis for phosphorus in some epilimnetic compartments before dealing with the phosphorus cycle within the epilimnion and within the whole lake.

As was mentioned above, the methods of compartmental analysis assume that the system under investigation is in a steady state. This condition is rarely, if ever, met in aquatic ecosystems, but during summer stagnation of temperate lakes a pseudo-steady state is sometimes achieved. Therefore it is not surprising that most tracer studies in lakes have been conducted in late summer. The model to be presented is for this period and thus is of limited value. Only when intensive investigations have been conducted throughout the year will we be able to pretend that we have a functional understanding of the phosphorus cycle in lakes. However, it is during the summer that primary production is highest, and nuisance blooms of algae develop. Consequently, a model covering only the summer, restricted as it is, may have some useful predictive value.

PHOSPHORUS COMPARTMENTS IN EPILIMNETIC WATER

Terminology

Since the validity of our chemical analyses has been questioned in the past and since the models to be presented in subsequent sections are based on the assumption that current analytical techniques do not necessarily measure definite phosphorus compartments that are equivalent to morphologically or chemically distinct components of lake water, I use herein an operational terminology to avoid attributing more meaning to chemical determinations than is inherent in them. The terminology of Strickland and Parsons (63), although criticized on etymological grounds by Olsen (46) and not strictly operational, is retained here since it is now widely used, and since further proliferation of terms would simply add to the terminological chaos that was documented by Olsen.

Soluble reactive phosphorus (SRP) refers to the value obtained when membrane-filtered water is analyzed by one of the variants of the molybdenum blue technique. This term implies neither that the orthophosphate measured was in solution before addition of the reagents nor that the intensity of the blue color is exclusively a function of orthophosphate concentration rather than that of interfering ions. When "orthophosphate phosphorus" (PO_4–P) is used it will *not* refer to the results of chemical analyses but to free orthophosphate in solution, the concentration of which is assumed to be as yet unmeasurable in the trophogenic zone of most lakes.

Soluble phosphorus (SP) refers to the value obtained when membrane-filtered (0.45 μ) water is analyzed after being digested with an oxidizing acid solution.

Soluble unreactive phosphorus (SUP) is the difference between SP and SRP.

Total phosphorus (TP) is obtained by analyzing whole lake water after acid digestion. It is assumed that the values obtained by this technique are indicative of the true phosphorus content of the sample.

Particulate phosphorus (PP) is the total phosphorus minus soluble phosphorus.

If the phosphorus fractions have been further subdivided, this is indicated in parentheses. For example, SUP (0.1 μ – 0.45 μ) refers to the difference between SUP in the filtrate of a 0.45-μ membrane filter and SUP in the filtrate of a 0.1-μ filter, and PP ($< 70\ \mu$) refers to particulate phosphorus that passes through a net with 70-μ apertures.

Orthophosphate Phosphorus

This compartment is central to any study of phosphorus cycling involving radioisotopes since it is the form in which isotopes such as ^{32}P and ^{33}P are normally purchased. It has been generally accepted that SRP is equivalent to PO_4–P and much research has been devoted to the elaboration of simple, sensitive chemical techniques for measuring SRP. [See Olsen (46) for a comprehensive review of these techniques.] Although suspicions that SRP might be greater than PO_4–P have been reported for some time (53, 63) no observations consistent with this suspicion were available until Kuenzler and Ketchum (36) reported their work on growth of *Phaeodactylum tricornutum* Bohlin in a chemically defined medium. They found that the maximum possible concentration of PO_4–P, calculated from the ratio of ^{32}P in solution to ^{32}P in cells, was 4% of the chemically determined SRP. Subsequently Rigler (57) suggested that anomalous uptake kinetics of $^{32}PO_4$ in natural lake waters could be explained by postulating a discrepancy of the same magnitude between SRP and PO_4–P in lake water as in Kuenzler and Ketchum's cultures. In both cases the discrepancy could have been much larger than that reported since Kuenzler and Ketchum's calculation assumed that all ^{32}P in filtrate was $^{32}PO_4$–P and Rigler's radiobiological assay simply estimated the maximum possible concentration of PO_4–P.

A gross error in the determination of PO_4–P could arise in one or all of three ways. First, lake water is generally filtered before SRP analyses are done. Filtration might damage delicate algal cells and release phosphate phosphorus or readily hydrolyzed phosphate esters into the filtrate. There is good evidence by Lasker and Holmes (39), Guillard and Wangersky (22), McAllister (44), Kuenzler and Ketchum (36), and Arthur and Rigler (2) that phytoplankton are damaged by filtration and that recently fixed ^{14}C may be released from the damaged cells. This suggests that PO_4–P might also be released. Finally, King (34) has shown that more PO_4–P is released by high-pressure than by low-pressure filtration, although his work suggested that the quantity released was small. A second possible source of error derives from the need, in all variants of the analytical technique, to acidify the sample with H_2SO_4 (usually 0.1 to 0.4 N). This could hydrolyze free phosphate esters (67) and release PO_4–P from fulvic acid–metal phosphates (40) or from colloidal iron phosphate. Evidence consistent with this suggestion was presented by Rigler (58) who showed that a partial purification of PO_4–P from lake and pond water on columns of hydrous zirconium oxide, which trapped PO_4–P efficiently but allowed all SUP to escape, significantly decreased the concentration of SRP. The inference here is that removal of acid-labile P before addition of acid reduced the amount of PO_4–P released into solution by acid hydrolyses of phosphate esters or insoluble metallophosphates. The next evidence was given by Chamberlain and Shapiro (9) who designed an elegant method to reduce the effective time of exposure to acid to 6 sec by adding excess acid, which prevents further formation of molybdophosphoric acid. Although they concluded that their work gave no evidence for or against the hypothesis that acid-labile phosphorus compounds are hydrolyzed during the analysis it is interesting to note that in the four lakes that had low concentrations of SRP (1 to 2 μg/l SRP by Harvey's method) and were not contaminated with excessive amounts of arsenate, the 6-sec method gave lower values for SRP than the more conventional methods with which it was compared. Furthermore, the 6-sec method gave SRP values close to the values for available P determined by a bioassay technique. These results are significant since Rigler's hypothesis was derived from results of experiments on waters in which SRP concentrations were 0.3 to 5 μg/l. Probably the discrepancy between SRP and PO_4–P will prove to be significant only in unpolluted lakes in which P is in short supply.

Of the elements that can interfere with colorimetic determination, and thus contribute to discrepancy between SRP and PO_4–P, arsenic may be the most important (25), although this possibility has generally been ignored. For example, Ambühl and Schmid (1) in a careful study of potential interference with the molybdenum blue reaction simply dismiss arsenic as being of no importance in lake waters. However, Chamberlain and Shapiro (9) have shown that arsenic does occur and must be considered as a potentially significant interfering substance.

In conclusion, there is considerable evidence suggesting that in the trophogenic zone of many lakes, chemically determined SRP is much greater than PO_4–P. In subsequent sections it is assumed that the PO_4–P compartment is very small and cannot be measured chemically. This imposes severe limitations on the interpretation of tracer results, since rate constants of PO_4–P uptake cannot be converted to P fluxes. Previously, PO_4–P was considered to be an excellent point of entry to the P cycle because it is a homogeneous and simple compartment. Once PO_4–P uptake rate was calculated, other fluxes could then be calculated. Having rejected the possibility of calculating PO_4–P uptake rate, I am forced to express other movements of P in terms of rate constants rather than rates.

Soluble Organic and Particulate Phosphorus

Although soluble organic phosphorus and particulate phosphorus continue to be treated as physically separable compartments, considerable evidence demonstrating the inadequacy of our current techniques for separating them has accumulated. It was first shown that different ratios of soluble organic P to particulate P reported by different workers, which were attributed by Hutchinson (29, 30) to varying amounts of colored material in the water, were not found when the same technique of separating these compartments was applied to lake waters ranging in color from 6 to 162 Pt units (56). Rigler also showed that the entire range of results obtained by several workers could be duplicated in one lake by applying the separatory techniques of those workers. Furthermore, filtration of lake water through 0.22- and 0.1-μ filters reduced the amount of "soluble" phosphorus below that found in the filtrate of a 0.45-μ filter. The occurrence of PP in the SUP fraction was more forcibly demonstrated by Chamberlain (8) who, recognizing the difficulty of obtaining precise analytical results and the possibility that fine-membrane filters adsorb organic P compounds, designed a test using water from a lake which had been labeled with ^{32}P. Beginning with the filtrate of a 0.45-μ filter he first refiltered it through a 0.1- and a 0.01-μ filter. Seventy-four percent of the ^{32}P in the original filtrate was removed on the former and 92% on the latter. If adsorption was not taking place, 92% of the soluble phosphorus (^{32}P) was associated with particles between 0.01 and 0.45 μ. To check for adsorption he ultracentrifuged another sample for times that should remove (*a*) particles larger than 0.065 μ and (*b*) those larger than 0.015 μ. In neither case was as much ^{32}P removed as by the corresponding filter but in (*b*) 80% of the ^{32}P was sedimented. Thus, although adsorption or filter clogging might interfere with results of differential filtration, it appears that at least 80% of the ^{32}P in the filtrate of a 0.45-μ filter was particulate.

All of the results above are consistent with the idea that methods allowing us to distinguish between dissolved organic P and PP have not been applied and reinforce the necessity of treating SUP and PP as operationally defined phosphorus compartments. The original assumption, implicit in the methodology of early workers, that all PP is associated with large plankton and tripton must be rejected. Instead, we must recognize that P is associated with particles ranging in size from the largest zooplankton down to colloids and that the choice of a 0.45-μ filter to separate PP and SUP, although convenient, is arbitrary. Much of the SUP is in particles smaller than 0.1 μ and much is colloidal, with perhaps only a small fraction in solution.

Zooplankton

Although it would be desirable to know what fraction of the limnetic phosphorus is in zooplankton, the difficulty of separating zooplankton from net phytoplankton has presumably inhibited investigations of this point. This is unfortunate since zooplankton sometimes have a higher biomass than phytoplankton and a significant fraction of the total P may be in this compartment at times.

Perhaps a more important consideration is that zooplankton are included in the total P of limnetic water, yet chemical sampling techniques do not adequately sample zooplankton. To obtain an adequate sample of macrozooplankton at a given depth it would be necessary to pool samples from different locations since zooplankton exhibit pronounced nonrandom horizontal distribution patterns (51, 38, 49). Also, because small-scale aggregations exist (11, 60), a single, small sample cannot be considered to be representative of the part of the lake from which it was taken. Since 50- or 100-ml samples are normally used for determination of total P in lake water it is not surprising that the total P in the trophogenic zone of lakes often appears to fluctuate wildly from day to day. The variability introduced by poor sampling was demonstrated by Chamberlain (8) who was able to attribute temporal fluctuations in total P in the epilimnion of a dystrophic lake to variance of PP ($>70\ \mu$). He also showed that one extra *Daphnia*, containing 0.19 μg P, in a 50-ml water sample, would increase the total P of the sample by 4 μg/l, a significant effect when total P averaged 14 μg/l.

Throughout this section I have stressed the inadequacies of chemical techniques rather than the advances that have been made over the last 50 years. This was motivated not by a destructive urge but by the recognition of the dependence of conclusions, to be drawn in the following sections, on chemical analyses as well as on interpretation of tracer kinetics. It is hoped that this emphasis will not only stimulate interest in improving existing techniques, but also encourage the reader to be critical of the conclusions that follow.

PHOSPHORUS CYCLING WITHIN THE EPILIMNION

It has been apparent for some time that in many lakes we cannot think of PO_4–P as a large phosphorus pool to be used by phytoplankton when needed and replaced only from paralimnetic sources. Rather it is an element in great demand, such that each atom is used and reused many times during its residence time in the epilimnion. Without some understanding of the kinetics of uptake and regeneration of available phosphorus, the forms of phosphorus available for use, and the details of the phosphorus cycle within the planktonic community, it would be very difficult to establish the role, if any, played by phosphorus in limiting production and determining the species composition of the phytoplanktonic community.

In discussing the phosphorus cycle within the epilimnion of lakes I restrict myself primarily to summer stagnation. In order to simplify the presentation, the epilimnion is treated as if it were a closed system in a state of equilibrium. This is not true since affluents, seepage, rain, and airborne allochthonous materials must continually contribute phosphorus. Losses of phosphorus from the epilimnion approximately equal inputs, but fluctuations in phosphorus content attest the fact that input rarely equals output. Furthermore, within the system, the biomass of each species fluctuates, often with startling rapidity. Therefore the epilimnion is really an open system that is rarely, if ever, in a steady state. However, if cycling within the epilimnion is rapid relative to movement through the epilimnion, and the time periods considered are short relative to those over which the biomass of individual species change, these oversimplifications will not seriously influence the interpretation of data.

I am taking PO_4–P as a starting point since it is the generally accepted source of phosphorus for phytoplankton and is the only clearly defined P compartment in the open water. When $^{32}PO_4$ is added to the surface water of lakes it is quickly removed from solution by seston. Rigler (54) inferred from serial filtration that bacteria and ultraphytoplankton were the organisms taking up the $^{32}PO_4$. Since the uptake of $^{32}PO_4$ by PP was prevented by heating the water to 85°C (53) or addition of metabolic inhibitors such as methylene blue or cyanide (50), it originally appeared to be indisputable that living organisms take up the $^{32}PO_4$. But if PO_4—P is as scarce as now postulated, the possibility that these treatments flooded the system with PO_4–P from killed organisms should have been considered. If this happened, any naturally occurring uptake of $^{32}PO_4$ by nonliving particles might be greatly reduced. Recently, Fuhs and Canelli (in press) have developed a radioautographic technique that should answer this question. Their preliminary results suggest that living and nonliving particles in lake water quickly become labeled with ^{33}P.

Since the kinetics of $^{32}PO_4$ uptake often conform to those predicted by a two-compartment exchange model in which ^{32}P in solution decreases exponentially to an equilibrium level, these results were explained by postulating a simple exchange of PO_4—P between solution and a solid compartment (54), and the turnover time of inorganic phosphate was calculated from them. The turnover time of PO_4—P is the time in which an amount of PO_4—P equal to that in the water is taken up by seston and the same amount released by seston. It is the reciprocal of the rate constant of loss of ^{32}P from solution. Although it has now been shown that the two-compartment model is inadequate (8), it is nevertheless reasonable to postulate that the PO_4—P compartment is continually replenished from some source and thus still acceptable to use turnover time as a measure of the rate of replacement of PO_4—P.

The turnover time of epilimnetic PO_4—P has now been measured in open-system, artificial ponds, temperate oligotrophic to extreme eutrophic, dystrophic, and bog lakes (8, 12, 56), and regardless of the trophic nature of the water body, it is generally between 1 and 8 min during summer stratification.

Similar turnover times can be expected during the productive period in all lakes that have a high ratio of PP to SRP. However, a few lake districts are characterized by lakes in which SRP contributes up to 80% of the total. The Polish lakes studied by Mackiewicz-Golachowska (41) are good examples of this type. In these lakes the turnover time of PO_4—P is probably extremely long. The only evidence concerning this point is from preliminary studies of two polar lakes (59). In Char Lake (unfertilized) turnover time in July averaged 63 min, longer than in temperate lakes, but not surprisingly so since the water temperature was 3°C. However, in a nearby sewage-enriched lake, where SRP equaled SP (V. Maseman, personal communication) and contributed 62% of the TP, uptake of $^{32}PO_4$ by plankton was unmeasurable. Thus a large PP/SRP ratio probably indicates a short turnover time of PO_4-P which, in turn, suggests an extreme shortage of available PO_4—P.

In winter, the turnover time of PO_4—P increases considerably, although less in bog lakes than others (56). This increase can probably be attributed to decreased temperature, increased concentration of PO_4—P, and to the reduced biomass of plankton which is characteristic of winter.

Uptake and Turnover of PO_4—P

Although it is obvious that PO_4—P is rapidly utilized and regenerated, and it appears that ultraplankton are primarily responsible for removing PO_4—P from solution, only a tentative model of phosphorus cycling within the epilimnion can be developed at the moment. Several observations are inconsistent with the two-compartment exchange model. These are considered and a provisional, simplified model consistent with these observations is presented.

If $^{32}PO_4$ is added to lake water in which the cycling of P is completely dominated by exchange of PO_4—P between solution and a particulate compartment, the loss of ^{32}P from solution will be described by the equation,

$$Y_t - Y_\infty = (Y_0 - Y_\infty)e^{-kt} \tag{1}$$

where Y_0 is the specific activity of PO_4—P at $t = 0$, Y_t is the specific activity of PO_4—P at any time (t) after the addition, and Y_∞ is the specific activity when equilibrium distribution of tracer has been attained. Thus a plot of $\ln(Y_t - Y_\infty)$ against time will yield a straight line, the slope of which (k) is the sum of the rate constants of PO_4—P movement from solution to solids and from solids to solution. Although the uptake of $^{32}PO_4$ added to lake water sometimes conforms to Eq. (1), Chamberlain (8) showed that the log plot of $Y_t - Y_\infty$ against time sometimes yields a curve which can be split by standard techniques (52) into two straight lines, and can be best described, by Eq. (2), as the sum of two exponentials

$$Y_t - Y_\infty = K_a e^{-\lambda_1 t} - K_b e^{-\lambda_2 t} \tag{2}$$

Uptake kinetics of this type are consistent with many models (65) but not with the two-compartment exchange model. The simplest way of making the model conform to these data is to add a third compartment, either a second particulate compartment (Fig. 1*a*) or a second solution compartment (Fig. 1*b*). The decrease of tracer in the SP fraction after addition of $^{32}PO_4$ would be described by Eq. (2) if either of the two models in Fig. 1 adequately described the system. If it is postulated that the size of compartment 1, or the rate constant of $2 \rightarrow 1$, varies temporally or from lake to lake, the fact that uptake kinetics conform sometimes to Eq. (1) and sometimes to Eq. (2) can be explained.

The functional difference between the two models in Fig. 1 is that compartment 1 is operationally defined as SP in Fig. 1*b* and as PP in Fig. 1*a*. In neither case is its chemical or biological nature defined, although Fig. 1*b* precludes the possibility of it representing a group of living organisms. Although it is intuitively more reasonable to postulate that different groups of organisms with different exchange rates and capacities exist than to postulate that a second solution compartment exists, several observations require the latter postulate. Initially Rigler (58) observed that $^{32}PO_4$ added

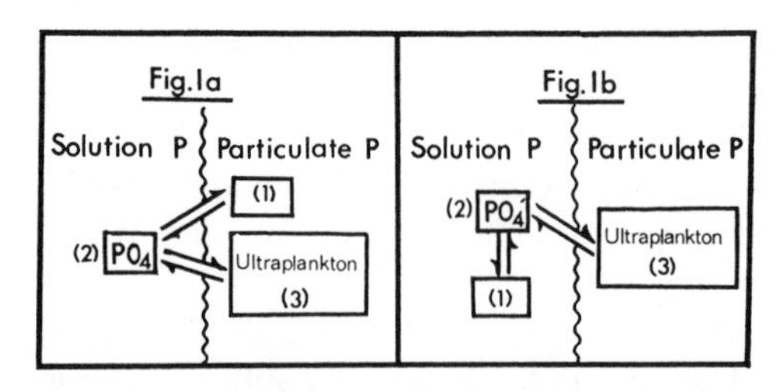

Figure 1. Two simple three-compartment models that are equally consistent with the observed kinetics of loss of $^{32}PO_4$ from solution in lake water: (*a*) contains two particulate compartments, (*b*) two solution compartments.

to membrane-filtered lake water was, at times, only partially removed on columns of hydrous zirconium oxide (HZO). Since this substance has a high affinity for PO_4^{3-} (35) and removed $^{32}PO_4$ from distilled water with over 95% efficiency, Rigler postulated the existence of a solution compartment that exchanged with inorganic phosphate. Subsequently, Chamberlain (8) showed that in lake water which had been equilibrated with ^{32}P in situ for many days, the SP compartment contained ^{32}P that on refiltration was removed and that could be removed by ultracentrifugation. He also showed by ion-exchange studies, and by measurements of uptake kinetics of the ^{32}P remaining in solution shortly after $^{32}PO_4$ was added to lake water, that within minutes some of the soluble ^{32}P is no longer PO_4—P.

Although Chamberlain's observations do not require postulation of an abiotic exchange, they are clearly inconsistent with the original two-compartment exchange model and also with that in Fig. 1*a*, since neither includes any solution compartment other than PO_4—P.

The best evidence for abiotic exchange was obtained by David Lean (personal communication) during a study of SUP in lake water. One of his results, presented in Fig. 2, showed that $^{32}PO_4$ added to membrane-filtered

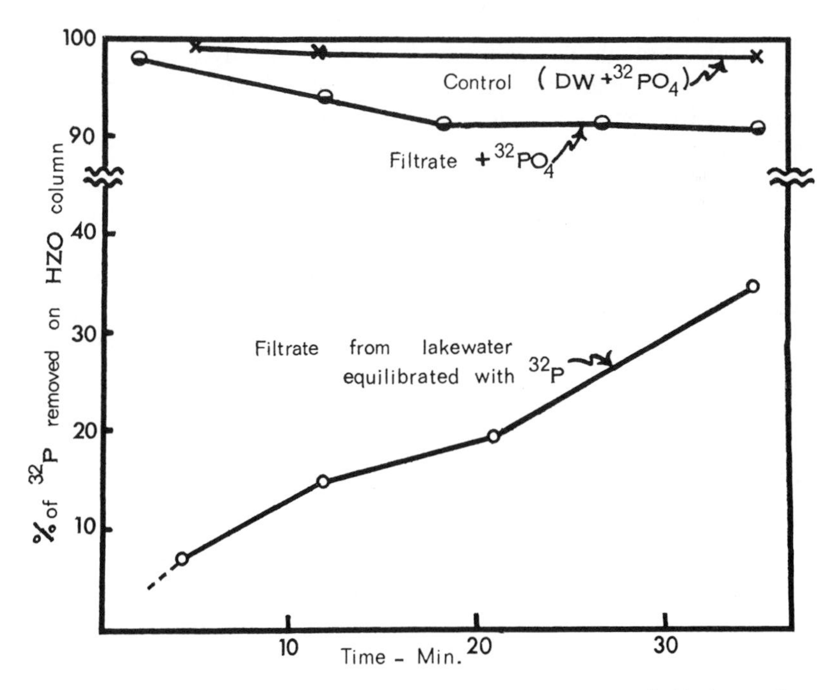

Figure 2. Unpublished results of David Lean suggesting existence of abiotic transformations of P in membrane filtrate of lake water.

lake water was initially removed almost quantitatively on a column of HZO, but after 35 min about 7% of the ^{32}P was in a form not taken up by HZO. On some occasions there was a negligible change, on others over 90% of the $^{32}PO_4$ was unreactive with HZO after 30 min. The converse experiment, in which the membrane filtrate of lake water equilibrated with $^{32}PO_4$ before being filtered was passed through a column of HZO, suggested that immediately after filtration very little of the soluble ^{32}P occurred as PO_4—P. Only 7% was removed on the column at first, but the amount steadily increased until at 35 min after filtration 35% of the filtrate ^{32}P was taken up. These results suggest that in membrane filtrate of lake water, PO_4—P is concurrently reacting with some unspecified substance and being released from some substance. Thus the model in Fig. 1*b*, although grossly oversimplified, is clearly more consistent with the observations on phosphorus cycling in epilimnetic water than is the model in Fig. 1*a*.

The most obvious oversimplification in Fig. 1*b* is that all PP is included in one compartment. This is inconsistent with the previously mentioned observation that $^{32}PO_4$ is taken up by ultraplankton, since it treats net phytoplankton and zooplankton as operationally indistinguishable from nanno- or ultraplankton. Another observation requiring subdivision of PP into two compartments was made by Chamberlain (8). He distributed $^{32}PO_4$ evenly through the epilimnion of a small, dystrophic lake and subsequently measured the specific activity of different limnetic P compartments. Five hours after addition of the $^{32}PO_4$ it was largely associated with PP ($<70\ \mu$), but over a period of about 10 days the net plankton and nannoplankton came into isotopic equilibrium (Fig. 3). Although the scatter of points precludes a detailed analysis, the turnover time of PP ($>70\ \mu$) is apparently much longer than the turnover time of PP ($<70\ \mu$). Thus we must recognize at least two particulate compartments, one comprising the smallest organisms through which P cycles rapidly, the other comprising the larger organisms through which P cycles slowly. Further insight about these compartments can be obtained by considering the biology of the limnetic system and regeneration of phosphorus.

Mechanisms Returning Phosphorus to Solution

Of the possible mechanisms returning phosphorus to the PO_4–P pool, three have been considered to be important. These are (*a*) direct release of phosphorus by the ultraplankton; (*b*) excretion of phosphorus by zooplankton; and (*c*) enzymatic hydrolysis of organic phosphorus compounds excreted by organisms or produced by autolysis or decomposition of dead plankton. Each has been stressed by one worker or another as the most important source of PO_4—P, but no systematic analysis of the relative importance of

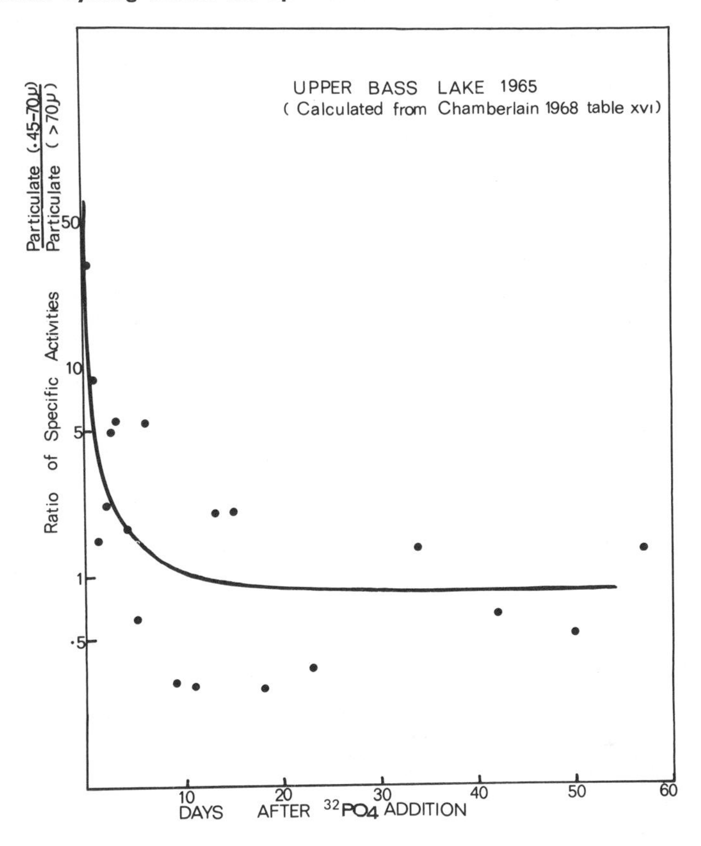

Figure 3. Relative specific activities of different fractions of particulate P in Upper Bass Lake as determined by Chamberlain (8).

each mechanism has been, or is, possible. However, it may be useful to discuss the limitations of previous conclusions and attempt a preliminary resynthesis based on recent data.

The conclusion easiest to negate is that of Rigler (55), who compared uptake kinetics of $^{32}PO_4$ by ultraplankton with release of PO_4—P by one zooplankter and concluded that release of P by zooplankton is negligible compared with direct release from ultraplankton. There were two errors in this approach, both of which minimized the role of zooplankton. First, as indicated by Johannes (33), the assumption that P excretion per unit weight is independent of body size was erroneous, since smaller organisms excrete more P per unit weight than large ones. This error, combined with the use of excretion rates measured on a crustacean (*Daphnia magna*) larger than limnetic zooplankters resulted in an estimate of P excretion by the zooplankton community that was much too low. Second, release of P from bacteria and algae was calculated on the assumption that it was approximately equal to uptake rate which, in turn, was equal to the uptake rate constant of $^{32}PO_4$ multiplied by the concentration of SRP. Since SRP probably exceeds PO_4–P by a factor of 100 in the lakes studied, the release of PO_4—P by ultraplankton would be correspondingly overestimated.

Until PO_4—P can be measured quantitatively another method must be used to estimate direct release of P by ultraplankton. I will use the results of unpublished experiments in which the rate of release of ^{32}P from seston was measured. Radioactive phosphate was added to a small sample of lake water and after 5 to 45 min the water was gently filtered through a 0.45-μ filter in a holder designed to allow nonradioactive, filtered lake water to flow over the cells and through the filter while a Geiger-Müller detector above the filter continuously recorded radioactivity of the layer of plankton. In this way, the washout of recently assimilated ^{32}P from ultraplankton was measured. The results are ambiguous since they usually showed a rapid initial loss for a few minutes, followed by a slow exponential loss until the experiments were terminated 3 hr later. There is as yet no basis for interpreting the rapid initial loss ($k = 6.8\ hr^{-1}$) which might be caused by (*a*) cell breakage, (*b*) loss from tripton, (*c*) desorption of ^{32}P from the membrane filter or, (*d*) loss from a quickly exchanging intracellular compartment. If the rapid initial loss is ignored and washout from a single compartment system is postulated to explain the subsequent slow loss, then the rate constant of loss calculated with Eq. (1) ranged from 0.006 to 0.046 hr^{-1} in different lakes, and averaged 0.019 hr^{-1}. This figure, although tentative, is used here as an estimate of the direct release of P as fraction $\cdot$ $time^{-1}$ (form unspecified) from the small phyto- and bacterioplankton and is compared with the release of P by zooplankton.

Two approaches can be used to measure regeneration of P by zooplankton. In the first, excretion by individual species of zooplankton is measured, either by confining the animals and chemically determining the increase of P in the ambient medium or by measuring the release of ^{32}P from uniformly labeled animals. The results of several studies that used these methods are plotted in Fig. 4. The various results are consistent only in showing that excretion per unit body weight decreases with increasing size of animal. This relationship holds both within (Peters, personal communication) or between species [Johannes (33), Barlow and Bishop (3), Hargrave and Geen (26). However, the results are of little value in a general model since they differ by more than two orders of magnitude, and there is as yet no basis for determining whether the differences are real, and reflect extreme variability of phosphorus excretion with environmental conditions such as

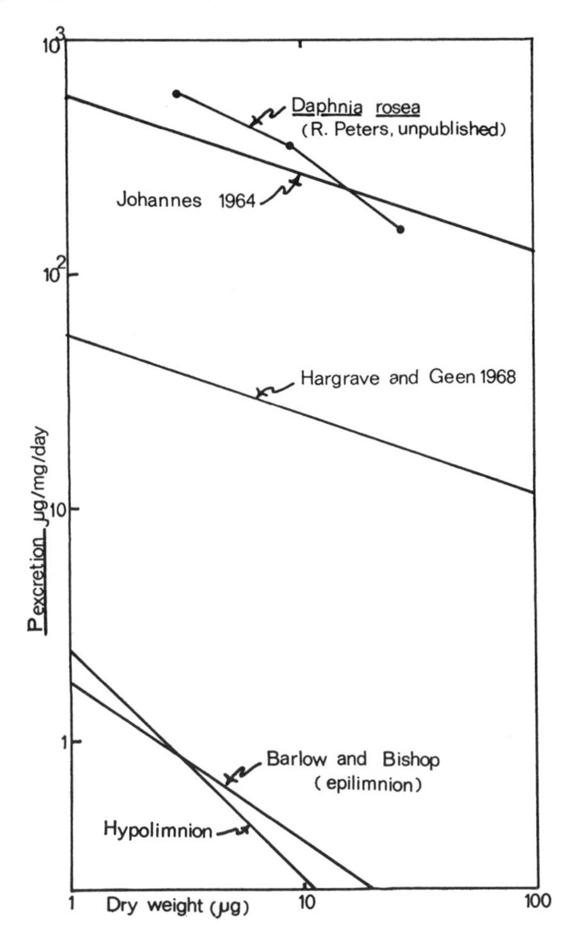

Figure 4. Excretion of P by zooplankton as measured by different authors. Only those results showing the effect of size are presented. Many additional individual values exist and these fall in the same (wide) range as the values given. Johannes' values were converted from body equivalent excretion times (his regression line, log BEET $= 0.33 \log W + 0.6$) on the assumption that the P content was 1% of the dry weight.

food supply or phosphorus content of food, or are produced by experimental artifacts. That experimental artifact could contribute to the difference is suggested by the fact that the most rapid rates were obtained from the isotope technique (Peters, Johannes) and the slowest (Barlow and Bishop, Hargrave and Geen) generally from chemical analyses. Incomplete labeling, in which the specific activity of a quickly exchanging compartment was higher than the specific activity of a slowly exchanging compartment, would introduce a positive error, and recycling of excreted P in the long-term experiments using chemical analyses would introduce a negative error. A further problem is that neither experimental variant has been applied to normal concentrations of animals feeding on a natural food supply.

For these reasons I reject the data in Fig. 4 and use an indirect method of calculating phosphorus regeneration from data on zooplankton grazing rates obtained in situ. It was shown above that the ultraplankton take up PO_4—P most rapidly. Since uptake of P by a population of a given species is related to the loss of P by that population and to the turnover of its biomass, it is not surprising that studies of the feeding behavior of zooplankton have shown that ultraplankton is intensively grazed, but net phytoplankton is not (6, 21, 24). In a steady-state system in which the biomass of ultraplankton and zooplankton is constant or changing at a very slow rate, one would expect all of the P assimilated by zooplankton to be excreted back into the water. Therefore a measure of grazing rate and assimilation rate of the zooplankton community on ultraplankton could be used to calculate the fraction of ultraplankton P returned to the water per unit time by herbivorous zooplankton. Recently, Haney (24) developed a rapid, in situ method of measuring the grazing rate of the entire zooplankton community (exclusive of animals smaller than 70 μ). He applied this method to a eutrophic lake and measured the rate at which particles such as bacteria (*Pseudomonas flourescens*), yeast (*Rhodotorula glutinis*), and small algae (*Chlamydomonas* sp.) were grazed. Although grazing rates fluctuated throughout the summer, the three sizes of food particle were grazed at roughly the same rate. The average grazing constant (fraction of small particles eaten per unit time) in the trophogenic zone during summer stratification was 0.033 hr^{-1}. Thus if zooplankton assimilated P from their food with an efficiency of 100%, the rate constant of loss of P from ultraplankton *due to grazing alone* would be 0.033 hr^{-1}. There are few measures of assimilation efficiency for P, but work by Marshall and Orr (43) and Corner *et al.* (16) on marine calanoids suggests that a value of 60 to 70% is not unreasonable. This is in accord with the value of 60% suggested by Conover (13, 14) as an approximation of the efficiency at which zooplankton assimilate carbon. If it is assumed that assimilation efficiency for P by the entire zooplankton community is 60%, then the rate constant of P loss due to grazing would be 0.02 hr^{-1}. Therefore, until the assumptions on which the calculations above depend have been tested, it can be tentatively concluded that direct release of P from ultraplankton (0.019 hr^{-1}) and excretion by zooplankton are equally important mechanisms regenerating P in the trophogenic zone of eutrophic lakes during summer stratification.

It should be noted that the rate constants cannot be converted to P fluxes since the size of the ultraplankton compartment is unknown. Also, the fate of unassimilated P is unknown. Presumably part of it is lost from the trophogenic zone in fecal pellets, part resuspended as detritus, and part as viable cells. Finally, there is no agreement on the form in which P is excreted by zooplankton even though studies have only attempted to show what fraction of the excreted P is SRP and SUP. Estimates range from 100% SRP (55) to 26.6% SRP (26), but the recent work of Butler *et al.* (7), in which care was taken not to contaminate excretion products with defecated material, showed that 71 to 87% of the excreted P was SRP. If Butler *et al.* are correct in implying that high SUP values are artifacts due to contamination of excreted P, then it is probable that most of the excreted P is readily available for reuptake by ultraplankton.

The third possible source of inorganic P which has attracted attention is hydrolysis of organic phosphorus compounds by phosphatases present in the water and at cell surfaces. Unfortunately the importance of this mechanism of regeneration relative to the other two cannot be evaluated. It was first demonstrated that phosphorus from killed zooplankton is rapidly returned to solution (15, 20). To evaluate the relevance of this observation we must know the natural mortality rate, exclusive of mortality due to predation, of zooplankton. Although Hall (23) and Tappa (64) have demonstrated that mortality of *Daphnia* may be as high as 0.25 day^{-1} during midsummer, they have not shown what fraction of this can be attributed to predation.

It has been shown that natural waters contain phosphatases produced by dead (62) and living (42) zooplankton as well as by some algae and bacteria (47). However, Rigler (55), although confirming excretion of phosphatase by zooplankton, was unable to demonstrate hydrolysis of naturally occurring phosphorus esters. More recent studies by Overbeck and Babenzien (48) and Berman (4) have clearly demonstrated phosphatase activity in whole lake water and cell-free filtrates of natural waters, but have given no information on the natural rate of regeneration of PO_4—P by these phosphatases. This is because all of Overbeck and Babenzien's experiments involved the addition of artificial substrates for the enzymes. Berman conducted some experiments without adding an artificial substrate, but in these he used whole lake water saturated with chloroform. In these experiments, the increase of SRP was probably due partially to hydrolysis of naturally occurring phosphate esters, but also to autolysis of organic phosphorus compounds from or within organisms killed by the chloroform.

Interpretation of these experiments is further confused by the work of Kuenzler and Perras (37) which showed that several marine algae possess phosphatases firmly bound to the cell surface. These enable the algae to utilize PO_4—P from phosphate esters, presumably without first releasing any of the PO_4—P into solution. Therefore we do not yet know the normal rate of regeneration of PO_4—P by phosphatases, although we do know that phosphatases capable of hydrolyzing a variety of organic P compounds exist in natural waters. Until more sophisticated experiments are conducted the importance of phosphatases cannot be evaluated relative to the other mechanisms previously discussed.

Simple Model of Phosphorus Cycle within the Epilimnion

The cycle of phosphorus within the epilimnion is summarized in Fig. 5. No rate constants are indicated because studies have been conducted in too few lakes to give any indication of the limits of variation. Also, in a multi-compartment system the method of calculating rate constants depends on the conceptual model of the system, and as yet this model is strictly provisional. However, the arrows representing P fluxes are drawn of varying thickness to suggest the movements of phosphorus which appear to dominate the phosphorus cycle. The active exchange of P pictured between the PO_4—P compartment and the soluble compartment A of unspecified composition is of variable importance. The rapid uptake of PO_4—P by ultraplankton

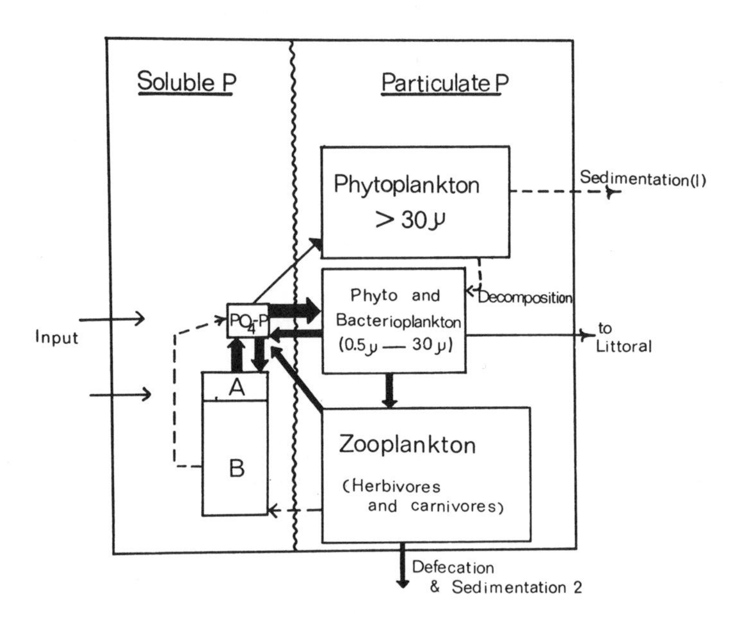

Figure 5. Diagrammatic representation of the phosphorus cycle in the epilimnion of eutrophic lakes. The thickness of arrows is intended to give an impression of the importance of various fluxes. Dashed arrows indicate fluxes of doubtful importance. Soluble P other than PO_4—P is subdivided to indicate that A, the part shown to be exchanging with PO_4—P, is only a small fraction of the total, and is probably different from the material released by zooplankton (B).

appears to be more general. Roughly equal amounts of the phosphorus taken up by ultraplankton are regenerated by direct release and through grazing by zooplankton. Most of the P excreted by zooplankton is ultimately regenerated as PO_4—P although a part of the P may be excreted in organic phosphorus compounds.

The relation of the largest fraction (B) of the SUP to the cycle is unclear although it appears that much of this material is colloidal or associated with small particles (0.1 to 0.45 μ). Phosphatases capable of hydrolyzing phosphorus esters are present in the water, but as yet the concentration of substrate and the rate of hydrolysis of organic phosphorus are unknown. Phytoplankton ($>30\ \mu$) are pictured as comprising a large P compartment through which P cycles relatively slowly and from which P is largely regenerated by decomposition.

Although this model has innumerable limitations only a few are mentioned here. It applies only to eutrophic lakes during summer stratification. The relative amounts of P in the various compartments are unspecified and in fact are unlikely to remain constant. The chemical and physical nature of the SUP is still largely unknown as is the position of this material in the phosphorus economy of the trophogenic zone. The value of the model is that it emphasizes the transient nature of the PO_4—P pool, the need to treat the phytoplankton as at least two compartments, and the possibility of interactions between PO_4—P and some components of the SUP pool.

PHOSPHORUS CYCLE WITHIN THE ENTIRE LAKE

Early Tracer Studies

In the previous section I treated the epilimnion as if it were a closed system, but this was done merely to simplify the presentation. Rapid movement of P out of the epilimnetic water was forcibly demonstrated by Einsele's (17) fertilization experiments in Schleinsee. At the same time Hutchinson (29) formalized the concept of the epilimnion as an open system by postulating a unidirectional movement of P from littoral sediments into the epilimnion, and a loss of P, mediated primarily by fecal pellets of zooplankton, from the epilimnion to the hypolimnion and deep-water sediments. Further understanding was not gained until Hutchinson and Bowen (31, 32) in their pioneering tracer experiments in Linsley Pond, introduced the technique that facilitated demonstration of phosphorus movements from one part of a lake to another and made it theoretically possible to elaborate and test models of the phosphorus cycle within lakes. However, the earliest studies

(10, 32, 27, 54) tended to confuse rather than clarify our concepts. I attempt therefore to evaluate these studies before presenting the more recent and more instructive work of Chamberlain (8).

Hutchinson and Bowen (32) analyzed their results by considering the lake as comprising three functional compartments: the epilimnion, to which the tracer was added; the littoral zone; and the water plus the sediments below the epilimnion. (They actually subdivided the deeper water into six strata, but found that this procedure did little to facilitate interpretation of the results.) Their results showed that between 1 and 4 weeks after addition of tracer to the epilimnion the amount of ^{32}P in the epilimnion decreased, but the ^{32}P in the open water of the whole lake increased by 20%. During this period there was definitely movement of phosphorus from the epilimnion into the deeper water. However, there was no evidence of movement out of the lake or into the littoral solids since there was no decrease of ^{32}P in the open water. In fact, the total appeared to increase during this period.

Serious difficulties are encountered when one attempts to reconstruct events during the first week of the experiment. Hutchinson and Bowen (32) argued that there was no significant loss of ^{32}P from the open water then, but this is not easily reconciled with their description of methods. The amount of ^{32}P added was 70 mCi, or one-fifth of an irradiation unit of 350 mCi. This amount should have yielded 1.3 to 3.9×10^9 dis/sec. The actual radioactivity cannot be given more precisely since shipments of radioisotopes at that time could have differed by as much as 50% from the quantity specified. Counts on uranyl acetate standards of the same weight as those used by Hutchinson and Bowen showed that their counting efficiency was at least 10%. Therefore, the minimum count their sample should have given on August 8 (one half-life after the date it was received) was $1.3 \times 10^9 \times 60 \times \frac{1}{10} \times \frac{1}{2} = 3.9 \times 10^9$ cts/min, whereas the authors stated that 1.35×10^9 cts/min was added. Consequently there is some uncertainty about the quantity of ^{32}P added and one cannot be certain that there was no loss of ^{32}P from the open water during the first week because no samples were taken in that time. If the amount actually added were larger than 1.35×10^9 counts/min, the observation that the quantity of ^{32}P transported down to the hypolimnion was five times greater during the first week than in successive weeks could be easily explained.

Coffin *et al.* (10) performed a similar experiment in a small (0.3 ha) acid-bog lake and subsequently Hayes *et al.* (27) reported another in an unstratified oligotrophic lake. Neither of these experiments added greatly to the model of P circulation between various communities because the method of compartmentalizing the lakes was inappropriate for this purpose. The compartments said to be considered were (*a*) SP in open water and (*b*) all other P in the system. The latter supposedly included all PP in the open water and all phosphorus associated with sediments and with littoral and benthic organisms in the lakes. If this were true, the kinetics should have been dominated by the rapid uptake of $^{32}PO_4$ by ultraplankton, and turnover times should have been of the order of 10 to 100 min rather than 7.6 days in the bog lake and 5.4 days in the unstratified lake as calculated by Hayes *et al.* (27).

This discrepancy is easily explained because the ^{32}P measured was not only that associated with SP, but also that in ultraplankton (e.g., bacteria) and ultratripton. This is not obvious from the publications. Hayes *et al.* (27) stated that their methods were as described by Coffin *et al.* (10), but this paper does not describe treatment of samples. However, Hayes *et al.* stated that "all [water] samples were filtered before analysis" (27, p. 205) and F. R. Hayes (personal communication) confirmed that filter paper was used for this purpose. Thus the compartment usually called "lake water" by these authors included not only SP but also PP small enough to pass through filter paper. The reader may be further confused because the authors sometimes write as if the movement of ^{32}P were from the open water to *Sphagnum* (27, p. 208) and at others as if it were from SRP to all PP (27, p. 214).

Clearly the methods used do not allow one to draw quantitative conclusions about movements of P between open water and littoral or between epilimnion and hypolimnion. Nevertheless these papers convey the impression that the phosphorus cycle is dominated by an equilibrium exchange of P between the open water and solids associated with the littoral and benthic zones. This model, whether or not it is justified by the data, is quite different from Hutchinson's.

Finally Rigler (54) reported an experiment conducted in a bog-dystrophic lake. He used Hutchinson and Bowen's (32) method of compartmentalizing the lake, but in his experiment ^{32}P lost in the outflow and to the sediments below the epilimnion was measured. The only unmeasured compartment was the littoral zone. After 4 weeks 88% of the ^{32}P had left the epilimnion, 1.2% through the outlet, 11% to the deep water, and 3% to the sediments below the epilimnion. The 74% not recovered presumably had gone to the littoral zone. Since total P of epilimnion increased during this period Rigler postulated an equilibrium exchange system between the mobile phosphorus of both the water and plankton in the epilimnion and phosphorus of solids in the littoral, and calculated that the turnover time of the mobile P pool in the epilimnion was 3.6 days. This model was very similar to that proposed by Hayes *et al.* (27), but like its predecessor was not justified by the experimental data. The most serious error in this work, subsequently pointed out by Chamberlain (8), was that the kinetics of tracer loss from the epilimnion were not analyzed, and were inconsistent with the two-compartment exchange model.

Thus by 1956 the literature was burdened with three different models, each of which derived from inadequate experimental evidence and/or faulty analysis of the results.

Qualitative Model of the Phosphorus Cycle

Chamberlain (8) conducted an experiment in which he spread 2 Ci of $^{32}PO_4$ through the epilimnion of a dystrophic lake (Upper Bass Lake) with an area of 5.8 ha and maximum depth of 9 m. This experiment was superior to previous ones in several respects. First, Chamberlain added enough tracer to facilitate accurate measurement of the amount remaining in the epilimnion for 2 months. Second, he demonstrated that the epilimnion was in a steady state with respect to P, and measured the specific activity of several operationally defined subcompartments within the epilimnion. Finally, he analysed his and previous results more rigorously than had been done previously.

His contribution is relevant in that he showed that the kinetics of loss of ^{32}P from the epilimnion were inconsistent with those predicted by the simple, two-compartment exchange model. By reanalyzing Rigler's (54) results he showed that these, contrary to Rigler's interpretation, were also inconsistent with this model. His results (Fig. 6) showed that (*a*) loss of ^{32}P could not be described by a single exponential term [e.g., Eq. (1)] but was biphasic [Eq. (2)], and (*b*) the slower exponential loss continued until the end of each experiment. Thus it was no longer necessary to postulate exchange of P between the epilimnion and a single littoral compartment, but it was necessary to postulate a system comprising more than two compartments. Again there were many possible ways of modifying the model to conform to the observed kinetics, but two observations suggest that Chamberlain's approach of subdividing the epilimnion was reasonable. First, an assumption of compartmental analysis is that the compartment to which tracer is added is uniformly labeled at $t = 0$. Rigler's (54) and Chamberlain's (Fig. 3) results show that this does not hold for ^{32}P added to the epilimnion. Initially almost all of the $^{32}PO_4$ enters PP ($<70\ \mu$) and only after 5 to 10 days is PP ($>70\ \mu$) in tracer equilibrium. This alone suggests that the epilimnion cannot be treated as one compartment. Since the period when PP ($<70\ \mu$) had a high specific activity corresponded to that when loss of ^{32}P was rapid, Chamberlain proposed that P moved out of the epilimnion as PP and that PP ($<70\ \mu$) was lost much faster than PP ($>70\ \mu$). Second, Confer (12), while studying the phosphorus budget of 200-liter open-system ponds, observed similar kinetics of $^{32}PO_4$ loss from the open water to those in Fig. 6. By adding a variety of tracer particles to the open water he demonstrated a rapid removal of particles by the littoral (in this case a layer of filamentous alga with

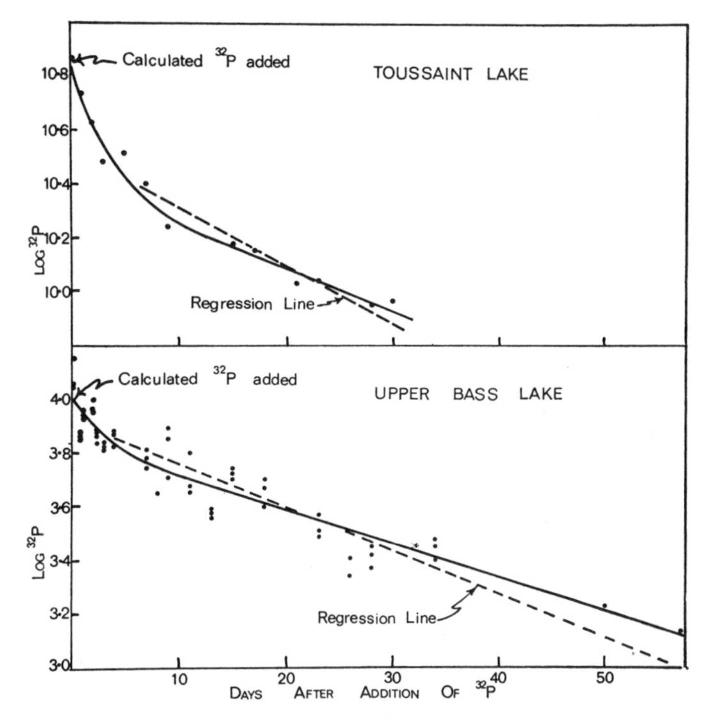

Figure 6. Loss of ^{32}P from the epilimnion of Toussaint Lake and Upper Bass Lake redrawn and modified from Chamberlain (8). The solid lines are fitted by eye. The regression lines are those calculated by Chamberlain, that from Toussaint Lake using only the data from day 7 to 30.

associated rotifers and protozoans), and found the highest instantaneous removal rate (3.85 day^{-1}) for the smallest particles.

In Fig. 7, Chamberlain's hypothesis concerning the gross circulation of phosphorus is shown along with the models of Coffin *et al.* (10), Hutchinson and Bowen (32), and Rigler (54). The littoral in Fig. 7*d* has been subdivided merely to indicate that there is no evidence for return of ^{32}P from the littoral to the epilimnion. However, exchange of P between the epilimnion and one littoral compartment is not ruled out, provided the amount of P in the exchanging littoral compartment is so much greater than that in the epilimnion that it would still have a negligible specific activity after 2 months. In this case return of ^{32}P to the epilimnion would also be unmeasurable.

Among the obvious flaws of the model in Fig. 7*d* is that it omits any flux of SP from the epilimnion to the littoral. This is omitted, not because the mechanism is improbable, but because evidence of this flux is missing. Certainly the work of Coffin *et al.* (10) and Hutchinson and Bowen (32)

A - Coffin et al, 1949

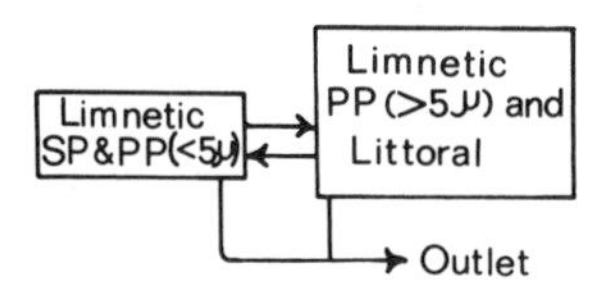

B - Hutchinson and Bowen, 1950

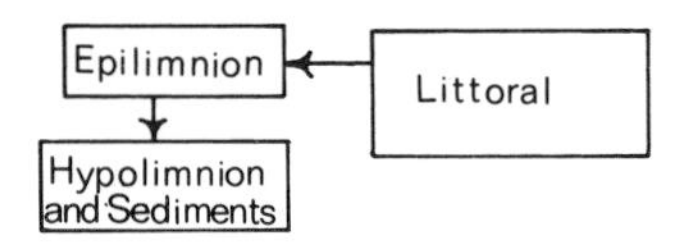

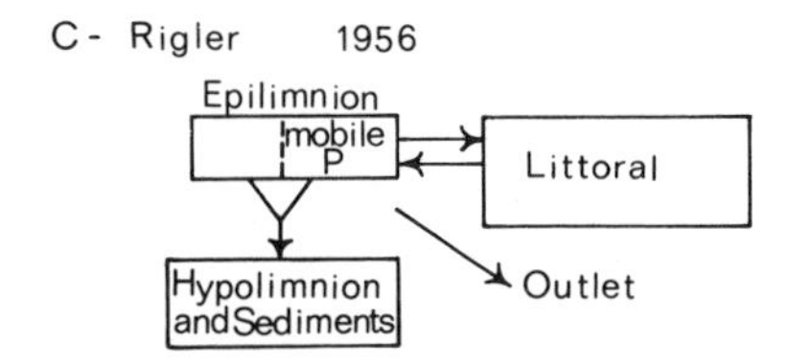

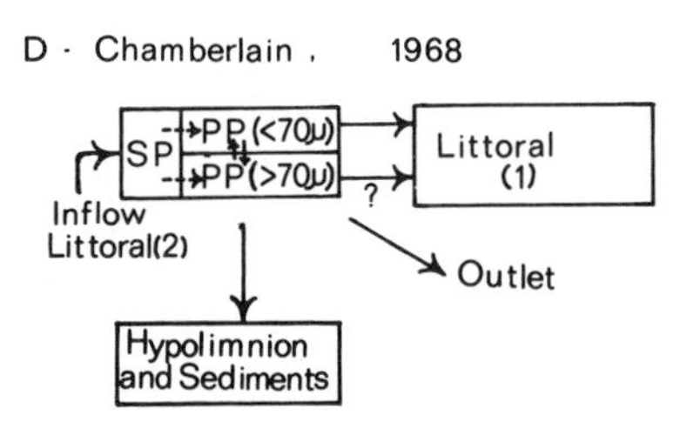

Figure 7. Four models that have been proposed for the circulation of phosphorus in lakes. Rigler's "mobile P" was, and remains, undefined. The cycling within the epilimnion in Chamberlain's model is much simplified. If an arrow does not touch a box, it indicated that the subcompartment from which the P comes is unknown. A second littoral compartment is included in Chamberlain's model although tracer data do not require it.

showed that littoral algae and macrophytes contained ^{32}P shortly after it was added to the water. Although these plants probably obtained the ^{32}P as SP there is as yet no way of distinguishing between direct uptake of limnetic SP and uptake of SP regenerated in the littoral from limnetic PP.

Another weakness of this model is that it does not specify the origin or nature of the P entering the epilimnion. Although there is almost no information on the nature of phosphorus entering seepage lakes,* an interesting

* Insufficient attention has been given to subsurface inflow to lakes generally. Brunskill and Ludlam (5) estimated that Fayetteville Green Lake with a drainage area of 4.3 km^2 receives more than 50% of its affluent by seepage. If the nature and quantity of inflowing P is to be evaluated, subsurface as well as surface affluents will need to be studied. Alternatively lakes in which paralimnetic conditions minimize the contribution of seepage will have to be chosen for study.

hypothesis was raised by Chamberlain to explain why the specific activity of SP ($<0.1\ \mu$) during his experiment in Upper Bass Lake was continuously much lower than that of other epilimnetic P compartments. He recognized that this discrepancy could be explained in several ways. There might be a systematic, positive error in the chemical analysis for SP ($<0.1\ \mu$), or much of the P in this compartment might not enter into the phosphorus cycle at all. However, he showed by a careful analysis of turnover times of P in seston and the epilimnion that a discrepancy equal to that observed could have been produced if all the P entering the epilimnion were unlabeled SP ($<0.1\ \mu$) exclusive of PO_4—P. This is an appealing hypothesis, but is, as yet, untested.

Movement of P between the epilimnion and deeper water is shown as being unidirectional. This is consistent with Hutchinson's model and the conclusions of McCarter *et al.* (45) who added ^{32}P to the hypolimnion of Teardrop Lake and found negligible quantities in the paper filtrate of surface water. However, a comparison of the results obtained by Coffin *et al.* (10), who had previously added the same quantity of ^{32}P to the surface of Teardrop Lake, with those of McCarter *et al.* (45), reveals some puzzling inconsistencies. If P flux through the metalimnion were unidirectional, one would expect no ^{32}P in epilimnetic organisms during the experiment of McCarter *et al.* (45). However, the experimental results did not fulfill this expectation. All epilimnetic organisms sampled were radioactive. Although sponges contained only 1/400 as much ^{32}P as during the experiment of Coffin *et al.*, *Sphagnum*, filamentous algae, and zooplankton contained about 1/14 as much. This suggests that a considerable amount of ^{32}P did reach the surface water during the experiment of McCarter *et al.*, perhaps in particulate form that would not be detected in filtered water. Unfortunately neither paper reported phosphorus content of the water during the experimental period. Consequently one cannot derive even a rough estimate of the quantity of P moving up and down. In the absence of good evidence it is still reasonable that more P moves down than up through the metalimnion. However, the unidirectional movement postulated by Hutchinson and depicted in Fig. 7*d* is not strictly correct.

Estimation of Rate Constants

One consequence of Chamberlain's work is that we must now treat a lake as a four-compartment system if we are to analyze the results of whole-lake tracer experiments with ^{32}P. Although the mathematics involved would presumably not be intractable, the existing data are inadequate to provide true rate constants of phosphorus movement out of the epilimnion to the littoral and hypolimnion. Furthermore, it appears that mere repetition of whole-lake tracer experiments will not provide these rate constants. Instead we need more studies designed specifically to elucidate the hypothetical pathways in the phosphorus cycle, to test existing models, and produce more realistic ones.

Although the rate constants of phosphorus loss from the epilimnion cannot be derived rigorously, it is useful to have rough approximations of them. Chamberlain attempted to obtain such values by making several simplifying assumptions. He first concluded, justifiably, that the turnover time of "mobile" phosphorus in the epilimnion, given as 3.6 days ($k_{out} = 0.28$ day^{-1}), is meaningless, and that the ultimate rather than the initial rate of tracer loss should be used to calculate k_{out}. In Table 1 Chamberlain's values for turnover time ($1/k_{out}$) of epilimnetic phosphorus are given, but the rate constants for loss to littoral and deeper water have been recalculated. The reasons for rejecting Chamberlain's figures and the oversimplifications required to derive the new values are given in the Appendix. Table 1 shows that the turnover time of P in the epilimnion ranges from 20 to 45 days. It is correlated inversely with area of lake and the subjective estimate of littoral vegetation. Unfortunately approximations of relative loss to the littoral and deeper water can only be made for two lakes, but these differ considerably. In Linsley Pond there was no loss to the littoral, and in Toussaint Lake an amount equal to 4% of the total P in the epilimnion moved into the littoral every day. Probably loss to the littoral is a function of the ratio between epilimnetic volume and biomass in the littoral zone. Confer's (12) observations concerning the phosphorus cycle in small (200 liter), artificial ponds are certainly consistent with this interpretation. His values for *net* daily movement of P from the open water to "littoral" were between 0.12 and 0.37, much higher than *gross* loss in Toussaint Lake. He also showed that loss of P to the sides increased as the biomass of attached, filamentous algae increased.

TABLE 1 Comparison of Three Lakes and the Calculated Rate Constants of P Transport in Them[a]

Lake and Source	Area (ha)	Littoral Vegetation (rank)	P Turnover Time (days)	Rate Constants		
				k_{out}	k_{EH}	k_{EL}
Toussaint (53)	4.7	1	20	0.05	0.01	0.04
Upper Bass (8)	5.8	2	27	0.04	—	—
Linsley Pond (32)	9.4	3	45	0.02	0.02	0

[a] The lakes are ranked in order of decreasing amount of littoral vegetation, subjectively estimated.

Downward movement of phosphorus, which is generally considered to be due to sedimentation of particulate matter such as algal cells and fecal pellets, was more similar in the two lakes. In Toussaint Lake the daily loss of P was 1% and in Linsley Pond it was 2%. These are similar to the values obtained by Bosch for Horew Bay of Lake Lucerne from analyses of the phosphorus content of sediment traps (19).* Bosch found that 1.4 to 2.5% of the P in the trophogenic zone was sedimented daily. It is interesting that the fraction lost daily to deeper water is so similar in different types of lakes, and furthermore that a loss of P proportional to the total amount present in the water was one of the assumptions made by Vollenweider (66) while developing his model for the phosphorus budget of lakes.

The similarity of the rate constants of P loss to the deeper water is also consistent with and might have been predicted from, the results obtained by Hutchinson (28). He showed that the areal hypolimnetic O_2 deficit in four lakes was directly proportional to the standing crop, as dry weight, of seston. This close correlation would only be expected if the fraction of seston daily sinking into the hypolimnion and decomposing were the same in lakes of different trophic status. Furthermore, Hutchinson calculated the rate constant of loss of seston, measured in calories, to be 0.031 and 0.033 day^{-1} in the two lakes for which adequate data existed. The similarity between these values and those calculated for phosphorus is certainly encouraging.

REFERENCES

1. Ambühl, H., and M. Schmid, "Die Bestimmung geringster Mengen von Phosphation im Wasser von Binnenseen," *Schweiz. Z. Hydrol.*, **27**, 172–183 (1965).
2. Arthur, C. R., and F. H. Rigler, "A Possible Source of Error in the ^{14}C Method of Measuring Primary Productivity," *Limnol. Oceanogr.*, **12**, 121–124 (1967).
3. Barlow, J. B., and J. W. Bishop, "Phosphate Regeneration by Zooplankton in Cayuga Lake," *Limnol. Oceanogr.*, *Suppl.*, **10**, R15–R24 (1965).
4. Berman, T., "Alkaline Phosphatases and Phosphorus Availability in Lake Kinneret," *Limnol. Oceanogr.* **15**, 663–674 (1970).
5. Brunskill, G. J., and S. D. Ludlam, "Fayetteville Green Lake, New York, I. Physical and Chemical Limnology," *Limnol. Oceanogr.*, **14**, 817–829 (1969).
6. Burns, C. W., "The Relationship between Body Size of Filter-Feeding *Cladocera* and the Maximum Size of Particle Ingested," *Limnol. Oceanogr.*, **13**, 675–678 (1968).
7. Butler, E. I., E. D. S. Corner, and S. M. Marshall, "On the Nutrition and Metabolism of Zooplankton, VI. Feeding Efficiency of *Calanus* in Terms of Nitrogen and Phosphorus," *J. Mar. Biol. Assoc. U. K.*, **49**, 977–1001 (1969).
8. Chamberlain, W. M., "A Preliminary Investigation of the Nature and Importance

* Gächter reported these results as percentage of PP. They were converted to the percentage of total P on the assumption that the PP/TP ratio was similar in 1966 and 1967.

of Soluble Organic Phosphorus in the Phosphorus Cycle of Lakes," Ph.D. Thesis, University of Toronto, Canada (Microfilm, National Library of Canada), 1968, 232 pages.

9. Chamberlain, W. M., and J. Shapiro, "On the Biological Significance of Phosphate Analysis; Comparison of Standard and New Methods with a Bioassay," *Limnol. Oceanogr.*, **14,** 921–927 (1969).
10. Coffin, C. C., F. R. Hayes, L. H. Jodrey, and S. G. Whiteway, "Exchange of Materials in a Lake as Studied by the Addition of Radioactive Phosphorus", *Can. J. Res. D*, **27,** 207–222 (1949).
11. Colebrook, J. M., "Some Observations of Zooplankton Swarms in Windermere," *J. Anim. Ecol.*, **29,** 241–242 (1960).
12. Confer, J. L., "The Inter-Relationships among Plankton, Attached Algae, and the Phosphorus Cycle in Artificial Open Systems," Ph.D. Thesis, University of Toronto, Canada, 1969, 129 pages.
13. Conover, R. J., "Food Relations and Nutrition of Zooplankton," in *Symp. Exp. Mar. Ecol.*, Occasional Publ. 2, University of Rhode Island, Kingston, Rhode Island, 1964, pp. 81–91.
14. Conover, R. J., "Assimilation of Organic Matter by Zooplankton," *Limnol. Oceanogr.*, **11,** 338–345 (1966).
15. Cooper, L. H. N., "The Rate of Liberation of Phosphate in Seawater by the Breakdown of Plankton Organisms," *J. Mar. Biol. Assoc. U. K.*, **20,** 197–200 (1935).
16. Corner, E. D. S., C. B. Cowey, and S. M. Marshall, "On the Nutrition and Metabolism of Zooplankton, V. Feeding Efficiency of *Calanus finmarchicus*," *J. Mar. Biol. Assoc. U. K.*, **47,** 259–270 (1967).
17. Einsele, W., "Die Umsetzung von zugeführten, anorganischen Phosphat im eutrophen See und ihre Rückwirkung auf seinen Gesamthaushalt," *Z. Fish.*, **39,** 407–488 (1941).
18. Fuhs, G. W., and E. Canelli, "Phosphorus-33 Autoradiography Used to Measure Phosphate Uptake by Individual Algae," *Limnol. Oceanogr.*, **15,** 962–967 (1972).
19. Gächter, R., "Phosphorhaushalt und planktische Primärproduktion im Vierwaldstättersee (Horwer Bucht)," *Schweiz. Z. Hydrol.*, **30,** 1–66 (1968).
20. Gardiner, A. C., "Phosphate Production by Planktonic Animals," *J. Cons. Int. Explor. Mer.*, **12,** 144–146 (1937).
21. Gliwicz, Z. M., "The Share of Algae, Bacteria and Trypton in the Food of the Pelagic Zooplankton of Lakes with Various Trophic Characteristics," *Bull. Acad. Pol. Sci.*, **17,** 159–165 (1969).
22. Guillard, R. L., and P. J. Wangersky, "The Production of Extracellular Carbohydrates by Some Marine Flagellates," *Limnol. Oceanogr.* **3,** 449–454 (1958).
23. Hall, D. J., "An Experimental Approach to the Dynamics of a Natural Population of *Daphnia galeata mendotae*," *Ecol.*, **45,** 94–112 (1964).
24. Haney, J. F., "Seasonal and Spatial Changes in the Grazing Rate of Limnetic Zooplankton," Ph.D. Thesis, University of Toronto, Canada, 1970, 176 pages.
25. Hansen, A. L., and R. J. Robinson, "The Determination of Organic Phosphorus in Sea Water with Perchloric Acid Oxidation," *J. Mar. Res.*, **12,** 31–42 (1953).
26. Hargrave, B. T., and G. H. Geen, "Phosphorus Excretion by Zooplankton," *Limnol. Oceanogr.*, **13,** 332–342 (1968).
27. Hayes, F. R., J. A. McCarter, M. L. Cameron, and D. A. Livingstone, "On the Kinetics of Phosphorus Exchange in Lakes," *J. Ecol.*, **40,** 202–216 (1952).

28. Hutchinson, G. E., "On the Relation between the Oxygen Deficit and the Productivity and Typology of Lakes," *Int. Rev. Hydrobiol.*, **36,** 336–355 (1938).
29. Hutchinson, G. E., "Limnological Studies in Connecticut, IV. Mechanism of Intermediary Metabolism in Stratified Lakes," *Ecol. Monogr.*, **11,** 21–60 (1941).
30. Hutchinson, G. E., *A Treatise on Limnology*, Vol. 1. Wiley, New York, 1957, 1015 pages.
31. Hutchinson, G. E., and V. T. Bowen, "A Direct Demonstration of the Phosphorus Cycle in a Small Lake," *Proc. Natl. Acad. Sci. Washington*, **33,** 148–153 (1947).
32. Hutchinson, G. E., and V. T. Bowen, "Limnological Studies in Connecticut, IX. A Quantitative Radiochemical Study of the Phosphorus Cycle in Linsley Pond," *Ecol.*, **31,** 194–203 (1950).
33. Johannes, R. E., "Phosphorus Excretion and Body Size in Marine Animals: Microzooplankton and Nutrient Regeneration," *Science*, **146,** 923–924 (1964).
34. King, P. H., "A Test of the Hypothesis that Vacuum Filtration of Lakewater Releases Orthophosphate," M.Sc. Thesis, University of Toronto, Canada, 1970, 82 pages.
35. Kraus, K. A., H. O. Phillips, T. A. Carlson, and J. S. Johnson, "Ion-Exchange Properties of Hydrous Oxides," in *Proc. Int. Conf. Peaceful Uses of Atomic Energy*, **28,** 3–16 (1958).
36. Kuenzler, E. J., and B. H. Ketchum, 1962. "Rate of Phosphorus Uptake by *Phaeodactylum tricornutum*," *Biol. Bull., Woods Hole*, **123,** 134–145 (1962).
37. Kuenzler, E. J., and J. P. Perras, "Phosphatases of Marine Algae," *Biol. Bull., Woods Hole*, **128,** 271–284 (1965).
38. Langford, R. R., and E. G. Jermolajev, "Direct Effect of Wind on Plantkon Distribution," *Verh. Int. Verein. Limnol.*, **16,** 188–193 (1965).
39. Lasker, R., and R. Holmes, "Variability in Retention of Marine Phytoplankton by Membrane Filters," *Nature*, **180,** 1295–1296 (1957).
40. Levesque, M., and M. Schnitzer, "Organometallic Interactions in Soils: 6. Preparation and Properties of Fulvic Acid–Metal Phosphates," *Soil Sci.*, **103,** 183–190 (1967).
41. Mackiewicz-Gołachowska, J., "Fosfór w wodzie jezior Polski postaci, ilośei i stratifikacja," *Zesz. Nauk. Wyzsz. Szk. Roln. Wrocławiv*, **15,** 191–205 (1968).
42. Margalef, R., "Role des entomostraces dans la regeneration des Phosphates," *Verh. Int. Verein. Limnol.*, **11,** 246–247 (1951).
43. Marshall, S. M., and A. P. Orr, "On the Biology of *Calanus finmarchicus*, VIII. Food Uptake, Assimilation, and Excretion in Adult and Stage V *Calanus*," *J. Mar. Biol. Assoc. U. K.*, **34,** 495–529 (1953).
44. McAllister, C. D., "Decontamination of Filters in the C^{14} Method of Measuring Marine Photosynthesis," *Limnol. Oceanogr.*, **6,** 447–450 (1961).
45. McCarter, J. A., F. R. Hayes, L. H. Jodrey, and M. L. Cameron, "Movement of Materials in the Hypolimnion of a Lake as Studied by the Addition of Radioactive Phosphorus," *Can. J. Zool.*, **30,** 128–133 (1952).
46. Olsen, S., "Recent Trends in the Determination of Orthophosphate in Water," in *Proceedings of an I.B.P. Symposium held in Amsterdam and Nieuwersluis*, 1967, pp. 63–105.
47. Overbeck, J., "Untersuchungen zum Phosphathaushalt von Grünalgen, II. Die Verwertung von Pyrophosphat und organisch gebundenen Phosphaten und ihre Beziehung zu den Phosphatasen von *Scenedesmus Quadricauda* (Turp.) Bréb," *Arch. Hydrobiol.*, **58,** 281–308 (1962).

48. Overbeck, J., and H. Babenzien, "Über den Nachweis von freien Enzymen im Gewässer," *Arch. Hydrobiol.*, **60**, 107–114 (1964).
49. Patalas, K., "Composition and Horizontal Distribution of Crustacean Plankton in Lake Ontario," *J. Fish. Res. Board Can.*, **26**, 2135–2164 (1969).
50. Pomeroy, L. R., "Experimental Studies of the Turnover of Phosphate in Marine Environments," in *Radioecology*, Schultz and A. W. Klement, Jr. (Eds.), Reinhold, New York, and A. I. B. S., Washington, 1963, pp. 163–166.
51. Ragotzkie, R. A., and R. A. Bryson, "Correlation of Currents with the Distribution of Adult *Daphnia* in Lake Mendota," *J. Mar. Res.*, **12**, 157–172 (1953).
52. Riggs, D. S., *The Mathematical Approach to Biological Problems*, Williams and Wilkins, Baltimore, Maryland, 1963, 445 pages.
53. Rigler, F. H., "The Circulation of Phosphorus in Lakewater," Ph.D. Thesis, University of Toronto, Canada, 1954, 117 pages.
54. Rigler, F. H., "A Tracer Study of the Phosphorus Cycle in Lakewater," *Ecol.*, **37**, 550–562 (1956).
55. Rigler, F. H., "The Uptake and Release of Inorganic Phosphorus by *Daphnia magna* Straus," *Limnol. Oceanogr.*, **6**, 165–174 (1961).
56. Rigler, F. H., "The Phosphorus Fractions and Turnover Time of Phosphorus in Different Types of Lakes," *Limnol. Oceanogr.*, **9**, 511–518 (1964).
57. Rigler, F. H., "Radiobiological Analysis of Inorganic Phosphorus in Lakewater," *Verh. Int. Verein. Limnol.*, **16**, 465–470 (1966).
58. Rigler, F. H., "Further Observations Inconsistent with the Hypothesis that the Molybdenum Blue Method Measures Orthophosphate in Lakewater," *Limnol. Oceanogr.*, **13**, 7–13 (1968).
59. Rigler, F. H., "The Char Lake Project. A Study of Energy Flow in a High Arctic Lake," in *Productivity Problems in Freshwaters*, Z. Kajak and A. Hillbricht-Ilkowska (Eds.), PNW, Warszawa, 1972, pp. 278–300.
60. Schröder, R., "Untersuchungen über die Planktonverteilung mit Hilfe der Unterwasser-Fernsehanlage und des Echographen," *Arch. Hydrobiol.*, *Suppl.*, **25**, 228–241 (1961).
61. Solomon, A. K., "Compartmental Methods of Kinetic Analysis," in *Mineral Metabolism*, Vol. 1 (A), C. L. Comar and F. Bronner (Eds.), Academic Press, New York, 1960, pp. 119–167.
62. Steiner, M., "Zur Kenntnis des Phosphatkreislaufes in Seen," *Naturwiss.*, **26**, 723–724 (1938).
63. Strickland, J. D. H., and T. R. Parsons, "A Manual of Seawater Analysis," *Bull. Fish. Res. Board Can.*, **125**, 1–185 (1960).
64. Tappa, D. W., "The Dynamics of the Association of Six Limnetic Species of *Daphnia* in Azicoos Lake, Maine," *Ecol. Monogr.*, **35**, 395–423 (1965).
65. Van Liew, H. D., "Semilogarithmic Plots of Data Which Reflect a Continuum of Exponential Processes", *Science*, **138**, 682–368 (1962).
66. Vollenweider, R. A., *Possibilities and Limits of Elementary Models Concerning the Budget of Substances in Lakes* (translated from German), Canada Sec. State-tr-0343, 1969, pp. 1–53.
67. Weil-Malherbe, H., and R. H. Green, "The Catalytic Effect of Molybdate on the Hydrolysis of Organic Phosphate Bonds," *Biochem. J.*, **49**, 286–292 (1951).

APPENDIX

It is not possible to calculate rigorously the rate constants of ^{32}P transport from the epilimnion to littoral and hypolimnion from the experimental results of Hutchinson (32), Rigler (54), and Chamberlain (8). In every case simplifying and probably unjustified assumptions must be made. Chamberlain (8) led the way in estimating these rate constants, but some of his methods do not seem to be the most desirable. The discussion herein shows why I have rejected some of his figures and the simplifying assumptions made in calculating new ones.

Although Linsley Pond appears to be the simplest case, Hutchinson and Bowen's (32) results present some interesting problems. First, there is uncertainty about the amount of ^{32}P added to the lake (see the subsection on early tracer studies), and second, the amount of ^{32}P in the open water should have decreased or remained constant; instead, it increased from August 1 to August 22. To overcome these problems I prefer to reject the results of the first week. The added advantage of this procedure is that the analysis is only applied to the period when the epilimnetic P subcompartments are likely to be uniformly labeled.

From August 1 to 22, epilimnetic ^{32}P first decreased and then increased. Since changes of this sort are not consistent with any steady-state model it is assumed that they were introduced by sampling errors and that it is more reasonable to assume that epilimnetic ^{32}P remained constant from August 1 to 22 at the average value of 614×10^6 counts/min. Since ^{32}P was increasing in the hypolimnion, I am essentially making the unlikely assumption that a movement of P, with the same specific activity as that in the epilimnion, into the epilimnion exactly balanced loss to the hypolimnion. Thus the exponential equations of compartmental analysis do not apply since ^{32}P in epilimnion remains constant and that in the hypolimnion will increase linearly. If Hutchinson and Bowen's value of 16×10^6 counts/(min)(day) is used as the rate of accumulation of ^{32}P in the hypolimnion, then the rate constant of transport to the hypolimnion is $16 \times 10^6 \div 614 \times 10^6 = 0.026$ day^{-1}.

This value differs from Chamberlain's (0.032 day^{-1}) because he assumed exponential loss of ^{32}P from epilimnion to hypolimnion, and included in his calculation the loss during the first week. Since total ^{32}P in the water did not change from August 1 to 22, Chamberlain concluded justifiably that the rate constants of transport to the littoral and through the outlet were both zero.

To calculate the rate constant for total transport of ^{32}P out of the epilimnion of Toussaint Lake and Upper Bass Lake, Chamberlain assumed that the slope of the ln-linear regression line for loss of ^{32}P from the epilimnion is equivalent to the rate constant. (For Upper Bass Lake he used all the points to calculate the regression line since the initial more rapid loss was not very obvious. For Toussaint Lake he used the data from day 5 to 30. I have used his regression lines but as indicated in Fig. 6 it might have been better to use Toussaint Lake data from day 7 to 30.) This assumption would be valid only if the epilimnion could be treated as a single compartment uniformly labeled from $t = 0$, and this is exactly what Chamberlain showed we can no longer do. Thus the calculated rate constant is only a first approximation at best, but until we have more comprehensive models of the phosphorus cycle and better data, a more rigorous treatment is impossible.

To estimate the rate constants of ^{32}P transport into the littoral and hypolimnion separately, Chamberlain assumed that transport of ^{32}P to the hypolimnion would be proportional to the concentration of ^{32}P in the epilimnion.

Thus

$$\frac{d{}^*P_H}{dt} = k_{EH}{}^*P_E \tag{1}$$

where k_{EH} is the rate constant of transport out of the epilimnion to the hypolimnion and ${}^*P_E$ and ${}^*P_H$ are the amounts of ^{32}P in the epilimnion and hypolimnion. (Actually the term "hypolimnion" is being used incorrectly and refers to all water and sediments below the epilimnion.) Since the epilimnetic ^{32}P, over that period during which it can be adequately described by the regression line, is given by

$$ {}^*P_{E(t)} = {}^*P_{E(0)}e^{-kt} \tag{2}$$

in which k is the rate constant of total transport of ^{32}P out of the epilimnion, we can substitute for ${}^*P_E$ in Eq. (1) to obtain

$$\frac{d{}^*P_H}{dt} = k_{EH}{}^*P_{E(0)}e^{-kt} \tag{3}$$

On integration with the condition that ${}^*P_H$ is zero at $t = 0$, Eq. (3) gives

$$k_{EH} = \frac{{}^*P_{H(t)}}{{}^*P_{E(0)}}\left(\frac{k}{1 - e^{-kt}}\right) \tag{4}$$

Thus k_{EH} can be obtained from the experimental data and the rate constant of transport from epilimnion to the littoral will be $k - k_{EH}$, provided loss of ^{32}P through the outlet is negligible.

To solve for k_{EH} in Eq. (4), Chamberlain used the $t = 0$ value from the regression line, and the amount of ^{32}P in the hypolimnion some weeks after addition of ^{32}P. I believe this procedure exaggerates the rate constant of transport into the hypolimnion since it assumes that rate of movement of ^{32}P will initially be proportional, not to the ^{32}P actually in the epilimnion, but to the lesser value given by the regression line. Also, it uses data from the period before the epilimnion is uniformly labeled. Since Hutchinson and Bowen (32) showed that five times as much ^{32}P accumulated in the hypolimnion of Linsley Pond during the first week as in any successive week it is probably advisable to use only data from the period after the PP ($>70\ \mu$) has become fully labeled.

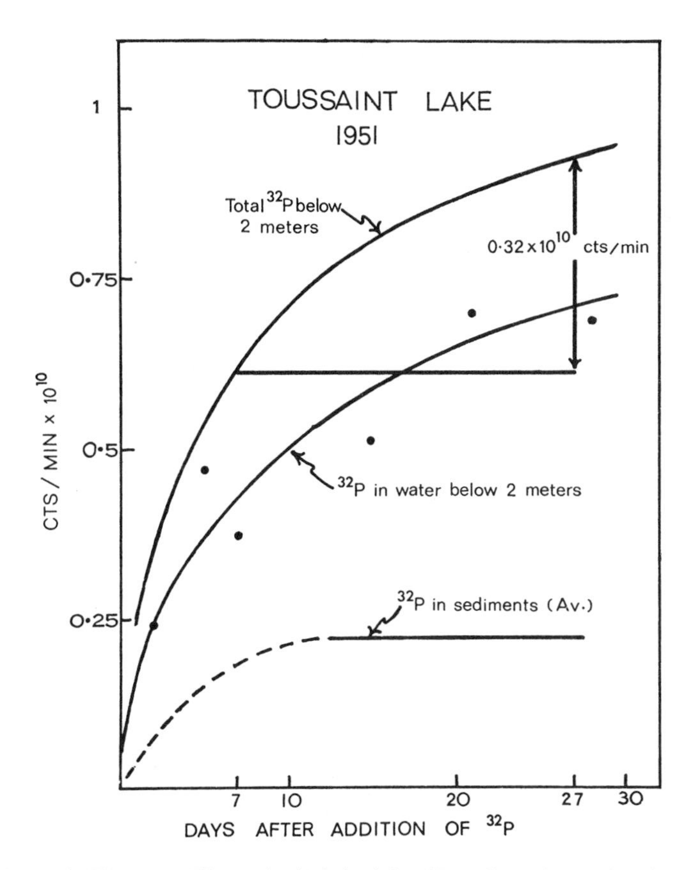

Appendix Figure 1. The method of obtaining $*P_{H(t)}$ from the results of Rigler's experiment in Toussaint Lake. The constant value for ^{32}P in sediments after day 12 is the average of values determined on three dates. The curvilinear increase to a maximum on day 12, the first day on which sediment was samples, is assumed. The upper curve is simply the sum of the lower two. The procedure of setting the ^{32}P in the hypolimnion at zero on day 7 is valid only if movement of ^{32}P between epilimnion and hypolimnion is essentially unidirectional.

For these reasons I prefer not to use Chamberlain's values. They were recalculated for Toussaint Lake as follows. The amount of ^{32}P in the epilimnion at $t = 0$ ($*P_{E(0)}$) was obtained from Chamberlain's regression line at day 7. Since his equation is log $*P_E = 10.53 - 0.021t$, $*P_{E(0)}$ is 2.44×10^{10} counts/min. Since the amount of ^{32}P in the hypolimnion at $t = 0$ must be zero, it is necessary to find the amount in the hypolimnion at $t = 0$ and subtract it from the amount there on the date chosen as t. This was done as shown in Appendix Fig. 1. Thus $*P_{H(t)}$ is 0.32×10^{10} counts/min, k is $0.021 \times 2.303 = 0.048$ day^{-1}, and

$$k_{EH} = \frac{0.32}{2.44}\left(\frac{0.048}{1 - e^{-0.048\times 20}}\right) = 0.0103 \text{ day}^{-1}$$

Since k is 0.048 day^{-1}, then k_{EL} is $0.048 - 0\,0103 = 0.038$ day^{-1}.

Since Chamberlain did not measure ^{32}P in the hypolimnion until near the end of his experiment, the calculations above cannot be applied to the data from Upper Bass Lake.

ACKNOWLEDGMENTS

This manuscript was prepared while I held a research grant from the National Research Council of Canada. I am grateful to Robert Peters and David Lean for suggestions and permission to use unpublished data. William Chamberlain and John Confer kindly read the manuscript, made many corrections, and clarified some of my most obscure passages.

31

Phosphorus in Lake Sediments

P. R. HESSE

Senior Technical Officer,
Food and Agriculture Organization of the United Nations, Rome, Italy

Although the purists among pedologists will disagree, lake sediments are merely special kinds of soil; the principal and most important differences from "ordinary" soils are that sediments are permanently waterlogged and are being continuously regenerated by deposition. In shallow and small lakes the sediments are formed largely from material brought in from the shoreline, surrounding swamps, and drainage basins. In deeper and large lakes the deposits are produced mainly internally from dead organisms and undigested organic material. Thus in very deep and extensive lakes the phosphorus content of the sediments will depend largely on a phosphorus cycle between mud, water, vegetation, and animals, whereas small and shallow lakes receive fresh supplies of phosphorus from external sources. The normal balance of biological processes in lakes is such as to immobilize phosphorus from solution faster than it is released by the decomposition of organic matter. Thus a deficiency of phosphorus can arise unless sufficient quantities—in inorganic forms—are supplied by incoming material. The consequence is that shallow and small lakes normally contain more available forms of phosphorus than do deep or very large lakes.

The nature and amount of phosphorus compounds in lake sediments will be influenced by the composition of the incoming materials, the climatic conditions, and the chemical and biological consequences of waterlogging.

INFLUENCE OF INCOMING MATERIAL

Effects of Parent Material

The composition of the incoming materials to a lake depends on the nature of their origin, that is, the geology, soils, and vegetation of the drainage basin. Whether the soils of the drainage basin are calcareous, saline, alkali, or acid will affect the nature of the lake sediments and, in particular, the nature of the phosphorus compounds introduced. For example, the bottom deposit of Lake Magadi in Kenya is not likely to receive much iron or aluminum phosphate from the surrounding calcareous soils, whereas such forms of phosphorus could be expected in a lake with a lateritic drainage basin. Organically bound phosphorus will be brought into lakes in soluble organic matter and in partially and undecomposed vegetable and animal material; most of this organic phosphorus will accumulate as such in the sediment. If the surrounding land is cultivated, this too will influence the composition of a lake mud and in general a cultivated drainage basin results in an increase in productivity of a lake; this increase is most often due mainly to phosphorus.

Effects of Lake Size

The depth and area of a lake influence the nature of its bottom deposit because heavier particles tend to settle in shallow water. Furthermore shallow, wide lakes are more subject to wind and wave effects, and as the sediment is carried by water movement, the distribution of particles will be different in lakes of different sizes and depths. Very shallow lakes usually have more vegetation around the edges than do deep lakes and this results in organic matter accumulation with a slow rate of decomposition; in extreme cases swamps and peat bogs can result. In contrast, the bottom deposits of very deep and large lakes are formed not so much by river or flood-borne materials but from dead plankton. These facts all affect the nature and distribution of phosphorus in lake muds.

The general experience of high productivity being associated with shallow lakes has been attributed to such factors as more light for algal growth, nutrient release during seasonal overturns, and the fertility of the drainage basins. In nearly every case, however, one of the main, immediate reasons for high productivity has proved to be an enrichment of the lake with phosphorus. Within shallow lakes the characteristics associated with high productivity have been found to increase with water depth. Thus concentrations of nutrients such as phosphorus and nitrogen occur in the center of a lake bed; this is further discussed in the section on inorganic phosphorus compounds in lake sediments.

INFLUENCE OF CLIMATE

Climate indirectly affects the nature of lake sediments in several ways. The composition of muds in tropical lakes can naturally be expected to differ somewhat from that of lakes in temperate or cold climates. Climate influences the nature of the incoming materials, particularly with regard to the effects of floods and erosion. Large and frequent floods result in modification of lakes by, inter alia, delta formation with the concomitant effects of excess silt and organic matter accumulation. As is discussed later in the chapter, different inorganic forms of phosphorus tend to be associated with certain-sized soil particles and thus particular forms may accumulate with silt. Organic phosphorus will accumulate as part of the organic matter and, quite possibly, will never reach the lake bed proper.

In lakes that experience seasonal changes of climate there is a cycle of water stratification and mixing. The surface water becomes less dense in the warm summer months and is mixed by wind action with the lower layers to give what is called the "epilimnion." The epilimnion is separated by a thermocline from the deeper, dense, and more cold "hypolimnion." During the ensuing autumn and winter months winds mix the epilimnion and the hypolimnion to give water of uniform temperature and oxygen content. This cycle, first described and investigated by Mortimer (11), has important effects upon, among other things, the phosphorus distribution in lakes. Decomposition of organic matter in the sediment during summer reduces the concentration of oxygen in the hypolimnion and so, in a shallow lake, carbon dioxide, methane, hydrogen sulfide, and other products of reduction accumulate; in deep lakes with thick hypolimnions, the oxygen removal is not so serious. On the other hand, in deep lakes there is a seasonal fluctuation in oxygen diffusion from the water to the mud surface. Oxygen entering the mud is rapidly consumed and a shallow depth of oxidized mud results. Biological activity in such a layer of oxidized mud can be very high due to a high oxygen uptake by products of anaerobic decomposition in lower layers diffusing upward, and this can affect the release and fixation of phosphorus. In some lakes the oxygen in the hypolimnion is depleted before it can diffuse into the mud and the oxidized-reduced boundary occurs in the water. Mortimer followed such changes by measurement of oxidation-reduction potentials and for the system he studied he was able to assign a definite *Eh* value of 0.2 V at pH 7 to the boundary.

In many lakes sedimentation of inorganic phosphorus has been found to occur only at the time of the autumn overturn when the hypolimnion and epilimnion are mixed and, moreover, the phosphorus so deposited is derived from that already in the mud by reduction during the period of stratification.

CONSEQUENCES OF PERMANENT WATERLOGGING

In an unsaturated soil the fine pores may or may not contain water but the large pores are filled with air. When both the fine and large pores are completely filled with water, the soil is said to be waterlogged. The chemical properties of a waterlogged soil depend mainly on the fact that it is oxygen deficient, which leads to the dominance of facultive and obligate anaerobes. The anaerobes utilize oxidized substances in the medium as hydrogen acceptors and the system rapidly becomes reducing. However, it is possible for a waterlogged soil to be oxidizing at its immediate surface if the overlying water is oxygenated, and this is the case with lake muds when the epilimnion and hypolimnion have been mixed. During the period of stratification and oxygen depletion of the hypolimnion, the oxidized surface of the mud is reduced and this results in the liberation of certain previously immobile elements and radicals.

Mortimer (11) found that reduction of a waterlogged mud surface liberated into solution ammonium, hydrogen sulfide, sulfides, manganese(II), iron(II), organic compounds, silicates, and phosphates. Most of these ions are normal products of reduction but silicate and phosphate are not, and Mortimer explained their liberation as being due to reduction of iron(III). Iron(II) ions diffuse upward through the mud surface where they are oxidized to form a thin layer of iron(III) hydroxide and iron(III) humate complexes. These compounds accumulate as a flocculent layer and strongly adsorb anions such as phosphate. When the complexes are destroyed by reduction of the iron, phosphate and other ions are released into the water and free exchange of ions between water and mud by diffusion recommences. Any hydrogen sulfide formed by reduction of sulfur compounds in the mud would also favor the solution of iron(III) phosphate. Calcium phosphates and aluminum–phosphorus complexes are not affected by reduction and so the phosphorus in a lake mud made available during periods of reduction is mainly that which is adsorbed onto iron complexes. It can be shown that when oxidation-reduction potentials fall below the limiting value of +200 mV, extractable phosphorus in lake sediments increases greatly and this redox value corresponds to the reduction of iron(III). In antithesis, there is some evidence that lowering the Eh value of calcareous and alkaline sediments reduces the solubility of phosphorus.

Sometimes an oxygen-depleted, waterlogged mud overlays an oxygenated mud, and in such cases soluble complexes of iron and manganese leached downward are precipitated; thus localized deposits of phosphate can be found in deep mud profiles. Normally, however, the phosphorus content of lake muds is remarkably constant with depth. Mineralization of organic phosphorus compounds is retarded by waterlogged conditions and, as most phosphorus compounds are almost insoluble, movement by leaching is small and far less important than in the case of nitrogen.

FORMS OF PHOSPHORUS IN LAKE SEDIMENTS

Inorganic Compounds

Phosphorus exists in all soils in different forms according to the pH, Eh, organic matter content, and so on. Inorganically, phosphorus is typically combined with iron, aluminum, calcium, and fluoride and is adsorbed as the anion onto clay minerals and organic complexes.

In acid sediments phosphorus is combined largely with iron and aluminum, and in neutral or alkaline sediments calcium phosphate predominates. Inorganic phosphorus added to lake waters rapidly appears as one or more of these forms unless it is taken up by organisms. Frink (4), investigating the forms and amounts of phosphorus in the sediment of a eutrophic lake, found that phosphorus combined with aluminum and iron increased in amount with depth of overlying water, whereas calcium phosphate decreased with water depth. This distribution was almost certainly a result of the differential deposition of the various mineral fractions of the incoming material. Aluminum and iron phosphates were associated with the finer soil particles and were thus carried by water movement to the deeper parts of the lake before settling; calcium phosphate was associated with the coarse particles and so was deposited in the shallow water.

Frink further found that there was an overall shift from aluminum phosphate to iron phosphate in the acid watershed soils, but that in the neutral lake deposit aluminum phosphate tended rather to convert to calcium phosphate. Frink thus deduced that not all the calcium phosphate found in the mud had been brought in as part of the sand fraction but that some was due to chemical transformation. The changes in form of phosphate occurred rapidly in the soils brought down from the watershed and in the phosphorus compounds released by biological activity. Similar results have been reported for other eutrophic lakes.

The chemical transformation of aluminum phosphate to iron phosphate in waterlogged soils is a common phenomenon (2, 7, 14) and has been tentatively explained as being due to the lower solubility product of iron phosphate. Conversion of aluminum phosphate to calcium phosphate is less well known and, in acid sediments, it is more likely that calcium phosphate would convert to iron phosphate (7). Calcium phosphates in nonacid lake muds most probably undergo hydrolysis and accumulate in the stable form of hydroxyapatite.

Hutchinson and Wollack (9) analyzed a deep core of mud taken from a lake in Connecticut and deduced that in early (postglacial) times the phosphorus must have entered the lake in association with aluminum in an easily eroded mineral. Later in the history of the lake and at depths of mud of less than about 13 m, the phosphorus content increased with nitrogen and was presumed to be organic in origin. Hutchinson found that a large part of the phosphorus assimilated by planktonic organisms was derived from the lake sediment and was returned to the sediment as organic phosphorus in the dead plankton. An hypothesis was advanced to explain the diffusion of phosphorus from the sediment by means other than molecular diffusion. The latter can occur only as phosphate when iron(III) is reduced, which with the pH of the lake in question was inevitably followed by precipitation of iron(III) phosphate again. Thus it was suggested that phosphorus could diffuse from the mud to the water by the metabolic activities of benthic organisms (e.g., *Tubifex* which regulated the metabolism of the lake) resulting in excretion of phosphorus into the water independently of iron.

Organic Compounds

Organic compounds of phosphorus originate in lake sediments partly from plant and animal remains and partly from microbial synthesis from inorganic phosphorus compounds. Organic matter, regardless of its nature, entering a lake will eventually deposit phosphorus largely as organic compounds that will mineralize very slowly. The organic phosphorus in peats and humus, for example, is highly resistant to microbial degradation; this may be due to complex formation, but little is known about the matter. Only the phosphorus in inorganic forms will enter, by microbial activity, the phosphorus cycle in a lake. Phosphorus thus differs from nitrogen in this respect. Plants normally contain little inorganic nitrogen and mineralization of organic forms is necessary for decomposition to proceed. With phosphorus, on the other hand, the inorganic forms present in the plant are usually adequate for microbial needs during decomposition and thus the organic matter can decompose independently of phosphorus mineralization. This has been demonstrated for aerobic decomposition (1) and there is no reason why it should not be valid for anaerobic decomposition. The important facts are the extent to which the inorganic phosphorus originally present in a plant is converted to microbial phosphorus and how far this is again recovered in inorganic forms.

Known organic compounds of phosphorus account for relatively little of the total organic phosphorus in sediments; the precise nature of organic compounds of phosphorus has yet to be determined, both in waterlogged soils and in normal soils. Undoubtedly present will be nucleic acids (mostly in plant remains and microorganisms), phospholipids (more in microorganisms than in plant material), and phytin. Phytin acts like the phosphate ion in that it combines with aluminum, iron, and calcium to forms salts, and these reactions help prevent microbial decomposition of the phytin as the salts are resistant to phytase.

SORPTION AND DESORPTION OF PHOSPHORUS IN LAKE SEDIMENTS

Inorganic Reactions

It is a general observation that soluble phosphorus compounds added to lake water rapidly disappear from solution. Although this can be explained partly by biological uptake, it is largely due to adsorption of the phosphate ion onto the lake sediment. Phosphorus adsorption increases with decreasing particle size due to specific surface effects and thus will be most marked in muds of the deeper parts of lakes (see the section on the effects of parent material). The adsorption of phosphorus can be demonstrated by adding soluble phosphorus in the radioactive form to suspensions of mud when the ^{32}P rapidly disappears from the water to appear in the mud. Experiments made in this manner (5) have shown a maximal adsorption of phosphate by muds at pH values between 5 and 7 when the $[H_2PO_4]^-$ ion is dominant. The adsorption is inorganic, reversible, and can be represented by the Freundlich equation $y = Kx^n$ where y is the phosphorus content of the water, x is the phosphorus adsorbed, and K and n are constants which differ according to the kind of sediment. Hayes found that organic matter depressed phosphorus adsorption, whereas iron increased it and that the adsorption capacity of sediments is directly related to the ratio of iron to organic carbon. The Freundlich equation has been found applicable also to the adsorption of phosphorus in strongly acid sediments (13).

Olsen (12) made similar experiments with radioactive phosphorus but worked at well-defined oxidation-reduction potentials and he determined simultaneously the two opposite processes of sorption and desorption between water and sediment. For the oxidized sediments investigated, the phosphorus equilibrium was described as the difference between a gross adsorption and a liberation. Mathematical descriptions were given of the liberated amount (b) of phosphorus from the mud to the water, $b = K^b \cdot C^{-w}$, and of the gross adsorbed amount (A) of phosphorus, $A = K \cdot C^v$. The latter expression follows the Freundlich adsorption isotherm. It was found possible to express the net amount of phosphorus adsorbed as the difference between the equations for the A and b curves. Figure 1 shows a typical result obtained by Olsen from analyses of a calcareous lake sediment. By plotting the concentration of radioactive phosphorus in solution against that in the sediment on a log/log basis straight lines were obtained for the gross amount adsorbed and for that exchanged. The difference between the equations of these lines gives, in the example quoted, the line $171.2C^{0.17} - 13.5C^{-0.5}$ and this represents the net adsorption of phosphorus by the sediment. The equilibrium concentration of phosphorus for which the net adsorption is nil can be found from the point where $A = b$, that is, where the two straight lines intersect. When experimenting with muds in the reduced state, the plotted lines gave a completely different picture due to iron(II) complicating the process, and a direct mathematical description was impossible to derive. From his experiments Olsen concluded that there was much fixation of phosphorus as calcium phosphate by the calcareous deposit at depth and that phosphate adsorption decreased with rising temperature.

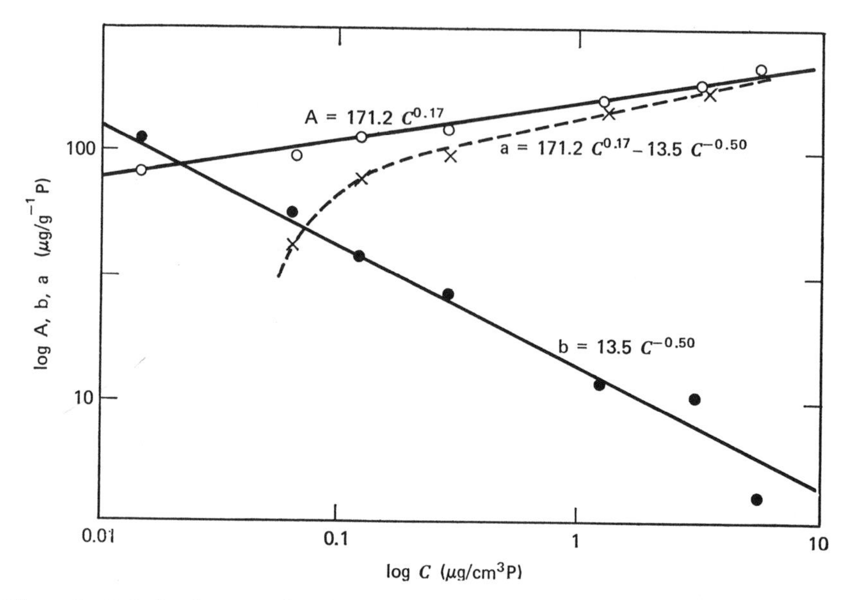

Figure 1. Relation between phosphorus in solution and amounts of phosphorus adsorbed and exchanged in a calcareous sediment (12).

The adsorption of phosphorus onto neutral muds will be affected by the presence of salts (e.g., potassium and sodium chlorides) and in some lakes this effect can be important and possibly limiting. It can also be demonstrated that calcium-rich deposits adsorb more phosphorus than can sediments saturated with ammonium ions and thus phosphorus adsorption is influenced by the nature and degree of organic nitrogen decomposition.

Organic Processes

The processes of phosphorus exchange at a mudwater interface are complicated by the presence of bacteria and plant life and by reactions involving organic complexes; in some, if not most, muds, bacterial activity dominates over inorganic chemical reactions.

Bacteria are active in removing inorganic phosphorus added to a lake, converting it to organic forms but, when the bacteria are killed (with antibiotics for example) a rapid exchange of phosphorus occurs with the mud. Although it is thought that dead plankton cannot itself adsorb phosphorus, addition of dead plankton to lake sediments results in a rapid loss of phosphorus from the water. This was shown by Hayes (5) to be due to an increase in bacterial activity at the mud surface from the utilization of readily decomposable organic matter. Hayes summarized his findings by concluding that the adsorption of phosphorus at a mud-water interface is controlled by two reactions—inorganic reactions influenced by oxidation-reduction potentials and organic reactions influenced by bacteria. The adsorption is stimulated by plankton fall-out and by oxygen diffusion. Findings by the International Rice Research Institute (10) that the oxidation-reduction potentials measured in waterlogged soils may well be "bacterial potentials" suggest that bacteria in fact influence both the organic and the inorganic adsorption processes.

It is possible that dead plankton could, in fact, be involved in the direct adsorption of phosphorus from lake waters. It has been suggested (6) that diatom frustules in Lake Victoria are responsible for the adsorption of sulfate ions—possibly because of the calcium present—and it would be of interest to investigate the effect upon phosphate ions.

PHOSPHORUS AND BIOACTIVITY IN LAKES

Many workers have reported the major importance of phosphorus as a nutrient in lakes. The nutrient elements in the muds of eutrophic lakes are the main source of food for algae and higher aquatic plants and, in most reported cases, phosphorus is the most important element. In Russia, silty lacrustine deposits are used to increase the fertility of sandy soils, and the effect has been attributed mainly to a high phosphorus content. Around the shores of parts of Lake Victoria in East Africa, the bottom deposits are pumped up to form highly productive market gardens, and phosphorus is the main deficiency of the soils of East Africa.

Also in Lake Victoria, the main source of food for fish is the phytoplankton *Melosira* and it has been shown (3) that the biological productivity of the lake is dependent on phosphorus. Addition of phosphate, even in small amount, gave significant increases in the growth of *Melosira*. An investigation of some of the shores of Lake Victoria has revealed the complete absence of phosphorus from the water until an area well into the papyrus zone is reached. The sudden appearance of phosphorus in the papyrus zone is accompanied by a sharp drop in the oxidation-reduction potential and the appearance of soluble iron. Thus in the open lake it appears that all available phosphorus is rendered unavailable by adsorption onto the sediment and only becomes soluble during the overturn when iron(III) is reduced. The fact that much of the phosphorus adsorbed on lake sediments becomes available when the mud is removed for agricultural use is explained by the fact that rooted plants are involved and that it is not a simple case of exchange with overlying water. High productivity per unit area in lakes is associated with either shallow depth or a base-rich medium, and commonly with both. Although several reasons for this have been advanced (11), the seasonal liberation of phosphate by reduction of the mud surface must surely be partly responsible.

SOME ANALYTICAL DIFFICULTIES IN THE DETERMINATION OF PHOSPHORUS IN MUDS

The determination of phosphorus in lake sediments presents difficulties not encountered when analyzing normal soils. First and foremost is the problem of obtaining the sample to be analyzed; the actual techniques of reaching and bringing up samples of mud are now fairly well established and are not discussed here, but there remains the question of how the sampling procedure will affect the chemical and biological nature of the sample.

Possibly the most important consequence of disturbing a waterlogged soil is the change in the state of its reduction. Even the insertion of electrodes into a mud causes fluctuations in the oxidation-reduction potential. Changes in redox potential of a lake mud brought to the surface and prepared for analysis are in the oxidizing direction and phosphorus will be affected principally as regards its form of combination with elements such as iron and manganese. Thus any phosphorus combined with iron(II) at the time of sampling may be converted to insoluble iron(III) compounds. Total amounts of phosphorus, whether inorganic, organic, or combined organic-inorganic, will not be affected. Experience has shown that well-poised lake muds resume their original oxidation-reduction level if stored underwater in the dark for some time and so more precise determinations can be made after such treatment. Ill-poised muds, however, may be irrevocably altered by sampling.

Treatment of the sample subsequent to sampling and prior to analysis can have far-reaching effects upon the determination of certain quantities, for example, iron(II), monsulfides, or reduced forms of nitrogen. For phosphorus, however, pretreatment of sediments is not so serious and it has been found that deliberate drying and oxidation of the sample is probably the most convenient procedure. This permits accurate subsampling and the application of methods of analysis used for ordinary soils (8). Frink (4) found that the various fractions of inorganic phosphorus in a lake sediment were unaffected by drying except for a slight increase in alkali-soluble phosphorus; he also reported no effect upon the sorption abilities of the muds. Frink, however, was dealing with a neutral sediment with little change in reaction when it was dried and his results would not apply to muds that undergo changes in pH value when dried and oxidized. For example, a sediment rich in oxidizable forms of sulfur could become extremely acid if dried and this would certainly influence the distribution of forms of phosphorus even if the total remained unaltered. Similarly the sorptive powers would be affected; in West Africa a peaty mud changed from pH 6.2 to pH 3.3 on drying and was found to adsorb 0.90 mg/g P in 30 min as compared with an adsorption of 0.55 mg/g P when wet (7).

If a sediment still in a reduced or partly reduced state is extracted with reagents for phosphorus determination by the usual molybdenum blue method, difficulties will arise from the reducing effect upon phosphomolybdate. The blue color produced on reduction of the molybdate complex varies in intensity with time and with concentration of reductant—even with kind of reductant (8). Thus the presence of unknown quantities of reductants of equally unknown and varied nature can make nonsense of the procedure.

Finally a word of caution regarding the use of radioactive phosphorus for sediment studies. As discussed earlier in the chapter, bacteria are active in removing soluble phosphorus from lake water. As bacteria do not differentiate ^{31}P and ^{32}P they will readily assimilate any ^{32}P added and so confuse the experimental results. Consequently before using radioactive phosphorus for inorganic adsorption experiments, bacteria should be killed, and this is best done with antibiotics. Sterilizing sediments by heat treatment should be avoided as certain, and probably relevant, physical-chemical properties of the sediment could be affected.

REFERENCES

1. Birch, H. F., *Plant Soil*, **15**, 347 (1961).
2. Chang, S. C., and W. K. Chu, *J. Soil Sci.*, **12**, 826 (1961).
3. Evans, J. H., *Nature, London*, **189**, 417 (1961).
4. Frink, C. R., *Soil Sci. Soc. Am. Proc.*, **33**, 269 (1969).
5. Hayes, F. R., *Oceanogr. Mar. Biol. Ann. Rev.*, **2**, 121 (1964).
6. Hesse, P. R., *Hydrobiol.*, **11**, 171 (1958).
7. Hesse, P. R., *Plant Soil*, **19**, 205 (1963).
8. Hesse, P. R., *A Textbook of Soil Chemical Analysis*, John Murray, London, 1971.
9. Hutchinson, G. E., and A. Wollack, *Am. J. Sci.*, **238**, 493 (1940).
10. International Rice Research Institute, The Philippines, *Ann. Rep.* (1963).
11. Mortimer, C. H., *J. Soil Sci.*, **1**, 63 (1949).
12. Olsen, S., *Verh. int. Verein. Limnol.*, **13**, 915 (1958).
13. Watts, J. C. D., *Malays. Agric. J.*, **47**, 187 (1969).
14. Yuan, T. C., *et al.*, *Soil Sci. Soc. Am. Proc.*, **24**, 447 (1960).

32

Phosphorus and Eutrophication

EUGENE A. THOMAS

Zürich University and Cantonal Laboratory, Switzerland

Literature concerning the eutrophication of lakes and streams regarding phosphorus and eutrophication, has greatly increased during the last decade. One work cited more than 500 references (43); a fundamental compilation of about 700 references has been published by Vollenweider (67). Interesting information can also be found in books giving general limnological descriptions of lakes as, for example, *Limnology in North America* by Frey (15).

Many different symposia have already been held on eutrophication, for example, *Biological Problems in Water Pollution* (1962) in Madison, Wisconsin as well as *Eutrophication: Causes, Consequences, Correctives* (1967), also in Madison. The problems of eutrophication are very varied according to local limnological and geographic circumstances. There is no attempt in the following work to present a complete compilation of existing literature. Examples from Switzerland and adjacent regions have been selected to show how incipient eutrophication caused by changes in the phosphate content is recognized and how its increase is accurately registered over the years.

Eutrophication of running water is only briefly referred to, its importance being ascribed to the origin of phosphates in the lakes. Lake morphology plays an important part as a sensitivity indicator in relation to the process of eutrophication as far as the characterization of lake eutrophication is concerned. Knowledge of the basic phosphate content is not sufficient for the determination of lake eutrophication. As is shown, the amount of phosphate fed daily into lakes is of greater importance. Phosphates can directly stimulate the growth of algae. It can also be observed, however, that free phosphates promote the division of bacteria and that many water bacteria produce growth factors for algae.

Special sections in the following work are devoted to measures that prevent eutrophication with special consideration of the depth at which sewage is introduced and the two particular aspects of lake amelioration: the immediate effect and the long-term effect. Reference is made to the special ratio between phosphate and nitrate within the sphere of eutrophication.

Fortunately examples of oligotrophication of lakes can also be mentioned whereby the conditions in Lake Washington and the Lake of Zürich are discussed. Oligotrophication has taken place in both these lakes as a result of measures which have been taken to protect the water.

CHARACTERISTICS OF LAKE EUTROPHICATION

There are no generally available definitions of the words "eutrophic" and "oligotrophic." We are not going to discuss the possible definitions of these words but it is necessary to state what we mean in this paper by "eutrophication."

Every observer of eutrophic lakes knows that most of these lakes have changed greatly during the last few years or decades. The most remarkable change consists of an extraordinarily rapid growth of algae in the surface water. Planktonic algae cause turbidity and floating films; shore algae cause ugly muddying and, very often, floating films and damage to reeds. Decay of these algae causes oxygen depletion in the deep water, in the thermocline layer, and in shallow water near the beach (52, 54, 55). This rapid growth of algae gives rise to numerous undesirable effects on treatment of potable water, fisheries, bathing sanitation, and recreation (tourism). This increase in algal growth is undoubtedly caused by increased fertilization of the lakes through the influence of man. Overfertilization results from the increase of wastewater in the lakes.

In Switzerland, the example of the Lake of Zürich was discussed years ago and the question of whether and how the eutrophication of the lake, which was already taking place, could be stopped and possibly reverted. Lauterborn observed in 1910 that human sewage sediments discharged into the Lake of Zürich was the reason for the vegetable planktonic organisms usually growing in the lake. They could suddenly begin to develop into enormous masses and often manage to remain in considerable quantities for years. Minder (33) expressed the opinion that the only hope of reducing strong development of vegetable planktons is by eliminating the supply of minimum substances. A "minimum substance" is any vital nutrient which, by its presence in the water in a small quantity, sustains the further development of plankton and algae. The importance of small concentrations of phosphorus in water was recognized by Yoshimura (71): "The limiting factor of the phytoplankton crop in Takasuka Pond may probably be due to phosphate at certain times and ammonia at other times. Phosphate seems to be the more important." Naumann (34) and Gessner (17) referred to the "physiological phosphate poverty" of Swedish lakes in the region of Aneboda, and Gessner (18) speaks of phosphorus as being the minimum factor limiting organism production in inland waters. Stangenberg (42), Thomas (46), and Ohle (37) also attached great importance to the supply of phosphate in phytoplanktonic production. Early indications about substances stimulating the growth of algae can be found in publications by Atkins (3) and Juday and Birge (23). Interesting recent work on this has been done by Provasoli (40).

PHOSPHORUS AND NITROGEN AS IMPORTANT FACTORS OF EUTROPHICATION IN LAKES

Not infrequently we find data in the literature where the author has reached no decision as to whether or not phosphate or nitrate, or the presence of another substance, is to be regarded as a factor limiting growth. Also no mention is made of whether or not great differences between lakes are to be expected. It is known that some taller water plants and types of algae need certain substances in considerable quantities for them to flourish. Among these substances are, for example, calcium, molybdenum, or boron. Other types of vegetation only thrive when calcium or other substances are present in only very small quantities. The demands of these specialists among the plants need not be considered. What is more important is whether or not the increased addition of phosphorus and nitrogen compounds to lakes is sufficient to bring about a corresponding increase in plankton production. It would then be necessary to add surplus quantities of nitrates and phosphates to the water from different lakes and to observe the response. This question can be dealt with experimentally in the laboratory by adding excess nitrates and phosphates to the water from these different lakes.

Tests were carried out on the surface water from 40 central European lakes, that is, in Switzerland, France, Germany, Austria, and Italy (47). One series of water samples was taken from the lakes in December ("winter water") which is during the circulation period when the nutrient content of the surface water is highest. The second series of test samples was taken at the end of the stagnation period in September ("summer water"), at which time the surface water contains the least nutrient material. When there was sufficient water from the individual lakes available, we filled two to six conical flasks from each sample with 300 ml of water which was sterilized after the flasks had been plugged with cottonwool. Ten milliliters of sterile nitrate and phosphate solution was then added. These solutions contained 823 mg/l $NaNO_3$ and 87.2 mg/l $NaH_2PO_4 \cdot H_2O$, respectively, so that the nutrient solutions in the conical flasks contained over 20 mg/l nitrate and over 2 mg/l phosphate, including the nitrates and phosphates already present in the lake water. The sterile nutrient solutions were then ready to be injected with plankton.

The injected material consisted of surface plankton from the mesotrophic Upper Lake of Zürich which we had collected in March for the winter-water tests and in April for the summer-water tests. This plankton proved to be suitable because it contained practically no crustaceans, very few rotatories, but did contain algae from very different groups such as diatoms, *Oscillatoria rubescens*, types of *Dinobryon*, *Peridineae*, *Mougeotia* (planktonic), and a few green algae. We determined the algal content per milliliter of this plankton concentration and added 1 ml of the injection material in the first test and 0.5 ml in the second one. After the injection, the nutrient solutions contained the following amounts per milliliter of the most abundant algae:

Winter water: *Tabellaria fenestrata*, 16 cells; *Asterionella formosa*, three cells; *Fragilaria crotonensis*, two colonies; *Synedra acus*, one cell; *Oscillatoria rubescens*, three threads.

Summer water: *Tabellaria fenestrata*, 18 cells; *Asterionella formosa*, 47 cells; *Synedra acus*, one cell; *Fragilaria crotonensis*, one colony; *Oscillatoria rubescens*, one thread.

The nutrient solutions injected in this way were left for 2 months in a diffused light at room temperature. At the end of the test period, it was found that the addition of abundant nitrate and phosphate to the water of the lakes tested was sufficient to obtain a plentiful development of planktonic algae after injection with a little plankton material in all the sterilized culture solutions. Tychoplankton algae, which were also in the water, sometimes developed too. The nutrient content diminished considerably when compared with the test samples that had not been injected.

In the tests with winter water, the original 20 mg/l NO_3^- was used up in the water from 25 of the 32 lakes, whereby only 0.5 to 4 mg/l remained in that from the other lakes. Of the original 2 mg/l PO_4^{3-}, 0.25 mg/l remained in only one case. The content generally dropped to 0.2 to 0.02 mg/l PO_4^{3-}, and in eight of the 32 cases it was even below 0.02 mg/l. In the tests with summer water, with one exception, the nitrate was about 98% consumed (less than 0.5 mg/l NO_3^-) and the phosphate content also generally sank to low values. Nitrite and ammonium ions were not formed in large amounts nor did they play a noteworthy part in these two tests for the nitrogen supply of the algae. No addition of ferric salt was necessary in these two tests with surface water.

In both series of tests, it was striking from the biological point of view that planktonic diatoms sometimes showed exceedingly strong development.

Green and blue algae also managed to dominate in some individual flasks and the appearance of *Oscillatoria rubescens* is worthy of note. A content of 219 threads/ml, as in the culture with summer water from the Lake of Constance, is just visible to the naked eye. Because these algae only flourished in a few flasks in both series of experiments, the conclusion may be drawn that nitrogen and phosphorus fertilizer alone is not decisive for the prevalence of recession of these blue algae.

The question on which the experiments were based can be answered as follows, according to the results obtained: With the addition of phosphates and nitrates, the content of all other vital substances in the lakes investigated is sufficient to cause increased production of plankton and to use up and convert into organic substances most of the added nutrient. A phosphate–nitrate ratio of $PO_4^{3-}/NO_3^- = 1:10$ proved to be favorable. A biochemical precipitation of lime, as in nature, can also occur. Any influence on the cultures by substances from the glass containers can be disregarded because in control tests with phosphates, nitrates, and the same injected material, but with distilled water instead of lake water, there was no growth. Therefore in these lakes, only the phosphorus and nitrogen compounds are to be considered regarding their importance as minimum substances for the general production of plankton.

INDICATIONS OF INCREASED PHOSPHATE TROPHY—EXAMPLES OF CENTRAL EUROPEAN LAKES

Some lakes in Central Europe have been regularly analyzed during the last few years for phosphate. We consider herein three lakes with different degrees of eutrophication: the Lake of Constance, the Lake of Zürich, and the Greifensee.

The Lake of Constance

Wagner (70) summarized the phosphate trophy of the Lake of Constance with a graphic illustration. The values given below expressed as mg/m³ of PO_4 ion, are taken from that illustration:

Year	PO_4-Ion Content	Year	PO_4-Ion Content
1940	0	1960	30
1950	5	1965	52
1955	12	1968	80
		1970	113

Wagner (69) states that phosphates (PO_4 ions) increase 3 to 6 mg/(m^3)(year). There was, however, no increase from 1964 to 1965. Wagner attributed this fact to climatic factors. The year 1965 was rainy. Only a little phosphate was washed from the ground but more phosphate was carried out of the lake through its outlet. I made similar observations on Swiss lakes.

Although the phosphate content of the Lake of Constance has increased considerably, phosphate is still a minimum factor in the epilimnion.

The Lake of Zürich

A short article about the phosphate trophy of the Lake of Zürich and a comprehensive work on the subject have been published (62, 63). In order to compare the Lake of Zürich with the Lake of Constance I have given some results illustrating the increase of phosphate content (PO_4 ions) in the Lake of Zürich (annual average of the content for the total volume). The values are expressed in mg/m^3.

Year	PO_4-Ion Content	Year	PO_4-Ion Content
1946	69.6	1964	269.0
1950	82.4	1965	231.6
1956	126.1	1966	234.8
1960	152.0	1967	230.2
1961	171.7	1968	240.9
1962	189.2	1969	261.3
1963	202.3	1970	251.2

After World War II, the phosphate content in the Lake of Zürich apparently increased sharply. Algal masses first appeared on the beaches in 1949. A very interesting fact is that the phosphate content was highest in a very dry year (1964), whereas in the very rainy years of 1965 and 1966 the phosphate content did not increase (there was no special washout of phosphates from agricultural areas but a decrease in PO_4^{3-} was observed). The clay particles washed into the lake from the shore seemed to bind some phosphate.

Nitrates played a role opposite to that of phosphates. The nitrate content of the Lake of Zürich was higher in the summer months in 1965 and 1966 when rains were heavy than in previous years. Heavy rains thus caused a decrease in phosphates and an increase in nitrates.

The Greifensee

Almost every year the Greifensee has a complete mixing, that is, a period of total circulation during which the phosphate content is uniform from the surface to the bottom. The mean annual phosphate content in relation to the total volume has not been calculated from the analyses that have been made for many years. The figures for the phosphate content are therefore the values at the time of total circulation. The values are expressed as mg PO_4 ions/m^3 H_2O.

Year	PO_4-Ion Content	Year	PO_4-Ion Content
1950	180	1961	1000
1951	220	1962	1000
1953	320	1963	1100
1954	550	1964	1300
1955	550	1965	1300
1956	600	1966	1000
1957	900	1967	1200
1958	800	1968	1320
1959	800	1969	1620
1960	900	1970	1450

Also in this lake, phosphates did not increase in the years with heavy rains but showed a marked decrease.

Since 1960, the purification plant at Uster has eliminated about 90% of the phosphates from the sewage water. Only a small proportion of the inhabitants in the catchment area of this lake live in Uster and there was still an increase in phosphates between 1950 and 1969. As of 1970, the phosphates were still not being eliminated in various communities in the catchment area of the lake. Because of the elimination of phosphates by the iron-return sludge process, the outflow of the Uster purification plant today contains less phosphates than the water of the Greifensee during total circulation.

PROBLEMS OF EUTROPHICATION THROUGH ADDITION OF PHOSPHATES

The Basic Value

Because of total circulation in holomictic lakes, all the water mechanisms are balanced from top to bottom. Lakes having no total circulation are also known, as shown by Bachmann (5): As shown by chemical analysis, the Lake of Ritom consisted of two totally different layers of water which were never mixed. The depth of the upper layer was 12 to 15 m. Findenegg (13) called such lakes "meromictic lakes." Some others, such as the Türlersee (45) and the Lake of Zürich (28) can be classified as "facultative meromictic." The state of maximum circulation should not be called "total circulation" but "main circulation" (46).

In all these types of lakes, the maximum quantities of phosphate ions are available to the phytoplankton of the epilimnion at the time of the spring turnover. We characterize this spring maximal value as the basic value. The available phosphate ions are quickly consumed at the time of increasing stratification and increasing water temperature.

On analysis of the samples, the values obtained are too high because, in the method used for analysis, some of the bound phosphates are freed. If phosphate addition from the catchment area of the lake were stopped, the growth of algae would stop as soon as the algae had consumed the basic phosphate content.

Relation between Basic Value and Continuous Addition of Phosphate

In the central European lakes, the basic value of phosphate ions decreases to traces or zero in the period from March to June. The significance of the *daily addition* of phosphate ions by sewage has been proved for the Lake of Zürich (64). Calculating the supply of phosphate to the vegetation zone (0 to 10 m depth in the Lake of Zürich), about two-thirds of the phosphate used in the epilimnion in one year comes directly from the sewage water (disregarding summer turbulence, zooplankton, empneuston, and agriculture) (Fig. 1).

Accurate data are available for the monthly addition of phosphate to the Greifensee in the publication by Pleisch (39) as well as from our own investigations into the phosphate concentration of the lake in other months. These values are compared in Fig. 2. This comparison also shows that the amount of phosphate added to the epilimnion is of great importance for eutrophication, compared with the basic value. The amount of phosphate added to the water layer which produces algae in the course of the year is about twice as much in the Greifensee as the basic value of this layer during main circulation. It is also possible for a lake to have a low basic value of phosphate ions but for it to show a trend toward eutrophication because of high daily additions of phosphate. We often find this phenomenon in large lakes to which a lot of phosphate has recently been added.

Phosphate Ions as a Stimulant for Growth of Bacteria

After Bosset (6) had shown that the bacterial content of drinking water increases greatly because of addition of phosphates, we investigated the effect of phosphate addition on bacteria in the waters of the Upper Lake of Zürich and the Lower Lake. Water samples were collected at the surface and at depths of 20 and 50 m. Analytical sodium phosphate was added to the

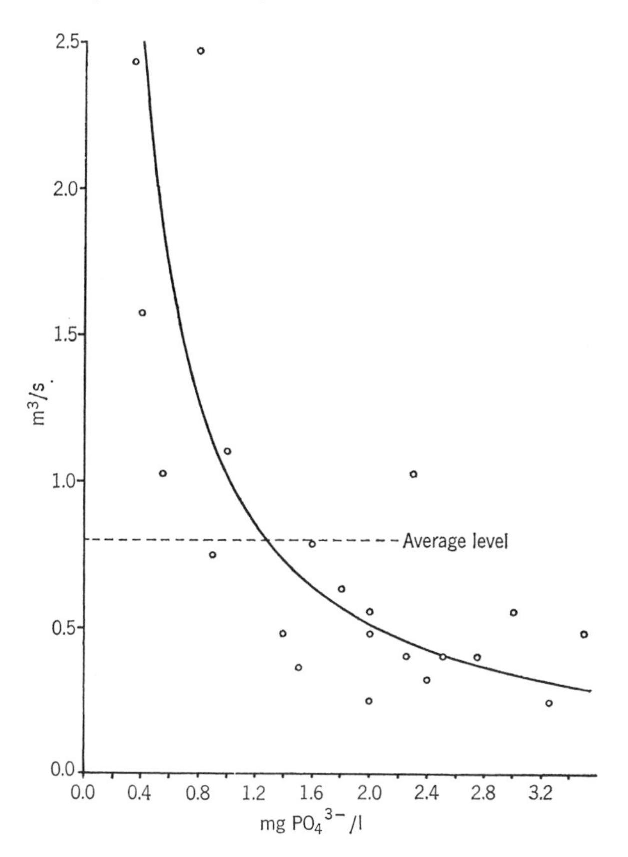

Figure 1*a*. Relation between the stream flow and the concentration of phosphates in the Mönchaltorfer River. The graph shows the theoretical dilution of the average phosphate load of the water in the stream. The dashed line represents the average level of the river [according to Pleisch (39)].

unsterilized lake water, which was kept in darkness, along with control water, at 20°C for 20 days. After 20 days the samples were examined for bacterial densities. Plate-count agar was used for a total count of bacteria. The results of the experiments (Table 1) show that increase in bacterial density is directly proportional to the phosphate content of the water in the Lake of Zürich.

Phosphate as a Stimulant for Bacteria for Production of Growth Factors for Algae

It is well known that most freshwater algae use growth factors for thriving. We have shown that for *Cladophora glomerata* there are numerous water

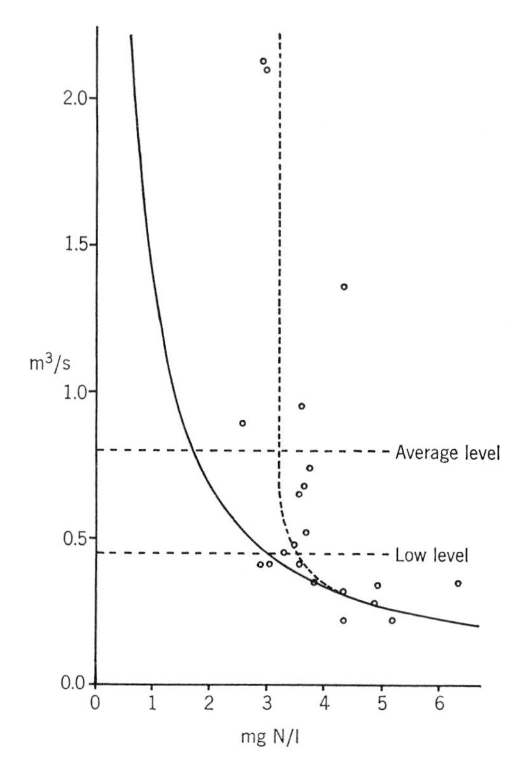

Figure 1*b*. Relation between the stream flow and the concentration of inorganic nitrogen compounds. The indicated graph is the theoretical dilution curve of the nitrogen load which is presumed to be constant. The dashed curve is the course of the graph of experimentally determined values [according to Pleisch (39)].

bacteria that are able to produce such growth factors (56). With our bacterial pure culture S_2 we sought to determine whether bacteria that produce growth factors are also stimulated by the addition of phosphate ions. Preliminary experiments show that the bacteria are so stimulated. Further tests are in progress to determine whether there is a simultaneous increase in production of growth factors and in growth of bacteria, as presumed.

Judging from our studies, increase addition of phosphates in the lake would bring about the following eutrophication mechanism:

increase in bacterial content,
increase in oxygen demand,
increase in production of growth factors for the algae,
increase in growth of algae.

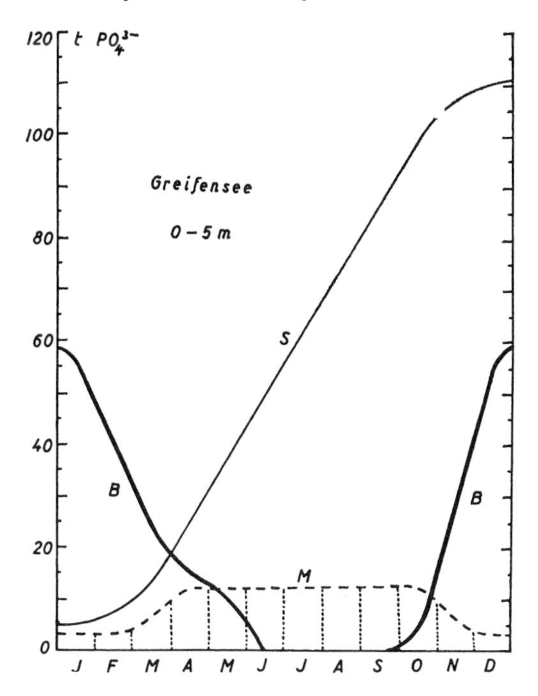

Figure 2. Relation between phosphate ions present in the epilimnion of the Greifensee (*B*: basic value) and the amounts of phosphate added monthly to the epilimnion by the discharge of sewage. The summation curve (*S*) was obtained by successive addition of the amounts of phosphate added monthly to the epilimnion (*M*: monthly amounts).

The phosphate hypertrophy of lakes allows the increased production of growth factors, and this production promotes the growth of algae. To avoid algal damage in lakes, we need to reduce the supply of phosphates.

THE INCREASE IN LAKE EUTROPHICATION CAUSED BY THE RETURN OF SEDIMENT PHOSPHATE

The phosphates absorbed by the phytoplankton are mostly deposited in the sediment when the plankton sinks. When an investigation is made on the sediment to determine how much of the bound phosphorus redissolves upon the decomposition of planktonic organisms, the results depend on the character of the lake. The organically bound phosphates are the ones which predominantly escape from the sediment (59). If the decomposition of the sediment takes place in an aerobic milieu, the release of sediment

TABLE 1 Increase in Bacterial Content of the Lake of Zürich 20 Days after Addition of 2 mg/l PO_4^{3-} (Average of Two Samples)

Sampling Location	PO_4^{3-} Content (mg/l)	Bacterial Colonies (total/m)
Upper Lake of Zürich[a]		
Surface		
Without addition	<0.02	7,315
With addition	1.7	25,025
20-m depth		
Without addition	<0.02	1,365
With addition	1.65	5,313
Lower Lake of Zürich[b]		
Surface		
Without addition	<0.02	3,653
With addition	1.65	32,300
50-m depth		
Without addition	0.25	1,773
With addition	2.1	5,225

[a] Middle of the lake at Bollingen.
[b] Middle of the lake at Thalwil.

phosphate is much less than it is in an anaerobic state. A release of phosphate of 1 to 2% was established in two lakes in the aerobic environment whereas in the anaerobic milieu in the same lakes it was between 8 and 28% and in extremely eutrophic lakes it was up to 44% for the same milieu (48, 56).

Part of these phosphates can return to the epilimnion again during the main circulation and lead to renewed development of plankton. Golterman closely observed the influence of mud on the chemistry of water in relation to productivity [Golterman and Clymo (21) cite additional references].

In a considerable number of lakes it has been observed that the process of eutrophication does not take place at a regular pace but in spasms, that is, it is not "rasant" but "saltant." If the addition of phosphate compounds to an oligotrophic lake is regular, as is diagrammatically shown in Fig. 3, eutrophication only progresses steadily until the oxygen above the mud on the bottom of the lake has been largely used up during the period of summer stagnation. When the consumption of oxygen above the bottom of the lake first takes place, the nutrients that have been retained in the sediment can return to the free water in large quantities and consequently into the

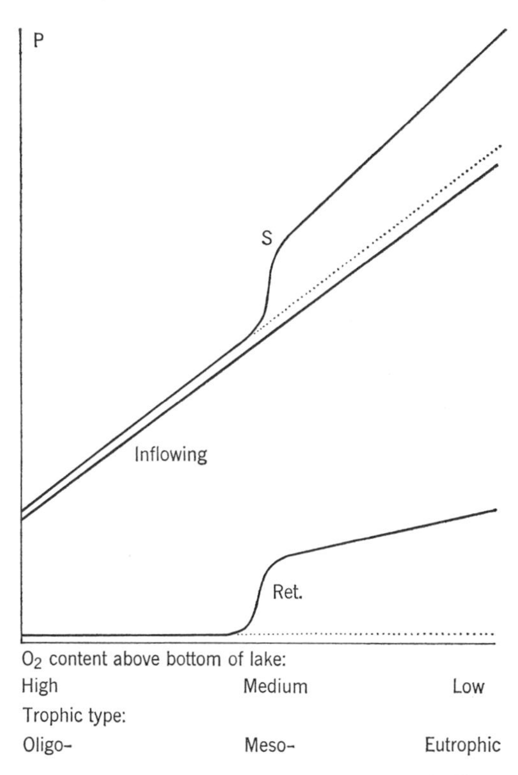

Figure 3. Diagram of lake eutrophication by leaps. Ret: organically combined phosphates of the sediment which return to the free water. *S*: sum of inflowing phosphorus compounds and those returning to the lake from the sediment.

circulation of material in the lake. The consumption of oxygen in the hypolimnion is therefore the cause that induces a sudden additional eutrophication of the lake. The flatter and more extensive the bottom of the lake is, the more evident is the sudden eutrophication due to the first appearance of deep water that lacks oxygen. If the hypolimnion has no oxygen every autumn, the sudden increase in eutrophication no longer takes place and a condition is reached in which eutrophication again progresses regularly, depending on the amount of nutrient substances that are added and those that return to the water from the sediment (Fig. 3). Nitrogen compounds return from the sediment into the free water in high percentages (49, 56).

Compared with the addition of phosphates from the inflows, the importance of the return of sediment during summer stagnation is slight but the basic value for phosphates rises (Fig. 2).

THE STRATIFICATION OF WATER FLOWING INTO A LAKE

The Stratification of Rhine Water in the Lake of Constance

In the colder part of the year, the vertical interchange currents in lakes are so intensive that, with the main circulation, there is a uniform mixing of the whole mass of water from the surface to the depths. At this time of the year, the water flowing into the lake is more or less uniformly mixed with the whole mass of water in the lake.

Conditions are different in summer when the surface water is warmer and the deep water colder than the inflowing water. Then the inflowing water finds its level at the depth of the lake which has the same specific weight. This layering process has been extensively studied in the Lake of Constance (24, 4). The water from the Rhine accordingly finds its place in summer at a depth of from 5 to 30 m in the Lake of Constance and flows through the lake along the whole length of the right side although it becomes increasingly mixed with the water of the lake.

Fertilizers such as phosphates and nitrates which are brought into the lake with this large, cool inflow are therefore not available to the surface plankton at first. Only the gradual mixing brings the fertilizers more and more into the trophogenic layer where they then lead to an increase in the plankton.

The Stratification of Sewage Water in the Lake of Pfäffikon during Summer Stagnation

Biologically purified sewage water was originally discharged into the Lake of Pfäffikon along an open channel without a tertiary purification stage. We tested how this water was stratified and took several samples, at different depths, along a line extending in the direction of the inflow as well as along the line bisecting the angle between the inflow line and the shore as indicated in Table 2. The spread of sewage water discharged into the lake could be clearly followed by chemical and especially by bacteriological methods (65).

The stratification of the sewage phosphates in the lake is of special interest to the question of eutrophication. Practically no increase in phosphate was observed in the surface water, even near the shore. On the other hand, increased phosphate values caused by the 6.4 mg/l phosphate flowing in with the sewage extended down the slope from the shore to a distance of 20 m and spread at a depth of 5 m to as far as 50 m from the shore. Outside the distance of 20 m from the shore, that is, at a depth of more than 5 m, the

TABLE 2 Stratification of Phosphates from the Discharge Point of the Pfäffikon Purification Plant into the Lake of Pfäffikon (Sept. 1953)

Depth	Sewage	2.5	5	7.5	10	15	20	35	50	100	500 m
		2.5	5		10		20		50 m (Left Radius)		
0.3	PO_4^{3-}	<0.2	<0.04		<0.02		<0.02		<0.02		
1.5	mg/l	**0.8**	**0.36**		—		—		—		
2.5					**0.24**		<0.02		<0.02		
5.0							<0.02		<0.02		
7.5									<0.02		
10.0									<0.02		
		Distance from Discharge of Sewage into the Lake (m)							(Center Radius)		
Depth	**Sewage**	**2.5**	**5**	**7.5**	**10**	**15**	**20**	**35**	**50**	**100**	**500 m**
0.3	**6.4**	<0.1	<0.04	<0.02	<0.02	<0.02	<0.02	<0.02	<0.02	<0.02	<0.02
1.5		**1.3**	**0.40**	—	—	—	—	—	—	—	—
2.5				**0.64**	<0.02	<0.02	<0.02	<0.02	<0.02	<0.02	<0.02
3.2					**0.5**	—	—	—	—	—	—
5.0						**0.16**	**1.04**	**0.36**	**0.12**	<0.02	<0.02·
6.0							**0.06**	—	—	—	—
7.5								<0.02	<0.02	<0.02	<0.02
10.0									<0.02	<0.02	<0.02
		2.5	**5**		**10**		**20**		**50 m (Right Radius)**		
0.3		<0.1	<0.04		<0.02		<0.02		<0.02		
1.5		**1.2**	**0.50**		—		—		—		
2.5					**0.48**		0.03		<0.02		
5.0							**0.16**		<0.02		

layers near the bottom of the lake were no longer influenced by sewage phosphate (Table 2).

In the preceding case, the sewage phosphates reached the aphotic zone in September so that they could not at first be utilized by the phytoplankton. This may be a main reason why, in the middle of summer, the production of plankton in such lakes generally recedes. The nutrients present in the surface water in spring have been depleted and the nutrients in the metalimnion only come up to the surface in autumn as a result of partial circulation. This explains the second maximum development of phytoplankton which usually occurs in autumn.

Choice of the Depth for Discharging Sewage Water into Lakes

In the case of the Lake of Pfäffikon, the discharged water is cooler in summer than the surface water of the lake so that it sinks to a deeper layer. Where bathing is allowed in a lake, it would undoubtedly be undesirable to discharge sewage water on the surface because currents caused by wind can prevent it from sinking to the deeper layers in the lake. On the other hand, the deeper layers of lake water must unquestionably be protected from discharged sewage so that the catchment of drinking water is not endangered.

There are numerous catchment points for drinking water in the Lake of Zürich at depths of between 30 and 40 m. The thermocline is at depths of from 5 to 15 m (28). A depth of from 3 to 6 m is chosen as the discharge depth for the outlets from the mechanical-biological purifying plants in which phosphates are also eliminated. These depths, at the upper limit of the thermocline layer, have proved to be good. The nutrients for algae, which still reach the lake in spite of purification of the sewage, are in a zone where the light is poor. The catchment of drinking water and bathing beaches are also protected. The discharge depths for sewage water must be correspondingly adjusted in lakes where the stratification of water is different.

PHOSPHATE AS A GROWTH-LIMITING FACTOR AND THE OLIGOTROPHICATION OF LAKES

By oligotrophication we mean the reduction of nutrients of algae connected with a reduction of algal production in the lake. The oligotrophication of eutrophic lakes is important for the removal of the undesirable effects on the treatment of water for drinking purposes and on fisheries, bathing sanitation, and recreation (tourism). Harmful algae are best controlled by limiting the inflow of phosphates for the following reasons:

1. Phosphate is present only in traces in oligotrophic lakes.
2. Natural tributaries running into these lakes contain very little phosphate as long as they are not subjected to pollution by the influence of man but they contain large quantities of nitrates.
3. Fewer phosphates than nitrogenous compounds are washed out of agricultural land.
4. Rainwater often contains large quantities of nitrogenous compounds that can be utilized by plants.
5. Bacteria and blue-green algae living in lake water are able to fix nitrogen or to produce the growth factors for algae.
6. The addition of only phosphate to lake water is sufficient to increase the growth of bacteria and blue-green algae.
7. Some blue-green algae produce toxins that are very toxic for warm-blooded animals.
8. Nitrogenous compounds from putrefied parts of organisms and sludge return to the biochemical cycle in larger quantities than phosphate compounds.
9. In eutrophic lakes, nitrates are eliminated from time to time by the process of denitrification.
10. It is cheap and easy to eliminate phosphates from sewage water (by alum or $FeCl_3$ in the activated sludge process).

PRECAUTIONS AGAINST EUTROPHICATION OF LAKES

Circumferential Pipelines

Circumferential pipelines are used to collect wastewater from the catchment area of a lake and to carry it to the central purification plant. The purified water is then diverted straight to the outflow of the lake. Such pipes are especially suitable for small lakes where installation does not involve extending the pipeline over too long a distance.

Such pipelines were installed some years ago around the Lake of Tegern (Germany), The Lake of Zell (Austria; advised by Dr. H. Liepolt), and the Lake of Hallwil (Switzerland). As early as 1912, two pipes, each about 6 km long, were installed around the lower part of the Lake of Zürich. One pipe extended from Kilchberg and the other from Zollikon to the outflow of the lake. Both of these cover a part of the city of Zürich. These pipes were installed to protect the drinking-water supply from this part of the lake and, undoubtedly, the installation of these pipes is a most effective precaution.

The high water level during rainy weather should be considered when such pipes are being installed. Whether pipes can be installed around large lakes depends on many local factors. Reports of the initial results are given by Liepolt (31) and Findenegg (personal communication).

Eliminating Phosphates from Sewage

There are many publications on the elimination of phosphates from sewage. I have 12 years' experience of phosphate elimination by the activated sludge process. Ferric chloride (10 or 15 mg/l of Fe^{3+}) is continuously added to the inflow or outflow of the aeration tank. About 90% of the phosphates can be eliminated in this way. The volume of sludge is not increased and the ferric phosphate sludge does not dissolve during decay.

In Switzerland, the cost of installations for eliminating phosphates is 1 to 2% of the total cost of mechanical and biological purification. The cost of chemicals used is $2 to $3/1000 m^3, or about $0.50 per person per year (60, 61).

In Switzerland there have been standards since 1966 for the condition of water discharged from purification plants. This cleaned sewage water, which is measured daily, must not contain more than 2 mg/l of PO_4^{3-}.

On June 19, 1967 the Swiss Federal Council requested all communities in the catchment areas of lakes to eliminate phosphates from sewage. Today there are more than 20 treatment plants working with phosphate elimination.

The elimination of nitrates or other ions is out of the question.

Prohibition of the Addition of Phosphates to Drinking Water

Phosphates are often added to drinking water but this greatly increases the content of bacteria in the distribution system and impairs the quality of drinking water (research by Dr. E. Bosset, Lausanne, 1965). For this reason, the Swiss Public Health Department in Bern has forbidden the addition of phosphates to drinking water (Circular No. 16 concerning the use of phosphates for the processing of drinking water, 1966). This precaution is very important for the protection of water. In this way it is possible to stop the extra increase of phosphates in sewage.

Phosphate in Washing and Cleaning Agents

It is well known that, since World War II, numerous washing and cleaning agents that contain a high percentage of phosphates that can be utilized by plants have been produced. After being used, such phosphates go into the sewage. Where this sewage water is connected to a central purification plant, they can be eliminated in a tertiary treating process before they reach lakes and streams where they cause damage.

There are, however, many isolated houses where the sewage water will not be connected to a central purification plant within a reasonable time. For this reason it would be better to use new detergents which contain no, or very little, phosphate. The use of precipitating material in the purification plants would then be correspondingly less and the running costs lower. New phosphate-free detergents must be well tested limnologically to safeguard water before they are put on the market, however.

Preventing Disposal of Animal Liquid Manure and Silo Waste

In some central European countries it is forbidden to discharge animal liquid manure and silo waste into bodies of water. Such agricultural sewage is better used in the field.

THE TWO ASPECTS OF LAKE SANITATION

The Immediate Effect

In many lakes, the area between 0 and 5 m in depth may be regarded as a special zone for plankton activity. This zone covers an area of about 350×10^6 m^3 in the Lake of Zürich; 750×10^6 mg phosphate/day, or 2 μg PO_4^{3-}/(l.) (day), or 0.02 mg/l. PO_4^{3-} are added to this zone. These are easily detectable quantities but they are continuously consumed by the phytoplankton in the lake. In other words, the daily addition of phosphate, especially from sewage, allows the increased phytoplankton to be continuously maintained during the summer. Because the winter supply of the surface water is depleted in May, the further addition of phosphate to the lake causes plankton growth in the remaining half of the summer.

These facts are of great importance for repressing eutrophication. If phosphate from purification plants is decreased 90% by eliminating the phosphates, the growth of algae must decrease to the same extent from about the middle of May. This means an improvement in the oxygen content of the deep water. If phosphate reduction in a lake is very well organized, there will be an immediate reduction of algal production the following summer.

The Long-Term Effect

With a sufficient decrease in the addition of phosphate, the basic value of phosphate in a lake will decrease instead of increasing. After many years, smaller quantities of phosphates will be offered to the algae before the beginning of the vegetative period and the trophic content of the lake will nearly approach its original state.

BIOASSAYS IN THE LAKE

The plankton test-tube method (51, 53, 55) is of special interest for testing the growth of phytoplankton organisms in lakes and the influence of phosphate and nitrate. Similar equipment was used by Goldmann (20) and Gächter (16). These tests showed the importance of phosphates for the growth of plankton algae. All algae, as was to be expected, do not have the same requirements regarding the phosphate concentration. It was proved in the plankton test tubes that the composition of plankton biocoenose changes quickly if the phosphate milieu factor is artificially changed. *Ankistrodesmus* and *Selenastrum*, as well as other green algae and diatoms, increased very rapidly with a high phosphate content (55). Gächter's investigations show that planktic primary production could clearly be increased from March to October by phosphate fertilization.

CHANGE IN THE CONTENT OF PHOSPHATE AND NITRATE BY FLOWING THROUGH THE SURFACE WATER THROUGH TWO LAKES

The Lake of Zürich offers an interesting insight into the reciprocal behavior of nitrate and phosphate. It receives the largest amount of water directly from the Upper Lake. This inflow of water has a high nitrate content and a low phosphate content. Sewage water rich in phosphate formerly flowed into this long, narrow lake from both sides. The basic value during the main circulation period was therefore distinctly less in the upper part of the lake than in the lower one. But, also during the summer, large amounts of phosphates constantly flowed into the epilimnion. These phosphate ions are quickly and completely consumed by the phytoplankton in the surface water in summer.

The influence of the addition of phosphate to the nitrate content is seen in Fig. 4. In the Upper Lake, the basic phosphate content and the daily addition of phosphate were still so limited that only a small part of the nitrates was used in April and the nitrates were not entirely consumed even in August. The basic phosphate content in the Lake of Zürich increased from Stäfa to Riesbach. The daily addition of phosphate in this part of the lake was then also high. As a result, the nitrate content dropped quickly in April 1962 and sank even toward zero near Wollishofen and Riesbach. In August

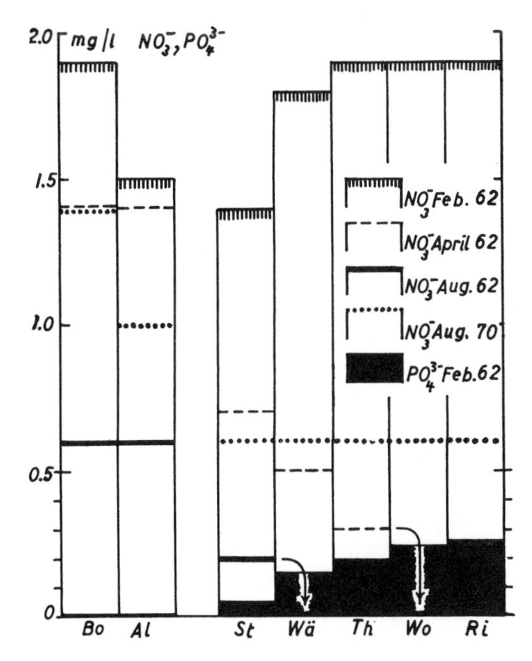

Figure 4. Behavior of nitrates and phosphates in water flowing through the Upper Lake of Zürich (Bollingen-Altendorf) and the Lower Lake of Zürich (Stäfa-Wädenswil-Thalwil-Wollishofen-Riesbach outlet) 1962 and 1970.

1962 there was only a small remainder of nitrate near Stäfa, and in Wädenswil the nitrate content sank toward zero as a consequence of the daily addition of phosphate.

The purification of sewage in the catchment areas of the Lake of Zürich was improved by the communities from 1967 to 1970; 90% of phosphates is removed from sewage in almost all purification plants today. The daily addition of phosphate has therefore been greatly reduced. At the same time, it can be seen that the nitrates in the surface water of the Lake of Zürich were no longer completely consumed in August 1969 and 1970. According to Fig. 4, 0.6 mg/l. NO_3^- remains in solution. On the basis of earlier investigations, it is to be assumed that phosphate is again the growth-limiting factor in the Lake of Zürich. Further investigations are being carried out.

CHANGE IN THE PHOSPHATE—NITRATE RELATION WITH INCREASING LAKE EUTROPHICATION

Observations on numerous Swiss lakes led us to draw up the scheme shown in Fig. 5. The scheme is concerned with the process of lake eutrophication

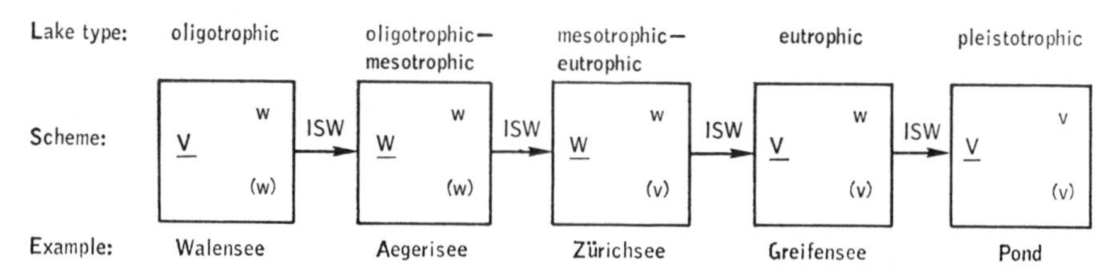

Figure 5. Scheme of eutrophication expressed through maximal values of surface water in summer. *V*: more than 0.2 mg/l NO_3^-; *W*: less than 0.2 mg/l NO_3^-; *v*: more than 20 mg/m³ PO_4^{3-}; *w*: less than 20 mg/m³ PO_4^{3-}; (*v*): more than 40 mg/m³ PO_4^{3-} after ignition; (*w*): less than 40 mg/m³ PO_4^{3-} after ignition; ISW: incoming sewage water.

and the behavior of phosphates and nitrates in the epilimnion in summer. As has been shown before, the limiting factor for algal growth in oligotrophic lakes on which man has had little or no influence is still phosphate. Free nitrate ions were present in these lakes throughout the year. Elimination of sewage nitrate in these cases would therefore be useless.

The daily addition of a great deal of sewage phosphate to an oligotrophic lake has the result that plankton algae completely consume the nitrates in the surface water. As all phosphate ions discharged into the lake are at first combined in an organic substance, the content of free phosphate does not increase in the surface water. If more and more phosphates and nitrates are discharged into a lake, these ions cannot always be completely utilized in the surface water. This leads to eutrophic and pleistotrophic types of lakes (Fig. 5).

In the case of the Lake of Zürich, the daily addition of phosphate has been reduced as a result of phosphate elimination in purification plants in recent years. Oligotrophication has taken place (Thomas, 1971).

SUMMARY

Phosphates, being fertilizers, contribute to the growth of plants in rivers and lakes. They particularly accelerate the growth of benthal and planktonic algae. Proliferation of algae may be designated as secondary pollution of the water. The algal masses die from time to time and their decomposition leads to tertiary pollution. Oligotrophic lakes with clear water contain very little phosphate but often considerable amounts of nitrate. Sewage phosphates then have a particularly unfavorable effect in such circumstances.

It has long been known that ground with a definite chemical composition produces the same type of plant communities under the same geographical conditions. Chalk-loving plants develop on ground that is rich in chalk and plants that dislike chalk flourish on ground where it is lacking. Similarly, a lack or an abundance of certain other substances such as molybdenum, copper, vanadium, cobalt, boron, arsenic, and so on, or of organic substances, has a selective effect on the development of certain types of algae and algal communities. Until today, no way that can be economically operated has been found by which the algal growth in water can be reduced with certainty by diminishing the addition of such substances. Iron is another metal that is also constantly being added to lake water because the ground contains rich iron compounds.

Apart from the phosphate from sewage, lakes receive few phosphate ions, so the question arises as to whether oligotrophication can be achieved in a lake by reducing the addition of phosphate. Measures taken in two large lakes provide answers to this question. The sewage that was previously discharged into Lake Washington is now cleaned in purifying plants and then pumped into the sea. After a short time, Edmondson (9) confirmed great improvements in the lake, which was then free of incoming phosphate.

The sewage from the catchment areas of the Lake of Zürich is now mechanically and biologically cleaned in 12 purification plants and freed of phosphates by treatment with $FeCl_3$ (66). Since this has been done, the water of the Lake of Zürich has been clear and greenish-blue again and also contains far fewer algae. The oxygen content of the lake has been more favorable in the last four years than in the previous four decades (Thomas, 1971). If these improvements can also be maintained during dry years, the amelioration of the lake may be regarded as very satisfactory. Even today, it may be stated with certainty that the treatment of sewage with $FeCl_3$ has had no unfavorable effects at all. The further course of developments in the lake will be tested.

REFERENCES

1. Ambühl, H., "Die Künstliche Belüftung des Pfäffikersees; Orientierung über die ersten Ergebnisse," *Verb. Schweiz. Abwasserfachleute, Ber. Nr.* 77/3, (1962).
2. Ambühl, H., "Die Nährstoffelimination aus der Sicht des Limnologen," *Schweiz. Z. Hydrol.*, **26,** 569–594 (1964).
3. Atkins, W. R. G., "The Phosphate Content of Fresh and Salt Waters in Its Relationship to the Growth of Algal Plankton," *J. Mar. Biol. Assoc. U.K.*, **13,** 119 (1923).
4. Auerbach, M., "Die Oberflächen-und Tiefenströme im Bodensee," *Deut. Wasserwirtsch.*, **34,** 193–202, 358–366 (1939).
5. Bachmann, H., "Temperaturen des Ritomsees," *Schweiz. Z. Hydrol.*, **2,** H.3, 21–28 (1924).
6. Bosset, E., "Incidences hygiéniques de la vaccination des eaux de boisson au moyen de polyphosphates," *Monatsbull. Schweiz. Ver. Gas Wasserfachmännern*, **45,** 146–148 (1968).

7. Edmondson, W. T., "Changes in Lake Washington Following an Increase in the Nutrient Income," *Verh. Int. Verein. Limnol.*, **14,** 167–175 (1961).
8. Edmondson, W. T., "Changes in the Oxygen Deficit of Lake Washington," *Int. Soc. Theor. Appl. Limnol., Proc.*, **16,** 153–158 (1966).
9. Edmondson, W. T., "The History of Lake Washington," in *Metro—The First Ten Years*, Municipality of Metropolitan Seattle, 1969, 39 pages.
10. Edmondson, W. T., "Eutrophication in North America," in *Eutrophication, Causes, Consequences, Correctives*, Proceedings of a Symposium, National Academy of Sciences, Washington, D.C., 1969, pp. 124–149.
11. Elster, H. J., and H. Lehn, *Bodensee-Projekt der deutschen Forschungsgemeinschaft*, Franz Steiner Verlag, Wiesbaden, 1963.
12. Eschmann, K. H., "Die Sanierung des Wilersees durch Ableitung des Tiefenwassers," *Gesundheitstech. (Zürich)*, **3,** *Nr.* 5, 125–128 (1969).
13. Findenegg, I., "Beiträge zur Kenntnis des Ossiacher Sees," *Carinthia* 2 (1934).
14. Forel, F. A., *Handbuch der Seenkunde; Allgemeine Limnologie*, Verlag J. Engelhorn, Stuttgart, 1901, 249 pages.
15. Frey, D. G., *Limnology in North America*, The University of Wisconsin Press, Madison, Wisconsin, 1963, 734 pages.
16. Gächter, R., "Phosphorhaushalt und planktische Primärproduktion im Vierwaldstättersee (Horwer Bucht)," *Schweiz. Z. Hydrol.*, **30,** 1–66 (1968).
17. Gessner, Fr., "Phosphat und Nitrat als Produktionsfaktoren der Gewässer," *Verh. Int. Verein. Limnol.*, **7,** 525–538 (1935).
18. Gessner, Fr., "Die Phosphorarmut der Gewässer und ihre Beziehungen zum Kalkgehalt," *Int. Rev. Hydrobiol.*, **40,** 197–207 (1939).
19. Gessner, Fr., *Hydrobotanik* II, VEB Deutscher Verlag der Wissenschaften, Berlin, 1959, 701 pages.
20. Goldman, C. R., "A Method of Studying Limiting Factors in situ in Water Columns Isolated by Polyethylene Films," *Limnol. Oceanogr.*, **7,** 99–101 (1962).
21. Golterman, H. L., and Clymo, R. S., "Chemical Environment in the Aquatic Habitat," *Proceedings I.B.P. Symposium*, N. V. Noord-Hollandsche Uitgevers Maatschappij, Amsterdam, 1967, 322 pages.
22. Grote, A., "Der Sauerstoffhaushalt der Seen," *Binnengewässer*, **14,** (1934).
23. Juday, Ch., and Birge, E. A., "A Second Report on the Phosphorus Content of Wisconsin Lake Waters," *Trans. Wis. Acad. Sci.*, **26,** 353 (1931).
24. Kiefer, F., *Naturkunde des Bodensees*, Jan Thorbecke Verlag, Lindau und Konstanz, 1955, 169 pages.
25. Kliffmüller, R., "Beiträge zum Stoffhaushalt des Bodensees; Stickstoff- und Phosphor-Haushalt," *Arch. Hydrobiol., Suppl. XXXV*, **3,** 309–371 (1969).
26. Kolkwitz, R., and Marsson, M., "Oekologie der pflanzlichen Saprobien," *Ber. Deut. Bot. Ges.*, **26,** 505–519 (1908).
27. Kolkwitz, R., and M. Marsson, "Oekologie der tierischen Saprobien," *Int. Rev. Ges. Hydrobiol. Hydrogr.*, **2,** 126–152 (1909).
28. Kutschke, I., "Die thermischen Verhältnisse im Zürichsee zwischen 1937 und 1963 und ihre Beeinflussung durch meteorologische Faktoren," *Vierteljahrsschr. Naturforsch Ges. Zürich*, **111,** 47–124 (1966).

29. Lauterborn, R., "Die Vegetation des Oberrheins," *Verh. Naturhist. Med. Ver. Heidelberg, N. F.*, **10**, 450–502 (1910).
30. Liebmann, H., "Tegernsee," *Münchn. Beitr: Z. Abwasser-, Fisch., Flussbiol*, Bd. 6, (1959).
31. Liepolt, R., "Die limnologischen Verhältnisse des Zellersees, seine Verunreinigung und Sanierung," Föderation Europäischer Gewässerschutz, Symposium Salzburg, Sept. 1966.
32. Mercier, P. and S. Gay, "Effets de l'aération artificielle souslacustre au lac de Bret," *Schweiz. Z. Hydrol.*, **16**, 248–308 (1954).
33. Minder, L., "Der Zürichsee im Lichte der Seetypenlehre," *Neujahrsblatt Naturforsch Ges. Zürich* (1943).
34. Naumann, E., "Grundzüge der regionalen Limnologie," *Die Binnengewässer*, Bd. 11, 1–176 (1932).
35. Nümann, W., "Die Möglichkeiten der Gewässerreinigung mit höheren Pflanzen nach den bisherigen Untersuchungsergebnissen und theoretischen Ueberlegungen," *Int. Rev. Ges. Hydrobiol.*, **55**, 149–158 (1970).
36. Odermatt, J., "Limnologische Charakterisierung des Lauerzersees mit besonderer Berücksichtigung des Planktons," *Schweiz. Z. Hydrol.*, **32**, 1–75 (1970).
37. Ohle, W., "Der labile Zustand ostholsteinischer Seen," *Der Fischwirt*, **12**, 7 p. (1951).
38. Olszewski, P., "Versuch einer Ableitung des hypolimnischen Wassers aus einem See," *Verh. Int. Verein. Limnol.*, **14**, 855–861 (1961).
39. Pleisch, P., "Die Herkunft eutrophierender Stoffe beim Pfäffiker- und Greifensee," *Vierteljahrsschr. Naturforsch Ges. Zürich*, **115**, 127–229 (1970).
40. Provasoli, L., "Algal Nutrition and Eutrophication," in, *Eutrophication, Causes, Consequences, Correctives*, Proceedings of a Symposium, National Academy of Sciences, Washington, D.C., 1969, p. 574–593.
41. Rivier, O., "Recherches hydrobiologiques sur le lac de Morat," *Bull. Soc. Neuchâteloise Sci. Nat.*, **61**, (1936). (Diss. 1937, Neuchâtel).
42. Stangenberg, M., "Zur Oligotrophie des Ohrid-Sees," *Arch. Hydrobio. Rybactwa, T. XII, Suwalki*, 153–178 (1939).
43. Stewart, K. M., and G. A. Rohlich, *Eutrophication—A Review*, California State Water Quality Control Board, Publ. No. 34. 1967, 188 pages.
44. Thienemann, A., "Der Sauerstoff im eutrophen und oligotrophen See," *Binnengewässer*, **4**, 1–175 (1928).
45. Thomas, E. A., "Limnologische Untersuchungen am Türlersee," *Schweiz. Z. Hydrol.*, **11**, 90–177 (1948).
46. Thomas, E. A., "Regionallimnologische Studien an 25 Seen der Nordschweiz," *Verh. Int. Verein. Limnol.*, **10**, 489–495 Cf. 479 (1949).
47. Thomas, E. A., "Zur Bekämpfung der See-Eutrophierung: Empirische und experimentelle Untersuchungen zur Kenntnis der Minimumstoffe in 46 Seen der Schweiz und angrenzender Gebiete," *Monatsbull. Schweiz. Ver. Gas Wasserfachmännern*, 25–32, 71–79 (1953).
48. Thomas, E. A., "Phosphatgehalt der Gewässer und Gewässerschutz," *Monatsbull. Schweiz. Ver. Gas Wasserfachmännern*, **35**, 224–231, 271–277 (1955).
49. Thomas, E. A., "Stoffhaushalt und Sedimentation im oligotrophen Aegerisee und im

eutrophen Pfäffiker- und Greifensee," *Mem. Ist. Ital. Idrobiol., Suppl.* **8,** 357–465 (1955); Colloque IUBS 19.

50. Thomas, E. A., "Ueber die Bedeutung der abwasserbedingten direkten Sauerstoffzehrung in Seen," *Monatsbull. Schweiz. Ver. Gas Wasserfachmännern,* **35,** 119–129 (1955).
51. Thomas, E. A., "Das Plankton-Test-Lot, ein Gerät zum Studium des Verhaltens von Planktonorganismen im See," *Monatsbull. Schweiz. Ver. Gas Wasserfachmännern,* (1958).
52. Thomas, E. A., "Sauerstoffminima und Stoffkreisläufe im ufernahen Oberflächenwasser des Zürichsees (Cladophora- und Phragmites-Gürtel)," *Monatsbull. Schweiz. Ver. Gas und Wasserfachmännern,* **40,** 140–147.
53. Thomas, E. A., "Vergleiche über die Planktonproduktion in Flaschen und im Plankton-Test-Lot," *Verh. Int. Verein. Limnol.,* **14,** 140–146 (1961).
54. Thomas, E. A., *The Eutrophication of Lakes and Rivers, Cause and Prevention, Biological Problems in Water Pollution,* Third Seminar, U.S. Department of Health, Education and Welfare, 1962.
55. Thomas, E. A., "Thermisch bedingte Horizontalzirkulationen, Wasserchemismus und Algenwucherungen an Zürichseeufern," *Hydrobiol.,* **20,** 40–58 (1962).
56. Thomas, E. A., "Versuche über die Wachstumsförderung von Cladophora- und Rhizoclonium-Kulturen durch Bakterienstoffe," *Ber. Schweiz. Bot. Ges.,* **73,** 504–518 (1963).
57. Thomas, E. A., "Massenentwicklung von Lamprocystis roseo-persicina als tertiäre Verschmutzung am Ufer des Zürichsees," *Vierteljahrsschr. Naturforsch. Ges. Zürich,* **109,** 267–276 (1964).
58. Thomas E. A., "Seetypen und Gewässerschutz," *Vierteljahrsschr. Naturforsch. Ges. Zürich,* **109,** 511–517 (1964).
59. Thomas, E. A., "Phosphat-Elimination in der Belebtschlammanlage von Männedorf und Phosphat-Fixation in See-und Klärschlamm," *Vierteljahrsschrift Naturforsch. Ges. Zürich,* **110,** 419–434 (1965).
60. Thomas, E. A., "Der Verlauf der Eutrophierung des Zürichsees," *Mittl. Oesterr. Sanit.,* **66,** (1965).
61. Thomas, E. A.," Der Pfäffikersee vor, während und nach künstlicher Durchmischung," *Verh. Int. Verein. Limnol.,* **16,** 144–152 (1966).
62. Thomas, E. A., "Die Phosphat-Hypertrophie der Gewässer; Notwendigkeit und technische Möglichkeit der Zufuhr-Drosselung," *Chem. Weekblad,* **63,** 305–319 (1967).
63. Thomas, E. A., "Die Phosphattrophierung des Zürichsees und anderer Schweizerseen," *Mitt. Int. Ver. Limnol.,* **14,** 231–242 (1968).
64. Thomas, E. A., "The Process of Eutrophication in Central European Lakes," in *Eutrophication,* Proceedings of a Symposium, National Academy of Sciences, Washington, D.C., 1969, pp. 29–49.
65. Thomas, E. A., "Untersuchungen über Auswirkungen und Folgen der Einschichtung von Abwässern in Seen," *Verh. Int. Verein. Limnol.,* **17,** 226–239 (1969).
66. Thomas, E. A., and H. Rai, "Betriebserfahrungen mit Phosphatelimination bei 10 kommunalen Kläranlagen im Kanton Zürich," *Gas, Wasser, Abwasser,* **50,** 179–190 (1970).

67. Vollenweider, R., *Scientific Fundamentals of the Eutrophication of Lakes and Flowing Waters, with Particular Reference to Nitrogen and Phosphorus as Factors in Eutrophication*, Organisation for Economic Co-operation and Development, September 1968, DAS/CSI 68. 27.

68. Vollenweider, R., *A Manual on Methods for Measuring Primary Production in Aquatic Environments*, Blackwell, Oxford and Edinburgh, 1969, 213 pages.

69. Wagner, G., "Beiträge zum Sauerstoff-, Stickstoff- und Phosphorhaushalt des Bodensees," *Arch. Hydrobiol.*, **63,** 86–103 (1967).

70. Wagner, G., "Die Zunahme der Belastung des Bodensees," *gwf-Wasser/Abwasser*, **111,** 485–487 (1970).

71. Yoshimura, S., "Seasonal Variation in Content of Nitrogenous Compounds and Phosphate in the Water of Takasuka Pond, Saitama, Japan," *Arch. Hydrobiol.*, **24,** 155 (1930).

Eutrophication and Biological Associations

KENNETH M. MACKENTHUN

Director, Division of Applied Technology, Environmental Protection Agency, Washington, D.C.

The enrichment of waters by nutrients through either man-created or natural means along with the attendant biological phenomena defines the term eutrophication. Present knowledge indicates that phosphorus and nitrogen are the chemical constituents usually responsible for the eutrophication phenomenon. Other elements are essential such as carbon, vitamins, and trace elements but often these are not limiting to nuisance biological development in natural lakes and streams.

Lund (48) in his thorough literature review stated that "Nitrogen and phosphorus can still be considered as two of the major elements limiting primary production. In some tropical and highly eutrophic temperate lakes, nitrogen may be a more important limiting factor than phosphorus. In many other lakes phosphorus is present in very low concentrations and seems to be the major factor limiting production. Evidence from the addition of fertilizers to fish ponds and from what is known about the eutrophication of lakes by sewage supports the view that phosphorus plays a major role in production." Carbon, as well as molybdenum, has been found to be limiting in particular natural waters (27, 42).

Evidence indicates that: (*a*) high phosphorus concentrations are associated with accelerated eutrophication of waters when other growth-promoting factors are present; (*b*) aquatic plant problems develop in reservoirs or other standing waters at phosphorus values lower than those critical in flowing streams; (*c*) reservoirs and other standing waters collect phosphates from influent streams and store a portion of these within consolidated sediments; and (*d*) phosphorus concentrations critical to noxious plant growths vary, and a given concentration will produce nuisance growths in one geographical area but not in another. Potential contributions of phosphorus to the aquatic environment have been indicated in the literature (Table 1).

The discharge of domestic sewage increases the concentration of phosphorus markedly. Organic phosphorus in the sewage and simple and complex

TABLE 1 Pounds of Phosphorus Contributed to Aquatic Ecosystems (51)

Major contributors
Sewage and sewage effluents: 3 lb/(capita) (year)[a]
Some industries, for example, potato processing: 1.7 lb/ton processed
Phosphate rock from 23 states (53)
Cultivated agricultural drainage: 0.35–0.39 lb/acre drained per year (24, 73, 86)
Surface irrigation returns, Yakima River Basin: 0.9–3.9 lb/(acre) (year) (81)
Benthic sediment releases
Minor contributors
Domestic duck: 0.9 lb/year (72)
Sawdust: 0.9 lb/ton (22)
Rainwater[b]
Groundwater, Wisconsin: 1 lb/9 million gal (40)
Wild duck: 0.45 lb/year (62)
Tree leaves: 1.8–3.3 lb/acre of trees per year (17)
Dead organisms; animal excretions

[a] Various researchers have recorded the annual per capita contribution of phosphorus in pounds from domestic sewage as 2 to 4 (15), 2 and 3 (56), 1.9 (61), and 3.5 (75).
[b] Influenced by pollution present in atmosphere "washed out" by rainfall.

phosphates from synthetic detergents are the principal contributions. Decomposition of the organic material, along with soluble phosphates, results in phosphorus concentrations in excess of the requirements for plant growth. The readily available soluble phosphorus often furnishes a food source for nuisance biological growths.

SEWAGE

The discharge of human wastes results in an abundance of nitrogen in all forms, causing an abrupt change in the nutrient balance of the stream. When untreated domestic sewage is discharged to a watercourse, organic nitrogen (proteins) and ammonia are the principal nitrogen constituents. In the water, nitrifying organisms decompose the organic materials and oxidize the ammonia to nitrite and nitrate. Since the nitrite ion is a transient form it is usually present in very low concentrations.

Treated sewage has undergone partial oxidation in the treatment process. Therefore the nitrite and nitrate forms are increased in well-treated sewage, while the organic nitrogen and ammonia are reduced.

Phosphorus is added to receiving waters principally as a component of pollution. Once added, it is combined with other constituents in populations of bacteria, algae, vascular plants, and fish and in benthic sediments. Once nutrients are combined within the ecosystem of the receiving waters, their removal is tedious and expensive; removal must be compared to inflowing quantities to evaluate accomplishment. In a lake, reservoir, or pond, phosphorus is removed naturally only by outflow, by insects that hatch and fly out of the drainage basin, by harvesting a crop, such as fish, and by combination with consolidated bottom sediments. Even should adequate harvesting methods be available, the expected standing crop of algae per acre exceeds 2 tons and contains only about 1.5 lb of phosphorus. Similarly, submerged aquatic plants approach at least 7 tons/acre (wet weight) and contain 3.2 lb/acre of phosphorus. Probably only half of the standing crop of submerged aquatic plants can be considered harvestable. The harvestable fish population (500 lb) from 3 acres of water would contain only 1 lb of phosphorus.

Sawyer (74) discussed factors that influence the development of nuisance algal growths in lakes. The surface area is important since the accumulations of algae along the shoreline of a large lake under a given set of wind conditions could easily be much larger than on a small lake, under equal fertilization per acre. The shape of the lake determines to some degree the amount of fertilizing matter the lake can assimilate without algal nuisances since prevailing winds blowing along a long axis will concentrate the algal production from a large water mass into a relatively small area. The most offensive conditions develop during periods of very mild breezes that tend to skim the floating algae and push them toward shore. Shallow lakes, too, respond differently than deep stratified lakes in which the deeper waters are sealed off by a thermocline. In the nonstratified waters all the nutrients dissolved in the water are potentially available to support an algal bloom. In stratified waters, only the nutrients confined to the epilimnion are available except during those brief periods when complete circulation occurs.

Chu (18) found that optimum growth of all organisms studied in cultures can be obtained in nitrate–nitrogen concentrations from 0.9 to 3.5 mg/l. and phosphorus concentrations from 0.09 to 1.8 mg/l., while a limiting effect on all organisms will occur in nitrogen concentrations from 0.1 mg/l. downward and in phosphorus concentrations from 0.009 mg/l. downward. The lower limit of optimum range of phosphorus concentration varies from about 0.018 to about 0.09 mg/l., and the upper limit from 8.9 to 17.8 mg/l. when nitrate is the source of nitrogen, while it lies at about 17.8 for all the planktons studied when ammonium is the source of nitrogen. Low phosphorus concentrations may, therefore, like low nitrogen concentrations, exert a selective limiting influence on a phytoplankton population. The nitrogen concentration determines to a large extent the amount of chlorophyll formed. Nitrogen concentrations beyond the optimum range inhibit the formation of chlorophyll in green algae.

SEDIMENTS

Keup (43), in flowing water studies, found that phosphorus is temporarily stored in bottom sediments or transported as a portion of the stream's bed load after its removal from the flowing water. Long-term storage is affected when the phosphorus is pooled in deltas or deposited on flood plains. Keup reviewed the literature on phosphorus discharges by specific streams (Table 2).

Sediments may serve only to support the water, or they may have a profound effect on the quality of the water that comes in contact with them. In a lake which man has not polluted seriously, the lake bed will resemble the soils of the surrounding land. As man "civilized" an area by plowing fertilized fields, and by discharging sewage and industrial wastes to the watercourse, lake-bed sediments assumed different characteristics because of the materials that became a part of them. Concentrations of certain materials in the sediments became greater, the soil chemistry more complex, and biological populations more numerous and specialized.

Matter that can settle may transport nutrients to the sediments by ion-exchange and sorption mechanisms. As coarser and denser materials settle rapidly, large quantities of nutrients may be effectively removed and buried. Thomas (84) observed a phosphorus reduction (as P) from 2000 to 150 μg/l. as particulate matter settled from water passing through a 25-mi-long reservoir that received 6,629,000 tons of sediments annually. The particle size of the suspended sediment was very small and was comprised of 54% clay, 40% silt, and 6% sand. Also, once a dissolved nutrient is incorporated into an organism, the tendency is for it to deposit as a solid. Metabolic cycling may delay settling of some elements such as nitrogen, carbon, or phosphorus; for other elements such as silicon, when "fixed" as a diatom valve, the fate of deposition is nearly assured.

The contribution of nutrients and phosphorus in particular from consolidated lake-bed sediments to the water's biodynamic cycle is a variable factor that depends to a great extent on the physical-chemical aspects of the environment. There is evidence to indicate that in an undisturbed mud-water system the amount of phosphorus released to the superimposed water is very small (30, 87).

Nutrients in the sediment have been found to be more important as a growth contributor for sago pond weed than nutrients in the water (64).

TABLE 2 Phosphorus Discharged by Selected North American Streams (43)

Principal Land Use	River	Number of Analyses	Season of Sampling	Drainage Area (mi^2)	Phosphorus (P) [lb/(annum) (mi^2)]	Population Density (mi^2)	Ref.
Forested	West Branch Sturgeon R. Mich.	27+	July	14	37	Sparse	8
	Pigeon, Minn.	4	Aug. and Sept.	600	28	Sparse	67, 68, 5
	Poplar, Minn.	4	Aug. and Sept.	114	21	Sparse	67, 68, 5
	Baptism, Minn.	4	Aug. and Sept.	140	42	Sparse	67, 68, 5
	St. Louis, Minn.	4	Aug. and Sept.	3430	58	Sparse	67, 68, 5
	Bois Brule, Wis.	4	Aug. and Sept.	113	97	Sparse	67, 68, 5
	Bad, Wis.	4	Aug. and Sept.	611	78	Sparse	67, 68, 5
	Montreal, Wis.	4	Aug. and Sept.	281	98	Sparse	67, 68, 5
	Black, Mich.	4	Aug. and Sept.	202	65	Sparse	67, 68, 5
	Presque Isle, Mich.	4	Aug. and Sept.	260	39	Sparse	67, 68, 5
	Ontonagon, Mich.	4	Aug. and Sept.	1290	44	Sparse	67, 68, 5
	Yakima, Wash.	?	Annual	182	473	Sparse	81
	Tieton, Wash.	?	7 months	237	492	Sparse	81
	Cedar, Wash.	?	Annual	125	204	Sparse	81
	Mulligan, Maine	12	4 seasons	21	4	Sparse	6
	Stetson, Maine	19	4 seasons	29	20	Sparse	6
	East Branch Sebasticook, Maine	56	4 seasons	56	128[a]	>63[b]	6
	Ellershe, Prince Edward Island	44	April–Dec.	10	113	Sparse	79
	Pigeon, N.C.	18	July	133	97	Light	This article
	Johnathans, N.C.	5	July	65	201	Light	This article
	Kankakee, Ind. and Ill.	6	June–Sept.	5280	139	28	34, 4
	Vermillion, Ill.	8	June–Sept.	1230	179	36	34, 4
	Fox, Ill. and Wis.	7	June–Sept.	2570	489	145	34, 4
Agricultural	Kaskaskia, Ill.	100	April–Dec.	5220	225	>174[b]	24
Urban	Streams near Madison, Wis.	?	?	?	235–262	?	73
	Du Page, Ill.	5	June–Sept.	325	18	380	34, 4
	Des Plaines, Ill. and Wis.						
	Above confluence with Chicago River	5	June–Sept.	635	570	1270	34, 4
	Total basin (includes Chicago River)	19	June–Sept.	2180	4020	2570	34, 4
	Chicago,Ill.	16	June–Sept.	810	6540	5650	34, 4

[a] One seasonal (9 months) industry contributes approximately 75%.
[b] Only sewered population known.

Plants with extensive root systems aid in recycling nutrients that have been buried below the interface and are otherwise unavailable to the overlying water.

BENTHIC ORGANISMS

Benthic organisms may transfer nutrients when that exchange is not reduced or prohibited by overlying materials. In a study on Connecticut lake sediments, Hutchinson and Wollack (36) found that diffusion of phosphorus from the mud may be aided by the metabolic activities of benthic organisms. Studies by Hooper and Elliott (33) on two species of protozoa indicated that organisms were capable of breaking down organic phosphates to inorganic phosphorus in aerobic conditions.

In addition to metabolic activities, benthic organisms may, through burrowing activities, resuspend or redeposit nutrients on the mud surface that would otherwise have been lost from the system. Aquatic oligochaetes may ingest quantities of material 2 to 3 cm below the interface, and midges may scrape up detritus from a depth of 5 to 10 mm (65). Aquatic organisms such as fishes also contribute to the overturn of bottom muds. In fish ponds located in Israel, phosphorus fixation was higher when mud was mixed with water by carp in the ponds (31). Other bottom-feeding fish such as catfish and bullheads probably contribute also to the overturn of bottom muds and the resultant release of nutrients as they disturb the bottom during feeding activities.

EFFECTS OF EUTROPHICATION

Jónasson (38) concluded that the bottom fauna fits into an ecological pattern set by primary production of algae, vertical distribution and abundance of macrophytes, dissolved oxygen, and nutrients. Increasing the supply of nutrients to the epilimnion causes increases in the standing crop and in the production of phytoplankton; transparency decreases; subsurface light dwindles; the macrophytes are excluded from deeper waters and eventually from the lake because of inadequate light; periods of dissolved oxygen deficiency become more prolonged; hypolimnionic pH decreases; and alkalinity increases. These environmental factors all have an adverse effect on benthos and may result in restricting the benthic inhabitants to a few midges and worms.

Larkin and Northcote (45) note that the eutrophication of lakes affects fish in many ways. These result primarily from the increase in production, the consequent deoxygenation of the hypolimnion and other waters, and the alteration of many other features of the biological environment that determine survival and abundance of various fish species. The abundance of food organisms caused by eutrophication may accelerate greatly the growth rate of the fish. On the other hand, eutrophic environments may force certain species such as ciscoes to live under undesirable conditions of temperature and dissolved oxygen, and they will fail to thrive even in the presence of abundant food (32).

When given the opportunity and because fish are mobile, they may respond to adverse environmental changes by moving from the area, to which they may return when conditions for existence become improved. On other occasions they may not be given the opportunity. Mackenthun *et al.* (52) reported an extensive mortality of fish resulting from the decomposition of algae that were flushed to the Yahara River through the control gates on Lake Kegonsa, Wisconsin. The lake was made eutrophic principally because of the inflow of treated sewage effluent. This, and particular climatological phenomena, resulted in a prolific algal growth that formed a thick scum several acres in area. When this decomposing mass was flushed to the river it eliminated the dissolved oxygen, and the water exhibited toxic properties.

As noted by Larkin and Northcote (45):

> More than 40 years ago, A. S. Pearse studied several lakes in Wisconsin, and his review on the ecology of lake fishes summarizes major differences in the quantity and species composition among the various lake types (63). Increasing eutrophy is associated with greater production. The largest oligotrophic lakes are dominated by salmonids and coregonines, whereas smaller oligotrophs support centrachids in abundance as well as coregonines. Such eutrophic lakes as Mendota, in Pearse's day produced large quantities of perch, largemouth bass, white bass, rock bass, carp, and buffalofish. The shallow Lake Wingra (maximum depth, 4.3 m) produced large quantities of carp, crappie, sunfish, dogfish, and perch. In the words of Pearse, 'Each lake presents a type in which one or more species of fishes may be at their best and become dominant.' It is scarcely surprising that with the changes attendant upon eutrophication, changes in fish populations should ensue.

Enrichment may cause both an increase and a decrease in fish growth in different stream sections. Environmental changes resulting from enrichment influence the total stream length inhabited by particular associations of fish. The coarse fishes normally associated with downstream reaches tend to move into the enrichment zone and often the finer fishes are reduced substantially or eliminated.

ASSESSMENT OF NUTRIENT PROBLEM

To assess a nutrient problem properly, consideration should be given to all of those sources that may contribute nutrients to the watercourse. These sources could include sewage, sewage effluents, industrial wastes, land drainage, applied fertilizers, precipitation, urban runoff, soils, and nutrients released from bottom sediments and from decomposing plankton. Transient waterfowl, falling tree leaves, and groundwater may contribute important additions to the nutrient budget. Flow measurements are paramount in a study to assess quantitatively the respective amounts contributed by these various sources during different seasons and at different flow characteristics. In the receiving lake or stream the quantities of nutrient contained by the standing crops of algae, aquatic vascular plants, fish, and other aquatic organisms are important considerations. A knowledge of those nutrients that are harvested annually through the fish catch, or that may be removed from the system through the emergence of insects, will contribute to an understanding of the nutrient budget.

The interaction of specific chemical components in water, prescribed fertilizer application rates to land and to water, minimal nutrient values required for algal blooms, vitamins required, other limiting factors, and the intercellular nitrogen and phosphorus concentrations are likewise important. Usually, it is necessary to determine that portion of the nutritive input attributable to man-made or man-induced pollution that may be corrected as opposed to that input that is natural in origin, and therefore usually not correctable. A nutrient budget is used to determine the annual input to a system, the annual outflow, and that which is retained within the water mass to recycle with the biomass or become combined with the solidified bottom sediments. The carbon, nitrogen, phosphorus, and their respective ratios are important values to aid in the identification of a material, to calculate the amount of major nutrients contained within a segment of the biomass or a stratum of sediment, and from which to judge the relative input of nutrients to the water mass when the ecosystem component undergoes decomposition, or natural chemical change (Table 3).

FIELD INVESTIGATIONS

The conduct of a field investigation to define the effects of eutrophication on the living aquatic resource involves a number of important sequential considerations. These considerations are formulation of objectives to define the problem and delimit the scope of the study; planning in detail the logical

TABLE 3 Carbon, Nitrogen, and Phosphorus in Freshwater Environmental Constituents (50)

Constituent	Standing Crop (lb/acre)		%C[a]	%N[a]	%P[a]	Ratio		Ref.
	Wet	Dry				C:N	N:P	
Phytoplankton	1,000–3,600	100–360						11
				6.8	0.69		10	26
			39	6.1	0.64	6.5	10	54
				9.0	0.52		17	11
Attached algae	2,000	200						59
				2.8	0.14		2	11
Vascular plants	14,000	1,800						69, 70
				1.8	0.18		10	28
Myriophyllum				3.2	0.52		6	11
Vallisneria				1.8	0.23		8	77, 78
Potamogeton				1.3	0.13		10	77, 78
Castalia				2.8	0.27		10	77, 78
Najas				1.9	0.30		6	77, 78
Myriophyllum				3.0	0.5		6	2
Bottom organisms								
Midges	200–400	40–80						21, 58
Chironomus				7.4	0.9		8	14
Hyalella				7.4	1.2		6	11
Hirudinea				11.1	0.8		14	11
Sialis				8.1	0.6		14	11
Fish	150–600							80
				2.5[b]	0.2[b]		10	9
				2.8	0.18–0.49			13
					0.19			46

TABLE 3 (continued)

Constituent	Standing Crop (lb/acre) Wet	Dry	%C[a]	%N[a]	%P[a]	Ratio C:N	N:P	Ref.
					0.20			55
					0.29			82
				2.6–3.3	0.18–0.24			37
Domestic wastes[a]					5.1–10.6[d]			24
				45[d]	8[d]		6	55
				20–40[d]	5.3–10.6[d]		4	15
				61.3[d]	10.7[d]		6	60
				18–28[d]	3.5–9.0[d]			7
Sediments								
Lake Tahoe			0.6–19.8	0.6–1.6		4–25		55
Wisconsin lakes			4.4–40.5	0.6–3.6	0.12–0.6	8–14	5–6	12, 41
Madison, Wisconsin lakes				0.7–0.9	0.1–0.12		6–9	76
Green Lake				0.6	0.17		4	82
Lake Sebasticook			10–34	0.3–1.8	0.06–0.16	8–44	5–16	54
Klamath Lake			8.6	1.2		7		Thomas, N. A., unpublished[f]
Boston Harbor			2.3–5.0	0.06–0.41				Stewart, R. K., 1968[e]
Organic river sediments			0.03	0.0027		12		25
Pulp and paper wastes in river			5.3	0.23		22		25

Untreated domestic wastes	3.54	0.3		12		25
Untreated chemical and fertilizers and domestic wastes	3.15	0.12		26		25
No tributary wastes	0.55	0.05		11		25
Sand: silt; clay; loam	0.4–2.1	0.02–0.10		20		Ballinger and McKee, unpublished[f]
Stable sludge; peat; organic debris	2.0–5.0	0.10–0.20		20–25		Ballinger and McKee, unpublished[f]
Paper mill wastes	6–15	0.10–0.30		50–60		Ballinger and McKee, unpublished[f]
Packinghouse wastes	2.8–4.3	0.30–0.50		8–10		Ballinger and McKee, unpublished[f]
Fresh sludge; decaying algae; sewage solids	5–40	0.70–5.0		7–8		Ballinger and McKee, unpublished[f]
Log pond bark	50.6	0.5	0.02	100	25	Thomas, N. A., unpublished[f]
Sewage sludge in river	5.8	0.28	0.18	21	2	Thomas, N. A., unpublished[f]
Algae; sawdust; sewage	14.6	0.93	0.11	16	9	Thomas, N. A., unpublished[f]

TABLE 3 (continued)

	Standing Crop (lb/acre)					Ratio		
Constituent	Wet	Dry	%C[a]	%N[a]	%P[a]	C:N	N:P	Ref.
Leaf litter			28.3	1.63	0.11	17	15	Warner, R. W., *et al.*, 1969[g]
Sand			0.2	0.02	0.005	10	4	Warner, R. W., *et al.*, 1969[g]
Loam			2.7	0.19	0.02	14	10	Warner, R. W., *et al.*, 1969[g]
Muck			7.3	0.52	0.04	14	13	Warner, R. W., *et al.*, 1969[g]
Floating waste wool			37–43	3.4–4.7	0.08–0.09	9–11	38–58	[h]

[a] As the total element in percentage of the dry weight, unless specified otherwise.
[b] Calculated on wet weight.
[c] Average sewage flow can be calculated at 100 gal per capita per day.
[d] mg/l.
[e] *Biological Aspects of Water Quality, Charles River and Boston Harbor, Massachusetts*, by R. K. Stewart, Technical Advisory and Investigations Branch, Cincinnati, Ohio, 1968.
[f] Technical Advisory and Investigations Branch, Cincinnati, Ohio.
[g] Analyses of soil types from *Black Water Impoundment Investigations*, by R. W. Warner, R. K. Ballentine, and L. E. Keup, Technical Advisory and Investigations Branch, U.S. Department of the Interior, Cincinnati, Ohio, 1969.
[h] *Fertilization and Algae in Lake Sebasticook, Maine*, Department of Health, Education, and Welfare, Technical Advisory and Investigations Activities, Cincinnati, Ohio, 1966.

events that will lead to a successful study and the many details necessary to ensure success in each phase of the investigation; data collection, which involves a selection of sampling sites, a judgment of the required number of samples, and a decision on the proper time, type, periodicity, and extent of sample collection; sample and data analyses and interpretation; and reporting of results with conclusions, recommendations, and predictions.

The first field study in the United States to address itself to the complex problem of determining a lake nutrient budget was that of Sawyer *et al.* (76). The essence of this report was later published (73). This 2-year study showed that Lake Waubesa, at Madison, Wisconsin, received at least 75% of its inorganic nitrogen and 88% of its inorganic phosphorus from sewage effluent. One facet of this study was historic because from it came the now famous and oft-quoted conclusion that a 0.30 mg/l. concentration of inorganic nitrogen (N) and a 0.010 mg/l. concentration of soluble phosphorus (P) at the start of the active growing season could produce nuisance algal blooms. This conclusion was based on the correlation of results of monthly nutrient and algal sample examinations from 16 southeastern Wisconsin lakes. Although these observations were confined to one geographical area, they have been substantiated reasonably well in subsequent field and laboratory studies on waters in which the total methyl orange alkalinity exceeds 40 mg/l.

PRESERVATION

To prevent biological nuisances in most waters, total phosphorus should not exceed 100 μg/l. P at any point within the flowing stream, nor should

TABLE 4 Total-to-Soluble Phosphorus Ratios in Water

Water	Total P to Soluble P	Ref.
Western Lake Erie	3.5	16
Detroit River mouth	5–7	PHS Detroit Project
Linsley Pond, Conn.	10.0	35
Northern Wisconsin lakes	7.0	40
Northeast Wisconsin lakes	2–10	39
Ontario lakes (8)	17	71
Southeast Wisconsin lakes (17)	9	Mackenthun, unpublished
Rock River, Wis.	2–15	Mackenthun, unpublished
Sebasticook Lake, Maine	2.8 Winter	54
	12.7 Spring	54
	7.0 Summer	54
	4.1 Fall	54

TABLE 5 Lake Nutrient Loadings and Retentions

Lake	State	Nitrogen (N)		Phosphorus (P)		Ref.
		Loading [lb/(year acre)]	Retention (%)	Loading [lb/(year acre)]	Retention (%)	
Washington	Wash.	280	—	12		1
Mendota	Wis.	20[a]	—	0.6[b]		3
Monona	Wis.	81[a]	48–70	7.5[b]	64–88	44
Waubesa	Wis.	435[a]	50–64	62.8[b]	−26–25	44
Kegonsa	Wis.	162[a]	44–61	35.9[b]	−21–12	44
Tahoe	Calif.	2	89	0.4	93	47
Koshkonong	Wis.	90	80	40	30–70	Mackenthun, unpublished
Green	Wash.	—	—	4.8	55	82
Geist	Ind.	440[a]	44	28	25	Mackenthun, unpublished
Sebasticook	Maine	—	—	2	48	Mackenthun, unpublished
Ross R. Barnett	Miss.	—	—	32	—	Mackenthun, unpublished

[a] Inorganic nitrogen only.
[b] Soluble phosphorus only.

50 μg/l. be exceeded where waters enter a lake, reservoir, or other standing water body (49). Those waters now containing less phosphorus should not be degraded because even lower concentrations may be critical in very low alkalinity waters. Adequate phosphorus controls must now be directed toward treatment of nutrient point sources and to wastewater diversion around the lake or dilution within the lake, where feasible.

PHOSPHORUS SOLUBILITY DISTRIBUTION

Total-to-soluble phosphorus ratios may vary from 2 to 17 or even 90%, dependent on the particular water, season, aquatic plant populations, and probably other factors (Table 4). These ratios are of value when they can be determined periodically within the same water body and changes in them correlated with volumetric response changes within the algal mass.

The nutrient loading to the lake on a unit basis gives some measure of comparability among various water bodies (Table 5). Likewise, a lake or reservoir usually retains a portion of those nutrients that it receives from its various sources. The amount or percentage of the nutrients that may be retained by a lake or reservoir is variable and will depend on (*a*) the nutrient loading to the lake or reservoir; (*b*) the volume of the euphotic zone; (*c*) the extent of biological activity; (*d*) the detention time within the basin or time allotted for biological activity; and (*e*) the level of the penstock or discharge from the basin.

Long-term remedial measures might be focused on reducing the nutrient concentration in troublesome areas or in altering some aspect of the topography that concentrates or fosters the development of nuisance algae or aquatic weeds. Such measures often involve costly physical modifications to correct existing conditions, as well as future planning to assure wise use of the area's natural aquatic resources.

REFERENCES

1. Anderson, G. C., "Recent Changes in the Trophic Nature of Lake Washington—A Review," *Algae and Metropolitan Wastes*, U.S. Public Health Service, SEC TR W61-3, 1961, pp. 27–33.
2. Anderson, R. R., R. G. Brown, and R. D. Rappleye, "Mineral Composition of Eurasian Water Milfoil, *Mypiophyllum Spicatum L.*," *Chesapeake Sci.*, **6** (1), 68–72 (1965).
3. Anonymous, *Report on Lake Mendota Studies Concerning Conditions Contributing to Occurrence of Aquatic Nuisances 1945–1947*, Wisconsin Committee on Water Pollution, Madison, 1949, 19 pages (mimeographed).

4. Anonymous, *Report on the Illinois River System, Water Quality Conditions*, Great Lakes Illinois River Basin Project, Department of Health, Education, and Welfare, 1963, Part II, Tables, Chapters 2 and 3 (not paginated).
5. Anonymous, *Compilation of Records of Surface Waters of the United States, October 1950 to September 1960*, St Lawrence River Basin Geological Survey Water Supply Paper 1727, Washington, D.C., 1964, Part 4, 379 pages.
6. Anonymous, *Fertilization and Algae in Lake Sebasticook, Maine*, Technical Advisory and Investigations Activities, Federal Water Pollution Control Administration, Cincinnati, Ohio, 1966, 124 pages (mimeographed).
7. Anonymous, "Sources of Nitrogen and Phosphorus in Water Supplies," Am. Water Works Assoc. Task Group Report 2610-P, *J. Am. Water Works Assoc.*, **59** (3), 344–366 (1967).
8. Ball, R. C., and F. F. Hooper, "Translocation of Phosphorus in a Trout Stream Ecosystem," in *Radioecology*, V. Schultz, and A. W. Klement, Jr. (Eds.) Reinhold, New York, 1963, pp. 217–228.
9. Beard, H. R., *Nutritive Value of Fish and Shellfish*, Report U.S. Commissioner of Fisheries for 1925, 1926, pp. 501–552.
10. Benoit, R. J., and J. J. Curry, "Algae Blooms in Lake Zoar, Connecticut," *Algae and Metropolitan Wastes*, U.S. Public Health Service, SEC TR W61-3, 1961, pp. 18–22.
11. Birge, E. A., and C. Juday, *The Inland Lakes of Wisconsin, The Plankton. I. Its Quantity and Chemical Composition*, Wisconsin Geological and Natural History Bulletin No. 64, Scientific Series No. 13, 1922, pp. 1–222.
12. Black, C. S., "Chemical Analyses of Lake Deposits," *Trans. Wis. Acad. Sci., Arts Lett.*, **24,** 127–133 (1929).
13. Borgstrom, G., Ed., *Fish as Food*, Academic Press, New York, 1961, 725 pages.
14. Borutsky, E. V., "Dynamics of the Total Benthic Biomass in the Profundal of Lake Beloie," *Proc. Kossino Limn. Sta. of the Hydrometeorological Service, U.S.S.R.*, **22,** 196–218 (1939), translated by M. Ovchynnyk, edited by R. C. Ball and F. F. Hooper.
15. Bush, A. F., and S. F. Mulford, *Studies of Waste Water Reclamation and Utilization*, California State Water Pollution Control Board, Sacramento, California, Publ. No. 9, 1954.
16. Chandler, D. C., and O. B. Weeks, "Limnological Studies of Western Lake Erie, V. Relation of Limnological and Meteorological Conditions to the Production of Phytoplankton in 1942," *Ecol. Monogr.*, **15,** 436–456 (1945).
17. Chandler, R. F., Jr., "Amount and Mineral Nutrient Content of Freshly Fallen Needle Litter of Some Northeastern Conifers," *Soil Sci. Soc. Am. Proc.*, **8,** (1943).
18. Chu, S. P., "The Influence of the Mineral Composition of the Medium on the Growth of Planktonic Algae, Part II. The Influence of the Concentration of Inorganic Nitrogen and Phosphate Phosphorus," *J. Ecol.*, **31** (2), 109–148 (1943).
19. Dean, J. M., "The Effect of Sewage on a Chain of Lakes in Indiana," *Hydrobiol.*, **24,** 435–440 (1964).
20. Deevey, E. S., "The Obliteration of the Hypolimnion," *Mem. Inst. Ital. Idrobiol., Suppl.*, **8,** 9–38 (1955).
21. Dineen, C. F., "An Ecological Study of a Minnesota Pond," *Am. Midl. Nat.*, **50** (2), 349–356 (1953).

22. Donahue, R. L., *Our Soils and Their Management*, The Interstate Printers and Publishers, Danville, Illinois, 1961, 568 pages.
23. Edmondson, W. T., "Eutrophication in North America," in *Eutrophication: Causes, Consequences, Correctives*, National Academy of Sciences, Washington, D.C., 1969, pp. 124–149.
24. Engelbrecht, R. S., and J. J. Morgan, "Land Drainage as a Source of Phosphorus in Illinois Surface Waters," *Algae and Metropolitan Wastes*, U.S. Public Health Service, SEC TR W61-3, 1961, p. 74.
25. Finger, J. H., and T. A. Wastler, "Organic Carbon–Organic Nitrogen Ratios of Sediments in a Polluted Estuary," *J. Water Pollut. Control Fed.*, **41** (2), R101–109 (1969).
26. Gerloff, G. C., and F. Skoog, "Cell Content of Nitrogen and Phosphorus as a Measure of Their Availability of Growth of *Microcystis aeruginosa*," *Ecol.*, **35** (3), 348–353 (1954).
27. Goldman, C. R., "Micronutrient Limiting Factors and Their Detection in Natural Phytoplankton Populations," *Mem. Ist. Ital. Idrobiol, Suppl.*, **18,** 121–135 (1965).
28. Harper, H. J., and H. R. Daniel, "Chemical Composition of Certain Aquatic Plants," *Bot. Gaz.*, **96,** 186 (1939).
29. Hasler, A. D., "Eutrophication of Lakes by Domestic Sewage," *Ecol.*, **28,** No. 4, 383–395 (1947).
30. Hasler, A. D., "Natural and Artificially (Air-Plowing) Induced Movement of Radioactive Phosphorus from the Muds of Lakes," *Int. Conf. on Radioisotopes in Scientific Res.*, UNESCO/NS/RIC/188 (Paris), **4,** 1 (1957).
31. Hepher, B., "On the Dynamics of Phosphorus Added to Fish Ponds in Israel," *Limnol Oceanogr.*, **3** (1), 84–100 (1958).
32. Hile, R., "Age and Growth of the Cisco, *Leucichthys artedii* (Le Sueur), in Lakes of the Northeastern Highlands, Wisconsin," *Bull. U.S. Bur. Fish.*, **48,** 211–217 (1936).
33. Hooper, F. F., and A. M. Elliott, "Release of Inorganic Phosphorus from Extracts of Lake Mud by Protozoa," *Trans. Am. Microscop. Soc.*, *LXXII*, **3,** 276–281 (1953).
34. Hurwitz, E., R. Beaudoin, and W. Walters, "Phosphates, Their "Fate" in a Sewage Treatment Plant-Waterway System," *Water and Sewage Works*, **112,** 85–89, 112 (1965).
35. Hutchinson, G. E., *A Treatise on Limnology*, New York, 1957, 1015 pages.
36. Hutchinson, G. E., and A. Wollack, "Studies on Connecticut Lake Sediments, II. Chemical Analyses of a Core from Linsley Pond, North Branford," *Am. J. Sci.*, **238** (7), 493–517 (1940).
37. Ingalls, R. L., *et al.*, *Nutritive Value of Fish from Michigan Waters*, Michigan State College Agricultural Experimental Station Bulletin No. 219, 1950, pp. 1–24.
38. Jonasson, P. M., "Bottom Fauna and Eutrophication," in *Eutrophication: Causes, Consequences, Correctives*, National Academy of Sciences, Washington, D.C., 1969, pp. 274–305.
39. Juday, C. E., *et al.*, "Phosphorus Content of Lake Waters of Northeastern Wisconsin," *Trans. Wis. Acad. Sci., Arts Lett.*, **23,** 233–248 (1927).
40. Juday, C., and E. A. Birge, "A Second Report on the Phosphorus Content of Wisconsin Lake Waters," *Trans. Wis. Acad. Sci., Arts Lett.*, **26,** 353 (1931).
41. Juday, C., E. A. Birge, and V. W. Meloche, "Chemical Analysis of the Bottom Deposits of Wisconsin Lakes, II. Second Report," *Trans. Wis. Acad. Sci., Arts Lett.*, **33,** 99–114 (1941).

42. Kerr, P. C., Federal Water Quality Administration, Athens, Georgia, personal communication, 1970.
43. Keup, L. E., "Phosphorus in Flowing Waters," in Water *Research*, Vol. 2, Pergamon Press, New York, 1968, pp. 373–386.
44. Lackey, J. B., and C. N. Sawyer, "Plankton Productivity of Certain Southeastern Wisconsin Lakes as Related to Fertilization, I. Surveys," *Sewage Works J.*, **17** (3), 573–585 (1945).
45. Larkin, P. A., and T. G. Northcote, "Fish as Indices of Eutrophication," in *Eutrophication: Causes, Consequences, Correctives*, National Academy of Sciences, Washington, D.C., 1969, pp. 256–273.
46. Love, R. M., J. A. Lovern, and N. R. Jones, *The Chemical Composition of Fish Tissues*, Department of Scientific and Industrial Research Special Report No. 69, H.M.S. Stationery Office, London, 1959, 62 pages.
47. Ludwig, H. F., E. Kazmierczak, and R. C. Carter, "Waste Disposal and the Future at Lake Tahoe," *J. Sanit. Eng. Div., Proc. Am. Soc. Civil Eng.*, **90** (SA3), Paper 3947, 27–51 (1964).
48. Lund, J. W. G., "The Ecology of Freshwater Phytoplankton," *Biol. Rev.*, **40,** 231–293 (1965).
49. Mackenthun, K. M., "The Phosphorus Problem," *J. Am. Water Works Assoc.*, **60** (9), 1047–1054 (1968).
50. Mackenthun, K. M., "Writing a Water Quality Report," *J. Water Pollut. Control Fed.*, **11** (1), 82–88 (1969).
51. Mackenthun, K. M., *The Practice of Water Pollution Biology*, U.S. Department of the Interior, Federal Water Pollution Control Administration, Washington, D.C., 1969, 281 pages.
52. Mackenthun, K. M., E. F. Herman, and A. F. Bartsch, "A Heavy Mortality of Fishes Resulting from the Decomposition of Algae in the Yahara River, Wisconsin," *Am. Fish. Soc. Trans.*, **75,** 175–180 (1948).
53. Mackenthun, K. M., and W. M. Ingram, *Biological Associated Problems in Freshwater Environments, Their Identification, Investigation and Control*, U.S. Department of the Interior, Federal Water Pollution Control Administration, 1961, 287 pages.
54. Mackenthun, K. M., L. E. Keup, and R. K. Stewart, "Nutrients and Algae in Lake Sebasticook, Maine," *J. Water Pollut. Control Fed.*, **40** (2), R72–R81 (1968).
55. McGauhey, P. H. *et al.*, *Comprehensive Study on Protection of Water Resources of Lake Tahoe Basin through Controlled Waste Disposal*, prepared for the Board of Directors, Lake Tahoe Area Council, Al Tahoe, 1963, 157 pages.
56. Metzler, D. F. *et al.*, "Emergency Use of Reclaimed Water for Potable Supply at Chanute, Kansas," *J. Am. Water Works Assoc.*, **50** (8), 1021 (1958).
57. Miller, W. E., and J. C. Tash, *Interim Report, Upper Klamath Lake Studies, Oregon*, Federal Water Pollution Control Administration, Pacific Northwest Water Laboratory, Corvallis, Oregon, 1967, 37 pages.
58. Moyle, J. B., *A Biological Survey of the Upper Mississippi River System* (*in Minnesota*), Minnesota Department of Conservation and Fisheries Inv. Report No. 10, 1940, 69 pages.
59. Neil, J. H., *Nature of Growth in a Report on Algae, Cladophora*, Report of Ontario Water Resources Commission, 1958, pp. 3–7.

60. Oswald, W. J., "Metropolitan Wastes and Algal Nutrition," *Algae and Metropolitan Wastes*, U.S. Public Health Service, SEC TR W61-3, 1960, pp. 88–95.
61. Owen, R., "Removal of Phosphorus from Sewage Plant Effluent with Lime," *J. Water Pollut. Control Fed.*, **25** (5), 548 (1953).
62. Paloumpis, A. A., and W. C. Starrett, "An Ecological Study of Benthic Organisms in Three Illinois River Flood Plain Lakes," *Am. Midl. Nat.*, **64** (2), 406 (1960).
63. Pearse, A. S., "Ecology of Lake Fishes," *Ecol. Monogr.*, **4,** 475–480 (1934).
64. Peltier, W. H., and E. B. Welch, *Factors Affecting Growth of Rooted Aquatic Plants*, T. V. A. Report, Division of Health and Safety, 1968, 45 pages (mimeographed).
65. Pennak, R. W., *Fresh-Water Invertebrates of the United States*, Ronald Press, New York, 1953, 769 pages.
66. Phinney, H. K., and C. A. Peek, "Klamath Lake, an Instance of Natural Enrichment," *Algae and Metropolitan Wastes*, U.S. Public Health Service, SEC TR W61-3, 1961, pp. 22–27.
67. Putnam, H. D., and T. A. Olson, *A preliminary Investigation of Nutrients In Western Lake Superior*, School of Public Health, University of Minnesota, 1959, 32 pages (mimeographed).
68. Putnam, H. D., and T. A. Olson, *Ar. Investigation of Nutrients in Western Lake Superior*, School of Public Health, University of Minnesota, 1960, 24 pages (mimeographed).
69. Rickett, H. W., "A Quantitative Study of the Larger Aquatic Plants Of Lake Mendota," *Trans. Wis. Acad. Sci., Arts Lett.*, **20,** 501–522 (1922).
70. Rickett, H. W., "A Quantitative Study of the Larger Aquatic Plants of Green Lake, Wisconsin," *Trans. Wis. Acad. Sci., Arts Lett.*, **21,** 381–414 (1924).
71. Rigler, F. H., "The Phosphorus Fractions and the Turnover Time of Inorganic Phosphorus in Different Types of Lakes," *Limnol. Oceanogr.*, **9** (4), 511–518 (1964).
72. Sanderson, W. W., "Studies of the Character and Treatment of Wastes from Duck Farms," *Proc. 8th Ind. Waste Conf., Purdue Univ. Ext. Ser.*, **83,** 170–176 (1953).
73. Sawyer, C. N., "Fertilization of Lakes by Agricultural and Urban Drainage," *J. New England Water Works Assoc.*, **61,** 109 (1947).
74. Sawyer, C. N., "Factors Involved in Disposal of Sewage Effluents to Lakes," *Sewage Ind. Wastes*, **26** (3), 317–325 (1954).
75. Sawyer, C. N., "Problem of Phosphorus in Water Supplies," *J. Am. Water Works Assoc.*, **57** (11), 1431 (1965).
76. Sawyer, C. N., J. B. Lackey, and R. T. Lenz, *An Investigation of the Odor Nuisances Occurring in the Madison Lakes, Particularly Monona, Waubesa and Kegonsa from July 1942–44*, Report of Governor's Committee, Madison, Wisconsin, two volumes, 1945 (mimeographed).
77. Schuette, H. A., and H. Alder, "Notes on the Chemical Composition of Some of the Larger Aquatic Plants of Lake Mendota, II. *Vallisneria* and *Potamogeton*," *Trans. Wis. Acad. Sci., Arts Lett.*, **23,** 249–254 (1928).
78. Schuette, H. A., and H. Alder, "Notes on the Chemical Composition of Some of the Larger Aquatic Plants of Lake Mendota, III. *Castalia odorata* and *Najas flexilis*," *Trans. Wis. Acad. Sci., Arts Lett.*, **24,** 135–139 (1929).
79. Smith, M. R., "Phosphorus Enrichment of Drainage Waters from Farm Lands," *J. Fish. Res. Board Can.*, **6,** 887–895 (1959).

80. Swingle, H. W., *Relationships and Dynamics of Balanced and Unbalanced Fish Populations*, Agricultural Experimental Station, Alabama Polytechnic Institute, Bulletin No. 274, Auburn, Alabama, 1950, pp. 1–74.
81. Sylvester, R. O., "Nutrient Content of Drainage Water from Forested, Urban and Agricultural Areas," *Algae and Metropolitan Wastes*, U.S. Public Health Service, SEC TR W61-3, 1961, p. 80.
82. Sylvester, R. O., and G. C. Anderson, "A Lake's Response to its Environment," *J. Sanit. Eng. Div., Proc. Am. Soc. Civil Eng.*, **90** (SA1), 1–22 (1964).
83. Thomas, E. A., "The Process of Eutrophication in Central European Lakes," in *Eutrophication: Causes, Consequences, Correctives*, National Academy of Sciences, Washington, D.C., 1969, pp. 29–49.
84. Thomas, N. A., "Biological Investigations of Tuttle Creek Reservoir, Kansas," 1969, manuscript.
85. Trelease, W., "The Working of the Madison Lakes," *Trans. Wis. Acad. Sci., Arts Lett.*, **7**, 121–129 (1889).
86. Weibel, S. R., personal communication, 1965.
87. Zicker, E. L., K. C. Berger, and A. D. Hasler, "Phosphorus Release from Bog Lake Muds," *Limnol. Oceanogr.*, **1** (4), 296 (1956).

34

Phosphorus and Ecology

CLAIR N. SAWYER, PhD

Consultant, Sun City, Arizona

"OHNE PHOSPHOR KEIN LEBEN" ANONYMOUS

ECOLOGICAL CONSIDERATIONS

In considering the ecological aspects of phosphorus, or any other element for that matter, it is well to remember that a finite amount of each exists and that the Law of Conservation of Matter applies to all, save the radioactive ones. Seldom are the elements per se of significance in environmental considerations, except for mining and refining operations involved in winning the elements from their ores. Of vastly greater importance are the compounds that are generated from the elements to meet the demands of our modern civilization. Although these compounds are usually widely disseminated throughout populated, and sometimes unpopulated, areas of the world, some of them, unfortunately, tend to become concentrated in certain areas. The soluble phosphate compounds are a classical example.

PHOSPHORUS DISTRIBUTION IN VARIOUS ECONOMIES

Agrarian

Cropping of land exerts a constant drain upon the phosphorus resources of the soil. Table 1 shows the phosphorus content of a wide variety of crops and food products derived from them. Continual removal of crops without recycling results in a depletion of available phosphorus in the soil, and crop yields eventually become limited by the amount released by natural weathering action of the soil.

TABLE 1 Phosphorus Content of Various Food Products[a]

		Crops	
Crop	Percent P	Crop	Percent P
Apples	0.011	Oats	0.318
Asparagus	0.055	Onions	0.039
Barley	0.343	Peanuts	0.394
Beans, green	0.050	Peas, green	0.124
Cabbage	0.038	Potatoes	0.053
Celery	0.101	Prunes, dried	0.068
Corn, field	0.110	Rice, whole	0.310
Figs, fresh	0.021	Soybeans	0.633
Grapes	0.018	Walnuts	0.309
Lentils, dried	0.392	Wheat	0.342

		Food Products	
Product	Percent P	Product	Percent P
Beef	0.198	Flour, white	0.096
Butter	0.004	Milk	0.088
Cheese, hard	0.547	Oatmeal	0.351
Chicken	0.218	Pork	0.262
Cream	0.048	Sardines	0.550
Eggs	0.111	Veal	0.235
Fish	0.221	Wheat bran	1.430

[a] Source: W. H. Petersen, J. T. Skinner, and F. M. Strong, *Elements of Food Biochemistry*, Prentice-Hall, Englewood Cliffs, N.J., 1949.

In spite of the fact that Oriental civilizations had practiced recycling of animal and even human wastes for centuries to maintain the fertility of the soil, European farmers of the early nineteenth century were careless in this regard and farmlands became seriously "run out." It was not until the midnineteenth century that the famous agricultural chemist, Justus von Liebig, established the fundamentals of plant fertilization and urged the farmers to conserve animal dung and return it to the fields.

The system of conservation and recycling shown in Fig. 1 served to maintain the fertility and, thereby, the productivity of farmlands in an agrarian society very well. Each farm was an ecosystem which, in order to remain viable, had to recycle the phosphorus and other fertilizers back to the fields. Such losses, as occurred due to soil erosion, were replaced to a large extent

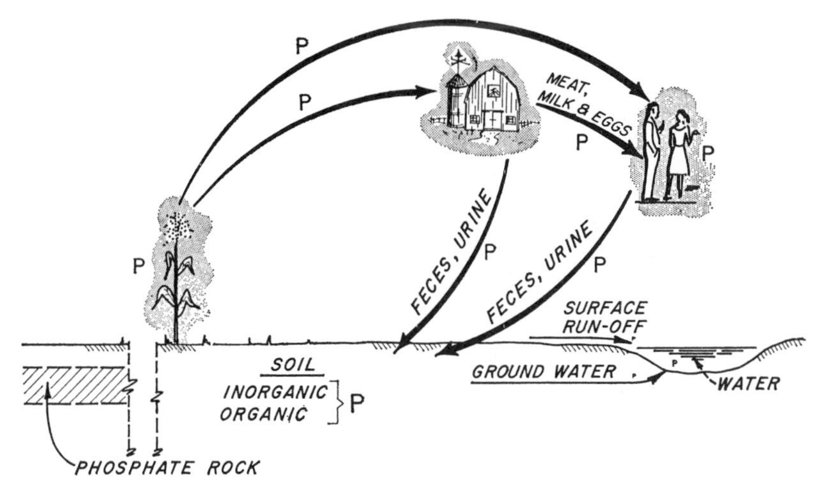

Figure 1. Phosphorus distribution in an agrarian economy.

through natural weathering of the rocks and soil, and each farm became a balanced ecosystem in itself, with productivity controlled by the character of the soil and the diligence with which conservation and recycling were practiced.

Simple Urban

With the industrial revolution, a marked change occurred in the living habits of the people. Many left the farms and rural areas to live in cities to be near the new jobs in the mills. All demanded food for their tables, however. The farmers now had a market for their produce and began to export certain crops, meat, and dairy products to the urban centers. Along with this marketable material goes much of the phosphorus and other fertilizing matter removed from the soil by the crops.

In the city, most of the food is consumed and, through human digestion and metabolic processes, a major part of the phosphorus, as well as other fertilizing elements taken from the soil, is released as waste products in human excrement. The amount of major fertilizing matters retained by the body is very small as shown in Table 2.

Because most of our cities and towns are sewered, nearly all of the phosphorus contained in food exported from our farms eventually finds its way into the sewers, as shown in Fig. 2, with little likelihood of ever being returned to the land to produce more crops.

The export of crops and animal products from farms soon led to depletion of even the richest soils, and productivity has been maintained only by

TABLE 2 Fate of Nutrients in Human Foods (9) (65-Year Life Span)

Element	Amount in Food[a] (lb)	Retained in Body at Death (lb)	Amount Excreted (%)
P	80	1.5	98.1
N	840	4.0	99.5
K	175	0.6	99.7

[a] Based on consumption of 17.5 tons of dry solids.

mining phosphate deposits to produce fertilizers to replace that "cropped" from the soil. As urban populations have grown, exports from the farms have increased, and the amount of phosphorus applied to farmlands in commercial fertilizers has increased, as shown in Fig. 3, in order to maintain fertility.

Modern Urban

Concurrent with our expanding urban populations and exportations of food from farm to city, there has been a marked expansion in the production of phosphatic compounds for industrial uses as shown in Fig. 4. Since 1948, the increase has been influenced greatly by the demand for complex phosphates in the formulation of synthetic detergents which have had wide application on the farm, in the city, and in industry.

Figure 5 shows the major avenues by which phosphorus moves through our modern economy and how a large part has only one useful life and is

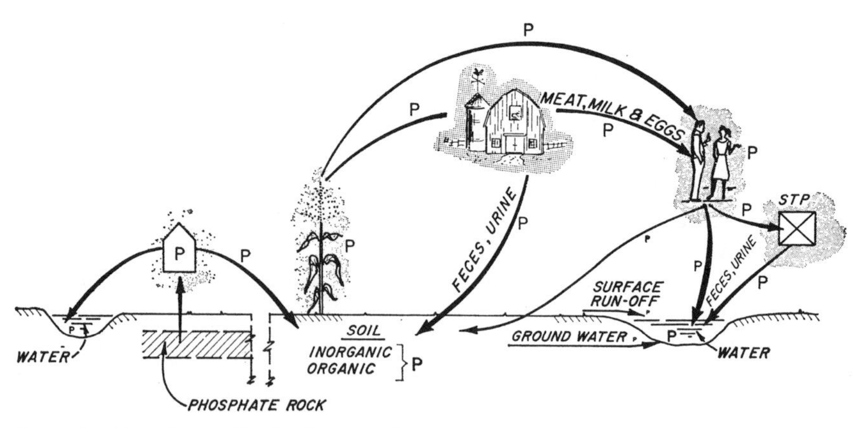

Figure 2. Phosphorus distribution in a simple urban economy.

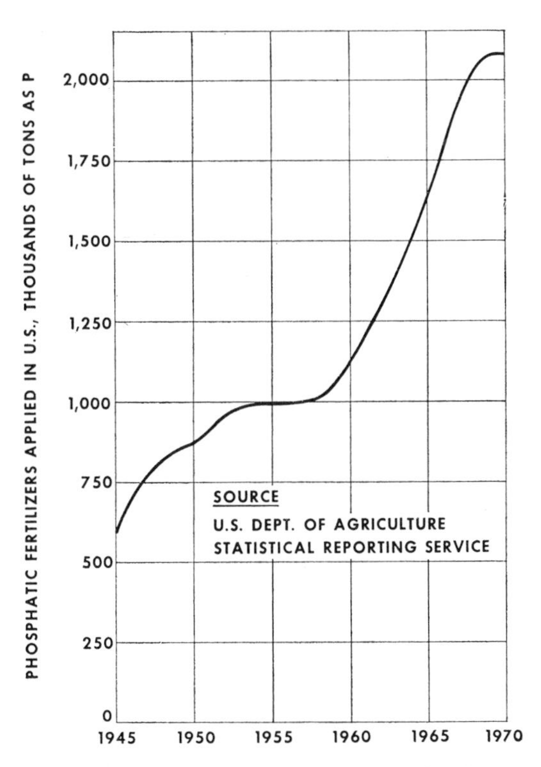

Figure 3. Phosphorus applied to farmlands in commercial fertilizers.

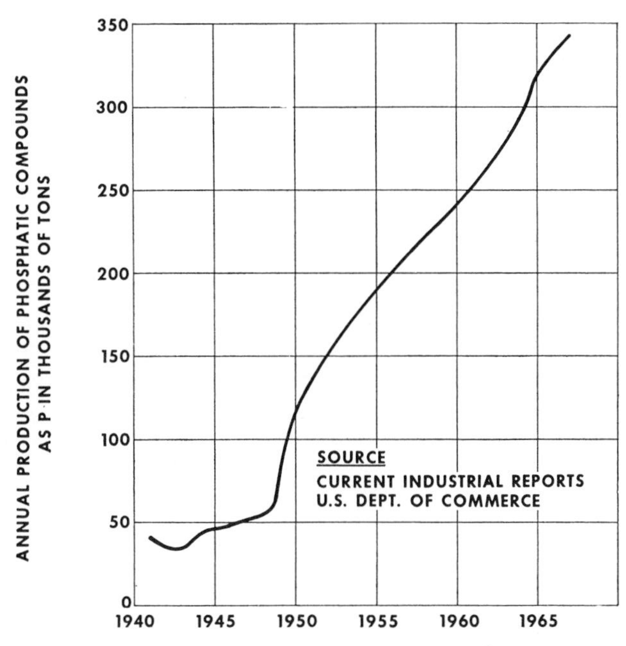

Figure 4. United States production of phosphatic compounds.

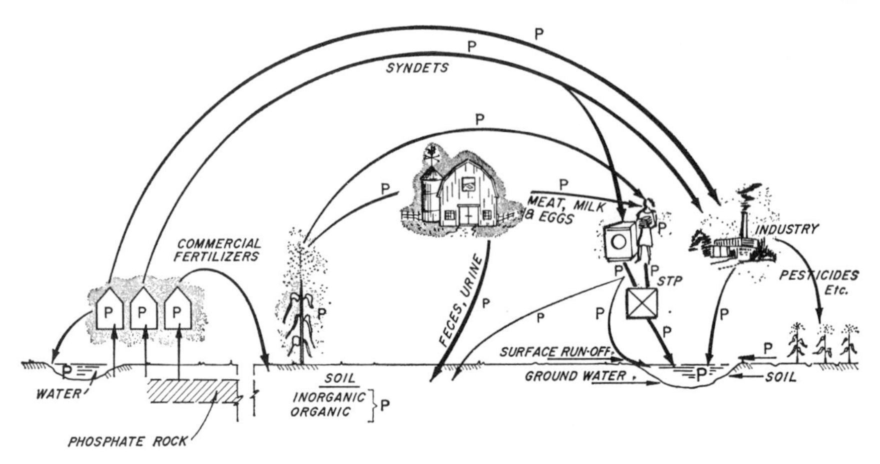

Figure 5. Phosphorus distribution in a complex urban economy.

then lost to the water environment via sewerage systems. It is this transfer of phosphorus from land-based ecosystems to aquatic ecosystems that makes this chapter worth writing. Once phosphorus enters sewer systems, man has lost control of it, except where recovery processes are employed. It is free to create distortions in the aquatic environment, since sewers are the recipient of most all wastewaters of domestic and industrial origin.

OTHER PATHWAYS OF PHOSPHORUS LOSSES FROM THE LAND

Agricultural

Although the export of farm products to urban areas constitutes a major avenue by which phosphorus escapes from agricultural areas, it is by no means the sole one. In spite of the fact that most soils have adequate exchange capacity to retain phosphate ions effectively to the point that very little phosphorus percolates to contaminate groundwaters, there are other losses due to wind and water erosion that must be recognized from the ecological viewpoint.

Losses through wind erosion have been reduced markedly since the historic dust storms of the 1930s through improved farming practices, particularly strip farming and the transfer of marginal lands from tilled to grassland. Nevertheless, significant losses do occur as is evidenced by analysis of snow and rain samples. Studies conducted in the Lake of the Woods area (12) have shown both rain and snow to contain about 0.006 mg/l. of phosphorus. Since the earth's surface is about 70% water, it is reasonable to assume that about 70% of the phosphorus losses resulting from wind erosion are deposited in the aquatic environment.

The major uncontrolled loss of phosphorus from agricultural lands is due to water erosion, or surface runoff. This is due to the actual physical suspension and transport of top soil and ground litter during periods of heavy rainfall. Natural litter, per se, is not a particularly important factor in loss of phosphorus because such matter is usually refuse such as straw, cornstalks, and the like, with little nutrient value. However, if the litter consists of animal manures, the loss can be considerable and the beneficiary is the receiving stream, lake, or reservoir.

Losses of phosphorus from farmlands, resulting from both percolation and surface runoff, have been reported as ranging from 160 to 262 lb/(year) (mi^2) (2, 8, 10). Minshall *et al.* (6) have reported losses of 60 lb/year in base flow. It is recognized that losses are influenced greatly by the slope of the land but improved farming practices, principally contour farming, have reduced losses materially. Of the losses that occur from agricultural areas, it is difficult to ascribe the fraction due to commercial fertilizers added, that due to recycling of manures, and that due to phosphorus native to the soil.

Forest Lands

There is a continual loss of phosphorus from forested lands because the root systems abstract phosphorus from the soil and transmit it to the living tree. Much of the phosphorus concentrates in the leaves. When the leaves fall and decompose, a large part of the phosphorus remains in the "duff" and gradually feeds back into the soil, with only minor amounts escaping in water running from the area. However, fires which burn trees, leaves, and "duff" release the phosphorus in soluble inorganic forms that can be extracted and carried away in relatively high concentration and amount.

Studies (8, 12) have shown that normal forest drainage can be expected to carry from 22 to 115 lb total P/(mi^2) (year), with only a minor part, 4 to 9 lb, being in a soluble form.

Urban Areas

The assessment of phosphorus losses from urban areas is fraught with many complications. The major one relates to whether storm drainage includes any domestic or industrial wastewater, and the second pertains to whether the people "curb" domesticated animals on grassed or paved areas. Studies

TABLE 3 Phosphorus Losses from Agricultural, Urban, and Forested Areas in Potomac Watershed

	Total P (lb/(mi^2)(year))
Agricultural	160
Urban	135
Forest	60

in the Potomac watershed (8) have indicated losses of the order shown in Table 3.

DOMESTIC WASTEWATERS

Domestic wastewaters vary considerably in composition and concentration, depending mainly on the habits and customs of the people, water consumption, and whether a separate or combined sewer serves the area. However, populations consume essentially the same amount of food per day, regardless of rainfall, consequently the amount of phosphorus and other nutrients discharged to the sewer in human excreta is quite constant. Total phosphorus, however, may vary markedly from day to day, depending on the habits of the people, particularly with respect to laundry and cleaning operations when synthetic detergents are used. In general, the nature of the sewer system has little bearing on the concentration of nutrients in domestic wastewaters under dry weather conditions. Under wet weather conditions, the concentration of nutrients decreases in combined sewer discharges because of dilution by surface drainage. Although, theoretically, the total quantity of nutrients should reach the treatment plants day by day, this seldom happens during severe or prolonged storms because of emergency overflows that bypass raw wastes en route. Some older separate sewer systems have many of the characteristics of combined systems because they are not watertight. Under dry weather conditions exfiltration occurs, and under wet weather conditions infiltration may be so severe as to result in bypassing of wastewater.

The composition of typical municipal wastewaters in terms of phosphorus and nitrogen content is shown in Table 4. A major point of interest is concerned with the $\pm 200\%$ increase in phosphorus that has occurred during the period 1945 to 1970, largely as a result of the use of synthetic detergents, and the effect it has had on the nitrogen–phosphorus ratio.

Numerous studies (8) during recent years on per capita contributions of phosphorus to municipal sewers have shown them to range from 2.90 to

TABLE 4 Phosphorus and Nitrogen Content of a Typical Municipal Wastewater

mg/l	1945	1970
Inorganic P	3	8
Organic P	1	1
Inorganic N	20	20
Organic N	15	15
N/P Ratio	9	4

7.90 lb/year. As a rule the higher values predominate in the hard-water areas where either larger amounts of synthetic detergents are used or products with higher percentages of phosphates are household favorites.

INDUSTRIAL WASTES

Many industrial wastes contain phosphorus because it is native to the product being processed, for example, all food products, particularly those rich in protein. The phosphorus content is often markedly increased by the use of phosphate-bearing detergents for sanitation purposes. Some industrial wastes that would normally be very lean in phosphates are phosphorus rich because of the use of detergents for washing purposes, such as those from the textile industry.

WASTEWATER DISPOSAL

We have seen that phosphorus gains access to the water environment, in unnatural manner, largely through discharges of domestic sewage, the use of phosphate-bearing detergents, food processing, and certain industrial wastes, all of which are associated with significant amounts of biodegradable organic matter. In addition, phosphorus gains access to the aquatic environment in winning it from its ores and in the production of fertilizer and chemicals. These are localized problems and have been considered in earlier chapters.

Most municipal and industrial wastes, because of their organic content, are given some degree of treatment before disposal. This has ranged from fine screening to extensive biological treatment. No matter what degree of treatment is provided, a major part of the phosphorus is carried along in the

TABLE 5 Phosphorus Removal by Various Methods of Wastewater Treatment

Method	Percent Removal	
	With Sludge Digestion	With Incineration
Screening		1
Primary settling	2–5	5–10
Biological—trickling filters	5–10	10–20
Biological—activated sludge		
Conventional	5–10	15–40
Extended aeration	—	2–5
Luxury uptake	—	40–90
[a]With Fe^{3+} or Al^{3+}	—	85–95
[a]Lime precipitation	—	85–95

[a] Special treatment methods.

aqueous phase as shown in Table 5, unless special efforts are made to remove it.

SIGNIFICANCE OF VARIOUS SOURCES OF PHOSPHORUS

In attempting an appraisal of the significance of the several sources of phosphorus upon the environment, it is well to remember that the major losses from land areas—farm, forest, and urban—occur in a sporadic manner and at times of high rainfall when the dilution factor is greatest. In addition, the losses occur at an infinite number of sources over a large geographic area. In contrast, phosphorus contributed by domestic and industrial wastewaters enters receiving waters in highly concentrated form at a finite number of point sources on a continuous basis, regardless of rainfall or stream flow. As a result, the worst possible conditions are created during drought conditions when dilution by receiving streams is at a minimum and the streams themselves are of the highest quality (6).

THE AQUATIC ENVIRONMENT

Ecological Aspects

From the discussion up to this point, it should be obvious that human populations, in their desire to concentrate in urban centers, have unwittingly but successfully siphoned a large part of the phosphorus and other nutrients involved in agricultural pursuits to the aquatic environment. In addition, the use of phosphate builders in synthetic detergents has increased the amount of phosphorus from two- to threefold in domestic wastewaters and up to 100-fold in some industrial waters; consequently, most of our surface waters which receive domestic wastewaters are abundantly rich in phosphorus in relation to other nutrients.

Now that we have established that many of our aquatic areas are phosphorus rich, we must admit that an unnatural stress has been placed upon the ecosystem and a response, either negative or positive, will occur.

Aquatic Plants

In view of all that has been said about losses of phosphorus and other plant nutrients to the aquatic environment, there would be little bewilderment if we were able to report that the yields of "aquatic alfalfa, clover, corn, soybeans, wheat, and so on" were increased. For what else would one expect under increased conditions of fertility? The aquatic equivalent of these crops, however, ranges from microscopic single-celled algae of many kinds to macroscopic rooted aquatic plants of considerable diversity, all bearing chlorophyll and capable of utilizing carbon dioxide and inorganic fertilizers, N–P–K, and a host of others, in creating new crops.

The amounts of the various major nutrients, commonly referred to as fertilizers, found in various aquatic plants are given in Table 6. Since these

TABLE 6 Analyses of Some Typical Aquatic Plants

	Percent			
	Carbon	Nitrogen	Phosphorus	N/P
Blue-green algae (14)				
Anabaena	49.7	9.43	0.77	12:1
Aphanizomenon	47.7	8.57	1.17	7:1
Microcystis	46.5	8.08	0.68	12:1
Green algae (14)				
Cladophora	35.3	2.30	0.56	4:1
Pithophora	35.4	2.57	0.30	8:1
Spirogyra	42.4	3.01	0.20	15:1
Rooted aquatics (3)				
Elodea	—	2.10	0.14	15:1
Lobelia	—	1.89	0.16	12:1
Potomogeton	—	3.19	0.30	11:1

analyses were obtained on specimens collected under bloom conditions, it may be assumed that they grew under luxury conditions in terms of nutrients and the results do not necessarily indicate limiting values. Gerloff and Skoog (4) have shown that *Microcystis aeruginosa* is able to grow at N/P ratios of 50:1.

From the data in Table 6, it will be noted that the blue-green algae are exceptional in the amounts of nitrogen which they contain. From this it may be inferred that they have a protein content much higher than the other plants. The phosphorus content is highly variable among and between the various forms. Also, the nitrogen–phosphorus ratio varies widely.

In considering the ecological aspects of aquatic plants, it is the single-celled and colonial, suspended, and free-swimming forms that are of prime concern, for they can proliferate anywhere within the euphotic zone. Rooted and attached forms are solely dependent on the water in the immediate area and bottom deposits within the littoral zone. Among the algae shown in Table 6, *Anabaena*, *Aphanizomenon*, *Microcystis*, and *Spirogyra* are suspended forms that can develop anywhere in the euphotic zone and are serious trouble makers because they tend to float at some stage in their life and can be concentrated and pushed about by gentle winds. It will be noted that all of these forms use much more nitrogen than phosphorus, since the N/P ratios range from 7:1 to 15:1. Growth of algae in waters with N/P ratios of 4 will be limited by nitrogen, and phosphorus will be in excess.

Ecosystem Dynamics

In any body of water having sufficient detention time for algae blooms to develop, die, settle, and decompose to feed nutrients back into the overlying waters, not only is the concentration of nutrients of great importance but the ratio of nitrogen to phosphorus is also important. Rivers and streams are normally exempt from these conditions.

From the data for N/P ratios shown in Table 6, we might conclude that if influent waters contained nitrogen and phosphorus in a ratio of 15:1 or greater, then both nutrients would be completely consumed and the productivity would be directly proportional to the nutrient input budget. This is an oversimplification of the situation, however, because many of the algae that grow in the water do not escape but die, settle to the bottom, decompose, and release both nitrogen and phosphorus to the water again. In several instances (7, 10) it has been demonstrated that the nutrient feedback from bottom deposits is considerably richer in phosphorus, N/P of 3:1 to 5:1. The release of nutrients with such a low N/P ratio from organisms known to contain much higher ratios remained an enigma until the concept of "luxury uptake" of phosphorus became established.

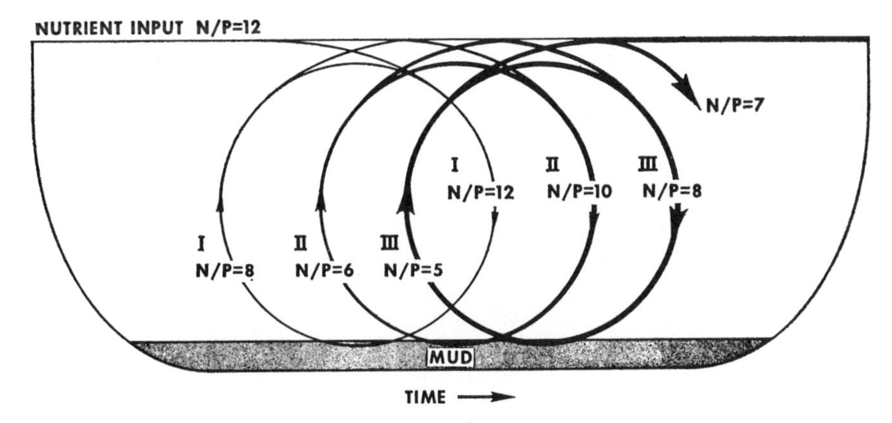

Figure 6. Author's concept of how luxury uptake and release of phosphorus affect phosphorus economy and eutrophication.

According to the concept of "luxury uptake," algae growing in the presence of excess phosphorus can store considerable amounts in excess of their actual needs. When death occurs, it is believed that the stored phosphorus is readily released and that it is this stored phosphorus that results in the low N/P ratios of nutrient feedback. It is this phenomenon that makes strict accounting of phosphorus effects so difficult.

Figure 6 presents the author's concept of how phosphorus recycles in a lake with respect to nitrogen. Briefly, a lake receiving a relatively high-quality water with an N/P ratio of 12 would produce algae with the same ratio. Upon death and decay, the algae would release nitrogen and phosphorus in a lower ratio of, say, 8. These recycled nutrients when mixed with the influent waters could develop a crop of algae with an N/P ratio of 10. The latter upon decomposition could release nutrients in a ratio of 6, and so on. It is this ability to conserve phosphorus while maintaining the capability of releasing it for further support of biological productivity that is believed to be a major factor in the natural eutrophication process in all bodies of water.

Edmondson (1) has shown how this trend to low N/P ratios has been radically reversed in Lake Washington by diversion of wastewaters.

Excess Phosphorus

All domestic wastewaters contain excess phosphorus in terms of algal requirements. Even before the advent of phosphate-built detergents, the N/P ratio of so-called natural sewage was in the range of 9:1 or 10:1. Studies on the Madison, Wisconsin lakes from 1942 to 1944 (5) showed

Figure 7. Nuisance bloom of blue-green algae.

levels of phosphorus in the lake receiving the sewage treatment plant effluent in excess of "luxury uptake" capabilities at all times.

Whenever stresses are placed upon natural systems, some adjustments are prone to take place. In most aquatic systems, it is believed that the stress of excess phosphorus is relieved in two ways. Both of these involve bottom deposits or muds serving as a "sink" or final resting place for the excess phosphorus.

Chemical Precipitation. In hard-water lakes and possibly some soft-water lakes with peculiar properties, excess phosphates are known to be precipitated as calcium salts. This reaction is favored by high pH levels which are created automatically where algae are actively growing and drawing upon the carbon dioxide reserves in bicarbonates and carbonates. Under the most favorable conditions, precipitation is seldom able to reduce the inorganic phosphorus level below 0.3 mg/l, which still leaves considerable excess phosphorus in the water.

Increased Biological Productivity. The other pathway by which excess phosphorus tends to isolate itself from an aquatic ecosystem involves increased biological productivity and, accordingly, it can result in distortions of the ecosystem that are highly objectionable. The phenomenon occurs

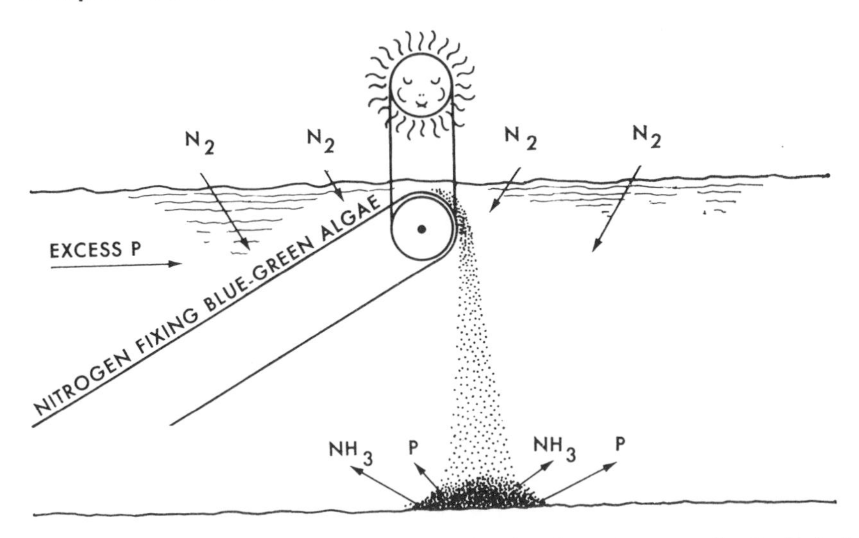

Figure 8. Schematic diagram depicting how excess phosphorus increases nitrogen budget of natural waters and hastens eutrophication.

because *certain* blue-green algae, species of *Anabaena*, *Gleotrichia*, *Aphanizomenon*, and *Nostoc*, are capable of reproducing rapidly in the absence of fixed forms of nitrogen through their ability to fix nitrogen from the gaseous atmospheric nitrogen dissolved in the water. It is often the late summer blooms of the nitrogen-fixing organisms that create conditions such as those shown in Fig. 7.

Blooms of nitrogen-fixing algae utilize phosphorus from the aquatic environment and immobilize it for the moment. Upon death, the organisms release the "luxury" phosphorus and through bacterial decomposition part of the biologically fixed phosphorus is released to the water and the remainder enters the bottom sediment "sink."

Removal of excess phosphorus by blooms of nitrogen-fixing algae is a highly hazardous and inefficient method of letting a problem solve itself. The distortions created in the ecosystem are several, as illustrated in Fig. 8. That part of the "fixed" nitrogen released to the water results in a net increase in the amount of available nitrogen in the ecosystem and serves to support abnormal blooms of ordinary non-nitrogen-fixing algae. The resultant growths cause radical diurnal variations in dissoved oxygen where algae are growing. Shoreline accumulations of algae are prone to create odorous conditions and may release toxins that are fatal to animals ingesting the water. The increased organic load (BOD) imposed upon hypolimnetic waters can cause the development of anaerobic conditions, killing of deep-water fishes, and a significant change in the entire ecology of a lake.

REFERENCES

1. Edmondson, W. T., "Phosphorus, Nitrogen and Algae in Lake Washington after Diversion of Sewage," *Science*, **169,** 690 (1970).
2. Engelbrecht, R. S. and J. J. Morgan, "Studies on the Occurrence and Degradation of Condensed Phosphate in Surface Water," *Sewage Ind. Wastes*, **31,** 458 (1959).
3. Gerloff, G. C., "Evaluating Nutrient Supplies for the Growth of Aquatic Plants in Natural Waters," in *Eutrophication—Causes, Consequences, Correctives*, Natural Academy of Sciences, Washington, D.C., 1969.
4. Gerloff, G. C., and F. Skoog, "Cell Contents of Nitrogen and Phosphorus as a Measure of Their Availability for Growth of *Microcystis Aeruginosa*," *Ecology*, **35,** 348 (1954).
5. *Investigation of Odor Nuisance Occurring in Madison Lakes*, Two Volumes, Governor's Committee Report, Madison, Wisconsin, 1942–1944.
6. Minshall, N., M. S. Nichols, and S. A. Witzel, "Plant Nutrients in Base Flow of Streams in Southwestern Wisconsin," *Water Resour. Res.*, **5,** 706 (1969).
7. *Occoquan Reservoir Study*, Report to Virginia State Water Control Board, Richmond, Virginia, by Metcalf and Eddy, Inc., Boston, Massachusetts 1970.
8. *Relative Contributions of Nutrients to the Potomac River Basin from Various Sources*, Technical Report No. 31, Chesapeake Technical Support Laboratory, FWQA, Department of the Interior, January 1970.
9. Sawyer, C. N., "ABC's of Cultural Eutrophication and Its Control," *Water and Sewage Works*, **118,** Two Parts, 278 and 322 (1971).
10. Sawyer, C. N., "Fertilization of Lakes by Agricultural and Urban Drainage," *J. New England Water Works Assoc.*, **61,** 109 (1947).
11. Sawyer, C. N., "Problem of Phosphorus in Water Supplies," *J. Am. Water Works Assoc.*, **57,** 1431 (1965).
12. *Shoal Lake Study*—1970, The Metropolitan Corporation of Greater Winnipeg, Winnipeg, Manitoba.
13. Sylvester, R. O., "Nutrient Content of Drainage Water from Forested, Urban and Agricultural Areas," in *Algae and Metropolitan Wastes*, U.S. Department of Health, Education and Welfare, Public Health Service, 1961.
14. *Water Quality Criteria*, FWPCA, Department of the Interior, 1968.

Phosphorus in Wastewater Treatment

JOHN B. NESBITT

Department of Civil Engineering, The Pennsylvania State University, University Park Pennsylvania

Phosphorus is an element that is both needed and undesirable in wastewater treatment. Its presence is absolutely essential to the proper functioning of all biological waste-treatment processes, but when present in excess it can create a pollution problem in the receiving body of water. In both of these instances it is merely serving its role as an essential element in the metabolism of organic matter by both plants and animals.

In biological treatment it must be present so that microorganisms can metabolize the organic matter and thus consummate the treatment process. If it is not there, or is there in insufficient amounts, as is the case with some industrial wastes, it will have to be added. If it is present in excessive amounts, as is the case with most municipal wastewaters, it may cause excessive growths of rooted and/or floating aquatic plants in the body of water receiving the wastewater flow. These plants (*a*) may cause diurnal fluctuations in the dissolved oxygen level of the water and thus discourage the growth of game fish; (*b*) can clog a stream or river making it undesirable for recreation and increasing the chances of flooding; (*c*) can cause taste and odor problems in water supplies; and (*d*) die and decay thus placing a continuing load on the oxygen resources of the stream and perhaps causing an air-pollution problem for the nearby population.

FORMS AND SOURCES OF PHOSPHORUS

In 1969 Nesbitt (1) discussed the forms and sources of phosphorus found in wastewaters. It was pointed out that phosphorus may occur in the form of organic phosphorus found in organic matter and cell protoplasm or as complex inorganic phosphates such as those used in agriculture, water-treatment processes, and in cleaning compounds, and as soluble inorganic orthophosphate (PO_4^{3-}), the final breakdown product in the phosphorus cycle, and the form in which phosphorus is most readily available for biological utilization.

Phosphorus may enter a wastewater treatment plant in all three of the above-mentioned forms. During the treatment process most of the organic and some of the complex phosphates are removed or decomposed to inorganic orthophosphate. According to Finstein and Hunter (2) the hydrolysis of the complex phosphates is accomplished by the biological flora present in the biological treatment section of the plant. In three activated sludge and three trickling filter plants they found that in the influent to the biological treatment units the complex phosphates constituted 15 to 75% of the total phosphorus while in the effluent from these units the percentages varied from 5 to 40%. In all the plants approximately half of the complex phosphorus was hydrolyzed to orthophosphate during biological treatment. Similar results were found by Engelbrecht and Morgan (3) in four Illinois wastewater-treatment plants and in a study (4) conducted at the treatment plant serving Pennsylvania State University and the Borough of State College, Pennsylvania.

In discussing the subject of phosphorus in wastewater some consideration should be given to whether the phosphorus is in the soluble or insoluble form. (In the following discussion, soluble phosphorus is defined as that which will pass laboratory filtration as used in a test for suspended solids.) In raw wastewater that has been treated by biological processes most of the phosphorus will be soluble although a small amount of insoluble organic phosphorus may be present in the form of cell protoplasm. In wastewater that has had the phosphorus removed by chemical precipitation most of the effluent phosphorus will be in the insoluble form (i.e., calcium, aluminum, or iron phosphates). Considerable difference of opinion exists as to whether these insoluble forms will dissolve in other units of the treatment plant or in the receiving body of water. Theoretically both the insoluble organic and inorganic phosphorus can dissolve. Under anaerobic conditions the cell protoplasm decomposes releasing the organic phosphorus as soluble inorganic phosphate, and this process has been observed by many researchers. The insoluble inorganic phosphates can become soluble and release their phosphorus under certain pH conditions. In practice, however, these conditions are not common and the few observations that have been made indicated the phosphorus was not released. These observations were made in the sludge digestion process for iron phosphate by Thomas (5) and aluminum phosphate by Barth and Ettinger (6) and O'Shaughnessy (7).

The quantity of phosphorus in municipal wastewaters has been increasing steadily through the years, but appears to have reached a plateau. In 1947 Rudolfs (8) reported that orthophosphate in the trickling filter effluent from 12 American sewage-treatment plants ranged from 0.27 to 2.37 mg PO_4/l.

and averaged 0.66 mg. In 1966 Nesbitt (9) showed how this value had increased steadily and finally stabilized at about 24 mg PO_4/l. although individual instances may show concentrations considerably above or below that amount.

PHOSPHORUS REQUIREMENTS IN WASTEWATER TREATMENT

In wastewater treatment phosphorus is an essential element only in biological treatment processes such as activated sludge, trickling filters, waste-stabilization ponds, and anaerobic fermentation.

The first three processes are aerobic, and phosphorus requirements for them are indicated by the work of Helmers *et al.* (10, 11) in studies of the nutritional requirements for the treatment of industrial wastes by the activated sludge process. For wastes devoid of phosphorus, they found that the minimum ratio of organic matter present as measured by the 5-day biochemical oxygen demand (BOD_5) test to the phosphorus that must be added (BOD_5/P) was approximately 100:1. It is expected that similar ratios would exist for both the trickling filter and waste-stabilization pond treatment processes. In a waste-stabilization pond, however, phosphorus requirements in the incoming waste could be considerably less because a portion of the phosphorus is recycled in a conventional pond via the anaerobic digestion process which occurs at the bottom of the pond.

Since municipal wastewater has an average BOD_5 of 250 mg/l., the 24 mg PO_4/l. (8 mg P/l.) in it is considerably in excess of the minimum phosphorus requirements for aerobic biological treatment.

Phosphorus requirements in anaerobic fermentation processes are similar to the requirements mentioned above for aerobic processes. Schroepfer and Ziemke (12) report that for an anaerobic contact process the raw waste feed should have a BOD_5/P ratio that does not exceed 100:1. In sludge digestion, an anaerobic fermentation process used in municipal wastewater treatment, the work of Sanders and Bloodgood (13) and Speece and McCarty (14) has indicated that the minimum carbon–phosphorus ratio for successful digester operation is approximately 100:1.

PHOSPHORUS REMOVAL IN WASTEWATER TREATMENT

Phosphorus can be removed from wastewaters both easily and efficiently. Before discussing these removal methods, however, some consideration should be given to a method for determining for each individual situation whether phosphorus removal will provide a solution to the pollution problem that exists, and if it will, just how much phosphorus should be removed. In 1969 Nesbitt (1) stated that this consideration should include such items as (*a*) will phosphorus removal control the nuisance growths and if so, what maximum concentration of phosphorus is permissible; (*b*) what other sources of phosphorus are in the drainage basin; (*c*) what is the background level of phosphorus in the receiving body of water; and (*d*) what dilution ratio is available for the plant effluent.

Item (*a*) poses the most difficult question and I believe it is safe to say that we do not know the answer to it. In 1966 Nesbitt (9) surveyed the literature on this subject and found that although much was written on it, few conclusions could be drawn from what was reported. The only positive statement was that of Sawyer (15) who concluded that any lake showing concentrations of inorganic phosphorus in excess of 0.01 mg P/l. and of inorganic nitrogen in excess of 0.30 mg N/l. at the time of the spring overturn could be expected to produce algal blooms of such density as to cause nuisance. Recently Dryden and Stern (16) in a study of some recreational lakes in California fed with sewage effluent only, concluded that phosphorus levels below 0.5 mg PO_4/l. would control nuisance growths of algae and that algae growth would almost stop at levels below 0.05 mg PO_4/l.

Item (*b*) can be determined, although perhaps not easily, by a survey of the drainage area. Reports of several such surveys are available in the literature and they show that phosphorus can come from domestic wastewater, agricultural drainage, and urban runoff. In many instances these reports show that domestic wastewater is the prime contributor to the phosphorus concentration; however, in a report on the nutrient sources for Lake Mendota, Wisconsin (17) it was estimated that only 36% of the phosphorus in the lake came from this source. Of the remaining, 43% came from rural runoff, 17% from urban runoff, and 4% from precipitation and groundwater. Therefore, such a survey definitely should be made before phosphorus-removal facilities are designed.

An analysis of quality and quantity data on the receiving body of water will provide data for items (*c*) and (*d*). When all the questions are answered, a required effluent level of phosphorus can be determined and a phosphorus-removal facility designed. As an illustration of what might be required, Eberhardt (18) calculated that, "Assuming a receiving water total available phosphorus concentration of 0.03 mg PO_4/l. (0.01 mg P/l.) or less is desired, a background level of 0.01 mg PO_4/l. exists, and a dilution of one part effluent to nine parts receiving water is available, an effluent total available phosphorus concentration of 0.2 mg PO_4/l. (0.065 mg P/l.) is required," It should be emphasized that this figure is an estimated one based on assumptions for a given situation and should not be taken as a goal for all phosphorus-removal procedures.

The discussion which follows is concerned only with those phosphorus-removal methods that can be included with a present wastewater-treatment plant either as a part of the primary or secondary treatment processes or as a tertiary treatment process. Processes such as electrodialysis and reverse osmosis which are designed to produce an effluent approaching drinking-water quality, or processes which have a limited geographical application, will not be included. Methods to be discussed include:

1. Physical treatment
   a. Sedimentation
   b. Flotation
   c. Filtration
2. Chemical treatment
   a. Precipitation with lime
   b. Precipitation with aluminum and iron
   c. Chemical-biological precipitation
   d. Ion exchange
3. Biological treatment
   a. Activated sludge
   b. Oxidation ponds

Removal by Physical Treatment Processes

The physical treatment processes of sedimentation, flotation, and filtration will remove only insoluble organic and inorganic phosphorus compounds. The efficiency of removal is related to the general ability of the process to remove suspended solids and to the fraction of the total phosphorus that is in these solids.

For the plain sedimentation process, Jenkins and Menar (19) have summarized the experience of most municipal wastewater-treatment plant operators when they studied the removal of phosphorus in the primary sedimentation units of several wastewater-treatment plants. Although such units can be expected to remove about 65% of the suspended solids, phosphorus removals varied from 5 to 15% and averaged 10%. Similar results should be obtained in the flotation process if it provides the same degree of solids removal.

Filtration through fine-grain media such as sand, anthracite coal, and diatomaceous earth when properly designed and operated is an extremely efficient process for removing suspended solids. For all practical purposes it can be expected to remove 100% of the insoluble phosphorus in a wastewater.

Sedimentation and filtration are also used to remove suspended solids and thus phosphorus following both chemical and biological treatment. Typical results for the combination of processes are presented subsequently.

Precipitation with Lime

A compilation of the various lime-precipitation schemes for the removal of phosphorus as presented by Nesbitt (1) is shown in Table 1 and Fig. 1. In Fig. 1, scheme 1 represents straight chemical treatment without any biological treatment, scheme 2 the common tertiary process, and scheme 3 the phosphate-extraction process (PEP) recently suggested by Albertson and Sherwood (20). Although not shown on any of the three schemes all of them can be followed by a filtration step to remove additional insoluble phosphorus. In schemes 1 and 2, optimum pH is about 11 and thus recarbonation may be required to prevent precipitation of calcium in many receiving waters and calcium deposition on subsequent process equipment. In scheme 3, recirculation of calcium phosphate sludge permits operation at about one pH unit lower, thus reducing chemical costs. Recarbonation and further removal of phosphorus are accomplished in the biological treatment units. No operating data on these processes are available.

Recently Buzzell and Sawyer (21) and Schmid and McKinney (22) reported results of laboratory studies which are similar to the PEP process and involve scheme 1 plus biological treatment by either activated sludge or oxidation ponds (or scheme 3 without lime recovery). They report phosphorus removals from 88 to 99% as shown in Table 1. A further advantage of these processes and scheme 3 is the reduction in size of the secondary treatment units and the reduced sludge handling costs because of the increased removal of BOD and solids in the primary treatment units. These schemes show much promise but no full-scale plants have yet been constructed. All of the lime-precipitation studies show good removals of phosphorus, but from the data in Table 1, it appears that a final filtration step is a prerequisite to producing an effluent having a minimum phosphorus concentration.

Precipitation with Aluminum and Iron

Figure 1 and Table 2 show the precipitation schemes and data from some of the tests which have been made on the precipitation of phosphorus from domestic wastewater with aluminum and iron. Other investigators have conducted additional studies using the iron coagulants ferric chloride and ferric sulfate on a jar test scale and reported results similar to those shown in Table 2. Coagulant doses for the iron coagulants were slightly less than those reported for aluminum. Again the results show good removals of phosphorus by coagulation and settling alone but minimum residuals still require filtration.

TABLE 1 Phosphorus and BOD Removal for Lime Precipitation

			Coagulant			Phosphorus Data		BOD
Ref.	Process Scheme[a]	Size of Test	Type	Dosage (mg/l.)	pH	Effluent (mg PO_4/l.)	Removal (%)	Removal (%)
23	2	0.77 mgd	CaO	545	11.0	5.2	79	94
23, 24	2 + Filtration	0.77 mgd	CaO	545	11.0	0.4	98	—
15	2	—	$Ca(OH)_2$	280	—	1.5	93	—
4	2	Jar	$Ca(OH)_2$	330	—	3.1	91	63[b]
25, 24	1	Jar	$Ca(OH)_2$	300	11.4	1.8	93	71
26, 24	2	10 gal/month	$Ca(OH)_2$	300	—	2.3	92	83
			Alum					
20	3	Jar	$Ca(OH)_2$	200	9.5–10	5[c]	—	70[c]
				20				
21	1 + Activated sludge	Laboratory	Lime	—	<11.0	0.5–4.2	88–96	87–93
21	1 + Oxidation ponds	Laboratory	Lime	—	<11.0	0.2–0.3	98–99	—
22	1 + Activated sludge	Laboratory	Lime	—	9.5	4.0	90	92
27	2 + Filtration	7 mgd	Lime	150	—	2.4	93	—
				200				

[a] Number refers to Fig. 1.
[b] Refers to additional removal obtained by lime precipitation.
[c] Includes only mixing, coagulation, and setting.

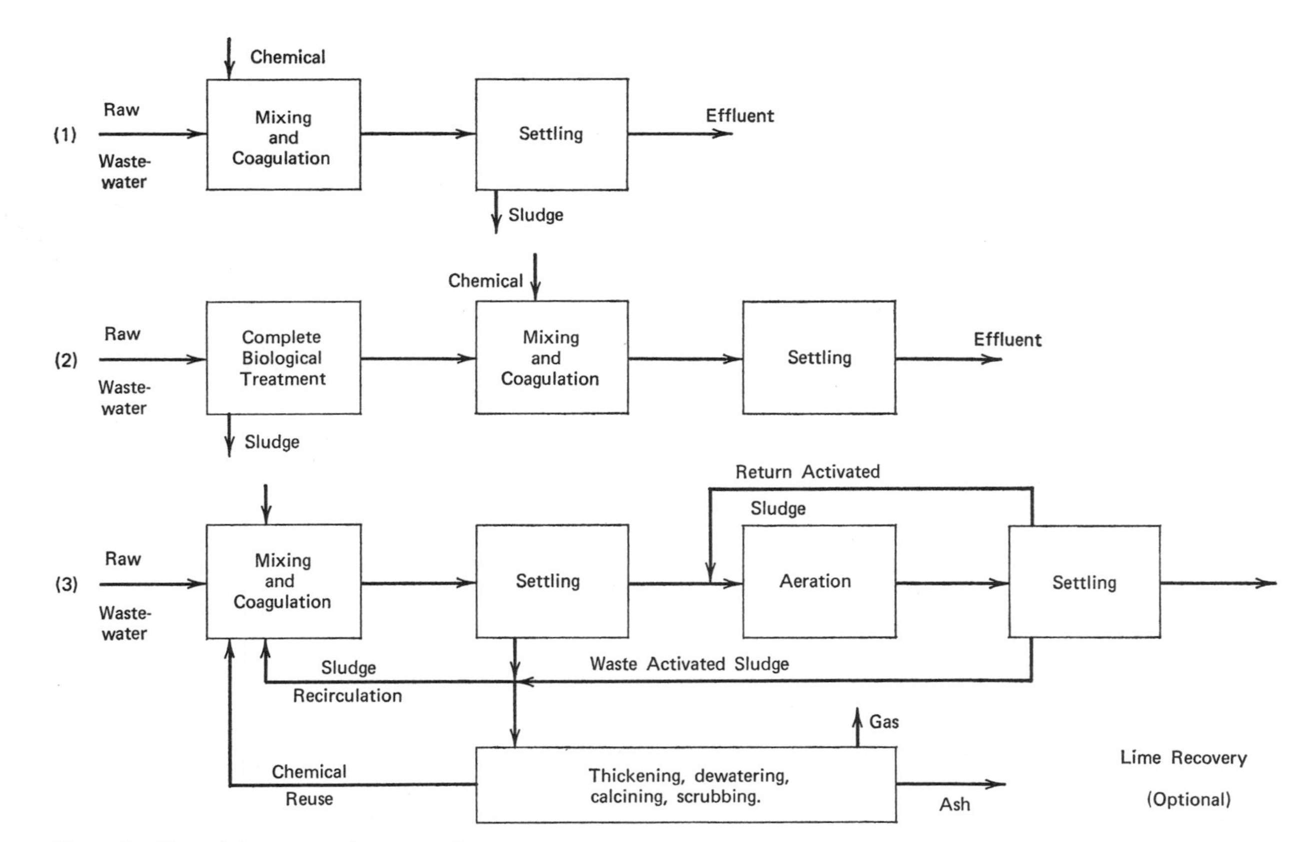

Figure 1. Phosphorus-removal process schemes.

TABLE 2 Phosphorus and BOD Removal for Aluminum and Iron Precipitation

Ref.	Process Scheme[a]	Size of Test	Coagulant: Type	Coagulant: Dosage (mg/l.)	Phosphorus Data: Effluent (mg PO_4/l.)	Phosphorus Data: Removal (%)	BOD Removal (%)	pH
2, 24	1	1.5 mgd	"Alum"[b]	94	3.15	81	70	—
			Act. sil.	3.4				
28, 24	1 + Filtration	—	"Alum"	94	0.33	98	—	—
			Act. sil.	3.4				
25, 24	1	Jar	"Alum"	400	0.8	97	73	—
29, 24	2	10 gal/month	Filter alum	200	2.33	85	95	7.1–7.7
29, 24	2 + Filtration	—	Filter alum	200	0.14	99	—	—
30, 24	2	10 gal/month	Filter alum	397	4.90	89	86	—
30, 24	2 + Filtration	10 gal/month	Filter alum	397	1.53	96	95	—
31, 32	2 + Filtration	2.5–4 mgd	Liquid alum	200	0.1–1.0	>95	96	—
33	2	Plant	$FeCl_3$	29	1.5	—	—	8.8
			$Ca(OH)_2$	200				
16	2	30 gal/month	Filter alum	300	0.5	99	—	6
16	2 + Filtration	30 gal/month	Filter alum	300	0.25	>99	>74[d]	6
31	1 + Trickling filter	0.5 mgd	$FeCl_2$	≃50	5.1	82	80	—
			polyel.	≃0.4				
			NaOH	≃24				
34	2	310 gal/month	"Alum"	101	7.0	80	72[d]	—
			Act. sil.	9.7				
34	2 + Filtration	310 gal/month	"Alum"	101	5.0	86	77[d]	—
			Act. sil.	9.7				

[a] Numbers refer tc Fig. 1.
[b] "Alum" refers to $Al_2(SO_4)_3 \cdot 18H_2O$.
[c] Filter alum refers to $Al_2(SO_4)_3 \cdot 14H_2O$.
[d] Tertiary treatment only.

Chemical-Biological Treatment

Chemical-biological treatment of wastewater for the removal of phosphorus is applicable only to the activated sludge treatment process and involves the addition of a metallic precipitant to the aeration-sedimentation units of the process. The process, which is shown schematically in Fig. 2, requires no additional treatment units other than facilities for chemical storage and addition and produces a combined chemical-biological sludge that is returned to the system with the excess sludge being wasted.

Present experience with the process indicates that its successful design and operation are dependent on (*a*) the form and amount of chemical precipitant used, (*b*) the pH of the mixed liquor solids age, and (*c*) the method of waste sludge disposal.

The choice of chemical precipitant and the operating mixed liquor pH are dependent on the compatibility of the mixed liquor microorganisms and the pH of minimum solubility of the metallic hydroxyphosphate which precipitates in the process. To a certain extent, the alkalinity of the wastewater may also affect this choice. These conditions limit the choice to compounds of aluminum and ferric iron. According to Stumm (35), the minimum solubility for $AlPO_4$ is pH 6.3 and for $FePO_4$, 5.3. However, most studies with these two metals have found that in domestic wastewater the pH of minimum solubility has a fairly broad range for both aluminum and iron and low effluent phosphate concentrations can be obtained at pH values between 5.5 and 7.0. This pH range is compatible with the mixed liquor microorganisms.

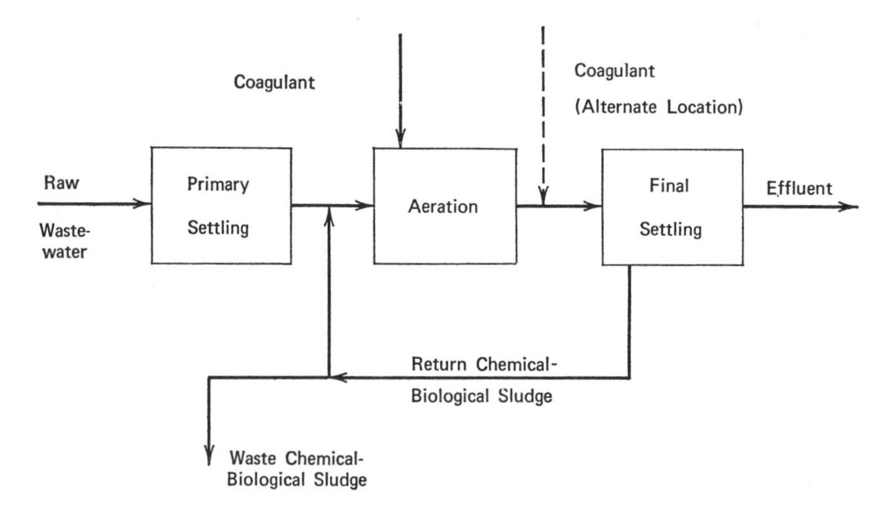

Figure 2. Chemical-biological phosphorus-removal process scheme.

The other possible metallic ions, calcium and ferrous iron, produce low phosphate concentrations only at higher pH values and are generally not compatible with biological treatment.

The aluminum ion has been fed successfully as granular aluminum sulfate [$Al_2(SO_4)_3 \cdot 18H_2O$] by Eberhardt and Nesbitt (24), liquid alum by Long and Nesbitt (36), and liquid sodium aluminate by Long and Nesbitt (36) and Barth and Ettinger (6). All of these chemical forms will provide the necessary aluminum ions that will precipitate the phosphate in the wastewater, and which one is chosen will depend on the type of chemical feed equipment available or desired, the alkalinity of the wastewater, and the cost of the chemical per unit volume of wastewater treated. Alum, either liquid or granular, will reduce the untreated wastewater pH and thus would be preferred for wastewaters having higher alkalinities and pH's, while liquid sodium aluminate (pH > 13) may be preferred in wastewaters having lower alkalinities and pH's. Aluminate also introduces no additional sulfate ions to the water, which may be a factor in waters containing considerable noncarbonate hardness.

Iron has been fed as ferric chloride by Thomas (5). The literature does not contain any reference to the use of ferric sulfate in a chemical-biological process; however, it has been used successfully in a tertiary treatment process and should be a satisfactory source of the ferric ion.

The quantity of chemical required to precipitate the phosphate ion appears to be dependent primarily on the amount of phosphate present, and to a lesser degree, on the pH of the mixed liquor after the precipitant has been added. In the pH range 5.5 to 7.2, a molar ratio of aluminum to phosphorus (Al/P molar) of at least 2:1 (this value is equivalent to an Al/P weight ratio of 1.74:1) is usually required to produce a low effluent phosphate concentration. According to Farrell *et al.* (37) there is some indication that with aluminum in low alkalinity waters, pH adjustment to about 6.3 could reduce this ratio to 1.5:1 *M* (1.3:1 weight). In alkaline waters pH adjustment should have little effect on the required ratio.

Eberhardt and Nesbitt (24) and Long and Nesbitt (36) have found that the point of chemical addition may have an effect on the clarity of the settled effluent from the process. From their results, it appears that the length of time between chemical addition and settling should be the minimum obtainable that provides adequate chemical mixing. Chemical addition at the effluent end of the aeration tank or at the influent end of the pipe or channel leading to the final settling tank is usually satisfactory. Prolonged aeration of the chemical-biological sludge may disperse a portion of the sludge and produce a milky-white colloidal precipitate in the settling tank effluent. The composition and reason for the formation of this colloidal precipitate is not yet known; however, it does contain insoluble phosphate and will pass through a single- or multiple-medium water-type filter very easily.

A reasonably low solids age is essential to the proper operation of the process. The longer the solids age, the higher the inorganic fraction of the mixed liquor suspended solids (MLSS). Solids age should probably be kept under 5 days and preferably around 3 days. With the solids ages realized in the extended aeration process, it is possible to lose all biological activity with all subsequent removals being strictly chemical in nature. Reducing the solids age may also help to control the formation of the colloidal precipitate referred to previously.

The chemical-biological sludge should settle more rapidly than straight biological sludge. Thus it is possible that although the sludge weight will exceed that of biological sludge, its volume may be only slightly greater. The sludge dewaters easily and may be disposed of directly or through aerobic or anaerobic sludge digestion.

The procedures used and results obtained by several investigators who have used the chemical-biological process are presented in Table 3. The data show that with Al/P molar ratios of approximately 2.4:1 effluents can be produced which have fairly low phosphate concentrations with the process producing removals of 90% or higher. In the table, lines 5 and 6 (Ref. 36 using $Na_2Al_2O_4$) represent a period when the effluent contained considerable colloidal precipitate. The data show that some of this precipitate was removed by a Reeve-Angel glass fiber filter (line 5) but that it passed easily through the dual medium filter (line 6). Lines 7 and 8 show the effect of a dual medium filter on the effluent when the colloidal precipitate is not produced. These results indicate that with this process most of the phosphorus leaving the final settling tank is insoluble suspended solids which, if properly coagulated, can be removed by a filtration step. Percentage removals are not calculated for the dual medium filter runs (lines 6 and 8) as the effluent phosphorus concentration was determined on a grab sample with the other data being determined from 24-hr composite samples.

Long and Nesbitt (36) have shown that the addition of liquid alum or sodium aluminate in the chemical-biological process also may reduce the effluent BOD and COD by a factor of 2 to 3 over that obtained when no chemical is added.

Ion Exchange

In 1966 Nesbitt (9) reported on the status of ion exchange as a method of phosphorus removal from municipal wastewaters. He concluded that the process would not be a practical one for the removal of phosphorus alone because of the extensive pretreatment necessary to prevent the organic

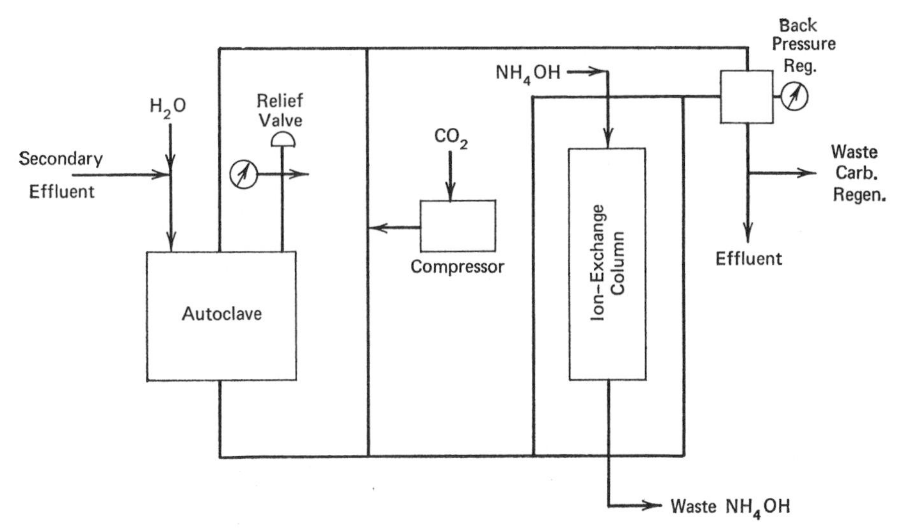

Figure 3. Ion-exchange process for phosphorus removal (38).

fouling of the resins, the problem of the disposal of used regenerant, and the cost of the regeneration chemicals.

The recent development of macroporous resins that will absorb much organic matter without fouling and the possibility of less expensive regeneration chemicals make this process worth consideration again. A flow diagram for such a process as suggested by Pollio and Kunin (38) is shown in Fig. 3. It uses a pressurized anion exchanger (Amberlite IRA 68) operating on the bicarbonate cycle and regenerated with ammonium hydroxide and carbon dioxide. The principal reactions are as follows:

Phosphorus removal (alkalization)

$$3(\mathrm{R{-}NH})\mathrm{HCO_3} + \mathrm{Na_3PO_4} \rightarrow (\mathrm{R{-}NH})_3\mathrm{PO_4} + 3\mathrm{NaHCO_3}$$

Regeneration

$$(\mathrm{R{-}NH})_3\mathrm{PO_4} + \overset{(1\,N)}{3\mathrm{NH_4OH}} \rightarrow 3(\mathrm{R{-}N}) + (\mathrm{NH_4})_3\mathrm{PO_4} + 3\mathrm{H_2O}$$

$$3(\mathrm{R{-}N}) + 3\mathrm{H_2O} + 3\mathrm{CO_2} \rightarrow 3(\mathrm{R{-}NH})\mathrm{HCO_3}$$

Similar reactions occur with other anions such as chloride and sulfate. During operation the secondary effluent enters the autoclave and is then forced upflow through the ion exchanger using a 50 psi CO_2 driving pressure. Upon exhaustion regeneration is first with 1 N NH_4OH operating by gravity in a downflow direction. The volume required for Haddonfield, New Jersey was 1% or less of the waste treated. To convert the resin to the bicarbonate

TABLE 3 Phosphorus Removal through Chemical-Biological Treatment

Line Number	Ref.	Size of Test	Type of Effluent Filtration Employed	Coagulant Type	Coagulant Dosage (mg/l.)	pH	Phosphorus Data (Total) Effluent ($mg\ PO_4/l.$)	Phosphorus Data (Total) Removal (%)
1	24	60 liters/day	None	Alum	335[a] (2.4:1)[b]	6.0–6.6	2.0[c]	94[d]
2	24	60 liters/day	Laboratory, Reeve-Angel glass	Alum	335[a] (2.4:1)[b]	6.0–6.6	0.05[c]	>99[d]
3	5	0.5 mgd	None	$FeCl_3$	29	7.1	1.85	94.2[d]
4	6	100 gal/day	None	$Na_2Al_2O_4$	61 (1.26:1)[b]	—	1.53	94.6[d]
5	36	1.0 mgd	Laboratory, Reeve-Angel glass	$Na_2Al_2O_4$	43 (2.5:1)[b]	7.3	1.41[c]	93.0[d]
6	36	1.0 mgd	Dual medium	$Na_2Al_2O_4$	43 (2.8:1)[b]	7.3	5.21[c]	—
7	36	1.0 mgd	None	Liquid alum	≈2.5:1[b]	7.1	3.83[c]	≈90[d]
8	36	1.0 mgd	Dual medium	Liquid alum	≈2.3:1[b]	7.1	0.31[c]	—

[a] Concentrations as $Al_2(SO_4)_3 \cdot 18H_2O$.
[b] Approximate Al/P molar ratio based on influent total soluble phosphorus.
[c] Total phosphorus by the binary acid wet-oxidation technique. Lines 6 and 8 are grab samples.
[d] Based on influent total phosphorus.

form, CO_2 was fed into the water-filled autoclave for 10 min at 75 psi. The carbonated water was then passed downflow through the exchange column using 50 psi CO_2 driving pressure.

Tests on chlorinated secondary effluent from Haddonfield, New Jersey which contained 21 mg PO_4/l. and 32 mg Cl/l. showed that after 226 bed volumes the phosphorus in the effluent was still less than 0.06 mg PO_4/l. Chlorides were essentially zero for 100 bed volumes and were 16 mg Cl/l. at 226 bed volumes. After 3000 bed volumes the resin still retained 99% of its initial exchange capacity with no downward trend in capacity visible. If further treatment is desired, the authors also present data on flocculation with bentonite and a polyelectrolyte followed by sedimentation for removal of organic matter, degasification for removal of CO_2, cold lime softening to remove some cations, and cation exchange for more cation removal including ammonia. In the phosphorus-removal process some disposal method will have to be found for the waste ammonium hydroxide regenerant which may amount to 100,000 gal/day for a 10-mgd treatment plant.

Activated Sludge

The sanitary engineering profession has not resolved the question as to how much phosphorus can be removed from wastewater by conventional biological treatment. The activated sludge process and the oxidation pond have been studied many times but the reported results are quite inconsistent.

Jenkins and Menar (19) operated a pilot-activated sludge plant at organic loadings up to ten times those normally encountered in practice and found that regardless of loading the weighted average phosphorus content of the cells as P was 2.62% of the volatile suspended solids (2.27% of the suspended solids) and ranged from 2.45 to 3.03%. Using growth data developed in the study they estimated that the maximum possible biological phosphorus removal would be 20%. Thus, including primary sedimentation, the maximum phosphorus removal that could be expected in an activated sludge treatment plant would be 28%.

In contrast to the studies mentioned above, Vacker *et al.* (39) reported on studies conducted at San Antonio, Texas, and Priesing *et al.* (40) at five other plants in the south-central region of the United States. They report the operation of activated sludge plants such that the removal of phosphorus reached 90% at an influent concentration of approximately 30 mg PO_4/l. The waste sludge at San Antonio contained up to approximately 7% P by weight. They emphasized that in order to obtain these results the following operating parameters must be strictly adhered to:

1. An optimum organic loading of 50 lb BOD/100 lb aeration tank solids must be maintained and should be held as constant as possible through

raw wastewater flow equalization tanks and control of return activated sludge.

2. Aeration tank DO should be 2 mg/l. at the midpoint of the tank and 5 mg/l. at the tank effluent.

3. The phosphate-rich waste sludge should be disposed of completely apart from the primary and secondary treatment processes in the plant.

An explanation of the different results obtained for phosphorus removal in the activated sludge process has been given by Bargman *et al.* (41) and Menar and Jenkins (42) in studies relating aeration, pH, and cation concentration in several activated sludge plants. These studies showed that the increased aeration required to produce high levels of dissolved oxygen in the aeration tank stripped carbon dioxide from the mixed liquor and thus raised the pH. At the higher pH values the cations in the wastewater precipitated the phosphorus in the proper mole ratios. Actual material balances of the precipitated compounds were close to those calculated from the combination with the available cations and the expected normal metabolic uptake of phosphorus.

According to Levin and Shapiro (43), Shapiro *et al.* (44), and Hennessey *et al.* (45) the biological uptake of phosphorus in an activated sludge plant can be increased considerably by stripping the phosphorus from the return activated sludge. This stripping can be accomplished by subjecting the sludge to controlled anaerobic or low pH conditions. Hennessey *et al.* (45) conducted experiments on this process with a pilot plant using pH control for phosphorus stripping and reported approximately 55% removal of total phosphorus from sewages having an influent concentration of approximately 30 mg PO_4/l. However, all the investigators of this process feel that 80 % removals can be achieved. A typical flow diagram for this process as proposed by Hennessey *et al.* (45) is shown in Fig. 4.

Oxidation Ponds

Because the algae that are indigenous to any oxidation pond are known to require more phosphorus in their metabolism than bacteria, this treatment process has received considerable attention as a potential phosphorus-removal process. In 1966 Nesbitt (9) reviewed the literature on this subject and concluded that the successful operation of an oxidation pond for phosphorus removal would require (*a*) proper design for the removal of organic matter, (*b*) operation in a warm climate to ensure continuous biological activity, (*c*) an artificial source of light to provide for continuous photosynthesis, and (*d*) some method harvesting the algae to provide a method for phosphorus to leave the system other than with the effluent.

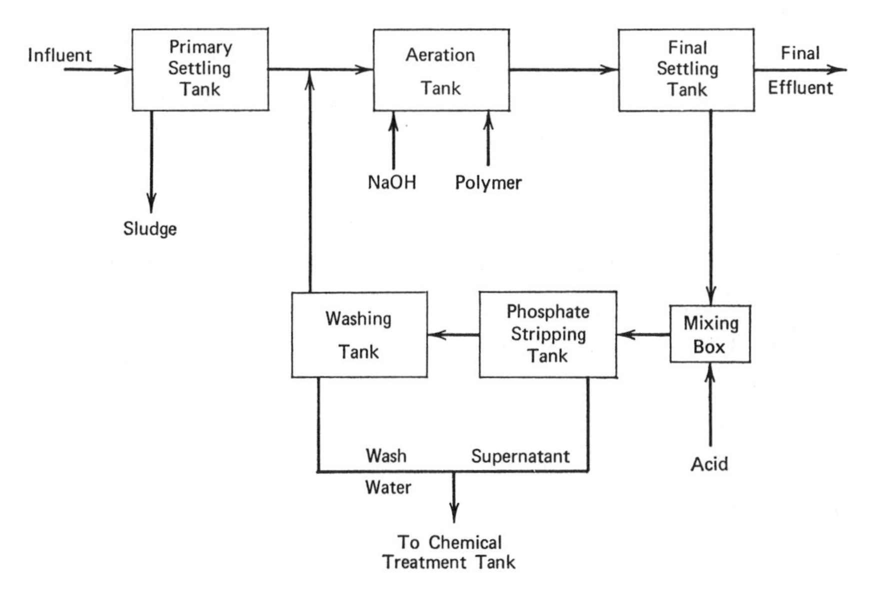

Figure 4. Flow diagram of pilot plant [after Hennessey *et al.* (45)].

An oxidation pond also can cause considerable phosphorus to be precipitated from wastewater as calcium phosphate. This precipitation occurs when the pH in the pond is raised because of the utilization of carbon dioxide by the algae culture. This removal is only temporary unless the precipitate is removed from the pond before it is redissolved under the anaerobic conditions existing at the bottom of the pond.

SUMMARY

In summary it should be repeated that in the proper proportions phosphorus is absolutely essential to the successful operation of a biological wastewater-treatment process. In municipal wastewaters, phosphorus is usually present in excess of this required amount and may thus contribute to pollution through the fertilization of aquatic growths in the receiving body of water. Although existing secondary treatment processes are not designed to remove this excess phosphorus, they can be modified or additional treatment units added so that most of this excess phosphorus can be removed.

REFERENCES

1. Nesbitt, J. B., "Phosphorus Removal—The State of the Art," *J. Water Pollut. Control Fed.*, **41**, 701–713 (May 1969).

2. Finstein, M. S., and J. V. Hunter, "Hydrolysis of Condensed Phosphates during Aerobic Biological Sewage Treatment," *Water Res.*, **1**, 247–254 (April 1967).
3. Engelbrecht, R. S., and J. J. Morgan, "Studies on the Occurrence and Degradation of Condensed Phosphates in Surface Waters," *Sewage Ind. Wastes*, **31**, 458–478 (April 1959).
4. Metcalf and Eddy, Engineers, *Disposal of Sewage and Industrial Wastes on the Spring Creek Watershed, Centre County, Pennsylvania*, Report to the Spring Creek Committee, March 10, 1961.
5. Thomas, E. A., "Phosphat—Elimination in der Belebtschlammanlage Von Mannedorf and Phosphat-Fixation in See- und Klarschlamm," *Viertel. Naturforsch. Ges. Zürich*, **110**, 419–434 (Dec. 1965).
6. Barth, E. F., and M. B. Ettinger, "Mineral Controlled Phosphorus Removal in the Activated Sludge Process," *J. Water Pollut. Control Fed.*, **39**, 1362–1368 (August 1967).
7. O'Shaughnessy, J. C., J. B. Nesbitt, D. A. Long, and R. R. Kountz, *Soluble Phosphorus Removal in the Activated Sludge Process, Part II, Sludge Digestion Study*, U.S Government Printing Office, Water Pollution Control Research Series, 17010 EIP 10/71, October 1971, 59 pages.
8. Rudolfs, W., "Phosphates in Sewage Treatment: I. Quantities of Phosphates," *Sewage Works J.*, **19**, 43–47 (January 1947).
9. Nesbitt, J. B., *Removal of Phosphorus from Municipal Sewage Plant Effluents*, Engineering Research Bulletin B-93, The Pennsylvania State University, February 1966.
10. Helmers, E. N., J. D. Frame, A. E. Greenberg, and C. N. Sawyer, "Nutritional Requirements in the Biological Stabilization of Industrial Wastes, II. Treatment with Domestic Sewage," *Sewage Ind. Wastes*, **23**, 884–899 (July 1951).
11. Helmers, E. N., J. D. Frame, A. E. Greenberg, and C. N. Sawyer, "Nutritional Requirements in the Biological Stabilization of Industrial Wastes, III. Treatment with Supplementary Nutrients," *Sewage Ind. Wastes*, **24**, 496–507 (April 1952).
12. Schroepfer, G. J., and N. R. Ziemke, "Development of the Anaerobic Contact Process, I. Pilot Plant Investigations and Economics," *Sewage Ind. Wastes*, **31**, 164–190 (February 1959).
13. Sanders, F. A., and D. E. Bloodgood, "The Effect of Nitrogen to Carbon Ratio on Anaerobic Decomposition," *J. Water Pollut. Control Fed.*, **37**, 1741–1752 (December 1965).
14. Speece, R. E., and P. L. McCarty, "Nutritional Requirements and Biological Solids Accumulation in Anaerobic Digestion," *J. Water Pollut. Control Fed.*, **34**, 229–230 (March 1962).
15. Sawyer, C. N., "Some New Aspects of Phosphates in Relation to Lake Fertilization," *Sewage Ind. Wastes*, **24**, 768–776 (June 1952).
16. Dryden, F. D., and G. Stern, "Renovated Waste Water Creates Recreational Lake," *Environ. Sci. Technol.*, **2**, 268–278 (April 1968).
17. *Report on Nutrient Sources of Lake Mendota*, Report of the Lake Mendota Problems Commission, Madison, Wisconsin, 1966, 41 pages (mimeographed).
18. Eberhardt, W. A., "Chemical Precipitation of Phosphorus within a High Rate Activated Sludge Process," Ph.D. Thesis, The Pennsylvania State University, University Park, Pennsylvania, December 1967.

19. Jenkins, D., and A. B. Menar, *The Fate of Phosphorus in Sewage Treatment Processes, I. Primary Sedimentation and Activated Sludge*, SERL Report No. 67-6, University of California, Berkeley, California, 1967.
20. Albertson, O. E., and R. J. Sherwood, "Phosphate Extraction Process," *J. Water Pollut. Control Fed.*, **41,** 1467–1490 (August 1969).
21. Buzzell, J. C., and C. N. Sawyer, "Removal of Algae Nutrients from Raw Wastewater with Lime," *J. Water Pollut. Control Fed.*, **39,** 10, Part 2, R16–R24 (October 1967).
22. Schmid, L. A., and R. E. McKinney, "Phosphate Removal by a Lime-Biological Treatment Scheme, *J. Water Pollut. Control Fed.*, **41,** 1259–1276 (July 1969).
23. Owen, R., "Removal of Phosphorus from Sewage Plant Effluent with Lime," *Sewage Ind. Wastes*, **25,** 548–556 (May 1953).
24. Eberhardt, W. A., and J. B. Nesbitt, "Chemical Precipitation of Phosphorus in a High Rate Activated Sludge System," *J. Water Pollut. Control Fed.*, **40,** 1239–1267 (July 1968).
25. Rand, M. C., and N. W. Nemerow, *Removal of Algae Nutrients from Domestic Wastewaters*, Report No. 9, Department of Civil Engineering, Syracuse University Research Institute, Syracuse, New York, 1965.
26. Garland, C. F., and G. L. Shell, *Integrated Chemical-Biological Wastewater Treatment*, AWTRA, FWPCA, U.S. Department of the Interior, November 1966.
27. Culp, R. L., A paper presented at a FWPCA sponsored workshop on phosphorus removal, Chicago, Illinois, June 26–27, 1968.
28. Neil, J. H., "Problems and Control of Unnatural Fertilization of Lake Waters," *Proceedings 12th Industrial Waste Conference*, Engineering Bulletin XLII, No. 3, Purdue University, Lafayette, Indiana, 1958, pp. 301–316.
29. Lea, W. L., G. A. Rohlich, and W. J. Katz, "Removal of Phosphates from Treated Sewage," *Sewage Ind. Wastes*, **26,** 261–275 (March 1954).
30. Bennet, G. E., R. Eliassen, and P. L. McCarty, *Reclamation of Reusable Water from Sewage*, Technical Report 58, Department of Civil Engineering, Stanford University, Stanford, California, October 1965.
31. "Phosphate Removal Processes Prove Practical," *Environ. Sci. Technol.*, **2,** 182–185 (March 1968).
32. Priday, W., H. E. Moyer, and R. L. Culp, "The Most Complete Wastewater Treatment Plant in the World," *Am. City*, **79,** 123–126 (September 1964).
33. Wuhrmann, K., "Stickstoff and Phosphorelimination: Ergebuisse von Versucken in Techmschen Masstab," *Schweiz. Z. Hydrol.*, **26,** 520–558 (1964).
34. Tossey, D., P. J. Fleming, and R. F. Scott, "Tertiary Treatment by Flocculation and Filtration," *ASCE J. Sanit. Eng. Div.*, **96,** 75–90 (February 1970).
35. Stumm, W., "Chemical Elimination of Phosphates as a Third Stage Sewage Treatment. A Discussion," *Proceedings of the International Conference on Water Pollution Research, London, September 1962*, Pergamon Press, New York, 1964, pp. 216–230.
36. Long, D. A., J. B. Nesbitt, and R. R. Kountz, *Soluble Phosphorus Removal in the Activated Sludge Process, Part I, Chemical Biological Process Performance*, U.S. Government Printing Office, Water Pollution Control Research Series, 17010 EIP 05/71, May 1971, 119 pages.
37. Farrell, J. B., B. V. Salotto, R. B. Dean, and W. E. Tolliver, "Removal of Phosphate from Wastewater by Aluminum Salts with Subsequent Aluminum Recovery," *Chem. Eng. Prog. Symp. Ser.*, **64,** 232–239 (1968).

38. Pollio, F. X., and R. Kunin, "Tertiary Treatment of Municipal Sewage Effluents," *Environ. Sci. Technol.*, **2,** 54–60 (January 1968).
39. Vacker, D., C. H. Connell, and W. N. Wells, "Phosphate Removal through Municipal Wastewater Treatment at San Antonio, Texas," *J. Water Pollut. Control Fed.*, **39,** 750–771 (May 1967).
40. Priesing, C. P., J. L. Witherow, L. D. Lively, M. R. Scalf, B. L. DePrater, and L. H. Myers, "Phosphate Removal by Activated Sludge Plant Research," paper presented at the 40th Conference of the Water Pollution Control Federation, New York, October 1967.
41. Bargman, R. B., W. F. Garber, and M. Spiegel, "Phosphate Removal by Activated Sludge Aeration," *ASCE J. Sanit. Eng. Div.*, **96** (SA1), 45–46 (February 1970).
42. Menar, A. B., and D. Jenkins, *The Fate of Phosphorus in Sewage Treatment Plant Processes, II. Mechanism of Enhanced Phosphate Removal by Activated Sludge*, SERL Report 68-6, University of California, Berkeley, California, August 1968.
43. Levin, G. V., and J. Shapiro, "Metabolic Uptake of Phosphorus by Wastewater Organisms," *J. Water Pollut. Control Fed.*, **37,** 800–821 (June 1965).
44. Shapiro, J., G. V. Levin, and G. H. Leo, "Anoxically Induced Release of Phosphate in Wastewater Treatment," *J. Water Pollut. Control Fed.*, **39,** 1810–1818 (November 1967).
45. Hennessey, T. L., K. V. Maki, and E. Y. Young, "Phosphorus Removal in Wastewater by a Modified Activated Sludge Process," a paper presented at a FWPCA-sponsored workshop on phosphorus removal, Chicago, Illinois, June 26–27, 1968.

Phosphorus as a Factor in the United States Economy

EDWARD D. JONES III

Research Economist, Central Intelligence Agency, Washington, D.C.

This chapter examines the economic importance of phosphorus and phosphorus products as produced and utilized in the United States. The phosphorus industry has entered the 1970s faced with several problems that could significantly alter the structure of the industry. The major problems that can be anticipated are greater foreign competition, overbuilt capacity, factor pricing, and, perhaps the most critical, ecological activism. This chapter attempts to put these problems in perspective by examining the entire phosphorus industry.

PHOSPHATE ROCK EXTRACTION

Phosphate rock is the primary raw material source for phosphorus chemical production. The United States is the largest producer in the world, accounting for 44.6% of the 1968 world production of 92.5 million tons (25). The next largest producers are the U.S.S.R. (including the production of apatite) and Morocco, accounting for 21.1 and 12.5%, respectively. Although there are a number of other countries producing phosphate rock, the three leading producers account for 78.2% of world production. In addition to being the largest producer, the United States is a major net exporter of phosphate rock. United States exports in 1969, 11.37 million tons, were 30.1% of marketable production. The primary recipients were Japan, Canada, Italy and West Germany. United States imports of phosphate rock in 1969 were 0.14 million tons, only 0.5% of apparent consumption, and came mainly from the Netherlands Antilles, Mexico, and Canada. These, due to the low fluorine content, were used primarily as animal feed supplements. In terms of the balance of trade in phosphate rock for 1969, the United States was in substantial surplus, exports amounting to $62.288 million, and imports $3.544 million.

The degree of growth of United States exports of phosphate rock in the 1970s will depend inversely on the rate of development and growth of foreign sources of rock closer to the primary United States export markets, and the development of indigenous rock sources in United States export markets for home consumption. As foreign competition becomes more efficient in production, transport costs will become a critical factor for the United States. For example, Morocco (23) and Spanish Sahara (25), a new entry in the phosphate rock market, both with rapidly developing phosphate rock industries, will probably garner a larger share of the European market as they are proximate to one of the major United States export markets. Indeed, there is speculation that one day Morocco may rival Florida production (15). Another example is Mexico which is currently exporting triple superphosphate to the United States which it derived from imported American phosphate rock (11). With the latest mining and processing technology available to it, Mexico could weaken American exports of phosphate rock by developing its home resources. When modern technology is augmented by the desire of such developing countries to utilize indigenous resources, especially when those resources are potentially major foreign-exchange earners, it is apparent that the United States will face much stronger foreign competition in the 1970s. Thus, the American phosphate rock industry will constantly have to improve and maintain high levels of technology and efficiency.

In 1969 37.725 million tons of marketable phosphate rock was produced from 62 mines operating in nine states centered in the southeast and west. Florida and North Carolina accounted for 79.3% of production, Tennessee and Alabama 8.8%, and the western states (Idaho, Montana, Utah, Wyoming, and California) 11.9%. The value of the rock sold in 1969 was $204,409,000 and the average price per ton was $5.57 (25). Of the total sold, 69.2% was for domestic use, and the remainder, 30.8%, for exports (Table 1).

TABLE 1 Phosphate Rock Production, by Uses and States, 1969, in Percent[a]

	Florida	Tennessee	Western States	United States, Total
Domestic, Total	62.5	100.0	88.7	69.2
Agriculture	97.0	—	25.0	73.0
Industry	3.0	100.0	75.0	27.0
Exports	37.5	—	11.3	30.8
Total	100.0	100.0	100.0	100.0

[a] Source: *Bureau of Mines Mineral Yearbook.*

Domestically, 73% went into agricultural uses, almost entirely fertilizer, and 27% into industrial uses. A breakdown of industrial consumption shows 26.0% for soap and detergents, 14.8% for animal feed, 11.1% for plating and polishing, and 48.1% for other industrial uses (11).

Regionally, utilization by source varies considerably. Tennessee rock is entirely for domestic use, and goes to industrial consumption via electric-furnace production. Seventy-five percent of the rock from the Western states goes into industrial uses. Florida, the national leader in production of phosphate rock for more than 75 years, channels all but 3% of its production into agricultural uses. In addition, 37.5% of Florida rock is exported. For the United States, the overall rock sold for industrial use by Florida, Tennessee, and the western states amounted to 8.1, 46.6, and 45.3%, respectively. A comparison of the statistics in Table 1 indicates that, absolutely, change in demand for exports and agricultural use of phosphate rock will have its greatest impact on Florida production, whereas a change in demand for industrial uses will have its impact on Tennessee and western states production. However, due to the relatively smaller scale of operations in Tennessee and the western states as compared with Florida operations, change in demand deriving from any source is likely to be a significant factor in those areas, directly or indirectly. In 1969, for example, Montana production fell 48%. This was due to weakened fertilizer demand which caused the phase out of inefficient plants that were previously operating on a marginal basis.

Total capacity of the phosphate rock industry is 51.165 million tons/year. Capacity utilization for 1969 was 73.7%, and an estimated 80.7% for 1970 (5, 16). The relatively large unused capacity is explained by the softening demand for fertilizer phosphates, the large number of obsolete operations in use, and the increase in plant capacity in the mid-1960s to meet the expected increase in fertilizer demand that did not materialize. Nonetheless, the increase in capacity utilization in 1970, taking into consideration a 0.3% decrease in production, reflects the closing of inefficient operations, namely seven plants, the closing of fertilizer phosphate plants and distribution facilities, and, in addition, operating the open plants more efficiently by cutting costs, for example, shortened work weeks and the postponement of some investment.

The importance of the phosphate rock industry is reflected in the 1967 Census of Mineral Industries (22). In the broad industry group "Chemicals and Fertilizer Minerals,"* phosphate rock ranked first in the number of establishments (69), employees and production workers (8000 and 5000,

* Includes barite, fluorspar, potash, soda, borate minerals, phosphate rock, rock salt, sulfur, and chemicals and fertilizer minerals, not elsewhere categorized.

respectively), man-hours worked (12 million), value of shipments ($297 million), and capital expenditure ($47 million). It ranked second in payroll ($59 million), wages ($35 million), and value added in mining ($199 million). Obviously, the phosphate rock industry is an integral part of the fertilizer and chemical mineral industry. To maintain this position, the phosphate rock industry cannot afford to let the ever-increasing foreign knowledge of mining and processing technology go unheeded. The next section outlines the industry to which phosphate rock is such an important input.

PRODUCTION OF PHOSPHORUS CHEMICALS

There are two distinct processes of manufacture currently used in the United States to produced phosphorus and phosphoric acid—the two building blocks for the production of all phosphorus compounds. First, there is the so-called wet-acid process which involves the treatment of phosphate rock with sulfuric acid to obtain a relatively impure phosphoric acid and a calcium sulfate precipitate. Due to high levels of impurities, phosphoric acid so derived usually requires purification prior to utilization. This is the primary source for normal superphosphate. Triple superphosphate is derived from treating phosphate rock with phosphoric acid. Although the triple superphosphate process is relatively more costly than the normal superphosphate process, certain economies are derived from the utilization of lower grade rock to produce the phosphoric acid to be used in the triple superphosphate process (21). In 1969 4.916 million tons of phosphoric acid (100% P_2O_5) was produced in the United States (24). Of this total 78.7% was produced by the wet-acid production process. In addition, 76.4% of the total went into fertilizer applications, indicating that nearly all of the phosphoric acid derived from the wet-acid process was for fertilizer use. Only slightly over 3% of the wet-acid went into nonfertilizer applications.

The other major production process is the electric-furnace method. Elemental phosphorus production by the electric-furnace process involves the input of phosphate rock, silica, and coke in an electric furnace, which, under a sufficiently high-temperature reaction, yields a white modification of elemental phosphorus, calcium silicate slag, and ferrophosphorus (26). The elemental phosphorus can then be converted to a highly purified phosphoric acid by oxidation and hydration in burning towers. In 1969 there were 13 plants in the United States producing elemental phosphorus with a total capacity of 760,00 tons/year (5). Production amounted to 610,000 tons in 1969, or 80.3% of capacity. By 1974 production is expected to increase to 740,000 tons, assuming, of course, the continued existence of detergent phosphates. The projected 1974 capacity could be handled with 97.4% of existing 1969 capacity. The United States accounts for approximately 60% of the world production of elemental phosphorus, but the share has gradually declined as other countries have begun domestic production. Finally, the electric-furnace technology is expected to remain the dominant process in phosphorus production into the 1980s (1).

A recent study has indicated that phosphate rock is a major cost item in elemental phosphorus manufacture. Specifically, phosphate rock accounted for 40 to 60% of total raw material cost, and was equivalent to 15 to 20% of total manufacturing cost (1). Illustrating total manufacturing cost, breakdown for an 80,000-tons P_4/year capacity with two MW electric furnaces was as follows: raw materials 35%, power 25%, capital 15%, maintenance and labor 15%, and other factors 10%. As indicated by the percentages, the sensitivity to price changes in power and phosphate rock costs is an important factor in the pricing of phosphorus. Although investment in electric-furnace technology is significantly greater than that of an equivalent wet-acid operation, lower grade rock utilization in the former offsets the difference somewhat.

Total phosphoric acid production (100% P_2O_5) in 1969 represented 84.7% of a total capacity of 5.807 million tons. Wet acid accounted for 78.7% of total production, and the remainder was produced by the oxidation and hydration of elemental phosphorus. The trend in recent years has been for the percentage of phosphoric acid derived from the wet-acid process to increase. For example, that derived from elemental phosphorus decreased from 27.8% in 1964 to 18.7% in 1968, while that derived via the wet-acid route increased from 72.2% in 1964 to 81.3% in 1968. The 2.6% drop in 1969 in wet-acid production resulted from the fall in fertilizer demand and the increased price of sulfur due to the sulfur shortage. Although the wet-acid process dominates due to the fertilizer usage requiring a relatively less pure acid than most industrial applications, factor pricing could affect the industry mix in the 1970s.

Perhaps the most critical factor is the cost of the two "critical inputs" in the wet-acid and electric-furnace processes, sulfur and electric power, respectively (1). An increasing ratio of cost of electric power to cost of sulfur has been partially responsible for a 2% decrease in 1970 of elemental phosphorus production. Electric power rates have increased substantially in the past 2 years while sulfur prices have declined to the point that they are only about 60% of the March 1969 price of $40.00/long ton. The problem is particularly acute in Tennessee following an increase in T.V.A. power rates in the fall of 1970 (6). If this trend continues, it will be economically advantageous for wet-acid producers to invest in developing new purification methods so that they can derive a larger share of the nonfertilizer market. It is also advantageous from the standpoint that, recently, fertilizer demand has been for higher analysis products, a trend that appears likely to continue, especially given the economies in distribution and shipping of the higher analysis products.

A trend in favor of electric-furnace production has been the shift of manufacturing plants away from raw material sources, and closer to ultimate market destination. The economic advantage derives from the fact that elemental phosphorus can be shipped as such and converted to phosphoric acid more cheaply than producing the acid at the elemental phosphorus plant and shipping the acid. This is one price advantage that the electric-furnace method enjoys over the wet-acid process. In addition, with the demand for higher analysis fertilizer products, concentrated superphosphate and diammonium phosphate, elemental phosphorus can expect increased demand, provided electrical power costs do not counteract the gains derived from increasing scale of operations, as occurred in 1969.

Of the top 50 volume chemicals produced in the United States in 1970, phosphoric acid (100% P_2O_5) ranked tenth with a production of 5.471 million tons, and elemental phosphorus ranked fiftieth with a production of 0.597 million tons (10). Over the period of the 1960s annual average growth has declined. For the periods 1960 to 1970, 1965 to 1970, and 1969 to 1970, growth rates were 10, 7, and 2% for phosphoric acid, and, for elemental phosphorus, 6, 4, and −2%. On the same period breakdown, growth rates for sodium tripolyphosphate, the primary detergent phosphate, were 6, 4, and −2%, in close accord with the rates for elemental phosphorus. Phosphoric acid production in 1968 failed to increase for the first time since 1942. Over the period 1942 to 1970 annual average growth of phosphoric acid was an extremely high 11.3% (12). Growth of phosphoric acid will depend on fertilizer demand and usage. The prospects for the 1970s, at least in the early part of the decade, indicate a growth rate less than the 7% sustained in the 1960s. Table 2 shows production of selected phosphorus compounds for the years 1954, 1963, 1967, and 1969. Of particular note is the growth in phosphoric acid production and the change in derivation by source.

FERTILIZER APPLICATIONS

Fertilizer applications accounted for almost 82% of phosphoric acid production in 1970. In 1969 slightly over 76% produced was for fertilizer applications. The major part of the increase in production of phosphoric acid can be accounted for by fertilizers between 1969 and 1970. In this period there was the continuation of the switch to higher analysis products, from normal superphosphate to triple superphosphate and diammonium phosphate. Comparing 1970 to 1969 crop years, normal superphosphate

TABLE 2 Production of Selected Phosphorus Compounds (Short Tons)[a]

	1954	1963	1967	1969
Calcium phosphate, dibasic 100% $CaHPO_4$	128,730	239,900	391,700	503,862
Phosphoric acid, 100% P_2O_5				
Total	1,138,026	2,904,800	5,066,400	4,915,534
From phosphorus	506,744	947,300	1,072,900	1,048,539
From phosphate rock and other	631,252	1,957,500	3,993,400	3,866,995
Phosphorus	266,887	488,100	587,000	624,460
Phosphorus oxychloride, 100% $POCl_3$	15,240	23,900	32,800	31,405
Phosphorus trichloride, 100% PCl_3	9,758	26,800	51,100	56,365
Tetrapotassium pyrophosphate, 100% $K_2P_2O_7$	—	37,400	54,900	41,467
Sodium phosphates				
Monobasic, 100% NaH_2PO_4	7,114	14,600	17,400	N.A.[b]
Dibasic, 100% Na_2HPO_4	26,604	23,700	22,600	N.A.
Tribasic, 100% Na_3PO_4	48,435	56,700	61,700	57,337
Meta, 100% $NaPO_3$	52,808	64,500	88,100	87,633
Tetra, 100% $Na_4P_2O_7$	108,989	109,300	108,800	84,039
Acid pyro, 100% $Na_2H_2P_2O_7$	14,238	20,400	25,200	N.A.
Tripoly, 100% $Na_5P_3O_{10}$	520,549	817,700	1,048,100	1,206,086

[a] Sources: U.S. Department of Commerce, Bureau of the Census, 1958 Census of Manufactures, Vol. II, Industry Statistics; 1967 Census of Manufactures, MC67(2)-28A, September, 1970; *Inorganic Chemicals Summary for 1969*, M28A(69)-13, July 20, 1970.

[b] N.A.—not available.

production declined 7%, domestic disappearance declined 10%, and inventories increased 7%. Triple superphosphate, on the other hand, showed production up 15%, domestic disappearance up 21%, and inventories down 1%. Prices are indicative of the trend. Normal superphosphate showed general price erosion in 1969, and a 13.1% drop in March 1970. Concentrated superphosphate also showed price erosion during 1969, but a strong uptrend in 1970 as fertilizer demand strengthened (27).

Phosphate fertilizer producers have begun phasing out inefficient operations, and investing in newer, more efficient plants and distribution facilities.

Capacity utilization has increased from between 60 and 65% in 1967 to 1968 to nearly 80% in 1970 overall (16). Triple superphosphate plants have a present capacity of 2.601 million tons/year. Ammonium phosphate and diammonium phosphate plants have a capacity of 3.169 million tons/year (5). Ammonium phosphates showed a 52.1% capacity utilization in 1969, while superphosphates, normal and concentrated, showed a 88.6% capacity utilization. Projections for 1974 show that ammonium phosphate production will be at 2.2 million tons, triple superphosphate will increase to 1.8 million tons, but normal superphosphate will decrease to 900,000 tons. Producers will have to adjust overcapacity to realistic estimates of demand. In addition, producers will be facing stronger foreign competition.

The United States is a major net exporter of phosphate fertilizers, exports accounting for 41.1% of domestic production in 1970. United States exports of superphosphates in 1969 were 847,000 tons valued at $33.922 million. Compared with 1968, volume and value of exports decreased 34.3 and 39.8%, respectively. Imports of phosphate fertilizers were 83,000 tons valued at $3.976 million in 1969 (25). Likely to affect United States phosphate fertilizer exports is the trend, especially in developing countries, of meeting fertilizer needs by home production. Demand in the developing countries for all fertilizers is expected to double the 1968–1969 consumption of 7.7 million tons by 1973, and reach 33.9 million tons by 1985 (8). One estimate suggests that in 1970 developing countries were able to meet 70% of all fertilizer needs by home production, and by 1972 will be able to meet 90% of phosphate fertilizer needs by home production (19). To meet the growing demand capital investment in the fertilizer industry in developing countries is expected to exceed $10 billion by 1985. These countries will need to import a good deal of raw materials for the fertilizer industry, especially phosphate rock. Combined with the initial capital investment required, the FAO concludes that, given the necessity to increase export earnings to justify increased imports of raw materials for home production, developing countries would be better off for the immediate future importing finished or semifinished fertilizers (8). Highlighting this conclusion is a recent study that offers a comparison of United States and Brazil superphosphate production costs (3). For granular triple superphosphate (46% P_2O_5) on a capacity basis of 154,000 tons/year, investment per annual ton of P_2O_5 capacity is $150 and $189 for the United States Gulf Coast and Brazil, respectively. In addition, when comparing operational requirement costs, although total fixed costs are only 1.24 times as great in Brazil, total variable costs are 1.56 times as great in Brazil. The major difference is in the cost of the phosphate rock, which, in terms of dollars per ton of P_2O_5, is 2.18 times as great in Brazil as in the United States. However, this study concludes that, even though home production in developing countries may be more expensive, local production may derive benefit over imports as local production exerts less stress on international finance requirements, especially if fertilizer raw materials are exploitable in the developing countries. Although at present developing countries may be better off importing finished fertilizer products, in the long run developing countries will benefit by establishing their own industries, especially if the inputs are indigenous. Korea and Taiwan, for example, already have become net exporters of fertilizers, and other developing countries will probably follow suit.

Two domestic factors that have affected fertilizer sales in the last several years have been poor weather in the spring planting season, and corn blight. Corn is the most widely planted crop in the United States and, as it requires heavy application of fertilizer to maximize yield, the fertilizer industry will be susceptible to changes in demand caused by these natural phenomena in the future.

SYNTHETIC DETERGENTS AND PHOSPHATES

Probably the most critical problem to face the phosphorus industry is the possible elimination of phosphates from synthetic detergents. In 1969 the soap and detergent industry in the United States produced 2.977 million tons of soap and synthetic detergents with sales of $1.563 billion. Synthetic detergents accounted for 84.2% of production and 77.0% of sales (9). Although synthetic detergents were first marketed in 1933, it was not until after World War II, and the introduction of phosphates on a large scale, that production and sales began to boom. In 1947 soap accounted for 90% of sales, and synthetic detergents accounted for 10%. By 1953 synthetic detergents had overtaken soap with 52% of total sales, and synthetic detergents have dominated ever since. Through the decade of the 1960s synthetic detergent sales were greater than 75% to total sales on an annual basis. Prospects for the 1970s, clouded by the possible elimination of phosphates from detergents, are uncertain.

The problem with phosphates in detergents is basically ecological, that is, allegedly, excessive concentrations of phosphates in United States waterways have caused severe eutrophication. However, eutrophication by phosphates is not isolated to that only from detergents. Soil runoffs of phosphate from fertilizer, animal wastes, plant wastes, and soil erosion also are responsible for concentrations of phosphates in waterways. The excessive phosphates in the waterways are responsible for upsetting the biological life cycle by causing algal blooms and depriving oxygen from the other living beings in the hydroenvironment. One estimate puts the amount of detergent phosphate in municipal sewage as high as 60% of total phosphate content (18). However, there is a wide variation of opinion concerning the ecological damage done by phosphates, whether phosphates are indeed the so-called limiting nutrient in eutrophication control, and whether detergent phosphates are the critical source. Currently, it would appear that the solutions to the problem are more emotionally charged and uncertain than scientifically based on concrete and irrefutable evidence. Although no Federal legislation has been passed to date, the Committee on Government Operations of the House of Representatives recommended on April 14, 1970 "that the phosphate content of all detergents be reduced in phases, starting immediately, and that all phosphorus be eliminated from detergents within two years" (18). Already a number of local communities have instituted restrictions on the phosphate content of synthetic detergents allowable in their areas, usually restricting the phosphorus content to a maximum of 8.7% of total content (14). To understand the magnitude of the problem, it is necessary to examine the use of phosphates in detergents and the alternative methods of correction of eutrophication, including substitutes for phosphates, such as nitrilotriacetate and polyelectrolytes, and advanced sewage-treatment facilities that would remove concentrations of all nutrients.

Approximately 80% of all sodium phosphates find use in detergent applications, with the remainder divided between water conditioning (10%), food and medicine (5%), and other uses (5%) (7). In turn, it was estimated in 1965 that 55% of all elemental phosphorus production found application in sodium phosphates (7). Sodium tripolyphosphate is the single largest input in the manufacture of synthetic detergents. In a typical heavy-duty detergent, it represented 50% of composition (18). In 1969 sodium tripolyphosphate production capacity was 1.2 million tons, with capacity utilization close to 100% (5). Projected production for 1974, ignoring the detergent phosphate problem, is estimated at 1.6 million tons. Other phosphates, namely trisodium phosphate, tetrasodium pyrophosphate, and tetrapotassium pyrophosphate, also are employed in the production of synthetic detergents. Tetrapotassium pyrophosphate is primarily used in the production of liquid detergents. Among the functions in detergents ascribable to phosphates are that they heighten the efficiency of the surface active agent, decrease the redeposit of dirt particles on the clothes, and soften the water through sequestration (17). In addition, phosphate has been shown to be nontoxic and safe for use. In 1967 the soap and detergent industry consumed the entire production of trisodium phosphate (61,683 tons), 64% of tetrasodium pyrophosphate (United States production: 108,850 tons), and 81% of sodium tripolyphosphate (United States production: 1,048,079 tons). Tetrapotassium pyrophosphate production in the same year was 54,877 tons, but the amount consumed by the soap and detergent industry is not available. Total phosphorus consumed in 1967 by soap and detergent industry was an estimated

243,000 tons, 41.5% of the phosphorus produced. In terms of elemental phosphorus content, sodium tripolyphosphate accounted for 88.2% of the total. Examining United States phosphate use in 1967 by the soap and detergent industry as a function of the total United States apparent domestic consumption, 6% was consumed. This represents a decline from 9% in 1963, but an increase of 15.6% in amount of phosphate consumed in weight. Although small in comparison to total phosphate production and usage, detergent phosphates represent an important consumption for one segment of the market, elemental phosphorus.

Although the detergent phosphate market has not been severely constricted as yet, there is a voluntary program instituted by detergent producers to reduce the phosphate content of detergents gradually (13). An estimate suggests that by late 1972 producers of detergents will have decreased phosphate usage by almost 36%, representing a then expected consumption of 642,500 tons of sodium tripolyphosphate. This would indicate an approximate decline in demand for total sodium tripolyphosphate production of 30%. Demand for elemental phosphorus would also be curtailed, perhaps, on the basis of this scenario, by as much as 15 to 20%. Of course, this will also affect demand for phosphate rock to be used in the production of elemental phosphorus. However, the severity of the constriction in demand will decrease sequentially going from detergent phosphates to elemental phosphorus to phosphate rock. In addition, repercussions will spread to other industries that are associated with the production of the aforementioned phosphorus products. The ultimate effect will be to increase the prices of phosphorus products, after an initial downward price tendency. The latter derives from the fact that producers will lower prices to liquidate stocks as they readjust to new, and lower, levels of demand. As producers adjust to lower levels of demand by producing less, they will be faced with higher levels of average fixed costs. These costs cover investment in buildings, equipment, land, and so on, that are being amortized, and are invariant with the level of output. Average variable costs, the costs incurred in the production process such as for labor, raw materials, and so on, do not generally hold to a hard-and-fast rule as do average fixed costs. Under the assumption of diminishing returns to scale, average variable costs generally decline to some point—a point of efficiency—and increase beyond that point. The rationale for this argument is that before and after reaching the efficiency point average variable costs generally rise as factors are employed less efficiently. With the projected decline in demand by the end of 1972, completely abstracting from the decline that would be attendant on a total ban, average variable costs, and thus average cost (i.e., average fixed cost + average variable cost), will probably have a tendency to increase. This observation is based on the fact that 1969 production involved almost total capacity utilization, and production at somewhat lower capacity utilization will increase the average costs of operation. Thus, after the initial adjustment of production levels to lower levels of demand, prices of detergent phosphates will have a tendency to increase. What this means is that competing uses of sodium phosphates will more than likely be faced with higher prices, and, consequently, those products that use sodium phosphates will be higher priced. Note that this is a tendency. Exact computation of effects will depend on a number of factors that are unclear at the moment. For example, not only would the extent of the ban be important, but also the timing. Gradual elimination of detergent phosphates would allow producers to adjust production with less disruption than if the ban were sudden and immediate. If the latter occurred, producers would drive the price of sodium phosphates very low to undercut competitors and liquidate stocks. This would have the possible effect of driving out of the market those producers who, had the ban been less drastic, ordinarily would have remained in the market. In the longer run, producers will adjust the scale of operations such that production will be efficient at lower levels of demand, but with generally higher average costs than had previously obtained.

ALTERNATIVE SOLUTIONS TO THE DETERGENT PHOSPHATE PROBLEM

There are several possible solutions to the detergent phosphate problem currently being considered. One approach, requiring the partial or total elimination of phosphates from detergents, is the substitution of nitrilotriacetate or polyelectrolytes in detergents to perform functions similar to the detergent phosphates. One study has estimated that, in addition to not cleaning as well as phosphate formulations, nonphosphate or low-phosphate products generally cost more per washload, even though, in some cases, input costs per pound of product are less (4). Although nonphosphate products are riding a crest of popularity due to concern for environment, marketing of nonphosphate products has been so swift that the safety of inputs alternative to phosphate have been questioned, for example, nitrilotriacetate (2). Although research is continuing to find a product similar to phosphate in cost, safety, and effectiveness, none has been found to date that meets these criteria and is readily substitutional.

Another approach is to provide wastewater-treatment facilities to negate eutrophication of waterways from a number of nutrient sources. However, the costs of providing advanced wastewater treatment are enormous. One estimate for the entire United States, based on a study of the Lake Erie-Lake Ontario drainage basins, puts the cost of providing advanced treatment at close to $10 billion (18). Others have estimated the costs to be substantially greater, perhaps three to five times greater. For the period 1971 to 1974 an estimated \$4 billion/year is needed in industrial and municipal outlays to control water pollution based on current requirements, with about half coming from each source for new investment and operating expenses (20). For fiscal 1972 the Federal contribution to pollution abatement is a budgeted \$2.014 billion in outlays with budget authority of \$3.127 billion (20).

The commitment to water-pollution control by all levels of government, industry, and the people, what can be called ecological activism, has grown substantially in the last several years. A concerted and well-designed program for the 1970s is a necessity in order to clean up the environment. If the available funds are less than that necessary to complete the task, there are several alternatives that present themselves. One, of course, is the realignment of priorities at all levels of activity. Another is the application of an excise tax on those products that pollute the environment. The argument for such a tax is that the market price of the product understates its "real" price, the latter including the externalized costs of repairing the environmental damage. An example is the newly enacted Mineral Severance Tax in Florida, the revenues of which are to be assigned partially to land reclamation (15). This might plausibly be an alternative to the removal of phosphates from detergents. Less precipitous than removal, it would discourage demand for phosphate detergents by increasing the price relative to nonphosphate detergents, would provide revenue to contribute to the clean up of the environment, and, as importantly, provide revenue to research in the quest for a suitable alternative to phosphates in detergents. However, before any action is taken, more research into the detergent phosphate problem is needed.

REFERENCES

1. Bryant, H. S., N. G. Holloway, and A. D. Silber, "Phosphorus Plant Design New Trends," *Ind. Eng. Chem.*, **62,** No. 4, 8–23 (April 1970).
2. Cohn, V., "Detergent Chemical Causes Defects in Rats," *Washington Post*, December 18, 1970, p. A1.
3. Development Centre of the O.E.C.D., *Supply and Demand Prospects for Fertilizers in Developing Countries*, Paris, O.E.C.D., 1968, pp. 192–202.
4. "Ecologic Detergents: Will the Bubble Burst?" *Chemical Week*, April 28, 1971, pp. 10–12.
5. *The Economics of Clean Water*, Vol. III, *Inorganic Chemicals Industry Profile*, U.S. Department of the Interior, Federal Water Pollution Control Administration, U.S. Government Printing Office, Washington, D.C., March 1970.
6. Emigh, G. D., "Phosphate," *Mining Engineering*, January 1971, pp. 59–60.
7. Faith, W. L., D. B. Keyes, and R. L. Clark, *Industrial Chemicals*, 3rd ed., Interscience, New York, 1965, pp. 699–706.

8. "Fertilizer Growth Seen Needing $11 Billion in Capital Investment For World's Developing Areas," *Oil, Paint and Drug Reporter*, February 23, 1970, p. 7.
9. "Good Growth in Sales Despite Slowing Economy," *Chemical and Engineering News*, September 7, 1970, p. 77A.
10. "Growth Sags for Top Fifty Volume Chemicals," *Chemical and Engineering News*, May 17, 1971. p. 15.
11. Guccione, E., "Phosphate," *Engineering and Mining Journal*, March 1971. pp. 136–137.
12. "Industrial Inorganic Chemicals, N.E.C.," *1967 Census of Manufactures, Industry Series*, Preliminary Report MC67(P)-28A-6, U.S. Department of Commerce, Bureau of the Census, U.S. Government Printing Office, Washington D.C., 1969.
13. "Industry Pollution Control Effort Begins to Make Impact on Nation," *Commerce Today*, December 14, 1970, pp. 31–32.
14. "New Phosphate Ban in New York State," *The Wall Street Journal*, June 9, 1971, p. 4.
15. "New Tax Hard to Dig," *Chemical Week*, June 30, 1971, p. 13.
16. "Phosphates are Moving Again," *Chemical Week*, March 10, 1971, p. 18.
17. "Phosphates in Detergents and the Eutrophication of America's Waters," *Hearings Before a Subcommittee of the Committee on Government Operations, House of Representatives, Ninety-First Congress, First Session, December 15 and 16, 1969*, House of Representatives, Committee on Government Operations, U.S. Government Printing Office, Washington, D.C., 1970.
18. "Phosphates in Detergents and the Eutrophication of America's Waters," *Twenty-Third Report by the Committee on Government Operations*, House Report No. 91-1004. House of Representatives, Committee on Government Operations, U.S. Government Printing Office, Washington, D.C., 1970.
19. "Plague of Overcapacity," *Chemical and Engineering News*, October 19, 1970, p. 25.
20. Scherer, J., "Pollution and Environmental Control," *Federal Reserve Bank of New York Monthly Review*, June 1971, pp. 132–139.
21. Shreve, R. Norris, *Chemical Process Industries*, 3rd ed., McGraw-Hill, New York, 1967, pp. 265–285.
22. *Statistical Abstract of the United States*, 91st annual ed., U.S. Department of Commerce, Bureau of the Census, U.S. Government Printing Office, Washington, D.C., July, 1970, pp. 652–653.
23. Stipp, H. E., "The Mineral Industry of Morocco," *Preprint from the 1969 Bureau of Mines Minerals Yearbook*, U.S. Department of the Interior, Bureau of Mines, U.S. Government Printing Office, Washington, D.C., 1970, pp. 1–7.
24. *Survey of Current Business*, U.S. Department of Commerce, Office of Business Economics, U.S. Government Printing Office, Washington, D.C., December 1970, p. S-24.
25. Sweeney, J. W., "Phosphate Rock," *Preprint from the* 1969 *Bureau of Mines Minerals Yearbook*, U.S. Department of the Interior, Bureau of Mines, U.S. Government Printing Office, Washington, D.C., 1970, pp. 1–14.
26. Van Wazer, J. R., *Industrial Chemistry and Technology of Phosphorus and Phosphorus Compounds—A Survey*, Interscience Reprint Series, Interscience, New York, 1969.
27. *Wholesale Prices and Price Indexes*, U.S. Department of Labor, Bureau of Labor Statistics, Monthly, U.S. Department of Labor, Washington, D.C., 1969 and 1970.

Environmental Phosphorus—An Editorial

EDWARD J. GRIFFITH

Monsanto Company, St. Louis, Missouri

The following editorial is an analysis of the current status of phosphorus in the environment. An attempt has been made to deal primarily with the purely scientific aspects of this emotion-packed subject. Unfortunately, only limited quantitative data exist concerning phosphorus in the environment which have not been obtained by extrapolation of a meager number of measurements that are often of questionable quality. Usually semiquantitative estimates of the behavior of phosphorus in the environment will suffice, but in special cases even carefully executed measurements made with the best available techniques may prove to be inadequate. The long-term changes of the concentration of phosphorus in the freshwater resources of the earth present such a special case.

In addition to the purely scientific reasons for desiring an in-depth understanding of the role of phosphorus in the environment, the current interest in pollution control and environmental protection also deserves consideration. Several questions are worthy of exploration:

1. Has the concentration of phosphorus in the freshwater resources of the earth changed appreciably during the last three decades?
2. How much data related to question 1 are available and how reliable are these data?
3. Is the concentration of phosphorus in the aquatic environment approaching a critical level?
4. What is a critical level?
5. Does environmental phosphorus create a potential hazard to the future of man?
6. Is man capable of general control of the concentration of phosphorus in the aquatic environment?
7. What new research should be in progress?

Satisfactory answers to most of these and similar questions do not exist today, but the questions should be reviewed.

THE CONCENTRATION OF PHOSPHORUS IN THE FRESHWATER RESOURCES OF THE EARTH

If only measured analytical data were accepted as justification for stating, "The concentration of phosphorus in the aquatic environment of the earth has changed significantly during the last three decades," the statement cannot be justified. Until the early 1960s the concentration of phosphorus in the freshwaters of the earth was considered by most investigators to be too insignificant to be worthy of analysis. In most cases even poor data do not exist. It is stated in the *U.S. Geological Survey Water Supply Paper No. 1473* (1959), "The determination of phosphate is not generally included in the chemical analysis of water. Except in unusual instances, the amount present is small and can be ignored without serious error." It is also stated, "So few data are available on the content of phosphate in natural water that little can be positively stated regarding the observed range of concentrations" (1). In the 1970 revision of the *Water Supply Paper 1473* the wording has been changed, but the conclusions are essentially the same. Too few data are available to assess the role of phosphorus in the aquatic environment. There is no question as to the validity of the conclusions presented above. Published analytical data either good or bad exist for but a very limited number of the earth's lakes and streams. Most of the data which do exist are of questionable value except for the specific reasons for which the measurements were made, and these were seldom obtained from experiments designed to yield a representative sampling of a total body of water.

It is impossible to conclude that the concentration of phosphorus has changed in the freshwater resources of the earth when no reliable datum point exists with which to compare current measurements. This is not to imply that the concentration of phosphorus has not changed in specific bodies of water. Neither does it imply that the total quantity of phosphorus entering the aquatic environment has remained unchanged. It does mean that extreme care must be exercised in rendering general conclusions regarding the status of phosphorus in the freshwater resources of the earth.

SCIENTIFIC STUDIES

Fortunately, during the last 10 years environmental phosphorus has received more attention than previously. It is unfortunate, however, that too many of the works concerned with the concentration of phosphorus in the aquatic environment have been more or less a side issue to support some preconceived concept with social or political overtones. Too few critical studies have been made that are directed specifically toward learning of the behavior of phosphorus in natural water systems. There are notable and commendable exceptions, but even these can hardly be considered adequate.

How reliable are the phosphorus measurements that have been obtained? Do they truly represent the systems from which the samples were taken? These questions are particularly disturbing when considering the numerous glib and unsupported statements made concerning the concentration of phosphorus in the aquatic environment. The Lake Erie studies are a prime example of both good and poor work, but, admittedly, the problems in assessing the status of phosphorus in Lake Erie are staggering.

Lake Erie has a surface area of 9940 mi^2 and has an average depth of 60 ft. What is the minimum number of samples that must be collected and analyzed at any one time to determine the "average" concentration of phosphorus in the total lake with an error no greater than $\pm 10\%$? How often must the samples be collected, and over how long a time span in order to know if long-term and short-term cycles are occurring which significantly bias any group of data? Does an "average value" for the concentration of phosphorus in a large dynamic heterogeneous body of water have a real meaning? A number of well-planned, statistically designed experiments could do much toward answering these and similar questions. The magnitude of the problem of obtaining a representative sampling of Lake Erie can be appreciated if one assumes that only one sample should be collected per square mile-foot (i.e., 1 sample/208,558,310 gal H_2O), a total of 600,000 samples should be analyzed. The samples should all be taken almost simultaneously because the phosphorus concentration in Lake Erie is reported to vary as much as 1200% at a single location in 3 weeks' time (2). Even in some of the more elaborate studies of Lake Erie, samples have been collected at less than 50 stations (3). After 10 years' study, little, if any, indisputable change has occurred in the quantity of phosphorus in the water of Lake Erie as assessed by current methods of sampling and analyzing for phosphorus when the 1962 data are compared with a more limited but more recent study (4).

Obviously, the current methods of collecting and analyzing grab samples for phosphorus are inadequate. Some form of a continuous instantaneous method of analysis for phosphorus must be developed if sufficient numbers of data are to be collected to yield reliable values. A specific ion electrode sensitive to dissolved phosphorus would be most helpful, but at this time reliable phosphorus electrodes are not available.

FACTORS WHICH GOVERN THE CONCENTRATION OF PHOSPHORUS IN A BODY OF WATER

Numerous variables determine the concentration of active phosphorus in a body of water. (Active phosphorus is defined as all phosphorus contained in a lake which is a component of or is capable of becoming a component of the biological cycles of a lake without suffering a change of state.) The size of the body of water is of prime importance. It has long been recognized that the solubility of calcium orthophosphate (apatite) is the governing factor for the concentration of active phosphorus in the seas and oceans of the earth at depths of more than about 50 m (6). Man has had no significant influence on the concentration of phosphorus in seawater, nor is it probable that he could influence the concentration measurably even if he chose to do so.

The smaller a body of water is, the greater influence man can exert upon it. In a small lake the quantity of phosphorus added to the lake, the type of phosphorus compounds added, the volume of the lake, the pH, solar radiation, the temperature, the water hardness of the lake, the terrain surrounding the lake, the composition of the bottom of the lake, the depth of the lake, and the quantity of water passing through the lake in a period of time all influence the concentration of active phosphorus in the lake. In some studies it has been tacitly assumed that the quantity of active phosphorus will continue to rise indefinitely as the quantity of phosphorus is increased in the water entering the lake (6). In small freshwater lakes the concentration of phosphorus may respond quickly to increased loads of phosphorus and the active phosphorus concentration may rise sharply when the load is increased. In large bodies of water an equilibrium will be established that will function to resist changes in the quantity of active phosphorus contained in the lake (1a). For example, if the total annual quantity of phosphorus added to Lake Erie is doubled or tripled over the next 20 years, this does not mean that the quantity of active phosphorus contained in the waters of the lake will double or triple. Most of the phosphorus will be precipitated as inactive phosphorus, becoming trapped in the sediments where it will ultimately become fluo- or hydroxyapatite, as has been the case for millions of years (5). Comparing the data presented in the 1967–1968 *Lake Erie Surveillance Data Summary* and the *Lake Erie Report* (1968), some interesting observations can be made which support these conclusions. Again the reader should not accept these data as more than indicative and the conclusions must not be considered definitive.

In the *Lake Erie Report* (1968) (6, p. 2), a graph implies that the input of phosphorus to Lake Erie increased about 12% between 1962 and 1968. In the *Lake Erie Surveillance Data Summary* the data of Table 7 (p. 39) report that the phosphorus in the sediments of Lake Erie increased roughly 39% for the total lake in the same period. These data *indicate* that phosphorus is precipitating at a more rapid rate as more phosphorus is being added to the lake. This is to be expected for a lake in which the calcium concentration is about 40 ppm (*Lake Erie Report*, Fig. 3.4, p. 33). The problems that arise with phosphorus enrichment occur because phosphorus is added to a lake at a greater rate than outflow and precipitation of phosphate can take place. Some investigators believe that the phosphate which precipitates as ferric phosphate may be converted to the more soluble ferrous phosphate under anaerobic conditions. Ultimately, even the solubilized ferrous phosphates will be converted to apatites, but it appears that most of the precipitated iron phosphates in large bodies of water will remain trapped in the muds.

It should also be mentioned that the phosphate precipitates also serve to remove other heavy metals and fluorides from the water resources of the earth (5, 7). Both heavy metals and fluorides are considered to be potential problems.

In order to better understand the role of phosphorus in our environment, more studies are needed to determine how changes of the quantity of phosphorus added to larger bodies will influence the quantity of active phosphorus contained in the body of water. The two extreme cases are known; in oceans there is no appreciable influence, in small bodies of water the corrective reactions do not have an adequate time to function, and the concentration of active phosphorus may become very high. More fundamental laboratory data must first be collected before truly meaningful studies can be made for larger bodies of water. At this time even the solubilities of most heavy-metal inorganic or organic phosphates have never been determined. Calcium pyro- and tripolyphosphates are known to be almost insoluble compounds, but their equilibrium solubilities are questionable. Moreover the longer chain phosphate, calcium hexapolyphosphate, is a remarkably insoluble substance (23) and would surely contribute no active phosphorus to a lake.

CRITICAL LEVELS OF PHOSPHORUS

Much of the current popular literature dealing with phosphorus in the aquatic environment leads the reader to two fallacious conclusions:

1. If the phosphorus concentration of the aquatic environment is not dramatically reduced in a short period of time, an impending disaster is imminent.
2. Most of the phosphorus found in the aquatic environment is a result of the widespread usage of synthetic detergents.

Neither of these implications can be ignored.

Phosphates are among the safest of all substances known to man. They are essential to man's nutrition. They are consumed in foods and soft drinks, taken in medicines and vitamin pills, and employed as toothpaste, all to the benefit of the user. No deaths have occurred from the phosphates contained in the aquatic environment and no deaths are anticipated. No illnesses have resulted from the phosphates contained in our water resources and no illnesses are anticipated.

Most of the concern for phosphorus in the environment has resulted because phosphorus is believed to be one of the nutrients for algae which may be controllable. It is no more nor less a nutrient than nitrogen or carbon or a host of other essential elements. It may be more controllable, but even algae cannot be considered anything more than a nuisance. No reputable scientist can in good conscience classify the growth of algae as a potential disaster! Mankind will continue to thrive on earth if absolutely nothing is changed with respect to the use of phosphorus, but life will be more pleasing if corrections are made. Even high-quality surface waters are not absolutely essential to man's survival. Over 80% of the freshwater resources of the United States are contained in groundwater, and man could have an adequate supply of freshwater even if he chose not to utilize surface water. Several large cities and hundreds of small cities already depend exclusively on wells as a source of water. The statement above in no manner condones the misuse of our water resources. It does emphatically refute the idea that human life on this planet is threatened by phosphorus, however.

The concept that most of the phosphate in the environment has occurred as a result of synthetic detergents is an example of neglect of the few facts that are known about environmental phosphorus. In 1970 6% of the phosphorus mined in the United States for use in the United States was used in the manufacture of detergents (see Chapter 36). Since at least 85% of the sewage effluents of the United States is deposited in receiving waters which terminate in the surrounding oceans, and about 30% of the population does not utilize sewers, less than 0.5% of the phosphorus mined in the United States flows into lakes as a result of detergent usage.

The areas of the United States in which sewage phosphates have created problems are mostly localized. In 1962 the Department of Health, Education and Welfare published a survey of the sewage disposal plants in the United States (18). Of the 14,549 plants that responded to the survey, 455 plants discharged sewage into 206 lakes. Eighty-two percent of these plants are located in nine states. Wisconsin, Michigan, Minnesota, New York, and Florida contribute 56% of the plants. No more than an equal number of plants employ streams that terminate in lakes, but it is noteworthy that today the problem areas of the United States are those areas that have traditionally employed lakes as receiving waters for sewage effluents. It is also noteworthy that these problem areas have existed for many years and have not suddenly occurred in the last two or three decades. Green Bay, Wisconsin, is a noteworthy example, having been named accordingly. Probably the primary reason it is generally believed that detergents contribute most of the phosphorus in the aquatic environment stems from sewage data collected in densely populated areas and the frequently quoted value for detergent phosphates in sewage. Doubtless in urban areas detergent phosphates constitute a large fraction of the phosphorus entering into sewage, but over 90% of the population of the United States lives on less than 6% of the land area. For 94% of the land area detergent phosphate plays a very minor role.

FERTILIZER PHOSPHATES

About 73% of the phosphorus mined in the United States is converted into fertilizers. Because phosphates interact strongly with some types of soil it is often stated that phosphates migrate too slowly in soils to consider fertilizers as a primary source of phosphorus in the aquatic environment. If nationwide as little as 0.4 lb/(acre)(year) (9) is considered a reasonable value for the quantity of phosphorus both eroded and leached from agricultural soil, the quantity of phosphorus from this source is comparable to the quantity of phosphorus employed in synthetic detergents each year. A loss of 0.4 lb/acre is equivalent to 250 lb/mi^2. Most estimates for the low-loss areas in the Lake Erie Basin are in this range. Souchelli estimated that 1800 million lb of phosphorus was eroded and leached from farms annually in the United States. This value is three times the quantity of phosphorus employed in synthetic detergents, and does not include a similar quantity of phosphorus which is cropped from the soil each year.

Lake Erie has received a major portion of the efforts expended on research for agricultural runoff of phosphorus. The Lake Erie Basin can hardly be considered typical of the United States. It is densely populated, highly industrialized, and the soil is of good quality compared to the sandy soils of other areas. Estimates for phosphorus inputs from municipal sources are very large compared to agricultural runoff in the Lake Erie Basin (22).

Of the total phosphorus mined in the United States for domestic use, a minimum of 12% finds its way into the aquatic environment via fertilizers within one year of the time it is applied. As time passes an even greater percentage of the fertilizer applied in previous years enters the aquatic environment. Within four years, as much as 40% of the phosphorus production of the previous four years can enter the aquatic environment as a result of fertilizer usage (10).

Fertilizers and the resulting food by-products account for the largest single source of man-induced phosphorus in the environment. A nearly steady state of input-output has developed. Detergent phosphates are the second largest source. Both of these sources are probably small compared to the quantities which enter the waterways by natural causes. If any single item on earth is essential to the future livelihood of mankind, adequate food supplies cannot be minimized in importance. The population of the earth can no longer be fed without the aid of fertilizer. As long as man must eat, a portion of the fertilizer will surely find its way into the aquatic environment. In one sense the choice is between people and algae.

NATURAL SOURCES

There is small doubt that the largest single source of phosphorus in the total freshwater environment comes as a result of natural causes, and yet this is an all but impossible thesis to support because reliable data are so limited. As a result of the interest in sewage in highly populated areas, the natural input of phosphorus has probably been underestimated by most of the authors attempting to assess the quantity of phosphorus entering the total aquatic environment. Conversely, Reid (12) has estimated that 68% of the phosphorus entering Lake Erie from the Maumee River occurs in January and February, but attributes the phosphorus to agricultural runoff. The *Lake Erie South Shore Tributary Loading Data Summary* (1967) (11) supports Reid's conclusions, but the 1967 *Report of the Lake Erie Enforcement Conference Technical Committee* minimizes runoff as does *Agricultural Pollution of the Great Lakes Basin* (22).

Trees contribute a large quantity of phosphorus to the aquatic environment each year. There are so many trees in the United States that no accurate number can be quoted. It is estimated that there are more the 450 billion ft^3 of lumber resources in the United States (13). Each tree contributes both leaves and pollen in very great quantities. Pollen is particularly rich in phosphorus (14), and the pollen from a single tree may contain one-half the weight of fat and one-tenth the weight of protein as a grain field covering the same area (15). Some forest streams become yellow with pollen each spring. Leaves are not as rich in phosphorus as is pollen, but the weight of the leaves is greater (16). Grasslands and flowers also contribute to the total load of phosphorus in the aquatic environment. Even wild-animal wastes may also become appreciable in some areas. Bird life alone is sufficient to cause enrichment of some Canadian lakes (17).

Two other major sources are natural leaching and erosions. Groundwater and spring water can be particularly rich in phosphorus (17). Eroded clay as seen in many muddy rivers following heavy rains is also rich in phosphorus (18). The abundance of phosphorus on earth is evidenced by the life and vegetation on earth. Where there is green vegetation and life, there also is phosphorus.

THE CONTROL OF PHOSPHORUS

The freshwater resources of the earth must be protected. This concept is too elementary to deserve debate. As the number of people on earth continues to increase, it will surely become more and more difficult to protect the quality of the aquatic environment. Sewage treatment is the logical first step toward improving and maintaining the water quality required for the future. The removal of phosphorus at the sewage-treatment plant in areas where phosphorus is a problem must be considered of immediate importance. The mandatory removal of phosphorus as a nationwide regulation is surely not advisable. The removal of phosphorus from sewage effluents of plants which utilize the Mississippi River, Chicago, for example, as receiving waters can only be a misdirected effort because the added phosphorus will hardly make a measurable change in the concentration of phosphorus in the river.

In some areas where excessive growth of aquatic vegetation is occurring and phosphorus is considered to be the limiting nutrient, the control of the influx of phosphorus and flushing the lake with water have proved to be helpful. In the construction of an artificial lake, a means of providing for bottom-water drainage would also be helpful. A means of providing flow from the bottom of a lake should do much to remove the nutrient-laden water from a lake as well as lessen the tendency for the silts to become anaerobic.

Whether or not man can control the input of phosphorus into a lake and continue to perform the essential activities of living depend on what levels he desires to control the "average" concentration of phosphorus. If the concentration must be lowered to 10 ppb of phosphorus at the beginning of the algal growing season to control algal blooms, the task could become very difficult for many lakes. In most instances lowering the concentration to 10 ppb will probably not be required, and adequate sewage treatment should improve the water quality if the lake is being abused with sewage waste. The disposal of only partially treated sewage into any lake cannot be considered responsible action.

Since phosphates are blamed for many of the problems associated with eutrophication, it is fortunate they are easily controlled in sewage effluents. It is questionable that the phosphates alone are responsible for many of the difficulties with which they are blamed, but the point is academic. It is self-evident that the freshwater resources cannot be expected to withstand the impact of a gigantic population unless they are protected with adequate sewage treatment. Mankind cannot prohibit phosphorus, he must use phosphorus, but he can control it when necessary.

NEEDED RESEARCH AND ANSWERS

Of all the scientific areas with a vested interest in phosphorus only chemistry has the charter to prepare new and useful products. Ten years ago numerous great scientific institutions throughout the world were engaged in basic and applied chemical research to find new and peaceful applications of phosphorus for the benefit of mankind. Within the last 10 years inorganic phosphate research as it existed 10 years ago has all but ceased to exist as a direct result of the concern for the protection of our environment. There is no way to begin to assess the meaning of this tragic loss to man's future. One need only review the impressive list of unique applications of phosphorus to grasp the impact of the loss of 10 years of research. Areas which have already suffered from this lost research are cancer detection (19), dentifrices and the control of caries (20), flame proofing and fire retardation, nutrition, agriculture, pesticides, sanitation and detergency, glass manufacture, water treatment, corrosion inhibition, and many others where new projects are no longer even considered because "the product contains phosphorus."

Some research has been undertaken to find better methods of removing phosphate from sewage or to improve the methods of analysis for phosphorus in the environment. This work is immensely important, but it does not fulfill many of the future requirements that must come from research directed toward new uses.

Future progress in the control of phosphorus in the environment has also suffered an untimely and critical blow. Without new chemical knowledge of phosphorus, the desired control of phosphorus in the environment is surely impeded if not made impossible. It is incredible that this loss has occurred at a time when millions throughout the world are hungry and phosphorus has long been recognized as "the key" to their survival (10). It is incredible that this loss has occurred because phosphorus supports life. It is incredible that *nuisance* blooms of algae can take precedence over all other important considerations. Our lakes and streams must be protected, but it must be done with insight and resourcefulness.

A mere listing of the information that is needed and the research that should be in progress is extensive. Only a few areas and problems are listed below, but this list spotlights some of the many needs.

1. The solubilities of most heavy-metal ortho- and polyphosphates are unknown. The few systems that have been studied quantitatively reveal that heavy-metal phosphates have very low solubilities and qualitative observations indicate that most, if not all, heavy-metal phosphates are very sparingly soluble. Quantitative data must be developed because these data

are indispensable in judging the behavior of phosphates and heavy metals in sewers, treatment plants, rivers, and lakes.

2. Additional work is needed to determine how phosphorus influences the rate of treatment in a sewage-treatment plant. What are the influences of phosphorus in excess of the biological demand? Does pure oxygen aeration and excess phosphorus provide a more favorable treatment time than sewage with lesser quantities of phosphate? Does it make any difference in rates and sludge quality if only organic phosphorus compounds or only inorganic phosphates are supplied in sewage?

3. About 30% of the phosphorus is removed from sewage in a typical treatment plant, but phosphorus from different sources has different properties. What percentage of the phosphorus that is found in the *effluent* is actually derived from detergents?

4. If the influent phosphorus to a treatment plant is reduced or increased by 10, 50, or 75%, how much is the effluent phosphate increased or reduced?

5. If phosphates are removed from detergents, will it then be safe to continue to put only partially treated sewage into lakes and streams? This does not seem to be logical. Generally scientists agree that if a lake is troubled with excessive quantities of phosphorus, the removal of only detergent phosphorus will accomplish nothing.

6. What is the optimum concentration level for phosphorus in a lake? Is it different from the concentration of phosphorus at which algae cannot bloom?

7. Does the optimum phosphorus level vary from lake to lake?

8. Does the particular use of a lake govern the optimum level of phosphorus in a lake? Does a fishing lake need more phosphorus than a boating or swimming lake, for example? A long range national zoning plan is needed to insure the proper utilization of our water resources.

9. Can fish flourish in a lake that is deficient in phosphorus? Some studies indicate that 40 lb of phytoplankton is required to support 1 lb of fish (21).

10. Is it possible to develop reliable data and methods of predicting runoff phosphorus values for soil? Few data are available and estimated runoffs vary from almost zero to as much as 11,000 lb of runoff phosphorus per square mile (9).

11. Polyphosphates of the types found in detergents prevent the growth of calcium and magnesium scale in boilers, hot-water heaters, and sand filters of water companies. If phosphates are removed from detergents and sodium carbonate washing powders or soaps are substituted, will feeder sewage lines grow scale as boilers do and increase sewer-line maintenance costs? The phosphates in human waste are not threshold agents or deflocculating agents.

12. What are the beneficial influences of blue-green algae in a lake?

13. What is the source of phosphorus for the algal blooms that occur at high elevations in the streams of the Rocky Mountains? Neither sewage, agriculture, nor detergents contribute to these blooms.

14. Core samples obtained by the Ohio Geological Survey to depths of 30 ft reveal that the phosphorus input to the sediments of Lake Erie has been rich in phosphorus and reasonably constant for at least the last few thousand years. Should not more deep core samples be obtained in order to better assess the past history of lakes?

15. What part is phosphorus playing in the precipitation of colloidal clays in lakes?

16. The ever-present need for better analytical methods is prominent. A method that continuously monitors and records total phosphorus as a function of depth in a lake is sorely needed.

Extensive new research must be initiated. The environmental problems of the next 50 years cannot be controlled by legislation, nor can the natural laws which govern be repealed or declared unconstitutional. Environmental phosphorus obeys only the laws of nature. To control phosphorus in natural bodies of water man must first learn the laws of nature.

ACKNOWLEDGMENT

The author wishes to express his gratitude to Dr. M. M. Crutchfield for his many helpful discussions and his suggested additions to and deletions from this manuscript.

REFERENCES

1. Hem, J. D., *U.S. Geological Survey Water Supply Paper* 1473, U.S. Printing Office, Washington D.C., 1959.

1a. Revised edition of *Water Supply Paper* 1473, 1970.

2. Chandler, D. C., and O. B. Weeks, *Ecol. Monogr.*, **15**, 437 (1942).

3. Federal Water Pollution Control Administration, *Lake Erie Surveillance Data Summary*, U.S. Department of the Interior, 1968.

4. Lange, W., *Water Res.*, **5**, 1031 (1971).

5. Rankama, K. R., and Th. G. Sahama, *Geochemistry*, The University of Chicago Press, Chicago, Illinois 1949.

6. Federal Water Pollution Control Administration, *Lake Erie Report*, U.S. Department of the Interior, 1968.

7. Curl, H., *Limnol. Oceanogr.*, **4**, 66 (1959).

8. Department of Health, Education and Welfare, *Inventory of Municipal Waste Facilities*, Publication 1065, 1962.

9. Lunin, J., *Advances in Environmental Science and Technology*, J. N. Pitts, and R. L. Metcalf, (Eds.), Interscience, New York, 1971.
10. Sauchelli, V., *Phosphorus In Agriculture*, The Davidson Chemical Corporation, Baltimore, Maryland, 1951.
11. Federal Water Pollution Control Administration, *The Lake Erie South Shore Tributary Loading Data Summary*, 1967.
12. Reid, G. K., *Ecology of Inland Waters and Estuaries*, Reinhold, New York, 1961, p. 187.
13. *American Encyclopedia*, Vol. 27, American Corporation, New York, 1956.
14. Sosa-Bourdourl, C., *C. R.*, **217,** 617 (1943).
15. Fribas, F., and H. Sagnomsky, *Biol. Zent.*, **66,** 129 (1947).
16. Steiner, M., *Biol. Zent.*, **67,** 84 (1948).
17. Clarke, F. W., *The Data of Geochemistry*, U.S. Geological Survey Bulletin 770, Washington, D.C., 1924.
18. Mattingly, G. E. G., and O. Talibudeen, *Topics in Phosphorus Chemistry*, Vol. 4, M. Grayson and E. J. Griffith (Eds.), Interscience, New York, 1967.
19. Subramanian, G., and J. G. McAfee, *Radiology*, **99,** 192 (1971).
20. Nizel, A. E., and R. S. Harris, *J. Dent. Res.* **43,** 1123 (1964).
21. Juday, C., *Trans. Wis. Acad. Sci.*, **29,** 1 (1942).
22. Environmental Protection Agency, *Agricultural Pollution of the Great Lakes Basin*, U.S. Government Printing Office, Washington, D.C., 1971.
23. Van Wazer, J. R., and S. Ohashi, *J. Am. Chem. Soc.*, **80,** 1010 (1958).

THE LAST PAGE

This book is dedicated to all students, young and old, who refuse to accept conjecture for fact while demonstrating the diligence to discern the difference. This book is dedicated to you who read the last page of this book.

EDWARD J. GRIFFITH

INDEX

Environmental Phosphorus Handbook has been compiled to define and clarify the extensive ramifications of phosphorus in the environment, and to provide a convenient (and badly needed) source of interdisciplinary information for the many people concerned with this important area of investigation. Because of the scope of the subject, the book has been edited by four experts, each of whom selected papers pertaining to his or her own areas of competence. As a result, *Environmental Phosphorus Handbook* covers an unusually varied range of topics, including the geochemistry of minerals containing phosphorus, the origin and fate of organic phosphorus compounds in aquatic systems, chromatographic analysis of oxoacids of phosphorus, gas chromatography detectors for phosphorus and its compounds, phosphorus in human nutrition, phosphorus nutrition of algae, the vertical movement of phosphate in freshwater, and phosphorus in wastewater treatment. Further, to present the clearest possible profile of phosphorus and its interactions, the 37 papers have been arranged according to a scale of reduction outlined in the Preface:

> "Phosphorus is viewed telescopically beginning with studies of gigantic systems followed by smaller and smaller subsystems, until the details of the minute microscopic and submicroscopic systems are explored. The manifestations of phosphorus are presented as they appear in meteorites and lunar samples, as in the geological formation of the earth, as in living organisms, and as in the molecules and atoms that constitute these systems. With this background the behavior and problems of phosphorus in lakes, streams, and sewage are reviewed. Then as a change of pace, phosphorus is treated as a factor in the economy of the United States. Finally, a simplified editorial overview of phosphorus in the environment completes the study."

The contrast and logical development of this approach afford a multi-perspective view of phosphorus that has never before been available in a single volume.

Environmental Phosphorus Handbook will be a valuable reference for anyone dealing with phosphorus or phosphorus-related problems. This includes scientists carrying out research in schools,